£3.00

The Dynamic Earth:
Textbook in Geosciences

The Dynamic Earth:
Textbook in Geosciences

Peter J. Wyllie

John Wiley & Sons, Inc.

New York · London · Sydney · Toronto

Printed in the United States of America.

10 9 8 7 6 5

Dedication

Dedicated to wives the world over who put up
with husbands who write textbooks when they
should be in bed and dedicated, in particular,
to Romy.

Preface

This book developed from a course called "The Solid Earth" that I organized at the University of Chicago in 1967. The aim of the course was to give first-year graduate students a birds-eye review of geology, and a global geophysical framework within which they could locate their specialized research topics in mineralogy, petrology, or geochemistry. There are many elementary texts in "Physical Geology" giving fine over-views of the Earth, but after a course using one of these, students usually take progressively more specialized subjects. By the time they reach the B.S. level they have learned enough geology to begin to appreciate the significance of the material that they had read with little comprehension a few years earlier. With this in mind, I prepared a more advanced course parallel in some respects to the elementary "Physical Geology" texts, dealing specifically with the major problems in earth sciences that the students might not otherwise meet in their subject-oriented Ph.D. courses.

The revolution in earth sciences was proclaimed in 1967, as I was preparing the course for the first time. The many new developments in global tectonic theory since then have convinced me that J. Tuzo Wilson is right when he states that geological curricula must be revised. I offer this text as one "revolutionary" design for geology courses, a design which emphasizes major units of the Earth, and major geological processes, without regard for the conventional subjects of curricula. The text is designed for graduate students in geology, geochemistry, and geophysics, but it could be used by well-prepared senior undergraduates either as the basis for a course, or as a supplementary text.

A single text incorporating all of the material that is appropriate for this volume would be enormous. I have selected and organized topics and material in what appears to me to be a useful sequence, knowing that no textbook is complete for a good teacher, and hoping that teachers will be able to use my selection of material as a basis for their courses, with the balance being supplied from their own experience, their favorite research topics, and the current literature. Specialists in most areas will probably find their subjects slighted by inadequate attention. For example, paleontologists, stratigraphers, and isotope geochemists will certainly feel this way, and with justification. Geophysicists may be disconcerted by the rather shallow treatment of physical topics but, on the other hand, geologists and geochemists may be relieved that the presentation is more comprehensible to them than that in the available texts on geophysics. The lack of reference to literature in languages other than English will be noted and deplored by some. For these and other shortcomings I can only apologize, plead that the life of an author is short, and claim that selectivity is essential.

I have adopted a didactic approach throughout. It is impossible for students to learn everything: the important thing is for them to learn how to examine evidence and ideas critically. For each major topic introduced I have tried to show how the associated ideas developed, and to present both sides

vii

of controversies where they exist. I made no attempt to review all pertinent literature. On the contrary, I made a special effort to keep literature citations down to a minimum, selecting specific papers for examination. Yesterday's theory is of historical interest only, and today's theory is tomorrow's history. I have therefore attempted to examine today's theories as part of the spectrum of history rather than as the definitive solutions so often claimed by enthusiastic researchers.

It is currently fashionable to denigrate advanced textbooks and monographs with the phrase: "It was out of date before it was published." I started the final draft of this volume in October 1969, and certainly by the time I submitted it to the publishers in August 1970, many of the facts recorded were already superceded by research papers in press. Minor revisions in November did not permit appreciable up-dating, despite my urge to include an additional chapter on the recent JOIDES results. This volume contains a record of the revolution in earth sciences during the 1960's, surely the most remarkable decade in the history of geology. I trust that this record of the accumulation of data and the development of ideas must have lasting value to students and teachers despite the flood of new data appearing annually. I anticipate also that many professional scientists, busily engaged in their own work during the 1960's, will find this volume a useful substitute for hours of library research. The historical account of the revolution will provide them with a sound background against which to view the new developments of the 1970's.

Among the many people who have helped me during the preparation of this text, I wish to thank in particular those who read parts of the manuscript at various stages. This does not make them responsible for the validity of my treatment and interpretations. Thanks to: A. L. Boettcher, F. Chayes, D. P. Gold, J. R. Goldsmith, D. H. Green, K. Ito, G. C. Kennedy, E. D. Jackson, I. B. Lambert, R. B. Merrill, A. A. Meyerhoff, R. C. Newton, M. J. O'Hara, J. K. Robertson, T. J. M. Schopf, A. M. Ziegler, and A. N. Onymous who reviewed the entire manuscript for the publisher. For their speedy, cheerful typing I must also thank Mrs. Irene Baltuska and Mrs. Glenda York.

University of Chicago, Illinois
February, 1971

PETER J. WYLLIE

Contents

1. INTRODUCTION: A GLOBAL APPROACH   1

2. SURFACE FEATURES OF THE EARTH   5
Introduction   5
Distribution of Continents and Oceans   5
Surface Relief of the Solid Earth   6
Major Structural Units and Physiographic Features   8
   Continents   9
   Ocean Basins   10
   Continental Margins   11
Tectonically Active Zones: Orogenic Belts and the World Rift System   12
   The World Rift System and Major Faults   12
   Distribution of Volcanoes   13
   Distribution of Earthquakes   13
   Global Tectonics of G. E. Rouse and R. E. Bisque   15

3. PHYSICAL PROPERTIES OF THE EARTH AND ITS INTERIOR   17
Introduction   17
Seismology and the Earth's Interior   17
   Seismic Waves   17
   Travel-time Curves   19
   Seismic Velocity Profiles and the Earth's Internal Structure   20
   Lateral Variations in the Upper Mantle   23
Properties of the Earth's Interior   26
   Density Distribution   26
   Elastic Properties, Pressure, and Gravity   29
   Temperatures within the Earth   30
   Viscosity   32
Magnetism, Gravity, and Heat Flow   33
   The Magnetic Field   33
   The Gravity Field   37
   Heat Flow   39

4. GEOLOGICAL PROCESSES AND THE GEOLOGICAL TIME SCALE   47
Introduction   47
Geological Cycles   48

Geological Processes   50
    Rock Materials   50
    Mineral Reactions   50
    Igneous Processes   55
    Sedimentary Processes   57
    Metamorphic Processes   58
Geological Time Scale   59
    Stratigraphic Classification   59
    Isotopic Age Determinations   59

5. THE NATURE OF THE CRUST–MANTLE BOUNDARY AND THE GABBRO–ECLOGITE PHASE TRANSITION   63
Introduction   63
Crust–Mantle Boundary in Tectonically Active Regions   65
Chemical Discontinuity or Phase Transition?   67
    The Moho as a Chemical Discontinuity   68
    The Moho as a Phase Transition   68
The Peridotite–Serpentinite Model of H. H. Hess   71
The Gabbro–Eclogite Transition as the Moho   72
    The Meteorite Analogy of J. F. Lovering   72
    G. C. Kennedy's Model   75
Experimental Studies on the Gabbro–Eclogite Phase Transition   76
    Instability of Crustal Silicates at High Pressures 76
    Reactions Among Mineral Assemblages   79
    Stability Limits of Gabbro and Eclogite   80
    The Transition of Gabbro to Eclogite   84
Development of the Phase Transition Hypothesis   86
    Revival by J. F. Lovering and G. C. Kennedy   86
    Modifications Based on Results of H. S. Yoder and C. E. Tilley   87
    Rejection by A. E. Ringwood and D. H. Green   89
    Revitalization by F. Press, K. Ito, and G. C. Kennedy   90
    Rejoinder by D. H. Green and A. E. Ringwood   90

6. THE COMPOSITION AND MINERALOGY OF THE EARTH'S MANTLE   93
Introduction   93
Extraterrestrial Evidence for the Composition of the Earth   94
    Cosmic and Solar Abundances of the Elements   94
    Classification and Chemical Composition of Meteorites   94
    Genetic Relationships Among Meteorites   101
    Origin and Differentiation of the Earth   102
    Compositions of the Core and Mantle   103
Ultramafic Rocks and the Upper Mantle   105
    Peridotite Mineralogy and Field Associations   105
    Petrogenesis of Ultramafic Rocks   106
    Extrapolation to the Upper Mantle   110

Composition of the Mantle 111
  Estimates Based on Terrestrial Rocks 111
  Hypothetical Peridotites 113
  Mantle Composition 114
  Chemical Variation Diagrams 116
  Trace Elements and Volatile Components 118
Phase Transitions in Mantle Peridotite and Eclogite 119
  Experiments on Peridotite Mineral Facies 119
  Effect of Water 120
The Olivine–Spinel Transition 123
  Phase Diagram for Fe_2SiO_4 123
  The Transition for Mg_2SiO_4 124
  The Transition for Olivine $Fo_{90}Fa_{10}$ 127
  The Transition Zone of the Mantle 400 to 1000 km Depth 129
Mineralogy and Petrology of the Mantle 131
  The Upper Mantle, the Transition Zone, and the Low-Velocity Zone 131
  The Lower Mantle 136

7. THE STRUCTURE, PETROLOGY, AND COMPOSITION OF THE EARTH'S CRUST 139
Introduction 139
Structure of the Crust 139
  Cross Sections Through Continents 140
  Cross Sections Through Oceanic Crust 143
  Cross Sections Through Continental Margins 147
  Schematic Structural Subdivisions of the Whole Crust 148
Petrology and Mineralogy of the Crust 150
Composition of the Crust 152
  Composition of Crustal Layers 153
  Composition of Crustal Units and of the Whole Crust 153
  Vertical Distribution of K, U, and Th in the Continental Crust 156
  Heat Production in the Crust 156
Deep Structure of Continents 159
  Concentration of Elements into the Crust 160
Ages of Continental Basements 161

8. MAGMA GENERATION 167
Introduction 167
Igneous Rock Associations 167
Development of Petrogenetic Theory in the Twentieth Century 168
  The First Half of the Century: Primary and Derivative Magmas 168
  Influence of High Pressure Experiments Since 1950 171
Rock-Water Systems as Guides to Magma Generation 173
  Water-Absent and Water-Excess Systems 173
  Water-Deficient Systems 176
  The Conditions $P_{H_2O} = P_{total}$ and $P_{H_2O} < P_{total}$ 178

Magma Generation in Crustal Rocks	181
Mineral Variation in Crustal Rocks	181
The Granite System	182
The Granodiorite System: Effect of Mineralogy	185
The System Granodiorite-Water	187
Anatexis in the Crust	189
Magma Generation in the Mantle	190
Mineralogy of Mantle and of Basalts	190
Melting Relationships of Peridotites and Basalts	192
Generation and Fractionation of Basaltic Magmas	196
Effect of Water	201
Basalt Petrogenesis	207
Generation of Batholiths and Andesites	208

9. GEOSYNCLINES AND THE OROGENIC CYCLE: CLASSICAL VIEWS — 211
Introduction	211
Classification of Geosynclines	212
Tectonic Elements	212
Metamorphism and Igneous Activity	216
Evolution of Geosynclines	217
Geosynclinal Couples	217
Geosynclinal Evolution According to J. Aubouin	219
"Pacific" Geosynclines of K. A. W. Crook	222
Evolution of Continents	223
Contemporary Geosynclines	225
Causes of Subsidence and Uplift	228
Subsidence of Geosynclines	228
Metamorphic Rocks as Guides to Geosynclinal Conditions	230
Theories of Global Tectonics	231

10. TECTONIC SIGNIFICANCE OF PHASE TRANSITIONS — 233
Introduction	233
Effect of Thermal Perturbations	235
Peridotite–Serpentinite Model of H. H. Hess	235
Gabbro–Eclogite Model of G. C. Kennedy	236
Effect of Pressure Perturbations	237
Sediment Deposition: Subsidence Followed by Uplift	238
Dynamics of Motion of a Phase Boundary after R. J. O'Connell and G. J. Wasserburg	238
The Models of G. J. F. MacDonald and N. F. Ness and G. W. Wetherill	239
Time-Dependent Solution of W. J. van de Lindt	241
Time-Dependent Solution of W. B. Joyner	243
Oscillatory Movements	245
Gravity and Phase Transitions: Centrifuge Models of H. Ramberg	246
Dome Models	247
Models of Subsiding Sheets	248

Phase Transitions and Mantle Convection: Analysis by J. Verhoogen   249
Metastable Phase Transitions Cause Earthquakes According to J. G. Dennis and
  C. T. Walker   252

11. CONTINENTAL DRIFT: DEBATE OF THE CENTURY   255
Introduction   255
The Debate Until 1950   256
  The Theory of Continental Drift   256
  Evidence Cited for Continental Drift   258
  Mechanisms Postulated for Drifting the Continents   259
Developments since 1950   261
  Geometrical Fit of the Continents   262
  Matching Age Provinces on Continental Reconstructions   263
Revolution in the Earth Sciences   266

**12. PALEOMAGNETISM, POLAR WANDERING, AND SPREADING SEA
FLOORS**   269
Introduction   269
The 1950's: Paleomagnetism and Polar Wandering   269
  Remanent Magnetism in Rocks   269
  Paleomagnetic Measurements   270
  Interpretation of Paleomagnetic Pole Positions   272
Paleomagnetism since 1960: Magnetic Reversals   277
  Polar Wandering and Continental Drift   277
  Paleomagnetism and the Earth's Radius   284
  Magnetic Reversals in Igneous Rocks   285
  Magnetic Reversals in Deep-Sea Sediments   292
The 1960's: Spreading Sea-Floor Concept   298
  The Bandwagon Began to Roll in 1960   298
  The Contributions of H. H. Hess and R. S. Dietz   298
  Some Other Convection Models   301

13. MAGNETIC ANOMALIES IN THE OCEAN BASIN   307
Introduction   307
1958–1968: Linear Magnetic Anomalies   308
  1958: *Discovery in the Pacific by R. G. Mason*   308
  1963: *Explanation by F. J. Vine, D. H. Matthews, and L. W. Morley*   313
  1965–1966: *Confirmation by F. J. Vine and J. T. Wilson*   315
  1965–1966: *Scepticism and Conversion of J. R. Heirtzler and Lamont Associates*   322
  1968: *Extrapolation by J. R. Heirtzler, G. O. Dickson, E. M. Herron, W. C. Pitman,*
   III, and X. Le Pichon   326
Problems and Interpretations since 1968   332
  Magnetization of Basalt   334
  Selected Interpretations   335
  A Note of Caution from N. D. Watkins and A. Richardson   337
  Near-Bottom Magnetic Results from the "Fish"   340

14. PLATE TECTONICS   343

Introduction   343

The Concept of Tectonics on a Sphere   344

Paving Stones of the North Pacific by D. P. McKenzie and R. L. Parker   345

Aseismic Crustal Blocks of the World by W. J. Morgan   345

Global Patterns of Surface Motion by X. Le Pichon   346

The New Global Tectonics by B. Isacks, J. Oliver, and L. R. Sykes   348

Small Crustal Plates by P. Molnar and L. R. Sykes   351

Evolution of Triple Junctions of Plates by D. P. McKenzie and W. J. Morgan   352

The Results of Deep Sea Drilling   352

Mechanism of Plate Tectonics   353

Midoceanic Ridges and Mantle Convection   356

Thermal Structure   357

Petrological Structure   357

Lithosphere Consumption Beneath Island Arcs   360

Thermal Structure   360

Magma Generation   361

Structures Associated with Trenches   364

Geosynclines, Mountain Building, and Sea-Floor Spreading   366

15. GLOBAL GEOLOGY IN THE 1970's   373

Introduction   373

Prognosis from the 1969 Penrose Conference and the 1970 Geodynamics Commission   374

V. V. Beloussov and R. W. van Bemmelen Prefer "Oceanization"   375

A. A. Meyerhoff Maintains that the Atlantic Ocean has been Open for 800 Million Years   376

Epilogue   377

REFERENCES   379

AUTHOR INDEX   393

SUBJECT INDEX   399

The Dynamic Earth:
Textbook in Geosciences

1. *Introduction: A Global Approach*

The Earth is a near-spherical body in space. It moves. It rotates about its own axis and it follows an elliptical path through the solar system about the sun, which is one of at least 100 billion stars rotating around the center of the Milky Way galaxy.

The structure of the Earth may be considered as a series of concentric shells with an inner core, an outer core, a mantle with several shells, and a crust. The hydrosphere, atmosphere, and magnetosphere form an envelope around the solid Earth and shield it from much of the radiation and many of the meteoritic particles that bombard the Earth from space. Most of the hydrosphere fills shallow depressions on the Earth's surface forming the oceans, but this shell overlaps into the atmosphere and into the crust. Movements occur within each of the shells.

Motions within the atmosphere such as wind are familiar; the circulation of the atmosphere on a global scale and its more localized rotary motions are essential ingredients of weather and climate at the surface of the Earth. Wind-driven surface currents in the oceans contribute toward the circulation of the oceans. The ocean currents carry heat energy from equatorial regions toward the poles, which influences climate and the energy balance in the hydrologic cycle. The hydrologic cycle involves the circulation of water through the hydrosphere, with evaporation of water from the oceans being followed ultimately by its precipitation from the atmosphere and by gravity-controlled flow over and through the crust back to the oceans. Weathering and downward erosion of the Earth's solid surface is accomplished largely by the moving atmosphere and by moving water within the hydrologic cycle.

Solar radiation provides the energy for the circulation of the atmosphere and the hydrosphere and thus for the erosion of the Earth's surface. Erosion would reduce the exposed solid surface to sea level within a very short period, in terms of the geological time scale, if there were no forces within the Earth causing uplift and repeated exposure of rocks to the action of solar radiation, the atmosphere, and the hydrosphere.

Geological study of rocks confirms that they have been folded and uplifted; local motions within the crust are thus established. Records of the Earth's former magnetic field from paleomagnetic studies and magnetic anomalies provide evidence that continental masses and oceanic crust have moved with respect to the poles. The existence of the Earth's magnetic field is explained by the dynamo theory which requires motions in the liquid outer core. These motions are dominated by the Coriolis force which arises from rotation of the Earth. Attempts to explain the origin and present distribution of the ocean basins, continents, and mountain ranges suggest that motions also occur in the upper mantle, possibly of a circulatory nature. The energy source for these motions is heat, probably derived from the decay of radioactive materials within the Earth.

It appears that the major features of the surface of the Earth are produced by the interactions of processes driven by two energy sources. The internal heat within the Earth causes motions in the mantle which influence the distribution and elevation of the continental masses. The external solar radiation drives the circulation of the atmosphere and hydrosphere, which produces the detailed sculpture of the solid Earth surface. The Earth's crust is a thin, brittle shell sandwiched between the Earth's mantle and the envelope comprising the atmosphere and hydrosphere.

Geology involves study of the physiographic features and rocks at the Earth's surface and deduction of the processes and history of their formation. The direct observations of this accessible part of the Earth lead to the conclusion that the present shapes and positions of the continents, distribution of features such as mountain ranges, as well as distribution of various rock types are second-order results of major processes occurring within the mantle. Thus, when we consider the Earth's surface, we find ourselves concerned with the Earth's interior, and hence with the origin of the Earth, which leads us to the origin of the solar system and the universe. These are complex problems and we cannot devote many pages to cosmology. Our attention will be concentrated on the crust of the Earth and the upper mantle which appears to dominate its behavior.

This book was conceived as a text for a course that provides an overview of the whole Earth, and simultaneously reviews the various subjects that are usually taught in geology and Earth sciences. Figure 1-1 is a "Tetrahedron of physical sciences," which includes all of the subjects applied to the study of the Earth. Geology remained largely descriptive and historical in its approach until recently when the application of physical and chemical methods and techniques led to the development of geophysics and geochemistry. We will be concerned for the most part with the front face of this tetrahedron and less with paleontology and the life sciences.

The tremendous advances made in Earth sciences have led in many universities to a proliferation of courses. Many specialized topics are now recognized as "subjects," each occupying a very small space in Figure 1-1,

but each with appropriate courses being taught and textbooks written. Students are burdened with more and more facts and more and more instruments for gathering data.

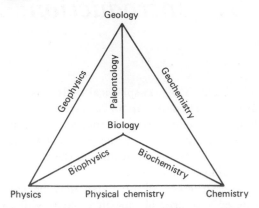

Figure 1-1. Tetrahedron of physical sciences. From "Introduction to Geophysics" by B. F. Howell. Copyright 1959. Used with permission of McGraw-Hill Book Company.

A. D. Stewart drew attention to this problem in a letter to *Nature* on March 9, 1968, entitled "Geology in British Universities." The fact that the article was reprinted in *Geotimes* eight months later indicates that the problem is not limited to Britain. Stewart pointed out that most university geology departments offer a range of semi-independent, specialized courses within the field from which the student is expected to construct his own view of the Earth. Departments are usually staffed with one teacher for each subject segment, with each segment completely separated from its neighbors. Standard segments include mineralogy, petrology, geochemistry, structural geology, stratigraphy, geophysics, sedimentology, and paleontology. The system tends to be self-perpetuating because as data accumulate original segments become split with the appointment of a new teacher. Paleontology, for example, may be split into macropaleontology and micropaleontology with an additional segment of vertebrate paleontology. What is needed, according to Stewart, is for teaching and research to be directed toward synthesis of geological information at levels above that of these segments. He proposed as an alternative a three-tier hierarchical scheme in which scale plays a key part, and teaching is process-oriented rather than descriptive. Level III is concerned with the standard subject segments and with techniques; level II is concerned with broader topics such as the Earth's core, mantle, and crust, the hydrosphere and atmosphere, and the evolution of life; level I is concerned with the Earth as a planet and with planetary studies.

Recent developments in the Earth sciences, especially those related to marine geophysics and paleomagnetism, have led to a great deal of synthetic research effort at level II, and this has certainly filtered through to undergraduate courses in universities. Several excellent introductory textbooks are available which present material more closely tied to the framework of level II than to the traditional segments. Unfortunately, however, after this there usually comes a series of successively more specialized courses. A presentation of material at level II should mean more to students when they have already learned something about geological processes and rock types than when they are first being introduced to the Earth sciences and that is what this volume attempts to provide. I have tried to present material appropriate for a student who is about to graduate with a bachelor's degree or who is beginning his graduate work in the Earth sciences.

We read a great deal today about the scientific revolution in the Earth sciences, and J. Tuzo Wilson considers that this revolution is shaking the very foundations of classical historical geology. In a revolutionary era the traditional subject-orientated instruction programs certainly require modification. Even if we are not living through a major revolution I consider it essential for students to be aware of a global, geophysical framework in which the various specialized research topics can be located, and to be familiar with the hypotheses

of world geology which embrace all of the subjects.

Geology is the science of the Earth, and the Earth is a chemical system subject to physical processes. I have therefore tried to incorporate both geophysics and geochemistry into consideration of geology on a global scale. The study can be pursued on three different scales.

1. We have the standard approach of regional geology, with rocks being examined in the range from local outcrops to mountains or other major tectonic units.

2. The history of such units is dependent on global geology, which is concerned with the size, distribution, and possibly the movements of continents and ocean basins.

3. The third scale relates to the Earth as a sphere, and is concerned with movements in the mantle and core which control the movements of the crust.

For complete understanding of geology we have to decipher the processes occurring within the Earth.

This volume is concerned mainly with the second scale, the development and distribution of the continents and ocean basins. The next six chapters review the chemistry, physics, and geology of the Earth's mantle and crust. The materials of the Earth are discussed in Chapter 4 together with a general review of geological processes. Most of these processes are involved in the mountain building cycle. The classical concepts of geosynclines and orogeny are reviewed in Chapter 9. Theories of the behavior of the Earth's interior must be capable of explaining the complexity of geosynclinal history.

Changes in temperature at depth in the mantle and crust cause changes of phase which may have significant effects at the surface and in Chapters 8 and 10 we consider the products and effects of partial melting at depth and the tectonic significance of solid state phase transitions. Vertical movements induced by deep-seated phase transitions may contribute to mountain building, but during the 1960's evidence supporting continental drift, in a modified guise of sea-floor spreading and plate tectonics, forced geologists to consider seriously the implications of global movements for the interpretation of mountain ranges and more localized problems. This is the stage at which revolution was proclaimed.

Chapters 11, 12, 13, and 14 include a history of the development of ideas related to continental drift and wandering poles, with special attention being paid to the concept of sea-floor spreading, the interpretation of linear magnetic anomalies of the ocean basins, and the formulation of the theory of plate tectonics. The new global tectonics involves the origin and evolution of the continents and of mountain ranges and the use of all of the traditional subject segments in geology. In Chapters 14 and 15 we examine the effects of revolution on reexamination and reinterpretation of classical geological concepts, and note the beginning of a counterrevolution by geologists unconvinced by the new evidence and ideas.

2. *Surface Features of the Earth*

INTRODUCTION

It is inevitable that a book written at the end of the geological revolution of the 1960's, which saw development of the global theories of sea-floor spreading and plate tectonics, must be greatly concerned with whether or not the oceans have grown and spread from a central rift, and whether or not the ancient continental masses have drifted with a young, spreading ocean floor. Nevertheless, we must begin with the state of the Earth as it is today.

Theories for the origin and history of the Earth must explain first the present distribution of the continents and ocean basins and second the major physiographic and geologic features of the interface between the Earth's solid surface and its envelope of atmosphere and hydrosphere. Interpretation of geophysical evidence, especially from paleomagnetic studies, linear magnetic anomalies of the ocean basins, and seismic studies, indicates that continents and ocean basins have undergone significant changes in relative position.

Evidence that relative motion may be in progress at the present time is provided by the observation that stable areas of the Earth's surface are traversed by elongated belts of instability. The distribution of volcanoes and earthquakes defines the active belts, and geological studies of folded orogenic zones define the location of earlier active belts on the continental masses. The existence of activity does not necessarily imply mobility and some geological evidence can be adduced to show that the relative positions of the continents and ocean basins have not changed for several hundred million years at least.

In this chapter we introduce the major physical features of the Earth's surface and the active belts that traverse them. These features are merely the superficial expressions of deep-seated processes, and therefore we will review the physical features of the Earth's interior in the following chapter. Together these two chapters provide the global framework for the topics discussed in subsequent chapters.

DISTRIBUTION OF CONTINENTS AND OCEANS

The world maps reproduced in figures through this book give a rather misleading picture of the relative areas and the distribution of the continents and oceans, and examination of a world globe is necessary for appreciation of the significance of the statistics reviewed. Figure 2-3 shows the distribution of the continents and oceans and Table 2-1 lists their areas.

Asia and Europe are often considered separately, but in a physiographic sense we must treat them as combined in the continent of Eurasia, which is almost twice as large as Africa, the next largest continent. Various authors subdivide the oceans in different ways; Table 2-1 lists areas for the Pacific, Atlantic, Indian, and Arctic Oceans. Separate areas are given in the table for each ocean and for each ocean plus marginal seas, gulfs, and straits. Other systems subdivide the Pacific and Atlantic into North and South

TABLE 2-1 Areas of Continents and Oceans and Mean Ocean Depths[a]

	Area (10⁶ km²)	Percent of land or ocean	Percent of world surface	Mean depth (km)
World surface	510	—	100	—
All continents	148	100	29.2	—
All oceans	362.0	100	70.8	3.729
Eurasia	54.8	36.8	10.8	—
Asia	44.8	29.8	8.7	—
Europe	10.4	7.0	2.1	—
Africa	30.6	20.5	6.0	—
North America	22.0	14.8	4.3	—
South America	17.9	12.0	3.5	—
Antarctica	15.6	10.5	3.1	—
Australia	7.8	5.2	1.5	—
Pacific Ocean	166.2	—	—	4.188
—with adjacent seas	181.3	50.1	35.4	3.940
Atlantic Ocean	86.6	—	—	3.736
—with adjacent seas	94.3	26.0	18.4	3.575
Indian Ocean	73.4	—	—	3.872
—with adjacent seas	74.1	20.5	14.5	3.840
Arctic Ocean	9.5	—	—	1.330
—with adjacent seas	12.3	3.4	2.4	1.117

[a] Data for oceans from Menard and Smith (1966).

Oceans at the equator, and include a Southern (Antarctic) Ocean arbitrarily bounded by the parallel 55°S.

More than 70% of the Earth's surface is covered by the oceans, and each of the three major oceans is larger than Eurasia. The Pacific Ocean constitutes just over half of all ocean surface. It is larger than all of the continents combined, and together with adjacent seas it covers 35.4% of the Earth's surface.

The continents are not distributed evenly over the Earth's surface. More than 65% of the land is in the northern hemisphere, with a concentration of continental material just south of the Arctic Circle. The continents of North and South America, Africa, and Asia with its Indian appendage, are roughly triangular in shape with their apices pointing south. About 81% of all of the land surface is situated on a land hemisphere with pole near Spain at 0° E-W, 38°N; this hemisphere contains 47% land and 53% sea. The opposite hemisphere with pole in New Zealand, the water hemisphere, contains 11% land and 89% sea. On the Earth as a whole about 45% of the surface has sea opposite sea, and only about 1.5% has land opposite land. Of the total land surface 95% is antipodal to sea. A striking example is provided by the opposite characters of the Arctic and Antarctic. The Antarctic continent is centered over the south pole and is surrounded by ocean, whereas the north pole lies in the Arctic Ocean which is almost completely surrounded by land.

SURFACE RELIEF OF THE SOLID EARTH

If the ocean water were removed from the surface of the Earth, we would see the continental masses rising abruptly from the ocean floor. The reason for the existence of the continents is one of the problems we seek to answer. Removal of the ocean water would also reveal the relief of the solid surface

beneath the oceans and expose the system of ridges and rises encircling the globe.

The distribution of the elevations of the surface of the solid Earth has been the subject of many studies since measurements of land heights and ocean depths became available. The distribution of elevations has been well

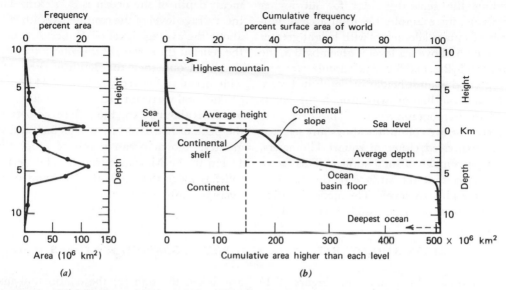

Figure 2-1. Distribution of areas of the solid Earth between successive levels. (*a*) Frequency distribution. (*b*) Cumulative hypsographic curve.

established for more than 50 years, but the number of depth soundings in the ocean basins has increased by many orders of magnitude during the last decade or two. Menard and Smith (1966) therefore re-examined the distribution of ocean depths using the modern bathymetric charts and considering specific physiographic and structural provinces.

Data for construction of the frequency distribution curve and the cumulative hypsographic curve of Figure 2-1 are listed in Table 2-2. The areas of the Earth's solid surface within 1 km intervals above and below sea level are given, as well as the percentage of the total surface area that is within each class interval. In Figure 2-1*a* these percentages are plotted at the midpoints of the level intervals, and the lines joining them show that the surface of the solid Earth is strongly concentrated at two levels: one corresponds to the continental platform and the other to the ocean basin floor.

The cumulative areas listed in Table 2-2 for total surface area above the lower limit of each height or depth interval are obtained by

TABLE 2-2 Hypsometry of the Earth's Surface[a]

Height or depth interval (km)	Area		Cumulative area (higher than lower limit of interval)	
	$\times 10^6$ km²	Percent of total	$\times 10^6$ km²	Percent of total
Above sea level: greatest height, Mt. Everest, 8.848 km; average height, 0.875 km				
> 5	0.5	0.1	0.5	0.1
4–5	2.2	0.4	2.7	0.5
3–4	5.8	1.1	8.5	1.6
2–3	11.2	2.2	19.7	3.8
1–2	22.6	4.5	42.3	8.3
0–1	105.8	20.8	148.1	29.1
Below sea level: greatest depth, Marianas trench, > 11 km; average depth, 3.729 km				
0–0.2	27.1	5.3	175.2	34.4
0.2–1	16.0	3.1	191.2	37.5
1–2	15.8	3.1	207.0	40.6
2–3	30.8	6.1	237.8	46.5
3–4	75.8	14.8	313.6	61.5
4–5	114.7	22.6	428.3	84.0
5–6	76.8	15.0	505.1	99.0
6–7	4.5	0.9	509.6	99.9
7–11	0.5	0.1	510.1	100.0

[a] Data for oceans from Menard and Smith (1966). For continents from Scheidegger (1963) after E. Kossina.

adding the separate areas for all higher intervals; for example, the total area in millions of square kilometers with height greater than 3 km above sea level is given by 8.5, the sum of 0.5, 2.2, and 5.8. The cumulative areas are plotted against height or depth in Figure 2-1b and the line drawn through the points gives the hypsographic curve. The curve gives directly the area of surface above any selected level, expressed either in square kilometers, or as percentage of total surface. This figure also shows that the surface of the Earth is dominated by two levels. The mean elevation of land above sea level is 0.875 km, and the mean depth of the ocean is 3.729 km. Thus the average level of the continents is 4.604 km above the average level of the ocean floor. If the relief at the surface of the solid Earth were smoothed out, then the depth of the world-wide ocean so formed would be 2.44 km.

The total vertical relief on the Earth's solid surface is about 20 km, the difference between the peak of Mt. Everest, at 8.85 km, and the greatest known ocean depth of more than 11 km in the Marianas trench. This vertical relief is more than half the thickness of the average continental crust.

MAJOR STRUCTURAL UNITS AND PHYSIOGRAPHIC FEATURES

The hypsographic curve in Figure 2-1b illustrates in a general way the major physiographic provinces. The two predominant structural units are (a) the continental platforms and (b) the ocean basin floors. These are linked through (c) the continental margins, which comprise the continental shelf and slope, and other provinces to be reviewed in more detail below. Figure 2-1b does not show the existence of (d) the elevated oceanic

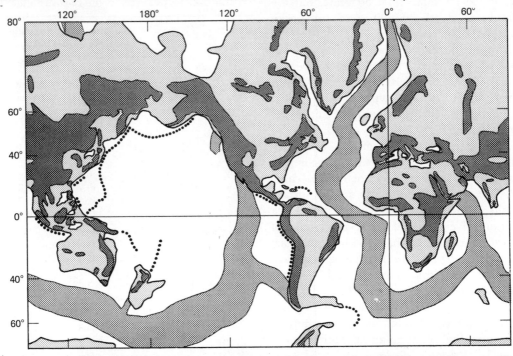

Figure 2-2. Major physiographic features of the world. Key: white, abyssal ocean floor; medium shading, oceanic ridge system; heavy dotted lines, oceanic trenches; light shading, continental platform and continental shelf; dark grey, mountains, intermontane basins, associated hills, and some elevated plateaus (various sources).

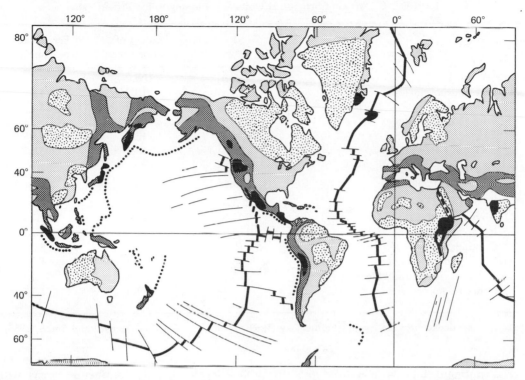

Figure 2-3. Major tectonic features of the world. Key: heavy lines, active rift systems of oceanic ridges; light lines, oceanic faults; dotted lines, oceanic trenches; light shading, continental platforms; ornamented, continental shields; dark grey, Tertiary folded mountain chains; black, Cenozoic volcanic regions (various sources).

ridges and rises, because their depths are included in the region labelled "continental slope." Small percentages of the oceanic and continental platforms are occupied by (e) ocean deeps and (f) high mountains respectively. The surface distribution of these six units is shown in figures 2-2 and 2-3, and surface areas are listed in Table 2-3. Note that although the maximum depths and maximum heights are at opposite ends of the diagram in Figure 2-1*b*, the ocean trenches are always located close to and almost parallel with land of high elevation such as an island arc or mountain range.

Continents

The position of the shore line between oceans and continents is not the boundary between the structural units. The continents are partly covered by water, and the flooded continental shelf and slope constitute 10.9% of the area of the whole Earth (Table 2-3). This is about one-quarter of the total area that is structurally continental. The sides of the continental slabs are represented by the continental slopes, which drop off abruptly from the edges of the continental shelves at angles of 2 to 3.5°, and continue down to the ocean basin floors lying in the depth range 3 to 6 km.

There is good evidence for worldwide changes in sea level, as well as areally restricted changes caused by local phenomena. Figure 2-1*b* shows that a small rise in sea level would produce a significant shift in the shore line position, an increase in area of continental shelf, and a decrease in area of sub-

TABLE 2-3   Areas of Structural Units and Physiographic Provinces[a]

	10^6 km²	Percent of land or ocean	Percent of world surface
Continent	149.0	100.0	29.1
(1) Precambrian shields	29.4	19.7	5.8
(2) Platforms	66.9	44.9	13.1
(3) Orogenic folded zones			
Riphean-Paleozoic	24.4	16.4	4.8
Mesozoic-Cenozoic	28.3	19.0	5.5
Submerged continent			
Continental shelf and slope	55.4	15.3	10.9
Ocean	306.5	84.6	60.0
(1) Abyssal ocean basin floor	151.5	41.8	29.7
(2) Ridge and rise	118.6	32.7	23.2
(3) Continental rise	19.2	5.3	3.8
(4) Island arc and trench	6.1	1.7	1.2
(5) Volcanic ridges, and volcanoes	5.7	1.6	1.1
(6) Other ridges and elevations	5.4	1.5	1.1

[a] Data for continents from Ronov and Yaroshevsky (1969). Data for oceans from Menard and Smith (1966).

aerial continent; for example, if the 25×10^6 km³ of ice stored in ice sheets and glaciers were to melt rapidly, sea level would rise by 50 to 70 m, before settling down at about 35 m after isostatic adjustment of land levels.

Figure 2-3 shows the main geological divisions of the continents.

1. The Precambrian shields.

2. The platform areas which are shields covered by a thin veneer of flat-lying younger sediments and the continental borderlands which are contiguous with the continental shelf.

3. The mountain ranges extending along orogenic belts of folded rocks.

The geologically young mountain chains follow two orogenic belts, each of which follows approximately a great circle. The circum-Pacific belt extends through the Philippines, Japan, Alaska, the Rocky Mountains, the Andes, and Antarctica. The Mediterranean-Asian belt follows the Alps, the Himalayas, Indonesia, New Guinea, and New Zealand. The continental areas with abundantly developed volcanic rocks of Cenozoic age are also indicated in Figure 2-3.

Ocean Basins

Table 2-3 lists six physiographic provinces for the oceans excluding the continental shelves and slopes because these, although covered by ocean, are not oceanic. Only three of these provinces occupy more than 2% of the ocean basin. (1) The abyssal ocean basin floor, and (2) the ridges and rises together cover more than 74%, and (3) the continental rise occupies 5.3%. The oceanic ridges and rises cover 32.7% of the ocean basin and 23.2% of the total Earth surface. This is much more than the folded orogenic belts of the continents. They form a worldwide system with many branches (Figures 2-2 and 2-3) and with mean depth of 3.970 km compared with a mean depth of 4.753 km for the ocean basin floor.

The continental rise, including partially filled sedimentary basins, consists of coalescing alluvial fans and piedmont plains deposited on the ocean floor beneath the continental slope, by turbidity currents and slumps. The continental rise, where recognized, is gently sloping or almost flat, and its characteristic features are similar to those of thick accumulations of sediments in partially enclosed basins such as the Gulf of Mexico.

Figure 2-4 shows the hypsometry of the world ocean basins and compares this with the depth distribution of the three major

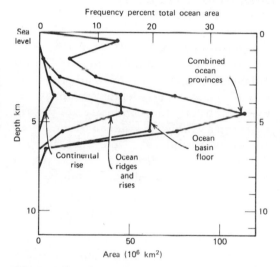

Figure 2-4. Hypsometry of all ocean basins for all physiographic provinces combined, and for individual major provinces. (From H. W. Menard and S. M. Smith, Jour. Geophys. Res., **71**, 4305, 1966, with permission).

physiographic provinces of the oceans; the ocean basin floor, the ridges and rises, and the continental rise. The areas of each province within the 1-km depth intervals can be read from the diagram, as well as the percentage of the total ocean area (compare Figure 2-1*a*). The combined curve for the oceans shows a double peak with maximum values at depths of 4.5 and 0.5 km, corresponding to the level of the deep ocean floor, and to the continental shelf and slope. Notice that the individual hypsometric curves for the ocean basin floor and for the oceanic

rises and ridges have similar shapes, which could be interpreted as indicating that an oceanic rise represents normal oceanic basin which has been elevated by an average of 1 km. Elevation or depression of the rises could affect sea level quite markedly.

Table 2-3 gives the areas of two more oceanic provinces and a third group of poorly defined features: (4) island arc and ocean trench, (5) composite volcanic ridges and individual oceanic volcanoes, and (6) long, narrow, steep-sided ridges not known to be volcanic, and poorly defined elevations.

Island arcs and trenches, including the whole system of low swells subparallel to the trenches, are considered as a single province covering only 1.2% of the Earth's surface, if continental equivalents or extensions of island arcs, such as Japan, are excluded. The median depth of 3.97 km is less than that for all ocean basins; the process which forms trenches and related features apparently elevates the oceanic crust on a regional scale.

The individual volcanoes and volcanic ridges rising from the ocean basin floor constitute a physiographic province distinct from the volcanoes associated with the oceanic rises and ridges. The volcanic ridges are formed from overlapping volcanoes. This province covers about the same surface area as the island arc and trench province, but it may be larger, because probably some of the remaining 1.5% of the ocean floor consisting of ridges and elevated regions not known to be volcanic will eventually prove to be of volcanic origin.

Continental Margins

The continental margin is the region between continent and ocean basin. The edge of the continental shelf is usually sharply defined by a marked increase in slope. The width of the continental shelf ranges from zero to 1500 km, with an average of 78 km, and the ocean depth at the shelf edge ranges from 20 to 550 m, with an average of 133 m. The continental margins are deeply incised by submarine canyons, which serve as routes for sediment transportation away from the continental shelves. The sediments are deposited in ocean trenches, or on the abyssal ocean floor, where they build up continental rises.

Three types of continental margin can be distinguished in Figure 2-3. The *Atlantic type* comprises the continental shelf, continental slope, and continental rise. The *Andean type* has a narrow continental shelf, with an oceanic trench below the continental slope. The *Island-arc type* consists of a volcanic island arc with oceanic trench, separated from the main continental mass by a small ocean basin. Some island arc systems are so far from continental masses, as much as 2000 km, that they can not be considered as features of continental margins.

TECTONICALLY ACTIVE ZONES: OROGENIC BELTS AND THE WORLD RIFT SYSTEM

The platforms of the continents and ocean basins (Figures 2-1*b* and 2-2) are stable, but the abrupt change in elevation at the continental margins makes them potential or actual sites of instability. The Earth is in a state of activity, as demonstrated by the occurrence of earthquakes and the eruption of volcanoes. The existence of high mountain ranges composed of intensely folded rocks confirms that the Earth has been active in the past. The unstable or active parts of the Earth occur along elongated belts bordering or traversing the stable platform areas. The distribution of orogenic belts has already been reviewed in Figures 2-2 and 2-3. The post-Precambrian orogenic belts can be subdivided into four regions with different periods of folding. Orogenic activity apparently is not continuous in any given location, but intermittent. Another major zone of activity follows the crest of the midoceanic ridge-rise system.

The World Rift System and Major Faults

During the last 20 years evidence has accumulated that a worldwide rift system follows the crest of oceanic ridges and rises and extends across parts of the continents. There is a striking rift valley in the center of the mid-Atlantic ridge, but there remains some uncertainty about the physiography of other rises. There appears to be nothing comparable to the Atlantic rift valley and associated ridges on the East Pacific Rise. Figure 2-3 shows the distribution of the rift system according to various authorities, although many more depth soundings are required before the continuity of the rift system on a worldwide scale can be considered absolutely established. Note that the rift system is not restricted to the oceans. It extends across the African continent, for example.

The map shows that the rift is discontinuous on a local scale because of displacement along fracture zones which were at first believed to be transcurrent faults, but which are now interpreted as transform faults (Chapter 14). In the East Pacific there is a series of major fracture zones about 100 to 200 km wide, approximately following great circles, which extend for a few thousand kilometers, with a vertical relief of a few kilometers. Individual ridges and troughs within the zones are several hundreds of kms long and a few tens of kilometers wide. Within the troughs occur the greatest ocean depths of the central Pacific, 6 to 7 km. Figure 2-3 shows that these enormous fracture zones are transverse to the crest of the East Pacific Rise, and it appears that they lie entirely on the broad flanks of the Rise.

Another group of fractures is the series of large transcurrent fault systems that parallel the margins of the Pacific. These are not shown in Figure 2-3 because it would complicate the diagram too much. Movements on these faults, such as the San Andreas fault in California, are of the order of 1 cm/year, which would provide long-term displacements large enough to be consistent with the hypothesis of continental drift. In the early 1960's many geologists considered trans-current faulting to be the dominant mode of tectonics in the circum-Pacific belt, with the evidence suggesting that the entire Pacific basin was rotating anticlockwise or that the whole basin was moving northwest. It now appears that the San Andreas fault is a trans-form fault (Chapter 14), and attention is directed toward sea-floor spreading (Chapter 12) rather than rotation of the sea floor.

Figures 2-2 and 2-3 show the distribution of major physiographic and geological features of the Earth's surface. The sites of current activity are defined by volcanoes in eruption and earthquakes.

Distribution of Volcanoes

Nearly 800 volcanoes are active today or known to have been active in historical times. More than 75% are situated in the circum-Pacific belt, which is known as the Ring of Fire. This belt coincides with the young mountain ranges of western America, and the volcanic island arcs fringing the north and western sides of the Pacific. The Mediter-ranean-Asian orogenic belt has volcanoes distributed sparsely, except for Indonesia and the Mediterranean where they are more abundant. The oceanic volcanoes have already been described as a physiographic province, and the remaining active volcanoes are strung along the oceanic ridges and associated with the African rift valleys, which both belong to the world rift system.

J. H. Tatsch presented a three-dimensional least-squares analysis of the distribution of volcanoes in 1964. He found that approxi-mately 93% of the active volcanoes appear to lie along belts defined by the traces of three mutually orthogonal planes passing through the center of the Earth. The circle traces of these planes intersect each other at the

points: lat. 5°N, long. 95°E; 30S-175W; 55N-165W; 5S-85W; 30N-5E; and 55S-15E.

Distribution of Earthquakes

Earthquakes are of three kinds, shallow, intermediate, and deep. Shallow earthquakes are those occurring at depths above 70 km. If the focus is between 70 and 300 km the earthquake is intermediate. Deep-focus earth-quakes are those originating at depths between 300 and 700 km. The earthquake belts of the Earth are defined in Figure 2-5a by the epicenters of all earthquakes recorded during the period 1961–1967. The intermediate- and deep-focus earthquakes are restricted in their distribution, as shown by Figures 2-5b, and 14-3.

The earthquake zones follow closely the distribution of volcanoes, although not necessarily in detail. More than 80% of the world's shallow earthquakes occur in the circum-Pacific zone. The same zone experi-ences about 90% of intermediate shocks and nearly all of the deep shocks. Other earth-quakes occur along the Mediterranean-Asian mountain system and on the world rift system. The relation of earthquakes to conti-nental margins and ocean trenches is shown in Figure 14-3.

Examination of Figures 2-5 and 14-3 shows that the deeper earthquakes of the circum-Pacific belt are displaced toward the conti-nents compared to the shallow-focus earth-quakes. According to H. Benioff, in 1955, these earthquake epicenters lie within fracture zones about 250 km wide, dipping beneath the continent or island arcs, and extending to depths of 650 to 700 km. He illustrated the two types of deep-seated seismic zones in Figures 2-6a and 2-6b. The oceanic fault zones extend downward at 61° from the ocean floor beneath the island arcs, some distance from the continental borders. The marginal or continental fault zones are more complex. Figure 2-6 shows an example extending down-ward from an ocean trench bordering a conti-nental mountain range. A shallow component

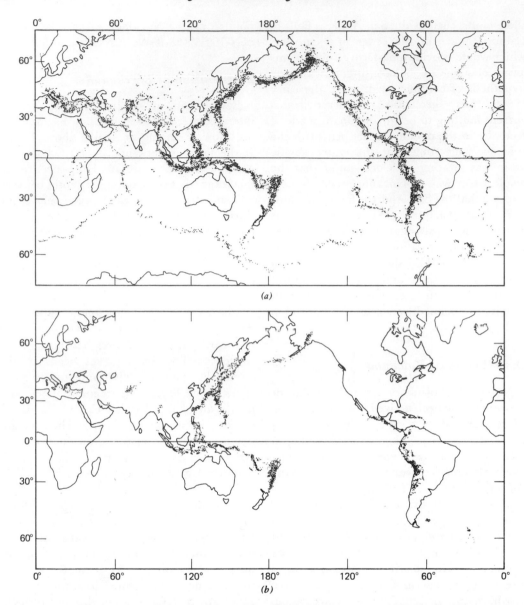

Figure 2-5. Seismicity of the Earth, 1961–1967. Plot of all earthquake epicenters recorded by U.S. Coast and geodetic Survey. (*a*) All epicenters. (*b*) Epicenters with depths between 100 and 700 km. (From M. Barazangi and J. Dorman, Bull. Seismol. Soc. Amer., **59**, 369, 1969, with permission.)

extends to a depth of 70 km, an intermediate component extends to about 300 km with an average dip of 32°, and a deep component extends from 300 to about 700 km with a dip of 60°. These seismic zones are now known as Benioff zones.

The use of digital computers in recent years

to redetermine the hypocenters of earthquakes in Benioff zones indicates that the earthquake foci in some island arc regions are confined to a zone only 50 to 100 km thick. It also appears that many deep earthquake zones extending under island arcs have dips of about 45°. Recent results show considerable

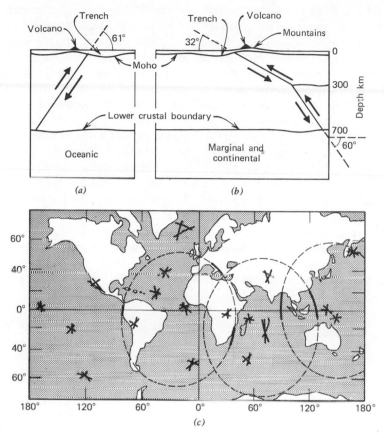

Figure 2-6. (*a*) and (*b*) Oceanic and marginal continental faults and crustal structure, showing deep fault zones marked by intermediate and deep-focus earthquakes (after Benioff, 1955, with permission of The Geological Society of America). (*c*) Three examples of the 21 belts that cover better than 90% of linear seismic zones longer than 700 km, according to Rouse and Bisque. These belts show the 19 multiple intersections plotted (after Rouse and Bisque, 1968, with permission of *The Mines Magazine*).

variation in the dips of Benioff zones from one location to another. L. R. Sykes concluded in 1966 that the shallow-, intermediate-, and deep-focus earthquakes associated with the Benioff zones dipping beneath island arcs are part of a continuous zone of tectonic activity. The earthquake foci under young folded regions are less deep than those under island arcs, and they are concentrated in relatively small areas. The Benioff zones constitute an important part of the New Global Tectonics reviewed in Chapter 14. They also form the basis of a global scheme presented by Rouse and Bisque in 1968.

Global Tectonics of G. E. Rouse and R. E. Bisque

There have been many attempts to explain the distribution of the active belts of the world within a comprehensive global tectonic system, and a recent example used Benioff zones as the basis for correlation. Rouse and Bisque noted that if a Benioff zone is extended through the Earth with an average dip of 60°, it turns out that it is in tangential contact with the surface of the core. This observation led them to examine the relationships of other tectonic features, including ocean ridges, island arcs, and major fault systems, to the circles described at the surface of the earth by planes tangential to the core-mantle interface.

They found that of the infinite number of planes possible 21 were sufficient to cover more than 90% of the known seismic belts. They defined the positions of 16 circle zones or belts by listing for each circle at least four major linear structural systems. A few circles are shown in Figure 2-6c. They reasoned that if the belts do have global structural significance then zones of multiple intersections might be of particular interest; they located 19 of them, as shown in Figure 2-6c. All but five of these do occur at zones of notable seismic or volcanic activity. Several of them occur on belts whose positions are defined by features in other quadrants of the globe, which suggests that the planes themselves do have global tectonic significance.

Rouse and Bisque proposed tentatively that the fluid motions in the core, which are invoked to explain the Earth's magnetic field, cause stress patterns in the mantle which are transmitted to the surface along planes tangential to the core-mantle interface. It was mentioned above, however, that the dip of many Benioff zones is now known to be 45° and not the 60° shown in Figure 2-6. Extrapolation of these zones through the Earth would provide no contact with the core-mantle interface. The significance of the multiple intersections plotted in Figure 2-6c therefore remains uncertain. What is certain, however, is that we must seek within the Earth for explanations of the active belts at the Earth's surface. Let us now turn to the physical features of the Earth's interior.

3. *Physical Properties of the Earth and its Interior*

INTRODUCTION

The Earth is a sphere in space, and physical properties of the sphere as a whole can be determined. The size, shape, and mass of the Earth have been measured with precision, and its magnetic and gravity fields and the geothermal flux from the sphere have also been measured. The Earth's interior is inaccessible, and we can only infer its chemical composition from indirect evidence (Chapter 6). However, the passage of earthquake waves through the Earth provides a kind of X-ray of its interior, and these waves bring to the surface information about the physical properties of material within the Earth. Interpretation of countless earthquake waves shows that the Earth has the concentric structure illustrated in Figure 3-1. The Earth's crust, ranging in thickness from about 5 km beneath the ocean to 35 km or more beneath the continents, is separated from the mantle by the Mohorovicic discontinuity (the Moho). Inside the mantle and separated from it by the Gutenberg-Wiechert discontinuity is the Earth's spherical core, with a radius of 3473 km.

This topic merits a whole book in itself. In this chapter we can only review broad outlines, with sufficient detail to demonstrate that interpretations have changed considerably within a decade. We should therefore be prepared to assume that further changes in interpretation will arise from new data gathered during the next decade.

SEISMOLOGY AND THE EARTH'S INTERIOR

The internal structure of the Earth depicted in Figure 3-1 is based on seismology, the study of earthquake waves. The use of underground nuclear explosions in the study of seismological problems is one of the outstanding advances of recent years. They have the advantage that both site and time of the event can be selected for the convenience of the observers.

Seismic Waves

An earthquake involves the sudden release of energy which is transmitted through the Earth in all directions, in the form of seismic waves. In order to track the paths followed by the waves and to determine their times of travel the precise location and time of the earthquake must be known. Because this information can only be derived from the seismic waves themselves, there is some uncertainty in the procedures. Underground nuclear explosions provide point sources of seismic energy with all ambiguities about the location of an earthquake focus removed.

Consider a nuclear explosion set off

17

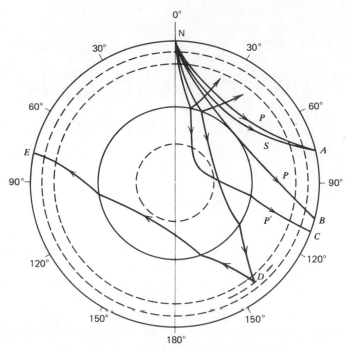

Figure 3-1. Cross section through the Earth, showing paths of seismic waves from earthquake epicenter, N.

precisely at the North Pole N in Figure 3-1. If the Earth were a homogeneous, elastic, isotropic sphere, elastic waves would spread in all directions through the Earth, following straight lines, until they reached the surface again. The velocity of the waves could easily be calculated from the time taken for a wave to reach any point on the surface.

We know, however, that the Earth is not homogeneous. Because the mean density of the Earth is 5.517 g/cm³, compared with densities of 2.6 to 3.0 for crustal rocks, part of the Earth's interior must be composed of material with density greater than 5.517 g/cm³. The moment of inertia of the Earth provides information about the mass distribution. For a sphere of uniform density of radius, r, and mass, m, rotating about an axis, the moment of inertia is $0.4\,mr^2$. The moment of inertia of the Earth, as determined from satellites in orbit, is $0.331\,MR^2$ (where M and R are the mass and the radius of the Earth), which indicates that mass is concentrated toward the

center of the Earth. Therefore the seismic waves are refracted as they pass deeper into the Earth, through material with different properties, and it has been established that they follow curved paths, or rays, like those shown in Figure 3-1. Where the property changes are sufficiently rapid, the waves may be reflected as well as refracted, and interpretation of the seismic records then reveals a surface of discontinuity at the depth where reflection occurred.

The release of energy at an earthquake focus or nuclear explosion produces several different seismic waves. There are body waves, which follow rays through the Earth like those shown in Figure 3-1, and surface waves, which travel around the outside of the Earth. The primary (P) wave is a compressional body wave, and it travels with about twice the speed of the secondary (S) wave, which is a transverse body wave. These waves are propagated because the Earth material is deformable, and their study thus depends upon elastic theory. For a perfectly elastic body, the velocities of P and S waves are given by:

$$V_P{}^2\rho = K + 4\mu/3 \qquad (3\text{-}1)$$

$$V_S{}^2\rho = \mu \qquad (3\text{-}2)$$

where ρ is density, K is bulk modulus or incompressibility, and μ is the modulus of rigidity. The consistency between these equations and seismic observations confirms that the theory of perfect elasticity is relevant to seismic wave propagation.

Surface waves (L waves) are strongly developed in the Earth. Rayleigh surface waves consist of motion in a plane perpendicular to the surface and parallel to the direction of transmission. Love surface waves are shear waves, with vibration perpendicular to direction of transmission and in a horizontal plane, which travel in a thin surface layer. The energy of surface waves is distributed through an appreciable depth into the Earth. Energy of long wavelength penetrates to greater depth than shorter wavelength energy, and the deeper energy travels faster. The variation of velocity with wavelength is called dispersion. Study of the dispersion of surface waves is yielding detailed information about the Earth's crust and upper mantle. Observations now extend out to the fundamental periods of free oscillations of the whole Earth.

Travel-time Curves

Consider again a nuclear explosion set off at the North Pole in Figure 3-1, at a precisely known time, and consider the transmission of seismic waves from this impulse to a series of points on the surface of the earth, whose positions lie on a great circle passing through the Pole. The position of each point is denoted by the angle subtended at the center of the Earth by the surface between the point and the energy source, as shown by the abscissa in Figure 3-2. In this example the angle is equivalent to degrees of latitude. At point A, denoted by the angle 75° subtended

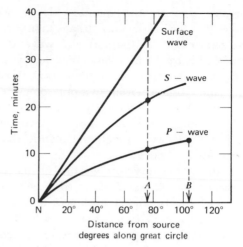

Figure 3-2. Travel-time curves for seismic waves emanating from epicenter N in Figure 3-1.

by the arc NA, energy is received first from the P wave, followed by the S wave; the waves follow the refracted rays shown passing through the mantle in Figure 3-1. At a later time surface waves which have travelled around the surface NA are received.

From the seismic records at the station A the time taken for each wave to reach A is determined and plotted on the travel-time diagram as shown in Figure 3-2. These points, along with similar sets of points for a series of recording stations between and south of NA, permit the construction of time-distance curves, or travel-time curves, for the different seismic waves. The graph obtained for the surface wave is a straight line, because the velocity does not change as the wave travels around the surface of the Earth. However, the graphs for the P and S waves are not linear; they have downward curvatures. This shows that with increasing distance of the recording station from the impulse source, and increasing length of the ray path within the Earth, the average velocities of the P and S waves increase. Figure 3-1 shows that this velocity increase is a function of depth of penetration of the rays into the Earth.

The ray NB in Figure 3-1 just penetrates to the core-mantle boundary and point B (103°) is thus a limit for seismic waves passing directly through the mantle. Figure 3-2 shows that for points more distant than *B* on the great circle *NB* the normal *P* waves disappear. Figure 3-2 also shows that the travel-time curve for the *S* wave terminates at about the same distance from *N*, but somewhat nearer to *N* because the *S* waves received at a given point such as *A* follow a somewhat deeper trajectory through the mantle than the *P* wave (Figure 3-1).

Many sets of travel-time tables giving the average times of travel along corresponding rays in terms of angular distances were prepared between 1920 and 1940, and by 1940 the errors for many distances were reduced from the order of a minute or more to only one or two seconds. In recent years major effort has been directed toward detecting differences in travel-time data for different environments.

Seismic Velocity Profiles and the Earth's Internal Structure

The velocities of seismic waves at depth can be calculated from the travel-time data, although the calculations are complex. The velocities of *P* and *S* waves have been determined with an accuracy of 2% at most depths. Figure 3-3 shows the velocities of *P* and *S* waves within the Earth, according to H. Jeffreys and B. Gutenberg. These curves agree closely, and they have been widely quoted for about 30 years. The principal uncertainties are in the velocity gradients at several depths in the Earth. Using Jeffreys' curves K. E. Bullen divided the Earth into a series of layers, with boundaries associated with levels where the velocities, or the velocity gradients, changed abruptly. These layers are designated *A*, *B*, *C*, etc., in Table 3-1 and Figure 3-3.

Figure 3-3 shows that at a depth of 2900 km, the *S*-wave velocity curve terminates, and there is an abrupt decrease in the velocity of the *P* wave. The maximum depth of penetration into the Earth for *P* waves following the ray *NB* in Figure 3-1 is 2900 km, and at

TABLE 3-1 First Approximation to the Earth's Internal Layering (Bullen, 1967)

Region	Range of depth (km)	Name	Characteristics of *P* and *S* velocities
A	0–33	Crust	Complicated
B	33–410	Upper mantle	Normal gradients
C	410–1000	Transition region	Greater than normal gradients
D'	1000–2700	Lower	Normal gradients
D''	2700–2900	Mantle	Gradients near zero
E	2900–4980	Outer core	Normal *P* gradient
F	4980–5120	Transition region	Negative *P* gradient
G	5120–6370	Inner core	Smaller than normal *P* gradient

distances between *N* and *B* a pulse is received at times corresponding to reflection from a surface at 2900 km depth. This surface is the boundary between the mantle and the core. At distances greater than *NB* (103°*C*), the *S* wave is not received at all, indicating that the core is composed of material that cannot maintain a shear stress. The core thus behaves like a liquid with respect to the transmission of elastic waves.

P waves are rarely observed at angles greater than 103°, and the area behind the core is thus called the shadow zone. *P* waves which reach the core are in part reflected and part refracted, as shown by the ray *NC* in Figure 3-1, and there is a focusing effect. The refracted *P* waves are slowed down. The material in the core has markedly different elastic properties than the material in the mantle. There are at least two layers in the core, the inner core probably being solid. Many more complex models for the core have been proposed since 1940.

Most of the discrepancies between the wave velocities of Jeffreys and Gutenberg are within the upper mantle and the transition zone between the upper and lower mantle (Figure 3-3), that is, in the depth interval of 40 to 1000 km. On the basis of Jeffreys' velocities,

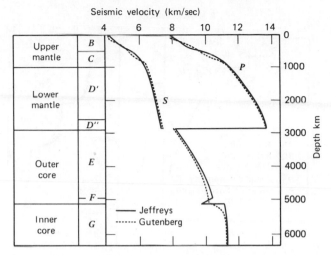

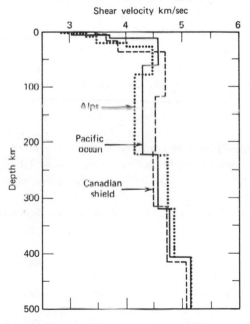

Figure 3-3. Velocities of *P*-waves and *S*-waves in the Earth according to Jeffreys and Gutenberg, showing Bullen's (1949) layers. See Bullen (1967) for review (with permission of Academic Press).

Bullen defined three subdivisions of the mantle, *B*, *C*, and *D*, with the latter subdivided into *D'* and *D"*. Neither of the boundaries *B-C* at 410 km and *C-D* at 1000 km appeared to be sharp, and several different depths have been assigned to them by various investigators. Gutenberg's velocity distributions in Figure 3-3 show a well defined low-velocity layer at 100–150 km depth, the existence of which he first proposed in 1926. Gutenberg has proposed that the *B-C* discontinuity is the outermost limit of the shadow zone caused by the low velocity layer.

The reality of a low-velocity zone is now established (Figures 3-4 and 3-5) although there is less certainty about its depth in relationship to tectonic environment. Surface wave studies have provided significant contributions. For the study of body waves computers have been a convenience but not a necessity. The study and interpretation of surface waves and free oscillations, on the other hand, could not have proceeded beyond the qualitative stage without the computer. Figure 3-4 compares layered shear velocity models depicting low-velocity zones in three different tectonic environments. These models were derived by analysis of Rayleigh wave dispersion.

The classic picture of the Earth's internal structure, shown in Figures 3-1 and 3-3 and Table 3-1, has been changed by the discovery of several discontinuities of first order in the mantle, and others which may be of first or second order. The depths to the discontinuities

Figure 3-4. Models for *S*-wave velocities in three different tectonic environments. (From J. Dorman, Geophysical Monograph **13**, 257, 1969, with permission of Amer. Geophys. Union.)

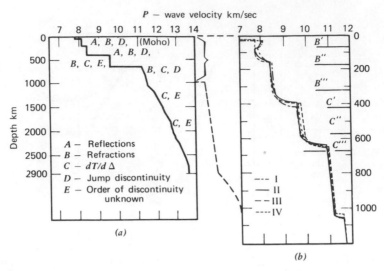

Figure 3-5. Models for *P*-wave velocity distributions in the mantle. (*a*) Schematic distribution after Knopoff (1967, with permission of Academic Press). The velocity gradients within the upper layers are not known and are shown as zero; there may be separate discontinuities at 640 and 800 km, (*b*) Models for four tectonic provinces of the United States. The most highly variable regions of the mantle are the *B* zones. Variations between the models in the *C* and *D* regions are not considered significant. The transition zones *C'* and *C'''* are well established, but the *D''* transition is questionable. I. East Basin and Range Province and North Rocky Mountains. II. Colorado Plateau and Rocky Mountains. III. Basin and Range Province. IV. Snake River plains and west Rocky Mountains. (From C. B. Archambeau, *et al.*, Jour. Geophys. Res., **74**, 5825, 1969, with permission.)

appear to vary in different parts of the world, and a schematic *P* wave velocity distribution is shown in Figure 3-5*a*. The discontinuities have been detected by various methods.

Figure 3-5*b* shows *P*-wave velocity profiles for the mantle within four different structural provinces of the western United States. In all provinces there are high-velocity gradients (sharp zones of transition) near 150, 400, 650, and possibly near 1000 km. The low-velocity zone is clearly defined beneath the Rocky Mountains and Colorado plateau. Beneath the Basin and Range province abnormally low velocities extend from near the base of a thin crust to depths of 150 km, but there is no lid with high velocities capping the zone. These detailed studies show that there is considerable variation in the properties of the upper mantle down to at least 150 or 200 km from one province to another and even within each province.

An example of the sensitivity of modern

seismic techniques is illustrated by the rays *NC* and *NDE* in Figure 3-1. The ray *NC* passes through the core, and we have already referred to the fact that part of the ray is reflected at the core-mantle boundary and recorded at a seismic station between *N* and *A*. The line at *D* represents the 650 km discontinuity shown in Figure 3-5*b*. After diffraction through the core, a portion of the ray *ND* is reflected from the underside of the discontinuity *D*, and this passes back through the core along the ray *DE*. Pulses received at *E* as a result of the seismic waves emanating from *N* can be detected and interpreted in terms of reflection from *D*.

Figure 3-6 shows the results of a detailed analysis of mantle structure by R. Z. Tarakanov and N. V. Leviy. They proposed that the upper mantle has a complicated layered structure like that of the crust, and they distinguished four layers of low velocity and low strength at depths of 60–90 km, 120–

160 km, 220–300 km, and 370–430 km. The model correlates the change with depth of magnitudes of earthquakes, velocities and amplitudes of P waves, and the derivative of an empirical travel-time curve.

Figure 3-6a shows the distribution of the maximum earthquake shock intensities plotted against the depth of the focus, for major earthquakes recorded throughout the world during 1896–1962. The maximum earthquake magnitudes fluctuate within the shaded band. There are significant minima in this profile, which are interpreted in terms of layers of reduced strength at these depths.

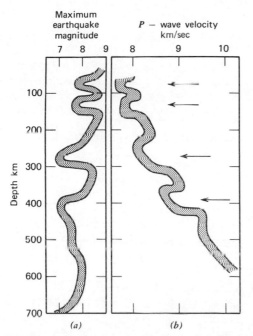

Figure 3-6. (a) Change of the maximum magnitude of earthquakes with depth. (b) The distribution of P-wave velocities with depth. Arrows show four low strength and low-velocity layers (after R. Z. Tarakanov and N. V. Leviy, Geophysical Monograph **12**, 43, 1968, with permission of Amer. Geophys. Union).

Empirical curves were constructed from data collected at more than 100 observation stations in the transitional zone from the Asiatic continent to the Pacific Ocean, across

the Kuril and Japanese Islands. Using a method devised by Gutenberg, they determined the P-wave velocity at the focal depths of earthquakes in the depth range from 30 to 680 km. The velocity profile so obtained is shown in Figure 3-6b, with the uncertainty in velocity determinations being shown by the shaded band. The four zones of reduced velocity are well correlated with the depths of weak layers shown in Figure 3-6a by the minima in the earthquake magnitude distribution.

Lateral Variations in the Upper Mantle

Figure 3-5b shows that there are lateral variations in the upper mantle. Figures 3-7 and 3-8 illustrate the pattern of these variations on a continental and a global scale.

The velocity of the P wave just below the Moho discontinuity, P_n, was generally taken to be near 8.1 km/sec until about 1960. By then it was discovered that the crust and upper mantle of the western United States is different from that of the eastern United States, and values of P_n were measured between 7.7 and 8.4 km/sec. Low values for P_n are also found beneath island arcs and midoceanic ridges, with the higher velocities under deep ocean basins (see Chapter 5). Figure 3-7a shows the estimated P_n velocity for the United States, based on data from deep seismic soundings, underground nuclear explosions, and earthquakes; P_n velocities are lower beneath the mountainous west than beneath the plains of the east.

The correlation of topography with P_n on a regional scale is matched approximately by crustal thickness, and by mean crustal seismic velocity. Figure 3-7b shows the variations in crustal thickness, with the depth to the Moho contoured at 10 km intervals, compared with the regions having P_n greater than, or less than 8 km/sec. The mean crustal P-wave velocity is also illustrated. The Rocky Mountain system appears to divide the United

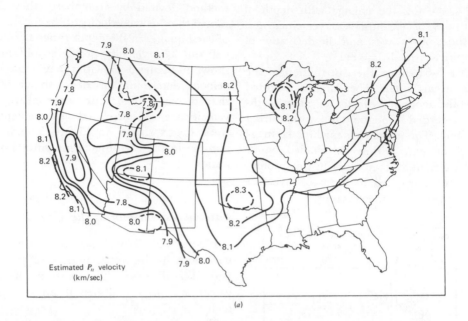

(a)

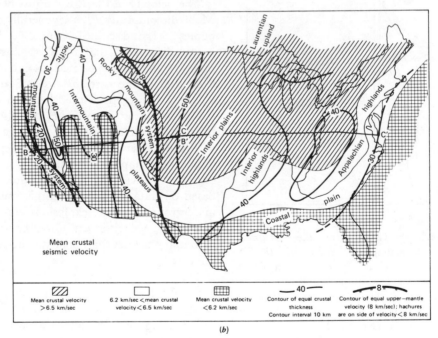

(b)

Figure 3-7. (*a*) Estimated P_n velocity for upper mantle in the United States, based on data from deep seismic soundings, underground nuclear explosions, and earthquakes (after E. Herrin, Geophysical Monograph **13**, 242, 1969, with permission of Amer. Geophys. Union). (*b*) Variations in crustal thickness, mean crustal velocity, and upper mantle velocity in the United States, showing locations of aeromagnetic profiles, *B-B′* and *C-C′* of Figure 7-2*a*. (From L. C. Pakiser and I. Zietz, Rev. Geophys., **3**, 505, 1965, with permission.)

States into two superprovinces involving both crust and upper mantle

1. In the eastern superprovince with lower elevation, P_n is greater than 8 km/sec, the mean crustal velocity is generally greater than 6.4 km/sec, and the crust is generally thicker than 40 km.

2. In the western mountainous superprovince P_n is less than 8 km/sec, the mean crustal velocity is generally less than 6.4 km/sec, and the crust is generally thinner than 40 km.

These correlations indicate that crust and upper mantle are closely linked in a tectonic sense, and that both have to be considered in isostasy and gravity calculations.

S_n is a short-period seismic shear wave that propagates in the uppermost mantle and does not penetrate the low-velocity zone. The efficiency of transmission of S_n on a worldwide scale is summarized in Figure 3-8. This shows that S_n propagates very efficiently across the stable continental shields and ocean basins. The areas for inefficient propagation include the oceanic ridges and rises, some orogenic belts, and the concave side of island arcs. Paths crossing the concave side of island arcs or the crests of oceanic ridges do not transmit S_n. The rigid outer shell of the upper mantle, part of the lithosphere, is assumed to correlate with low attenuation of S_n (or high Q), and the results then imply that the uppermost mantle is weaker beneath the island arcs and oceanic ridges than elsewhere. These locations can be interpreted as discontinuities in the lithosphere.

There is considerable current interest in the hypothesis of mantle convection (Chapters 12 and 14). If flow has occurred in the upper

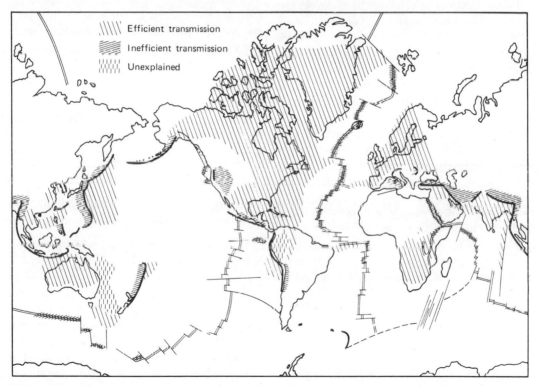

Figure 3-8. Summary of world regions where S_n waves propagate inefficiently in the upper mantle. Island-arc structures are represented by bold dark lines; crests of midocean ridges by double lines; and fracture zones by single lines. (From P. Molnar and J. Oliver, Jour. Geophys. Res., **74**, 2648, 1969, with permission.)

mantle this could have produced anisotropy in a horizontal plane. Recent seismic velocity measurements have therefore been directed toward identification of anisotropy in the uppermost mantle beneath the oceans. The seismic results were reviewed by H. G. Ave'Lallemant and N. L. Carter (1970) in their discussion of modes of flow in the upper mantle, based on experimental syntectonic recrystallization of olivine. The metamorphic texture and fabrics produced in the recrystallized grains are similar to those occurring in some peridotites and dunites believed to be mantle-derived tectonites; these include nodules in kimberlites and basalts, and orogenic peridotites (see Chapter 6).

PROPERTIES OF THE EARTH'S INTERIOR

The study of seismic waves yields the distribution of seismic wave velocities within the Earth, and several properties of the Earth's interior can be calculated from these profiles, the average density of the Earth, and its moment of inertia. The distributions of ρ, μ, k, g, and pressure, p, within the Earth are closely interlocked, and Equations 3-1 and 3-2 show that the velocities of body waves are expressed in terms of the ratios of density and elastic properties. Most investigators have concentrated on deducing a consistent density distribution first and deriving the other property distributions from this.

The temperature distribution within the Earth is not well known, and no unique solution is possible for the geothermal gradient at depth. Knowledge of the thermal state of the Earth and its variation with time, however, is of fundamental importance for geophysical and geological problems, and there have been many estimates for the positions of geotherms in various tectonic environments.

Other properties of the Earth's interior which have been estimated as a function of depth include viscosity and electrical conductivity.

Density Distribution

The general method for deducing the density distribution involves many assumptions. It requires that the Earth be treated as a series of concentric layers, each assumed to have homogeneous composition with no abrupt changes in physical properties. If m is the mass of material within a sphere of radius r, and G is the universal constant of gravitation, then the value of g at distance r from the Earth's center is

$$g = \frac{Gm}{r^2} \qquad (3\text{-}3)$$

and since the stress in the earth's interior is essentially equivalent to hydrostatic pressure, the variation of p with depth is given by

$$\frac{dp}{dr} = g\rho = -\frac{Gm\rho}{r^2} \qquad (3\text{-}4)$$

The incompressibility is linked with the density and the pressure by

$$k = \rho\left(\frac{dp}{d\rho}\right) \qquad (3\text{-}5)$$

These equations, together with 3-1 and 3-2, give the Williamson-Adams equation:

$$\frac{d\rho}{dr} = -\frac{Gm\rho}{r^2}\left(V_p^2 - \frac{4V_s^2}{3}\right) \qquad (3\text{-}6)$$

This cannot be integrated directly, but it can be applied to each of the assumed homogeneous layers, giving the rate of change of density at a particular depth. If a value for the density of the upper mantle is assigned to material beneath the crust, its density deeper in the layer can be calculated. A new value for $d\rho/dr$ is calculated for the new density, and the calculation is repeated, layer by layer, until a discontinuity in the wave velocity

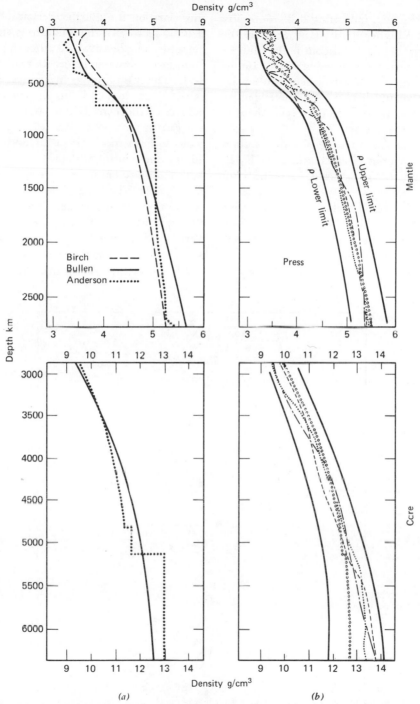

Figure 3-9. Density distributions in the mantle. (*a*) Recently proposed density models (1961, 1965) compared with the standard Bullen $A(i)$ model, as compared by Anderson (1967, with permission of Academic Press). (*b*) Three plausible, random density distributions obtained by the Monte Carlo procedure, compared with the Bullen model (after Press, 1968, with permission from Science. Copyright 1968 by the American Association for the Advancement of Science). Bullen curve, open circles.

profile is reached, indicating a discontinuity in density. A higher density is assumed for the material just below the discontinuity, and the successive calculations are continued. Limits are placed on the density discontinuities by the requirement that the complete density distribution must give the correct moment of inertia and average density.

Figure 3-9*a* shows the density distribution derived in a series of calculations by K. E. Bullen since 1936, assuming a density of 3.32 g/cm³ for the material immediately below the Moho (Figure 3-10). This is representative of all models based on Equation 3-6. The results shown for the core represent only one possible hypothesis, and there is still no generally accepted density distribution for this part of the Earth. Perhaps the most striking feature of the diagram is the abrupt increase in density at the mantle-core boundary, between layers *D* and *E* (Table 3-1). The assumptions necessary for these calculations depend upon preconceived models of the composition and structure of the Earth. It would obviously be advantageous to have more direct methods of determining the density at depth.

Two other density models are shown in Figure 3-9*a*. That of F. Birch is based on an assumption of linear relationship between density and the *P*-wave velocity. D. L. Anderson has shown how long-period surface wave dispersion and free oscillations of the Earth can yield densities in a direct manner,

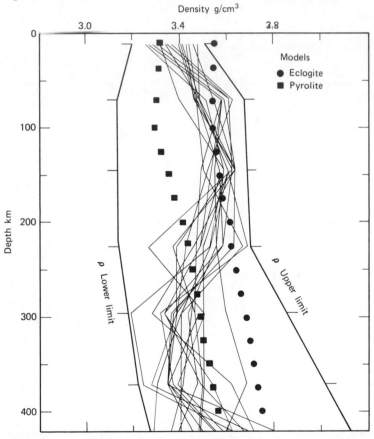

Figure 3-10. Successful density models for suboceanic upper mantle using Monte Carlo method; bounds define range permitted in selection. Points show density values according to Clark and Ringwood (1964) for "pyrolite" and eclogite models (after Press, 1969, with permission from Science. Copyright 1969 by the American Association for the Advancement of Science).

if the seismic wave velocities are completely and accurately known from body wave studies. One of his models is shown in Figure 3-9a. Note the low-density zones and the two regions of very rapid increase in density, compared with the seismic velocity profiles in Figure 3-5.

A completely different approach was adopted by F. Press, involving the Monte Carlo statistical procedure, which is quite independent of any of the usual assumptions. Approximately five million Earth models were randomly generated by feeding into the computer values for Earth parameters including P-wave velocity, S-wave velocity, density distribution in the mantle (which is limited between upper and lower bounds), core density, and core radius (limited within the range 3473 $\pm$ 25 km). These models were tested against available geophysical data including the mass of the Earth, its dimensionless moment of inertia, travel times of seismic waves, and Earth eigenperiods for free oscillation modes. Only six of the five million models satisfied the known geophysical data, and only three of these were plausible.

Every successful model required an increase in the radius of the Earth's core by 18 to 22 km. Figure 3-9b compares Bullen's standard density distribution with the three random distributions found to be plausible by Press's method. Inner core densities are significantly higher than in the standard model. In the mantle deeper than 1000 km the successful solutions have a surprisingly narrow spread of densities within the permitted range. Within the transition zone of the mantle, between about 400 and 1000 km depth, the density distributions exhibit higher density gradients than the standard model. Within the upper mantle the successful models show large variability, with surprisingly large density fluctuations, and with one or two density minima. The range of acceptable upper mantle densities, 3.34 to 3.54 g/cm³, corresponds to the range covered by peridotite to eclogite. Applying these randomly generated density variations to existing concepts of

mantle composition leads to the hypothesis that the upper mantle is laterally and radially variable, with extensive zones, or layers, of eclogite within a peridotite mantle. The results also have implications for the composition of the core and mantle, which will be considered in Chapter 6.

A similar study gave the first independent determination of density in the suboceanic upper mantle, as shown in Figure 3-10. The successful models show that in the vicinity of 100 km the densities occupy a narrow band between 3.5 and 3.6 g/cm³, in the upper part of the permissible range. In the depth range 250 to 400 km the models have densities reduced to the range 3.3 to 3.5 g/cm³. The figure also shows densities computed by S. P. Clark and A. E. Ringwood for petrological models of mantle composed of either eclogite, or of a hypothetical peridotite, termed "pyrolite" (Chapter 6). The eclogite model alone is consistent with the results between 80 and 150 km, and the pyrolite model is favored in the region near 300 km. Press found a low velocity zone for shear waves in all successful models, with center between 150 and 250 km, beneath a lid about 100 km thick. He concluded that the mantle between 80 and 150 km, including the lower part of the lithosphere, consists of peridotite with about 50% eclogite.

Elastic Properties, Pressure, and Gravity

From the seismic wave velocities (Figures 3-3 and 3-5) and the density distribution (Figure 3-9), all of the elastic constants may be calculated. Their variation with depth is shown in Figure 3-11a. From the density distribution, the mass m of material within a sphere of radius r can be calculated, and this permits determination of the acceleration due to gravity at any level within the Earth from Equation 3-3, and it also gives the pressure distribution within the Earth, as in Equation 3-4. The pressure distribution is compared with the elastic constants in Figure 3-11a and with the gravity distribution in Figure 3-11*b*. Pressure increases continuously with depth. Gravity tends to increase below the surface of the Earth, but there is little change to a depth of 2400 km. It increases slowly from here to the mantle-core boundary, and within the core it decreases steadily to zero at the Earth's center. The figure also shows one estimate of the temperature distribution within the Earth, and therefore provides values for pressure and temperature at any selected depth.

The results shown in Figures 3-3, 3-5, 3-9, 3-10, and 3-11, depicting the variation with depth of seismic wave velocities, density, elastic properties, gravity, and pressure, include all that is known with any degree of certainty about the physical properties of the Earth's deep interior. No hypothesis concerning the mantle and core can be considered acceptable unless it is consistent with these facts.

Temperatures within the Earth

There is great uncertainty about the temperature distribution within the Earth, as shown by the selected estimates of geotherms for the outer mantle in Figure 3-12. The present thermal state of the Earth depends upon its thermal history. Hypotheses for the thermal history depend critically upon assumptions about the origin and early history of the Earth, the chemistry of material composing the Earth, and the physical properties of this material. Estimated geotherms therefore depend on these same assumptions.

Limits are placed on the temperature distribution by several factors. Surface heat flow measurements relate to the thermal gradients only within the upper few tens of kilometers. Inferences about temperature distributions at greater depths are based on observed seismic velocity profiles and variations in electrical conductivity, but these depend on assumptions about the physical properties of postulated mantle material at high pressures and temperatures. Seismic data show that the mantle is essentially crystalline and that the outer core is liquid. The temperature within the mantle is therefore below the solidus curve for mantle

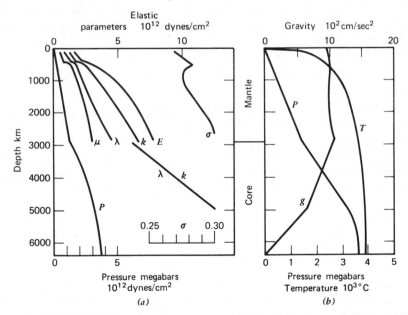

Figure 3-11. Properties of the Earth's interior. (*a*) Variation of pressure and elastic properties with depth. (*b*) Variation of pressure, gravity, and temperature with depth. (Various sources; summaries in Howell, 1969, and Jacobs *et al.*, 1959.)

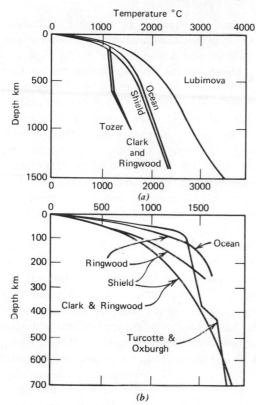

Figure 3-12. Estimates of temperature distribution with depth in the Earth.

material, but the fact that volcanoes exist and have erupted at frequent intervals and in most locations through the long span of geological time shows that the solidus temperature is locally exceeded. Therefore the geotherms probably remain close to the solidus for the upper mantle at least.

The present temperature distribution within the Earth depends upon its temperature when formed, the amount of heat generated as a function of depth and time, and the rate of outward flow of heat. It seems probable that significant cooling from the Earth's surface has not occurred much below 100 km. The Earth's initial temperature is not known, but current hypotheses favor accretion of a cold body (Chapter 6). Sources of internal heat include:

1. Gravitational energy converted to heat during formation of the Earth's core; heat

from gravitational energy during accretion of the Earth was probably generated near the surface and radiated away into space.

2. Energy of friction arising from Earth tides, dissipated as thermal energy probably in zones of reduced strength, such as the low-velocity zone.

3. Decay of radioactive elements; this is the principal contribution to the Earth's thermal budget and the present heat flow.

Processes for the outward flow of heat from within the Earth include (a) conduction, (b) radiation, and (c) convection. The transfer of heat by thermal conduction is a very slow process in silicate rocks, but it is the main process in the outer layers of the Earth. At higher temperatures, possibly at depths of 150 km and more, radiation with exciton and photon transfer becomes a dominant mode of heat transfer. This could easily increase the effective conductivity by a factor of 10, but the magnitude of the effect is largely conjectural. Convection, if it occurs in the mantle, is an efficient mechanism for transferring deep-seated heat to the surface, and it provides a large increase in effective thermal conductivity. Heat is also transported directly to the surface by magmas and hydrothermal solutions, but this appears to represent a small fraction of the thermal budget.

Four different approaches have been used to estimate the temperature distribution in the Earth:

1. Methods based on deduction of the thermal history of the Earth.

2. Methods based on variations in physical quantities such as seismic wave velocities and electrical conductivity.

3. Methods based on geochemical and petrological data and inferences.

4. Methods based on convection models.

Figure 3-12 shows the range of temperature profiles deduced in recent years by some of these approaches.

The geotherm calculated by E. A. Lubimova in Figure 3-12a is based on conduction and the Earth's thermal history. Figure 3-11b shows the temperature distribution extended down to the Earth's center.

S. P. Clark and A. E. Ringwood developed a petrological model involving the formation of continents by vertical differentiation of the upper mantle. Because of the approximate equality of heat flow from oceans and continents, the temperature beneath the oceans is higher, at a given depth, than that beneath the continents, as shown in Figure 3-12*a*. The two geotherms must converge at depth. For consistency the model requires a very high thermal conductivity at high temperatures, and they concluded that this is produced by radiative transfer. Clark and Ringwood presented curves to 400 km and D. C. Tozer extended them to 1400 km. At 1000 km depth these curves give temperatures about 800°C lower than the conduction geotherm of Lubimova. Ringwood later modified these conduction-radiation geotherms by changing the contribution of radiative transfer. This had the effect of raising the geotherms by about 250°C at 250 km depth, as shown in Figure 3-12*b*. These are the geotherms used in the following chapters.

The effect of convection superimposed on the conduction-radiation model of Clark and Ringwood was examined by D. C. Tozer, with the results shown in Figure 3-12*a*. He first extended the geotherms from 400 to 1400 km, and concluded that for these temperature distributions there was a viscosity increase by a factor of 10^4 at a depth of 600 km, and that the mantle was convectively unstable above this depth. He therefore proceeded on the assumption that the lower boundary of a convecting layer occurred at 600 km; the upper boundary was taken as 50 km beneath the oceans and 150 km beneath the continental shields. The effect of convection is to lower the temperature of the bottom of the layer by about 600°C, according to Figure 3-12*a*. Below 600 km he calculated a new conduction equation solution producing the change in geothermal gradient shown in the figure. He found that the temperature distribution in the convecting region was very

insensitive to the heat source distribution, which is why the geotherms for oceanic and shield regions are so similar.

The temperature distribution for another convection model, in the oceanic regions, is given in Figure 3-12*b*. D. L. Turcotte and R. Oxburgh considered a model for mantle convection using boundary-layer theory, with rates of flow estimated from magnetic anomaly data (Chapter 13). The temperature profile through the cold conduction boundary layer above the convection cell, at about 100 km from an oceanic ridge, was estimated by comparing measured values of heat flow as a function of distance from midoceanic ridges with theoretical models calculated using assumed mantle properties. This extends down to about 50 km. Below 150 km they concluded that the thermal gradient would be close to the adiabatic value, and they adopted an earlier proposal of B. Gutenberg that the geotherm in the upper mantle would follow a 1400°C adiabatic gradient. They matched the estimated boundary layer profile to a 1400°C adiabatic profile through the zone of main horizontal flow and into the slow moving core of the convection cell. The rather abrupt increase in temperature at about 400 km depth is due to the increase in adiabatic gradient through a phase transition zone (Chapter 6). At 350 km depth the estimated temperature is about 350°C higher than Tozer's convection model, and very close to the conduction-radiation model of Clark and Ringwood.

Viscosity

Estimates of the viscosity of the upper mantle have been based on the rates of uplift of large crustal regions, such as Fennoscandia after removal of its ice sheet. Values lie in the range 10^{21} to 10^{22} poises, increasing to about 2×10^{22} at a depth of about 1000 km. Two different interpretations for the cause of the nonequilibrium figure of the Earth give estimates for the average viscosity of the mantle as either 10^{26} poises, or 5×10^{22}

poises. Seismic anelasticity measurements of the mantle give estimates for the average viscosity of 2.4×10^{22} poises, with 8×10^{21} poises for the upper mantle and 10^{23} poises for the lower mantle.

There has been considerable discussion in the literature about whether the mantle behaves as a plastic body with a yield strength which would involve a power law or exponential dependence of strain rate on stress or whether it behaves as a viscous fluid with linear or nonlinear (Newtonian) relationships between stress and strain. Diffusion creep is likely to be a significant factor for deformation in the Earth's mantle at high temperatures, and assuming that diffusion creep alone occurs, Turcotte and Oxburgh used their temperature profile for the mantle (Figure 3-12b) to estimate the dependence of viscosity with depth. They found a distinct minimum in viscosity of 1.09×10^{21} poises near 100 km depth. A second minimum of

3.10×10^{21} poises at 425 km is due to the temperature increase associated with the phase transition. Other values tabulated include: 49 km, 3.9×10^{22} poises; 300 km, 7.75×10^{21} poises; 700 km, 1.18×10^{23} poises.

In 1970 N. L. Carter and H. G. Ave'Lallemant experimentally deformed dunite and peridotite at mantle pressures and temperatures and at constant strain rates. They found that the steady-state deformation is best fit by a power-creep equation, and extrapolation gives estimates of 10^{20} to 10^{21} poises for the viscosity over most of the upper mantle.

Uncertainties in estimates of mantle viscosity are confirmed by the range of estimates listed above. There is increasing evidence, however, for the existence of a low viscosity layer at depths corresponding to the zone of low seismic velocity in the upper mantle, and many estimates for this layer are about 10^{21} poises.

MAGNETISM, GRAVITY, AND HEAT FLOW

We have discussed the physical properties of different parts of the Earth's interior, and now we can review some physical properties of the Earth as a whole, which are measured at the surface. There is a magnetic field associated with the Earth which approximates to that of a dipole magnet with an axis slightly offset compared to the rotation axis. This is illustrated in Figure 3-13a, which shows also the distribution of the magnetic lines of force near the Earth's surface. The study of paleomagnetism and anomalies in the present magnetic field has contributed greatly to current concepts of global geology. The Earth has an external gravity field which arises from the distribution of mass within the Earth. Gravity variations or anomalies therefore provide information about the density in different parts of the Earth. Most major geophysical and geological processes are controlled by, or at least influenced by, the generation, distribution, and transfer of heat

within the Earth. Measurement of surface heat flow thus provides a boundary condition of fundamental significance.

Measurements of the magnetic field, the gravity field, the heat flow, and similar compilations for the travel times of seismic waves, crustal thickness, and the surface topography of the Earth, can be expanded in terms of spherical harmonics. Then the correlation coefficients between the different sets of data can be compared with their harmonic coefficients. This provides the means for investigation of the structure and specifically the inhomogeneities in the Earth's mantle, as reviewed in 1969 by M. F. Toksöz, J. Arkani-Hamed, and C. A. Knight.

The Magnetic Field

At any point on the surface the Earth's magnetic field is defined by its strength and direction, which are usually expressed in

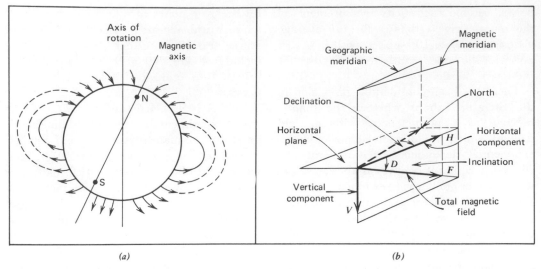

(a)   (b)

Figure 3-13. (a) Earth's magnetic field relative to axis of rotation and magnetic axis. (b) Horizontal and vertical components of the Earth's total magnetic field, resolved in the vertical plane through the magnetic meridian.

terms of the magnetic elements shown in Figure 3-13b. The total magnetic field is represented by a vector F making an angle of dip D (the inclination) with the horizontal. The vector is resolved into a horizontal component H and a vertical component V. The local magnetic meridian is the vertical plane containing F and H. The angle measured from the geographic meridian to the magnetic meridian is the declination or variation.

Isomagnetic maps show contours connecting points on the Earth's surface with equal values of the magnetic elements. Isogonic maps show declinations, isoclinic maps show inclinations, and isodynamic maps show magnetic intensity, either the total field or a resolved component. Figure 3-14a is an isoclinic map. The zero isocline, called the magnetic equator, is a curve quite close to the geographic equator. The angle of dip becomes vertical at two magnetic dip poles, presently located in northern Canada (73°N, 100°W) and Antarctica (68°S, 143°E); their positions change with time. Each dip pole is about 2300 km from the antipodal point of the other, and therefore a line joining the poles does not pass through the center of the Earth. The

magnetic field is strongest near the dip poles (0.7 oersted in the south) and weakest near the magnetic equator (about 0.3 oersted). The magnetic dip poles are not coincident with the geomagnetic poles which are defined below.

Magnetic anomalies. The irregularity of contours in isomagnetic maps shows that the concept of a single dipole (Figure 3-13a) is too simple. The observed magnetic field varies from place to place owing to the permeability and magnetization of rocks beneath the surface and to the ionosphere. The difference between the simple dipole field and the observed field is called the nondipole field or the magnetic anomaly. Anomalies are measured in gammas (γ), where $1\gamma = 10^{-5}$ oersted.

Small-scale anomalies are caused by near-surface features, and mapping local magnetic anomalies is a useful prospecting technique for ore bodies. In addition to these local anomalies there are large-scale anomalies extending over thousands of square kilometers. The anomaly of the vertical intensity V (Figure 3-13b) is shown in the isodynamic map of figure 3-14*b*. These anomalies must be caused by the properties of materials deep within the Earth.

The regular field which best approximates the Earth's field is obtained by spherical harmonic analysis of the observations for any year from all parts of the Earth, which resolves the field into components from internal (94%) and external sources. The potential of the total magnetic field is expanded in spherical harmonics. This leads to representation of the observed field by a number of magnetic dipoles, each with a different orientation, distributed at the center of the Earth as required by the coefficients of the harmonic analysis. The first and most powerful, which accounts for the regular dipole part of the Earth's field, is called the

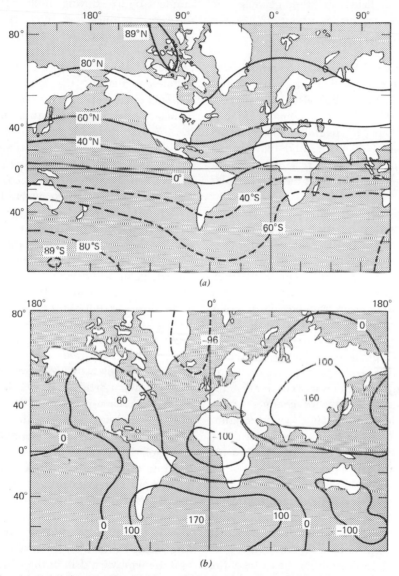

(a)

(b)

Figure 3-14. (*a*) Isoclinic map showing lines of equal magnetic inclination in 1945 (from Vestine *et al.*, 1947, courtesy of Carnegie Institution). (*b*) Map showing lines of equal anomalies of vertical magnetic intensities (*V* in Figure 3-13*b*) in 10^{-3} gauss (after Bullard *et al.*, 1950, with permission of author).

inclined geocentric dipole. The geomagnetic axis of this dipole makes an angle of 11.5° with the rotation axis and intersects the surface at the geomagnetic poles. The positions of these poles, unlike those of the dip poles, have changed little in the past century: 79°N, 70°W, and 79°S, 110°E. Latitude relative to this axis is called the geomagnetic latitude. The higher harmonics constitute the irregular or nondipole field, or the geomagnetic anomaly. Its strength is about 5% of the dipole field.

Variations with time. The intensity and direction of the magnetic field not only varies from place to place, but also from time to time. Direct records of the geomagnetic field have been kept at magnetic observatories during the last 400 years, and two distinct types of change have been recognized: short-term fluctuations and long-term or secular variations.

Short-term periodic variations with durations of hours, days, or years can be explained by electric currents in the mantle and ionosphere; intensity variations are of the order of 10 to 100γ. Irregular, transient fluctuations which may exceed $10^3\gamma$, called magnetic storms, probably result from bombardment of the ionosphere by solar radiation because they usually occur at times of sunspot activity.

Secular changes include changes in declination, inclination, and intensity. Isoporic maps show the distribution of isopors or lines joining places with equal rates of changes of the magnetic elements during a given time interval. Secular variations in intensity are in the range of several to tens of gammas per year. There appear to be cyclic changes in the direction of the field with periods of several hundreds of years, which produce a range of field directions of 10 to 20° on either side of the average. The magnitude of the change varies from one magnetic observatory to another, which indicates that the cause of this secular variation is regional rather than planetary. In addition to these directional changes of the geomagnetic field, the patterns illustrated by isomagnetic maps drift westwards. This effect is produced because the nondipole field drifts westward over the regular dipole field at a rate of 0.18° per year. If this rate remains constant, the map of intensity anomalies shown in Figure 3-14b would travel right around the world in 2000 years. Another type of secular variation is shown by the decrease of the Earth's magnetic moment by 5% during the last 100 years. If this rate continues, then the Earth's magnetic field would vanish in about 2000 years. The evidence indicates, however, that the field is probably oscillating rather than reversing.

Paleomagnetism, the study of fossil or remanent magnetism in rocks, provides the means of extending observations farther back in time than the 400 years available in direct records from observatories. Paleomagnetic studies show that the Earth's magnetic field has vanished often in the geological time scale, on each occasion when the Earth's polarity underwent a complete reversal. Changes in the attitude of the geomagnetic axis, loosely termed polar wandering, have also been recognized from the paleomagnetic record. Polar wandering and the correlation of magnetic reversals with linear magnetic anomalies parallel to the oceanic rifts are largely responsible for the revival of the concept of continental drift. We shall return to these topics in Chapters 12 and 13.

The only plausible explanation for the origin of the Earth's magnetic field involves electric currents in the fluid core, which are probably induced and maintained by some magneto-hydrodynamic phenomenon. The speed of the secular variations supports this dynamo theory. It is hardly conceivable that such rapid changes could occur in the solid mantle, but the times appear reasonable for convective motions in a fluid core. The secular variations may be considered as resulting from large-scale eddies near the surface of the core, and the westward drift of the whole magnetic pattern could possibly be explained by variable coupling between the core and the mantle. The problem is extremely complex.

The Gravity Field

The computed variation of the acceleration due to gravity within the Earth was shown in Figure 3-11b, and in this section we will consider variations in gravity as measured at the surface. The major component of gravity is the attraction between the Earth and a body at the surface and this is dominated by the whole Earth. There are, however many effects modifying the measured values. If the Earth were a perfect sphere whose density varied only as a function of distance from its center, gravity at its surface would be uniform, as given by Equation 3-3. The force of gravity would be directed toward the center of the Earth and perpendicular to its surface, which is therefore an equipotential surface—one that is everywhere horizontal. The gravity potential at the surface would be $-GM/R$, where M and R are, respectively, the mass and radius of the Earth. For this hypothetical Earth the geoid would be the spherical equipotential surface.

Because of the Earth's daily rotation, however, there is a centrifugal force accelerating the matter of the Earth in opposition to gravity, and this force is greater nearer the equator. This accounts for the equatorial bulge and the flattening of the poles. The figure of the Earth is approximated by a spheroid or an ellipsoid, and the sea level value of gravity varies from 978.049 cm/sec² (gals) at the equator to 983.221 cm/sec² at the poles. The theoretical value of gravity as a function of latitude (θ) on the ellipsoid is given by an International Gravity Formula:

$$g = 978.049(1 + 5.2884 \times 10^{-3} \sin^2 \theta - 5.9$$
$$\times 10^{-6} \sin^2 2\theta) \text{ gal} \quad (3\text{-}7)$$

Satellite studies have provided more precise values than surface gravity surveys, and following is the revised geodetic standard:

$$g = (978.03090 + 5.18552 \sin^2 \theta$$
$$- 0.00570 \sin^2 2\theta) \text{ gal} \quad (3\text{-}8)$$

Variations or anomalies in gravity measurements are measured in milligals (1 mgal = 0.001 gal) or the gravity unit: 1 g.u. = 0.1 mgal.

The surface of the oceans is almost an equipotential surface, with minor departures arising from the attraction of the water toward the continents. The geoid is the equipotential surface which most nearly approximates mean sea level, and this is the reference level for all gravity measurements. Figure 3-15 shows that the geoid is slightly distorted compared to the equilibrium ellipsoid.

The value of gravity itself is not constant on the equipotential geoid, because it varies with latitude according to Equation 3-8. Data for the equilibrium figure of the Earth have been obtained from geodetic measurements, and location of the geoid by gravity measurements provides another method.

Gravity corrections and anomalies.

The results of gravity measurements vary not only with position on the geoid, but also because the surface of the Earth is not everywhere at the same altitude. Equation 3-3 gives the variation of gravity with distance from the center of the Earth and, using this equation, measured values of gravity can be corrected for the altitude effect by changing the result to what the value would have been if the measurement were taken at sea level, considered equivalent to the geoid. This free-air correction is about 0.3086 mgal/m, although it varies with latitude and altitude. This correction takes no account of the nature of the material between the geoid and the point of measurement. If the material were air, the free-air correction should correct the gravity measurement to the value appropriate for this point on the geoid. The difference between the theoretical value from Equation 3-7 and the corrected value is called the free-air anomaly.

A part of the free-air anomaly is caused by the attraction of land above sea level, or by the comparative lack of attraction of sea

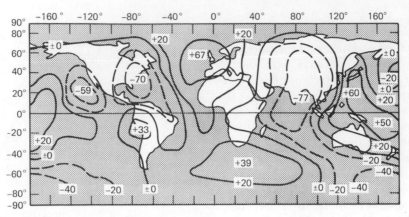

Figure 3-15. Satellite geoid map with contours in meters, representing the zonal and non-zonal gravity harmonics up to order 8. (From W. H. Guier and R. R. Newton, Jour. Geophys. Res., **70**, 4613, 1965, with permission; as presented by Schmucker, 1969.)

water in ocean basins. The Bouguer correction allows for the effect of the material between the elevation of the measuring station and the geoid, and additional topographic corrections can be made to allow for the effect of neighboring mountains and valleys. For rocks with average density 2.67 g/cm³, the Bouguer correction is 0.1119 mgal/m. The free-air and Bouguer corrections are always of opposite sign. The combined correction for an increase in elevation of 1 m is a reduction of the measured value by 0.1967 mgal.

The free-air anomaly is a measure of the effects of mass excesses and deficiencies within the Earth. Subtraction of the topographic correction and the Bouguer correction eliminates that part of the free-air anomaly due to elevation and topography. The remaining Bouguer anomaly, the difference between the corrected measurement of gravity and the value given by the International Gravity Formula (Equation 3-7), represents the effect of buried rocks of greater or less than average density, or of geological structures involving rocks of contrasting densities. Study of these anomalies has proved very useful in determining crustal structure, although no unique solution can be obtained from gravity data alone, because an infinite number of mass distributions at depth are capable of producing the same anomaly at the surface.

Consideration of Bouguer anomalies on a worldwide scale indicates that gravity is less than average in high areas and greater than average over oceanic areas, which indicates that rocks beneath high areas have relatively low density compared to rocks beneath oceanic areas. This relationship is explained by the law of isostasy, which can be stated thus: all large land masses on the Earth's surface tend to sink or rise so that, given time for adjustment to occur, their masses are hydrostatically supported from below, except where local stresses are acting to upset equilibrium. We can suppose that there is the same weight of material in each vertical column of rock above some level of compensation within the Earth. When the rock is lighter the surface is higher as for the continents, and when the rock is denser the surface is lower as in the ocean basins. Airy's theory of isostatic compensation is illustrated schematically in Figure 10-1, with the level of compensation situated at some depth beneath the deepest "crustal" column. Using either Airy's theory or Pratt's theory isostatic corrections for gravity measurements can be calculated. Subtraction of these from the Bouguer anomalies leaves a remainder which is called the isostatic anomaly; this is usually much smaller than the free-air and Bouguer anomalies. The assumption of hydrostatic equilibrium at some depth, of the order of 100 km, thus appears to be justified; but large isostatic anomalies do exist, indicating that portions of the Earth's crust are not in isostatic equilibrium.

The gravity anomalies discussed above are differences between corrected gravity measurements and the theoretical gravity calculated from the International Gravity Formula. This assumes that the geoid is a regular ellipsoid. In fact the geoid is slightly distorted. The amount of distortion is very small and it appears certain that the distortions are caused by density variations in the mantle rather than in the crust. Crustal effects have been compensated for by the free-air, Bouguer, and isostatic corrections. At present surface gravity measurements can tell us nothing about the deeper mantle.

Geoid heights from satellite data. Since the first artificial satellite was launched in 1957 there have been many determinations of the Earth's geoid by satellite observations. A satellite samples all longitudes and a range of latitudes as the Earth rotates beneath it, and its path is affected by the irregularities in the gravity field produced by irregular mass distributions within the Earth. Satellite data provide no information about variations of gravity on the scale associated with physiographic features of the Earth's surface but they do detect large-scale anomalies in the Earth's gravity field.

The expression for the gravitational potential outside of the Earth is usually written in terms of spherical harmonics, and the equation includes coefficients describing the gravity field of the Earth. The satellite data reveal departures from the gravity field that would be associated with hydrostatic equilibrium within the Earth. It appears that the sources of some of the observed harmonics, and perhaps of most, must lie within the mantle.

The satellite data may be expressed interchangeably in terms of gravitational potential, surface gravity, or geoid shape. Figure 3-15 shows a convenient representation with the surface of the geoid contoured at intervals in meters above and below the standard ellipsoid, or theoretical geoid. Various authors have used different methods for determination of the geoid from satellite data; results have been obtained which differ by 40 m or more in some locations, but we can expect an accuracy to within 10 m in future work. Combinations of satellite and surface gravimetric data are now being used to obtain a more complete picture of the gravitational field of the Earth than has been obtained from satellite data alone.

Depressions of the geoid surface are regions of negative free-air gravity, and elevations of the geoid surface are regions of positive free-air gravity. These anomalies are probably produced by lateral density differences within the mantle; they bear no direct relation to the distribution of continents and oceans. One objective of the study of gravitational evidence is to set limits to the depth within the Earth that lateral density variations do occur. The density variations could result from lateral changes in composition or in temperature. Interpretations of the geoid distortions, and the probable density variations in the mantle, require comparison of the geoid with the results of other physical measurements of the Earth, such as heat flow, travel-time anomalies for seismic waves, and spherical harmonic analysis of major tectonic features.

Heat Flow

The present rate of heat loss from the Earth's surface is about 2.4×10^{20} cal/year. In terms of energy this outflow of heat is the most impressive of geophysical processes; the more spectacular energy loss involved in earthquakes and volcanic activity is orders of magnitude less.

Methods of measurement. Despite its significance, satisfactory procedures for the measurement of heat flow have been developed only recently. In 1945 there were about

20 measurements, all on land. The design of apparatus for measuring heat flow through the ocean floor, together with the increased effort in marine geophysics, improved the situation greatly. By 1960 there were about 200 measurements, by the end of 1964 there were at least 1000 measurements published, and by July 1968 about 2600 results were available, with new values being reported at the rate of 500 per year. About 90% of all measurements were made at sea. Taking into consideration the areas of land and ocean (Table 2-1), this is equivalent to about three times more data per unit area of ocean than of land. Unfortunately the measurements on land are very unevenly distributed, and there are large gaps in continental and high-latitude regions.

The measurement of heat flow through the ocean floor, initiated by E. C. Bullard, is now a routine procedure, but the measurement of heat flow through land is still difficult. It is necessary to drill holes about 300 m deep, which is a very expensive business. Critical studies are required to find the optimum depths, hole size, horizontal spacing, instrumentation, and procedure for large-scale surveys of heat flow on continents. We can expect major efforts by groups such as the International Upper Mantle Committee to provide many more data in continental areas in the future, data which are critically needed to test geophysical hypotheses.

If the Earth's surface layer were an ideal uniform rock, the geothermal flux q would be provided by a measurement of the thermal gradient in the rock (two temperature measurements at distance z apart in a vertical direction), and the thermal conductivity K.

$$q = K \left(\frac{\partial T}{\partial z} \right) \qquad (3\text{-}9)$$

There are many factors that may affect underground temperatures: the circulation of underground water, including the migration of interstitial water in porous sediments on the ocean floor; vertical and lateral variations in thermal conductivity; the effects of topography, uplift and erosion, and recent glaciations; the effects of magmatic events; the effect of variable movements of water over the ocean floor; disturbances produced by drilling the hole on land or by penetration of the probe into ocean-floor sediments and by ventilation cooling in mines. Corrections may be applied for some of these effects.

On land the first measurements should be at a depth of at least 30 m in order to avoid the effect of annual temperature variations at the surface. The most difficult problem is determination of a good mean value for the thermal conductivity of the rocks within the vertical interval. The conductivity is low, and it may vary by tens of percent from one sample to another taken from an apparently uniform rock. Errors in individual heat flow measurements under favorable conditions at sea are usually no more than 10%, but they can reach 20%. The quality of heat-flow measurements on land varies from crude estimates to elaborate determinations. If the mean heat flow of a region is computed from more than 10 observations, it is believed that heat-flow variations among regions are significant if they exceed 0.2 μcal/cm^{2}sec.

Review of data. The heat-flow data listed in Tables 3-2 and 3-3, and plotted in Figures 3-16, 3-17, 3-18, and 3-19 are derived largely from three reviews by W. H. K. Lee and S. Uyeda in 1965, R. W. Girdler in 1967, and R. P. Von Herzen and W. H. K. Lee in 1969. Table 3-2 gives the data for oceans and continents and shows that statistical analyses of 2584 selected heat-flow measurements in 1969 do not differ significantly from the analyses of fewer than half this number of measurements in 1965. Table 3-3 shows the values and statistics for major physiographic provinces.

Figure 3-16 compares the histograms of heat-flow measurements for continents and oceans. This demonstrates that the heat flows in these two environments are essentially the same (Table 3-2), with arithmetic means

TABLE 3-2 Selected Heat-Flow Values for Oceans and Continents: Individual Values and Grid Averages (9 × 104 square nautical miles per grid element)[a]

Region	Lee and Uyeda, 1965			Von Herzen and Lee, 1969 (for July, 1968)		
	N	$\bar{q}$	S.D.	N	$\bar{q}$	S.D.
All continents	131	1.43	0.56	255	1.49	0.54
All oceans	913	1.60	1.18	2329	1.65	1.14
Atlantic	206	1.29	1.00	406	1.43	1.07
Indian	210	1.47	0.89	331	1.44	1.09
Pacific	497	1.79	1.31	1232	1.71	1.24
Arctic				29	1.23	0.33
Mediterranean seas	[b]			71	1.33	0.89
Marginal seas	[b]			260	2.13	0.63
				Girdler, 1967		
Africa	13	1.20	0.20	13	1.20	0.21
Asia[c]	37	1.49	0.58			
Japan				38	2.21	2.73
Australia	19	1.75	0.62	20	1.76	0.62
Europe	22	1.62	0.60	31	1.91	1.70
N. America	40	1.19	0.44	44	1.26	0.57
	Grid Averages			Von Herzen and Lee, 1969		
All continents	51	1.41	0.52	79	1.45	0.47
All oceans	340	1.42	0.78	577	1.46	0.78
Atlantic	65[d]	1.21	0.64	127	1.32	0.56
Indian	94[d]	1.35	0.67	110	1.36	0.82
Pacific	181[d]	1.53	0.87	305	1.50	0.78

[a] N = number of observations. $\bar{q}$ = arithmetic mean of heat flow in $\mu cal/cm^2sec$ (HFU). S.D. = standard deviation in HFU (heat-flow units).
[b] The few data available were included under oceans.
[c] Almost all data from Japan.
[d] Includes data from Mediterranean and marginal seas.

differing by less than 0.2 HFU (1 HFU = 1 heat flow unit = 1 $\mu cal/cm^2sec$). Despite the absence of measurements from large areas, it appears that more than 99% of the Earth's surface can be considered normal, with heat flow varying around 1.5 HFU; the remainder represents thermal areas with heat flux many times greater. This is indicated by the positive skew distributions for the histograms in Figures 3-16a and c. The mode for each of these two distributions is 1.1 HFU. Mass transfer of material bringing energy from deep sources is required to explain the localized thermal areas. The arithmetic mean of the world heat-flow measurements is 1.58 HFU, with standard deviation 1.14, and a mode for the histogram of 1.1.

For comparing heat-flow over large regions it is useful to represent the measurements in such a way that the effect of localized high values is reduced. The logarithmic heat-flow values give distributions closer to normal frequency curves, as shown in Figures 3-16b and d. Girdler suggested that geometric

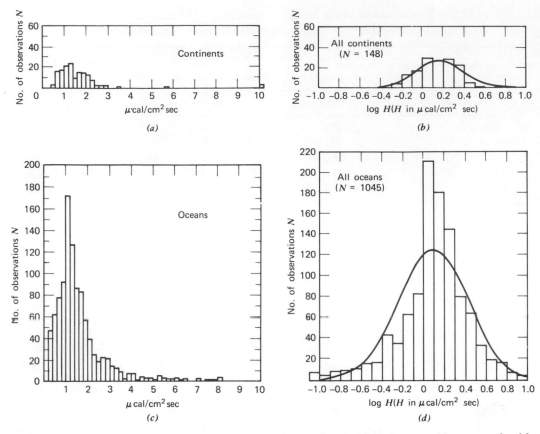

Figure 3-16. Histograms of heat flow observations for continents (*a*) and oceans (*c*) compared with histograms and normal frequency curves for the logarithms of heat flow values for all continental areas (*b*) and all oceanic areas (*d*) (after Girdler, 1967, with permission of Interscience).

means of heat-flow values should be considered as well as arithmetic means before making deductions concerning heat flow over large areas of the Earth's surface. The geometric means are smaller than the arithmetic means.

Lee and Uyeda introduced another method for reducing the bias of localized high heat-flow values, by working out the average heat flow for grid elements of equal area. Table 3-2 shows the statistics of averages for grid elements of 9×10^4 square nautical miles, or $5 \times 5°$ at the equator. The arithmetic mean and the standard deviation for the grid averages are smaller than for the original data, but the averages confirm the equality of heat flow through continent and ocean

floors. As noted in connection with Figure 3-12, this appears to require a different vertical distribution of radioactive materials within the mantle and crust beneath continents and oceans, and implies deep-seated differences in temperature between the two environments; unless convection in the mantle equalizes the temperature distributions.

Statistics of heat-flow values for the continents are listed in Table 3-2, but the small numbers and the irregular geographical distributions of the measurements permit no conclusions to be drawn. Similar statistics for the various oceans are also tabulated. The mean and scattering of the values increases from the Atlantic to the Indian to the Pacific, as shown by the histograms presented by Lee and Uyeda. These differences could be related to the different structures and histories of the oceans, but they could also be caused by nonuniform sampling of measurements.

Lee and Uyeda also presented histograms for heat-flow values associated with major geological features on land and beneath the ocean and found good correlation of heat flow with tectonic environments. The statistics are given in Table 3-3. Heat-flow values for shield areas are uniform and low, in accord with their long history of stability. The average heat flow in post-Precambrian orogenic and nonorogenic areas is essentially the same, but values in the orogenic areas are more scattered. If data from the Cenozoic volcanic areas were included with the orogenic areas, the average orogenic heat flow would be higher. Heat-flow values in the ocean basins are fairly uniform and low, with a mean of 1.28 HFU, whereas those in the ridges and other areas are higher and more scattered. The crestal zones of ridges are characterized by high heat flow and large variations, whereas relatively low values are observed over the flanks. This is shown in Figure 3-17a.

The high values for Japan represent the only extensive study among island arc areas, and the large standard deviation indicates the wide dispersion of values. The average heat flow in oceanic trenches is only 0.99 HFU, but the number of measurements is very small. The profiles of heat-flow values across island arc and ocean-trench systems are illustrated in Figures 3-17b and c. The lowest values are near the trench axis, and the high values are on the continental side of the arcs. These profiles correlate with the distribution of epicenters of earthquakes. The shallow focus earthquakes are concentrated in the belt of low heat flow, and the heat flow is high where the earthquakes have focal depths greater than about 150 km. This is coincident also with the oceanward limit of the active volcanoes.

Global analysis. A set of heat-flow data, in which each measurement point is expressed in terms of its latitude and longitude, can be fitted by a least squares method to a spherical harmonic expansion. This procedure provides an objective contour map of isoflux that best fits the observations. A global representation

TABLE 3-3   Statistics of Selected Heat-Flow Values for Major Physiographic Provinces.[a]
(Lee and Uyeda, 1965)

Province	N	$\bar{q}$	S.D.	S.E.	Modes
Ocean basins	273	1.28	0.53	0.03	1.1
Ocean ridges	338	1.82	1.56	0.09	1.1
Ocean trenches	21	0.99	0.61	0.13	1.1
Other seas	281	1.71	1.05	0.06	1.1
Precambrian shields	26	0.92	0.17	0.03	0.9
Phanerozoic nonorogenic	23	1.54	0.38	0.08	1.3
Phanerozoic orogenic areas[b]	68	1.48	0.56	0.07	1.1
Paleozoic orogenic areas	21	1.23	0.40	0.09	1.1
Mesozoic-Cenozoic orogenic	19	1.92	0.49	0.11	1.9, 2.1
Island arc areas	28	1.36	0.54	0.10	1.1
Cenozoic volcanic areas[c]	11	2.16	0.46	0.14	2.1

[a] N = number of observations. $\bar{q}$ = arithmetic mean of heat flow in $\mu cal/cm^2 sec$ (HFU). S.D. = standard deviation in HFU (heat-flow units). S.E. = standard error in HFU.
[b] Excluding Cenozoic volcanic areas.
[c] Excluding geothermal areas.

of the heat-flow field up to third-order spherical harmonics is given in Figure 3-18. Contour lines over regions in which no data exist are dashed. The analysis averages out the small-scale variations in the heat-flow field, and for a third-order spherical harmonic analysis variations on a scale less than 2000 km do not appear. The contour map shows a general heat-flow pattern, with low values in the central Pacific and in the Atlantic and high values in the eastern Pacific and east Africa; the African high is uncertain because there are few measurements there. The heat-flow measurements are so un-

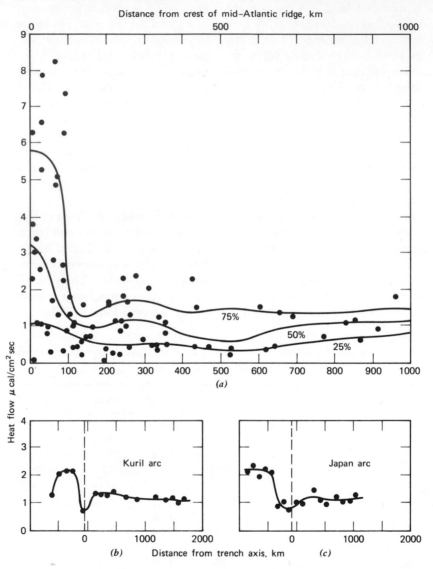

Figure 3-17. (*a*) Heat flow values versus distance from the crest of the mid-Atlantic Ridge; 75,- 50-, and 25-percentile lines are given. The 50-percentile line separates half the data points above and half the data points below it (after W. H. K. Lee and S. Uyeda, Geophysical Monograph **8**, 1965, with permission of Amer. Geophys. Union). (*b*) and (*c*) Profiles of heat flow values across island arcs, averaged in 100-km intervals (after Vacquier *et al.*, 1966; from Geophysical Monograph **12**, 349, 1968, with permission of Amer. Geophys. Union).

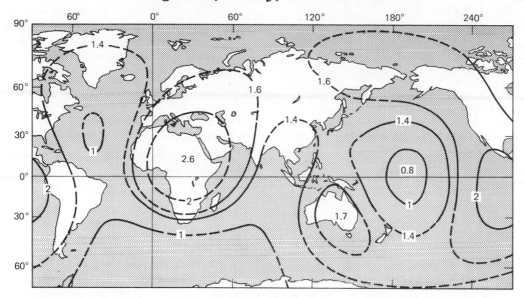

Figure 3-18. Orthogonal function representation (to third order spherical harmonics) of 987 heat flow values. Contour lines are in μ cal/cm² and are dashed over regions where no data exist (after W. H. K. Lee and S. Uyeda, Geophysical Monograph **8**, 1965, with permission of Amer. Geophys. Union; see Toksöz *et al.*, 1969, Figure 3*b*, for more recent map of heat flow variations).

evenly distributed across the Earth's surface that the use of spherical harmonic analysis could be misleading. There is a tendency for anomalies to be produced when there are no observations, as indicated by the large positive anomaly over Africa; much of this region is Precambrian shield, where low heat flow might be expected (Table 3-3).

The heat-flow map with all small-scale anomalies removed is related to properties within the mantle in a fashion similar to the map of the geoid in Figure 3-15. It has been proposed that a correlation exists between these two maps, with heat flow being low where gravity is high and vice versa. Girdler examined this proposition by assigning heat-flow measurements directly to regions of positive and negative gravity contoured on the geoid map (Figure 3-15). For each region he prepared a histogram of logarithmic heat-flow values to reduce the effects of local anomalies, and he compared the arithmetic and geometric means of each histogram with the world means. The overall results are plotted in Figure 3-19. The arithmetic mean heat flow for all four regions of negative

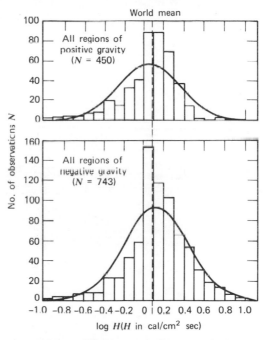

Figure 3-19. Histograms and normal frequency curves of the logarithmic values of heat flow for all regions of negative gravity and all regions of positive gravity for Guier and Newton's satellite geoid of figure 3-15 (after Girdler, 1967, with permission of Interscience).

geoid height is 1.67 HFU and that for all four regions of positive geoid height is 1.33 HFU. Girdler concluded that these results tend to support the hypothesis that undulations of the geoid are due to temperature differences in the mantle. The correlation is not well established, however, because of the uncertainty of our knowledge of both the heat-flow and gravity fields. Others have concluded that heat flow variations are not at all correlated with geopotential variations (e.g., Toksöz et al., 1969). We should also note that the gravity field represents the present mass distribution, whereas the surface heat-flow field may lag millions of years in indicating the temperature distribution at some depth in the Earth's interior because of slow thermal conduction. The thermal time scale for a layer 500 km thick is of the order of 10^9 years or longer. Convection in the mantle would reduce the time scale.

4. *Geological Processes and the Geological Time Scale*

INTRODUCTION

The aim of geology, and of the geophysical sciences in general, is to determine how and why the physical world became what it is today. We seek answers to questions such as these:

1. How did the continents form?
2. Why do they have their present distributions?
3. How is it that they persist?

In order to answer these questions we have to study the major features of the Earth's crust and to consider the geological processes involved in their formation and development. We have to examine carefully the rocks to see what clues they provide about their origins, and therefore we have to study the minerals composing the rocks. The study of minerals requires consideration of matter on an atomic scale. We discover that crustal processes have sources within the mantle, and that these processes can be related to mineralogical and phase changes occurring on an atomic scale and driven by thermal energy. Therefore, in order to comprehend the processes involved in large scale phenomena such as the formation of the great mountain ranges and the eruption of lavas from volcanoes and in floods, we have to study the materials of the Earth on all scales from the large dimensions considered in Chapter 2 down to atomic dimensions.

In Chapter 2 we examined the major features of the Earth as they exist at the present time. We know that despite their apparent permanence on the time scale of human life and civilizations, these are only transient features when considered in terms of the geological time scale. The mightiest mountains have been uplifted from beneath the oceans, and the low-lying continental shields were once traversed by rugged mountain ranges. Geology is a historical subject, and its study involves elucidation of the processes operating now, deduction of the processes operating during the past, determination of the effects of these processes operating over long periods of time, and their correlation with the stages of Earth history indicated in the stratigraphic record. This chapter is concerned with geological processes, and most of them are introduced and related to each other within the framework of the Rock Cycle, the Hydrologic Cycle, and the Tectonic Cycle. These cycles, combined within the dimensional framework of the geological time scale, provide an outline review of geology, and they introduce most of the subjects usually presented as separate courses in standard geological curricula. Subsequent chapters are more concerned with major physical units of the Earth or with major processes than they are concerned with conventional subjects.

GEOLOGICAL CYCLES

The major features of the Earth's surface are produced by the interactions of processes driven by two energy sources: the heat within the Earth is an internal source and the solar radiation is an external source. The internal source provides the energy for igneous, metamorphic, and tectonic activity, and the external source provides the energy dominating the circulation of the atmosphere and hydrosphere, the processes of weathering and erosion, and the formation of sedimentary rocks. These processes are connected through the Geological Cycle whose main features are summarized in Figure 4-1. Igneous, sedimentary, and metamorphic processes are reviewed in more detail in Figures 4-5, 4-6, and 4-7.

The Geological Cycle really consists of a number of interrelated cycles. (a) The Rock Cycle traces the relationships among igneous,

sedimentary, and metamorphic rocks; but this relationship exists only because of the processes involved in (b) the Hydrologic Cycle which governs erosion and sedimentation, and those involved in (c) the Tectonic Cycle—specifically, subsidence and uplift. (d) The Geochemical Cycle is concerned with the migration of elements within the concentric shells of the Earth, from one shell to another and from one cycle to another.

Figure 4-1 shows that the Geological Cycle is not closed. New material is added by uprise of magma generated in the mantle at depths of 50 km or more. The additional material provides silicates for the Rock Cycle, as well as water and other volatile materials for the Hydrologic Cycle. There is evidence that the Earth's crust, hydrosphere, and atmosphere may have formed progressively, as a result of

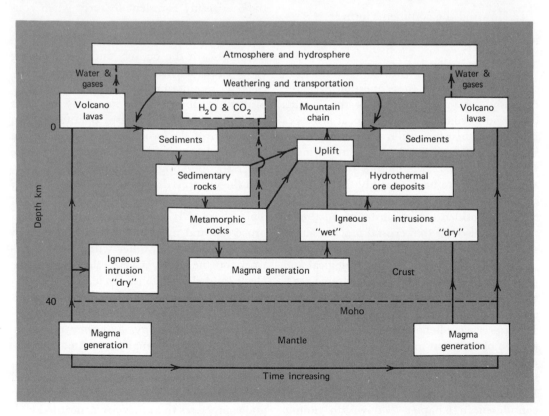

Figure 4-1. The geological cycle.

repeated magmatic activity, which leads to outward migration of the more fusible elements. The process is referred to as degassing or defluidization of the Earth.

We consider the Geological Cycle and the Rock Cycle to begin with the eruption of mantle-derived lava to form volcanoes, as shown in the left-hand side of Figure 4-1. Addition of water and other gases to the atmosphere and hydrosphere is indicated. Note also that some of the magma may be trapped in the crust to form a relatively dry igneous intrusion. Some of the water given off by volcanoes is undoubtedly juvenile, carried upward from the mantle, but at least a portion of it must be meteoritic, derived from water in the crust or from ocean water in the case of submarine eruptions and oceanic volcanoes.

As soon as a volcano emerges it is attacked by the atmosphere and hydrosphere. A series of physical and chemical processes disintegrates the exposed rocks, and the products of weathering are transported to lower altitudes where eventually they are deposited as sediments. Organisms and material of the biosphere may be involved in sediment formation in several different ways.

With continuation of the cycle sediments may become deeply buried and converted to sedimentary rocks by compaction, cementation, and recrystallization. If the Tectonic Cycle operates, the rocks are carried to deeper levels within the crust, where they are subjected to crushing, deformation, folding, recrystallization, and possibly reconstitution by metasomatism. Reactions occurring during progressive metamorphism cause evolution of carbon dioxide and water which migrate toward the surface. Regional metamorphism may culminate in partial fusion or anatexis of the rocks with the formation of migmatites and magmas. These magmas probably contain more dissolved water than mantle-derived magmas, and they are distinguished as "wet" intrusions in Figure 4-1. Crystallization of either "wet" magmas, or of relatively "dry" magmas from the mantle, leads eventually to the evolution of a dilute aqueous gas phase

which condenses at lower temperatures to a hydrothermal solution. The pneumatolytic gas phase and hydrothermal solution are products of an efficient geochemical process that concentrates certain elements. Precipitation of these elements in appropriate physicochemical environments may yield ore deposits.

The Tectonic Cycle continues with uplift and the formation of mountains composed of the folded and metamorphosed sediments and igneous intrusions. Figure 4-1 shows that the simple cycle of subsidence followed by uplift may be interrupted by periods of uplift and erosion at any stage in the cycle; these interruptions are represented in rock records by unconformities. The cycle is repeated with the deposition of sediments whenever significant uplift occurs, although the deposition of sediments is not always followed by deep burial and metamorphism. An Orogenic Cycle includes the whole sequence of cycles involved in the formation of a mountain chain (Chapter 9).

Normally the Hydrologic Cycle is considered with respect to the migration of meteoritic water through the atmosphere, the hydrosphere, and the uppermost part of the crust. A representation of the processes involved is shown in Figure 4-2. As shown in the figure, however, this cycle has direct connections with the deeper part of the Rock Cycle. Not only is juvenile water added from magmas of mantle origin but also meteoritic water is carried out of the main cycle for long periods of time, when sediments containing hydrous minerals and pore fluids are carried deep into the crust during a Tectonic Cycle. The water may return to the near-surface cycle via different routes. One route is simply by uplift and exposure through erosion of the overlying rocks. Progressive metamorphism of the buried sediment causes dehydration, and some of the released water migrates back to the surface. If anatexis occurs, the water dissolves in the magma and it may then be returned to the surface by volcanic activity or by upward migration of hydrothermal fluids emanating from the igneous intrusions.

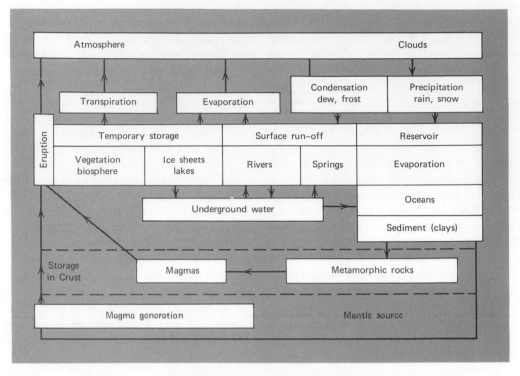

Figure 4-2. The hydrologic cycle.

GEOLOGICAL PROCESSES

Geological processes form rocks. Rocks are chemical systems adjusting to their physical environment in attempts to attain a state of thermodynamic equilibrium. As a basis for review of geological processes we should therefore examine the chemistry of the materials composing rock systems, the range of physical conditions to which they are subjected in the Geological Cycle, and the types of chemical reaction that result.

Rock Materials

Representatives of the main rock types involved at various stages of the Rock Cycle are listed in Table 4-1. The rocks are divided into four groups. The first constitutes the mantle material from which the original magma is derived, and the second includes the igneous rocks derived from the parental magma or magmas. Sediments constitute the third group,

and our main interest is in those occurring in orogenic environments, which subsequently become metamorphosed to form the fourth group. The compositions of metamorphic rocks cover the range of igneous and sedimentary rocks. Representative minerals composing these rocks are listed in the table.

For review of the types of chemical reaction experienced by rock systems we can group the minerals into three classes: (a) anhydrous silicate minerals stable only at high pressures (e.g. pyrope, jadeite); (b) anhydrous silicate minerals typical of crustal rocks (e.g. pyroxenes, feldspars, quartz); (c) hydrated and carbonated minerals (e.g. clay minerals, micas, hornblende, calcite).

Mineral Reactions

Magmas are generated in the mantle at depths to 100 km or more, corresponding to

TABLE 4-1 Selected Materials Involved in the Rock Cycle

	Rocks	Representative Minerals	
		Anhydrous	Hydrated and Carbonated
Upper mantle source	Peridotite	Olivine	Amphibole
	Eclogite	Pyroxenes	Phlogopite
		Garnet	Clinohumite
		Spinel	
Igneous rocks	Peridotite	Olivine	Amphibole
	Gabbro	Pyroxenes	Biotite
	Diorite	Feldspars	Muscovite
	Granite	Quartz	
Sedimentary rocks	Sandstone	Quartz	Kaolin
	Arkose	Feldspars	Clay minerals
	Greywacke	Salts	Chlorite
	Shale		Carbonates
	Limestones		Salts
	Evaporites		
Metamorphic rocks	Schists	Quartz	Muscovite
	Gneisses	Feldspars	Chlorite
	Marbles	Garnet	Serpentine
		Pyroxenes	Biotite
		Aluminosilicates	Epidote
		Cordierite	Staurolite
			Hornblende

pressures up to and greater than 30 kb and temperatures of the order of 1500°C. Conditions are less extreme in the Rock Cycle itself, which is confined to the range of crustal conditions with pressures up to 10 kb and temperatures rarely reaching 1000°C. There are three types of reactions illustrated in Figure 4-3 which occur within this range of pressures and temperatures: (a) polymorphic transitions and solid-solid reactions; (b) dissociation reactions: dehydration and decarbonation; (c) Melting reactions, either dry or in the presence of water under pressure.

Many simple mineral reactions are univariant, but in complex rock systems the reactions may be multivariant. These various types of reactions are represented on PT projections by curves or bands with different attitudes in terms of the P and T axes.

Figure 4-3a for the one-component system SiO_2 illustrates two kinds of polymorphic transition. The transition of quartz to coesite is pressure sensitive, being induced by increased pressure. In contrast the transition of α-quartz to β-quartz is temperature sensitive, the temperature of the transition being not greatly affected by change in pressure.

Pressure-sensitive reactions such as those illustrated in Figures 4-3b and 5-13 justify our division of anhydrous silicate minerals into groups (a) and (b) in the preceding sections. Plagioclase feldspar is a dominant constituent of crustal rocks, but Figures 4-3b and 5-11 show that it breaks down at high pressures as shown by univariant curves for the reactions:

$$NaAlSi_3O_8 \rightleftharpoons NaAlSi_2O_6 + SiO_2 \quad (4\text{-}1)$$

albite (Ab)   jadeite (Jd)   quartz (Qz)

$$3CaAl_2Si_2O_8 \rightleftharpoons Ca_3Al_2Si_3O_{12} + 2Al_2SiO_5$$

anorthite (An)   grossularite (Gr)   kyanite (Ky)

$$+ SiO_2 \quad (4\text{-}2)$$

quartz (Qz)

Figures 4-3*b* and 5-11 also show that the magnesian garnet, pyrope, is stable only at

mantle pressures; the reaction plotted shows its formation from aluminous enstatite (En), sapphirine (Sa), and sillimanite (Si). In a complex rock system similar reactions occur through a pressure interval as shown in Figure 5-15, where the low pressure silicate assemblage of gabbro (augite + plagioclase) is converted to the high pressure assemblage of eclogite (omphacite + pyrope-rich garnet).

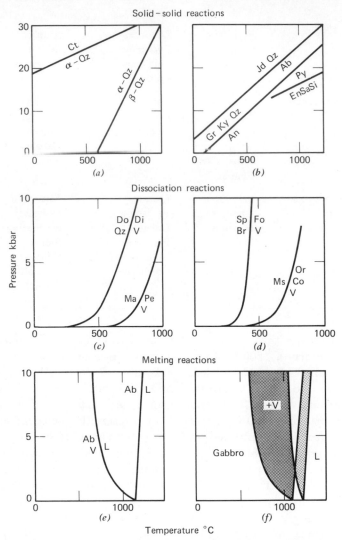

Figure 4-3. Types of reactions occurring in metamorphic and igneous processes. See text for details of reactions and abbreviations. Sources of data: (*a*) Coesite, Boettcher and Wyllie, 1968; β-quartz, Cohen and Klement, 1967. (*b*) Pyrope, Boyd and England, 1962; anorthite, Newton, 1966; jadeite, Boettcher and Wyllie, 1968. (*c*) Diopside, Turner, 1968, p. 144; periclase, Goldsmith and Heard, 1961. (*d*) Serpentine, Scarfe and Wyllie, 1967, and Johannes, 1968; muscovite, Evans, 1965, and Lambert *et al.*, 1969. (*e*) Boettcher and Wyllie, 1969, for data and review. (*f*) See Figure 6-12.

See also Figure 6-10 for transformations in peridotite.

The hydrated and carbonated minerals are abundant in sediments and many metamorphic rocks. During the progressive metamorphism of sedimentary rocks water and carbon dioxide are driven off by successive dehydration and decarbonation reactions as illustrated in Figures 4-3c and d. A dissociation reaction may involve the breakdown of a single mineral as illustrated by magnesite and muscovite:

$$MgCO_3 \rightleftharpoons MgO + CO_2 \quad (4\text{-}3)$$

magnesite (Ma) periclase (Pe) vapor (V)

$$KAl_3Si_3O_{10}(OH)_2 \rightleftharpoons KAlSi_3O_8$$

muscovite (Ms) orthoclase (Or)

$$+ Al_2O_3 + H_2O \quad (4\text{-}4)$$

corundum (Co) vapor (V)

In rock systems, however, it is more usual for reactions to involve at least one additional mineral among the reactants, as for the following reactions plotted in Figures 3-3c and d:

$$CaMg(CO_3)_2 + 2SiO_2 \rightleftharpoons CaMgSi_2O_6$$

dolomite (Do) quartz (Qz) diopside (Di)

$$+ 2CO_2 \quad (4\text{-}5)$$

vapor (V)

$$Mg_3Si_2O_5(OH)_4 + Mg(OH)_2$$

serpentine (Sp) brucite (Br)

$$\rightleftharpoons 2Mg_2SiO_4 + 3H_2O \quad (4\text{-}6)$$

forsterite (Fo) vapor (V)

For similar reactions involving minerals that exhibit solid solution the univariant reactions could be replaced by reaction bands.

Other dissociation reactions may involve the evolution of both water and carbon dioxide as in the univariant reaction:

$$Ca_2Mg_5Si_8O_{22}(OH)_2 + 3CaCO_3 + 2SiO_2$$

tremolite (Tr) calcite (Ca) quartz (Qz)

$$\rightleftharpoons 5CaMgSi_2O_6 + 3CO_2 + H_2O \quad (4\text{-}7)$$

diopside (Di) vapor (V)

Such reactions may proceed through divariant intervals in rock systems, the precise nature of the reaction being controlled largely by the composition of the pore fluid in the rock and the ease with which the vapor can escape from the region of the reaction.

Figures 4-3c and d show that dissociation reactions tend to be temperature-sensitive, except at very low pressures, but at higher pressures corresponding to upper mantle conditions, hydrated minerals may break down by pressure sensitive reactions. This is illustrated in a general way for amphibole in Figures 6-11 and 6-12. At pressures above 15 kb the slope of the reaction (dP/dT) becomes negative.

Magma is formed by melting reactions, and the typical shape of a fusion curve for a silicate is shown in Figure 4-3e by:

$$albite (Ab) \rightleftharpoons liquid (L) \quad (4\text{-}8)$$

The corresponding melting interval for a mineral aggregate or rock is shown by the narrow shaded band in Figure 4-3f.

Solution of a small amount of water under pressure in the silicate liquid has a dramatic effect in lowering the melting temperatures of silicates, as shown in Figure 4-3e by the curve for:

$$NaAlSi_3O_8 + H_2O \rightleftharpoons liquid \quad (4\text{-}9)$$

albite (Ab) vapor (V) (L)

The solubility of H_2O in the silicate liquid and the solubility of the solids in the gas phase increase with increasing pressure. The effect of water on a rock is to increase the melting interval as shown in Figure 4-3f for a gabbro, because the liquidus curve is depressed to a lesser extent than the solidus curve. In the presence of excess water the subsolidus gabbro is converted to amphibolite.

This pattern of melting reactions in the presence of excess water is modified when high pressure minerals are produced, as illustrated in Figures 6-12, 8-2 and 8-3. The slope of the solidus is reversed at pressures above about 15 kb, where the plagioclase of the gabbro breaks down to yield jadeite, grossularite, kyanite, and quartz (Figure 4-3b). The temperature of melting then increases with increasing pressure, and the solidus curve is then approximately parallel with the curve for the dry rock. The involvement of hydrous minerals in melting reactions is considered in Chapters 6 and 8.

The *PT* conditions for the occurrence of these various types of mineral reactions, involving materials of different composition, can be determined experimentally in the laboratory. The composition of rock material available for reaction varies, however, with depth and geological environment and the range of temperatures occurring within the Earth at any given depth (pressure) is also limited. For the experimentally determined mineral reactions to have relevance to geological processes, therefore, we must consider the appropriate combinations of rock composition, pressure, and temperature.

Figure 4-4 shows the pressure existing at each depth within the outer part of the Earth. The two dotted lines are geotherms showing estimates for the temperature distribution at depth in continental shield regions, and in oceanic regions (Figure 3-12b). The geotherm could rise to higher temperatures in active orogenic areas. The composition of the mantle

and crust will be considered in detail in Chapters 6 and 7, but for present purposes let us adopt a conventional model of a crust composed essentially of granitic rocks overlying gabbroic rocks, and a mantle composed of peridotite with layers and lenses of eclogite.

Figure 4-4 shows the range of *PT* conditions occupied by the facies of regional metamorphism. The facies are separated, for the most part, by dissociation reactions (Figures 4-3c and d), but the boundary between the eclogite facies and the other facies involves the formation of high-pressure phases (Figure 4-3b). See Figure 9-9 for details.

If water is available in the crust as is usually assumed, then magma generation can be expected in the crust if conditions correspond to the shaded bands shown in Figure 4-4 for granite-water and gabbro-water (Figure 4-3f). If water is available in the mantle, then traces of liquid may develop at temperatures down to the dashed line extending the solidus for the system gabbro-peridotite-water to pressures greater than 10 kb (see also Figures 4-3f and 6-13). Magma

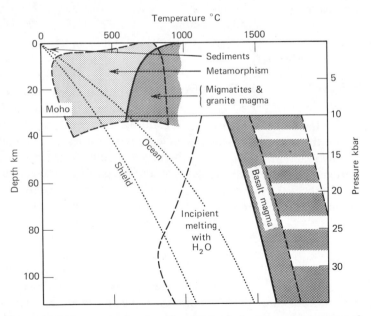

Figure 4-4. Depth-temperature ranges of major geological processes, compared with estimated temperatures beneath continental shields and oceans (see Figure 3-12). See Figure 9-9 for details of metamorphic facies.

generation in the mantle, however, is usually assumed to occur under essentially dry conditions, and the melting intervals for dry peridotite and eclogite are indicated by the alternating wide and narrow shaded bands.

Most reactions occurring in the formation of sedimentary rocks involve solution chemistry, and they occupy a very small range of pressures and temperatures near the origin of Figure 4-4. For discussion and illustration of sedimentary reactions, other diagrams, such as *Eh* versus *pH* diagrams, are more useful than the *PT* projections of Figures 4-3 and 4-4.

Figures 4-3 and 4-4 show the equilibrium positions of selected mineral reactions, as determined under static conditions in the laboratory. It does not necessarily follow that identical reactions will occur under identical pressure-temperature conditions in the Earth. Rock systems strive to attain a state of thermodynamic equilibrium but they do not always achieve this. Many rock materials include metastable ingredients, and these participate in metastable reactions. Many geological processes occur under irreversible conditions. Thus, although reaction curves determined experimentally in the laboratory provide a guide and possibly limits to the depths and temperatures of formation of rocks, they should be applied to rock systems only with due caution.

The occurrence of chemical or mineralogical reactions in rocks should not be considered separately from the many physical parameters that affect the rocks. In the formation of most sedimentary rocks, for example, the physical aspects of transportation and deposition of solid particles are at least as significant as the chemical reactions. In magmatic processes the temperature, pressure, and phase diagram suffice to define the conditions for the onset of partial fusion, but the extent of melting is controlled largely by the amount of heat available. This depends on the physical conditions in the crustal megasystem. Migration of magma from its place of origin introduces dynamic

aspects which form no part of the static phase diagram. It is quite evident from their mineralogy and texture that many igneous rocks did not crystallize under equilibrium conditions.

The interrelationships of chemical reactions and physical controls are clearly displayed in metamorphic rocks. Metamorphic rocks can be formed at depth and subsequently uplifted to the surface with little or no mineralogical change; this fact proves that simply changing the pressure and temperature of a rock mass within the crust may not be sufficient to make a reaction proceed. Mineral reactions in metamorphic rocks are strongly influenced by the interstitial solutions filling pore spaces or absorbed as surface films on the mineral grains. The pore fluid also provides a medium for mass transport of material in response to temperature, pressure, or chemical potential gradients, and the metasomatism so produced is not represented on the equilibrium phase diagrams for closed systems. Equally important in the influence and control of chemical reactions in metamorphic rocks is the effect of stress. Folding, deformation, recrystallization, and mineral reactions in metamorphic rocks are all interdependent.

Igneous Processes

The location and source of igneous rocks within the mantle and crust, and with respect to the Rock Cycle, is illustrated in Figure 4-1. Figure 4-5 summarizes in schematic form the interrelationships among the major processes involved in the generation and emplacement of magmas. Some of these are discussed in more detail in Chapters 6 and 8.

Figure 4-5 shows the three types of rock material from which magma can be generated by partial fusion. These are (a) dry assemblages of anhydrous minerals, (b) dry assemblages of anhydrous and hydrous minerals, (c) assemblages of minerals with an aqueous pore fluid. The presence or absence of water, and its state of combination if present, are factors of fundamental importance for the

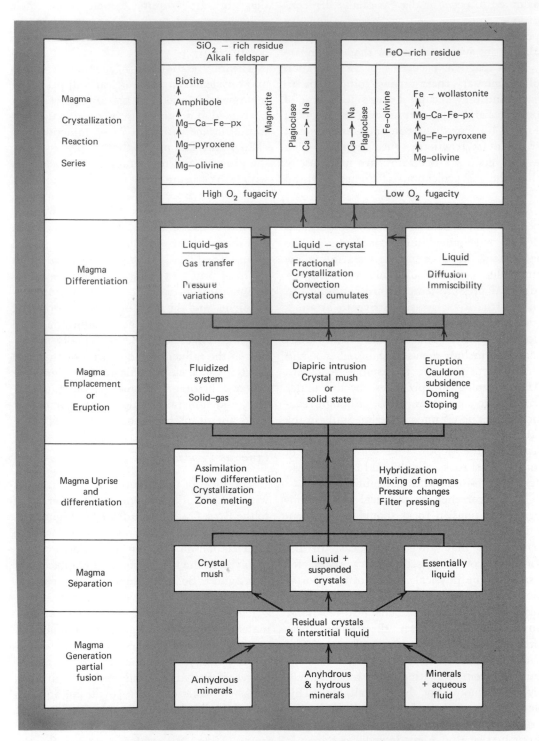

Figure 4-5. Schematic representation of igneous processes.

generation of magmas. The figure then outlines the major processes involved in (a)
magma separation, (b) uprise of magma, (c)
magma emplacement or eruption, (d) magma
differentiation either during uprise or during
solidification, and (e) the continuous and
discontinuous reaction series which dominate
differentiation in crystal-liquid systems.

Sedimentary Processes

The relationship of sedimentary rocks to
others in the Rock Cycle was illustrated in
Figure 4-1. The processes involved in their
formation are summarized in schematic form
in Figure 4-6. The main subdivision is into
the processes of (a) weathering, which begins
as soon as the rock is exposed to the hydrosphere, (b) transportation, (c) deposition, and
(d) lithification. Extending below the part of
the diagram showing transportation and
deposition in the oceans is a path leading to a
large box, which shows in more detail the
various routes that materials derived by
physical and chemical disintegration of
exposed rocks can follow into the ocean. It
also illustrates the processes leading to the
formation of the various types of sediment
occurring in different oceanic environments.

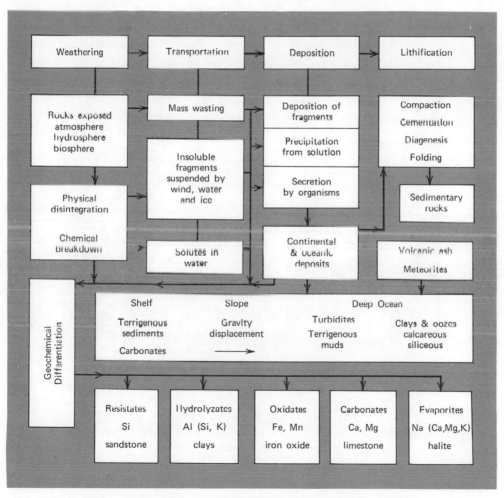

Figure 4-6. Schematic representation of sedimentary processes.

Extending from weathering boxes and transportation is a box labelled geochemical differentiation, which shows the chemical separation that occurs through sedimentary processes, and which classifies the sediments in these terms.

Metamorphic Processes

Figure 4-1 indicated that any rock could become metamorphosed if it were buried within the crust through subsidence. Subsidence is therefore labelled as process (1) in Figure 4-7, which summarizes in schematic form the major processes occurring during metamorphism. Deep burial need not be sufficient cause for regional metamorphism to occur, and box (2) indicates the possible effect of energy and material contributions from the mantle in active, orogenic belts. Up-

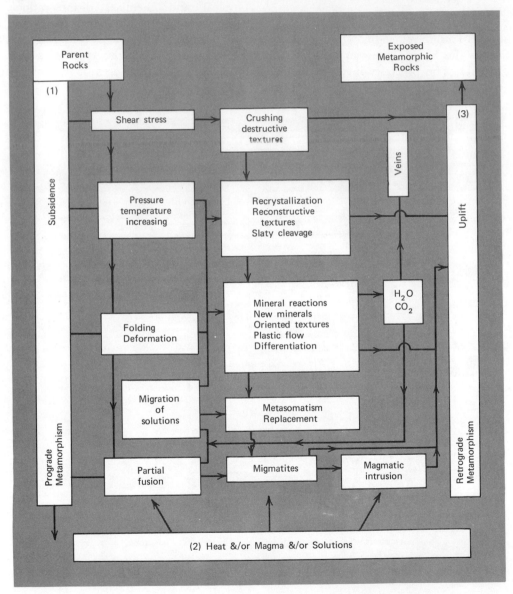

Figure 4-7. Schematic representation of metamorphic processes.

lift, which makes the examination of metamorphic rocks possible, is identified as major process (3). The main agents of metamorphism during subsidence are illustrated along with their effects upon the rocks, such as recrystallization, the formation of new minerals, and the development of textures peculiar to metamorphosed rocks.

GEOLOGICAL TIME SCALE

Geology is a historical subject, and its study can only be successful if sequences of geological events can be calibrated against a time scale. At any point on the surface of the Earth geological history is divided into (a) periods when rocks are forming either by sedimentation or by intrusion or extrusion of magma, and (b) periods of erosion when rocks are being destroyed and transported to other sites. The order and succession of rock units in time can be determined for a particular point or area, and this forms the stratigraphic column for that area. This column represents a finite time interval, including periods of rock formation, and unconformities representing periods of erosion. Lithological and biostratigraphical correlation of stratigraphic columns from different areas leads to the development of a generalized geological column representing the sequence of events in geological history without reference to any specific area. The subdivisions of this column, and the physical and biological events that they represent, provide an uncalibrated, relative time scale. Analysis of radioactive isotopes and their daughter elements permits calibration of the time scale.

Stratigraphic Classification

Two of the early attempts to subdivide the geological column into significant units are shown in Table 4-2. Charles Lyell's classification of 1833, incorporating the best of previous concepts, established the pattern to be followed for the next century. Many modified classifications were proposed. Continued study of the rocks in each group, their faunas, and the positions of major unconformities, led to the development of a catastrophic philosophy. Geological history was supposedly punctuated by repeated, synchronous changes in both physical and biological conditions. Diastrophism, represented by widespread unconformities, became the basis for correlation of stratigraphic columns and for the subdivision of the geological column. This was so in the text of T. C. Chamberlin and R. D. Salisbury in 1905, by which time classification had evolved into an essentially modern state (Table 4-2). The trend of the last fifty years has been toward a simplified classification, because detailed classifications developed to fit the stratigraphy of one area created additional problems in application to other areas. The idea of worldwide, synchronous orogeny and its relation to stratigraphy has undergone critical reexamination.

At the present time the science of stratigraphy is in a state of flux. The attention of stratigraphers in the petroleum industry has been directed by their search for oil and gas to tangible assemblages of strata that are potential structural traps, rather than to subdivisions representing portions of geological time. This led to the development of a dual classification, involving on the one hand mappable rock units, and on the other groupings of strata differentiated with respect to their position in time. This concept has now gained general acceptance, and Table 4-3 is the classification presented by W. C. Krumbein and L. L. Sloss in their text book of 1963.

Isotopic Age Determinations

All efforts to calibrate the geological time scale were failures until minerals and rocks were analyzed for radioactive isotopes and

TABLE 4-2 Stratigraphic Classification (modified from Krumbein and Sloss, 1963, Table 2-1)

Lyell, 1833		Chamberlin and Salisbury, 1905		Present Usage		Age at Base of Period, M.Y.	Duration M.Y.
Recent (Tertiary Period)	Newer Pliocene / Older Pliocene	Cenozoic	Present / Pleistocene	Cenozoic	Quaternary	1.5–2	1.5–2
	Miocene / Eocene		Pliocene / Miocene / Oligocene / Eocene		Tertiary — Neogene / Paleogene	65	63–63.5
Secondary Period	Cretaceous wealden	Mesozoic	Cretaceous Comanchean	Mesozoic	Cretaceous	136	71
	Oolite or Jura limestone group		Jurassic		Jurassic	190–195	54–59
	New red sandstone group		Triassic		Triassic	225	30–35
		Paleozoic	Permian	Paleozoic	Permian	280	55
Carboniferous Group — Coal measures			Coal measures or Pennsylvanian		Pennsylvanian (U. Carboniferous)	345	65
Carboniferous Group — Mountain limestone			Subcarboniferous or Mississippian		Mississippian (L. Carboniferous)		
Carboniferous Group — Old red sandstone			Devonian		Devonian	395	50
Carboniferous Group — Grauwacke and Transition limestone			Silurian		Silurian	430–440	35–45
Primary Period			Ordovician		Ordovician	~500	70
			Cambrian		Cambrian	570	70
		Proterozoic	Keweenawan Animikean Huronian	Precambrian	Regionally defined systems	4600	85% of Earth's history
		Archeozoic	Archean complex				

TABLE 4-3　Classification of Time and Rock Units (from Krumbein and Sloss, 1963, Table 2-3)

A. Observable units
　　1. *Rock-stratigraphic (lithostratigraphic) units*
　　　　a. Formal rock units
　　　　　　supergroup (rarely applied)
　　　　　　group
　　　　　　subgroup (rarely applied)
　　　　　　formation
　　　　　　member, tongue, lentil
　　　　　　bed (rarely applied as formal unit)

　　　　b. Informal rock units
　　　　　　sequence
　　　　　　bed (oil sands, quarry layers, key
　　　　　　　and marker beds)
　　　　　　heavy mineral zone, insoluble
　　　　　　　residue zone
　　　　　　electric-log zone, radioactivity
　　　　　　　zone, velocity zone
　　　　　　"marker-defined" unit

　　2. *Biostratigraphic units*
　　　　a. Assemblage zone
　　　　　　subzone
　　　　　　zonule

　　　　b. Range zone
　　　　　　local range zone

　　　　c. Concurrent-range zone

B. Inferential units
　　1. *Time-stratigraphic (chronostratigraphic) units*
　　　　―――――――――――――
　　　　―――――――――――――
　　　　system
　　　　series
　　　　stage

　　Geologic time units
　　　　eon
　　　　era
　　　　period
　　　　epoch
　　　　age

　　2. *Ecostratigraphic units*
　　　　a. Ecozone

　　　　b. Geologic-climate unit
　　　　　　glaciation
　　　　　　stade
　　　　　　interstade
　　　　　　interglaciation

---

their daughter elements. Because radioactive isotopes decay at a constant rate, these analyses permit calculation of the time interval since the mineral or rock was formed, provided that the amount of daughter element was zero at the time of formation, and provided that elements were neither lost nor gained during this interval. If these conditions are satisfied the calculated time interval is the age of the mineral or rock. Alternately the calculated time may date an event such as metamorphism, during which the radioactive clock was reset by the migration of elements. If the assumptions and conditions are not satisfied the decay time calculated may have no geological significance. Since it is often difficult to decide whether or not the conditions were satisfied, it is usual to refer to the times determined as "apparent ages". Table 4-4 shows the isotopes most often used for age determinations of minerals and rocks, their half-lives, and the dating range for each.

A 1964 version of the ages of the boundaries between geological periods, based on radioactive isotopes, is given in Table 4-2. The age of the Earth is about 4.6×10^9 years. The Cambrian period began 570×10^6 years ago. Precambrian time thus spans more than 85% of the Earth's history. Figure 4-8 shows the durations of the geological periods plotted against a linear time scale. It also shows the extent of the time scale for calibrated magnetic reversals, compared with the extrapolated geomagnetic time scale based on magnetic anomalies in the oceanic regions (Chapters 12 and 13).

TABLE 4.4   Radioactive Decay Schemes and Age Determination Methods

| Isotope | | Half-life, | |
Parent	Daughter	years	Effective dating range
^{40}K	^{40}Ca	1.47×10^9	
	^{40}Ar	1.19×10^{10}	10^4 years to Earth formation
^{87}Rb	^{87}Sr	$4.7, 5.0 \times 10^{10}$	10^7 years to Earth formation
^{232}Th	^{208}Pb	1.39×10^{10}	10^7 years to Earth formation
^{235}U	^{207}Pb	7.13×10^8	10^7 years to Earth formation
^{238}U	^{206}Pb	4.51×10^9	10^7 years to Earth formation
^{14}C	^{14}N	5.57×10^3	0 to 5×10^4 years
Fission tracks from U decay			0 to 1×10^8 years

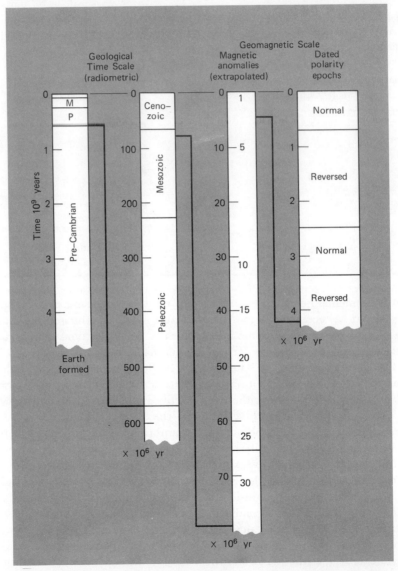

Figure 4-8.   Radiometric geological time scale compared with geomagnetic time scale. The scale for each successive column is increased by a factor of 10. For the geomagnetic scales, see Figures 12-13, 12-21, and 13-15.

5. The Nature of the Crust-Mantle Boundary and the Gabbro-Eclogite Phase Transition

INTRODUCTION

The term "crust" has been used in various ways. It was for many years applied to the relatively strong, brittle, outermost shell of the Earth, about 100 km thick, which extended down to the asthenosphere or weaker shell of the interior. As more was learned about the nature of the outer layers of the Earth, the word crust became more widely used in a petrological sense, to refer to the rocks occurring at and near the surface. It was only in 1955 that the crust was limited specifically to the rock material occurring above the Mohorovicic discontinuity. The asthenosphere is now usually equated with the low-velocity zone, and the term lithosphere (formerly stereosphere) is used for the relatively rigid shell above the low-velocity zone, which includes the crust and part of the upper mantle. This can be treated as a single tectonic unit despite the abrupt change in seismic velocities occurring within it.

For fifty years since its discovery the Moho was identified as a rather sharp boundary at which the velocity of compressional seismic waves, P-waves, increased from the range 6.8-7.2 km/sec to the range 8.0-8.2 km/sec, characteristic of the upper mantle. It was considered to be a worldwide discontinuity as shown in Figure 5-1, a compilation by A. Holmes of data available to the end of 1962. Figure 5-1 shows the contrast between conti-

nents and oceans. The average depth to the Moho in stable continental regions is 35 km, whereas in the ocean basins the average depth is 11 km, with about 1 km of sediment above a crust 5.3 km thick and the rest ocean. The oceanic crust was thought to thicken beneath some, but not all, oceanic ridges and rises. Beneath the mid-Atlantic ridge, the Moho was reported at a depth of about 25 km, but the East Pacific Rise is underlain by crust of normal thickness.

Figures 5-1b and c show the Moho extending down to depths of more than 60 km beneath high mountain ranges forming mountain roots. In general higher continental elevations are accompanied by thicker crusts in agreement with Airy's principle of isostasy, but Holmes mentioned exceptions like the Colorado Plateau and some parts of the Andes. The cross sections through island arcs and ocean trenches in Figures 5-1b and c show that the Moho is lowered to depths of about 30 km, producing considerable crustal thickening in a narrow belt.

This simple picture has been revised since 1960, largely through the results of explosion seismology, because it is now known that the upper mantle is heterogeneous and P_n is variable immediately beneath the crust (Figure 3-7a). There are also lateral variations in P for rocks in the lower crust (Figure

3-7*b*). Figure 3-7*a* shows that in the eastern part of United States P_n exceeds 8.3 km/sec, but beneath the mountainous part of western United States P_n falls as low as 7.7 km/sec. Low values of P_n have also been measured beneath other tectonically active regions, including island arcs and midoceanic ridges. The picture is further complicated in some tectonically unstable environments by the occurrence of layers of rock with *P*-wave velocities between 7.2 and 7.7 km/sec, which are intermediate between normal values for lower crust and upper mantle. In these environments the Moho ceases to give clear refractions and the terms Moho and P_n are not unambiguously defined. This led to some discussion in the early 1960's as to whether the material with intermediate *P*-wave velocities should be called anomalous lower crust, anomalous upper mantle, or mantle-crust mix.

At the present time the crust is usually defined as the outer shell of the Earth above the level where the *P*-wave velocity increases to more than 7.7 km/sec, and the Moho is then defined as the layer within which the *P*-wave velocities increase rapidly or discontinuously from crustal values to values above 7.7 km/sec. The sharpness of this layer is poorly defined, with estimates ranging from 0.1 km below parts of the Pacific Ocean through about 0.5 km in some stable continental regions, to several kilometers in other regions.

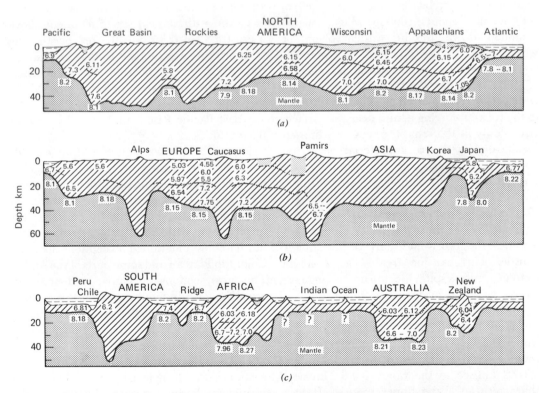

Figure 5-1. Composite seismic sections through characteristic parts of the continental and oceanic crusts. Figures in italics are the velocities of *P*-waves in km/sec. The levels of discontinuities are indicated where they have been recorded. Certain thick layers of sediments are shown by fine dots. Compiled from the data available to the end of 1962. (A. Holmes, *Principles of Physical Geology*, 2nd edition, The Ronald Press Company, New York, Thomas Nelson, London. Copyright 1965.)

CRUST-MANTLE BOUNDARY IN TECTONICALLY ACTIVE REGIONS

The layers of rock with *P*-wave velocities between 7.2 and 7.7 km/sec were earlier called anomalous, but it has now been established that they are a constant feature of tectonically active regions. For this reason, and because both crust and mantle are finely stratified, Kosminskaya and Zverev (1968) drew attention to the need for determining not only the depth to the Moho, but also the deep structure in the transition from the crust to the mantle and the layers within the mantle itself. A summary and an averaged representation of many Soviet results obtained by deep seismic soundings (DSS) are given in Figure 5-2. Crustal and upper mantle structures are subdivided into four types: continental and oceanic, with the continental margins including subcontinental and suboceanic types. The seismic velocity sections illustrated represent average characteristics for the four types. The Moho is readily identified in all sections except the subcontinental, which includes a great thickness of material with seismic velocity between 7.6 and 7.8 km/sec. Material with velocities in this range does not occur in the other three sections.

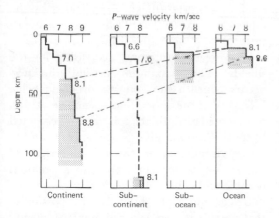

Figure 5-2. Velocity sections of the crust and upper mantle for different types of crust, averaged from data collected from many locations (after I. P. Kosminskaya and S. M. Zverev, Geophysical Monograph **12**, 122, 1968, with permission of Amer. Geophys. Union).

Figures 5-3 and 5-4 show similar data plotted in different ways by Drake and Nafe (1968) who have examined the variation of *P*-wave velocities with depth in different geological environments throughout the world. Figure 5-3*a* shows all available *P*-wave velocities from the Atlantic Ocean basins. The velocities are plotted against the depth of their measurement, depth below the solid surface being recorded rather than depth below sea level in order to reduce the scatter. The data points for mantle and crustal velocities are rather narrowly confined to the boxes in Figure 5-3*a*, although the data for the oceanic basement, layer 2, are more scattered compared to the box which encloses the most typical values. The fourth box shows the range of velocities recorded for the ocean basin sediments.

Similar diagrams were prepared for velocity data available in the various geological environments shown in Figure 5-4, and results are compared in Figure 5-3*b* for the stable ocean basins, the stable Precambrian shields with elevations of less than 500 m, and continental margins. The data for the Pacific and Indian Ocean basins fit into the boxes of Figure 5-3*a* for the Atlantic Ocean basin, so the same boxes represent all ocean basins in Figure 5-3*b*; there are distinct velocity gaps between the four boxes representing mantle, crust, layer 2, and oceanic sediments. Data for the low-lying continental shields occupy the three boxes shown; there is a distinct velocity gap between the boxes representing mantle and crust, but there is an overlap in velocities for the boxes representing lower and upper crust; velocities in the lower crust occupy the same range as those in the oceanic crust. Velocities measured beneath continental margins occupy the whole range from sediments to mantle, with a general trend of increasing velocity with increasing depth; there is no gap between the mantle velocities and the velocities characteristic of the crust beneath the stable continental

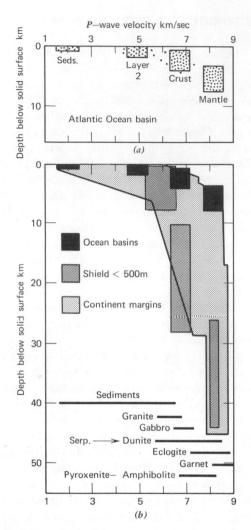

Figure 5-3. Seismic *P*-wave velocities located with respect to their depth beneath the solid surface. (*a*) Data points for the Atlantic ocean basin are concentrated into four distinct ranges, for mantle, crust, layer 2, and sediments. (*b*) Results based on world wide data are shown for three types of crust: data points for all ocean basins occupy the same ranges as in *a*; data points for the shields occupy three distinct ranges in terms of velocities and depths; data points for continental margins overlap with the specific ranges for other environments and occupy the values between as well. Velocities for rocks measured in the laboratory are plotted (after C. L. Drake and J. E. Nafe, Geophysical Monograph **12**, 174, 1968, with permission of Amer. Geophys. Union).

regions and ocean basins. Figures 5-3 and 5-2 both show that material with *P*-wave velocities between 7.2 and 7.7 km/sec is present beneath continental margins and absent beneath stable continents and ocean basins.

Figure 5-4 summarizes the data of Figure 5-3 in a different way and includes results from additional geological environments. For each environment there are four columns. Each column represents a certain range of *P*-wave velocities; the ranges 4.5–6.5 km/sec and 6.5–7.2 km/sec correspond to the crustal boxes in Figure 5-3. The range 7.2–7.7 km/sec corresponds to the velocity gap between crust and mantle boxes in Figure 5-3, and the range 7.8–8.5 km/sec corresponds to the mantle boxes in Figure 5-3. The vertical extent of the columns in Figure 5-4 shows the depth interval below the solid surface within which these velocities are encountered. This figure shows ranges of velocity in various environments, and it must not be confused with specific velocity profiles like those given in Figure 3-5*b*.

The central part of Figure 5-4 shows a progression from stable ocean basins, across continental margins and orogenic belts, to the low-lying, stable continental shields. The minimum crustal thickness, indicated by the top of the mantle column, indicates a depth to the Moho increasing from ocean basins to older orogenic belts and decreasing for the shields. The maximum crustal thickness occurs beneath the younger orogenic belts. Layers with the anomalous seismic velocities, represented by the black column, are restricted to continental margins, orogenic belts, and to the environments on the left and right of Figure 5-4, which correspond to anomalous oceanic and continental regions respectively.

Drake and Nafe discussed the possibility that the rock material with anomalous velocity is of a transient nature, representing additions to the crust from the mantle during orogenic processes. Production or disappearance of this material in response to changing conditions at depth could be responsible for up-

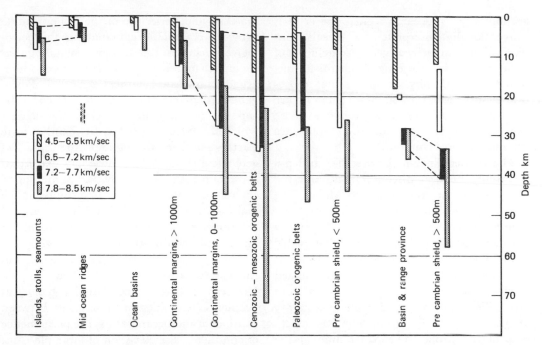

Figure 5-4. Depth intervals for *P*-wave velocities in four ranges, as they are encountered in various structural provinces; this is a summary of data gathered on a world-wide basis. The four velocity ranges correspond to (1) the upper crust, (2) the stable lower crust, (3) anomalous crust or mantle in tectonically active regions, and (4) upper mantle. See Figure 5-3 (after C. L. Drake and J. E. Nafe, Geophysical Monograph **12**, 174, 1968, with permission of Amer. Geophys. Union).

lift or subsidence and for changes in crustal thickness. Figure 5-3 includes the known seismic velocity ranges of a number of rock types and of particular interest are those for gabbro, eclogite, amphibolite, and dunite (or peridotite). These are the rock types with velocities appropriate for those measured in the lower crust and upper mantle. The most likely interpretation of the transient material discussed by Drake and Nafe is that this is related in some way to the gabbro-eclogite phase transition.

Conventionally the Moho has been regarded as a level marking a chemical discontinuity between two different types of rock. In more recent years the hypothesis that the Moho is a phase boundary between gabbro of the crust and chemically equivalent eclogite of the upper mantle has received serious consideration, and the consequences of the hypothesis have been examined in detail because of their tectonic implications. Changes of pressure or temperature at depth in the region of such a phase transition would result in uplift or subsidence at the surface of the Earth.

CHEMICAL DISCONTINUITY OR PHASE TRANSITION?

The rock materials comprising the upper mantle include peridotite, eclogite, and dunite as described in Chapters 3 and 6. Models for the lower continental crust reviewed in Chapter 7 involve gabbro, amphibolite, or rocks of intermediate composition in the high pressure granulite facies. The upper continental crust corresponds in composition to a granodiorite. In oceanic regions the crust is conventionally interpreted as basalt, but serpentine is an alternative. Arrangements of these materials providing a chemical discontinuity with physical properties appropriate for the Moho are summarized in Figure 5-5, and arrangements providing a phase transition appropriate for the Moho are summarized in Figure 5-6.

The dashed lines in Figures 5-5 and 5-6 represent chemical discontinuities, and the heavy solid lines represent the Moho. The chemical discontinuities indicated in these figures are schematic; although a boundary between upper and lower crust is detectable in many localities by geophysical methods, it is not universally present and the generalized model of a two-layer crust is passing from favor. There is also considerable uncertainty about, and possibly variation in, the thickness of the Moho. The Moho line in the figures is 1 km thick.

The Moho as a Chemical Discontinuity

Figure 5-5 shows four models for the Moho as a chemical discontinuity based on current information about the upper mantle and crust. These are:

1. Peridotite-gabbro (or basalt).

2. Peridotite-amphibolite.

3. Peridotite-granulite of intermediate composition.

4. Eclogite-granulite of intermediate composition.

In model (4) the upper mantle of gabbroic composition is composed of eclogite, because the overlying crustal material is in the eclogite facies (although of different composition). Note that the eclogite of the upper mantle does not extend indefinitely downwards, but it is terminated eventually by a chemical discontinuity against peridotite.

The conventional view that the Moho is a chemical discontinuity between basic crustal material and peridotite of the upper mantle provides reasonable explanations for petrological processes. Reasons for concluding that the upper mantle is of ultrabasic composition are discussed in Chapters 3 and 6. Petrogenesis demands that mantle material be capable of supplying the basaltic magma which has been erupted so frequently at the Earth's surface throughout the span of geological time. Furthermore it is likely that this magma is produced by partial melting of crystalline material rather than by complete melting, which is what would be required if the upper mantle were composed mainly of basaltic material in the form of eclogite.

There are difficulties facing this model, and these were discussed in 1959 by G. C. Kennedy with reference to major geological problems including the differences between continents and ocean basins, the permanence of continents, the elevation of large plateaus, the formation of geosynclines, and the origin of mountain belts. The present rate of heat loss from the Earth requires that radioactive material be concentrated in the upper mantle and crust. Because of the high concentration of heat-producing radioactive elements in rocks of the continental crust, and their low concentration in known ultrabasic rocks, we might expect higher values for heat flow from the continents than from the thin crust of the oceanic regions. The average heat flow from the continents, however, appears to be approximately the same as the average heat flow from the ocean basins (Chapter 3). This is more readily explained if the upper mantle beneath the oceans is of basaltic composition than if it is a peridotite. Finally there is the question of whether or not a chemical discontinuity between basalt and peridotite could persist for periods of the order of 10^9 years without being smoothed out by geological processes.

The Moho as a Phase Transition

Figure 5-6 illustrates two models for interpretation of the Moho as a phase transition

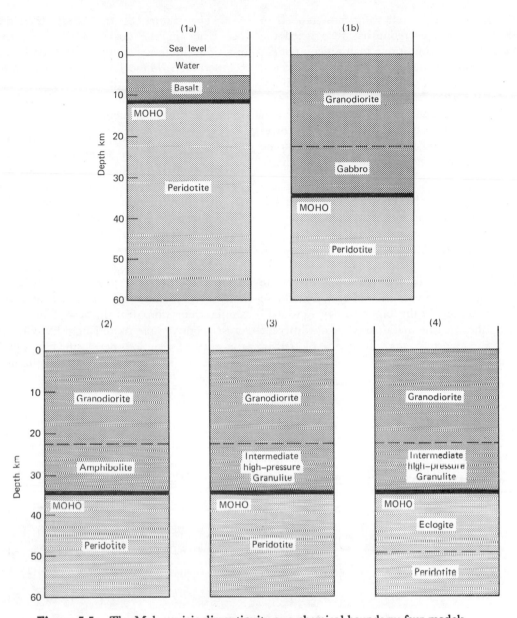

Figure 5-5. The Mohorovicic discontinuity as a chemical boundary; four models.

These are (5) peridotite-serpentinite in oceanic regions, and (6) eclogite-gabbro (basalt). Note that the eclogitic upper mantle is replaced at deeper levels by a peridotite. The Moho in model (5) is represented by a reaction involving the hydration of peridotite. Migration of water is therefore involved if it changes its position and the reaction is not an isochemical phase transition. Model (6) does involve an isochemical phase transition; the term eclogite has been used for rocks in the eclogite facies with a fairly wide range of compositions, but in this context eclogite is considered to be the chemical equivalent of the overlying basalt. Although the phase transitions are indicated in Figure 5-6 as sharp boundaries, following most models in the literature, everyone has recognized the fact that the transitions in multicomponent rocks would occur through a finite thickness, which could extend through several kilometers of rock.

Before we can decide whether or not a phase transition occurs at the depths corresponding to the Moho, in different tectonic environments, we must know the following:

1. The composition of the rock in the region of the Moho. Figure 5-5 indicates that there are different interpretations available.

2. The distribution of temperature with depth in any locality under consideration. The uncertainties here have been discussed in Chapter 3, and Figure 5-10 will illustrate the kind of problem encountered because of these uncertainties. In the following figures we will use the geothermal curves for different tectonic environments which were plotted in Figure 3-12b.

3. The nature of the phase transitions occurring in the rocks of (1) and their positions in terms of pressure (depth) and temperature. These have to be determined experimentally and critical evaluation of the phase transition hypothesis depends upon review of the experimental data.

If it appears that a phase transition layer does correspond to the Moho, then the kinetic response of the transition to changes in pressure and temperature or the availability of water (for serpentinite) must be determined for evaluation of the tectonic influence of the phase transition layer. We might expect that phase transitions near the base of the continental crust would migrate readily in response to temperature and pressure perturbations but, even with geological time available, similar migrations near the base of the oceanic crust might be far more sluggish. Whatever the rate of reaction peridotite can not be serpentinized if there is no water available.

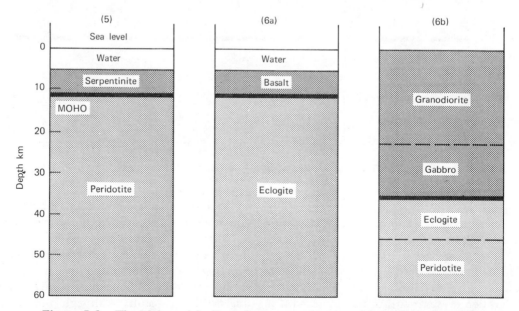

Figure 5-6. The Mohorovicic discontinuity as a phase transition zone; two models.

THE PERIDOTITE-SERPENTINITE MODEL OF H. H. HESS

Hess proposed in 1959 that layer 3 of the oceanic crust (Figure 7-8) might be serpentinite rather than the generally accepted basalt, and he has developed the idea in several papers (Chapter 12). Model (5) of Figure 5-6 is illustrated in more detail in Figure 5-7. Hess had earlier discussed the tectonic implications of serpentinization of peridotite in the upper mantle beneath the continents and Figure 5-7 provides the background for this process. Detailed examination of the process, however, will be reserved for our discussion of Figure 10-1.

Figure 5-7 is made up of three diagrams. The *P-T* diagram shows the experimentally determined univariant reactions involving the formation of serpentine from olivine:

$$Mg_3Si_2O_5(OH)_4 \quad + \quad Mg(OH)_2$$
$$\text{serpentinite (Sp)} \qquad \text{brucite (Br)}$$

$$\rightleftharpoons 2Mg_2SiO_4 \quad + \quad 3H_2O \quad (5\text{-}1)$$
$$\text{forsterite (Fo)} \qquad \text{vapor (V)}$$

$$5Mg_3Si_2O_5(OH)_4 \rightleftharpoons 6Mg_2SiO_4$$
$$\text{serpentinite (Sp)} \qquad \text{forsterite (Fo)}$$

$$+ \ Mg_3Si_4O_{10}(OH)_2 \ + \ 9H_2O \quad (5\text{-}2)$$
$$\text{talc (Ta)} \qquad \text{vapor (V)}$$

We will let reaction (5-2) represent the transition of serpentinite to peridotite by dehydration. The other two diagrams correspond to crustal models (5) in Figure 5-6 for the oceanic environment and (1*b*) in Figure 5-5 for an average continental crust.

The geotherm for continental regions on the *P-T* diagram intersects reaction (5-2) at point *A*, and that for oceanic regions intersects it at point *B*. The intersection *A* corresponds to a depth of about 55 km in the continental model, in the peridotite of the upper mantle. Given equilibrium conditions in the system peridotite-H_2O, and sufficient water, the upper mantle between the Moho and a depth of 55 km should be composed of serpentinite, with peridotite occurring only below 55 km. If this situation existed the abrupt change in physical properties corresponding to the Moho would occur at the depth of 55 km instead of the measured depth of 35 km. This indicates that a

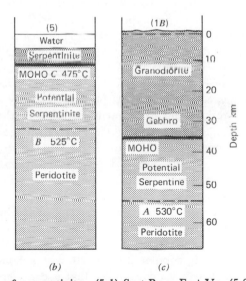

(a)   (b)   (c)

Figure 5-7. Mantle peridotite and the stability of serpentinite. (5-1) Sp + Br = Fo + V; (5-2) Sp = Fo + Ta + V (Kitahara *et al.*, 1966; Scarfe and Wyllie, 1967). Hess' proposal that the oceanic crust is composed of serpentinite, or hydrated mantle, requires that at one time the geotherm was in position *D*. Note the extent of mantle peridotite that is in the stability field of serpentinite, should water become available. See Figure 10-2.

significant layer of the upper mantle in stable continental areas (20 km in model (1*B*)) is potential serpentinite, should water become available to hydrate the mantle peridotite.

The intersection *B* in Figure 5-7 corresponds to a depth of about 32 km, 20 km below the Moho in the oceanic model (5). Therefore the Moho is definitely not an equilibrium reaction boundary between serpentinite and peridotite. An equilibrium reaction between peridotite and serpentinite would correspond to conditions given by point *C* on reaction curve (5-2) in Figure 5-7, and this requires that the geotherm passes through *C* following a curve such as *CD*. Hess proposed that generation of new ocean basin floor at the midocean ridges was accompanied by a high geothermal gradient such as *CD*. This produced an isothermal surface for 475°C at a depth of about 12 km. Water emanating from the Earth, or percolating downward from the oceans, hydrated that portion of the

mantle above the isotherm, converting it to serpentinite. After the emanation of water from the mantle ceased the geotherm migrated toward the present oceanic geotherm in Figure 5-7 and the 475°C isotherm moved to deeper levels in the Earth. The boundary between serpentinite and peridotite remained to mark the position of a fossil 475°C isotherm.

This hypothesis modifies the conventional model of the oceanic basaltic crust extending beneath the continents to form the lower continental crust (models (1*a*) and (1*b*) in Figure 5-5). The 5 km-thick layer 3 of the oceanic crust is here regarded simply as hydrated mantle, with only a very thin veneer of basalt and deep-sea sediments representing crust in the normal sense, that is material above the mantle. This is an integral part of Hess' concept of sea-floor spreading, and the proposal is reviewed in some detail in Chapter 12.

THE GABBRO-ECLOGITE TRANSITION AS THE MOHO

The suggestion that the upper mantle is composed of eclogite appears to have originated with L. L. Fermor in 1913. The idea was explored in the 1920's by V. M. Goldschmidt and A. Holmes, but the concept of a peridotite mantle overlain by a basaltic layer was preferred by most petrologists. Following theoretical and experimental studies in the 1950's, this conventional view was challenged by proposals that the Moho represent a phase transition from gabbro to eclogite, its dense chemical equivalent. The first detailed discussions of the geological and geophysical implications of this hypothesis appear to be those of J. F. Lovering in 1958 and G. C. Kennedy in 1959.

Lovering compared the Earth with achondrite meteorites concluding that a considerable portion of the Earth's upper mantle is of basaltic composition and that the Moho is therefore a phase change from basalt to eclogite. Kennedy cited several major geo-

logical observations which have been inadequately explained by the traditional view that the Moho is a chemical discontinuity. He then proceeded to develop plausible explanations for these observations by assuming that the Moho is represented by a phase transition from gabbro to eclogite. These two papers have been superseded by more recent work, but their historical significance and the lessons to be learned from them justify a rather detailed examination and explanation.

The Meteorite Analogy of J. F. Lovering

In 1958 Lovering outlined evidence supporting his model of a differentiated parent meteorite body, with 95% by volume of its silicate mantle being composed of achondrite material. He compared this model with a layered Earth, and concluded by analogy that at least the outer 100 km of the Earth is of basaltic composition. Extraterrestrial evi-

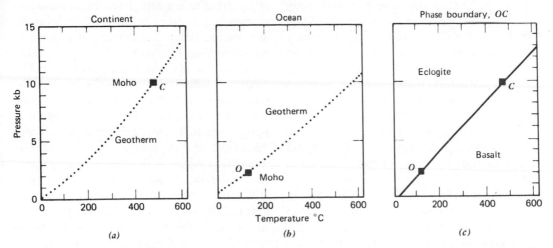

Figure 5-8. J. F. Lovering's hypothesis that the Moho is a phase transition boundary. (*a*) Given the calculated geotherm and the known depth of the Moho beneath continents, point *C* lies on the basalt-eclogite transition boundary. (*b*) Given the calculated geotherm and the known depth of the Moho beneath oceans, point *O* lies on the basalt-eclogite transition boundary. (*c*) According to the assumptions made, the basalt-eclogite transition boundary is the line through *OC*. (After J. F. Lovering, Trans. Amer. Geophys. Union, **39**, 947, 1958, with permission.)

dence for the composition and mineralogy of the Earth's mantle is reviewed in Chapter 6, and the analogy with chondrites is generally preferred over the analogy with achondrites. Lovering (1962) himself has since modified his assumptions about the relationships among chondrites and achondrites, but there does remain significant geochemical evidence that the Earth's chemistry may be closer to that of basaltic achondrites than to chondrites.

If the upper mantle is of basaltic composition then the Moho can be explained only as a phase change from basalt to eclogite. Therefore, Lovering reasoned, if the pressure and temperature conditions were known for the base of the crust beneath the oceans and the continents, this would provide two points on the basalt-eclogite phase transition curve. Temperatures in these two environments were estimated from surface heat flux and an assumed typical distribution of radioactivity and resulting geotherms are plotted in Figure 5-8. They are very similar to the curves that we have adopted in this text. A depth of 35 km to the Moho in continental regions corre-

sponds to a pressure of about 10 kb. This provides the point *C* in Figure 5-8a as a point on the basalt-eclogite phase transition curve at 480°C. Similarly the point *O* represents a depth of 11 km below sea level in oceanic regions, where the temperature is 123°C according to the calculated geotherm and this is a second point on the phase transition curve. In Figure 5-8c the phase transition curve has been drawn through these two points. Lovering noted that the position of this estimated curve was in reasonable accord with the available theoretical and experimental data, and he pointed out that the use of a univariant line for the complex rock reaction is a simplification.

Given the estimated phase boundary of Figure 5-8, and assuming that the Moho is everywhere a transition from basalt to eclogite, a knowledge of the depth of the Moho in any locality would be sufficient to provide the temperature at the Moho. Similarly for a given temperature distribution with depth, the position of basalt and eclogite layers within the Earth and the boundaries between them can be read directly from diagrams like those in Figure 5-8. When this is done for the conditions used in Figure 5-8 to derive the phase transition, the system works well for the continental region but it provides incorrect results for the oceanic region as shown in Figure 5-9.

Figure 5-9a reproduces from Figure 5-8 the continental geotherm and the phase transition curve. Figure 5-9b illustrates the crustal section used by Lovering for the temperature distribution calculation, with about 10 km of granodiorite overlying basalt, the Moho, marking the transformation of basalt to eclogite, occurs at the depth provided by the intersection point C, and the resulting section corresponds to that illustrated in (6b) of Figure 5-6. The oceanic geotherm and the

phase transition curve from Figure 5-8 are reproduced together in Figure 5-9c, with the intersection point O occurring at a pressure corresponding to the depth of the Moho, beneath 5 km of ocean and 6 km of crust. The relative positions of the geotherm and the transition boundary in Figure 5-9c, however, indicate that the material above the Moho is eclogite and that below the Moho is basalt, and this, of course, is the wrong way round (see section (6a) in Figure 5-6). This anomaly in Lovering's derivation was pointed out two years later by Harris and Rowell (1960), who noted that only if the continental thermal gradient is steeper than the oceanic one can the Moho coincide with the eclogite transformation in both environments. This is the model used by Kennedy as illustrated in Figure 5-10(a).

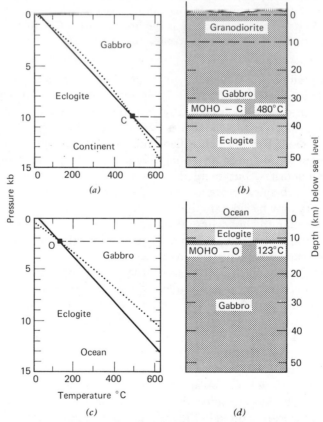

Figure 5-9. J. F. Lovering's model from Figure 5-8 expressed as crustal sections. (a) Combination of Figure 5-8a and 5-8c for the continents. (b) Crustal section for continents derived from a; 10 kms granodiorite assumed by Lovering in heat flow calculations. (c) Combination of Figure 5-8b and 5-8c for the oceans. (d) Crustal section for oceans derived from c.

G. C. Kennedy's Model

Lovering adopted a model for the mantle requiring the existence of a phase transition, and deduced where the phase transition curve would be on the basis of the known Moho depths and calculated temperature distributions. Kennedy (1959) adopted the phase transition hypothesis and selected a phase transition curve and geotherms which provided intersections at pressures corresponding to the Moho depths in different environments. His scheme is presented in Figure 5-10a, and what the intersections mean in terms of crustal sections are illustrated in Figures 5-10b, c, and d. The oceanic geotherm intersects the phase transition curve at point A, providing the crustal section shown in Figure 5-10b, and similarly the intersections B and C in Figure 5-10a provide the crustal sections in

Figure 5-10c and 5-10d respectively. These correspond to sections (6a) and (6b) in Figure 5-6.

The main difference between the models of Lovering and Kennedy is in the positions of the geotherms for oceanic and continental regions. Kennedy's model certainly explains the variation of depth to the Moho in different tectonic environments, but unfortunately it does not fit with current opinions for temperature distributions at depth. In Figure 5-10e the same phase transition curve is compared with the two geotherms adopted in Figure 3-12b. Intersection of the oceanic geotherm at the point E in Figure 5-10e provides the crustal section shown in Figure 5-10f, with a depth to the phase transition (the Moho) of about 25 km, which is too deep to satisfy geophysical measurements (Figure 5-5). Similarly the intersection of the

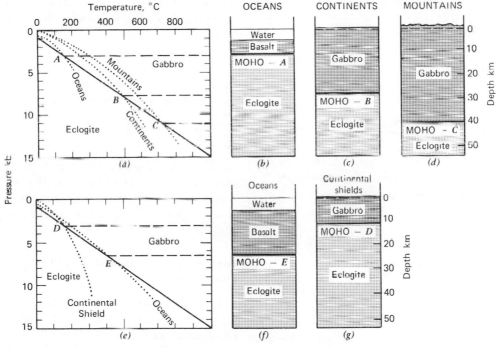

Figure 5-10. G. C. Kennedy's hypothesis that the Moho is a phase transition boundary. (a) Assumed gabbro-eclogite transition boundary, and geotherms for different environments. (After Kennedy, 1959, with permission, *American Scientist*, Journal of the Society of the Sigma Xi). (b), (c), and (d). Crustal sections derived from a. (e) Assumed gabbro-eclogite transition boundary from a with geotherms from Figure 3-12. (f) and (g) Crustal sections derived from e.

continental shield geotherm at the point *D* provides the crustal section in Figure 5-10*g* with the phase transition occurring at a depth of only 12 km, which is too shallow for continental regions. It is also too shallow for the occurrence of a basaltic layer according to most crustal models (Figures 5-5 and 5-6). Thus the arrangements of geotherms following the current best estimates do not support the occurrence of a basalt-eclogite phase transition beneath both continents and oceans, using Kennedy's estimate for the phase transition boundary as a basis for analysis.

Figure 5-10 shows that quite small changes in the positions of either the phase transition boundary or of the geotherms would cause significant variation in the depths to the phase transition in crustal sections. In order to evaluate the hypothesis, therefore, it is most important to know precisely the temperature distributions and position of the phase transition. The problems in connection with the geotherms have already been discussed so let us now examine the experimental evidence for the phase transition. More is available than when Lovering and Kennedy presented their papers.

EXPERIMENTAL STUDIES ON THE GABBRO-ECLOGITE PHASE TRANSITION

In this context eclogite is considered to be the high pressure, dense chemical equivalent of a gabbro. The olivine, pyroxene, and plagioclase of the gabbro are replaced by omphacite and pyrope-rich garnet. The transformation of gabbro to eclogite involves a series of complex reactions which can be oversimplified as follows:

$3Mg_2SiO_4$ + $3CaMgSi_2O_6$
olivine (Ol)   pyroxene (Px)

  + $[3CaAl_2Si_2O_8$ + $2NaAlSi_3O_8]$
plagioclase (Pl)

    $\rightleftharpoons 3CaMg_2Al_2Si_3O_{12}$
garnet (Ga)

    + $[2NaAlSi_2O_6 + 3CaMgSi_2O_6]$
omphacite (Om)

      + $2SiO_2$ (5-3)
quartz (Qz)

In addition to the formation of pyrope and jadeite there are reactions involving increased solubility in pyroxenes of Al_2O_3 from the anorthite component of the plagioclase and the formation of almandine and grossular components in garnet at lower pressures than the pyrope-rich garnet which characterizes eclogite.

There are three experimental approaches providing information about the location of

the gabbro-eclogite phase transition in terms of pressure and temperature:

1. Locate experimentally the positions of the reaction curves for the conditions of breakdown or formation of individual minerals, and of more complex reactions involving the stability of solid solutions.

2. Locate the positions of univariant reactions involving mineral assemblages with bulk compositions approaching that of the rock.

3. Study the transition interval in natural or synthetic rock samples.

We will examine these experiments in some detail to provide examples of the difficulties involved in obtaining conclusive results in laboratory phase studies of complex systems.

Instability of Crustal Silicates at High Pressures

The breakdown of plagioclase to yield jadeite is essential for the formation of omphacite, and the two appropriate univariant reactions are plotted in Figure 5-11:

$NaAlSi_3O_8$ + $NaAlSiO_4$
albite (Ab)   nepheline (Ne)

      $\rightleftharpoons 2NaAlSi_2O_6$ (5-4)
jadeite (Jd)

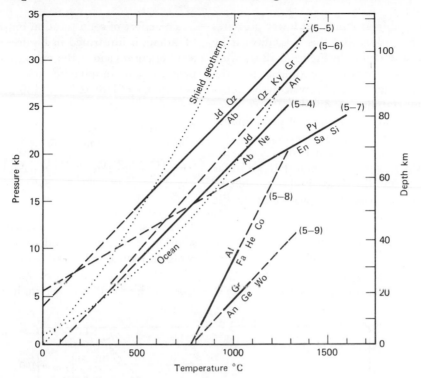

Figure 5-11. Mineral reactions related to the gabbro-eclogite transition. Solid lines show the limits of experimental measurement; dashed lines are extrapolations. (5-4) $Ab + Ne \rightleftharpoons Jd$; (5-5) $Ab \rightleftharpoons Jd + Qz$ (Boettcher and Wyllie, 1968); (5-6) $An = Gr + Ky + Qz$ (Newton, 1966, Boettcher, 1970); (5-7) $En + Sa + Si \rightleftharpoons Py$ (Boyd and England, 1962); (5-8) $Al \rightleftharpoons Co + He + Fa$ (Yoder, 1955, Hsu, 1968); (5-9) $Gr \rightleftharpoons An + Ge + Wo$ (Newton, 1966, Boettcher, 1970).

$$\underset{\text{albite (Ab)}}{NaAlSi_3O_8} \rightleftharpoons \underset{\text{jadeite (Jd)}}{NaAlSi_2O_6} + \underset{\text{quartz (Qz)}}{SiO_2} \quad (5\text{-}5)$$

In their reports of the first experimental studies of portions of these reaction curves in 1957 and 1960 Robertson, Birch, MacDonald, and LeComte noted with due caution that, although the results give an indication of the position and slope of the basalt-eclogite transition, they are insufficient to determine whether or not the Moho was a phase change.

The anorthite component of plagioclase also breaks down under similar pressure conditions, yielding grossular garnet and other minerals as shown in Figure 5-11:

$$\underset{\text{anorthite (An)}}{3CaAl_2Si_2O_8} \rightleftharpoons \underset{\text{grossular (Gr)}}{Ca_3Al_2Si_3O_{12}}$$

$$\underset{\text{kyanite (Ky)}}{2Al_2SiO_5} + \underset{\text{Quartz (Qz)}}{SiO_2} \quad (5\text{-}6)$$

The stability field of pyrope, bounded on its low pressure side by dense silicate minerals, was located by Boyd and England as shown in Figure 5-11:

Aluminous enstatite (En) + sapphirine (Sa)

+ sillimanite (Si) $\rightleftharpoons$ pyrope (Py)   (5-7)

They suggested that the pyrope and jadeite stability curves could together be taken as an approximation to the basalt-eclogite transition. Yoder's preliminary curve for the stability of almandine in Figure 5-11:

$$\underset{\text{almandine (Al)}}{5Fe_3Al_2Si_3O_{12}} \rightleftharpoons \underset{\text{Fe-cordierite (Co)}}{2Fe_2Al_4Si_5O_{18}}$$

$$+ \underset{\text{hercynite (He)}}{FeAl_2O_4} + \underset{\text{fayalite (Fa)}}{5Fe_2SiO_4} \quad (5\text{-}8)$$

shows that the Fe-bearing garnet in eclogite

would be stable at somewhat lower pressures than pure pyrope, and this is confirmed by the schematic curves in Figure 5-12 showing the pressure conditions required for the formation of garnets with given pyrope contents. The stability conditions of garnet are further clouded by the effect of CaO. Figure 5-11 shows that at high pressures grossular is produced along with other phases by the breakdown of anorthite, but that grossular also has a stability field similar to that of almandine; it decomposes according to the reaction:

$$2Ca_3Al_2Si_3O_{12} \rightleftharpoons CaAl_2Si_2O_8$$

grossular (Gr)   anorthite (An)

$$+ \ Ca_2Al_2SiO_7 + 3CaSiO_3 \quad (5\text{-}9)$$

gehlenite (Ge)   wollastonite (Wo)

The effect of solid solution on the stability of jadeite is illustrated in Figure 5-12, where a set of curves shows the pressure required to produce an omphacite with a specified percentage of jadeite in solid solution according to the reaction:

$$CaMgSi_2O_6 + NaAlSi_3O_8$$

diopside (Di)   albite (Ab)

$$\rightleftharpoons [CaMgSi_2O_6 + NaAlSi_2O_6] + SiO_2$$

omphacite (Om)   quartz (Qz)

$$(5\text{-}10)$$

These results indicate that considerable solid solution of diopside in jadeite is required in order to reduce the pressure of formation by more than a kilobar or two and the same is true for the solution of acmite in jadeite.

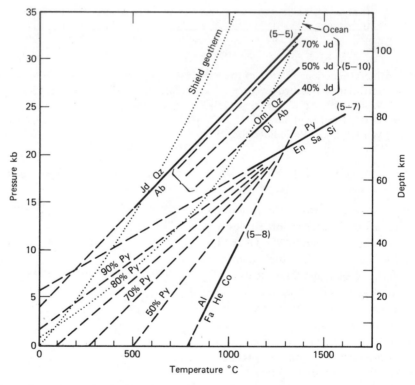

Figure 5-12. Mineral reactions related to the stability of omphacite and pyrope solid solutions. Solid lines show the limits of experimental measurement; dashed lines are extrapolations or estimates. See Figure 5-11 for reactions (5-5), (5-7), and (5-8). (5-10) Di+Ab ⇌ Om+Qz (Kushiro, 1969); the lines show pressure required for formation of omphacite with specific jadeite content. The dashed lines between reactions (5-5) and (5-8) show the pressure required to form a garnet with specific pyrope content (based on *PX* and *TX* diagrams of Yoder and Chinner, 1960).

The intersections of the two geotherms with the reaction curves in Figures 5-11 and 5-12 indicate that crystalline gabbro would be converted to eclogite at no great depth within the Earth, but there are large margins of uncertainty regarding the depth at which the change occurs. The depth would obviously be strongly dependent upon the composition of the gabbro and in particular upon the degree of silica saturation as indicated by the two curves for jadeite in Figure 5-11. Figure 5-12 demonstrates the depth dependence imposed by other compositional variables.

Reactions Among Mineral Assemblages

Experimental results for the stability of mineral assemblages are plotted in Figure 5-13. The reaction:

$$4MgSiO_3 + MgAl_2O_4$$
enstatite (En)   spinel (Sp)

$$\rightleftharpoons Mg_3Al_2Si_3O_{12} + Mg_2SiO_4 \quad (5\text{-}11)$$
pyrope (Py)   forsterite (Fo)

indicates that in a silica-undersaturated environment, a higher pressure is required to produce pyrope along with forsterite than to produce pyrope alone (compare Figure 5-12). The other reactions plotted in Figure 5-13, most of which involve the formation of a garnet, indicate the complexities involved in a four-component system, $CaO\text{-}MgO\text{-}Al_2O_3\text{-}SiO_2$, which is simple compared with a whole rock composition.

Kushiro and Yoder (1966) studied reactions occurring between anorthite and forsterite and between anorthite and enstatite. The mixtures used approached gabbro in composition. Univariant reactions encountered

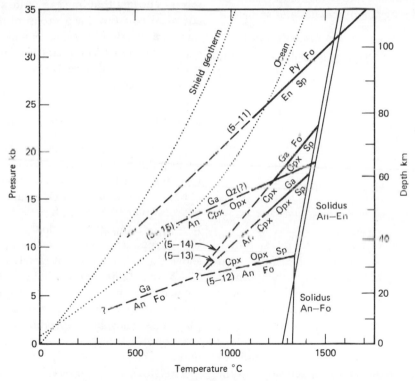

Figure 5-13. Mineral reactions in the system $CaO\text{-}MgO\text{-}Al_2O_3\text{-}SiO_2$ after Kushiro and Yoder (1966): (5-12) $An + Fo \rightleftharpoons Cpx + Opx + Sp$; (5-13) $An + Cpx + Opx + Sp \rightleftharpoons Ga$; (5-14) $Cpx + Opx + Sp \rightleftharpoons Ga + Fo$; (5-15) $An + Cpx + Opx \rightleftharpoons Ga + Qz$ (?). Reaction (5-11) is from MacGregor (1964): $En + Sp \rightleftharpoons Py + Fo$. Solid lines show the limits of experimental measurements; dashed lines are extrapolations. (With permission from Journal of Petrology, Clarendon Press, Oxford.)

involve the formation of aluminous pyroxenes and of garnets with compositions represented by the formulas $CaMgSi_2O_6.xCaAl_2SiO_6$, $MgSiO_3.xMgAl_2SiO_6$, and $CaMg_2Al_2Si_3-O_{12}$:

anorthite (An) + forsterite (Fo)
$$\rightleftharpoons \text{aluminous diopside (Cpx)}$$
+ aluminous enstatite (Opx) + spinel (Sp)

$$(5\text{-}12)$$

and at a higher pressure any remaining anorthite is used up by the reaction:

anorthite (An) + aluminous diopside (Cpx)
+ aluminous enstatite (Opx) + spinel (Sp)
$$\rightleftharpoons \text{garnet (Ga)} \quad (5\text{-}13)$$

the garnet having a composition near 75% pyrope, 25% grossular (Py_3Gr_1). Additional garnet is produced at a higher pressure by reaction among the pyroxenes and spinel:

aluminous diopside + aluminous enstatite
+ spinel $\rightleftharpoons$ garnet + forsterite   (5-14)

Reaction between anorthite and enstatite to produce aluminous pyroxenes is rather similar to reaction (5-12), except that quartz is produced instead of spinel. At a higher pressure anorthite reacts with the aluminous pyroxenes to yield garnet:

anorthite + aluminous diopside
+ aluminous enstatite = garnet
+ quartz(?)   (5-15)

This reaction occurs at a higher pressure than the similar reaction (5-13), indicating that the formation of garnet with composition close to Py_2Gr_1 takes place at higher pressures in silica-saturated environments than in silica-undersaturated environments. A similar statement is true for the formation of jadeite as shown by reactions (5-4) and (5-5) in Figure 5-11. Note also that the slope of the garnet-producing reactions (5-13) and (5-15) changes markedly with silica-saturation.

It appears from these results that mineral assemblages of anorthite with enstatite and forsterite are transformed through a transition interval characterized by aluminous pyroxenes, with garnet becoming stable on the high pressure side. Extrapolation of reactions (5-12), (5-13), and (5-14) to lower temperatures would probably produce an invariant point pinching out the aluminous pyroxene field, with a reaction involving the direct formation of a garnet from anorthite and forsterite at temperatures below about 750°C. None of the reactions in Figure 5-13 were determined at temperatures below 1100°C and it is a long extrapolation from the determined curves to the geotherms. The dependence of the positions and slopes of the garnet-producing reactions upon the chemical composition of the reacting system confirms the necessity for examining the gabbro-eclogite phase transition in systems approaching more closely the compositions of the rocks themselves. The fact that one can write a balanced chemical equation for a group of minerals is no guarantee that the reaction will occur, and even less of a guarantee that the reaction will occur in a complex rock system containing additional components.

Stability Limits of Gabbro and Eclogite

Figure 5-14*a* shows the preliminary diagram published by Yoder and Tilley in 1962 together with the definitive runs on samples of basalt and eclogite. Their results just below the solidus at 10 and 20 kb provide a reversed bracket for the gabbro-eclogite transition at high temperatures but, as the plotted runs show, the transition is not closely defined. At 20 kb and 1200°C, an olivine tholeiite was converted completely to an eclogitic assemblage (clinopyroxene + garnet) and at 10 kb, plagioclase feldspar was produced from partial reaction of natural eclogite in runs at 900°C and above.

The slope of the interval is based on the results of Kennedy (1959) at 500°C. Using a basaltic glass (fused olivine tholeiite) as starting material, he reported that plagioclase grew at 10 kb, decreased in amount with increasing pressure, and in later

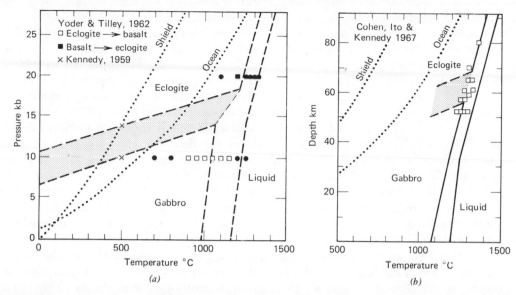

Figure 5-14. The gabbro-eclogite phase transition. Solid lines show the limit of experimental determination; dashed lines are preliminary estimates. Compare Figure 5-15. (*a*) After Yoder and Tilley (1962). Runs at 10 and 20 kbar were made with natural rock samples. Kennedy's preliminary runs were made with a basaltic glass. (With permission from Journal of Petrology, Clarendon Press, Oxford.) (*b*) After Cohen *et al.* (1967). Runs were made with a basaltic glass. (With permission from American Journal of Science, Yale University.)

experiments reported to Yoder and Tilley (pp. 472 and 504) was replaced by jadeitic pyroxene by 14 kb. Yoder and Tilley retained reservations about these results because garnet, essential for eclogite, was not produced. In a more recent paper Kennedy emphasized the problem of nucleation of garnet (Cohen et al., 1967, p. 514). In the absence of other information, however, Figure 5-14a was presented as the best estimate available for the gabbro-eclogite transition. This preliminary phase diagram has been widely reproduced in the literature, and applied to problems related to the crust and upper mantle. It is testimony to the difficulty of obtaining satisfactory results in subsolidus mineral reactions that, despite the great significance of this transition, it was not until five years later that any more results were published in detail. A preliminary report by Ringwood and Green (1964) on the re-crystallization of quartz tholeiite glasses was

followed in 1966 by a comprehensive discussion of the geophysical applications of their results, and the detailed experimental data were published in 1967 (Green and Ringwood). In the same year Cohen, Ito, and Kennedy (1967) described results obtained with the olivine tholeiite glass used by Kennedy in his earlier experiments at 500°C.

Cohen, Ito, and Kennedy completed determination of the melting relationships of their basalt up to 40 kb pressure, with the results shown in Figure 5-14b; the only runs plotted are those related to the gabbro-eclogite phase transition. Their results show that the phase transition at the basalt solidus, in the temperature interval 1250 to 1300°C, lies between 16 and 21 kb with the minimum possible width being between 17.2 and 19.9 kb. Ten exploratory subsolidus runs at temperatures between 1000 and 1150°C were made in an attempt to locate the slope of the transition interval. The samples were usually inhomogeneous and results were inconsistent. The authors concluded that development of other techniques would be necessary for location of the gabbro-eclogite phase transition at lower temperatures. The difficulty of garnet nucleation, noted by all previous investigators of reactions involving pyrope, becomes an increasingly serious problem with decreasing temperature and pressure. Details of Kennedy's earlier experimental data at 500°C do not appear to have been published and his results were not used by Cohen, Ito, and Kennedy to estimate the position of the transition zone at lower temperatures. They did, however, discuss the gradient required for extrapolation of their results in such a way that the transition would occur in the region of the continental Moho, and they noted that relevant reactions such as the formation of pyrope, reaction (5-7) in Figure 5-11, and the reaction between anorthite and forsterite, reaction (5-12) in Figure 5-13, do have appropriate gradients. Such an extrapolation would be consistent with Kennedy's 500°C results and would thus be rather similar to the zone indicated by Yoder and Tilley.

Green and Ringwood (1967) used a glass prepared from a natural quartz tholeiite as their main starting material, and by adding oxides and minerals in required proportions they obtained seven other glasses representing a range of basaltic types. Their results for quartz tholeiite B are shown in Figure 5-15a. Only the definitive runs are plotted, those which locate the solid lines bounding the transition interval between gabbro and eclogite. The lower line represents the appearance of garnet, and the upper line represents the disappearance of plagioclase; the transition interval was located between 1000°C and the solidus at about 1250°C. The transitional rocks were called garnet granulites in contrast to the gabbros or pyroxene granulites. The corresponding diagram of Green and Ringwood (Figure 7) includes additional runs on other quartz tholeiite compositions, and it shows the extent of the shifts produced in the boundary positions by minor changes in rock composition.

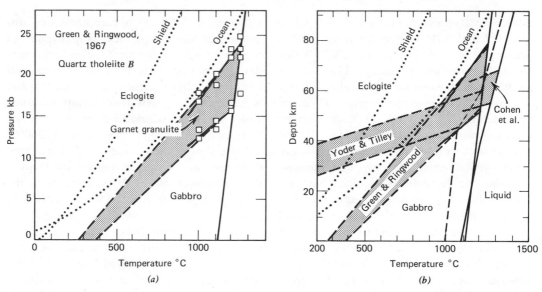

Figure 5-15. The gabbro-eclogite phase transition. Solid lines show the limit of experimental determination; dashed lines are extrapolations or estimates. Compare Figure 5-14. (*a*) After Green and Ringwood (1967). Runs were made with basaltic glass. (From Geochimica Cosmochimica Acta, **31**, 1967, with permission. Copyright Pergamon Press.) (*b*). Comparison of transition intervals from *a* and Figures 5-14*a* and *b*.

Most runs using the other basaltic compositions were limited to 1100°C. Each basalt glass crystallized in the same three principal mineral stability fields and at 1100°C the transition interval between the low pressure and high pressure assemblages varied in width from 3.5 to 12 kb. The positions of the transition intervals for two of the basalts are shown in Figure 5-16.

As Ringwood and Green pointed out (1966, p. 399) extrapolation of the boundaries of the garnet granulite field to lower temperatures has an appreciable uncertainty. The average gradient of the two experimental boundaries in Figure 5-15a is 21 bar/°C, which is similar to the slope of the jadeite reactions and several others plotted in Figures 5-11, 5-12, and 5-13. This slope was used for the extrapolation shown by the dashed lines in Figure 5-15a, with the width of the transitional zone made proportional to the absolute temperature.

The three interpretations for the position of the basalt-eclogite transition zone, based on experimental studies with natural rock compositions, are compared in Figure 5-15b. The subsolidus runs of Yoder and Tilley, the reversals with crystalline rock samples plotted in Figure 5-14a, are consistent with the results obtained by Cohen et al. with olivine tholeiite glass (Figure 5-14b). Also the subsolidus results of Green and Ringwood for the beginning of the transition zone marked by the appearance of garnet, using quartz tholeiite glass (Figure 5-15a), are consistent with the other two sets of results. There is a significant difference in the pressure required for the elimination of plagioclase, the curve of Green and Ringwood indicating higher pressures at temperatures near the solidus. With this exception, which could be the effect of compositional differences among the starting materials, the results are in reasonable agreement at temperatures near the solidus. The reaction curves in Figure 5-13 for the appearance of garnet and the disappearance of anorthite in the system $CaO-MgO-Al_2O_3-SiO_2$ are also consistent with the rock results at near-solidus temperatures.

The gradient for the transition interval presented tentatively by Yoder and Tilley depended upon Kennedy's observation that basaltic glass yielded plagioclase feldspar at 10 kb and 500°C, and this result conflicts with the extrapolation of Green and Ringwood's results also obtained with glass starting material. The pressure difference of more than 5 kb at 500°C is too large to be explained by uncertainties in pressure measurement in different apparatus and the discrepancy probably relates to metastable crystallization.

Many experimenters have found that subsolidus runs of short duration using glass as a starting material have yielded metastable products (Fyfe, 1960) and it is well known that metastable plagioclase is produced readily from glass, whereas magnesian garnet is extremely reluctant to nucleate from glass. Another complication is the tendency of metastable pyroxene to form first in basaltic glasses. Cohen, Ito, and Kennedy did not report the results of their subsolidus experiments at temperatures below 1200°C because they obtained inconsistent results. Green and Ringwood described experiments at 1100°C using the quartz tholeiite composition which were designed to investigate the attainment of equilibrium and reversibility of the plagioclase boundary, and they concluded that "there is no evidence that the use of glassy starting material causes growth of long-persisting metastable phases or nonnucleation of other phases at least for the compositions and experimental procedures used in this study".

The work of Green and Ringwood (Figure 5-15a) on subsolidus relations in basaltic compositions is the most detailed study available, but I would be surprised if the last word has been written on the position of the gabbro-eclogite phase transition at low temperatures. The difficulty of locating subsolidus reactions is indicated by the history of successive reports for the experimental determination of transitions among the polymorphs of Al_2SiO_5, most reports differing from the others and many with confident claims that reversibility was achieved. Location of the curve bounding the transition interval on its high pressure side (Figure 5-15a) is based on the limit of x-ray detection of plagioclase. In these multiphase assemblages such a procedure is not likely to provide a reliable slope for long extrapolation to low temperatures. Furthermore the effect of bulk composition on the slope of the transition zone remains to be determined. Figure 5-11 illustrates the effect of silica saturation on the pressure of formation of one eclogitic component, jadeite. The positions of reactions (5-13) and (5-15) in Figure 5-13 demonstrate that the degree of silica saturation may have a marked effect on the other eclogitic component, pyrope-rich garnet, not only with respect to the pressure of formation but also with respect to the slope of the garnet-producing reaction.

The Transition of Gabbro to Eclogite

Green and Ringwood studied the phase relationships for a number of basaltic compositions at 1100°C, and illustrated in diagrammatic form the nature of the mineralogical changes occurring through the transition zone for quartz tholeiite *B* and for alkali olivine basalt. The proportions of minerals present were estimated by comparison of powder patterns and diffractometer records of runs with specially prepared standards. Their figure incorporated the results of runs and interpretations of inferred reactions occurring among minerals; a somewhat simplified version is given in Figure 5-16. Note that the transition zone for the alkali olivine basalt is wider than that for the quartz tholeiite and it occurs at a lower pressure. With increasing pressure the anor-

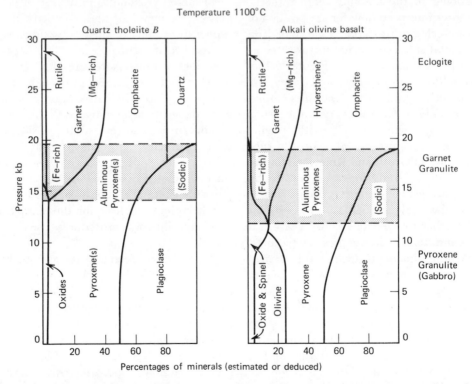

Figure 5-16. The mineralogical changes occurring within the basalt-eclogite transition for two different rocks at 1100°C, after Green and Ringwood (1967) and Ringwood and Green (1966). This is a simplified version. Note that the pressure of the phase transition interval varies with rock composition. (From Geochimica Cosmochimica Acta, **31**, 1967, with permission. Copyright Pergamon Press.)

thite component of the plagioclase reacts with the pyroxenes and olivine to yield aluminous pyroxene—see reaction (5-12)—with garnet being formed at a slightly higher pressure as in reactions (5-13) and (5-15). The pyroxenes probably contain Al_2O_3 in the form of Tschermak's molecule with the jadeite content remaining low until the upper part of the transition zone. Within the transition interval the mineralogical changes are gradual with steady decrease in plagioclase and increase in garnet. The garnet becomes richer in pyrope through the interval, and the plagioclase becomes richer in albite until this breaks down to yield the jadeite component for omphacite.

Other inferred mineralogical changes were discussed in detail by Green and Ringwood.

Ringwood and Green discussed the transition through garnet granulite in some detail in their 1966 paper with specific reference to Figure 5-16. They concluded that to a first approximation, but beyond reasonable doubt, both density and seismic velocity change regularly throughout the garnet granulite interval, from gabbro with density of 3.0 g/cm^3 to eclogite with average density 3.5 g/cm^3. Their measured results for the alkali olivine basalt at 1100°C are given in Figure 5-17. Note the regular density varia-

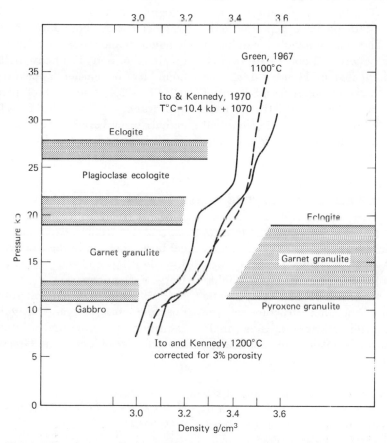

Figure 5-17. Experimentally measured variation in density as gabbro is transformed to eclogite (see transition interval in Figure 5-16). According to Green (1967) the density changes continuously through the transformation interval, but according to Ito and Kennedy (1970 and personal communication) there are two distinct density changes at the beginning and near the end of the transformation interval. (Based on preprint figure kindly supplied by K. Ito and G. C. Kennedy.)

tion through the garnet granulite interval after the first change.

In 1970 Ito and Kennedy reported experiments showing sharp density changes within the transition interval:

1. From gabbro with density near 3.0 g/cm³ to garnet granulite with density near 3.2 g/cm³.

2. From garnet granulite to eclogite with density 3.4–3.5 g/cm³.

They used the same olivine tholeiite glass previously studied with Cohen (Figure 5-14*b*). Runs were made between 6 and 35 kb at points on a line believed to be 20–50°C below the solidus. The density variation curve through their data points is reproduced in Figure 5-17. The change for the incoming of garnet occurs between 12 and 14 kb, or 5 kb lower than indicated in Figure 5-14*b*. The earlier results were too high because garnet nucleates with difficulty; the 1970 results were obtained by seeding the glass starting material with powdered garnet granulite of the same bulk composition.

Ito has since informed me that the runs for the 1970 results were actually just above the solidus, so they repeated the work at 1200°C with no glass present. Figure 5-17 compares their revised density curve, corrected for 3% porosity in the run charges. Two sharp density changes near 12 and 21 kb are still present, and there is also a less marked density change near 27 kb where plagioclase disappears. According to these results it appears that the gabbro-eclogite transition could be considered as composed of three transitions; two of these are associated with density changes sufficient to cause seismic discontinuities.

Plagioclase feldspar persists in plagioclase-eclogite for a pressure interval above the plagioclase-out boundary determined by Ringwood and Green for the alkali olivine basalt. Ito and Kennedy determined the boundary of garnet appearance as $P = 0.014T - 5.4$, and that of plagioclase disappearance as $P = 0.020T + 4.0$, with reactions being reversed at 800°C and higher. These boundaries should be added to Figure 5-15*b*: an exercise for students.

DEVELOPMENT OF THE PHASE TRANSITION HYPOTHESIS

Within the space of a few years the status of the phase transition hypothesis as an explanation for the Moho has changed several times. It is clear from the preceding review that neither the structure of the crust-mantle boundary, nor the fine structure of the gabbro-eclogite phase transition is as simple as it appeared to be in the late 1950's when the hypothesis was developed by Lovering and Kennedy.

We can review the changing status of the phase transition hypothesis and the development of ideas about the nature of the Moho, against a series of historical markers.

1. The revival of the hypothesis in the late 1950's by Lovering and Kennedy.

2. The preliminary experimental work on the transition by Yoder and Tilley in 1962.

3. More detailed experimental studies on gabbroic rocks published between 1966 and 1968 by Ringwood and Green; and by Cohen, Ito, and Kennedy.

4. Experimental determination in 1970 of the fine structure of the gabbro-eclogite transition by Ito and Kennedy, which evoked a rejoinder from Green and Ringwood in 1971.

Revival by J. F. Lovering and G. C. Kennedy

The models presented by Lovering and Kennedy have been described and reviewed in Figures 5-8, 5-9, and 5-10. Figure 5-9 shows that Lovering's model is internally inconsistent, and Figure 5-10 shows that Kennedy's model is inconsistent if currently accepted geotherms are employed. It works only if temperatures beneath the oceans are lower than those beneath the continents at moderate depths. Several papers published during the following years pointed out that the Moho beneath both continents and oceans could not readily be explained as the result of the same phase transition.

One of the difficulties facing the phase transition hypothesis relates to surface heat flow. If the Moho is a phase transition variations in surface heat flow, which reflect variations in the positions of geotherms at depth, should be accompanied by changes in the thickness of the crust. Wide variations in heat flow measurements from the Pacific Ocean are not accompanied by changes in crustal thickness. Observations of this kind were cited as evidence against the phase transition hypothesis in oceanic regions. Hess's proposal that the Moho beneath the oceanic crust represents a fossil isotherm for the reaction between serpentinite and peridotite, according to model (5) in Figure 5-6, is independent of present values of heat flow.

Modifications Based on Results of H. S. Yoder and C. E. Tilley

The phase diagram published by Yoder and Tilley in 1962, and illustrated in Figure 5-14a, supported many proposals that the Moho beneath the continents, at least in some environments, could be caused by the gabbro-eclogite phase transition, whereas the Moho beneath the oceans could be caused either by a chemical change from basalt to peridotite or by some other phase transition, such as serpentinite-peridotite (models 5 and 6b in Figure 5-6).

Observations on surface heat flow and crustal thickness do not conflict with the phase transition hypothesis if the model presented by Yoder and Tilley (Figure 5-14a) is correct, because of the shallow gradient of the phase transition, about 7 bar/°C. Cohen, Ito, and Kennedy also seem prepared to accept a shallow gradient of about 10 bar/°C. Green and Ringwood, on the other hand, presented experimental evidence that the slope of the phase transition zone was much steeper, 21 bar/°C (Figure 5-15a), which cannot be reconciled with the phase transition hypothesis because it would produce large variations in crustal thickness with surface heat flow.

Another difficulty introduced by the work of Yoder and Tilley is the indication that the gabbro-eclogite phase transition is spread through a width of several kilobars (Figure 5-14a), which represents even more kilometers within the Earth. We know that the Moho discontinuity itself may be of appreciable width, but it is not as wide as the interval in the phase diagram. Yoder and Tilley suggested that the part of the reaction producing garnet might make the largest contribution to the seismic velocity change, so that seismic techniques would sense an effective change through a much smaller depth interval. This was denied by Ringwood and Green who concluded from their experiments (Figure 5-16) that garnet increases gradually through the transition interval. They concluded further that if the Moho were a gabbro-eclogite phase transition, then the seismic velocity distribution in the transformation zone would be smeared out to such an extent that there would be no discontinuity. The more recent data of Ito and Kennedy, however, suggest that the formation of garnet does make a large contribution to the density (Figure 5-17) and therefore to the seismic velocity.

The preliminary phase diagram of Yoder and Tilley (Figure 5-14a) was widely used and applied to many petrological and geophysical problems; for example Pakiser (1965) presented a model for the crustal structure of the United States based on geothermal gradients, observed seismic discontinuities, and the basalt-eclogite transformation similar to that given in Figure 5-14a. He made excellent use of all available data, and concluded that a wide transformation zone

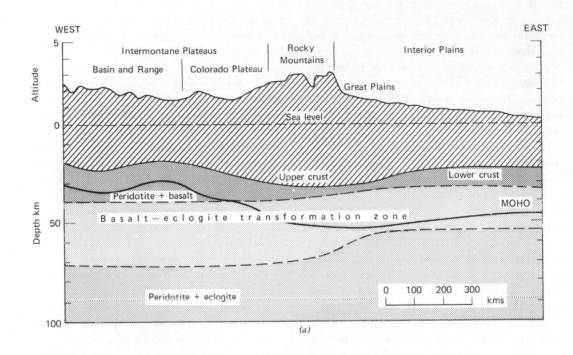

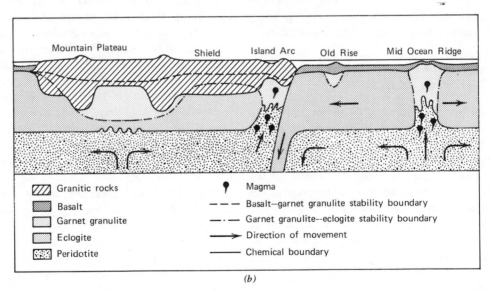

Figure 5-18. (*a*) Section through the crust from central Nevada to southeastern Nebraska showing the crustal and upper mantle structure from the Basin and Range province to the Great Plains. The inferred basalt-eclogite transformation is based on the data available in 1965 Figure 5-14*a* (after Pakiser, 1965 survey. Reprinted with permission from U.S. Geological Professional Paper 525–B). (*b*) Hypothetical crust-mantle structure based on a gabbro-eclogite phase transformation with two density contrasts bordering the garnet granulite as in Figure 5-17 (after Ito and Kennedy, 1970, with permission of Amer. Mineral. Society).

occurred mainly above the Moho in much of the eastern United States and entirely beneath the Moho in much of the western United States, as shown in Figure 5-18a. This model requires that the Moho is a chemical discontinuity separating basic crustal rocks from ultrabasic mantle rocks. It also requires considerable modification if the more recent results of Green and Ringwood are used (Figure 5-15a).

Rejection by A. E. Ringwood and D. H. Green

The hypothesis that the continental Moho is caused by an isochemical transformation from gabbro to eclogite was rejected by Ringwood and Green in 1966 on the basis of their experimental results illustrated in Figures 5-15a, 5-16, and 5-17. According to these results, and their extrapolation to lower temperatures, eclogite is thermodynamically stable through most of the normal continental crust (Figure 5-15a). This conclusion invalidates the gabbro-eclogite phase transition as an explanation of the Moho in normal continental regions. It also requires that the base of the continental crust is not of basaltic composition, but presumably of an intermediate composition occurring in the eclogite mineral facies. This produces a crustal model of type (3) or (4) in Figure 5-5, model (3) being favored.

Extrapolation of the experimental results to low temperatures is subject to uncertainties, and the position and slope of the phase boundaries may vary considerably with composition. Alternatives to Figure 5-15a therefore should not be dismissed until more data are available. The phase transition could be located at depths appropriate for the Moho in stable continental regions if extrapolations of high temperature results to low temperatures adopted a more gentle gradient for the transition as indicated by Yoder and Tilley's zone in Figure 5-15b and as discussed by Cohen, Ito, and Kennedy. This, however, seems unlikely.

Ringwood and Green reviewed the relationship between temperature at the Moho beneath the continents and the thickness of the crust, and they noted that although the surface heat flux varies by a factor of two or more in stable continental regions there is no apparent correlation between heat flow and crustal thickness.

They also argued that an eclogite layer (density 3.5 g/cm³) immediately below the continental crust passing downwards into peridotite (density 3.3 g/cm³) would be gravitationally unstable; furthermore the density of the upper mantle was generally believed at that time to lie between 3.3 and 3.4 g/cm³. The gravitational instability of eclogite overlying peridotite had been mentioned in 1960 by Harris and Rowell, who stated that the phase transition hypothesis might therefore endear itself to those seeking a new cause for mantle convection. By 1966 many geophysicists were actively seeking mechanisms for mantle convection, and Ringwood and Green did develop this theme of gravitational instability in a tectonic scheme for the orogenic cycle.

Despite the arguments listed by Ringwood and Green against the phase transition hypothesis for the Moho beneath normal continental areas, they allowed the possibility that conditions for the phase transition might occur *within the crust* in tectonically active areas characterized by high heat flow. Such areas include midoceanic ridges, island arcs, continental rifts, and continental areas that have undergone recent mountain building. These are the areas where the *M* discontinuity beneath the crust is poorly defined, and where there appears to be a layer with seismic velocities intermediate between those regarded as characteristic of the crust and those characteristic of the upper mantle (Figure 5-4). Figure 5-15a shows that a geotherm tracing higher temperatures than that beneath the oceanic regions would intersect the phase transition zone; and that a phase transition possibly from gabbro to eclogite, and certainly from garnet granulite to eclogite, could occur within the crust. The similar gradients of the geotherms and the phase transition zone would cause the transition phase to be spread out through several kilometers, which is an argument against the occurrence of the phase transition in continental regions where the Moho is sharply defined, but which is quite consistent with the occurrence of the phase transition in regions characterized by intermediate crust-mantle seismic velocities or by abnormally low upper mantle velocities.

Revitalization by F. Press, K. Ito, and G. C. Kennedy

One of the major points put forward by Ringwood and Green against the existence of an eclogite layer in the upper mantle is its high density. Despite the apparent gravitational instability of eclogite overlying peridotite, however, this is the structure of the suboceanic mantle deduced by Press in 1969 from independent geophysical evidence. Press introduced the concept of gabbro crystallizing from magma beneath the midoceanic ridges and becoming converted to eclogite as a result of cooling upon lateral migration away from the ridge (Figure 14-11).

Ringwood and Green also concluded that the seismic velocity distribution within the gabbro-eclogite transition interval would be smeared out to such an extent that occurrence of the transition within the Earth could not produce a seismic discontinuity. The results of Ito and Kennedy (Figure 5-17) introduce two abrupt changes in density associated with the transition, each of which would be capable of producing a relatively sharp discontinuity in seismic velocity. They related the garnet granulite interval to the layers with P-wave velocities in the range 7.4–7.8 km/sec which occur beneath some tectonically active areas (Figure 5-4). In some of these areas there are double discontinuities associated with the crust-mantle boundary, which could correspond to (a) the phase boundaries above and below the garnet granulite zone, or to (b) a combination of a chemical change from crust of acid or intermediate composition to underlying garnet granulite, and a phase change where the garnet granulite is replaced at a deeper level by eclogite.

Ito and Kennedy suggested that eclogite could be stable at the Moho beneath shield areas where the temperature is about 400°C at 35 km depth, but that in active regions with higher heat flow garnet granulite would be stable beneath the crust. There is no evidence for a garnet granulite layer where the temperature is low as at the suboceanic Moho and they discussed various possibilities to account for this. Great flexibility is introduced by the combination of two discontinuities associated with the gabbro-eclogite phase transition and the various chemical discontinuities possible among granitic rocks, peridotite, and rocks of gabbroic composition. Their sketch of hypothetical crust-mantle structures demonstrating some of these combinations is shown in Figure 5-18*b*.

Rejoinder by D. H. Green and A. E. Ringwood

This story of the development of a hypothesis is not closed. In the 1971 preprint from which Figure 5-17 was taken, Ito and Kennedy concluded that their results on the stability field of eclogite measured at temperatures down to 800°C contrast with those published by Green and Ringwood (Figure 5-15*a*). Green and Ringwood respond that this statement is not true. Green kindly gave me a 1971 preprint in which he and Ringwood concluded that the new work by Ito and Kennedy is in excellent agreement with their own earlier work and conclusions. They described the contribution as an improved determination of the gabbro-eclogite transition for a specific rock composition which adds little information not already demonstrated in their earlier work (Figures 5-15*a* and 5-16).

Green and Ringwood reviewed the experimental data and the range of uncertainty in the determination of densities for the Ito and Kennedy results in Figure 5-17, and they concluded that there is no experimental justification for the stepped density versus pressure curve. In their view, as stated in 1966, the gabbro-eclogite phase transition may certainly be involved in active regions with high heat flow, but it is not the cause of seismic discontinuities. Teachers will continue to find excellent material for seminars in future publications related to this controversy.

6. The Composition and Mineralogy of the Earth's Mantle

INTRODUCTION

The core and mantle together essentially determine the bulk composition of the Earth. The mantle constitutes 67.2% of the total mass of the Earth and 90% of its volume. The crust, hydrosphere, and atmosphere, which are so significant in geological and geochemical theories, together amount to less than 0.5% by weight of the whole Earth. Geophysical measurements have provided reasonable physical models for the Earth (Chapter 3), but the development of chemical models which correlate with the variation of physical properties within the Earth is hindered by lack of direct information about the chemistry of the Earth's interior.

We have four main approaches for estimating the composition and mineralogy of material within the earth:

1. The use of data from extraterrestrial sources and the formulation of physical and geochemical models for the origin of the solar system and the Earth. Sources include the stellar and solar abundances of the elements and the compositions of meteorites. The lunar rocks have been strongly fractionated, so they provide little direct information about early planetary processes and compositions.

2. The study of ultramafic and basaltic igneous rocks derived from the mantle. The rocks selected as representatives of the upper mantle must have concentrations of K, Th, and U adequate to account for surface heat flow.

3. Comparison of the variations of density, elastic properties, and other properties in the mantle measured by geophysical methods, with experimentally determined values for appropriate materials at high pressures and temperatures. This involves the results of static compression and shock wave experiments.

4. Given chemical compositions deduced from the above approaches the mineral phase assemblages and phase transitions at various depths are estimated by phase equilibrium studies at high pressures and temperatures using whole rocks, silicate systems, or germanate model systems.

We cannot infer the bulk composition of the Earth from terrestrial rocks, because the rocks exposed in the crust have been too strongly fractionated by various geological processes, but examination of igneous rocks derived from the mantle does place limits on the composition of the uppermost mantle. For the Earth as a whole we have to rely on interpretation of the chemistry of extraterrestrial bodies.

EXTRATERRESTRIAL EVIDENCE FOR THE COMPOSITION OF THE EARTH

In this section we consider only enough of the voluminous literature on the chemistry of extraterrestrial materials to permit comprehension of the basis for estimates of core and mantle composition. We start with the cosmic abundances of the elements which are generally assumed to be approximated by the solar abundances and stellar abundances, neither of which have been adequately determined. Meteorites, the oldest rocks known to us, provide samples of extraterrestrial material that can be analyzed in exhaustive detail. Much effort has been directed to recognition among the meteorites of the most primitive, least differentiated material. Comparison of the chemistry, mineralogy, and petrology of the meteorites with terrestrial igneous rocks should eventually yield an acceptable model for Earth formation and differentiation, although at present the application of meteorite analogies to quantitative calculations of Earth compositions must be regarded as conjectural.

Cosmic and Solar Abundances of the Elements

The term "cosmic" abundances refers to elements in the solar system and the stars of our galaxy. The composition of the sun is central to all discussions relevant to the bulk composition of the Earth, but even during the past decade there have been large changes in the estimates of the solar abundances of some elements. Stellar spectroscopic abundance measurements extend the range of elements measured compared to those measured in the sun, without improving the precision of abundances. Table 6-1 compares estimates of cosmic abundances of the elements with the solar abundances and the abundances of the elements in selected meteorites. The values are listed in cosmic abundance units (cau), which give the number of atoms present per 10^6 Si atoms.

Suess and Urey refined the subject of cos-

mic abundances in the 1950's and their interpretive approach used data on meteoritic and terrestrial abundances as well as solar and stellar abundances. Table 6-1 lists their estimates published in 1956, and compares them with the solar abundances adopted by Goles after a careful review of available data in 1969 and with abundance data for two types of carbonaceous chondrites obtained for the most part after 1956. The fifth column gives a 1967 estimate of cosmic abundances obtained by normalizing the recent solar and meteoritic abundances and by averaging the abundance ratios of 10 nonvolatile elements. The sixth column gives Goles' adopted abundances for the solar system and this bears comparison with the solar abundances. Comparison of the meteorite data with the solar and cosmic abundances indicates that there are significant differences among them. Let us therefore examine meteorites in somewhat more detail.

Classification and Chemical Composition of Meteorites

Meteorites differ from each other in terms of chemistry, mineralogy, structure, and color, and all of these properties have been used as criteria for classification. Figure 6-1 is a classification scheme illustrating the main chemical and mineralogical features. The four major groups in the left portion of the diagram are the irons, the stony-irons, and the stony meteorites, the stones comprising the chondrites and the achondrites. These are distinguished from each other on the basis of their proportions of metal (iron-nickel alloy) to silicate, which is shown by the horizontal axis. Troilite, FeS, occurs in all except the achondrites. The right-hand portion of Figure 6-1 shows the subclassification of the major classes.

Of all meteorites known, 61% are stones, 35% are irons, and only 4% are stony-irons; but of all the meteorites that were actually

TABLE 6-1   Comparisons of Cosmic, Solar, and Meteoritic Abundances of the Elements (from Goles, 1969) in Cosmic Abundance Units (cau, atoms per 10^6 Si atoms; to two significant figures)

Element	Cosmic (Suess and Urey, 1956)	Solar (Goles, 1969)	Carbonaceous chondrites Type I	Carbonaceous chondrites Type II	Cosmic (Cameron, 1968)	Solar System (Goles, 1969)
H	4.0×10^{10}	4.8×10^{10}	5.5×10^6	3.0×10^6	2.6×10^{10}	4.8×10^{10}
Li	100	1.7	50	16	45	16
Be	20	11	—	0.81	0.69	0.81
C	3.5×10^6	1.7×10^7	8.2×10^5	4.5×10^5	1.4×10^7	1.7×10^7
N	6.6×10^6	4.6×10^6	4.9×10^4	2.6×10^4	2.4×10^6	4.6×10^6
O	2.2×10^7	4.4×10^7	7.7×10^6	5.5×10^6	2.4×10^7	4.4×10^7
Na	4.4×10^4	9.1×10^4	6.0×10^4	3.5×10^4	6.3×10^4	3.5×10^4
Mg	9.1×10^5	7.4×10^5	1.1×10^6	1.0×10^6	1.1×10^6	1.0×10^6
Al	9.5×10^4	6.9×10^4	8.5×10^4	8.4×10^4	8.5×10^4	8.4×10^4
Si	$\equiv 1.0 \times 10^6$	$\equiv 1.0 \times 10^6$	$\equiv 1.0 \times 10^6$	$\equiv 1.0 \times 10^6$	$\equiv 1.0 \times 10^6$	$\equiv 1.0 \times 10^6$
P	1.0×10^4	1.9×10^4	1.3×10^4	8100	1.3×10^4	8100
S	3.8×10^5	8.0×10^5	5.1×10^5	2.3×10^5	5.1×10^5	8.0×10^5
K	3100	2200	3200	2100	3200	2100
Ca	4.90×10^4	6.0×10^4	7.2×10^4	7.2×10^4	7.4×10^4	7.2×10^4
Sc	28	30	31	35	33	35
Ti	2200	1800	2300	2400	2300	2400
V	220	630	300	590	900	590
Cr	7800	3800	1.3×10^4	1.2×10^4	1.2×10^4	1.2×10^4
Mn	6900	3000	9300	6200	8800	6200
Fe	6.0×10^5	2.5×10^5	9.0×10^5	8.3×10^5	8.9×10^5	2.5×10^5
Co	1800	2400	2200	1900	2300	1900
Ni	2.7×10^4	2.3×10^4	4.9×10^4	4.5×10^4	4.6×10^4	4.5×10^4
Cu	210	100	590	420	920	420
Zn	400	250	1500	630	1500	630
Ga	11	20	46	28	40	20
Ge	50	16	130	76	130	76
Rb	6.5	10	6.0	4.1	6.0	4.1
Sr	19	25	24	25	58	25
Y	8.9	80(?)	4.6	4.7	4.6	4.7
Zr	55	20	32	23	30	23
Nb	1.0	10	—	—	1.2	0.90
Mo	2.4	10	—	—	2.5	2.5
Ru	1.5	3	1.9	1.8	1.6	1.8
Rh	0.21	1	—	—	0.33	0.33
Pd	0.68	1	1.3	1.3	1.5	1.3
Ag	0.26	0.4	0.95	0.33	0.5	0.33
Cd	0.89	3	2.1	1.2	2.1	1.2
In	0.11	1	0.22	0.10	0.22	0.10
Sn	1.3	6(?)	4.2	1.7	4.2	1.7
Sb	0.25	0.1(?)	0.40	0.20	0.38	0.20
Ba	3.7	16	4.7	5.0	4.7	5.0
Yb	0.22	8(?)	0.21	0.22	0.21	0.22
Pb	0.47	4	2.9	1.3	2.9	1.3

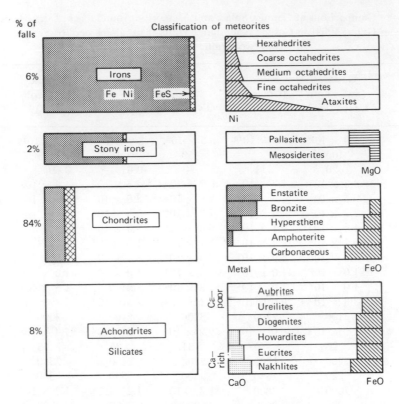

Figure 6-1. Meteorites are divided into four broad classes (left) on the basis of their ratio of metal (shaded) to silicate (white). Further subclassification (right) is based on various compositional or structural criteria (after E. Anders. Reprinted from Accounts of Chemical Research, October 1968, p. 289. Copyright American Chemical Society, reprinted by permission).

observed to fall, over 92% are stones and of these 84% are chrondrites. This has led many investigators to conclude that outer space is populated preponderately by chondritic meteorites. A disproportionate number of iron meteorites has been collected, which certainly results from their unusual and distinctive appearance. Stony meteorites are easily overlooked among terrestrial materials unless they are observed to fall.

Iron meteorites. The iron meteorites consist essentially of iron-nickel alloy distributed between two phases, kamacite with composition $Fe_{93}Ni_7$ and taenite with composition $Fe_{65}Ni_{35}$, together with troilite. The absolute value for the nickel content of all iron meteorites is about 11% by weight; no

irons contain less than 4% Ni, and a very few contain more than 20% Ni. There are small included crystals containing P, S, Cr, and C and, rarely, magnesian olivine and enstatite. The average composition of the metal is listed in Table 6-2.

Figure 6-1 shows the three main types of iron meteorites, the hexahedrites, the octahedrites, and the ataxites which are classified on the basis of their nickel content as shown and on their structure. Octahedrites are the commonest class and only one ataxite fall has been observed.

Iron meteorites are extremely fractionated compared with other meteorites and the sun (Tables 6-1 and 6-2). Such an efficient fractionation could have been achieved by a process of melting and separation of the

TABLE 6-2   Compositions of Meteorites and Meteoritic Material (weight percent)

	Iron meteorites		Average Bronzite Chondrite				Average Carbonaceous Chondrite	
	(1) Average Metal	(2) Metal	(3) Silicate	(4) Chondrite	Oxide	(5) Chondrite	(6) Type I	(7) C-, S- and H_2O-free
O			43.7	33.24	SiO_2	36.57	22.63	33.32
Fe	90.78	90.72	9.88	27.79	MgO	23.69	15.96	23.50
Si			22.5	17.10	FeO	9.67	10.42	35.47
Mg			18.8	14.29	Al_2O_3	2.30	1.64	2.41
S			—	1.93	CaO	1.77	1.56	2.30
Ni	8.59	8.80	—	1.64	Na_2O	0.86	0.74	1.10
Ca			1.67	1.27	Cr_2O_3	0.42	0.34	—
Al			1.60	1.22	MnO	0.33	0.22	—
Na			0.84	0.64	P_2O_5	0.26	0.35	—
Cr			0.38	0.29	K_2O	0.10	0.07	—
Mn			0.33	0.25	TiO_2	0.10	0.07	—
P			0.14	0.11	CoO	—	0.06	—
Co			—	0.09	NiO	—	1.29	1.90
K			0.11	0.08	H_2O	—	19.29	—
Ti			0.08	0.06	C	—	3.97	—
FeS			(5.3)	—	Organ.	—	5.53	—
	100.00	100.00	100.03	100.00	Fe	16.90	—	100.00
					Ni	1.64	—	
					Co	0.09	—	
					FeS	5.30	16.73	
						100.00	100.87	

(1) Average composition of metal from iron meteorites. Brown and Paterson, 1947.
(2)–(5) Average bronzite chondrite. B. Mason, 1965.
(2) Average metal from bronzite chondrite.
(3) Average silicate from bronzite chondrite.
(4) Average bronzite chondrite in terms of elements.
(5) Average bronzite chondrite in terms of oxides.
(6) Average carbonaceous chondrite Type I. Wiik, 1956.
(7) Analysis (6) recalculated on C-, S-, and H_2O-free basis. A. E. Ringwood, 1966.

molten alloy, presumably by gravity settling, from a complementary silicate rock or immiscible silicate liquid. Study of the textures and phase relationships of the kamacite and taenite indicate that the cooling rate between 600 and 400°C was very slow, in the range 1 to 10°C per million years, which is consistent with the rates expected if the material were situated near the center of an asteroidal body with radius 70 to 200 km.

Achondrites. The achondrites are stony meteorites without chondrules amounting to only 8% of observed falls. They are very similar to terrestrial igneous rocks in texture and mineralogy and they exhibit a wide range of chemical composition and mineralogy. Nickel-iron alloy is rarely present. Figure 6-1 shows their subclassification into calcium-rich and calcium-poor types, and the positive correlation between calcium content and the ratio of FeO to MgO. These are usually regarded as differentiated materials which have crystallized from a silicate melt in a meteorite parent body. Measured ratios of K/Rb, Rb/Sr, $^{87}Sr/^{86}Sr$, and K/U in meteorites and terrestrial rocks have been used to support the thesis that the composition of the mantle is closer to that of Ca-rich achondrites than to that of chondrites. Implications for interpretation of the Moho, given an Earth model with outer layers composed of achondritic material, were reviewed in Chapter 5.

Chondrites. Chondrites are characterized by the presence of chondrules, small rounded bodies consisting largely of olivine or pyroxene or both, with an average diameter of about 1 mm. The textures of chondrules indicate an igneous origin: they crystallized from drops of silicate liquid. Because it is generally agreed that of all meteoritic material the chondrites represent most closely the composition of the primordial dust of the solar nebula, they have been subjected to intensive chemical and mineralogical study. This has established the existence of significant variations among them.

In terms of their elemental compositions chondrites are very similar to one another, but Figure 6-1 shows a subdivision into five classes on the basis of degree of oxidation of their iron. The more oxidized chondrites contain less metal and FeS, and more FeO in the silicates. There are also differences in the absolute abundances of iron. There are distinct chemical differences between each group of chondrites, as shown in Figure 6-2, which plots both the oxidation state of the iron and the abundance of iron. The dashed lines with 45° slope represent constant total iron contents, and the areas covering analyses for the different groups lie on at least two lines.

The carbonaceous chondrites are the most primitive types among the chondrites. Their mineralogy is not well known because of their fine grain size and the presence of opaque

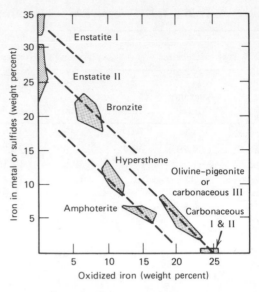

Figure 6-2. Relationship between oxidized iron and iron as metal or sulfide in chondritic meteorites. Dashed lines are lines of constant total iron content. Based on data from B. Mason and H. Craig (after Kaula, 1968, with permission of John Wiley, New York).

carbonaceous material which hinders microscopic examination. They are characterized by the presence of hydrated silicate material, including minerals such as serpentine and organic compounds in amounts up to 10%. There is little or no nickel-iron alloy because the iron is usually fully oxidized. Types I, II, and III are recognized on the basis of volatile element content: Type I does not contain chondrules. These volatile-rich chondrites obviously formed at low temperatures and they cannot have been subjected to temperatures any higher than a few hundred degrees. It has been suggested that they may be more abundant in space than the ordinary chondrites (bronzite and hypersthene chondrites), despite the fact that the latter constitute more than 90% of chondrite falls. The carbonaceous chondrites are so friable that most of them may break up when they enter the atmosphere. Analysis of the first sample of lunar soil returned from the Apollo 11 voyage in 1969 supports this suggestion; the chemistry of the soil indicates that it contains an admixture of about 2 wt% of carbonaceous chondrites (Keays, *et al.*, 1970).

The other classes of chondrites have mineralogy consistent with the attainment of high temperatures during at least one stage of their history. The mineralogy of the various classes varies around an average of 40% olivine, 30% pyroxene, 10% oligoclase, 10–20% nickel iron alloy, and 5–15% troilite. Note the sodic character of the plagioclase; in terrestrial igneous rocks containing 70% of magnesian olivine and pyroxene, plagioclase would be much richer in the anorthite component. Table 6-2 lists the average compositions of the metal and silicate fractions in chondrites. The metal fraction is very similar to the average composition of the metal in iron meteorites. The textures of many chondrites indicate that they have been heated and metamorphosed. Those with sharply defined chondrules have not been metamorphosed and the mineral compositions indicate lack of equilibrium. In others recrystallization has caused blurring of the chondrule boundaries, and the minerals approach equilibrium compositions appropriate for the temperature experienced. The degree of metamorphism thus provides

another criterion for the classification of chondrites.

The chemical compositions of carbonaceous chondrites are compared with the cosmic and solar element abundances in Table 6-1 and with the bronzite chondrites in Table 6-2. The abundances of most of the nonvolatile elements in the chondrites agree with those of the sun, which lends support to the development of a chondritic Earth model. It has also been shown that the rate of heat production in an Earth of chondritic composition is consistent with the known heat flow through the surface of the Earth, provided that the radioactive elements are strongly concentrated towards the surface. This is not a unique argument for a chondritic model, however, because the same heat flow can be produced using other models.

Abundance patterns in chondrites. The detailed chemical study of meteorites has revealed the pattern of element abundances depicted in Figures 6-3 and 6-4. The abundances of 48 elements, mostly in the transition groups of the periodic table, remain similar to solar and cosmic abundances in all classes of meteorites, but the shaded elements in Figure 6-3 are depleted in many meteorites.

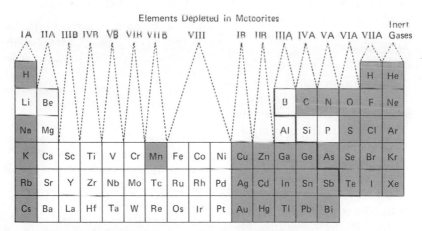

Figure 6-3. Elements in meteorites (after Anders, 1968). With the exception of Mn, all depleted elements (shaded) are situated in the main groups of the periodic table. Their only common property is volatility. (Reprinted, by permission, from Accounts of Chemical Research, October 1968, p. 289. Copyright American Chemical Society.)

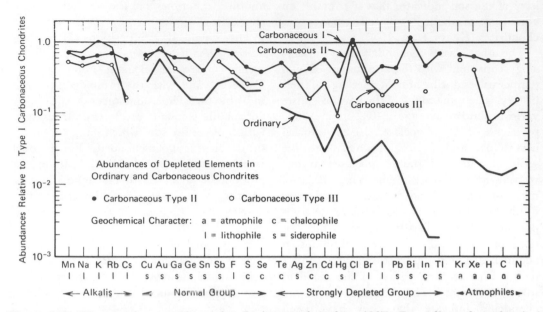

Figure 6-4. Elements in meteorites (after Larimer and Anders, 1967). Regardless of geochemical character all depleted elements are underabundant in Type II and III carbonaceous chondrites by nearly constant factors: 0.6 and 0.3 respectively. Apparent exceptions occur at Cl and Bi, but the parallelism of the curves suggests that the fault lies with the Type I data used for normalization. In ordinary chondites elements of the "normal" group are likewise depleted by a constant factor of ∼0.25, while elements of the "strongly depleted" and "atmophile" groups are depleted by progressively smaller factors. Alkalis other than Cs are not depleted. (From Geochimica Cosmochimica Acta, **31**, 1967, with permission. Copyright Pergamon Press.)

Type I carbonaceous chondrites approximate most nearly the solar abundances (Table 6-1), and elements which are strongly depleted in ordinary chondrites may be only slightly depleted in other carbonaceous chondrites and enstatite chondrites. Figure 6-4 compares the depletion in the various chondrites. The more volatile, depleted elements decrease in abundance consistently from carbonaceous chondrites Types I, II and III in ratios of 1.0 to 0.6 to 0.3, and enstatite chondrites of Type I fit into the same pattern.

It has been shown by Larimer and Anders (1967) and Anders (1968) that the abundance patterns of trace elements in chondrites can only be explained if the meteorites are a mixture of two types of materials:

1. Material that lost most of its volatile components during a high-temperature event.

2. A low-temperature material that retained its volatile elements.

The observed trace element depletion factors can be explained if we postulate that carbonaceous chondrites of Type I contain 100% of the low-temperature material, and Type III contains only 30% of the low-temperature material. There is a second fractionation pattern evident in the ordinary chondrites and Type II enstatite chondrites in which some elements show even further depletion.

The trace element fractionation patterns shown in Figure 6-4 strongly support a two-component origin for the chondrites. Loss of the depleted elements from the high-temperature fraction could possibly have occurred (a) during accretion of the meteorite parent bodies from the solar nebula, or (b) during a later reheating of these bodies. Larimer and

Anders concluded that the fractionations must have occurred in the solar nebula as it cooled from high temperatures, and that they could not have occurred in meteorite parent bodies. Some other properties of meteorites probably were established in parent bodies. The evidence suggests that most meteorites accreted at a temperature of 520–680°K and carbonaceous chondrites at ≤ 400°K.

Genetic Relationships Among Meteorites

One aim of meteorite studies is to determine the petrogenetic relationships among them and to apply these relationships to the history of planetary origin and evolution. The meteorites were classified in Figure 6-1 and their chemistry has been illustrated in Figures 6-2 and 6-4. Carbonaceous chondrites of Type I are of special interest because their element abundances agree so well with the solar abundances. Ringwood (1966) pointed out that, with very few exceptions, the compositions of all chondrites could be derived solely by the removal of appropriate amounts of trace and minor elements from the Type I carbonaceous chondrites. Figure 6-5 is Ringwood's interpretation of the genetic relationships among the principal groups of meteorites, showing some of the major chemical variations and the processes required for their derivation from the primitive carbonaceous chondrite.

There are three main theories for the origin of meteorites, each with many variants. They begin with the condensation from the primeval cloud, the solar nebula, of material similar to chondrites. Planetary models assume that all meteorites are fragments from a single disrupted planet. Serious objections to this theory are overcome in the second theory by the assumption that meteorites are derived from two successive generations of bodies; the primary objects of lunar dimensions which were subsequently fragmented by mutual collisions and the secondary objects of asteroidal size which accumulated from the debris about 4.3×10^9 years ago and which were in turn broken into fragments to produce the meteorites. This is a complex theory and other investigators prefer

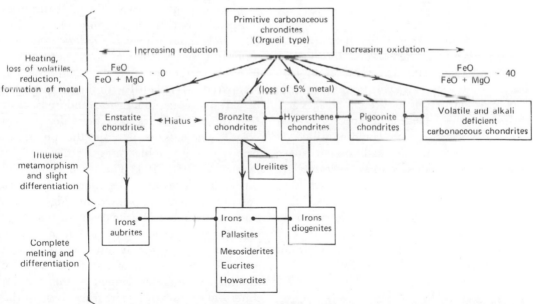

Figure 6-5. Genetic relationships among principal groups of meteorites (after Ringwood, 1961, see also Ringwood 1966. From Geochimica Cosmochimica Acta, **24**, 1961, with permission. Copyright Pergamon Press).

a third theory proposing that meteorites were formed in asteroidal bodies or planetismals which are presumably easier to break up than a single body of planetary size.

Whatever the precise sequence of events Figure 6-5 indicates that all of the chondrites could be derived from carbonaceous chondrite, Type I, by selective loss of volatile elements (Figure 6-4) and by varying degrees of reduction (Figure 6-2). Figure 6-2 also shows that hypersthene chondrites must also be depleted in about 5% of metallic iron, if derived from carbonaceous chondrites. The iron meteorites and the achondrites can then be derived from specific chondrites by melting and differentiation as depicted in Figure 6-5.

Larimer and Anders (1967) emphasized that the observed uniform depletion factors (Figure 6-4) cannot be attained in a simple heating process, and they concluded that the only feasible alternative was a two-component model with chondrites being mixtures of an undepleted fraction (A) and a depleted fraction (B). They identified fraction A as ^{18}O-rich matrix, and fraction B as ^{18}O-poor high-temperature material comprising the chondrules-plus-metal.

Apparent ages reported for meteorites lie consistently within the range 4.4–4.7×10^9 years, and these date the event resulting in the differentiation of meteoritic material into irons and achondrites. Gas retention ages give a wider spread, including an apparent age at 600×10^6 years, which could possibly represent a catastrophic collision in space between parent bodies. Studies of excess ^{129}Xe concentrations in meteorites, derived from the extinct radionuclide ^{129}I, indicate that the history of the solar system extended no further back than 200×10^6 years beyond the time of cooling the meteorite parent bodies about 4.5×10^9 years ago.

Origin and Differentiation of the Earth

Theories for the origin of the Earth begin with nucleosynthesis in the Galaxy, about 10×10^9 years ago, when various nuclear processes formed more complex elements from the primeval hydrogen. Matter initially dispersed through the rotating galaxy gathered together into more dense clouds, and the increased gravitational attraction caused these to collapse inward to form protostars. Gravitational potential energy was converted to heat and the gas pressure increased. The solar nebula produced in this way consisted of a large protosun, at high temperature, and a thin disc-shaped nebula of gas and dust particles, with temperature decreasing away from the protosun. Several lines of evidence suggest that the solar system was formed about 4.6×10^9 years ago, either by the action of some external force which produced the planets catastrophically from the sun, or more likely by the local condensation of gas and aggregation of particles within the rotating nebula to form small solid bodies which accreted to form protoplanets and planets. It is usually assumed that meteoritic material was formed during this stage of accretion of the tenuous solar nebula into solid bodies about 4.55×10^9 years ago.

The sun contains over 99.6% of the mass of the solar system and its composition is thus effectively that of the whole system, both now and probably when the solar nebula was forming. An estimate of its composition is compared in Table 6-1 with the compositions of carbonaceous chondrites, and it is clear that these meteorites have been depleted in volatile elements compared to the sun; H, C, N, O, and the noble gases are depleted by factors of 10 to 100. The achondrites and other chondrites are even more deficient in volatile elements as shown by Figure 6-4. Some of the more volatile elements that were lost during accretion of the parent meteorite bodies were presumably lost from the Earth as well. On the other hand the relative abundances of many elements within the sun differ only by small factors from those in the chondrites. The abundances of the non-volatile lithophile elements Mg, Al, Ca, Sc, Ti, Sr, and Ba are about the same in the sun and carbonaceous chondrites as are those of Co and Ni. It is therefore assumed that the abundances of the nonvolatile elements in the sun and carbonaceous chondrites provide a good estimate of element abundances in the Earth. There is one possible exception: Table 6-1 shows a considerable excess of Fe in the meteorites compared to the sun. Ringwood maintains, however, that when realistic estimates of uncertainties are made, it is not justifiable to conclude that solar Fe abundance is significantly less than that in meteorites.

According to most theories the Earth was formed by the cold accretion of particles of metal, troilite, and silicates with bulk composition approximated by chondrites (Table 6-2). Mason, Anders, and especially Urey have developed models involving several stages. Fractionation of iron with respect to silicate may have occurred in the solar nebula. At least one period of high temperature is required during the preterrestrial stage, and this may have developed in the dust cloud itself, or within earlier generations of bodies of asteroidal or lunar size that were subsequently fragmented. A rather complex series of events is required, causing the loss of volatile elements from the early accreted material believed to be similar to carbonaceous chondrites (Figure 6-4), the reduction of iron oxides to yield metal (Figure 6-2), and the formation of chondrules. Once the Earth had formed, slowly enough that conversion of gravitational potential energy to heat did not cause melting, then heating by short-lived radioactive elements would cause melting of the dispersed iron within about 600×10^6 years. The heavy molten iron, incorporating nickel and troilite (Table 6-2), would become concentrated towards the center of the Earth, while the lighter silicate fraction would be displaced outward. According to some models the silicate fraction also was partially fused, and in other models the fractionation was achieved by convection without melting of the silicate phases. The mantle and core material remain in chemical

equilibrium with each other during the differentiation.

Ringwood concluded that these multistage theories lacked plausibility mainly because of their complexity, and he proposed a single-stage theory involving the processes shown in Figure 6-5. The Earth formed directly by accretion from the primitive oxidized dust in the solar system, which he considered chemically equivalent to carbonaceous chondrites Type I with a slightly higher H/C ratio (Tables 6-1 and 6-2). As a direct result of the accretion process, and simultaneously with it, there occurred (a) the loss of volatile elements (Figure 6-4), (b) reduction to metal or iron oxides (Figure 6-2), nickel oxide, and some silica, (c) melting of both metal and silicate phases, and (d) a major differentiation of the Earth into metallic core and silicate mantle, followed by (e) further differentiation within the silicate mantle. Ringwood argued against an earlier silicate-iron fractionation within the nebula, claiming in effect that the solar Fe/Si ratio is in error (Table 6-1). He maintained also that although the mantle and core are separately in equilibrium, they are not in chemical equilibrium with each other as required in the multistage theories.

Compositions of the Core and Mantle

The masses of the core and mantle are known from geophysical measurements, and the bulk composition of the Earth is estimated by assigning compositions to the core and mantle and combining these in the appropriate proportions. The results obtained thus depend upon the model adopted for the origin and differentiation of the Earth. An early estimate by Washington is compared in Table 6-3 with recent estimates derived from the multistage and single-stage theories.

Washington's estimate is based on a model with a central core composed of material equivalent to iron meteorites, subsequent layers composed of stony-iron and chondritic material, and a dominant peridotite shell

TABLE 6-3   Estimated Compositions of the Whole Earth, the Core, and Mantle Using Extraterrestrial Information (weight percent)

	Earth			Core		Mantle			
	Washington 1925	Mason 1966	Ringwood 1966	Mason 1966	Ringwood 1966	Mason 1966	Oxide	Mason 1966	Ringwood 1966
O	28	30	30			44	SiO_2	48	43
Fe	40	35	31	86	84	9.9	MgO	31	38
Si	15	15	18	—	11	23	FeO	13	9.3
Mg	8.7	13	16	6.0		19	Fe_2O_3	—	—
S	0.64	1.9	—	7.4		—	Al_2O_3	3.0	3.9
Ni	3.2	2.4	1.7		5.3	—	CaO	2.3	3.7
Ca	2.5	1.1	1.8			1.7	Na_2O	1.1	1.8
Al	1.8	1.1	1.4			1.6	Cr_2O_3	0.55	—
Na	0.39	0.57	0.9			0.84	MnO	0.43	
Cr	0.20	0.26	—			0.98	P_2O_5	0.34	
Mn	0.07	0.22	—			0.33	K_2O	0.13	
P	0.11	0.10	—			0.14	TiO_2	0.13	
Co	0.23	0.13	—	0.40	—				
K	0.14	0.07	—			0.11			
Ti	0.02	0.05	—			0.08			

with the composition of achondrites; the outer shells of basalt and granite amount to a small fraction.

The second column of Table 6-3 gives Mason's estimate of the Earth's composition based on the average composition of the bronzite chondrites as listed in Table 6-2. He used Birch's model of a mantle plus crust making up 67.6% and a core 32.4% the mass of the Earth. The calculation assumes that the mantle and crust together have the same composition as the silicate portion of the bronzite chondrites, and that the 5.3% of troilite in the bronzite chondrites is incorporated in the core, the remaining 27.1% of which is nickel-iron of the same composition as in bronzite chondrites. The calculated compositions of the core and of the mantle plus crust are also listed in the table, the mantle plus crust composition being given in terms of elements and also of oxides. Mason emphasized that the assumptions that he made are oversimplifications, but he con-

sidered the results to be encouraging when compared with other estimates (Tables 6-3 and 6-6).

Ringwood adopted a single-stage model for the origin of the Earth mainly because of the complexity of the multistage models. But in order to estimate the composition of the Earth and its major parts from meteoritic compositions he had to make a more complex series of assumptions than did Mason. Ringwood took as his starting material the carbonaceous chondrite, Type I, which is listed in Table 6-2 and the last column shows the principal components recalculated on a C-, S-, and H_2O-free basis. He assumed that this composition approximates closely to the primitive, accreting, oxidized Earth in terms of nonvolatile elements. He concluded that geochemical and geophysical evidence indicates that the $FeO/(FeO + MgO)$ molecular ratio of the mantle is between 0.1 and 0.2, and he adopted a value of 0.12. The chemical analysis in Table 6-2 was rearranged to give this ratio by reducing all of the nickel and an appropriate amount of iron, and by transferring the reduced nickel and iron to the core. This provided a core amounting to only 26.48% of the mass of the Earth, and in order to increase the mass of the core to its correct value of 31% more material must be transferred from the silicate portion of the devolatilized and partly reduced carbonaceous chondrite. Ringwood concluded that SiO_2 is the common oxide that would be most readily reduced to metal, and in order to obtain an Earth model from the primitive composition, he therefore transferred 11 wt/% of silicon (20 atomic %) from the mantle to the core; this provided the required mantle/core mass ratio. The resulting estimate of the Earth's composition is shown in the third column of Table 6-3.

The core composition estimates by Mason and Ringwood are compared in Table 6-3. The density of the outer core shown in Figure 2-5 is at least 10% lower than the probable density of iron at these conditions, which indicates that the core contains a low density element in addition to iron. The only likely elements appear to be sulfur or silicon. Mason prefers an Earth model with core and mantle in equilibrium, which denies the existence of free silicon in the core if the mantle contains appreciable amounts of FeO. He concluded that the addition of sulfur to the core satisfied both geochemical and geophysical requirements. Ringwood, on the other hand, concluded that most of the primordial sulfur was lost from the Earth by

volatilization either before or during accretion, and that the separation of the mantle from the core occurred in such a manner that the compositions of the core and mantle are not in chemical equilibrium. Silicon is then considered to be the most likely extra component of the Earth's core.

Table 6-3 also compares the estimated compositions of mantle plus crust according to Mason and Ringwood. These analyses are reproduced in Table 6-6 along with estimates of mantle composition derived from terrestrial evidence, and these will be discussed together in a later section.

There is geochemical evidence that the abundances of many nonvolatile trace elements, including U, Th, Ba, Sr, Ta, Zr, Hf, and the rare earths, are strongly concentrated in the upper mantle and crust. Ringwood concluded that their upward concentration occurred at a very early stage in the Earth's history, about 4.55×10^9 years ago, when all or most of the mantle was subjected to partial or complete melting. This coincides with the single decisive high-temperature event which caused loss of volatile elements, reduction of oxides, and formation of the core. These trace elements are incompatible with the major mineral phases of the mantle because of their sizes and ionic charges, and therefore they would have migrated upward with the silicate liquid, forming a protocontinental layer. The continents, hydrosphere, and atmosphere were then derived later by fractional melting of the protocontinent and subsequent magmatic processes.

ULTRAMAFIC ROCKS AND THE UPPER MANTLE

The estimates of mantle composition in Table 6-3 are equivalent to ultramafic rocks, and we turn therefore to examination of ultramafic rocks exposed at the Earth's surface. Ultramafic rocks could possibly have been derived directly from the mantle or indirectly from other mantle-derived material such as basaltic magma. The petrological problem is to determine which rocks, if any, represent mantle

material and to deduce the processes which led to their emplacement into the crust.

Peridotite Mineralogy and Field Associations

There are many different rock types in the ultramafic clan and these occur in a variety of field and petrographic associations. The most abundant minerals in rocks with compositions similar to those in Table 6-3 are olivine and pyroxenes, and combinations of these minerals in different proportions form the various peridotites. The alumina is distributed among the pyroxenes and accessory minerals such as plagioclase, spinel, and garnet. Small amounts of water are likely to be accommodated in amphibole, phlogopite, or serpentine, and given sufficient water at crustal temperatures, peridotite may be converted to serpentinite.

For petrogenetic discussions relevant to the composition of the upper mantle it is convenient to consider four associations of ultramafic rocks which are summarized in Table 6-4.

1. The layered, stratiform, and other intrusions involving gabbro or diabase together with accumulations or concentrations of mafic minerals. These occur in varied tectonic environments, but they are rarely affected by contemporaneous orogeny.

2. The alkalic rocks of stable continental regions including kimberlites, mica peridotites, members of ring complexes, and ultrabasic lava flows. These usually occur in stable or fractured continental regions, and their distribution in belts appears to be controlled by deep-seated tectonics with linear trends.

3. The several serpentinite-peridotite associations of the orogenic belts that have been classified together as alpine-type intrusions. These include large and small bodies distributed along deformed mountain chains and island arcs, usually along with gabbro or basic volcanic rocks. Relationships among rocks in this association are often complicated by metamorphism and metasomatism. The orogenic peridotites are subdivided into the high-level ophiolites, and the "root-zone" peridotites which occur in crystalline rocks.

4. Serpentinites and peridotites of the oceanic regions. These have been identified on midoceanic ridges, within deep ocean trenches, and where small scarps expose material below the cover of basalts forming the ocean floor.

5. In the present context we must also consider peridotite and eclogite nodules occurring in alkali olivine-basalts and kimberlites.

Petrogenesis of Ultramafic Rocks

Important factors to be considered in petrogenesis include the source of the material, its variation in physical state and temperature between its source and its posi-

TABLE 6-4   Classification of Main Ultramafic Rock Associations

Ultramafic Associations	Tectonic Environments	Source of Rock
1. Layered intrusions	Nonorogenic Varied	Basalt magma from mantle
2. Alkalic complexes, Kimberlites	Cratonic	Magmas from mantle
3. Alpine-type intrusions (a) ophiolites (b) root-zone	Orogenic Island arcs	Magmas and Solids from mantle
4. Oceanic rocks	Ocean trenches Ocean floors Oceanic ridges	Mantle or oceanic crust
5. Nodules (a) alkali olivine-basalts (b) kimberlites	Varied	Mantle, crust, or magma concentrates

tion of intrusion, and its postintrusion history. The more complex the postintrusion history of an ultramafic rock the more difficult is its petrogenetic interpretation. The following discussion is a brief outline of a very complicated subject which I have reviewed elsewhere in some detail (Wyllie, 1967, 1969, 1970). Figures 6-6 and 6-7 provide a schematic representation of the processes involved in the petrogenesis of ultramafic rocks.

Layered intrusions. There is general agreement that the parent for association (1) in Table 6-4 is basaltic magma derived by partial fusion of the upper mantle, and that gravity settling of early-formed crystals is the dominant process. Figure 6-6 depicts magma generation beginning with diapiric uprise of mantle peridotite to a level where partial melting produces interstitial basaltic

liquid. The crystal mush continues to rise, and in the left-hand diagram basaltic magma separates from the mush within the mantle and rises into a large reservoir or magma chamber within the crust. Crystal settling from this basic magma produces a series of ultramafic cumulates. The magma flows over the cumulates and upwards out of the chamber, possibly producing dykes, sills, volcanic necks, volcanoes, and lava flows. Crystal settling, flow differentiation, and other processes of crystal concentration may yield ultramafic rocks in the general locations shown in the figure.

Alkalic complexes, and kimberlites. The parent magmas proposed for the ultrabasic rocks of group (2) in Table 6-4 include undersaturated alkalic basalts and alkalic ultrabasic liquid magmas, evidence for the

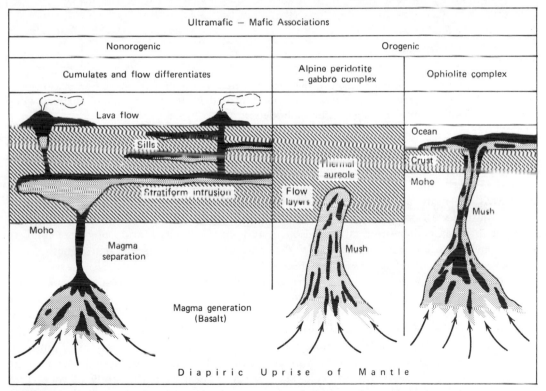

Figure 6-6. Schematic representation of the processes involved in the origin and emplacement of associations of ultramafic and mafic rocks. See text for explanation. Stipple—crystalline ultramafic material; black associated with stipple—interstitial basic liquid or crystalline basalt or gabbro (after Wyllie, 1970, with permission, Amer. Mineral. Society).

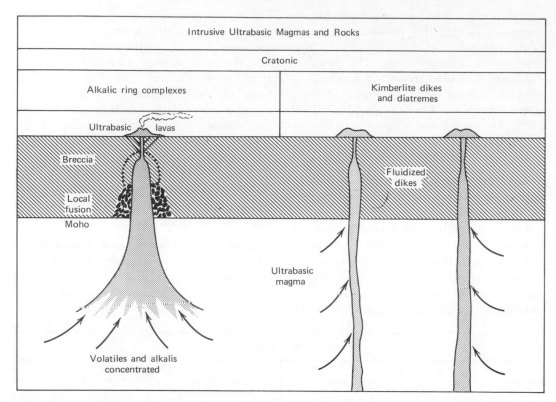

Figure 6-7. Schematic representation of the processes involved in the origin and emplacement of intrusive ultrabasic magmas and rocks. See text for explanation. Stipple—ultrabasic material (after Wyllie, 1970, with permission, Amer. Mineral. Society).

latter including the existence of alkalic ultrabasic lavas. The upper mantle is usually considered to be the source of these magmas, as illustrated schematically in Figure 6-7, but contamination with crustal material is often invoked to explain their unusual chemistry. Alkalis and volatiles are concentrated with the generation of ultrabasic magma which rises in dyke or pipe-like form into the crust and, eventually, ultrabasic lavas are erupted at the surface. The alkalis and volatiles become concentrated in residual liquids in the intrusions, facilitating differentiation. Exsolution of the volatiles causes brecciation, and explosive eruption of fragmental volcanic rocks. This effect is well illustrated by kimberlites: the deep-seated massive rock crystallizes from an ultrabasic magma, and higher level rocks are emplaced in diatremes as

fluidized solid-gas systems, xenocrysts, and xenoliths of wall rocks being mixed with fragments of altered kimberlite. The explosive effects and the metasomatic effects of alkalis and volatiles in the residual magmas and exsolved gases and solutions, tend to obscure the early stages of petrogenesis.

Ultramafic rocks of the orogenic belts (alpine-type). There is no general agreement about the petrogenesis of these rocks, and there is little doubt that the association includes rocks with a variety of origins and histories. If we are to extrapolate down to the upper mantle from the study of rocks now exposed at the surface, it is necessary to decipher these histories. Figure 6-6 illustrates schematically some of the processes involved in the emplacement of these rocks.

The middle diagram represents the root-zone peridotites, group (3b) in Table 6.4. The ultrabasic crystal mush rises into the crust under conditions permitting little physical separation of basaltic magma. Some gravity layering may develop within the mush within the upper mantle, and this is further complicated by the development of flow layering during emplacement, with characteristics different from that of ultramafic cumulates in stratiform intrusions. A similar diagram would represent another process often invoked, the diapiric intrusion of solid mantle material. A major question is the extent to which ultramafic rocks of cumulate origin, depicted in the left-hand diagram of Figure 6-6, have become involved in the orogenic process and thus converted into rocks which merit the appellation "alpine-type". The origin of ultramafic rocks enclosed in crystalline rocks of the orogenic belts has been explained by five types of interpretation: (a) intrusion of ultrabasic magma, (b) intrusion of crystal mush of partially melted mantle, (c) diapiric or tectonic intrusion of solid mantle peridotite or of peridotite plus gabbro, (d) reintrusion of peridotite or serpentinite from any source, and (e) metasomatism and metamorphism of basic lavas and dolomites.

The right-hand diagram of Figure 6-6 represents the processes involved in the petrogenesis of the ophiolite suite, with the crystal mush breaking through the oceanic crust. There are four main types of interpretation for the origin of the associated ultramafic and mafic rocks. The first three involve the extrusion of very large volumes of material in submarine eruptions, forming a thick peridotite-gabbro-basalt complex, with later gravity slides occurring on regional scales to produce chaotic structures.

1. Extrusion of essentially liquid basaltic magma which undergoes differentiation on the sea floor,

2. Extrusion of a partially molten diapir of rising mantle with partial separation of basic magma during and after uprise and continued differentiation of the massive pile on the sea floor (Figure 6-6),

3. A two-stage process with solid intrusion of peridotite and gabbro from the mantle during a pregeosynclinal period being followed by intrusion and extrusion of basaltic magma at intervals through the whole geosynclinal cycle,

4. Basaltic lavas and sediments are metasomatically transformed into peridotites and serpentinites.

Ultramafic rocks of the oceanic regions. During the last few years serpentinites and peridotites have been dredged from many sites in the Atlantic and Indian Oceans especially near fracture zones although very few have been found in the Pacific Ocean. Several recent interpretations of ultramafic rocks of the ophiolite suite involve the supposition that these represent ocean floor or suboceanic mantle in eugeosynclines, which has been incorporated into the orogenic belt by thrusting or gravity sliding. The structure of midoceanic ridges appears to be similar to that of some chromitite-bearing alpine peridotite-gabbro complexes of continental environments, which implies that gabbro interlayered with peridotite is an essential constituent of the upper mantle. According to Hess's sea-floor spreading hypothesis (Chapter 12) the oceanic crust is composed of serpentinized mantle peridotite, migrating laterally away from midoceanic ridges. The global scheme of plate tectonics (Chapter 14) introduces the prospect that a midoceanic ridge may be overridden by a continent. These ideas imply that the ultramafic rocks of the orogenic belts, island arcs and ocean trenches, the midoceanic ridges, and the ocean floor between all have common petrogenetic links with the upper mantle.

Ultramafic nodules. The processes involved in the formation of the nodules of group (5) in Table 6-4 can be related to the left-hand diagram of Figure 6-6 and to Figure 6-7. Nodules in basalts may represent fragments of the mantle or fragments of crystal cumulates or flow concentrates from a magma chamber or the walls of a conduit or volcanic neck (Figure 6-6). The nodules in kimberlites may represent fragments carried upward from various levels in the mantle or from metamorphic rocks of the deep crust (Figure 6-7). There is good evidence that in a single locality, for either basalt or kimberlite host, the nodules present may be of several types with distinct origins. Pyroxene peridotite nodules of uniform composition and mineralogy occur in both basalts and kimberlites, and garnet peridotites with a wide variation in mineralogy occur only in kimberlites. Eclogitic nodules, ranging in composition from basalt to picrite basalt, may accompany peridotite nodules in kimberlites and rarely in basalts. For interpretation of nodule suites it is necessary to determine both the depth of origin and process of formation: processes may be elucidated by recognition of cumulus, tectonite, or metasomatic textures in the rocks.

Extrapolation to the Upper Mantle

Review of the petrogenesis of the ultramafic rocks in groups (3), (4), and (5) of Table 6-3 indicates that some of these rocks represent mantle material, but it also posts clear warnings that extrapolation to the upper mantle should be made with caution. Rocks of group (1), in stratiform or layered intrusions or extrusions, are produced by the concentration of mafic minerals crystallized from basaltic magma. They can provide only indirect evidence of mantle chemistry. The alkalic rocks of the cratonic belts in group (2) do not provide direct representatives of the mantle because they have been strongly fractionated. The high concentrations of volatile components and alkalis obscure many details of their petrogenesis, and their chemical relationships to the upper mantle.

The ultramafic rocks of the orogenic belts probably include representative mantle material as depicted in Figure 6-6, and these rocks have often been cited as guides to mantle composition. Unfortunately the effects of metamorphism and reintrusion tend to blur the petrogenesis of these associations, and peridotites of any origin that have become involved in the orogenic process may be difficult to distinguish from those of alpine-type that are contemporaneous with the orogenic cycle. Certainly petrologists must exercise careful judgement before equating any specific orogenic peridotite with the upper mantle.

The ultramafic rocks of the oceanic regions, especially those near the center of the mid-oceanic ridges, may provide samples of the upper mantle, but more data are needed about the rock association and the geological structures before extrapolation can be made with confidence.

The ultramafic nodules may provide the most direct evidence of mantle chemistry and mineralogy, assuming that distinction can be made among those that are crystal cumulates or concentrates. The problems of interpretation were summed up by Jackson in 1969:

"Trying to reconstruct the depth of origin and process of formation of xenoliths in any one individual tuff or flow is akin to trying to map the geology of an inaccessible highland area by looking at the boulders in the bed of an emergent stream. Multiple origins and differences in distances of transport of fragments, even those deposited side by side, might as well be assumed in both cases, and generalizations ought to be made with extreme care."

Once the petrogenesis of an ultramafic rock (or of an ultramafic-mafic rock association) has been unravelled and an ultramafic rock has been traced back to a mantle source, its composition and mineralogy have still to be related to the history of that portion of the mantle from which it was derived. Petrologic considerations support the geophysical evidence that the upper mantle is heterogeneous (Chapter 3). The upper mantle will include material ranging in composition from original, primitive, undifferentiated peridotite to residual peridotite or dunite from which a basaltic or picritic liquid fraction has been partially or completely removed. It will include peridotite precipitated from picritic or basaltic magma as it fractionated en route to the surface (Chapter 8). It will include layers and lenses of eclogite crystallized from a basaltic or picritic liquid that failed to reach the crust. It may also include portions of peridotite somewhat enriched in eclogitic components by percolation of basaltic liquid

and its local concentration in dispersed form in favorable sites.

Ultramafic rocks, properly selected, should thus provide information about the composition and mineralogy of the upper mantle, but the petrogenetic processes involved between mantle source and crustal exposure are varied and complex. For satisfactory estimates of mantle composition based on terrestrial ultramafic rocks the field association and the petrogenesis of the rocks should be thoroughly investigated before extrapolation to the mantle.

COMPOSITION OF THE MANTLE

The general idea of a peridotitic mantle as a source for basaltic magmas by partial melting dates at least as far back as Bowen's treatise on the igneous rocks in 1928, but it was Ringwood who formalized the concept in 1962. Ringwood is largely responsible for directing the attention of petrologists to model compositions for the upper mantle. He postulated the existence of a primitive mantle material which was "defined by the property that on fractional melting it would yield a typical basaltic magma and leave behind a residual refractory dunite-peridotite of alpine type." This he called "pyrolite" (pyroxene-olivine-rock). He referred to the valid objections to inventing new names for hypothetical materials and defended this one mainly on the grounds that the trace element chemistry of "pyrolite" does not match that of natural peridotites. Rocks are named on the basis of their mineralogy, however, not on their trace element content; and I see no need to add yet another to the names that already exist. Most petrologists concerned with upper mantle constitution refer to known varieties of peridotites although geophysicists tend to use "pyrolite".

Among the first attempts to correlate the composition of the upper mantle with specific mantle-derived ultramafic rocks are those of Hess and Green both in 1964. Hess reviewed the chemistry of oceanic serpentinites, and

Green proposed that the average composition of the orogenic peridotite at Lizard, England represented undepleted upper mantle material. Since 1967 many accounts of the chemistry of ultramafic rocks in a variety of environments have included discussions of mantle chemistry. Petrologists have been concerned with identification of ultramafic rocks representing (1) undepleted mantle, and (2) residual mantle after removal of a basic liquid fraction; they have also attempted to identify (3) the rock produced by crystallization of the liquid fraction (Chapter 8). In this section we examine first (Table 6-5) the compositions of selected ultramafic rocks which have been cited as representatives of the upper mantle and then the calculated compositions of hypothetical mantle peridotites. Estimated mantle compositions based on extraterrestrial evidence are then compared with estimates based on terrestrial rocks and the hypothetical rocks (Table 6-6). Some aspects of the chemistry are compared in chemical variation diagrams, which include not only the rock compositions cited in this chapter but many others.

Estimates Based on Terrestrial Rocks

The Lizard peridotite of Cornwall, England, is one of a distinctive group of orogenic ultramafic intrusions, which Green interprets as a portion of the upper mantle mobilized by partial melting (middle diagram of Figure 6-6). The average composition of this intrusion, listed in column (1) of Table 6-5, was presented by Green (1964) as equivalent to undepleted mantle peridotite.

Garnet peridotites occur in the root-zones of orogenic belts and as nodules in kimberlites. Carswell (1968a) studied the garnet peridotites enclosed in gneiss at Ugelvik, Norway, and concluded that they represent primary mantle material; their average composition is listed

in column (2) of Table 6-5. He also reviewed other occurrences of garnet peridotite in basement rocks of Czechoslovakia and Switzerland and in kimberlites. All of these rocks prove to be similar mineralogically, occupying restricted chemical ranges, and he therefore concluded that they too probably represent primary upper mantle material. Points for these rocks are plotted in Figure 6-8.

Hess (1964) reviewed the mineralogy and chemistry of the serpentinite cores from the drill hole near Mayaguez, Puerto Rico where

TABLE 6-5   Compositions of Mantle-Derived Ultramafic Rocks

Weight percent	Continents		Oceans		Nodules		Hypothetical	
	(1)	(2)	(3)	(4)	(5)	(6)	(7)	(8)
SiO_2	44.77	44.65	39.82	43.56	44.18	41.10	42.71	40.3
MgO	39.22	41.66	48.60	41.53	40.95	46.33	41.41	32.7
FeO	8.21	6.81	7.86[a]	7.77[a]	7.34	9.31	6.51	7.1
Fe_2O_3	—	—	1.00[a]	1.00[a]	1.16	1.24	1.57	1.8
Al_2O_3	4.16	3.50	0.87	2.36	2.81	0.56	3.30	3.7
CaO	2.42	2.02	0.37	2.51	2.49	0.17	2.11	2.1
Na_2O	0.22	0.23	0.37	0.32	0.22	0.23	0.49	0.5
K_2O	0.05	0.04	[b]	[b]	0.04	0.03	0.18	0.0(2)
Cr_2O_3	0.40	0.59	0.46	0.40	0.3	0.35	0.45	0.3
NiO	0.24	0.29	0.46	0.34	0.27	0.44	0.42	0.2
CoO	—	—	—	—	—	—	0.02	—
MnO	0.11	0.14	0.10	0.10	0.14	0.15	0.13	0.1
P_2O_5	—	—	0.08	0.07	—	—	0.06	0.1
TiO_2	0.19	0.08	0.01	0.04	0.09	0.08	0.47	0.4
H_2O+	—	—	—	—	—	—	0.17	9.7
CO_2	—	—	—	—	—	—	—	0.8
Cl	—	—	—	—	—	—	—	0.2
Total	99.99	100.01	100.00	100.00	99.99	99.99	100.00	100.00

[a]Ferrous ferric ratio adjusted so that Fe_2O_3 is 1%.
[b]means less than 0.005.

(1) Green (1964). Average composition for the Lizard peridotite.
(2) Carswell (1968a). Mean of three garnet peridotites from Ugelvik, Norway.
(3) Hess and Otalora (1964). Average (D + E)-type serpentinite, recalculated water free, residual type.
(4) Hess and Otalora (1964). Average C-type serpentinite, recalculated water-free.
(5) Harris et al. (1967). Mean of five high calcium, high aluminum olivine nodules.
(6) Harris et al. (1967). Average of three olivine nodules with CaO and Al_2O_3 contents less than 1%, residual nodules.
(7) Green and Ringwood (1963). Pyrolite with 4:1 of, respectively, average anhydrous dunite and the mean of Nockold's (1954) average normal tholeiite and normal alkali basalt.
(8) Nicholls (1967). Composition of volatile-rich parts of the upper mantle, such as may occur beneath the midoceanic ridges (Nicholls, analysis 3, table 9).

the exposed serpentinite appears to be similar to rocks dredged from the ocean bottom near the island. The serpentinite forms the basement beneath Puerto Rico, and is considered by Hess to represent oceanic crust or hydrated mantle. Several distinct types of serpentinite are represented, and columns (3) and (4) in Table 6-5 give average compositions involving three types recalculated on an anhydrous basis. Hess proposed that the average in column (3) might be the residue after extraction of basalt from original mantle material with average composition (4). These analyses are similar to those of serpentinized peridotites from midoceanic ridges as shown by the recent review of 23 specimens by Miyashiro, Shido, and Ewing (1969). The peridotite mylonites from St. Peter and St. Paul Rocks in the mid-Atlantic Ocean include several varieties, and they appear to represent rocks which crystallized at different levels within the mantle. They therefore have bearing on the heterogeneity of the upper mantle according to Melson, Jarosewich, Bowen, and Thompson (1967). Individually the rocks analyzed are not considered equivalent to the mantle, but an estimated average composition of St. Paul's Rocks which is plotted in Figure 6-8 is in accord with suggested mantle compositions.

Figure 6-8 includes seven compositions specified as upper mantle material on the basis of study of peridotite nodules in kimberlites and basalts. Two of these are listed in Table 6-5. Harris, Reay, and White (1967) derived columns (5) and (6) from the chemistry of 27 peridotite and three garnet-peridotite nodules. The average of five selected analyses of peridotite nodules with CaO and Al_2O_2 contents between 2 and 3% is given in column (5), and this they considered to represent mantle material that has undergone only limited or no depletion in fusible components. Column (6) is the average of three peridotite nodules with CaO and Al_2O_3 each less than 1%. This they considered to represent samples of residual mantle left after extraction of basaltic magma

or samples of cumulate olivine precipitated from magma rising through the mantle.

Hypothetical Peridotites

The definition of Ringwood's "pyrolite" was given above. Ringwood and Green have emphasized flexibility in their model, and between 1963 and 1967 they have presented four pyrolite analyses. All four are plotted in Figure 6-8 and two of them are listed in column (7) of Table 6-5 and column (5) of Table 6-6. The ratio of peridotite:basalt may vary within the limits 1:1 to 4:1, with 3:1 being the most likely. Column (7) is based on a 4:1 mixture of average anhydrous dunites and an average of tholeiite and alkali basalt. Other model compositions included a primitive Hawaiian olivine tholeiite composition and a hypothetical peridotite with composition calculated from analyzed minerals separated from alpine-type peridotites. Column (5) in Table 6-6 is based on the 3:1 mixture.

Nicholls (1967) approached the problem of upper mantle composition by assuming that it was composed of (a) a volatile fraction, consisting of compounds that have been liberated from volcanoes to the atmosphere and hydrosphere, (b) a basaltic lava fraction derived by partial melting of the mantle, and (c) a residual fraction remaining after depletion in volatiles and basalt. Nicholls examined the possible compositions of each of these fractions to define their limits and then evaluated the proportions in which they might be combined in undepleted mantle material. He concluded that the preferred composition for the volatile fraction is 91% H_2O, 7% CO_2, 2% Cl and traces of other components. A basaltic fraction was calculated, and an estimate of the residual fraction was based on oceanic serpentinites. Column (8) in Table 6-5 is Nicholl's estimate of undepleted upper mantle material, still rich in volatile components, which he suggests may exist beneath the midoceanic ridges: notice the high water content. Column (6) in Table 6-6 represents mantle material from which

TABLE 6-6   Estimates of Primitive Mantle Compositions

Weight percent	Extraterrestrial		Terrestrial		Hypothetical	
	(1)	(2)	(3)	(4)	(5)	(6)
SiO_2	48.09	43.25	44.5	44.2	43.95	45.1
MgO	31.15	38.10	41.7	41.3	39.00	36.7
FeO	12.71	9.25	7.3	7.3	7.50	7.9
Fe_2O_3	—	—	1.5	1.1	0.75	2.0
Al_2O_3	3.02	3.90	2.55	2.7	3.88	4.1
CaO	2.32	3.72	2.25	2.4	2.60	2.3
Na_2O	1.13	1.78	0.25	0.25	0.60	0.6
K_2O	0.13	—	0.015	0.015	0.22	0.0(2)
Cr_2O_3	0.55	—	—	0.30	0.41	0.3
NiO	—	—	—	0.20	0.39	0.2
MnO	0.43	—	0.14	0.15	0.13	0.2
P_2O_5	0.34	—	—	—	—	0.1
TiO_2	0.13	—	0.15	0.1	0.57	0.5
Total	100.00	100.00	100.36	100.02	100.00	100.0

(1) Mason (1966), Table 6-3.

(2) Ringwood (1966), Table 6-3.

(3) White (1967). Upper mantle composition estimated from frequency histograms of 168 ultramafic rocks. Total iron is divided arbitrarily between FeO and Fe_2O_3. Na_2O and K_2O are from Stueber and Murthy (1966) and Stueber and Goles (1967).

(4) Harris et al. (1967). Estimated upper mantle based on analysis (3), and on analysis (5) for nodules in Table 6-5.

(5) Green and Ringwood (1967). Synthetic peridotite used in experimental phase studies; designated pyrolite II. Ratio of dunite: basalt is 3:1.

(6) Nicholls (1967). Estimated mantle material from which volatile components have been removed but not the basaltic components (Nicholls, analysis 1, table 9).

the volatile fraction has been removed, but which still retains the basaltic fraction; this is the analysis plotted in Figure 6-8 along with the residual peridotite.

Mantle Composition

Table 6-6 compares estimates of primitive, undepleted mantle compositions based on extraterrestrial evidence, terrestrial rocks, and theoretical models. Columns (1) and (2) are the estimates of Mason and Ringwood, determined as illustrated in Tables 6-2 and 6-3. The hypothetical compositions were determined as discussed in the preceding section. Column (3) is White's (1967) estimate of mantle peridotite based on frequency histograms for the chemistry of 84 peridotites,

53 serpentinites, and 31 peridotite nodules. White argued that the mode or most abundant class of exposed ultrabasic rock should represent original mantle peridotite. He plotted separately the analyses for rocks with between 1 and 5% CaO plus Al_2O_3 to eliminate the residual dunitic and more basaltic analyses, so that the mode for unfractionated mantle peridotite would not be masked. Harris, Reay, and White compared the dominant ultramafic rock in column (3) with the average compositions of peridotite nodules (Table 6-5) and presented a preferred composition for original or undepleted mantle based on these analyses and other considerations. This is column (4) in Table 6-6.

Element abundances in suites of ultramafic rocks have been studied by activation analysis and spectrochemical analysis, which make possible the accurate determination of elements previously difficult to measure in ultramafic rocks. Fisher, Joensuu, and Bostrom (1969) analyzed 58 ultramafic rocks from diverse sources. Among the rocks they could find no chemically coherent type that might represent the upper mantle. Their average values agree well with estimates for upper mantle compositions (as in Tables 6-5 and 6-6), but the wide spread observed casts doubt on the validity of averages based on only a few samples. Average values of some elements with their ranges are as follows: $Al = 1.6 \pm 1.6$; $Cr = 0.3 \pm 0.3$; $Ni = 0.16 \pm 0.16$. They concluded that either the

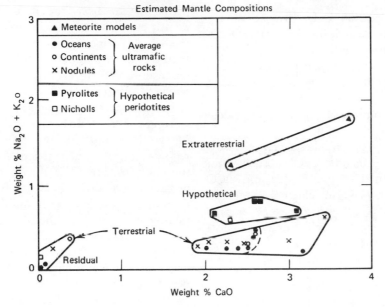

Figure 6-8. Chemical variation of estimated mantle conditions. See Figure 6-9 (after Wyllie, 1970, with permission, Amer. Mineral. Society).

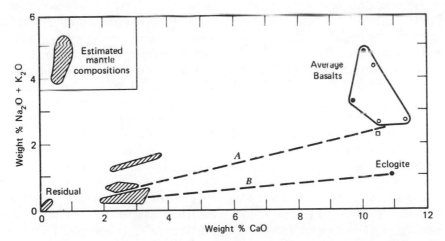

Figure 6-9. Expanded version of Figure 6-8, comparing mantle estimates with average basalts and a picritic eclogite (after Wyllie, 1970, with permission, Amer. Mineral. Society).

mantle is heterogeneous or that none of the rock types included in their study is representative of primitive mantle material.

Despite the different models adopted by Mason and Ringwood for their extraterrestrial estimates and the different approaches used for the other estimates in Table 6-6, these proposed mantle compositions agree with respect to the following:

1. More than 90% by weight of the mantle is represented by the system FeO-MgO-SiO_2 and no other oxide exceeds 4%;

2. Together the components Na_2O-CaO-Al_2O_3 lie within the range 5 to 10%, and more than 98% of the mantle is represented by the six components (with Fe_2O_3 calculated as FeO);

3. No other oxide reaches a concentration of 0.6% of the mantle. FeO substitutes for MgO in most magnesian minerals, and Na_2O substitutes for CaO in plagioclase at low pressures, and in jadeitic pyroxene at high pressures. Therefore the mineralogy of mantle material can be represented in simplified form in the quaternary system CaO-MgO-Al_2O_3-SiO_2.

Chemical Variation Diagrams

The estimated mantle compositions listed in Tables 6-5 and 6-6, together with other estimates that I have reviewed elsewhere (Wyllie, 1970), are compared in Figure 6-8, a standard diagram showing variation in alkali content of rocks with variation in lime content. Distinct areas with no overlap are occupied by each of the four groups of analyses, (a) extraterrestrial, (b) terrestrial undepleted, (c) terrestrial depleted, and (d) hypothetical. This is also true in two other chemical variation diagrams not reproduced here. The first shows the variation in Al_2O_3 using a triangular diagram for CaO-Al_2O_3-alkalis; the CaO/Al_2O_3 ratios for the mantle estimates occupy a limited range with Al_2O_3 being stored relative to lime plus alkalis in the residual types. The second shows the mantle estimates plotted against two standard differentiation indices, $100(Na_2O + K_2O)/(CaO + Na_2O + K_2O)$, and $100(FeO + Fe_2O_3)/(MgO + FeO + Fe_2O_3)$.

In Figure 6-8 the terrestrial estimates are identified as rocks from oceanic or continental environments or as nodules. The estimates of undepleted mantle based on ultramafic rocks are grouped together closely; only three points depart from the main group. The mantle as a whole, according to the meteorite models, contains significantly more alkalis than the upper mantle according to estimates based on observed terrestrial rocks. Also the observed ultramafic rocks contain less alkalis than the hypothetical models.

Figure 6-9 is an expanded version of Figure 6-8 comparing the mantle estimates with average basalts taken from various sources. The line A is drawn through the origin, which corresponds closely to residual dunite, in a position separating the terrestrial mantle estimates from the hypothetical peridotites. The average basalts lie above this line, on the same side as the hypothetical peridotites, confirming that they can be derived by subtraction from the hypothetical peridotite leaving residual dunite. The position of the terrestrial mantle estimates below line A, however, demonstrates that the average basalts cannot be derived by direct subtraction from these known ultramafic rocks if the residual product is to be equivalent to those peridotites and dunites normally regarded as residual types.

The hypothetical models assume that the average basalts are primary magmas derived from the mantle by partial melting, uprise, and eruption. If this is correct then the terrestrial ultramafic rocks are too low in alkalis to provide estimates for the upper mantle. In two stimulating reviews, however, O'Hara (1965, 1968) pointed out that the concept of primary magmas is only one limiting case for magma generation (Chapter 8). He argued that basaltic liquids normally regarded as primitive types could not have been in equilibrium with mantle material at the depth where partial melting occurred. The low pressure basaltic compositions erupted at the surface are the end products of a sequence of events acting upon the original high pressure liquid formed at depth. Experimental studies indicate that this liquid is picritic rather than basaltic, and recent petrological studies suggest that deep-seated liquid compositions cluster along the line *B* in Figure 6-9.

Carswell (1968b) examined in detail the garnet peridotites at Kalskaret in Norway and discovered a linear compositional trend between two end members, one the eclogite in Figure 6-9 and the other a dunite plotted with the residual rocks in Figure 6-8. The line *B* connecting these two end members passes through the group of undepleted mantle estimates based on ultramafic rocks. Carswell interpreted the field and textural relationships to indicate tectonic emplacement of the garnet peridotites as relatively cold intrusions from the mantle in the solid state. He concluded that within the upper mantle a picritic partial melt fraction was not completely liberated from a dunitic residual fraction of the original mantle material, but became trapped and mixed with the dunite; there it crystallized as eclogite. The mixed rocks which plot close to the line *B* were subsequently emplaced into the crust.

The high temperature orogenic peridotite masses of Beni Bouchera, Morocco, the Sierrania de la Ronda, Spain, and Etang de Lers, France, have been interpreted by Kornprobst (1969) and Dickey (1970) in terms of upward movement of mantle peridotite. This caused partial fusion (central diagram in Figure 6-6), and the small proportion of liquid later crystallized in magmatic layers forming various pyroxenites with or without garnet. Crystallization occurred within the mantle at high pressure, and tectonic movements emplaced the solid masses into the orogenic root zones. The peridotites in the Beni Bouchera massif plot in the area of terrestrial mantle estimates of Figures 6-8 and 6-9 or very close to it, and

the magmatic pyroxenite layers plot in a narrow zone between 10.5 and 15.5% CaO, which forms a continuation of the line *B*.

Additional evidence for the existence of picritic liquids is provided by a series of minor intrusions in Skye, Scotland, and Ubekendt Ejland in west Greenland studied by Drever and Johnston (1967). These intrusions are termed "picritic" owing to their richness in olivine, but their conventional explanation as simple mixtures of basaltic liquid plus suspended olivine crystals is inadequate to account for the data presented. Chemical analyses of the picritic and related calcic rocks when plotted on Figure 6-9 lie quite close to the line *B* and its extension, between 5.5 and 13.9% CaO. Drever and Johnston suggested that these materials may represent liquids formed at high pressures by partial fusion of an ultrabasic source rock. This is consistent with the interpretations for the other rocks with compositions close to the line *B*.

In view of the evidence that liquid fractions produced by partial melting of mantle peridotite at depth have compositions near the line *B* in Figure 6-9, it seems that this is the material that should be combined with residual peridotite or dunite in order to obtain a hypothetical mantle peridotite and not an average basaltic composition as employed in the models of Ringwood and Nicholls. This procedure does yield a composition corresponding to the terrestrial ultramafic rocks of Figures 6-8 and 6-9 at least with respect to alkali content. Possibly, therefore, the estimated mantle compositions based on selected ultramafic rocks may provide reasonable indications of some peridotite compositions occurring within the upper mantle, whereas the hypothetical peridotites in Figure 6-8 may be too high in alkali content.

Possible explanations for the difference in alkali content between the extraterrestrial estimates and the terrestrial estimates in Figure 6-8 include: (a) the lower mantle retains more alkalis than the upper mantle, (b) the accreting Earth lost more alkalis and other volatile elements than the meteorite parent bodies, (c) the chondrite models for the origin of the Earth are invalid.

Trace Elements and Volatile Components

The concentrations of major elements in the mantle listed in Table 6-6 appear to be reasonably well established, but large uncertainties remain about the concentrations and distribution of trace elements and volatile components. Related to the trace elements is the question of whether the composition of the mantle and crust is approximated more closely by chondritic meteorites or by Ca-rich achondrites. Arguments about the relative merits of chondritic and achondritic models have been concerned with concentrations in the crust, mantle, and meteorites, of the elements K, U, Th, Rb, Cs, Ba, Sr; and with the ratios Rb/Sr, $^{87}Sr/^{86}Sr$, K/Rb, and K/U. The arguments have not been resolved.

It is difficult to measure accurately the abundances of trace elements in mantle-derived ultramafic rocks, and there may be marked variation in trace element concentrations from one specimen to another in the same ultramafic body. Trace element distribution may be controlled to a large extent by the mineralogy. The concentrations of Rb, Sr, and K, for example, are very sensitive to the distribution of small proportions of amphibole. Secondary alteration which may appear quite minor in a petrographic sense can cause serious contamination with respect to whole rock values for elements such as K, Rb, Cs, Sr, and the rare earth group. In some mantle-derived ultramafic rocks U is distributed homogeneously in clinopyroxenes, but in other similar rocks most of the U occurs as contaminants along cracks, grain boundaries, or in microinclusions. There have been many proposals that the upper mantle has trace element concentrations different from the whole mantle arising from differentiation of the mantle at an early stage in Earth history and from the continued depletion of the

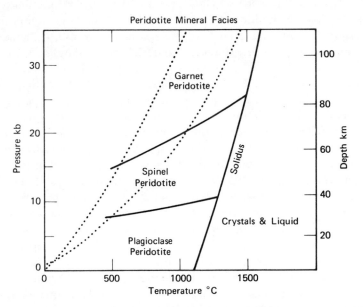

Figure 6-10. Peridotite mineral facies. Generalized phase diagram for an estimated mantle peridotite using data from various sources (after Wyllie, 1970). The lines indicate the solidus of the peridotite and the positions of divariant reaction intervals separating facies for spinel-peridotite from plagioclase-peridotite and from garnet-peridotite. Dotted lines represent geotherms for oceanic and shield regions (Figure 3-12). (With permission, Amer. Mineral Society.)

upper mantle in basaltic components. This makes a difficult problem even more complex.

The regime of volatile components in the Earth is of fundamental significance for many problems. It is generally held that the crust, hydrosphere, and atmosphere have been produced by defluidization of the Earth. The petrographic study of mantle-derived ultramafic nodules and rocks provides evidence that phlogopite and amphibole are primary constituents of the mantle, at least locally, suggesting that some H_2O remains in the mantle. The presence of CO_2 is inferred from the existence of CO_2-filled inclusions in olivines derived from upper mantle sources. The gases given off during volcanic eruptions provide data on volatiles in the mantle and H_2O is the dominant juvenile component. Interpretation is complicated by contamination with meteoritic solutions and by chemical reactions during uprise and eruption of the magma. The proportion dissolved in most magmas is small but uncertain.

Nicholls (1967) suggested that in certain environments, such as beneath the mid-oceanic ridges, the mantle peridotite might contain as much as 9.7% H_2O and 0.8% CO_2 (analysis (8) in Table 6-5). The peridotite mylonites of St. Peter and St. Paul rocks apparently contained an abundant fluid phase during intrusion from the mantle. The proportion of H_2O and other volatile components in the mantle is probably variable both laterally and vertically and probably no more than 0.1% except in special environments. Even a small proportion of volatiles, however, may have a very pronounced effect on the behavior of mantle rocks with respect to (a) physical strength and (b) temperature for beginning of melting.

PHASE TRANSITIONS IN MANTLE PERIDOTITE AND ECLOGITE

Evidence from extraterrestrial sources, petrology, and geophysics indicates that the mantle is composed of peridotitic material, and that in the upper mantle this is mixed with dunite and eclogite. Experimental studies on these compositions at appropriate pressures and temperatures delineate the mineral facies occurring within the upper mantle. The transition of gabbro to eclogite was reviewed in detail in Chapter 5, and the behavior of eclogite at depths greater than 100 km is probably similar to that of garnet peridotite.

Experiments on Peridotite Mineral Facies

There has been much speculation about the phase diagram for peridotite, and Figure 6-10 shows the general pattern of the phase changes for lherzolite, a peridotite consisting of olivine, clinopyroxene, and orthopyroxene together with an aluminous phase. The aluminous phase is plagioclase feldspar in the low-pressure plagioclase-peridotite facies.

With increasing pressure the plagioclase reacts with olivine to yield aluminous pyroxenes and spinel producing a spinel peridotite. At higher pressures garnet-peridotite is produced by reaction of the spinel with orthopyroxene yielding pyrope-rich garnet and olivine. The boundaries between the subsolidus facies are plotted as univariant lines but in fact they are reaction intervals.

Experimental approaches and problems involved in locating subsolidus phase boundaries for rocks were discussed in Chapter 5, and several of the mineralogical reactions related to peridotite in the system CaO-MgO-Al_2O_3-SiO_2 were reviewed and illustrated in Figures 5-11, 5-12, and 5-13. Figure 6-10 is based on these results and on experimental studies above 1100°C by Green and Ringwood using the synthetic pyrolite II listed in Table 6-6. A different synthetic peridotite, with lower MgO/SiO_2 ratio, introduced an additional phase field just below the solidus; this is pyroxene-peridotite without spinel. With increasing temperature within the spinel-peridotite field the spinel dissolves in the aluminous pyroxenes and solution is completed at about 1250°C. The generation of garnet from the spinel-free peridotite then occurs at a higher pressure about 30 kb on the solidus instead of 25 kb.

The general pattern of phase relationships illustrated is correct, although the positions of the boundaries may change considerably with variation in the content of CaO, and minor trivalent oxides such as Al_2O_3, Fe_2O_3, and Cr_2O_3. The geotherms for continental and oceanic regions indicate that at depths greater than 55 or 70 km the mantle consists of garnet peridotite (possibly with dunite and eclogite). Plagioclase peridotite can have only very limited distribution in the mantle.

Estimates of the mineralogy of mantle spinel-peridotites and garnet-peridotites are compared in Table 6-7. J. L. Carter examined the modal abundances of 150 nodules from five basaltic host rocks in terms of the fayalite content of their olivine. He found continuity except for a small gap between 12–14% fayalite, and he argued that this gap represents primitive upper mantle material. The corresponding mode is listed. H. Fujisawa considered the compositions of olivines in mantle-derived nodules and ultramafic rocks and concluded in 1968 that the olivine in the upper mantle contains 10% fayalite. The other assemblages listed in Table 6-7 are estimated upper mantle compositions recast into norms appropriate for the two facies, using analysis 4 in Table 6-6 and pyrolite III similar to analysis 5 in Table 6-6.

Effect of Water

Addition of water to peridotite causes depression of the melting temperature and introduces three hydrous minerals; serpentine, amphibole, and phlogopite. The melting curve and curves for the breakdown of the hydrous minerals are shown in Figure 6-11. The curve for phlogopite gives its maximum temperature stability, with the phlogopite melting incongruently to a vapor-absent assemblage, forsterite plus liquid. In the presence of excess vapor phlogopite melts together with forsterite at a lower temperature about 1200°C. The curve for serpentinite gives its maximum temperature of stability in the presence of excess water vapor. It would dissociate at lower temperatures if vapor were absent. The curve for hornblende is based on results in the presence of a liquid phase with excess water vapor. With water pressure less than load pressure, as in a vapor-absent system, the amphibole would be stable to higher temperatures and pressures. Notice the opposite effects of decreasing water pressure relative to load pressure for (a) a subsolidus dehydration reaction as for serpentine, and (b) a dissociation reaction involving a liquid phase as for phlogopite and amphibole.

TABLE 6-7   Estimates of Mantle Mineralogy

	Spinel Peridotite		Garnet Peridotite	
	Carter 1966	Harris et al., 1967	Harris et al., 1967	Ringwood, 1969
Olivine	55 ± 10	65.3	67	57
Orthopyroxene	27 ± 5	21.8	12	17
Clinopyroxene	14.5 ± 3.5	11.3	11	12
Spinel	3 ± 1	1.5	—	—
Garnet	—	—	10	14
Total	99.5	99.9	100	100

Given a peridotite containing just a trace of water we see that this could be stored in phlogopite, with no free vapor phase, until the temperature exceeded the phlogopite curve. Then a small amount of liquid would be produced by the incongruent melting of phlogopite.

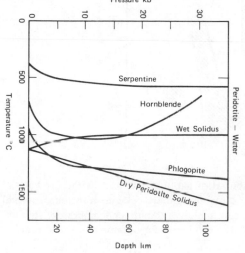

Figure 6-11. Experimentally determined reactions involving peridotite and water, based on various experimental studies (after Wyllie, 1970). Curves are plotted for the solidus of peridotite dry (Ito and Kennedy, 1967), and in the presence of excess water (Kushiro et al., 1968); other curves give the upper temperature stability limits for serpentine (Kitahara and Kennedy, 1966; Scarfe and Wyllie, 1967), hornblende (Kushiro, 1970), and phlogopite (Yoder and Kushiro, 1969). (With permission, Amer. Mineral. Society.)

Addition of water to gabbro causes depression of the melting temperature and introduces one hydrous mineral, an amphibole. Figure 6-12 compares the melting intervals for a dry gabbro, with the position of the gabbro-eclogite transition interval indicated below the solidus (Figures 5-14 and 5-15), and for the gabbro in the presence of excess water. The slope of the solidus (dP/dT) reverses at 15 kb, where plagioclase feldspar is replaced by jadeitic pyroxene. The curve for the breakdown of hornblende is indicated, and the gabbro-eclogite transition is masked

by the hornblende. At pressures above 25 kb, the subsolidus assemblage is quartz eclogite. For a gabbro undersaturated in silica to such an extent that it yielded olivine eclogite, or a simple garnet-pyroxene eclogite, we can expect that the solidus would increase in temperature rather sharply within a short pressure interval above 15 kb where plagioclase disappears; the mineralogy would then be similar to that of the high pressure peridotite (Figure 6-11).

Given a gabbro containing only 0.1% of water we can deduce the general pattern of phase relationships from Figure 6-12 and the resulting isopleth is shown in Figure 6-13. The liquidus temperature is depressed only slightly compared to the dry liquidus. Within the stability field of hornblende all of the water is fixed in the hornblende and no melting occurs. There are two vapor-absent fields for hornblende-gabbro and hornblende-eclogite with a transition interval indicated between them. The breakdown curve for hornblende in Figure 6-13 is located at somewhat higher temperatures than in Figure 6-12 because of the vapor-absent conditions.

At pressures above the stability of hornblende in the subsolidus region the water exists as an interstitial film in eclogite. The curve for the beginning of melting is the same as that for melting with excess water as in Figure 6-12, but for an eclogite without quartz this curve would be at a higher temperature. Because of the very small amount of water present only a trace of liquid develops.

Below about 27 kb pressure the solidus is the breakdown curve for hornblende. The subsolidus vapor-absent assemblages are stable until hornblende releases water, which then passes directly into a vapor-absent liquid phase. The light-shaded area represents gabbro or eclogite with a trace of interstitial liquid, undersaturated with water; its composition is intermediate rather than basic. The liquid content increases within the dark-shaded band, which corresponds closely to the melting interval in the dry system (Figure 6-12).

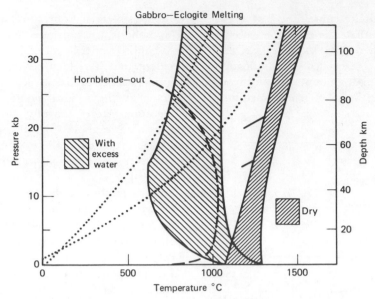

Figure 6-12. Generalized diagram for the melting interval of gabbro, dry, and in the presence of excess water (after Lambert and Wyllie, 1968, 1970). Note the gabbro-eclogite transition interval indicated just below the solidus for dry gabbro. The dashed line gives the upper temperature stability limit of hornblende with excess vapor. The dotted curves are geotherms for oceanic and shield regions. (With permission, Amer. Mineral. Society, Wyllie, 1970).

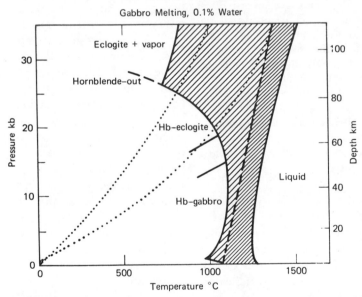

Figure 6-13. Estimated isopleth for the phase fields intersected by material of gabbroic composition in the presence of 0.1% water, based on Figure 6-12 (after Wyllie, 1970; Lambert and Wyllie, 1970). The solidus corresponds to the hornblende-out curve in Figure 6-12, but it is at a higher temperature because of water deficiency. Note the transition interval between gabbro and eclogite just below the solidus. Only a trace of liquid, probably of intermediate composition, is present through the light-shaded area; basic liquid is produced only within the dark shaded interval, which corresponds closely to the dry melting interval in Figure 6-12. (With permission, Amer. Mineral. Society.)

THE OLIVINE-SPINEL TRANSITION

Estimated compositions of the upper mantle (Table 6-6) imply that olivine is the dominant mineral (Table 6-7). When H. Jeffreys first presented evidence for the second order discontinuity in 1936 J. D. Bernal suggested, in the ensuing discussion, that the discontinuity might be accounted for by transition of the orthorhombic mineral olivine into a denser cubic form, spinel. He cited the earlier observation of V. M. Goldschmidt that the analogous compound, Mg_2GeO_4, was known to exist in two forms, one isomorphous with olivine and the other cubic. It has since been claimed several times that the spinel polymorph of Mg_2GeO_4 could not be synthesized.

In 1952 F. Birch published a detailed paper on the elasticity and constitution of the Earth's interior, in which he proposed that within the transition zone of the mantle there occurred a series of phase changes leading to closer packing, and that at depths greater than 1000 km the elastic properties of the mantle were consistent with those of close-packed oxide phases. This was the time when apparatus for the simultaneous development of high pressures and temperatures was being designed, and it was not long before many investigators were attempting to locate phase transitions which might occur within the Earth's interior. An early success was the reproducible formation of the spinel polymorph of Mg_2GeO_4 in the presence of water vapor under pressure by D. M. Roy and R. Roy in 1954; the phase transition curve is shown in Figure 6-14. F. Dachille and R. Roy then approached the problem of locating a corresponding transition in Mg_2SiO_4 by studying the solid solution and transition in the system Mg_2GeO_4-Mg_2SiO_4 with results published in 1956 and 1960.

The greatest contribution to this problem has been made by A. E. Ringwood. In a series of three papers on the constitution of the upper mantle in 1958 he published the results of thermodynamic studies of the olivine-spinel transition based on experimental studies in various model systems at 1 bar and at high pressures. Evidence for the chemical composition of the mantle was reviewed. Relating the composition to the phase transitions anticipated at high pressures and temperatures within the Earth, he set up a specific model for the mantle, in which the transition zone of the upper mantle between 400 and 1000 km (Figure 3-3) was explained by the olivine-spinel transition modified by solid solution effects. In these three papers and following papers on the same topic Ringwood made many predictions about the upper mantle. Ten years later, from his review of the problem at the International Conference on Phase Transformations and the Earth's Interior, held in Canberra in February 1969, it is clear that his percentage of successful predictions is remarkably high.

Between 1956 and 1966 the existence of an olivine-spinel transition was confirmed in many substances, including Fe_2SiO_4, and there were several estimates made for the conditions of transformation of Mg_2SiO_4 based on rather long extrapolations. The development of new types of high pressure, high temperature apparatus in 1966 led to rapid advances in study of the transition in the system Fe_2SiO_4-Mg_2SiO_4. In 1970 Ringwood and Major demonstrated that for compositions near Mg_2SiO_4 the transition is more complicated than previously supposed. Some of the results published between 1956 and 1969 are reviewed below and summarized in Figures 6-14 and 6-15.

Phase Diagram for Fe₂SiO₄

The upper mantle olivine is generally believed to have a composition near 90% forsterite, 10% fayalite and attention was directed early to the transitions in Fe_2SiO_4. The phase diagram for Fe_2SiO_4, published by S. Akimoto and E. Komada in 1967, is

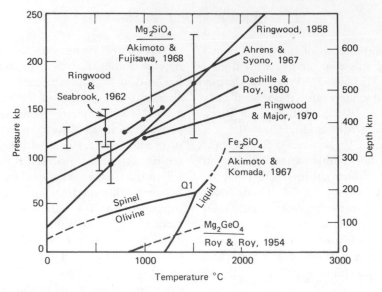

Figure 6-14. Experimental and theoretical results related to the olivine–spinel transition in the upper mantle. Mg_2GeO_4: olivine-spinel transition. Fe_2SiO_4: olivine-spinel transition and melting curves meeting at triple point Q_1. Mg_2SiO_4: olivine-spinel transition, see text and Figure 6-15.

shown in Figure 6-14. The olivine-spinel transition has a slope similar to that illustrated for the same transition in Mg_2GeO_4. The melting curve for fayalite meets the olivine-spinel transition curve at a triple point, Q_1, at 62 kb and the fusion curve for spinel extends upward in pressure after a sharp inflection at Q_1. This shows the pattern of phase relationships sought for the composition Mg_2SiO_4.

The Transition for Mg_2SiO_4

Figure 6-15 summarizes the attempts to determine the conditions for the breakdown of forsterite by extrapolation from within binary systems. Curves through individual points estimated in this way are given in Figure 6-14.

Mg_2GeO_4-Mg_2SiO_4. The univariant curve for the olivine-spinel transition in Mg_2GeO_4 is shown in Figure 6-14. The effect of adding Mg_2SiO_4 as a component was determined by Dachille and Roy in isobaric and isothermal sections through the *PTX* model

required to represent the binary system completely. In 1956 they reported the synthesis of single spinel phases at 20 kb with about 18% Mg_2SiO_4 in solution. Detailed results were published in 1960. Figure 6-15*a* shows the two-phase field for coexisting olivine and spinel solid solutions determined up to 65 kb pressure at 542°C. Extrapolation of the measured solid solution loop to a closure point (the dashed lines) for the composition Mg_2SiO_4 provided an estimate of 100 kb for the olivine-spinel transition in forsterite at this temperature. This corresponds to one point on the univariant curve for the transition. Dachille and Roy estimated that the slope of the curve passing through this point would be between 13°C/kb and 25°C/kb and an average value is given in Figure 6-14.

In the second of his three papers on the constitution of the mantle Ringwood in 1958 reported data on the extent of solid solution at 660°C and 30 kb between Mg_2GeO_4 and Mg_2SiO_4. From this he calculated the free energy of the transition in forsterite and used this to calculate the pressure required to

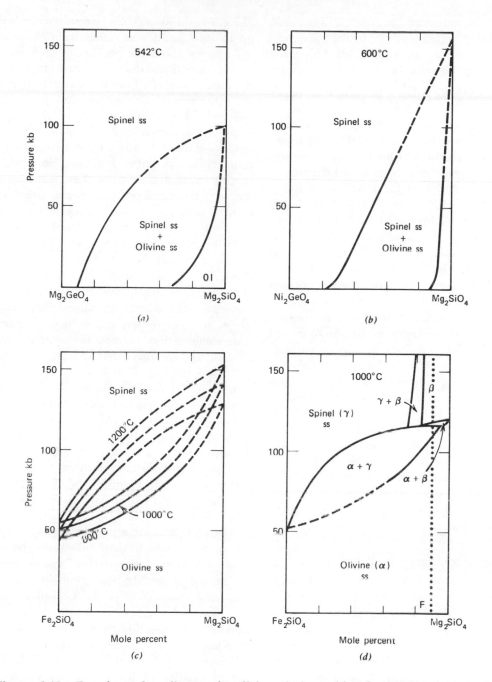

Figure 6-15. Experimental studies on the olivine-spinel transition for Mg_2SiO_4. (*a*) Isothermal extrapolation from Mg_2GeO_4–Mg_2SiO_4 by Dachille and Roy (1960). (*b*) Isothermal extrapolation from Ni_2GeO_4–Mg_2SiO_4 by Ringwood and Seabrook (1962). (*c*) Isothermal extrapolation from Fe_2SiO_4–Mg_2SiO_4 by Akimoto and Fujisawa (1968). (*d*) Direct determination of phase relationships by Ringwood and Major (1970). Olivine with composition near Mg_2SiO_4 transforms not to spinel but to a spinel-like beta-phase. Composition F is used in Figures 6-16 and 6-17. (With permission, American Jour. Science, Jour. Geophys. Research, Physics Earth and Planet. Interiors.)

cause the transition at 660°C. The most probable pressure was 88 kb, the maximum 116 kb, and the minimum 72 kb. This result is shown in Figure 6-14.

Ni₂GeO₄-Mg₂SiO₄. In the first paper of the 1958 set Ringwood presented a phase diagram for the solvus at 1 bar pressure between forsterite and Ni_2GeO_4 which has a spinel structure. From the solid solution relationships he calculated that the pressure for the transition of forsterite to the spinel structure at 1500°C was 175 ± 55 kb. In his second paper Ringwood drew the univariant curve through this point and the 660°C point estimated from the Mg_2GeO_4 system as shown in Figure 6-14. In an appendix to this paper he reported reconnaissance results up to 70 kb pressure, at temperatures between 560°C and 660°C, on the effect of pressure on the solid solubility of Mg_2SiO_4 in the spinel Ni_2GeO_4. A linear extrapolation indicated that at 600°C the spinel form of Mg_2SiO_4 should be stable at 125 kb (a maximum value). This result is not plotted in Figure 6-14, but the revised value of 130 ± 20 kb at 600°C determined in 1962 by Ringwood and Seabrook is shown. This point was located by the extrapolation shown in the isotherm for the binary system in Figure 6-15b. The extrapolated pressure shown is 155 kb, but this was corrected downward to 130 kb because of a revision in the high pressure scale.

Fe₂SiO₄-Mg₂SiO₄. There have been several studies of the solid solution of Mg_2SiO_4 in the spinel form of Fe_2SiO_4 and Figure 6-15c shows three isotherms located between 43 and 96 kb at 800, 1000, and 1200°C, by S. Akimoto and H. Fujisawa in 1968. Extrapolations of the two-phase fields to closures, as shown by the dashed lines in the figure, give for points on the forsterite-spinel transition the values 127 kb at 800°C, 140 kb at 1000°C, and 151 kb at 1200°C. These three points and the line connecting them are plotted in Figure 6-14. An earlier preliminary report gave an estimate of 150 kb at 800°C.

Figure 6-14 shows that the results described up to 1968 indicate a rather wide range of uncertainty about the position of the transition in pure forsterite. Calculated curves for the transition suffer from the uncertainties in the experimental results. For example T. J. Ahrens and Y. Syono calculated the slope of the curve shown in Figure 6-14, with its range of error, from the thermochemical and compressibility data available in 1967. Having calculated the slope they then passed the curve through a preliminary experimental point at 150 kb and 800°C, which was the following year moved downward through 33 kb by Akimoto and Fujisawa. They gave as the equation for the transition curve: $P = 110 \pm 11$ kb $+ 0.050\,T°C$. In the same year, using essentially the same data, D. L. Anderson calculated an almost identical curve, but noted that the revised pressure scale would reduce the pressure estimates by 20–30% and the curve that he used therefore had the equation: $P = 74$ kb $\pm 0.053\,T°C$. This is not plotted in Figure 6-14, but it would be quite close to the curve of Dachille and Roy.

Ringwood and Major reviewed in detail the experimental and theoretical data in this system prior to 1969 and presented the 1000°C isotherm shown in Figure 6-15d. This shows that the previous extrapolations were based on an incorrect assumption. Pure forsterite does not transform to a spinel but to a spinel-like material which they called the beta-phase. Previous claims that spinel was synthesized from Mg_2SiO_4 probably arose from misidentification of the beta-phase or of a hydrated magnesium silicate. No more than 80% of Mg_2SiO_4 dissolves in the spinel (γ) phase at 1000°C below 150 kb.

Figure 6-15d shows the first direct experimental location of the breakdown of pure forsterite with the beta-phase produced at 120 kb and 1000°C. This point is shown on Figure 6-14 and Ringwood and Major assumed that the slope of the curve for this transition was similar to that previously determined for the transition of fayalite to spinel.

The Transition for Olivine Fo$_{90}$Fa$_{10}$

It is generally assumed that the olivine in the upper mantle has a composition close to 90% forsterite. Figure 6-15d shows the changes (under equilibrium conditions) that occur in an olivine of this composition (F) with increasing pressure at 1000°C. At about 108 kb the olivine begins to transform to spinel (γ) with only 48% Mg$_2$SiO$_4$ in solution. With increased pressure both the spinel and the remaining olivine become enriched in magnesium, and at 116 kb their compositions become olivine with 94% Mg$_2$SiO$_4$ and spinel with 75% Mg$_2$SiO$_4$. These phases react at constant pressure to yield the beta-phase with 81% Mg$_2$SiO$_4$ in solution until all of the spinel is used up. With pressure increasing the olivine is transformed into the beta-phase, and the beta-phase composition reaches that of the original olivine at 119 kb just as the last trace of olivine disappears. The olivine is thus transformed to a more dense phase through a divariant transition interval of 11 kb with a univariant (isothermal invariant) reaction occurring within the interval.

The position of this transition interval as a function of pressure and temperature can be estimated from the schematic PTX model for the system Mg$_2$SiO$_4$-Fe$_2$SiO$_4$ which is given in Figure 6-16. The front face of the model is the PT diagram for the system Fe$_2$SiO$_4$ with three unary curves meeting at the invariant point Q$_1$ (Figure 6-14). Extension of the liquid volume and adjacent phase fields into the binary system are shown for completeness, but they need not concern us here. The back face of the model shows only the transition of forsterite to spinel as anticipated in Figures 6-14 and 6-15c. The phase fields for the beta-phase required by Figure 6-15d have been omitted for geometrical clarity.

The transition interval in which olivine and spinel coexist is given by the intersection of the constant composition plane for 90% Mg$_2$SiO$_4$ with the two-phase space in Figure 6-16. The isopleth showing the estimated transition band is given in Figure 6-17. The width of the interval is based on the experimental value of 11 kb at 1000°C in Figure 6-15d. According to Figure 6-16 the transition interval is a simple two-phase field between olivine and spinel but Figure 6-15d shows that the interval is divided into two parts by the univariant reaction:

$$\text{forsterite}_{ss} + \text{spinel}_{ss} \rightleftharpoons \text{beta-phase}_{ss} \quad (6\text{-}1)$$

This boundary is added as a dashed line in Figure 6-17, and the high pressure field is labelled beta-phase according to Figure 6-15d instead of spinel as indicated by Figure 6-16. Ringwood and Major pointed out that magnesian spinel and the beta-phase had almost identical densities at run conditions.

The geotherm in Figure 6-17 is Ringwood's estimate for shield regions extrapolated from below 250 km (Figure 3-12b) along a curve 250°C higher than the geotherm calculated earlier by Clark and Ringwood (Figure 3-12a). If the upper mantle were composed only of olivine with the composition represented in this diagram, Fo$_{90}$, it would be progressively converted to a spinel phase between A and B; at B the spinel phase would be replaced by the beta-phase; and between B and C, the remaining olivine would be progressively converted to the beta-phase. In the depth interval 305 to 430 km the olivine would be converted completely to the beta-phase with a density increase of 10.6%.

The uncertainties about estimating the temperature at depth within the Earth were discussed in Chapter 3, and the position of the transition interval in Figure 6-17 is only schematic. Figure 6-14 illustrates the uncertainties about the experimental data on which the position and slope of the transition interval is based. As all investigators have concluded, however, the seismic discontinuity at a depth of 400 km does correspond remarkably well with the transition of mantle olivine into a more dense phase.

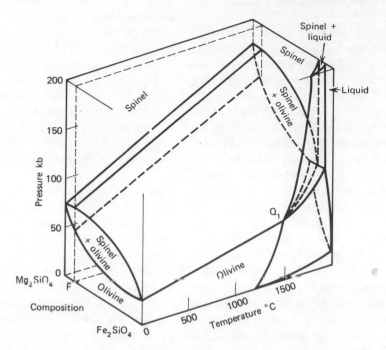

Figure 6-16. Schematic phase diagram in PTX space for the system Mg_2SiO_4–Fe_2SiO_4 using model corresponding to Figure 6-15c, omitting the beta-phase (6-16d) for clarity. The vertical plane through composition F (Figure 6-15d) intersects the transition interval as shown in Figure 6-17.

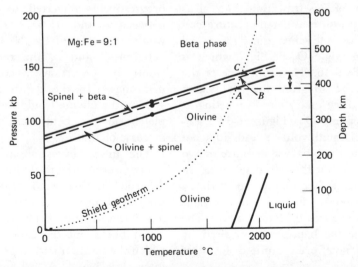

Figure 6-17. Schematic isopleth for the composition F (Figures 6-15d and 6-16) in the system Mg_2SiO_4–Fe_2SiO_4, showing detail through the transformation interval of olivine to beta-phase (Figure 6-15d). This is related to transitions within the mantle by the depth scale and the shield geotherm from Figure 3-12 (Ringwood 1966). Mantle material of this composition would undergo the transition in the depth interval *A–C*.

The Transition Zone of the Mantle 400 to 1000 km Depth

When experimental and theoretical investigations of the olivine-spinel transition began in the 1950's the aim of the exercise was to explain the seismic wave velocity distribution within the transition zone of the mantle, between 400 and 1000 km, as shown in Figure 3-3. More recent interpretations of the seismic velocity distribution (Figure 3-5) show that the transition zone includes three seismic discontinuities at depths near 400, 650, and 1050 km.

The estimated position of the transformation interval for mantle olivine, as shown in Figure 6-17, is a satisfactory explanation for the sharp discontinuity near 400 km but before this had been detected, the problem was to find some way to extend the effect of the olivine-spinel transformation in order to explain the whole transition zone between 400 and 1000 km.

In 1958 Ringwood suggested that the olivine-spinel transformation in the mantle could occupy a wide depth range if (a) the transformation interval had a much steeper slope than that illustrated in Figure 6-17, and (b) there were a marked increase in the Earth's temperature gradient between the two discontinuities. In papers between 1958 and 1966 Ringwood developed models in which the smoothed velocity distributions from 400 to 1000 km depth (Figure 3-3) were explained by the smearing-out effects associated with solid solution. According to these models the transition zone in garnet-peridotite is dominated by the olivine-spinel transformation at the beginning, but with an additional series of reactions, including the breakdown of orthopyroxene, leading to the progressive increase in density of the material. Solid solution effects spread the individual transitions over substantial depth ranges. In early models the heterogeneous upper mantle garnet peridotite passed through a wide transition interval until at a depth of 1000 km it consisted of a homogeneous rock composed of disordered, defect spinel in which all the other minerals had dissolved. An alternative explanation is that the minerals break down into close-packed oxides. In later models Ringwood suggested that spinels would transform to denser phases with silicon in octahedral coordination.

In his 1969 review Ringwood revised the model taking into account new experimental data and the changed picture of seismic velocities in the mantle (Figure 3-5), and he noted that the effects of solid solution did not have the result previously attributed to them. The apparent success of his previous models emphasizes a point made by Ringwood: however plausible a model may appear the acquisition of data usually requires that it be modified. It is certainly true that limited data can be extrapolated to fit observations far more readily than detailed data.

Figure 6-17 shows that some explanation other than the olivine-spinel transition is required to explain the changes in the seismic velocity profile near 650 and 1000 km (Figure 3-5). Details of Ringwood's 1969 model are shown in Table 6-8.

In 1967 D. L. Anderson had assumed that the spinel broke down to a "postspinel" phase composed of MgO, FeO, and SiO_2 with silicon in octahedral coordination and with the same thermochemical properties as the oxides. He calculated the univariant curves and transformation loops for the spinel-postspinel reaction in the system Mg_2SiO_4-Fe_2SiO_4, and used them in conjunction with similar olivine-spinel intervals based on extrapolations of the type in Figure 6-15c. The depth of a transformation interval such as AC in Figure 6-17 changes as a function of composition (Figures 6-15 and 6-16). By varying the composition Anderson fitted the intervals for both the olivine-spinel and spinel-postspinel transformations to the depths of the seismic transition zones near 400 and 650 km. The transition regions are consistent with temperatures near 1500°C at 365 km, and 1900°C at 620 km (compare Figure 3-12). He concluded that the seismic

TABLE 6-8   Mineralogy of the Mantle Peridotite as a Function of Depth: 1969 Model of A. E. Ringwood[a]. Compare Figure 3-5*b*

Depth km	Mineral assemblages and transformations	Weight percent mineral	Coordination of elements	Zero pressure density
To 80	*Plagioclase peridotite* *Spinel peridotite*	Table 6-7		
80–350	*Garnet peridotite* olivine orthopyroxene clinopyroxene garnet	 57 17 12 14	Si-4 Mg, Fe, Ca, -6, 8	3.38
350–450	Pyroxene → garnet Olivine → spinel[b]			
450–600	Spinel[b] Garnet Jadeite	57 39 4	Si-4, 6	3.66

600–700	Spinel[b] → Sr_2PBO_4 structure
	  or MgO + ilmenite structure
	Garnet → ilmenite + perovskite
	  structures
	Jadeite → calcium ferrite
	  structure			

700–1050	Ilmenite solid solution Strontium plumbate structure Perovskite, $CaSiO_3$ Calcium ferrite, $NaAlSiO_4$	36 55 6.5 2.5	Si, Mg, Fe -6	3.99 -4.03
1050–1150	Transformations into phases denser than isochemical oxide mixture			
1150–2900	Speculative. Depends on germanate analog systems and shock wave studies		Si, -6 Mg, Fe 6	7% higher than mixed oxides

[a] See also Ringwood, 1970.   [b] Includes beta-phase, Figure 6-15d.

velocities and the depths to the transition zones are consistent with a mantle composition between about 80 and 60 mole-percent Mg_2SiO_4; the Fe_2SiO_4 content appears to increase with depth from 500 to 700 km. Estimating compositions and temperatures in the mantle by correlating geophysical data with laboratory measurements points the way to future advances. The reliability of the available experimental and theoretical data on phase transitions in mantle material, however, remains rather uncertain as shown by Figures 6-14 and 6-15 for the simple olivine composition itself. According to Table 6-7 the upper mantle may contain only 55–67% olivine. The effects of solid solution have not yet been evaluated.

MINERALOGY AND PETROLOGY OF THE MANTLE

High pressure, high temperature apparatus provides, in effect, an indirect experimental probe into the Earth's interior for comparison with the properties measured by geophysical methods. Before 1966 the experimental probe was limited to less than 200 km in depth. Since 1966 the development of new apparatus has permitted the laboratory reproduction of conditions corresponding to 600 km. Many new phase transformations in silicates have been discovered in particular by A. E. Ringwood and associates. The study of germanate systems, which serve as analogs for silicate systems at higher pressures, has revealed additional phase transformations that may be experienced by the silicates at considerably greater depths. Shock wave techniques are capable of developing transient pressures corresponding to conditions in the lower mantle and phase transformations in silicates have been discovered in this way. Shock wave studies provide information for the formulation of an equation of state for mantle materials.

Ringwood's 1969 correlation of phase transformations in mantle peridotite (pyrolite model) with the seismic properties of the mantle is summarized in Table 6-8. He assumed that the molecular ratio of Fe/(Fe + Mg) remained constant at 0.11 throughout the mantle.

The Upper Mantle, the Transition Zone, and the Low-Velocity Zone

The mineralogy of the mantle at any depth can be estimated by following geotherms through experimentally determined mineral facies in material of appropriate composition. Estimates of mantle composition and the experimentally determined phase diagrams have been reviewed. Combining and extrapolating these data to pressures corresponding to depths of 600 km provides Figures 6-18 and 6-19 which illustrate schematically the phase relationships in probable mantle

materials dry or in the presence of water traces. A guide to the petrology of the mantle is given by the geotherm which is the oceanic geotherm from Figure 3-12, and a corresponding sequence of rock types with depth through the mantle is shown in Figure 6-20. Estimates of temperature distribution in the Earth depend critically upon assumptions regarding the relative significance for heat transfer of conduction, radiation, and convection as discussed in Chapter 3. The combination of extrapolated experimental data and arbitrarily selected geotherms to provide estimated mantle cross sections is intended only to show patterns for the petrology of the mantle; specific temperatures and depths can be modified as required by the acquisition of extended and improved experimental data.

The upper mantle is composed of peridotite and eclogite (Figure 5-18). The density distributions determined by F. Press using Monte Carlo methods (Figures 3-9b and 3-10) suggest that the upper mantle is chemically and mineralogically zoned, laterally as well as vertically, and that there may be as much as 50% eclogite in the suboceanic mantle between depths of 80 and 150 km. In addition to these rocks there is residual peridotite or dunite and precipitated peridotite resulting from the extraction of picritic or basaltic magma and its fractional crystallization during uprise. Superimposed on this heterogeneous mantle is a concentric succession of mineral facies.

Consider first a dry mantle. Experimentally determined mineral facies for peridotite and eclogite down to about 100 km are shown in Figures 5-15 and 6-10, and these are reproduced in Figures 6-18a and 6-19a. The general pattern of transitions for peridotite or eclogite in the mantle according to Figures 6-18a and 6-19a can be seen in Figure 6-20. Figure 6-10 shows that plagioclase peridotite can exist only in oceanic regions or in high-temperature regions with a thin crust. At

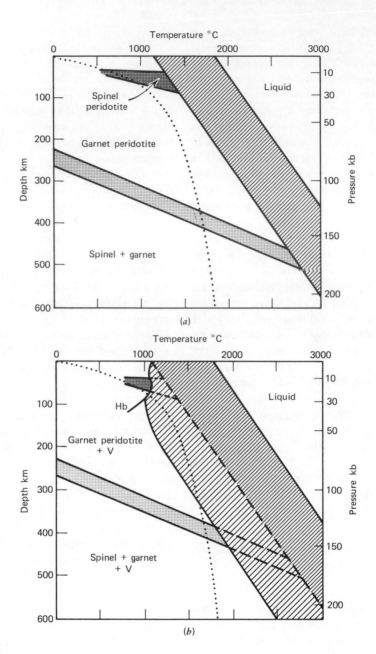

Figure 6-18. Schematic phase diagrams for peridotite and peridotite-water extrapolated to pressures corresponding to 600 km depth in the mantle (after Wyllie, 1971). (*a*) Dry peridotite, based on Figures 8-14*a*, 6-10, and 6-17. (*b*) Isopleth for peridotite with 0.1% water, based on Figures 6-18*a*, 6-11, and the procedure used in construction of Figure 6-13. The melting interval consists of two parts: the light-shaded band represents incipient melting, and the dark-shaded band is almost equivalent to dry melting. The depth scale and the geotherm (dotted) relate the phase diagrams to upper mantle of specific compositions. (With permission, Jour. Geophys. Research.)

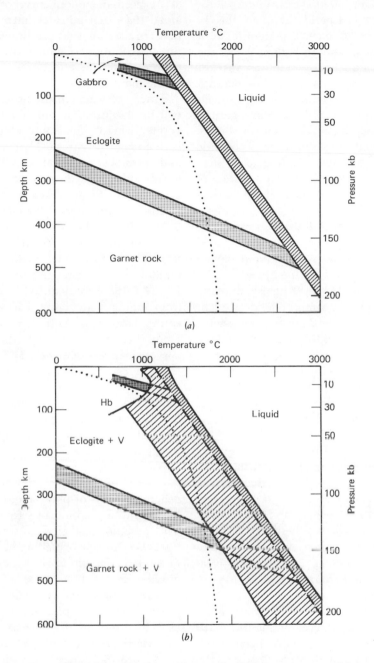

Figure 6-19. Schematic phase diagrams for gabbro and gabbro-water compositions extrapolated to pressures corresponding to 600 km depth in the mantle (after Wyllie, 1971). (*a*) Dry gabbro, based on Figures 8-14, 5-15, and analogy with 6-18*a*. (*b*) Isopleth for gabbroic material with 0.1 % water, based on Figures 6-18*b* and extrapolation of Figure 6-13. The melting interval consists of two parts: the light-shaded band represents incipient melting, and the dark-shaded band is almost equivalent to dry melting. The depth scale and the geotherm (dotted) relate the phase diagrams to upper mantle material of specific compositions. (With permission, Jour. Geophys. Research.)

depths of about 30 km this undergoes a transition to spinel-peridotite, and this in turn transforms to garnet peridotite at a depth of about 70 km. In regions where the geotherm is high spinel-peridotite can exist to a maximum depth of about 85 km. Beneath the continental shields a thin layer of spinel-peridotite would be transformed into garnet-peridotite at depths of 55 to 60 km. Figure 5-15 shows that crystallization of basaltic magma in the upper mantle can yield gabbro, but in cooling to the temperature of the normal geotherm the gabbro should transform to eclogite.

According to Table 6-8 the transition zone begins with a seismic discontinuity produced by the breakdown of pyroxene and olivine in the garnet peridotite of the upper mantle. The pyroxene dissolves in the garnet already present producing a complex solid solution. The formation of the β-phase from olivine is illustrated in Figures 6-17 and 6-18 and the formation of garnet in Figure 6-19. Because of the similarity in physical properties and transition parameters Ringwood found it convenient to treat the β-phase and spinel together as spinel. This reaction zone is shown near 400 km depth in Figure 6-20. Accounting for the seismic discontinuity near 650 km is the breakdown of all three minerals, spinel, garnet, and jadeite producing new phases with silica in six-fold coordination. The transition zone ends near 1050 km with a series of small discontinuities; these are interpreted in terms of transformations into phases which are denser than an isochemical oxide mixture and characterized by Mg coordinations higher than six.

Partial melting occurs if the temperature exceeds the solidus curves in Figures 6-18a and 6-19a. These curves have been determined experimentally down to depths of 100 to 150 km (Figures 5-15, 6-11 and 6-12), but results from different laboratories for different compositions vary, and there is a large uncertainty in extrapolation of the slopes of the melting intervals from 100 to 600 km. None of the calculated geotherms shown in Figure

3-12 exceeds the solidus curves of Figures 6-18a and 6-19a which indicates that magma generation requires unusual conditions of temperature distribution (Chapter 8). It also leads us to consider the effect of water on melting in the mantle.

Traces of water in the uppermost mantle will be stabilized in amphibole and phlogopite where potassium abundances are high enough (Figures 6-11, 6-12, and 6-13), and possibly in titanoclinohumite. The hornblende breakdown curves in Figures 6-18b and 6-19b are taken from Figures 6-11 and 6-13. The interval for incipient melting of eclogite in the presence of 0.1% water shown in Figure 6-19b is extrapolated directly from Figure 6-13. The melting relations for peridotite in the presence of 0.1% water in Figure 6-18b were deduced in the same way from the data in Figure 6-11 and the same extrapolation was made. It is experimentally established that the hydrothermal solidus curves change slope (dP/dT) from negative, to positive at pressures between 15 and 30 kb, and it appears from the eclogite-water system that the slope becomes similar to that for the dry rock (Figures 6-12 and 6-13). There is a large uncertainty in the assumption adopted in Figures 6-18b and 6-19b that the solidus curves in the presence of water become parallel with the dry solidus curves, but the general pattern of phase relationships is not likely to differ significantly from that shown. The oceanic geotherm passes through the interval of incipient melting in Figures 6-18b and 6-19b. Therefore, given this temperature distribution and a trace of water in a mantle composed of peridotite and eclogite, a trace of interstitial silicate liquid will exist in the depth zone where the geotherm exceeds the solidus temperature as indicated in Figure 6-20 (see discussion of Figure 6-13).

The existence of a low-velocity zone in the upper mantle for most tectonic environments is now established for both S-waves and P-waves (Figures 3-4 and 3-5b). There have been many attempts to explain this:

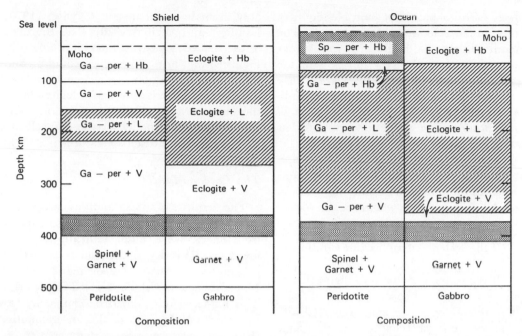

Figure 6-20. Schematic sections through the upper mantle in two different tectonic environments, for mantle material composed of either peridotite or eclogite, in the presence of traces of water (after Wyllie, 1971). These were determined by following geotherms (Figure 3-12) through the mineral facies in Figures 6-18*b* and 6-19*b*. The thickness of the zone of incipient melting depends upon both environment (geotherm) and composition. (See Anderson, 1970). (With permission, Jour. Geophys. Research.)

1. By departure of the geotherm from the critical temperature gradient for which the seismic velocity remains constant,

2. In terms of chemical or mineralogical zoning.

Although these factors may contribute to the low velocities they have not proved adequate to account for the properties of the zone. Geophysicists now conclude that partial melting is required to explain their observations, and the assumption has been usually implicit that melting occurred because the geotherm rose above the dry solidus of peridotite or eclogite. This introduces a number of problems, and it requires that the geotherm be rather sensitively controlled and constrained close to the solidus lest too much liquid is produced. A more likely explanation is that incipient melting occurs in the presence of traces of water as suggested by D. L. Anderson and C. G. Sammis (1969).

The pattern of phase relationships for material of gabbroic composition, containing only 0.1% water, is shown in Figures 6-13 and 6-19*b*. I. B. Lambert and I suggested in 1968 that the low-velocity zone corresponds to the level where amphibole becomes unstable because at greater depths traces of water in eclogite would cause the formation of interstitial magma. At the same time I. Kushiro published the solidus curve for peridotite-water in Figure 6-11 and suggested that water in mantle peridotite could be responsible for partial melting in the low-velocity zone. The geotherms in Figures 6-13 and 6-19*b* show that variation in temperature at depth would cause the boundary between vapor-absent hornblende eclogite and partially melted eclogite to migrate up or down, and

Figure 6-18*b* shows that hornblende may play a similar role in peridotite mantle. The amount of interstitial liquid produced is very small and almost a direct function of the water content; it does not vary significantly for wide variations in temperature and depth. For a given water content the amount of liquid produced in peridotite is much less than in eclogite.

If the low-velocity zone is caused by incipient melting in the presence of water its downward termination could be due to several effects. There would be no liquid below the zone:

1. If the water content became vanishingly small.
2. If the water was stored in high-pressure hydrous crystalline phases.
3. If the bottom of the zone coincided with the lower limit of eclogite; for the same amount of water there would be very much less liquid in the underlying peridotite than in the eclogite-bearing peridotite.
4. If the geotherm passed through the solidus into the subsolidus region. This is the condition illustrated in Figures 6-18*b*, 6-19*b*, and 6-20.

Figure 6-20 shows the patterns for the petrology of hypothetical mantle sections composed of either peridotite or gabbro with a trace of water present, in two different tectonic environments. These are derived by following the continental shield and oceanic geotherms from Figure 3-12 through the schematic phase diagrams of Figures 6-18*b* and 6-19*b*. Note that the water is contained either in (a) hydrous minerals in vapor-absent zones, (b) hydrous silicate liquid in vapor-absent zones, or (c) aqueous vapor phase. Note that changing the position of the geotherm in Figure 6-18*b* could introduce a zone containing vapor between the hornblende peridotite and the peridotite with liquid. The presence of carbon dioxide and other volatile components in the mantle would require modification of this scheme in detail, but the general pattern would prob-ably remain unchanged (Wyllie, 1970; Lambert and Wyllie, 1970). The thickness of the shaded zone with interstitial liquid in Figure 6-20 varies with the geotherm and therefore with tectonic environment, and the distribution of peridotite and eclogite within the mantle also affects its position. The versatility of this model for the low-velocity zone is one of its attractive features.

The Lower Mantle

The densities of model mantle materials at lower mantle pressures, derived principally from shock wave data, indicate that the lower mantle is denser than the hypothetical peridotite with six-fold coordination of Si, Mg, and Fe, and with $Fe/(Fe + Mg) = 0.11$. This higher density could be caused by (a) increased Fe/Mg ratio in the lower mantle, (b) phase transformations to denser structures with Mg and Fe in coordination higher than six, or (c) a combination of these factors. Ringwood presented evidence suggesting that the required density increase could be caused by phase transformations without significant change in composition. F. Press and D. L. Anderson concluded that a change in composition was required to explain the geophysical data.

The successful Earth models selected by the Monte Carlo procedure (Figure 3-9*b*) were compared by Press with laboratory measurements on rocks and minerals that are candidate constituents for the mantle. He concluded that the transition zone is a compositional boundary as well as a region of phase changes. He estimated that the inhomogeneous mantle is consistent with an increase in the molar FeO/(FeO + MgO) ratio by a factor of two. The increase corresponds to a change from $(Mg_{0.8}Fe_{0.2})_2SiO_4$ to $(Mg_{0.6}Fe_{0.4})_2SiO_4$. Using a completely different procedure reviewed a few pages back Anderson concluded that the Fe_2SiO_4 content of the mantle appeared to increase with depth from 500 to 700 km. He re-examined the problem in more detail the following year, 1968, using a combination of several different types of geophysical evidence, and concluded that if the FeO/MgO ratio in the upper mantle is 0.1 then it is 0.27 in the lower mantle.

If the lower mantle is enriched in iron compared with the upper mantle this has significant geochemical and geophysical implications. It has a bearing on hypotheses of differentiation of the Earth into core and mantle and on mantle convection hypotheses. Concentration of FeO relative to MgO at depth is stabilizing with respect to convection and makes the case for mantle-wide convection more difficult. Convection would then be restricted to the upper mantle.

7. The Structure, Petrology, and Composition of the Earth's Crust

INTRODUCTION

The structure of the Earth's crust can be determined from seismic, gravity, and magnetic measurements. Seismic studies provide data about the physical properties of rocks in the crust and the location of discontinuities between rock units with different properties. The seismic velocities can be compared with velocities measured for rocks in the laboratory at known pressures and temperatures (Figure 5-3). Gravity surveys cannot provide unique crustal sections because the anomalies are capable of explanation by more than one model, but with seismic measurements used as a control in interpretation gravity surveys provide significant data about the density distribution in crustal sections. Magnetic anomalies recorded near the Earth's surface are strongly influenced by local rock masses, but aeromagnetic surveys make it possible to distinguish major crustal units such as shields and folded belts of various ages. During the 1960's there were remarkable advances in technique and interpretation with national and international programs bringing all three methods to bear on specific regions.

The rock types composing the structural units of the crust can be estimated from the distribution of rocks at the surface and in drill cores of limited depth, from deductions about the geological history and petrogenesis of exposed rocks, from geochemical considerations, and from the physical properties of rocks at depth determined by geophysical methods. The composition of the crust can then be calculated from the known average compositions of rocks assigned to the various crustal units and layers.

STRUCTURE OF THE CRUST

Figure 5-1 outlines the main features of the crust in the major tectonic environments: the continents, ocean basins, and continental margins including island arcs and trenches. It also depicts the change in crustal thickness associated with the active belts: the oceanic ridges and the folded mountain chains. Selected velocities for P-waves indicate a sharply defined boundary between crust and mantle and a poorly defined boundary between upper and lower crust. The boundary between the crust and mantle was reviewed in detail in Chapter 5 and, in this context, we also examined the variation and distribution of seismic velocities as a function of depth in various tectonic environments. Figures 5-2, 5-3, and 5-4 show average P-wave velocities in different depth intervals.

Conventionally seismic data have been interpreted in terms of a two-layered crust divided into a "granitic" layer and a "basaltic" layer by the Conrad discontinuity and this is the general pattern depicted in Figure 5-1; the profiles of average seismic velocities in Figures 5-2, 5-3, and 5-4 are consistent with a layered crust. Soviet seismologists have largely abandoned this model in favor of more complex models that are more realistic geologically. A velocity profile for a layered crust is shown in Figure 7-1*a*, and this differs from the conventional model only by the presence of two velocity reversals, or low-velocity layers, within the crust. Figure 7-1*b* shows another layered model with the conventional downward sequence of increasing velocities interrupted by layers with high seismic velocities; this fits some observed travel-time curves. A crust composed of blocks rather than layers would give velocity profiles like those shown in Figure 7-1*c*.

The distribution of the major physiographic features of the Earth's surface was reviewed in Chapter 2, and these were described as tectonically stable or unstable. In Chapter 3 we noted a good correlation of heat flow with the tectonic environment, and

in Chapter 5 we correlated seismic velocities for the crust and upper mantle in these same environments. These geophysical parameters, along with gravity anomalies (Chapter 3) lead to the division of the Earth's crust into the types listed in Table 7-1.

The crustal structures for these types suggest a classification based on both crust and upper mantle. There are two crustal types, continental and oceanic, and two mantle types, stable and unstable. Areas of the Earth's surface can then be identified as one of the four types in Table 7-1. The unstable types include the active tectonic belts reviewed in Chapter 2: the folded mountain chains, the midoceanic ridges, and the island arc systems of the continental margins.

Cross Sections Through Continents

Figure 7-2 shows the results of a transcontinental geophysical and geological survey across the North American continent along a great circle passing approximately through Washington, D.C., Denver, and San Francisco. The schematic geological cross section

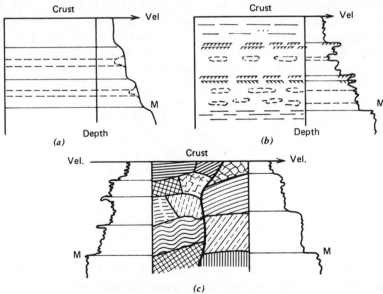

Figure 7-1. Crustal models, after Kosminskaya et al., (1969). (*a*) Continuous boundaries. (*b*) Piecewise continuous boundaries corresponding to piecewise continuous travel-time curves. (*c*) Assumed block-layered crust corresponding to the geophysical and geological points of view. (From Geophysical Monograph **13**, 195, 1969, with permission of Amer. Geophys. Union.)

TABLE 7-1   Tectonic Classification of Earth's Surface (Brune, 1969)

Crustal type	Tectonic characteristic	Crustal thickness (km)	P_n (km/sec)	Heat flow HFU	Bouguer Anomaly (mgal)	Geologic features
Continental crust overlying stable mantle						
Shield	very stable	35	8.3	0.7–0.9	−10 to −30	Little or no sediment, exposed batholithic rocks of Precambrian age. Moderate thicknesses of post-Precambrian sediments.
Midcontinent	stable	38	3.2	0.8–1.2	−10 to −40	
Continental crust overlying unstable mantle						
Basin-range	very unstable	30	7.8	1.7–2.5	−200 to −250	Recent normal faulting, volcanism, and intrusion; high mean elevation.
Alpine	very unstable	55	3.0	variable 0.7–2.0	−200 to −300	Rapid recent uplift, relatively recent intrusion; high mean elevation.
Island arc	very unstable	30	7.4–7.8	variable 0.7–4.0	−50 to +100	High volcanism, intense folding and faulting.
Plateau	Not adequately studied					
Oceanic crust overlying stable mantle						
Ocean basin	very stable	11	8.1–8.2	1.3	+250 to +350	Very thin sediments overlying basalts, linear magnetic anomalies, no thick Paleozoic sediments.
Oceanic crust overlying unstable mantle						
Ocean ridge	unstable	10	7.4–7.6	high and variable 1.0–8.0	+200 to +250	Active basaltic volcanism, little or no sediment.
Ocean trench	Not adequately studied					

in Figure 7-2*a* was deduced from seismic measurements. Values for the velocities in the upper mantle and lower crust can be read from Figure 3-7 which also shows the approximate line of the traverse. Contrast the position of the Moho in Figure 7-2 with its position in Figure 5-1*a*, which was based on data available as recently as 1962.

We noted in connection with Figure 3-7 that the Rocky Mountain system appears to divide the United States into two super-provinces involving both crust and upper mantle. This is confirmed in Figure 7-2 by the contrast in magnetic character west and east of the Rocky Mountains; the amplitudes and wavelengths of the anomalies differ. The

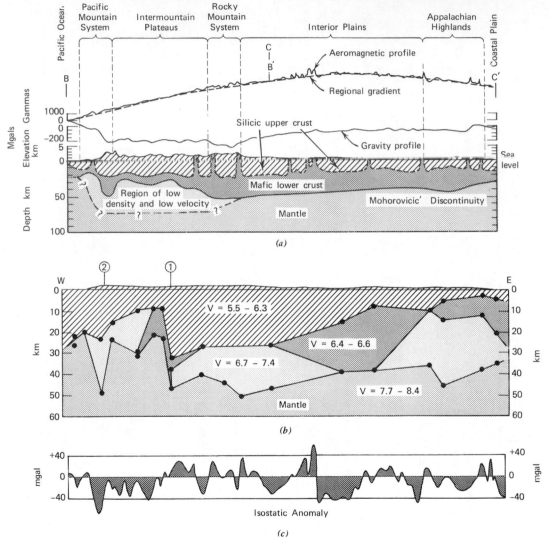

Figure 7-2. (*a*) Transcontinental geological and geophysical cross section of the earth's crust and upper mantle. Indicated distribution of mafic material in the silicic upper crust is schematic only. Sedimentary rocks overlying crystalline basement are not shown. See Figure 3-7*b* for location of sections (after Pakiser and Zietz, 1965). (*b*) Composite seismic cross section through the crust not far removed from section in *a*, with *P*-wave velocities. (*c*) Isostatic anomalies for the section in *b*. (*b* and *c* after Woollard, 1968.) (From Rev. Geophysics, **3**, 505, 1965; Geophysical Monograph **12**, 312, 1968; with permission of Amer. Geophys. Union.)

magnetic profile may be correlated with heat flow. In regions of low heat flow the geotherms are depressed and there is a greater thickness of rocks below the Curie temperature which produces more intense anomalies. The gravity profile in Figure 7-2a also shows the contrast between the eastern and western provinces of the continent.

Figure 7-2b is a composite seismic cross section through the crust along a profile from Chesapeake Bay, Maryland, to central California; this is not far removed from the great circle profile of Figure 7-2a. Units of the crust with similar ranges of seismic velocities are shown (compare Figure 3-7b), and isostatic anomaly values are given. The striking features of this profile at the Moho are the abrupt transition in structure of the Wasatch Mountain front (location 1) and the local root and negative isostatic anomaly beneath the Sierra Nevada (location 2). The seismic velocities indicate that the predominantly granitic crust of the west is replaced by a predominantly basic crust beneath the Appalachians and eastern seaboard, and that a thick intermediate layer beneath the eastern plains is replaced by granitic material beneath the western plains.

Figure 7-3 shows a crustal cross section through the Russian platform including the Ukrainian shield and the Dnieper graben. The crust is a layered structure with many seismic interfaces. In some areas the seismic velocities indicate that the Moho may be a layer up to 5 km thick between the crust and

mantle. The main difference between this section and those shown in Figure 7-2 is the existence of many deep fractures which extend right through the crust and into the mantle breaking the continuity of the Moho. These fractures break up the continental crust into separate tectonic blocks.

Cross Sections Through Oceanic Crust

The properties of the oceanic crust are often more uniform than those of the continental crust. The median values for normal oceanic crust are shown in Figure 7-8. Three layers can be distinguished: Layer 1 (the sedimentary layer), Layer 2, and Layer 3 (the oceanic layer). The median values for these layers are: (1) Layer 1, 0.3 km thick, with seismic velocity varying between 1.5 and 3.4 km/sec; (2) Layer 2, 1.4 km thick, with velocity between 3.4 and 6.0 km/sec, median near 5.1 km/sec; (3) Layer 3, 4.7 km thick, median velocity 6.8 km/sec.

These features are illustrated by Figure 7-4a, a seismic refraction profile across the Atlantic Ocean between Sierra Leone and Brazil (Figure 9-10). The distribution of seismic velocities as a function of depth is shown in Figure 5-3a. The thickness of the oceanic crust remains quite constant both in the Atlantic and other oceans (Figure 5-3b). Consistent regional variations are associated with the midoceanic ridges (Figure 7-5). Approaching the continental margins abrupt changes in crustal thickness occur (Figures

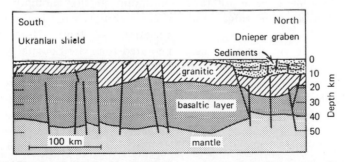

Figure 7-3. Crustal cross section through the Russian continental platform, showing in particular the deep fractures separating main regions and individual blocks (after Sollogub, 1969. From Geophysical Monograph **13**, 189, with permission of Amer. Geophys. Union).

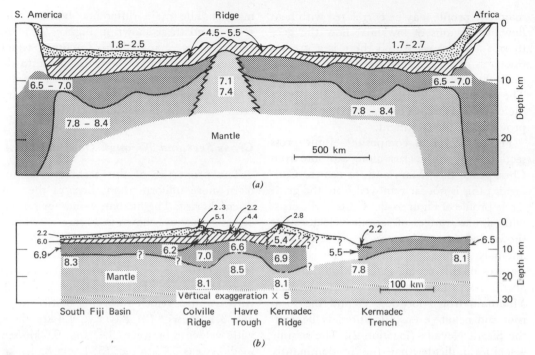

Figure 7-4. (*a*) Schematic cross section across the equatorial Atlantic Ocean, showing *P*-wave velocities (from various sources). (*b*) Crustal section across the Kermadec arc system based on seismic refraction profiles, with *P*-wave velocities (after D. E. Karig, 1970, Jour. Geophysical Res., **75**, 239, with permission).

5-4, 7-4*a*, and 7-6); the sedimentary layer is thicker in the continental rises near the continent.

Figure 7-4*b* shows the oceanic crust in the Pacific Ocean north of New Zealand with a section across the Kermadec-Tonga island arc system. This system includes an oceanic trench and a series of ridges and basins far removed from a continent. The interarc basin, consisting of a series of ridges and troughs with relief of 1000 m, is a region of high heat flow underlain by oceanic crust. The crust thickens to 15 km beneath the major ridges bordering the interarc basin and becomes more complex in structure.

The oceanic ridges form one of the tectonically active regions with anomalous upper mantle where both mantle velocities and mantle depths are less than normal in oceanic regions (Figures 5-4 and 7-8). Figure 7-4*a* shows Layer 2 directly overlying the anomalous mantle material of the mid-Atlantic ridge which extends through the oceanic Layer 3. In a few locations normal mantle with seismic velocity 8.3 km/sec instead of the usual 7.3 km/sec material has been identified.

Figure 7-5*a* shows the Bouguer gravity anomaly across the mid-Atlantic ridge. The minimum in the anomaly shows that the ridge is compensated isostatically by density variations at depth. If it were not compensated then the Bouguer anomalies would have remained constant at 350 mgal across the ridge. The three structure sections shown in Figure 7-5 satisfy both the seismic and the gravity data. The steep slope of the Bouguer anomaly requires that most of the compensation must lie at shallow depths and it is difficult to provide compensation for the ridge flanks except by the density reversal shown. The layer of anomalous mantle de-

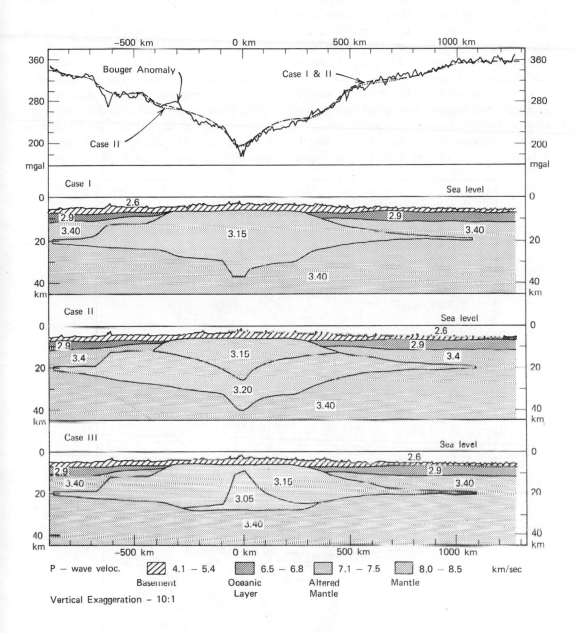

Figure 7-5. Three possible crustal models across the north mid-Atlantic ridge which satisfy gravity anomalies and are in accord with seismic refraction data. In all three models the anomalous mantle found seismically under the crest of the ridge is assumed to underlie the normal mantle under the flanks of the ridge. In Case I the anomalous mantle is assumed to have a uniform density; in Case II its density is assumed to increase downward, and in Case III the material constituting the anomalous mantle is assumed to be lighter near the axis of the ridge (after Talwani et al., 1965). See Figure 14-11. (From Jour. Geophys. Res., **70**, 341, 1965, with permission.)

picted in Figure 7-5 would not be detected by the usual seismic refraction techniques, because it underlies a higher-velocity layer of normal mantle. The configuration of this layer suggests that it may have been converted from normal mantle by a phase transition. Note that the density distributions in the anomalous layer for the three models provide considerable variety for interpretation in geological and petrological terms.

The structure of the East Pacific rise is similar to that of the mid-Atlantic ridge

(Figure 7-8) in that it is underlain by anomalous mantle and it is isostatically compensated. It differs from the mid-Atlantic ridge because the oceanic Layer 3 is continuous across the rise, with the anomalous mantle beneath the crest rising to somewhat higher levels than that beneath the flanks. The compensating layer of anomalous mantle may extend deeper than beneath the mid-Atlantic ridge.

The oceanic crust has not yet yielded all of its secrets. In January 1969 improved refraction measurements in the deep-ocean

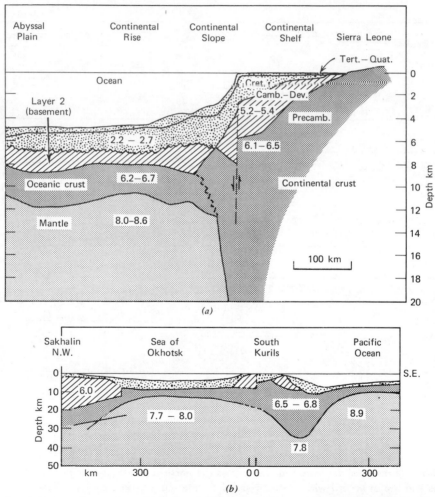

Figure 7-6. Seismic cross sections through continental margins and transitional zones, showing *P*-wave velocities. (*a*) Sierra Leone continental margin, vertical exaggeration 20 : 1 (after R. E. Sheridan et al., 1969, Jour. Geophys. Res., **74**, 2512, with permission). (*b*) Continental margin adjacent to the Sea of Okhotsk, and the South Kurils island arc (from I. P. Kosminskaya and S. M. Zverev, Geophysical Monograph **12**, 122, 1968, with permission Amer. Geophys. Union).

region between the Marshall and the Hawaiian Islands revealed a previously un-recorded basal layer between the mantle and the normal Layer 3. The layer has seismic velocities between 6.9 and 7.6 km/sec with an average of 7.3 km/sec (contrast Figures 7-4, 7-6, and 7-8). G. L. Maynard reported in 1970 that this basal layer may be widely present in the deep Pacific Ocean, and he outlined its implications for interpretation of gravity anomalies and the dispersion of seismic surface waves.

Cross Sections Through Continental Margins

The abrupt changes in the elevation of the Earth's solid surface and in the depth to the Moho in the transitional zones between continents and ocean basins are illustrated in Figure 5-1. Variations in seismic velocities for the continental margins in general have been reviewed in Figures 5-2, 5-3, and 5-4. Figures 7-4a and 9-10 show crustal sec-

tions through the continental margins bordering the Atlantic Ocean, and Figure 7-6a shows a more detailed cross section through one margin. Note the thick accumulations of sediment on the continental shelf and the continental rise. This represents the Atlantic type of continental margin (Chapter 2). The structure of the crust beneath the sediments in the continental margins is complex, and a significant feature is the existence of buried ridges which act as dams for the sediments. Figure 7-7 summarizes several ways in which such ridges may be formed based on interpretations of many crustal sections. Major fault systems appear to parallel many continental edges.

Figure 7-6b represents the Island arc type of continental margin (Chapter 2). The Kurils island arc is separated from the Asian continent by the Sea of Okhotsk which is a small ocean basin. This is comparable with the section through Asia and Japan in Figure 5-1 and fairly typical of continental margins in the north and west Pacific Ocean. The

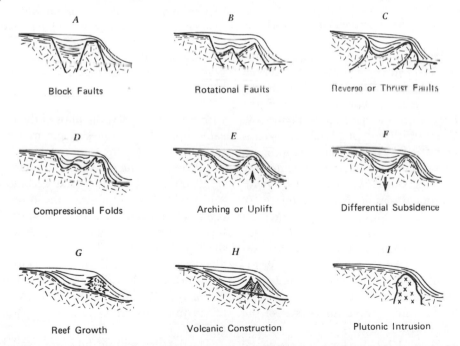

Figure 7-7. Possible origins of buried ridges within continental margins. Other causes can also be imagined, and any particular ridge may have resulted from a combination of processes (after Burk, 1968. With permission of The New York Academy of Sciences.)

depth to the Moho is similar beneath both the Pacific Ocean and the Sea of Okhotsk, but the latter has a thick accumulation of sediments over Layer 3. Sediments of considerable thickness are also shown in the trench bordering the island arc. The crust is greatly thickened beneath the arc-trench system, and the upper mantle here has anomalous seismic velocities. The crust is thicker than beneath the Kermadec-Tonga island arc system (Figure 7-4*b*).

Schematic Structural Subdivisions of the Whole Crust

Figure 7-8 shows a selection of individual crustal sections from various tectonic environments many of which were included in the cross sections in Figures 7-2, 7-3, 7-4, 7-5, and 7-6. Of particular interest are the sections indicating tectonic environments with large thicknesses of sediments. Standard sections for the average oceanic and continental crusts are given for comparison. Figure 7-8 may be compared with Figure 5-4 which provides similar information averaged for many seismic profiles.

Individual crustal sections such as those in Figure 7-8, and crustal cross sections such as those in Figures 7-2 to 7-6, provide the data for schematic representations of the whole Earth's crust as depicted in Figure 7-9. These diagrams were prepared as a basis for computation of the average composition of the crust. Figure 7-9*a* was presented by A. Poldervaart in 1955, and in 1969 with much additional seismic data available A. B. Ronov and A. A. Yaroshevsky presented an improved version Figure 7-9*b*. There are significant differences between the two models in the areas assigned to oceanic crust, continental crust, and the continental margins (transitional, suboceanic, or subcontinental) and in the average thicknesses of the crust in the various regions.

Tables 2-3 and 7-2 give for Figure 7-9*b* the estimates of areas of the structural units, average thicknesses and volumes of the layers

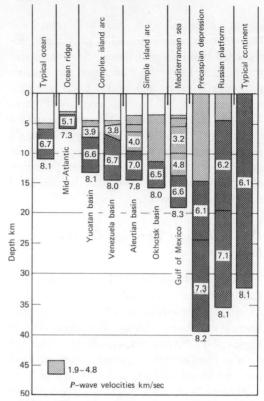

Figure 7-8. Crustal sections for different tectonic environments compared with typical oceanic and continental sections (after Menard, 1967, and Kosminskaya et al., 1969. With permission Amer. Geophys. Union.)

within the crust, and the mass of the layers in each unit. The volumes of the major structural units are (a) the total continental crust, 6,500 km³; (b) the total subcontinental crust, 1,540 km³; and (c) the total oceanic crust, 2,170 km³. The volumes of the three major layers of the crust are (a) the sedimentary layer including volcanic rocks, 985 km³, (b) the "granitic" layer, 3,590 km³; and (c) the "basaltic" layer, 5,635 km³. The composition of the crust can be calculated from Figure 7-9 if the compositions of the different crustal layers and units can be estimated (Tables 7-4 to 7-7).

The detailed seismic data now available for the United States permits a more precise calculation of the volumes of different layers

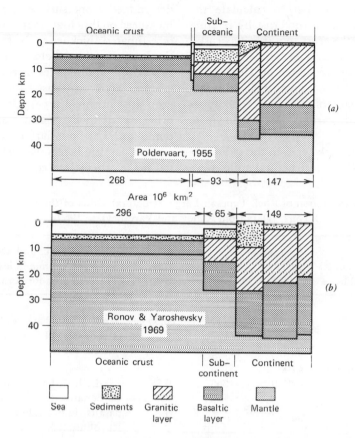

Figure 7-9. Schematic subdivisions of the Earth's crust used as basis for calculating the composition of the crust. Main subdivisions are into (1) sedimentary layer, (2) "granitic" layer, and (3) "basaltic" layer. See text for details. (*a*) Scheme of Poldervaart (1955, with permission The Geol. Soc. Amer.). (*b*) Scheme of Ronov and Yaroshevsky (1969. With permission of Amer. Geophys. Union, from Geophysical Monograph **13**, 37).

and their variations from one part of the continent to another. Figure 7-2 shows that there are differences from one tectonic environment to another. L. C. Pakiser and R. Robinson in 1966 estimated the volumes of the upper low-velocity unit (composition assumed to be silicic) and the deeper high-velocity unit (composition assumed to be mafic) within each of ten regions covering the conterminous United States. The percentages of mafic rocks range from a low of 25% in the California coastal region to a high of 77.7%

in the Columbia plateaus region with an average of 53.7% for the total crust considered. The regions were then grouped into an eastern superprovince and a western superprovince divided by the Rocky Mountains following the conclusions arising from evaluation of Figure 7-2. The average volume percentages of mafic crust calculated for these superprovinces are 57.4% for the east and 43.6% for the west. Average compositions calculated according to these estimates are listed in Tables 7-5 and 7-7.

PETROLOGY AND MINERALOGY OF THE CRUST

Poldervaart was the first to calculate the average composition of the crust by relating the petrology of the crust to a specific crustal model, Figure 7-9*a*, and essentially the same approach was followed later by Ronov and Yaroshevsky. Table 7-2 gives their estimates of the percentages of different rock types occurring in the various layers shown in Figure 7-9*b*.

Volumes of the rock types in the sedimentary layers in different structural units were determined from surface areas of outcrops measured on lithologic maps and on measured or estimated thicknesses. Note that 25.3% of the layer in folded belts consists of volcanic rocks. The abundances of rock types in the granitic layer are based on measured outcrop area in shield areas. Granites, granodiorites, and their gneissic equivalents are dominant. The boundary between the granitic and basaltic layers was arbitrarily chosen to make them of equal thickness. The

TABLE 7-2   Distribution of Rock Types in Large Structural Units of the Crust and the Crustal Layers. (Data from Ronov and Yaroshevsky, 1969)

Crustal layers in different crustal units	Average thickness (km)	Volume (km³)	Mass (10^{24}g)	Types of rocks and abundances. Percent volume of layer, except percent area for oceanic Layer 1
Continental Platform				
Sedimentary	1.8	135	0.35	Sands, 23.6. Clays, 49.5. Carbonates, 21.0. Evaporites, 2.0. Basalts, 3.9
Continental Geosynclinal Folded Belts				
Sedimentary	10.0	365	0.94	Sands, 18.7. Clays and Shales, 39.4. Carbonates, 16.3. Evaporites, 0.3. Basalts, 12.6. Andesites, 10.2. Rhyolites, 2.5
Subcontinental, Shelf and Slope				
Sedimentary	2.9	190	0.48	Similar to above groups
Continental and Subcontinental				
Granitic		3590	9.81	Granites, 18.1. Granodiorites, 19.9.
continental	20.1			Syenites, 0.3. Gabbro, 3.7.
subcontinental	9.1			Peridotites, 0.1.
				Gneisses, 37.6. Schists, 9.0.
				Marbles, 1.5. Amphibolites, 9.8
Basaltic		3760	10.91	Acid igneous and metamorphic
continental	20.1			rocks, 50.0. Basic igneous and
subcontinental	11.7			metamorphic rocks, 50.0
Oceanic				
Layer 1 Sedimentary	0.4	120	0.19	Terrigenous, 7.3. Calcareous, 41.5. Siliceous, 17.0. Red Clays, 31.2
Layer 2	0.6	175	0.44	Sediments, 50.0.
	0.6	175	0.52	Basalts, 50.0
Layer 3	5.7	1700	4.92	Oceanic tholeiitic basalts, 99.0. Alkaline differentiates, 1.0

constitution of the "basaltic" layer is not known. There are two extreme possibilities; (a) it is petrologically similar to the granitic shell or (b) it is composed essentially of basalt or gabbro. Ronov and Yaroshevsky assumed that in continental and sub-continental regions the layer changed gradual-ly from the granitic rock assemblage to gabbroic material near the Moho. Table 7-2 indicates a shell composed of 50% acid rocks and 50% basic rocks, including plutonic rocks, their metamorphic equivalents, and also paragneisses. In the oceanic crust Layer 3 is assumed to be oceanic tholeiite with minor differentiates. The constitution of Layer 2 is not known, and it is assumed to consist of 50% consolidated sediments equivalent to those in Layer 1, and 50% basic rocks equivalent to those in Layer 3.

The information about distribution of rock types in Table 7-2 was reorganized in order to show the abundances of the main rock types in the crust as a whole with the results shown in Table 7-3. This confirms that the most abundant rock types in the crust are (a) granites, granodiorites, and diorites, (b) basalts, gabbros, and amphibolites, and (c) their metamorphic equivalents. The clays and shales are dominant in the generally thin veneer of sediments at the surface of the Earth.

From the abundance of rock types, and the known mineralogy of the rocks, Ronov and Yaroshevsky worked out the abundance of minerals in the crust. Table 7-3 shows that quartz and feldspars together constitute 63% of the crust. The mafic anhydrous minerals amount to 14% and the hydrated silicates, mica, amphibole, clays and chlorite, amount

TABLE 7-3 Abundances of Main Rock Types and Minerals in the Crust (Data from Ronov and Yaroshevsky, 1969)

Rocks	% Volume of Crust	Minerals	% Volume of Crust
Sedimentary		Quartz	12
Sands	1.7	Alkali feldspar	12
Clays and shales	4.2	Plagioclase	39
Carbonates (including salt-bearing		Micas	5
deposits)	2.0	Amphiboles	5
Igneous		Pyroxenes	11
Granites	10.4	Olivines	3
Granodiorites, diorites	11.2	Clay minerals (+ chlorites)	4.6
Syenites	0.4	Calcite (+ aragonite)	1.5
Basalts, gabbros, amphibolites,		Dolomite	0.5
eclogites	42.5	Magnetite (+ titanomagnetite)	1.5
Dunites, peridotites	0.2	Others (garnets, kyanite, andalu-	
Metamorphic		site, sillimanite, apatite, etc.)	4.9
Gneisses	21.4		
Schists	5.1		
Marbles	0.9		
Totals		*Totals*	
Sedimentary	7.9	Quartz + feldspar	63
Igneous	64.7	Pyroxene + olivine	14
Metamorphic	27.4	Hydrated silicates	14.6
		Carbonates	2.0
		Others	6.4

to 14.6%. Carbonates amount to only 2% and other minerals are even less abundant. This shows the dominance of quartz and feldspars in crustal rocks in sharp contrast with the mineralogy of the underlying mantle (Table 6-7).

The assumption that the "basic" layer contains at least 50% of basic rocks is rooted in the historical development of petrological theory, and recently it has been supported by laboratory studies which confirm that basic rocks have seismic velocities corresponding to those measured in the lower crust. Other rocks have similar seismic velocities, however, and there are now good reasons to conclude that the lower continental crust is not composed largely of basic rocks.

According to Figure 5-15 the pressure-temperature conditions appropriate for the lower crust are inappropriate for the existence of gabbro. Material of gabbroic composition, if present, should exist as eclogite, but the density of eclogite is too high to fit the measured seismic properties of the lower crust. For this reason A. E. Ringwood and

D. H. Green proposed in 1966 that we should abandon the notion of a basic lower crust and consider instead an assemblage of acid-intermediate rocks in the eclogite facies. Ringwood and Green suggested that as an alternative to this dry assemblage basic rocks could exist in the lower crust as amphibolites in environments where the tectonic evolution had been simple enough that the water was not driven off. Amphibolites are stable in the lower crust and they have appropriate seismic velocities.

The composition of Layer 3 of the oceanic crust is also a matter for debate. The conventional interpretation of a basaltic composition (Table 7-2) is challenged by the hypothesis of sea-floor spreading (Chapter 12). According to this Layer 3 is composed of serpentinite, or hydrated mantle, as shown in Figure 5-6 (model 5) and in Figure 12-23a. The discovery of metamorphosed basalts in dredge hauls from the mid-Atlantic ridge provides some support for an alternative proposal that Layer 3 is composed of greenstone or amphibolite.

COMPOSITION OF THE CRUST

The many estimates of the composition of the crust follow one of five approaches, involving:

1. Averages of available analyses of rocks, such as the average of all igneous rocks or averages of igneous and sedimentary rocks in proportion to their abundances;

2. Averages of rocks weighted in proportion to their occurrence, using geological maps as a basis;

3. Indirect methods such as the analysis of glacial clays which represent a sample from a large area covered by a continental ice sheet;

4. Indirect computations based on various combinations of average granitic and basaltic rocks, with combination ratios being selected to best explain the compositions of sediments

derived by weathering of the crust, or the rare earth abundance patterns in sedimentary rocks;

5. Averages of rocks assigned to crustal models such as those in Figure 7-9. This is the only approach which has taken into consideration the composition and volume of the oceanic crust.

Despite the diversity of approaches most averages are similar in composition to an intermediate igneous rock, and the early estimates have not been greatly revised by later estimates. The uncertainty about the petrology of the lower continental crust, however, and the possibility that much of the oceanic crust might be composed of serpentinite rather than basalt (Chapters 5 and 12) leaves us with uncertainty about the

TABLE 7-4 Average Chemical Composition of Major Layers of the Crust in Weight Percent
(After Ronov and Yaroshevsky, 1969)

| Type of Crust | Continental and Subcontinental | | | | Oceanic | | | |
Layer	Sediment	Granitic	Basaltic	Total continental[a]	Layer 1	Layer 2	Basaltic	Total oceanic
SiO_2	50.0	63.9	58.2	60.2	40.6	45.5	49.6	48.7
TiO_2	0.7	0.6	0.9	0.7	0.6	1.1	1.5	1.4
Al_2O_3	13.0	15.2	15.5	15.2	11.3	14.5	17.1	16.5
Fe_2O_3	3.0	2.0	2.9	2.5	4.6	3.2	2.0	2.3
FeO	2.8	2.9	4.8	3.8	1.0	4.2	6.8	6.2
MnO	0.1	0.1	0.2	0.1	0.3	0.3	0.2	0.2
MgO	3.1	2.2	3.9	3.1	3.0	5.3	7.2	6.8
CaO	11.7	4.0	6.1	5.5	16.7	14.0	11.8	12.3
Na_2O	1.6	3.1	3.1	3.0	1.1	2.0	2.8	2.6
K_2O	2.0	3.3	2.6	2.9	2.0	1.0	0.2	0.4
P_2O_5	0.2	0.2	0.3	0.2	0.2	0.2	0.2	0.2
C	0.5	0.2	0.1	0.2	0.3	0.1	0.0	0.0
CO_2	8.3	0.8	0.5	1.2	13.3	6.1	—	1.4
S	0.2	0.0	0.0	0.0	—	—	0.0	0.0
Cl	0.2	0.1	0.0	0.1	—	—	0.0	0.0
H_2O +	2.9	1.5	1.0	1.4	5.0	2.7	0.7	1.1

[a] Total subcontinental is very similar.

average composition of the whole crust. The abundances and distribution of most minor elements within the crust are not well known.

Composition of Crustal Layers

From the abundances and distribution of rocks listed in Table 7-3, and from the known average compositions of these rock types, Ronov and Yaroshevsky calculated the average chemical compositions of the major layers depicted in Figure 7-9*b* for continental and oceanic environments. The results are given in Table 7-4. The composition of the sedimentary layer includes a significant proportion of volcanic rocks as indicated in Table 7-2. Despite this there exist distinct differences between the composition of the sedimentary layer and the granitic and basaltic layers of the continents. The sedimentary layer is higher in the volatile components H_2O and CO_2 and considerably higher in CaO. Poldervaart pointed out that the average composition of sediments does not correspond at all to the average igneous rock of intermediate composition.

Composition of Crustal Units and of the Whole Crust

According to Table 7-4 the average composition of the continental and subcontinental crust (79% of the volume) differs from that of the oceanic crust (21% of the volume). Tables 7-5 and 7-6 show that there are considerable differences among various estimates for the same structural units; these depend largely upon the petrology and composition assumed for the lower crust and upon the values used for areas and thicknesses of layers and units. The differences are carried through to the average compositions of the whole crust above the Moho which are compared in Table 7-7.

TABLE 7-5   Estimates for the Chemical Composition of Structural Units of the Continental Crust (Weight Percent)

Type of Crust	Shield		Young Folded Belts	United States			Total Subaerial Continental
				West	East	Average	
Author[a]	P	R & Y	P	Pakiser and Robinson			R & Y
SiO_2	59.8	66.0	58.4	60.0	57.1	57.9	60.2
TiO_2	1.2	0.6	1.1	1.1	1.3	1.2	0.7
Al_2O_3	15.5	15.3	15.6	15.1	15.2	15.2	15.2
Fe_2O_3	2.1	1.9	2.8	2.3	2.3	2.3	2.5
FeO	5.1	3.1	4.8	4.9	5.7	5.5	3.8
MnO	0.1	0.1	0.2	0.1	0.2	0.2	0.1
MgO	4.1	2.4	4.3	4.5	5.6	5.3	3.1
CaO	6.4	3.7	7.2	6.3	7.5	7.1	7.1
Na_2O	3.1	3.2	3.1	3.2	3.0	3.0	3.0
K_2O	2.4	3.5	2.2	2.4	2.0	2.1	2.9
P_2O_5	0.2	0.2	0.3	0.2	0.3	0.3	0.2
CO_2	—	—	—	—	—	—	1.2
$H_2O +$	—	—	—	—	—	—	1.4

[a] *P*, Poldervaart (1955); *R & Y*, Ronov and Yaroshevsky (1969); Pakiser and Robinson (1966).

TABLE 7-6   Estimates for the Chemical Composition of Oceanic and Suboceanic (Subcontinental) Crust (Weight Percent)

Type of crust	Oceanic		Suboceanic or Subcontinental	
Author[a]	P	R & Y	P	R & Y
SiO_2	46.6	48.7	49.5	59.5
TiO_2	2.9	1.4	1.9	0.7
Al_2O_3	15.0	16.5	15.1	15.1
Fe_2O_3	3.8	2.3	3.4	2.5
FeO	8.0	6.2	6.4	3.9
MnO	0.2	0.2	0.2	0.2
MgO	7.8	6.8	6.2	3.2
CaO	11.9	12.3	13.2	5.9
Na_2O	2.9	2.6	2.5	2.9
K_2O	1.0	0.4	1.3	2.8
P_2O_5	0.3	0.2	0.3	0.3
CO_2	—	1.4	—	1.5
H_2O+	—	1.1	—	1.4

[a] *P*, Poldervaart (1955); *R & Y*, Ronov and Yaroshevsky (1969).

Table 7-5 shows that the shield composition estimated by Ronov and Yaroshevsky is more silicic than that estimated by Poldervaart, but the difference is reduced for their estimates of the total continental crust. Poldervaart's estimate of suboceanic crust (Table 7-6) is more mafic than that of Ronov and Yaroshevsky for the subcontinental crust which refers to similar major structural units (Figure 7-9). The estimate of continental crust composition by Pakiser and Robinson, which is based on detailed structural sections similar to those in Figure 7-2, is more mafic than those based on the crustal models of Figure 7-9 and more mafic than estimates obtained by other methods. In contrast Ringwood and Green concluded that the average chemical composition of the crust is more acidic than the estimates reviewed here because of the instability of gabbro in the lower crust. The new seismic results used by Pakiser and Robinson reverse the conclusion of Poldervaart that the folded belts are less silicic than the shields. Table 7-5 shows that the tectonically active western superprovince of the United States is more silicic than Poldervaart's young folded belts (his analogous crust), and the eastern superprovince is more mafic than Poldervaart's shields (his analogous crust).

The differences in estimated compositions of oceanic crust in Table 7-6 result from the composition adopted for the basalt of Layer 3. Poldervaart used a value for average basaltic rocks. Since 1955 more attention has been paid to the study of basalts dredged from the ocean floors and the estimate of Ronov and Yaroshevsky is based on the compositions of oceanic basalts.

Table 7-7 compares the two estimates for the average composition of the whole Earth's crust above the Moho based on the crustal models of Figure 7-9. Ronov and Yaroshevsky included in their contribution a minor revision which led to insignificant changes in the analysis listed; they reported estimates for CO_2 and H_2O+ of 1.40% and 1.37%, respectively.

The average composition of the Earth's crust is very similar to the average composition of andesites as shown by McBirney's averages

listed in Table 7-7. McBirney distinguished between andesites of island arcs and continental margins.

Markhinin studied the distribution and composition of volcanic rocks erupted from the Kuril Islands, an island arc system extending northward from Japan. A cross section through the Kurils is shown in Figure 7-6b. The volcanic rocks included basalts, andesites, and rhyolites, but the average composition listed in Table 7-7 is very similar to an average andesite. Markhinin pointed out that the entire observable geological sequence of the Kuril Islands consists either of volcanic material or of the products of reworking of the volcanic material. He estimated that since the beginning of the Cretaceous, about 6.5×10^6 km^3 of volcanic material, mostly pyroclastic, had been erupted from the volcanoes of the Kuril Islands, and he concluded that this volume of material was sufficient to convert an original oceanic crust into a crust of continental type. He reviewed the thesis

TABLE 7-7   Estimates for the Average Chemical Composition of the Earth's Crust Compared with Average lavas (Weight Percent)

	Earth's Crust		Andesites		Kuril Islands[c]
			Island arcs[a]	Continental Margins[b]	
Author	Poldervaart 1955	Ronov and Yaroshevsky 1969	McBirney 1969		Markhinin 1968
SiO_2	55.2	59.3	58.7	58.7	58.1
TiO_2	1.6	0.9	0.8	0.8	0.7
Al_2O_3	15.3	15.9	17.3	17.4	17.1
Fe_2O_3	2.8	2.5	3.0	3.2	3.4
FeO	5.8	4.5	4.0	3.5	4.1
MnO	0.2	0.1	0.1	0.1	0.1
MgO	5.2	4.0	3.1	3.3	3.4
CaO	8.8	7.2	7.1	6.3	7.1
Na_2O	2.9	3.0	3.2	3.8	2.8
K_2O	1.9	2.4	1.3	2.0	1.2
P_2O_5	0.3	0.2	0.2	0.2	—

[a] Average of 89 calcic andesites..
[b] Average of 29 calc-alkaline andesites.
[c] Average of 427 analyses of volcanic rocks erupted in Kuril Islands.

that the Earth's continental crust was formed progressively through geological history by the eruption and reworking of volcanic material on island arcs and related structures (see Figure 4-1 for an outline of the Rock Cycle).

Vertical Distribution of K, U, and Th in the Continental Crust

The average chemical composition of the crust changes with depth from the sedimentary layer, through the granitic layer, and through the basaltic layer. According to Table 7-4 an increase in Fe, Mg, and Al is accompanied by a decrease in the amount of combined H_2O and CO_2. There is an increase of Si, Na, and K from sediment to granitic shell and then Si and K decrease with depth. The ratios K/Na, Ca/Mg, and Fe_2O_3/FeO decrease, and Al/Si increases with depth.

The abundances and distributions of trace elements are less well known than those of the major elements. I. B. Lambert and K. S. Heier have detected trends in the abundances of some trace elements with depth in the continental crust. Of particular significance for problems related to heat generation, heat flow, and orogenic processes are the distributions of K, U, and Th. In 1967 Lambert and Heier compared the abundances of these elements in rocks of similar bulk composition in three groups:

1. Igneous and sedimentary rocks of the Paleozoic folded belts of eastern Australia;
2. Intrusive granites and metamorphic greenstones and gneisses of the shield regions in western Australia;
3. Medium to high pressure pyroxene granulite subfacies rocks of the shield regions.

This represents a sequence corresponding to greater depths in the continental crust. The relatively rare outcrops of pyroxene granulite provide samples from the deepest levels of the crust that are available in any areal extent at the surface.

In all three groups there is a wide and similar range of K. The overall average of Th is similar in groups 1. and 2., although the metamorphic shield rocks of group 2. are somewhat depleted in U. The pyroxene granulites of group 3. are depleted in U and Th by factors of five and nine respectively, compared with the concentrations in the lower grade shield rocks of group 2., and the average K content is also lower. The decrease in U does not strictly parallel that of Th, but major depletion of both U and Th occurs between the shield rocks of group 2. and the pyroxene granulites of group 3. Values of U, Th, Th/K, and U/K in the mafic rocks studied showed no significant variation with metamorphic grade.

Lambert and Heier concluded that metamorphic processes involving movement of a vapor phase and partial melting contribute towards upward migration of granitophile elements including K, Th, and U. They suggested that the pyroxene granulite facies rocks may represent the residue remaining after partial melting and upward migration of an interstitial magma, transporting Th and U, which leaves the deep-seated rocks depleted in the heat-producing radioactive elements.

Heat Production in the Crust

If radioactive heat production in the crust is concentrated into upper levels, then erosion through a long period may be capable of removing enough of the upper layer that heat production in the crust and heat flow at the surface could be strongly affected. The results of Lambert and Heier confirmed that the overall average contents of the radioactive elements in the surface rocks of the western Australian basement shield are similar to those in the Paleozoic rocks in eastern Australia; yet the heat flow from the shield is lower than that from the younger orogenic region in the east. Lambert and Heier suggested that the higher heat flow in the east is accounted for because the upper layer with concentrated radioactive elements is thicker in this region than in the shield region. In the shield areas the radioactivity of the surface rocks is representative of only a thin layer resting on crust that has been depleted in Th and U.

The concentrations of K, Th, and U in large granitic plutons at 38 localities in the United States were measured by R. F. Roy, D. D. Blackwell, and F. Birch in 1968, and these were converted into heat productivity values expressed in units of HGU (1 HGU = 10^{-13} cal/cm^{3}sec). Heat flow measurements at each locality were recorded in HFU (1 HFU = 10^{-6} cal/cm^{2}sec). The variations in heat flow, when plotted against the heat generation of the surface rocks at each site, lie close to three lines of the form $Q = a + bA$, where Q is the surface heat flux in HFU and A is the heat generation in HGU. The three lines, shown in Figure 7-10, define three heat flow provinces, the eastern United States, the Sierra Nevada, and a zone of high heat flow in the western United States which includes the Basin and Range province. Figure 7-10 shows that there is lateral variation in heat flow within each province and that the ranges of heat flow overlap from one province to another. The distinctive character of each province becomes clear only when the heat flow is plotted against the heat productivity of the surface rocks.

Figure 7-10 shows that the lateral variation in heat flow within a province is a linear function of the heat productivity of the surface rocks. The linear relation is consistent with only a few classes of radioelement distributions in the crust. The limiting cases are (a) nearly constant heat production to the lower limit of the plutons, and (b) an exponential decrease in heat production with depth. In both cases the variability of heat flow over large areas is caused by variations in a near-surface layer.

The linear relationship is most simply interpreted if there is a constant heat flux from the upper mantle and lower crust, corresponding to the intercept a on the ordinate,

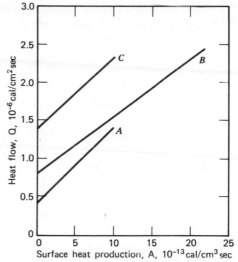

Figure 7-10. Summary of relationship between heat flow and surface heat production in the Sierra Nevada (A), eastern United States (B), and Basin and Range province (C). (After Roy, et al., 1968.) See also Lachenbruch (1970) and Tilling et al. (1970). (With permission of North-Holland Publishing Co.)

to which is added the radioactive heat generation within a surface layer of depth b. It is assumed that the radioactivity measured at the surface is constant from the surface to the depth b; the variations from place to place within the province produce the range of heat flow values. The values of a and b determined from the lines plotted in Figure 7-10 are as follows:

A. Sierra Nevada, $a = 0.40$ HFU, $b = 10.1$ km;

B. Eastern United States, $a = 0.79$ HFU, $b = 7.5$ km;

C. Basin and Range province, $a = 1.4$ HFU, $b = 9.4$ km.

These results are consistent with the conclusion of Lambert and Heier based on the geochemistry of metamorphic rocks, but whereas Lambert and Heier relate them to variation in thickness of the surface layer, the United States measurements are interpreted in terms of variations in heat productivity of the layer from place to place. The thickness (*b*) appears to remain fairly constant within the range 7 to 11 km.

The heat flow from beneath the plutons consists of a component from the upper mantle and a component from the lower crust. These must be insignificant over large areas of the United States, which confirms the conclusion of Lambert and Heier that there is an abrupt decrease in heat production a few kilometers below the surface of the crust. The similar slopes for the lines in Figure 7-10 suggest that the thickness of the heat producing layer varies little, and therefore the different regional values must arise largely from variations in the heat flux from the lower crust and upper mantle.

The curve for the eastern United States (the Central Stable Region) appears to apply also to the Australian shield; this is considered to be the reference curve for normal continental heat flow. Heat flow from the upper mantle is estimated as 0.4 HFU. If the lower crust in the Basin and Range province has a heat productivity similar to that inferred for

the Central Stable Region, then it has approximately 1.1 HFU flowing from the mantle. These values are summarized in Figure 7-11 for the three provinces. From the values of heat flow from the mantle and heat productivity in lower and upper crust the temperature-depth curves for the three provinces were calculated assuming steady state conditions (compare Figure 3-12). There are only small temperature differences within each province, the heat flow variations being produced by heat production only in the surface layer, but there are large temperature differences from one province to another because of the variations in heat flow from the mantle.

The heat flow from the normal, stable continental regions has a mode of 1 1–1.2 IIFU, and this is apparently controlled by a distribution of static heat sources. Roy, Blackwell, and Birch pointed out that the coincidence of this mode with the average heat flow in ocean basins (Table 3-2) "revives the argument for equality of radioactive sources beneath the continents but with different vertical distributions."

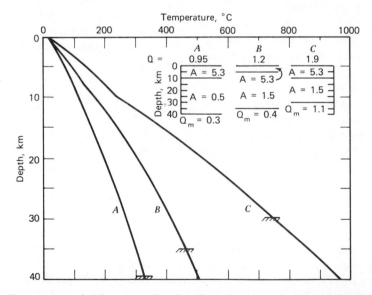

Figure 7-11. Temperature-depth curves for the three heat-flow provinces for the models shown (assumes steady state). The thermal conductivity is 6.5×10^{-3} cal/cm sec °C for the upper layer and 5.0×10^{-3} cal/cm sec °C for the lower layer of the crust. (After Roy et al., 1968.) Compare Hyndman et al. (1968). (With permission of North-Holland Publishing Company.)

DEEP STRUCTURE OF CONTINENTS

In a series of papers in the first half of the 1960's G. J. F. MacDonald reviewed the evidence that vertical segregation had been the dominant feature in the process of continent formation. This implies that the observed differences between continental and oceanic crust (Figures 5-1, 7-3, 7-4, and Tables 7-5 and 7-6) must extend into the mantle to depths of several hundreds of kilometers. MacDonald's most detailed review on the deep structures of continents was published in 1963.

According to MacDonald's arguments, large scale lateral convective movements have not been involved in the formation of continents. If the continental structure extends downwards to depths of 500 km or so, then severe restrictions are placed on theories of continental drift (Chapter 11). Lateral movements would have to involve blocks of crust and mantle with a thickness of 500 km, which implies that any convection in the mantle must occur at greater depths than this. Since about 1966 other arguments supporting seafloor spreading and plate tectonics (Chapters 13 and 14) appear to have received more wide-spread credence than those of MacDonald. We saw in the preceding section, however, that the relationship between heat flow and heat production in surface rocks revives the arguments for different vertical distributions of radioactive elements beneath the continents and the oceans, which is one of the major arguments for the deep structure of continents.

The evidence considered by MacDonald to indicate that continental structure extends to depths of about 500 km involves observations on (a) heat flow, (b) gravity, (c) surface waves in the upper mantle, and (d) concentration of deep-focus earthquakes along continental borders. The rocks composing the continental crust contain greater concentrations of radioactive elements than those composing the oceanic crust, and the continental crust is much thicker than the oceanic crust. The continental crust therefore contributes much more thermal flux to the measured value at the surface than the oceanic crust. In fact it appears that more than half the heat flow from the continents is generated within the crust and probably within a rather thin layer of the crust at the surface (Figures 7-10 and 7-11). We saw in Chapter 3 that the average values for heat flow through the continents and ocean floors are approximately the same. This observation requires that there is a greater contribution from the mantle beneath the oceanic crust than from beneath the continental crust. MacDonald concluded that this could be explained only if the abundance of radioactive elements in the suboceanic mantle was greater than that beneath the continents, which implies considerable vertical differentiation of elements and major differences in chemical composition between the suboceanic and subcontinental regions of the mantle.

Others have proposed that the thermal flux contribution from the suboceanic mantle is augmented by large-scale convection and that this is sufficient to equalize the heat flow from oceans and continents, but MacDonald presented additional arguments against the feasibility of mantle convection based on viscosity estimates. Roy, Blackwell, and Birch showed that the variations in heat flow from the surface within a continental heat flow province appear to be controlled by a distribution of static heat sources (Figures 7-10 and 7-11). They noted that if the oceanic heat flow were controlled by dynamic processes such as convection it would be surprising if this were equal to the average from static heat sources in the continental crust.

Artificial satellites have improved our knowledge of the Earth's gravity field (Chapter 3) and confirmed that there are regional variations which are of such extent that they cannot be accounted for in terms of crustal structures. This indicates lateral variations in mantle structure which could be explained in terms of temperature variations or compositional variations.

The present equatorial bulge is larger than that calculated for hydrostatic equilibrium. The present bulge corresponds to the equilibrium figure 10^7 years ago, assuming that the current rate of deceleration of the Earth remained constant through that interval. If we accept a model for the Earth with an elastoviscous mantle the interval of 10^7 years can be interpreted as a relaxation time which led MacDonald to estimate an average viscosity for the mantle of 10^{26} poises. This would prohibit convection. Some lower estimates of mantle viscosity are reviewed in Chapter 3.

The study of surface waves has revealed significant differences in the structure of the upper mantle beneath the oceans and continents. This is shown by the velocity profiles in Figure 3-4. Figure 3-5b shows that there are differences in wave velocity profiles in the mantle beneath different tectonic environments of the continents as well.

The association of deep-focus earthquakes with continental margins (Figures 2-5 and 2-6) can be explained in several ways (Chapters 12 and 14). MacDonald interpreted the deep zones of weakness as due to the effects of thermal stress which he considered to be consistent with his conclusions based on heat flow, gravity, and surface wave observations. The equality of heat flow from oceans and continents requires that at a given depth the temperature below the oceans is greater than that below the continents if convection is prohibited by the high viscosity deduced from gravity observations. MacDonald estimated temperature differences of 50 to 150°C. The temperature differences, and the induced zone of thermal stresses, are assumed to disappear at a depth of about 700 km corresponding to the deepest earthquake foci. MacDonald estimated that this pattern of thermal differences beneath oceans and continents would be produced by differ-

ences in radioelement content extending down to about 500 km in the mantle.

The evidence for the deep structure of the continents supports the models for vertical segregation of mantle material in the development of continents, and MacDonald outlined the following scheme. The early protocontinents were localized by restrictive fracture in the cool outer layers of an Earth initially expanding through radioactive heating. The continents were produced by uprise of basaltic and andesitic magmas including concentrations of the radioactive elements. Depletion of the subcontinental mantle in radioactive elements produced different thermal structures beneath the continents and oceans and caused the localization of deep fractures in the zones of thermal stress beneath the protocontinental margins. Further growth of the continents thus tended to be concentrated around the margins.

This is the hypothesis for the growth of continents by marginal accretion. The other major hypothesis for the origin and evolution of the continents is that a thin granitic crust formed very early in the Earth's history (either from the mantle or from extraterrestrial material) and was subsequently broken up into continents, recycled repeatedly, and resorbed by the mantle in a process of oceanization. In order to establish the pattern of continental evolution we need to know the distribution in the crust and mantle of the radioactive elements and their daughter products. Geochronological and isotope tracer studies should permit us to delineate continental age patterns and to determine the incidence of orogenic disturbances.

Concentration of Elements into the Crust

S. R. Taylor has calculated the percentage concentration of the elements in the crust relative to the whole Earth using a model of type I carbonaceous chondrite for the Earth (Chapter 6) and an overall andesitic crustal composition (Table 7-7) derived by vertical

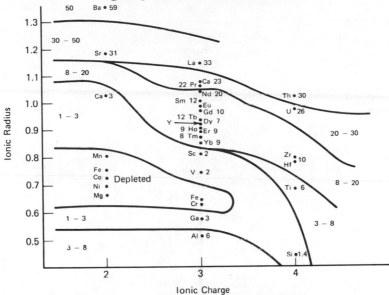

Figure 7-12. Concentration of chemical elements in the continental crust assuming an andesitic crustal composition and a chondritic Earth model. The numbers represent the percentage concentration of the elements in the crust on a vertical segregation model. The effect of ionic radius in fractionating elements of the same valency (e.g. rare earths) is striking (after Taylor, 1968, with permission of Pergamon Press).

segregation. The nonvolatile elements are plotted in Figure 7-12 on the basis of their crystal chemical properties of valency and ionic size. The boundaries between groups of elements with the same ranges of calculated concentrations show the strong control of crystal chemistry on the crustal enrichment.

The elements such as Mn, Fe, Co, Ni, Mg, and Cr are retained in the major minerals of the mantle (Chapter 6) and consequently depleted in the crust. Elements with size or valency different from these elements are strongly concentrated in the upper mantle or crust. These include K, Rb, Cs, Ba, U, and Th. The concentrations of the rare earth elements into the crust show a striking dependence upon ionic size. Figure 7-12 shows that there has been a very strong upward concentration of certain elements within the Earth and this conclusion remains valid whether or not the dominant process is vertical segregation.

AGES OF CONTINENTAL BASEMENTS

The tectonic structure of the North American continent has fostered the concept of the growth of continents by marginal accretion, and a compilation of apparent ages from radiometric data by A. E. J. Engel in 1963 strengthened this concept. Figure 7-13 shows the broad pattern of ages and geological provinces in North America as defined by major granite-forming, mountain-building events. When stripped of its thin veneer of younger sediments the continent has an ancient core, six times older than its margins and a zonal pattern with age increasing from the core to the margins.

Each broad province in Figure 7-13 is composed of a series of overlapping volcanic-sedimentary and granite-forming cycles. The granite-forming events tend to be localized

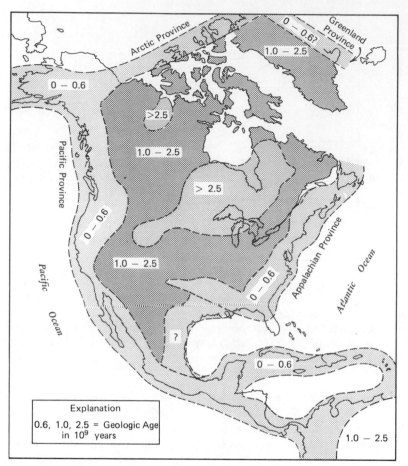

Figure 7-13. Gross patterns and ages of geologic provinces in North America as defined by major granite-forming, mountain building events dated by 1963 (after Engel, 1963). Compare Figure 7-14. (From Science, with permission. Copyright 1963 by the American Association for the Advancement of Science.)

along the sites of maximum crustal instability delineated at the time of intrusion by sinuous mountain belts. Each province overlaps pre-existing provinces by 20 to 60%. It has been estimated that the Appalachian province (0.2–0.8×10^9 years) overlaps the Grenville province (about 1×10^9 years) by about 60%, and in western North America there is near obliteration of older provinces by a succession of younger provinces dated 0.1, 0.4, 1.0, and 1.8 billion years ago. Engel illustrated the extent of overlap in another figure which also showed that the 1.0–2.5×10^9 year zone consists of two rather distinct parts:

basement rocks in the north and west are in the range 1.8–2.5×10^9 years and those in the south and east are 1.0–1.8×10^9 years.

The oldest rock complexes in the core of North America are composed of rocks characteristic of present day island arcs and continental margins. The geology of the younger provinces is well known; they form elongate sheaths to the continent built in part on preexisting older provinces but including constituents of fringing island arcs. The successive volcanic-sedimentary belts involving the adjacent continental crust have become folded, partially melted, and stabilized by welding onto the continental crust. The liquid formed by partial melting forms the granitic intrusions which engulf much of the geosynclinal pile. This is the model reviewed by Engel.

Many more measurements were available for a similar study by W. R. Muehlberger, R. E. Denison, and E. G. Lidiak in 1967 and their results are summarized in Figure 7-14. They studied buried basement rocks from more than 3000 drill-holes and scattered outcrops in the continental interior and combined the petrographic data with isotopic ages to outline a series of geological units and their geological histories.

They presented four maps showing the areas known to be underlain by rocks with isotopic ages equal to or greater than 2.5, 1.7, 1.35, and 1.0 × 10⁹ years respectively and on each map they also outlined the area showing

the probable minimum size of the continental crust at the stipulated time. They showed that more than 50% of the present continent was in existence 2.5×10^9 years ago and concluded that lateral continental accretion is less important than usually assumed. They stated that Engel's diagram (Figure 7-13) showed not continental accretion but the percentage of the continent that was stabilized by a particular time *and* was not involved again in a major granite-forming orogenic event. Engel did point out that much of the younger crust was built of reworked older crust, and certainly he implied that the original 2.5×10^9 year old continent was larger than the 16% of the present continent shown in Figure 7-13, but the case was made that the oldest rocks formed a core for the continent as a whole.

Figure 7-14 shows the outcrop areas of

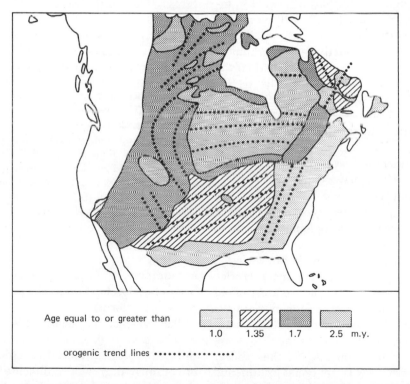

Age equal to or greater than

1.0 1.35 1.7 2.5 m.y.

orogenic trend lines ••••••••••••••

Figure 7-14. Generalized orogenic trend lines and ages of geologic provinces in North America according to Muehlberger et al. (1967). Each region delineated is known to be underlain by rocks giving isotopic ages equal to or greater than that indicated. Compare Figure 7-13.

successively younger continental belts and
their generalized orogenic trend lines. This
shows that the 1.0–2.5 × 10⁹ years belt is
more complex than indicated in Figure 7-13.
The older tectonic trends are truncated by
the younger trends. Many of these rocks have
had a multiple history that isotopic studies
may elucidate.

Two years later, in 1969, P. M. Hurley and
J. R. Rand surveyed the available age data
for the basement rocks of North and South
America, Africa, Europe, India, Australia,
and Antarctica. The results extended over
two-thirds the land area of the Earth.
Examination of the natural grouping of the
apparent ages led to consideration of three
groups of rocks, (a) older than 1.7×10^9
years, (b) in the range 0.8–1.7×10^9 years,
and (c) younger than 0.8×10^9 years. There
seems to have been a universal quiet interval
between 0.7 and 0.9×10^9 years ago, and
there is a slight dip in the world-wide histogram of reported ages at 1.7×10^9 years.

For comparison of the data from all con-
tinents they were plotted on a geographical
reconstruction of the continents based on the
hypothesis of continental drift (Chapter 11,
Figures 11-1 to 11-4). The result is shown in
Figure 7-15. Compare the age provinces on
North America in this figure with those in
Figures 7-13 and 7-14.

The continental basement complexes older
than 1.7×10^9 years occupy two areas as
shown in Figures 7-15 and 11-4, and these are
transected by younger geologically active
belts. Older relict ages are found in many of
the younger transcurrent zones. The two
areas enclosing the oldest rocks are encircled
almost entirely by younger continent. Hurley
and Rand proposed that the two areas of old
rocks represent former continental nuclei
surrounded by successively younger belts.
The continental age patterns on the pre-
continental drift reconstruction thus appear
to be consistent with marginal continental
accretion in two supercontinents, neither of
which was broken up prior to the onset of
drift about 200 m.y. ago. Note, however, that

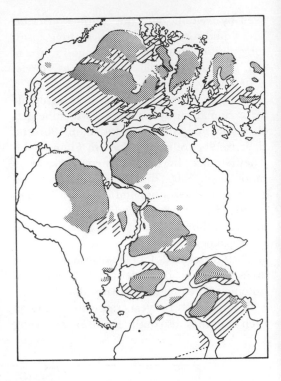

Figure 7-15. Continents reassembled in a predrift reconstruction—see Chapter 11. Lighter
hatching, regions underlain by rocks having apparent ages in the range 800 to 1700 million
years; heavier hatching, regions having apparent
ages >1700 million years. It appears that there
are two (or one) central regions of older rocks
transected and totally surrounded by belts of
younger rocks. This suggests that there was no
significant fragmentation or scattering of the continental nuclei prior to the last great drift episode.
(After Hurley and Rand, 1969. From Science,
with permission. Copyright 1969 by the American
Association for the Advancement of Science.)

in addition to marginal activity orogenic
belts also developed within the continental
cores or platforms.

The evidence outlined in this section supports the view that continents have grown
through geological time. Estimating the rate
of continental growth is difficult because of
the problem of deciding how much of an
orogenic belt of a given age represents new
accreted material and how much represents

reworked older continental crust. In the early 1960's Hurley and his associates and Engel showed that for North America, to a first approximation, there has been a linear growth rate. This is indicated by a plot of the areal extent, including known overlap, of each age province against the age interval. Equal surface areas of crust appear to have been generated in equal time intervals. Comparison of the ratio of $^{87}Sr/^{86}Sr$ with the age of granitic rocks led to an estimate of an average growth rate of about $7 \times 10^3 km^2/10^6$ years. In 1969 Hurley and Rand developed a more complex model for estimating the rate of generation of new continental crust using the partition of Rb into the crust relative to Sr as a criterion. K-Ar ages were calibrated against whole-rock Rb-Sr isochron ages and then used to estimate the distribution of area relative to age. Under the terms of their definitions and assumptions it appears that starting about 3.8×10^9 years ago there has been an accelerating generation of new crustal

material, with a rate of increase of $20 km^2/10^6$ years/10^6 years, or about $600 km^3/10^6$ years/10^6 years.

In order to trace the evolution of the crust and its relationships with the mantle through geological time it is necessary not only to map the apparent ages of crustal rocks but also to work out the strontium isotope and lead isotope evolution trends. Strontium studies have been applied to the problem for more than a decade but the data are still sparse. The available lead isotope data are still insufficient to resolve such fundamental questions as whether:

1. The continental crust has grown continually through geological time by addition of new material from the mantle,

2. The formation of continental crust was largely completed during the period $2.5-3.5 \times 10^9$ years ago with younger crust representing regeneration of the initial crust.

8. *Magma Generation*

Magma is a mobile assemblage of rock matter with the essential ingredient a silicate liquid; it usually contains suspended crystals and sometimes a separate gas phase. Magma is generated whenever conditions become appropriate for partial fusion of rocks within the Earth. Because of its lower density a magma tends to rise and differentiation of the liquid phase occurs in response to the changing conditions. Magmas may break through the crust to produce extrusive volcanoes or lava floods, or they may crystallize at depth within the mantle or crust as intrusive plutonic rocks. Since early Earth history repeated magma generation and uprise has caused progressive defluidization of the mantle.

The Earth's crust is composed essentially of igneous rocks and their metamorphic equivalents (Tables 7-2 and 7-3); it is thus the product of volcanism and plutonism. Its average chemical composition is very similar to that of andesites or the average of the whole suite of volcanic rocks building the island arcs (Table 7-7). It is through the

study of igneous rocks and processes that we hope to gain some insight into the question of whether (a) the continents have grown continuously through geological time by addition of new material from the mantle or (b) younger crust represents regeneration of older, initial crust, by magmatic processes. The formation of magma also provides a guide to temperatures within the Earth.

There are many different igneous rock types and rock associations, and classification of these provides the essential basis for petrogenetic theories. The problem is complex as illustrated by our brief review in Chapter 6 of the occurrences, distribution, and petrogenesis of ultramafic rocks; Table 6-4 lists five major associations occurring in different environments. Despite the complexity in detail, if we are concerned with major Earth processes such as the origin of continents and mountain ranges (Chapters 7 and 9), we may concentrate our attention on the most abundant rocks: the extrusive (a) basalts, (b) andesites, and (c) the intrusive diorites, granodiorites and granites of batholiths

IGNEOUS ROCK ASSOCIATIONS

It is in the oceanic tectonic provinces that we seek samples of magmas generated in the mantle and uncontaminated by crustal materials. Here we find two varieties of basalt with minor differentiates and ultramafic rocks. Subalkaline basalt, saturated or nearly saturated with silica and with the rather

unfortunate name "tholeiite," is the dominant basalt of the ocean floors and midoceanic ridges and the oceanic volcanic islands. Alkaline basalts, undersaturated in silica, occur in oceanic islands. Ultramafic rocks occur on the midoceanic ridges, and their relationship to the ophiolite complexes and other ultramafic rocks of the orogenic belts was reviewed in Chapter 6 and will be mentioned again in Chapter 14.

Tholeiitic basalts and alkaline basalts are also widely distributed on the continents with minor associates of andesite and rhyolite or trachyte and phonolite. The ultramafic rocks of the large layered intrusions (Table 6-4) are produced by crystal accumulation from a parent subalkaline basaltic magma. The rifted portions of otherwise stable continental platform areas are characterized by the rarer highly alkaline lavas. Plutonic equivalents of these lavas form usually small feldspathoidal intrusions which may be associated with ultramafic rocks (Table 6-4) and carbonatites.

The calc-alkaline andesites with associated basalts and rhyolites occur characteristically in orogenic belts of the continents and in the volcanic island arcs of the continental margins. They are interspersed with sediments in geosynclinal piles, and they have been repeatedly erupted during and after the folding and uplift of the geosynclinal rocks.

Batholiths are emplaced in orogenic belts and in some respects they appear to be the plutonic equivalents of the andesites. Their field relationships and petrology have often been adduced as evidence that the plutonic rocks were derived by reworking of crustal material by metamorphism and metasomatism without the intervention of magmatic activity.

DEVELOPMENT OF PETROGENETIC THEORY IN THE TWENTIETH CENTURY

Figure 4-5 summarizes the main processes in petrogenesis beginning with magma genesis and outlining various sequences of events that occur between melting and emplacement or eruption of the magma. The composition of the liquid which finally crystallizes as a plutonic intrusion, or erupts from a volcano, is the product of a very complex history. Petrogenesis is concerned with unravelling that history. The problem has been to examine the rocks, the end-products of the history, and to construct a petrogenetic scheme by inductive reasoning. During the past twenty years new approaches to the subject have made it easier to devise tests for petrogenetic theories. These include (a) geophysics of the solid Earth, (b) high pressure experimentation, and (c) geochemistry.

Improved techniques in geophysics have advanced our knowledge of the Earth's interior to such an extent (Chapters 3 and 6) that petrogenetic theory must now be based on the geophysical data. The development of apparatus for high pressure, high temperature studies makes it possible to reproduce crustal and mantle conditions in the laboratory, and experimental melting of

minerals and rocks places rather close restrictions on the conditions of magma generation. Improved analytical techniques for measuring isotopes and trace element concentrations in igneous minerals and rocks have led to a wealth of new data, much of which can be interpreted in terms of the composition of a mantle source or the extent of contamination by crustal material. These results provide limits for petrogenetic theories.

The First Half of the Century: Primary and Derivative Magmas

The concept of primary magmas has held a dominant position in petrogenesis. A primary magma is one formed either by partial or complete fusion of crystalline rock at depth or, according to older concepts, by tapping a body or layer of liquid or glassy magma persisting at depth from some early stage of the Earth's development. Basaltic and granitic magmas are the most abundant primary magmas. Figure 8-1 illustrates some of the models proposed for magma generation.

Until the mid-1930's many petrologists accepted a hypothesis, promulgated by R. A. Daly, that a worldwide substratum of basaltic

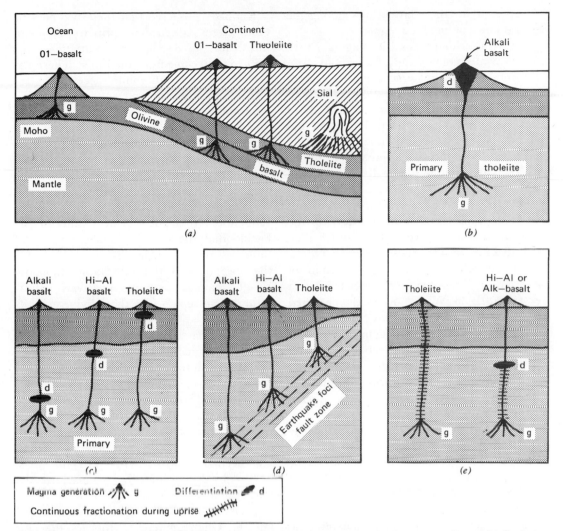

Figure 8-1. Various models for magma generation and fractionation in the crust and mantle. See text for review.

magma existed beneath the lithosphere. Daly subsequently abandoned the idea when it appeared that neither basaltic magma nor glass had suitable seismic properties, and he then favored fusion of basaltic rocks near the base of the crust. The concept of a worldwide subcrustal magma zone with composition equivalent to an olivine-rich basalt, however, was retained by A. Rittmann in 1962; he maintained that at high pressures magma viscosity would become high enough for the transmission of shear waves because the periods of the seismic transverse waves are shorter than the relaxation time of the magma. In the second edition of his petrology text T. F. W. Barth (1962) cites Rittmann and tentatively reverts to the hypothesis of a glassy basaltic substratum beginning at a depth of 50 to 70 km beneath the surface. Rittmann proposed that this extended to depths of at least 1400 km and perhaps to more than 2000 km. The hypothesis of a liquid substratum has not supplied explanations for either the properties of the mantle

(Chapter 6) or the relationships among the several kinds of basaltic magma.

Given a primary basaltic magma N. L. Bowen argued convincingly, from about 1915, that most igneous magmas could be derived from this parent by fractional crystallization. He based his arguments mainly on study of the crystallization of silicate liquids in the laboratory, but he did not neglect the standard petrological approaches. Other processes are listed in Figure 4-5.

During the 1920's the detailed study of basalts in the Tertiary province of Scotland led E. B. Bailey and others to develop the concept of magma types. Both subalkaline and alkaline basalts occur, and each type is associated with a characteristic group of differentiates representing distinct magma series. In 1933 W. Q. Kennedy proposed that each of these basaltic magma types was an independent primary magma, and he gave them the names tholeiite (subalkaline) and olivine basalt (alkaline). He examined the distribution of similar basalts on a worldwide basis and presented the petrogenetic scheme illustrated in Figure 8-1*a*. There are two basaltic crustal layers to provide for the independent formation of the two primary magmas with the subalkaline tholeiite layer being present only beneath the continents. Kennedy distinguished between volcanic igneous associations developed by differentiation of the primary basalts and plutonic igneous associations developed by fusion of sialic material of the crust above the basaltic layers. Figure 8-1*a* shows explicitly the development of batholithic magmas by fusion of the crust. Present interpretations of temperature distribution at depth do not permit the generation of basaltic magmas by fusion of basaltic layers above the Moho (Chapters 3 and 7). As long ago as 1928 before these models were developed Bowen had proposed that basaltic magmas were generated by partial fusion of feldspathic peridotite or eclogite beneath the Earth's crust, and this forms the basis of most recent models for the generation of primary basalts.

Kennedy proposed that these two basaltic types were independent primary magmas, but others have sought differentiation processes to derive one from the other. There has been incredible confusion in successive chemical, petrographical, and mineralogical descriptions and definitions of these two magma types, as well as uncertainty about their tectonic environments, which were reviewed in a delightful paper by F. Chayes in 1966.

Kennedy made a distinction between the olivine basalt and the tholeiite on the presence or absence of olivine. C. E. Tilley in 1950 noted that the olivine basalts of the Pacific had the chemical characteristics of tholeiites despite their olivine content. He therefore defined the subalkaline tholeiitic basalt magma as one relatively silica-saturated in such a way that olivine exhibits a reaction relationship with the liquid during crystallization. The alkaline olivine basalt, on the other hand, is relatively silica-undersaturated and the olivine exhibits no reaction relationship. A chemical distinction is that the tholeiitic rocks have hypersthene in the norm. In his review Chayes recognized the real existence of tholeiitic basalts and alkali olivine basalts but, because these two names had been used in so many different ways, he proposed that they be replaced by the terms subalkaline basalt and alkaline basalt respectively which are unambiguous and well known. Unfortunately it appears that his plea will be ignored and "tholeiites" may be with us for evermore. Tilley also recognized the existence of basalts unusually rich in alumina, and high-alumina basalt is now regarded as a third primary magma.

The celebrated "Granite Controversy" raged bitterly through the 1930's and 1940's. According to Bowen's thesis the diorites, granodiorites, and granites comprising the batholiths, as well as the eruptive andesites, were derived by fractional crystallization of primary basalts. Others suggested that the enormous volumes of granitic magma were augmented by contamination of the parent basalt by assimilation. Many petrologists, especially those who had worked in Pre-cambrian gneissic terranes, were convinced that the batholiths were formed by metasomatism of preexisting rocks in the solid state. The necessary components were introduced and subtracted from the original rock either by diffusion of ions through crystal lattices or by the flow of interstitial pore fluids. "Magmatists" and "transformists" had hardly a civil word for each other.

At the turn of the century J. J. Sederholm had systematically studied metamorphic rocks that had undergone anatexis, or "re-melting," with the formation of migmatite. This is a mixed rock part metamorphic and part magmatic. Figure 8-1a shows Kennedy's interpretation of the plutonic igneous associations being produced by fusion of the crust. These ideas eventually were formulated into experimentally based anatectic models, and the bitterness of the debate subsided during the 1960's as many petrologists found at least partial satisfaction in the model that produced magmatic rocks by partial fusion at the culmination of metamorphism.

Influence of High Pressure Experiments Since 1950

When Tilley's presidential address was published at the midpoint of the century this coincided with a turning point in petrology. His review of the primary basalts brought reasonable order into the problem. At the same time O. F. Tuttle designed a simple pressure vessel which permitted the routine study of mineral and melting reactions in the presence of water vapor under pressure. For years petrologists had expressed the need for such experiments. Now here was an apparatus that could be used with ease by students.

The first results were published in 1950 by Bowen and Tuttle on the feldspar system, $KAlSi_3O_8(Or)$-$NaAlSi_3O_8(Ab)$-H_2O. In 1958 Tuttle and Bowen published a monograph on the "Origin of granite in the light of experimental studies in the system $NaAlSi_3O_8$-$KAlSi_3O_8$-SiO_2-H_2O" in which they developed an anatectic model. They showed how partial melting of crustal rocks in the presence of small amounts of water offers a mechanism for producing large batholithic masses of granite. Similar conclusions were reached by H. G. F. Winkler and associates in a series of melting studies on sedimentary rocks and metamorphic rocks.

The Tuttle vessel and others utilizing a fluid medium to transmit the pressure to the sample are effectively limited in their pressure range to conditions in the Earth's crust. For the reproduction of mantle conditions an apparatus compressing a solid pressure medium is required. In the 1950's L. Coes was the first to synthesize minerals in this type of apparatus. Many solid-pressure designs have been used but most of them are elaborate and rather difficult to operate. There is one design, a single-stage piston-cylinder apparatus, which is now functioning routinely in many laboratories reproducing conditions in the mantle corresponding to depths of 100 km and more. This was described in 1960 by F. R. Boyd and J. L. England and developed also by G. C. Kennedy and associates. Since 1960 experimental data yielded by this apparatus have strongly influenced petrogenetic theory.

In the early 1960's most piston-cylinder experiments involved single minerals or simple assemblages but in 1962 H. S. Yoder and C. E. Tilley published a pioneering study on the effect of water on the melting of various basalts up to 10 kb pressure, together with exploratory work on eclogite under mantle conditions (the latter with the aid of S. P. Clark, F. R. Boyd, and J. L. England). From their study it became clear that the depth of magma generation and fractionation would have a marked effect on magma composition. They introduced the basalt tetrahedron, Figure 8-12d.

In 1965 A. E. J. Engel and associates concluded that in the ocean basins the only primary magma is a low-potassium tholeiitic basalt derived by partial fusion of the peridotite mantle. Various arguments led them to conclude that the alkaline basalts capping the oceanic islands were formed by differentiation processes occurring near the surface. Yoder and Tilley had shown that this process could not be crystallization differentiation because of the low-pressure thermal divide between these two liquids (Figure 8-12d), so Engel invoked other processes such as gas transfer for the conversion of the oceanic tholeiites into alkaline basalts. The model is illustrated in Figure 8-1b.

Yoder and Tilley showed that shifting thermal divides could cause a given basaltic composition to follow quite different fractionation paths depending on the pressure. These observations led them to the model illustrated schematically in Figure 8-1c with tholeiitic basalt being the only primary magma developed in the mantle. If the basalt rises directly to the surface and undergoes low pressure fractionation only a tholeiite and its usual derivatives can be erupted. On the other hand if the primary tholeiite undergoes a period of differentiation at depth within the mantle this causes formation of a derivative alkali olivine basalt. High-alumina basalt is produced by differentiation of the primary tholeiite at a high level within the mantle or possibly deep within the crust.

H. Kuno was the first, in 1960, to explain the different basalts by pressure effects; his model includes three primary magmas. Kuno has shown that the three types of basaltic magma occupy successive belts across island arcs, and he assumed that the position of the Benioff zone of earthquake foci gives the sites of magma generation. Figure 8-3d shows that the depths of magma generation are successively deeper beneath volcanoes erupting tholeiites, high-alumina basalts, and alkali olivine basalts. Hence Kuno's conclusion that the compositions of the three magma types are controlled by depth of generation, and their classification as separate primary magmas.

M. J. O'Hara published a stimulating review in 1965, revised and expanded in 1968, in which he pointed out that the concept of primary magmas was only one extreme model for magma generation. Implicit in the model is the assumption that a magma is generated at depth and then rises rapidly to the surface with little or no change of composition. It is the frequent eruption of tholeiite basalt according to the concept which permits one to recognize magma of this composition as being the primary product of melting at depth; Figure 8-1c shows two variations of the model which involve the derivation of other magmas from a primary tholeiite by a period of high pressure differentiation. O'Hara pointed out that enough experimental information was available on the compositions of liquid fractions in equilibrium with peridotites at high pressures to indicate that these compositions did *not* correspond to the lavas erupted at the surface and interpreted as primary magmas. Therefore he concluded that a new model for magma generation and fractionation is required.

The other extreme model is one where a magma is generated by partial fusion of mantle peridotite, and this experiences continuous fractionation from source to surface eruption. This is illustrated in Figure 8-1e which should be contrasted with the primary magma models of Figures 8-1b and d. O'Hara also discussed models between these two extremes as being somewhat more reliable than either extreme and an example is shown in Figure 8-1e.

In order to develop this type of model in detail it is necessary to work out the effect of bulk composition, pressure, and temperature on the composition of the liquid coexisting with residual peridotite. A rising magma can follow an infinite number of different paths through a system with these variables. Most evidence indicates that basaltic magma contains only traces of juvenile water although additional water may be dissolved during transit through the crust.

Even traces of water have significant effects, however, so we turn next to the general pattern of phase relations in rock-water systems. This provides the framework for more detailed examination of magma generation from materials of the crust and mantle based on high pressure experimental studies published since 1967.

ROCK-WATER SYSTEMS AS GUIDES TO MAGMA GENERATION

Basaltic magmas are produced by melting of the mantle and granitic magmas may be produced by melting of crustal material; andesitic magmas may be produced by differentiation of basaltic magmas or by melting in the mantle or crust. As shown in Figure 4-5 magmas may be generated by partial fusion of three types of assemblages: (a) anhydrous minerals, (b) anhydrous and hydrous minerals, and (c) minerals in the presence of a pore fluid usually considered to be aqueous, although carbon dioxide and other components are expected. For models of magma generation in the mantle we therefore consider the experimental results in the systems peridotite-gabbro, peridotite-water, and gabbro-water and for the crust we consider the systems gabbro-water and granodiorite-water.

Most magmatic processes occur with no vapor present and with water pressure less than total pressure: these are water-deficient conditions which have received little experimental attention. The dry and water-excess experimental results provide limits for the natural conditions, and the general pattern in water-deficient systems can be interpolated between these limits. C. W. Burnham reviewed the conditions for anatexis in the crust and mantle in water-deficient systems in 1967.

Water-Absent and Water-Excess Systems

The availability of piston-cylinder apparatus during the 1960's permitted the measurement of melting curves for dry minerals and rocks at pressures greater than 10 kbars. The effect of pressure is to raise the melting temperature as shown for various minerals and rocks in Figures 4-3, 4-4, 5-13, 5-15, 6-10, 6-11, 6-12, 6-14, 6-17, 6-18a, and 6-19a.

Chapter 4 included a review of melting reactions in silicate-water systems, Figures 4-3e and 4-3f. Figure 6-11 shows the effect of excess water on the solidus of peridotite, and Figures 6-12 and 8-18 compare the melting interval for gabbro composition dry and with excess water. The initial study of Bowen and Tuttle in 1950 on the system $NaAlSi_3O_8$-$KAlSi_3O_8$-H_2O was followed by many others. By 1967 the curves for the beginning of melting for many silicate assemblages in the presence of excess water had been measured to 10 kbars; the Tuttle pressure vessel and internally-heated hydrostatic vessels of the type used successfully by H. S. Yoder and C. W. Burnham yielded these results. In 1967 A. L. Boettcher and I used the piston-cylinder apparatus to extend the melting curves in feldspar-quartz-water and rock-water systems to pressures greater than 10 kb in order to evaluate the effects of water on magma generation in the mantle. By 1970 melting curves in the presence of excess water for individual feldspars for most feldspar-quartz combinations and for most major rock types had been located through a wide range of pressures. Selected results are shown in Figures 8-2 and 8-3.

Figure 8-2a shows the univariant reaction curves for individual feldspars and for quartz in the presence of excess water. The effect of water under pressure is to lower the melting temperature. The curve for orthoclase continues to pressures greater than 20 kb, but the curves for albite and anorthite terminate at invariant points where the feldspars break down to yield dense minerals such as jadeite, zoisite, and kyanite (Figures

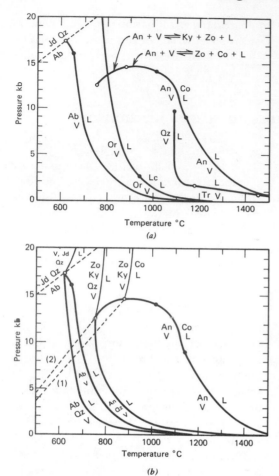

Figure 8-2. Solidus curves (univariant) in feldspar-quartz-water systems with excess water, compiled by Merrill et al. (1970). Heavy lines are solidus curves; light lines are solidus curves at pressures above the stability limit of feldspars; light dashed lines are subsolidus reactions; open circles are invariant points; closed circles are singular points. An = anorthite, Ab = albite, Or = Sanidine, Qz = quartz, Tr = tridymite, Lc = leucite, Co = corundum, Zo = zoisite, Ky = kyanite, Jd = jadeite, L = liquid, V = vapor.

(a) The system SiO_2-H_2O, and reactions involving single feldspars.

(b) Univariant reactions in the system $CaAl_2Si_2O_8$–$NaAlSi_3O_8$–SiO_2–H_2O.

Reaction (1): $An + Co + V \rightleftharpoons Zo + Ky$

Reaction (2): $An + V \rightleftharpoons Zo + Ky + Qz$

(Sources of data: Qz, Kennedy et al., 1962; Kushiro, 1969. An and An–Qz, Stewart, 1967;

4-3b, 5-11, and 5-12). The curve for quartz terminates at a second critical end-point.

The melting curves for plagioclase feldspars and plagioclase feldspar-quartz mixtures in the presence of excess water are shown in Figure 8-2b. The light dashed lines show the subsolidus reactions which limit the high pressure stability of the plagioclase feldspars, and the light solid lines indicate the hydrothermal melting curves which replace the feldspar and feldspar-quartz curves at higher pressures. Involvement of the dense mineral phases in the assemblage reverses the slope of the solidus so that with increasing pressure in the presence of excess water the temperature of beginning of melting increases just as in the dry systems.

The mineralogy of the major igneous rocks of the Earth's crust is dominated by feldspars and quartz (Table 7-3), and the similarity of the curves for the beginning of melting of rocks in the presence of excess water to pressures of 15 kb shows this influence (Figure 8-3). The effect of water on the melting temperature of peridotite is much less than on the crustal rocks that include feldspars and quartz. Curves for granite, tonalite, and gabbro, representing acid, intermediate, and basic compositions in the calc-alkaline suite are roughly parallel with each other; they are separated by only 30°C between 8 and 15 kb. At pressures above 15 kb the solidus curves reverse slope and become similar to those for dry silicates and rocks (see also Figures 6-18b and 6-19b).

Determination of the solidus curves for the beginning of melting of rocks is only the beginning of elucidation of magma generation. In order to examine the products of partial melting it is necessary to know the phase relationships through the melting

Figure 8-2 continued

Boettcher, 1970. Or and Or–Qz, Lambert et al., 1969. Ab and Ab–Qz, Boettcher and Wyllie, 1969; Tuttle and Bowen, 1958). (From Jour. Geology with permission. Copyright 1970 by The University of Chicago Press.)

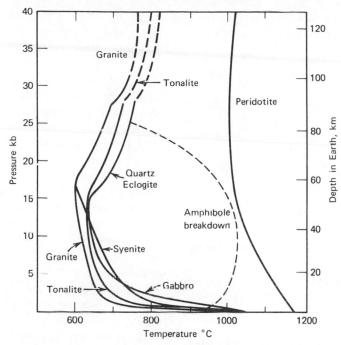

Figure 8-3. Solidus curves for rocks in the presence of excess water, compiled by Merrill et al. (1970). Granite-H$_2$O (rhyolite-H$_2$O) after Piwinskii (1968), Boettcher and Wyllie (1968). Tonalite-H$_2$O (andesite-H$_2$O) after Piwinskii (1968), Lambert and Wyllie (1970). Gabbro-H$_2$O (basalt-H$_2$O) after Lambert and Wyllie (1968, 1970), Hill and Boettcher (1970), Tuthill (1969) at 5 kb, Holloway and Burnham (1969) at 2, 5 and 8 kb. Syenite-H$_2$O after Merrill et al.(1970). Peridotite-H$_2$O after Kushiro et al. (1968). (From Jour. Geology, with permission. Copyright 1970 by The University of Chicago Press.)

interval of the source rock. The first detailed experimental work of this kind was the 1962 study of Yoder and Tilley on basalt compositions with excess water to 10 kb and dry at greater pressures. Additional studies of dry basalt compositions and peridotites began to appear in 1967 by A. E. Ringwood, D. H. Green, and associates at Canberra, and by K. Ito, G. C. Kennedy, and L. H. Cohen at Los Angeles. From Chicago A. J. Piwinskii and I in 1968 reported phase relationships at crustal pressures through the melting intervals of granites, granodiorites, and tonalites in the presence of excess water, and in 1968 and 1970 I. B. Lambert and I published similar results for gabbro and tonalite at mantle pressures between 10 and 25 kb.

Figures 6-12 and 8-18a for the composition gabbro-water illustrate the main differences in phase relationships produced at pressures up to 10 kb, if excess water is added to dry rock:

1. The solidus and liquidus temperatures are lowered.

2. The temperature interval between liquidus and solidus is increased.

3. Hydrous minerals are stabilized and they become involved in reactions including liquid.

Two other factors are introduced at higher pressures corresponding to mantle conditions:

1. The slope of the solidus is reversed so that with increasing pressure in the presence of excess water the temperature of beginning of melting increases.

2. At high pressures the slope (dP/dT) for the breakdown of hydrous minerals changes

from positive to negative so that with increasing pressure the dehydration temperature becomes lower.

For amphibole this effect becomes marked at about 15 kb, as shown in Figure 8-3, and for the gabbro composition the amphibole breaks down below the solidus at pressures greater than about 25 kb. An amphibole stability curve intersects the peridotite-water solidus between about 8 and 17 kb (Figures 6-11 and 6-18).

Water-Deficient Systems

Terms such as vapor-absent and water-deficient have been widely used without adequate definitions probably because experimental studies on melting relationships in water-deficient systems were not reported until the late 1960's. In 1971 J. K. Robertson and I prepared a consistent set of definitions for four types of subsolidus assemblages in silicate-water systems, providing a basis for models of magma generation. These types also have relevance for processes of magma crystallization:

Type I: Water-absent. An assemblage of anhydrous silicate minerals with no aqueous vapor phase.

Type II: Water-deficient and vapor-absent. An assemblage of silicate minerals including hydrous minerals but with no aqueous vapor phase.

Type III: Water-deficient and vapor-present. An assemblage of silicate minerals with or without hydrous minerals and with an aqueous vapor phase present. There is insufficient water present to saturate the liquid when the crystalline assemblage is completely melted at the existing pressure.

Type IV: Water-excess. An assemblage of silicate minerals with or without hydrous minerals and with more than sufficient water to saturate the liquid when the crystalline assemblage is completely melted at the existing pressure.

A system at a given pressure is water-deficient if it contains less water than necessary to saturate the liquid formed by complete melting (Type III) or less water than that required to fully hydrate the subsolidus assemblage (Type II). Water-deficient systems are characterized by vapor-absent conditions through at least one temperature interval. It is convenient for petrogenetic models to distinguish between the two types of water-deficient systems. Figure 8-4 illustrates the pattern of melting for all four types of systems.

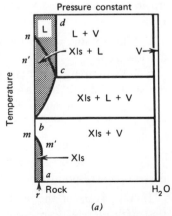
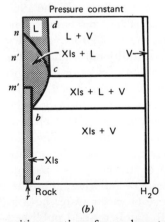

Figure 8-4. Schematic isobaric temperature-composition sections for rock–water systems. Shaded areas are vapor-absent; *b–c* is saturation boundary; *m–m′* is dissociation interval of hydrous minerals. See text for significance of lettered points. (*a*) Hydrous minerals dissociate below solidus temperature. (*b*) Hydrous minerals remain stable to temperatures above vapor-present solidus (after Robertson and Wyllie, 1971, from Amer. Jour. Sci., with permission).

Figure 8-4 shows a rock composition, r. If this rock is completely dehydrated it becomes a water-absent system. The anhydrous rock melts in the temperature interval $n'-n$. If the rock is held at some temperature below the solidus in the presence of excess water vapor under pressure, additional hydrous minerals may be formed changing the rock composition to point a. The shaded area within the saturation boundary $a-b-c-d$ is vapor-absent. The line above a marks the limit of water-deficient systems of Type II which are shown by the shaded area for "crystals." Subsolidus assemblages between a and the H_2O axis include systems of Types III and IV. The point m' gives the temperature where dissociation of the hydrous minerals begins.

In Figure 8-4a dissociation is completed in the temperature interval $m'-m$, before the solidus temperature is reached, and the Type II assemblages therefore do not melt; for all compositions, melting begins in the presence of excess vapor. The point c gives the amount of water required to saturate the liquid when the rock is completely melted. Therefore for temperatures above point m the compositions between b and c are water-deficient Type III and the others are water-excess Type IV.

In Figure 8-4b, compositions in the shaded area to the left of line $a-b$ are water-deficient Type II, compositions between points b and c are water-deficient systems of Type III, and those on the H_2O side of point c are water-excess Type IV. Dissociation of the hydrous minerals does not begin until the solidus temperature for Types III and IV has been exceeded, and Type II compositions therefore do melt. As soon as dissociation begins, at m', a small amount of H_2O-undersaturated liquid is produced.

The shape of the saturation boundary between $b-c$ is different for every rock. The boundary gives the amount of water required to saturate the assemblage of crystals + liquid. Its position at each temperature is therefore given by adding (a) the

water content of the crystals to (b) the product of the amount of liquid present with the solubility of water in the liquid. $P_{H_2O} = P_{total}$ for all assemblages with vapor and $P_{H_2O} < P_{total}$ in vapor-absent regions.

Figure 8-4 permits comparison of the melting relationships in systems of Types I, II, III, and IV. Note the high temperature and narrow temperature interval, $n'-n$, for xls + L in Type I. The solidus temperature for Type II, if this assemblage does melt as in Figure 8-4b, coincides with the temperature of beginning of dissociation of the hydrous minerals and is thus quite independent of Types I, III, and IV; the liquid produced is undersaturated with water and $P_{H_2O} < P_{total}$. The solidus temperatures for the vapor-present Types III and IV are identical and much lower than for Type I; the melting temperature is independent of the amount of water and the initial liquid is saturated with water, $P_{H_2O} = P_{total}$. The melting pattern of water-deficient Type III assemblages is initially the same as for water-excess Type IV assemblages. Melting of Type IV assemblages is completed with excess vapor present throughout, but for Type III assemblages the vapor is dissolved where the temperature reaches the saturation boundary, $b-c$; further increase in temperature yields a vapor-absent assemblage with the liquid becoming progressively more undersaturated with water, and with P_{H_2O} decreasing. The temperature of the saturation boundary $b-c$ for Type III assemblages depends sensitively upon the water content and the mineralogy of the rock; with increasing water content it migrates from solidus to liquidus, b to c, providing the transition from Type II to Type IV. The liquidus temperature in the vapor-absent region is successively lowered with increasing water content from Type I to Type IV.

Figure 8-4b is the normal model for magma generation in the crust. The pattern in Figure 8-4a would prevail for crustal rocks such as granodiorite at pressures below a few tenths of a kilobar (Figure 8-10a), for the system gabbro-water at pressures below about 1 kbar (Figure 6-12), and for peridotite-water below about 8 kbars (Figures 6-11 and 6-18*b*). The same pattern would be repeated for gabbro(eclogite)-water at pressures above 25 kbars (Figures 6-12, 6-19*b*, and 8-18*a*) and for peridotite-water at pressures above about 16 kbars (Figures 6-11 and 6-13*b*). Thus Figures 8-4*a* and 8-4*b* are the two basic models for magma generation in the mantle, the appropriate model being governed by composition and depth.

For details of the composition and amount of liquid produced during progressive fusion we need detailed phase relationships within the melting interval in the water-deficient assemblages for specific rocks. Experimental data are not yet available but interpolation between experimental results for dry Type I systems and for water-excess Type IV systems locates the positions of phase boundaries in the water-deficient region with little variation possible.

The Conditions $P_{H_2O} = P_{total}$ and $P_{H_2O} < P_{total}$

P_{total} refers to the applied pressure in experiments or to the lithostatic pressure resulting from the superincumbent rocks. P_{H_2O} is the partial pressure of water or the pressure that water alone would have if it occupied the volume of the system. Most experimental results using water have been in water-excess systems, and they have often been presented as a function of P_{H_2O}; there has been a tendency in the literature to apply the term P_{H_2O} only to vapor-present systems. Therefore it may be worthwhile to emphasize that the use of P_{H_2O} need not imply the existence of a vapor phase. We have assumed that $P_{H_2O} = P_{total}$ in all vapor-present assemblages (Types III and IV), and noted that $P_{H_2O} < P_{total}$ in vapor-absent assemblages including hydrous minerals or undersaturated liquids (Types II and III).

Solution of solids in the vapor phase

reduces P_{H_2O} compared with P_{total}; the effect is small in most silicate-water systems at crustal pressures but at mantle pressures this becomes a factor to consider. For most magmatic processes $P_{H_2O} < P_{total}$, first, because a vapor phase is not invariably present. Second, when a vapor phase is present it contains components additional to water. Therefore even if pore fluid pressure equals P_{total} the P_{H_2O} is less. Third, rocks may exist with pore fluid pressure less than P_{total}, with the differential pressure being supported by grain-to-grain contacts; under these conditions P_{H_2O} in the multicomponent pore fluid is even lower.

In vapor-absent assemblages P_{H_2O} at a given pressure and temperature is a function of water content: it is a dependent variable. In vapor-present assemblages at a given pressure and temperature P_{H_2O} is a function of the water content of the vapor phase. In experiments and in the Earth vapor phase components can be transferred in or out of the system so that P_{H_2O} can be varied independent of P_{total}. If there are two independent pressure variables a system gains an extra degree of freedom. A univariant melting reaction such as $Ab + Qz + V \rightleftharpoons L$ in Figure 8-5 becomes divariant. From the univariant melting curve for the condition $P_{H_2O} = P_{total}$ there extends a divariant four-phase solidus surface referred to three orthogonal axes, P_{total}, P_{H_2O}, and T. Figure 8-5*a* shows contours for the divariant surface projected onto the P_{total} T plane.

P_{eH_2O} is used in Figure 8-5*a* instead of P_{H_2O} because it is a term more amenable to rigorous definition. It was introduced and defined by H. G. Greenwood in 1961 specifically in connection with subsolidus reactions. The equilibrium pressure of water in a system at a given temperature, P_{eH_2O}, is the pressure of pure water that would be in equilibrium with the system through a membrane permeable only to water. The contours for constant P_{eH_2O} in Figure 8-5*a* are based on the first quantitative experimental results for $P_{eH_2O} < P_{total}$ with a liquid phase

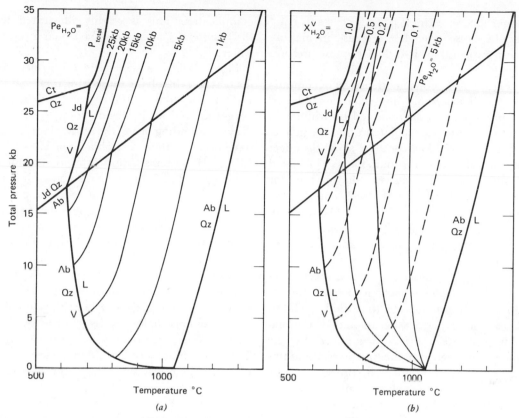

Figure 8-5. Univariant melting reactions for mixtures of albite and quartz both dry and in the presence of excess water, with the jadeite formation curve and coesite transition curve. Estimated values for divariant surfaces are long extrapolations from experimental data from below 6 kb. (*a*) Each light line (schematic) gives the locus of a melting reaction in the presence of a vapor phase with P_{eH_2O} maintained constant at the specified value. These curves are contours for the divariant solidus surfaces represented in terms of three independent variables, temperature, P_{total}, and P_{eH_2O}; the surfaces connect the dry curves and the water-saturated solidus reaction curves. (*b*) Each light line (schematic) gives the possible locus of a melting reaction in the presence of a vapor phase with $P_{vapor} = P_{total}$, and with the water proportion in the vapor phase fixed at the specified value. These curves are contours on the divariant solidus surface of (*a*). The dashed lines, transferred from figure (*a*), show the general relationship of the composition contours to the water pressure contours.

present. These were obtained by G. Millhollen from experiments on the effect of H_2O-CO_2 mixtures on the solidus curve for a nepheline syenite up to 6 kb. The slope of the projected contours shows that with P_{eH_2O} constant the melting temperature increases with increasing pressure. The limiting curve is the dry melting curve with P_{eH_2O} constant at zero. At a given total pressure the melting temperature increases with decreasing P_{eH_2O}.

The solidus curve changes slope at 17.5 kb where albite breaks down to yield jadeite and quartz. The family of contours is constrained to remain between the dry melting curve and the solidus with excess water. The P_{eH_2O} contours cross the jadeite reaction curve with a slight change of slope. At pressures above the jadeite curve the geometrical constraints indicate that the P_{eH_2O} contours become more closely spaced near the solidus than they were at lower pressures. This means that a larger decrease in P_{eH_2O} at constant P_{total} should cause a smaller increase in melting temperature than it did in the low pressure range.

Figure 8-5b shows another way of contouring the divariant melting surface again based on numerical results obtained by Millhollen up to 6 kb and extrapolated up to the jadeite reaction curve. The contours show the effect of mixed volatile components of fixed composition; the curves shown are for mixtures with $X^V_{H_2O} = 0.1$, 0.2, and 0.5, the balance being composed of CO_2 in Millhollen's experiments. These contours also must change slope at the jadeite reaction curve and, because of the crowding together of the P_{eH_2O} contours, it seems likely that the $X^V_{H_2O} = 0.5$ contour and possibly the 0.2 contour would migrate toward the solidus. These curves need experimental determination, but the restrictions imposed by the known geometry of the phase relationships suggest that in the high pressure range the aqueous vapor phase can be diluted considerably without producing much increase in the melting temperature. This kind of effect was reported by R. E. T. Hill and A. L. Boettcher in 1970 in experiments with gabbro-H_2O-CO_2, but they attributed it to increased solubility of CO_2 in the basaltic liquid at high pressures. Figure 8-5b offers an alternative or a contributory explanation.

We can conclude from Figure 8-5 that although additional components such as CO_2 may have a marked effect on the temperatures of melting and crystallization under crustal conditions this effect may be reduced under mantle conditions.

The effect of lowering P_{H_2O} at constant P_{total} by diluting the vapor with another component is to increase the solidus temperature. Similarly the dissociation temperature of a hydrous mineral is lowered if P_{H_2O} is reduced relative to P_{total} (Figure 8-6a). The effect of multicomponent pore fluids on conditions of melting in systems including Types II and III assemblages is therefore complicated. This is readily illustrated by the system muscovite (Ms)-quartz (Qz)-water.

Figure 8-6a is a schematic PT projection for reactions involving muscovite, quartz, and H_2O vapor showing the relevant univariant curves around an invariant point, Q. Other curves extending from the univariant point have been omitted. Assemblages of Ms + Qz (Type II systems) dissociate at pressures below Q and melt at pressures above Q. The dissociation curve is unchanged if vapor is added. The melting curve for the vapor-absent Ms + Qz assemblage is at a higher temperature than that for the vapor-present assemblage (Type III). The high temperature melting curve for Ms + Qz is a Type II reaction equivalent to m' in Figure 8-4b. The effect of reducing P_{H_2O} below P_{total} on the two reactions below Q is shown by the dashed lines, contours for constant P_{H_2O}. These contours are considered here as if they are identical with P_{eH_2O} contours (see Figure 8-5).

Figure 8-6b shows the phase fields for a specific assemblage of Ms + Qz + V where $P_{H_2O} = P_{total}$ up to 3 kb, but with P_{H_2O} remaining constant at 3 kb as P_{total} increases further. This is not entirely an arbitrary choice because D. S. Korzhinskii believes that during regional metamorphism, $P_{fluid} = P_{total}$, but P_{H_2O} reaches a maximum value of only 1.5 or 2 kb through the Earth's crust. The Ms + Qz + V assemblage dissociates along the contour for $P_{eH_2O} = 3$ kb which is situated at a temperature below Q. Therefore muscovite dissociates below the solidus; melting begins at the solidus for the Qz-Or-Si-V assemblage. At a depth corresponding to the invariant point, Q, the estimated temperature difference between dissociation and beginning of melting is about 150°C; with $P_{H_2O} = P_{total}$ Ms + Qz + V melts at Q without dissociation.

Figure 8-6b also shows the curve for Ms + Qz transferred as a dashed line from Figure 8-6a. The vapor-absent assemblage remains stable right up to the dashed line. This is at a

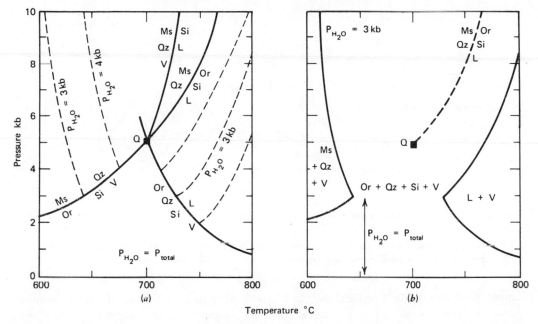

Figure 8-6. Dissociation and melting reactions involving muscovite, quartz, and aqueous vapor. (*a*) Condition $P_{H_2O} = P_{total}$, in part schematic (based on Lambert et al; 1969). Dashed lines (schematic) contour divariant surfaces for $P_{H_2O} < P_{total}$ (compare Figure 8-5). (*b*) Condition for excess vapor, with $P_{H_2O} = P_{total}$ up to 3 kb, and P_{H_2O} constant at 3 kb for higher values of P_{total}. Dissociation temperatures are lowered and melting temperatures are raised. The dashed line, transferred from figure (*a*), shows the melting temperature for muscovite-quartz mixtures with no vapor phase present. New mineral abbreviations: Ms = muscovite, Si = sillimanite.

lower temperature than the melting curve for the vapor-present assemblages with $P_{H_2O} < P_{total}$. We are accustomed to thinking that the effect of aqueous pore fluids is to lower the melting temperatures of rocks, but if $P_{H_2O} < P_{total}$ because of a multicomponent pore fluid Figure 8-6*b* shows that a pore fluid can cause melting to occur at a temperature higher than it would if there were no fluid phase present at all.

MAGMA GENERATION IN CRUSTAL ROCKS

The variables affecting the temperature of melting of crustal rocks and the composition of the liquid produced are (a) mineralogy especially the Ab/Or ratio in alkali feldspar, the An/Ab ratio in plagioclase, and the feldspar/quartz ratio, (b) total pressure, (c) P_{H_2O}, and (d) the amount of water. If no pore fluid is present the temperature of melting depends on the dissociation temperature of the hydrous minerals (Type II system).

Mineral Variation in Crustal Rocks

The greater part of the Earth's crust is composed of the calc-alkaline rock series or their metamorphosed equivalents (Tables 7-2 and 7-3). The initial liquids produced by their partial fusion approach compositions in the system $NaAlSiO_4(Ne)$-$KAlSiO_4(Kp)$-$SiO_2(Qz)$ which is illustrated in Figure 8-7*a*. This was termed the Residua System by Bowen because fractional crystallization of

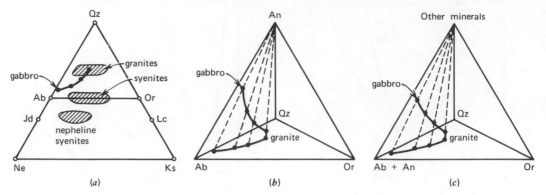

Figure 8-7.   Mineral variation (normative) in dominant crustal rocks, the calc-alkaline series gabbro-diorite-granodiorite-granite and their metamorphic equivalents. (*a*) Plotted in the Residua System. (*b*) Plotted in the Granodiorite System, and projected down onto the Granite System, a portion of the Residua System. (*c*) Plotted in terms of feldspars, quartz, and all other minerals.

magmas produces residual liquids approaching compositions in this system. The alkali feldspar join is a thermal divide separating SiO_2-saturated liquids of granitic composition from SiO_2-undersaturated liquids of the nepheline syenite clan. The compositions of syenites and trachytes, expressed in terms of their normative components, project close to the alkali feldspar join. The SiO_2-saturated part of the Residua System has been known as the Granite System since the work of Tuttle and Bowen on the system Ab-Or-Qz-H_2O.

In the calc-alkaline series from granite, granodiorite, tonalite (quartz diorite) to gabbro there is a progressive increase in the anorthite component of the plagioclase feldspar. Addition of anorthite (An) to the Granite System produces the Granodiorite System shown in Figure 8-7*b*. The four representative rock types are plotted in this tetrahedron in terms of their normative feldspars and quartz recalculated to 100%. They lie close to a curved line within the tetrahedron, and if they are projected down through the An apex on to the base, as illustrated, the result is a triangular diagram showing the rock series in terms of normative plagioclase-Or-Qz. The rock series is shown projected in just this way on the base of Figure 8-7*c*, where all other normative

components of the rocks (the mafic minerals) are represented by the apex of the tetrahedron. This shows the increasing content of mafic minerals in rocks ranging from granite through to gabbro.

The phase relationships in the system in Figure 8-7 can be determined by three approaches:

1. Experimental study of synthetic systems in the presence of water under pressure (dry reactions are too sluggish).

2. Experimental study of rocks and rock series occupying different positions within the tetrahedron.

3. Petrographic and mineralogic study of plutonic and volcanic rocks in which the compositions of phenocrysts and quenched liquid or groundmass can be determined.

Approaches 1. and 2. are considered in this chapter. I. S. E. Carmichael in 1963 presented a detailed study involving the third approach with respect to the Granodiorite System.

The Granite System

The phase relationships in the Granite System are sufficiently well known that it can be used to illustrate the effects of the variables influencing melting temperatures and liquid compositions.

Figure 8-8*a* shows a schematic version of the system Ab-Or-Qz-H_2O at 1 kb pressure. There are no hydrous minerals represented and so no Type II assemblages. The approximately triangular surface *efdgc* is the vapor-saturated liquidus surface giving the compositions of silicate liquids saturated with H_2O. Starting assemblages with compositions above *efg* are Type IV and those below *efg* are Type III. For all subsolidus assemblages and for all vapor-present hypersolidus assemblages $P_{H_2O} = P_{total}$, and for all vapor-absent hypersolidus assemblages $P_{H_2O} < P_{total}$. The surface *ijh* gives the compositions of vapors coexisting with liquids on the surface *efg*. The volume between these surfaces is a miscibility gap between silicate liquids and aqueous vapors.

There are temperature minima, *m* and m_1, on the phase boundaries for the system Ab-Or-H_2O. These minima are reflected in the dry system Ab-Or-Qz by the minimum *M* on the field boundary *ab* which separates fields for the primary crystallization of alkali feldspar and a silica phase, and on the vapor-saturated quaternary liquidus surface by the minimum M_1 on the field boundary

cd. The surface *abcd* separates the phase volumes for the primary crystallization of alkali feldspars and a silica phase. There is a temperature trough, or minimum, on this surface connecting the points *M* and M_1 (actually this dotted line need not begin and end at these two points but near enough for our purposes).

Effect of mineralogy and water content.

The composition of the first liquid depends on the ratio of Ab/Or. For quartz and a feldspar with composition between Ab-Or the first liquid has a composition on the boundary *cd* and approaching M_1 for a wide range of feldspar compositions; it is identical with M_1 only for one particular feldspar composition. For most Type IV assemblages a considerable amount of liquid with composition on the boundary cM_1d forms within a narrow temperature interval. When all of the feldspar or quartz dissolves the liquid composition changes across the vapor-saturated liquidus surface towards the bulk composition of the mixture with a considerable increase in temperature producing much less liquid. The feldspar/quartz ratio determines whether later liquids are in area *cdg* or *cdef*.

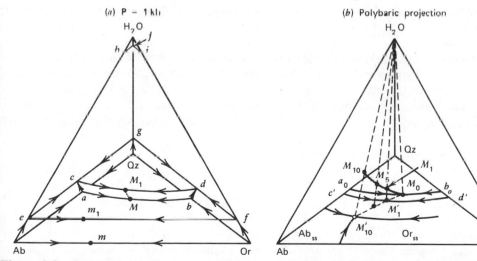

Figure 8-8. Schematic liquidus phase relationships in the Granite System, Ab-Or-Qz-H_2O. (*a*) Polythermal isobar at 1 kb. (*b*) Polythermal polybaric projection showing compositional variation of specific points as a function of pressure. *c'-d'* is a projection of *c-d* in figure *a*.

For a water-deficient Type III mixture with bulk composition below the surface efg the composition of the first liquid developed is also on the boundary cM_1d, close to M_1 for most bulk compositions. Fusion continues until all of the H_2O available has been used to saturate the liquid. Then with continued fusion the liquid composition migrates downwards across the surface $cdab$ in the vapor-absent region with P_{H_2O} decreasing. Further changes in composition within the H_2O-undersaturated volume beneath surface efg depend upon the bulk composition of the starting mixture.

Whatever the water content of a mixture the first liquids developed are H_2O-saturated with compositions along cd and clustering around M_1. They are independent of the amount of H_2O and dependent only upon the feldspar composition. For the H_2O-deficient systems, however, the later liquids developed are constrained to follow paths on the surface $cdab$.

At any fixed pressure compositions on the boundary cd, when expressed in terms of anhydrous components, differ from those of the boundary ab for the dry system Ab-Or-Qz. Therefore as liquid compositions migrate down the surface $cdab$, with increasing temperature and decreasing H_2O content, their compositions change with respect to compositions on the boundary cd. The compositions of H_2O-undersaturated liquids coexisting with alkali feldspar and quartz are thus dependent not only upon the Ab/Or and feldspar/quartz ratios but also upon the amount of H_2O present in the mixture.

Effect of total pressure and of P_{H_2O}.
With increasing pressure more water dissolves in the silicate liquids, more solids dissolve in the vapor phase, and the surfaces efg and ijh therefore approach each other. With increasing pressure the boundary cd of Figure 8-8a migrates towards the side ef, and the position of the quaternary minimum M_1 migrates towards the corner e. In the dry system similarly with P_{H_2O} remaining zero

the position of the boundary ab changes with changing pressure but to a lesser extent.

Figure 8-8b shows selected composition changes caused by increasing total pressure from 1 bar to 10 kb. For vapor-present assemblages the change is caused in effect by increasing P_{H_2O} from zero to 10 kb. The minimum temperature on the boundary a_0b_0 at 1 bar is at the point M_0. If P_{H_2O} is increased to 1 kb the boundary moves to cd in Figure 8-8a with minimum M_1. Figure 8-8b shows both the position of M_1 within the tetrahedron, and that of the 1 kbar boundary in terms of its anhydrous components by its projection $c'M_1'd'$.

The effect of increasing total pressure to 1 kb with P_{H_2O} remaining at zero changes $a_0M_0b_0$ (Figure 8-8b) to aMb (Figure 8-8a). The boundaries a_0b_0, ab, and $c'd'$ (projected) do not coincide. In particular the lines M_0M_1' (projected) and M_0M do not coincide. Therefore the effect of P_{H_2O} on the anhydrous composition of the quaternary temperature minimum, M_0M_1', differs from that of total pressure, M_0M.

Figure 8-8b shows the positions of the quaternary temperature minima for P_{H_2O} zero (M_0), 1 kb (M_1), 5 kb (M_5), and 10 kb (M_{10}), and their projections on to the anhydrous base. At a pressure of about 5 kb the solidus temperature is lowered enough that two feldspars are stable with liquids on the vapor-saturated surface, and at higher pressures the minimum is replaced by a quaternary eutectic (e.g. M_5 and M_{10}) but this does not affect the present discussion. The change in pattern of field boundaries is shown by the projections for 10 kb meeting at the projected eutectic M_{10}'.

Consider an assemblage containing only 1% water at 10 kb with feldspar compositions such that the first liquid formed by partial fusion is close to the eutectic M_{10}; this contains about 15% dissolved water. All of the available water is concentrated in this first liquid, and further melting is accomplished by the liquid changing composition with decreasing water content in the vapor-absent region; P_{H_2O} decreases. The liquid path extends downward from the vicinity of M_{10}, following a path corresponding closely to M_1M in Figure 8-8a. The path probably follows the same general trend as $M_{10}M_0$ in Figure 8-8b but it is certainly not identical with it. Thus with progressive fusion of a Type III assemblage an originally H_2O-saturated, albite-rich liquid (near M_{10}) becomes deficient in water with P_{H_2O} decreasing as the liquid becomes more under-saturated and with the liquid composition becoming enriched in quartz/feldspar. The later stages of fusion are controlled by the original quartz/feldspar ratio.

These few examples are sufficient to demonstrate that the compositions of liquids developed by partial fusion in silicate-water systems are dependent not only upon the silicate mineralogy but also upon the total pressure, and independently upon P_{H_2O}, if (a) this becomes less than P_{total} because of low water content, or (b) if the vapor phase is diluted by nonreacting volatile components.

The Granodiorite System: Effect of Mineralogy

The Granite System provides a model of limited applicability because the plagioclase feldspar in a rock has a marked influence not only upon the temperature and composition of the initial liquid, but also on the composition of liquids developed with progressive fusion.

Figure 8-9 shows the phase relationships in the Granodiorite System in the presence of excess water at 1 kb pressure with liquid compositions expressed in terms of anhydrous components. No vapor-absent relationships are represented. The complete tetrahedron corresponds to the vapor-saturated surface *efg* in Figure 8-8a. The phase diagram for Ab-Or-Qz on the base of the tetrahedron is the surface *efg* of Figure 8-8a projected downward radially through the apex H_2O onto the anhydrous base; the compositions of liquids cM_1d in Figure 8-8a are thus

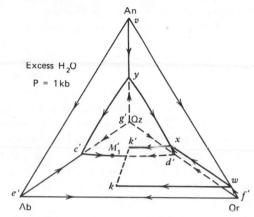

Figure 8-9. Schematic isobaric liquidus phase relationships with excess vapor at 1 kb pressure for the Granodiorite System (Figure 8-7b).

plotted in terms of their anhydrous components as $c'M_1'd'$ from Figure 8-8b. The other triangular sides of the tetrahedron in Figure 8-9 are similar projections. Rising from the field boundary $c'M_1'd'$ to the edge An-Qz is a surface separating the volume for the primary crystallization of quartz (in the presence of excess vapor) from the volumes for the primary crystallization of feldspar. The surface *wxkk'* separates the small alkali feldspar volume from the large plagioclase feldspar volume.

The field boundary xk' gives the compositions of liquids coexisting with vapor and with three crystalline phases: an alkali feldspar, a plagioclase feldspar, and quartz. It extends from x towards the temperature minimum M_1' with a very small content of An. The three surfaces emanating from this field boundary give the compositions of liquids coexisting with vapor and two of the crystalline phases; and the volumes give the range of liquid compositions coexisting with vapor and one of the crystalline phases. At 1 kb the boundary xk' terminates at the critical point k' where the two coexisting feldspars become coincident at the top of the solvus. Similarly the quinary surface *wxkk'* terminates at the critical line kk' situated just above the base of the tetrahedron.

Figure 8-7*b* shows that the compositions of calc-alkaline igneous rocks, represented in terms of normative quartz and feldspars, project into the plagioclase volume of Figure 8-9. Consider a granodiorite composed of plagioclase feldspar, alkali feldspar, quartz, and mafic minerals in the presence of excess water (Type IV system). The mafic minerals can be neglected in the early stages of fusion. The composition of the rock in Figure 8-9 is represented by a point on the triangle produced by joining the composition of the quartz and two feldspars. The composition and temperature of the first liquid developed for this rock, and for any other rock composition represented in the same triangle, is given by the apex of the four-phase tetrahedron plagioclase + alkali feldspar + quartz + liquid; the liquid lies on the field boundary *xk'*.

With increasing temperature, the liquid composition changes up the boundary towards *x*, with the plagioclase feldspar and alkali feldspar changing composition simultaneously, until one of the crystalline phases is dissolved; this is probably alkali feldspar for a granodiorite. Further fusion causes the liquid to follow a curved path across the surface *c'k'xy*, with simultaneous solution of quartz and plagioclase until all of the quartz is dissolved. Then the liquid follows a curved path through the plagioclase volume towards the bulk composition of the rock, while the remaining plagioclase becomes further enriched in anorthite.

For a given plagioclase feldspar composition the composition of the coexisting alkali feldspar is fixed; it does not change much with change in plagioclase composition. The first liquid produced does not have a composition at M_1', the minimum liquidus temperature in the system, nor at the composition *k'*, the minimum temperature on the liquidus boundary for the coexistence of three crystalline phases. The first liquid lies on the boundary *k'x*, and it is sensitively dependent upon the composition of the plagioclase. The effect of increasing An content in original plagioclase

is to increase the Or content of the first liquid developed. In the limit for a plagioclase of composition An the first liquid is *x*. Liquid *k'* would develop from a three-phase assemblage with a specific albite-rich plagioclase.

A small temperature change while the liquid composition changes along the boundary *k'x* produces a high proportion of liquid. Less liquid is produced within similar temperature intervals after the liquid leaves this boundary. Thus, the products of partial melting of crustal rocks containing the components of the minerals orthoclase, plagioclase and quartz tend to be concentrated along the field boundary *k'x*, which remains quite close to the Residua System so that the liquids are essentially granitic. A high proportion of granitic liquid is produced within a small temperature interval. The temperature of beginning of melting increases for rocks with plagioclase feldspars richer in anorthite (Figures 8-3 and 8-7*b*).

Alkali feldspar is less widely distributed in crustal rocks than quartz and plagioclase, but there are many rocks not containing free alkali feldspar which nevertheless contain the Or component in other minerals. Plagioclase-quartz-muscovite gneisses, for example, undergo reactions involving the breakdown of muscovite and the release of the component Or at temperatures close to those for the beginning of melting (Figure 8-6). Similarly breakdown of biotite can yield the Or component for melting. The addition of hydrous minerals to assemblages in Figure 8-9 introduces Type II systems as shown for the system granodiorite-H_2O in Figure 8-11.

Now let us consider briefly how the illustrated phase relationships might change under other conditions. With increasing P_{H_2O}, the positions of the surfaces will change, as illustrated by the migration of the boundary *c'd'* shown in projection in Figure 8-8*b*. If $P_{H_2O} < P_{total}$ with no change in P_{total} all temperatures are increased (Figure 8-5): the positions of the surfaces are changed in the same sense as but not coincident with the changes that are produced by decreasing P_{H_2O} with the condition $P_{H_2O} = P_{total}$. If the amount of water available for a mineral assemblage is less than that required for saturation of the liquid, at a given confining pressure, then the first liquids developed are H_2O-saturated on the line $k'x$ until all the water is used up, and then the H_2O-undersaturated liquids ($P_{H_2O} < P_{total}$) follow paths through narrow volumes close to the surfaces shown in terms of anhydrous components but not coincident with them.

The System Granodiorite-Water

Figure 8-10 shows partly schematic phase relationships for a granodiorite. Figure 8-10b shows the Type I system, dry, and Figure 8-10a shows the effect of excess water in the Type IV system. The Type IV results show a standard pattern for calc-alkaline igneous rocks of intermediate composition, although in some rocks pyroxene coexists with hornblende down to subsolidus temperatures. The minerals quartz, potash-feldspar, and the sodic portion of the plagioclase yield a granite liquid within a narrow temperature interval above the solidus, and biotite and hornblende persist to temperatures where they dissociate. The hornblende and the more lime-rich plagioclase coexist with granite liquid through a considerable temperature interval. Piwinskii and I reviewed these results in 1970.

The equivalent of Figure 8-4b for granodiorite-water provides the basis for models of magma generation. The assemblages of Types II and III up to the saturated liquid at c are significant. Figure 8-11 shows an enlarged portion of Figure 8-4b constructed from available data for the system granodiorite-water including Figure 8-10. The temperatures of phase boundaries in the vapor-present fields were located from the

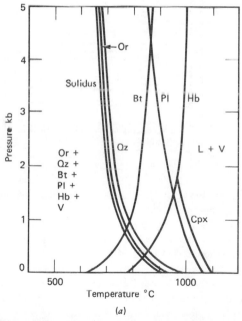
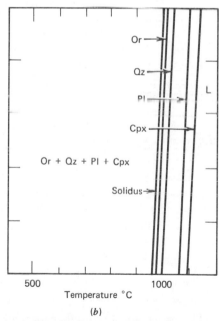

Figure 8-10. P-T projections contrasting water-excess and dry melting relationships for a granodiorite, representative of the dominant rocks of the Earth's crust (after Robertson and Wyllie, 1971). (a) Water-excess, extrapolated from data by Pinwinskii and Wyllie (1968). (b) Inferred melting relationships for the anhydrous equivalent of granodiorite. New mineral abbreviations: Bt = biotite, Hb = hornblende, Pl = plagioclase, Cpx = clinopyroxene. (From Amer. Jour. Sci., with permission.)

boundaries at 2 kb in Figure 8-10a. Isothermal boundaries for the solidus, liquidus, and the upper limit of each mineral extend into the water-deficient region as far as the saturation boundary, b–c. The shape of the saturation boundary was determined from estimates of the percentage of liquid at each temperature between the solidus and liquidus with excess vapor and the known solubilities of water in similar silicate liquids at each temperature. The dry rock at this pressure melts between the points n' and n taken from solidus and liquidus in Figure 8-10b.

Field boundaries in the vapor-absent region of Figure 8-11 were interpolated from the water-absent points between n'–n to the corresponding points on the saturation boundary between b–c. There is very little variation possible from the estimated curves shown for feldspars and quartz. More variation is possible for biotite and hornblende, because they have no stability on the water-absent axis, and their dissociation temperatures in the vapor-absent region have not been measured experimentally. There are two alternatives: congruent or incongruent dissociation. Figure 8-11 illustrates the former; each hydrous mineral phase boundary reaches a temperature maximum, m, at the boundary between Types II and III compositions or within the Type II range, and with further decrease in water content the stability curve is lowered to a point, p, and then to the anhydrous axis.

The major changes in phase relationships in Figure 8-11 caused by passing from water-excess to water-deficient conditions with progressive decrease in water content include the following:

1. The liquidus temperature increases.

2. The temperature interval between solidus and liquidus increases.

3. The temperature stability of hydrous minerals (biotite and hornblende) increases.

4. The temperature interval for the coexistence of two feldspars with liquid increases.

5. Quartz is stable with liquid through a significant temperature interval.

6. The amount of liquid produced within a given temperature interval above the solidus decreases.

Similar schematic diagrams have been constructed for the system gabbro-water at 10 and 20 kb in Figure 8-18.

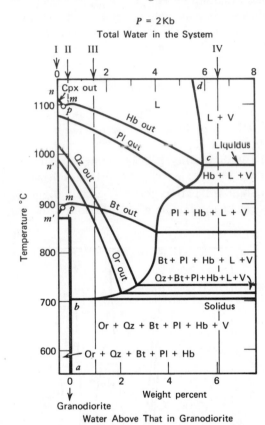

Figure 8-11. Schematic isobaric *T–X* section for granodiorite-water at 2 kb pressure. Composition I is for the dry dehydrated rock (Figure 8-10b), and composition IV corresponds to the water-excess rock (Figure 8-10a). Phase relationships are vapor absent to left of line a–b–c–d, and compositions II and III are thus water-deficient at this pressure. (After Robertson and Wyllie, 1971.) This corresponds to the generalized Figure 8-4b. Congruent reaction is indicated for Bt and Hb (m); incongruent reactions provide alternative arrangements. (With permission from Jour. Geology. Copyright 1971, University of Chicago Press.)

Anatexis in the Crust

Figure 8-11 provides a guide to the pattern of magma generation from the dominant calc-alkaline rock types of the continental crust. The confining pressure of 2 kb corresponds to a depth of about 8 km; the requisite temperatures for fusion would not be attained at this depth but the essential pattern remains unchanged for greater depths. There are two models corresponding to fusion paths for assemblages III and II.

The Type III composition represents an intermediate rock or its metamorphic equivalent with 1% pore fluid, which is probably a generous estimate for the fluid content of a metamorphic rock in the lower crust. Melting begins with the fusion of quartz and two feldspars and the solution of vapor, producing a water-saturated liquid of granite composition. The solidus temperature is controlled by P_{H_2O} and is independent of the amount of water. The amount of water-saturated liquid produced is controlled by the water content. A small increase in temperature above the solidus produces much additional water-saturated liquid. The pore fluid disappears within a few degrees of the solidus, except for components other than water which are insoluble in the silicate liquid or which have very low solubilities; carbon dioxide is such a component but we need not consider its effects (Figure 8-5). The temperature interval within which the water-saturated liquid exists (the temperature interval between the solidus and the saturation boundary b–c) increases with water content but it remains small for any reasonable amount of pore fluid.

With further increase in temperature the liquid becomes progressively more unsaturated with water. Within the vapor-absent region the percentage of liquid developed as a function of temperature is less than that for excess vapor. For a wide temperature interval, about 200°C in Figure 8-11, the undersaturated granite liquid coexists with quartz, alkali feldspar, plagioclase feldspar, and more refractory minerals until the quartz and alkali feldspar dissolve. Then through another wide temperature interval, the liquid composition becomes closer to that of a granodiorite as the granite liquid dissolves the more refractory minerals. The rock is not completely melted until a temperature of nearly 1100°C is attained.

In deep crustal environments where a granitic liquid has formed and migrated upward carrying with it the original pore fluid the remaining rocks are Type II systems. Subsequent anatexis can then occur only at considerably higher temperature, corresponding to m' in Figures 8-4h and 8-11 where water is released by dissociation of hydrous minerals. The liquid so produced is undersaturated with water, and its composition is different from that produced in Type III models. There may be major differences in the products of anatexis of crustal rocks depending upon whether or not a pore fluid is present.

Water-saturated granite liquids produced by anatexis in the crust can exist only for a few degrees above the solidus. It appears that the normal product of anatexis is a mush composed of crystals and water-undersaturated granite liquid. The amount of pore fluid controls the amount of liquid generated from the granitic minerals. Water pressure for a liquid within the vapor-absent region may be considerably less than the load pressure and therefore upward migration of such a liquid, or liquid-crystal magma, can proceed without excessive crystallization until the load pressure is decreased to a level approaching the water pressure in the undersaturated liquid.

High temperatures are required for complete melting of a granodiorite, about 1000°C with excess water and nearly 1100°C for more reasonable crustal models (Types III and II), even under the most favorable conditions with water being the only volatile component. Such high temperatures make it improbable that liquid magmas of intermediate composition can be generated by anatexis in the crust.

For consideration of the partial fusion of

basic rocks in the crust the experimental determination of diagrams such as Figure 8-18b is required.

MAGMA GENERATION IN THE MANTLE

The basaltic magmas erupted at the Earth's surface were derived originally by partial fusion of the mantle but, as shown by Figure 4-5, many things may have happened between magma generation and eruption. Figure 6-18b shows the general pattern for melting of mantle peridotite in the presence of traces of water. The melting interval consists of two parts: the low temperature band shows where traces of liquid are produced by water, and the upper portion above the dashed line shows where significant melting occurs producing basaltic or picritic liquids. Factors affecting the composition of the liquid at its source include:

1. The mineralogy which varies as a function of pressure and temperature.

2. The depth (pressure).

3. The temperature interval above the solidus which controls the percentage of melting.

4. The amount of water present and its physical state (in pore fluid or combined in hydrous minerals).

Other factors affecting its composition during transit from source to surface include: (a) the speed of movement towards the surface, (b) fractionation during uprise, (c) fractionation at specific depths due to interruption of uprise, (d) changes in P_{H_2O} and f_{O_2} during crystallization.

A major objective of petrology is to determine the composition of liquids developed by partial fusion of mantle peridotite, as a function of pressure (depth), temperature (and percentage melting), and composition including especially the water content; and to determine the changes in composition of

these liquids during crystallization at various pressures. This is an enormous experimental task.

Mineralogy of Mantle and of Basalts

We saw in Chapter 6 that the peridotite of the upper mantle consists of olivine (Ol), orthopyroxene (Opx), clinopyroxene (Cpx), and an aluminous mineral, which is plagioclase (Pl), spinel (Sp), or garnet (Ga), depending upon the depth (Table 6-7, Figure 6-10). For some bulk compositions the spinel in spinel peridotite dissolves in the pyroxenes before melting begins, leaving an assemblage of olivine plus aluminous pyroxenes (A-px). The system CaO-MgO-Al_2O_3-SiO_2 contains representatives of all these minerals and has been widely used to illustrate mantle reactions.

Figure 8-12a shows the compositions of the mantle minerals plotted on the ternary systems bounding the quaternary tetrahedron. Figure 8-12b shows the garnet solid solution series, Py-Gr, on the plane Wo-En-Al_2O_3 extending through the tetrahedron. Another geometrical feature to note is that the garnet join intersects the mid point of the An-Fo join at the composition Gr_1Py_2 (Figure 8-12d). Partial fusion of the assemblage Fo + En + Di + (An, Sp, or Ga) in the system CaO-MgO-Al_2O_3-SiO_2 yields a liquid representing basalts, the isobaric invariant liquid at A in Figure 8-13. Figure 8-12c shows the relationship of the inner tetrahedron Fo-En-Di-An to the other aluminous minerals, garnet solid solution and calcium-tschermak's molecule (CaTs), and to quartz. This is analogous to the simple basalt tetrahedron of Figure 8-12d introduced

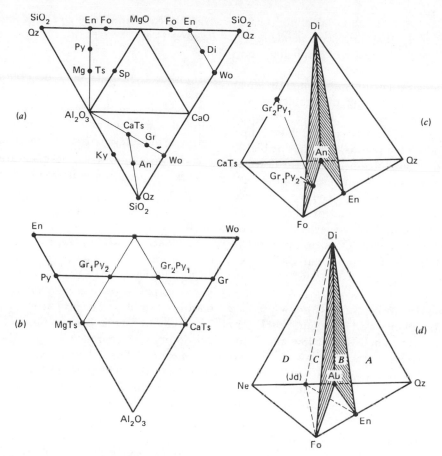

Figure 8-12. Model mantle systems, showing mantle mineralogy. (*a*) System CaO-MgO-Al$_2$O$_3$-SiO$_2$; the tetrahedron has been opened up at apex SiO$_2$ for clarity. (*b*) Garnet and Tschermak's molecule joins on the plane Wo-En-Al$_2$O$_3$ through the tetrahedron represented in (*a*). (*c*) Relationship of garnet join to join Fo-An in the basaltic compositional region of (*a*). (*d*) Simplified basalt tetrahedron (after Yoder and Tilley, 1962); compare (*c*). Major basaltic types occupy volumes *A, B, C,* and *D.* The planes separating these volumes are replaced by other planes at pressures where jadeite becomes stable at the expense of albite. (From Jour. Petrology, with permission.)

by Yoder and Tilley in 1962 as a means for classifying the major basaltic types in terms of their normative mineralogy.

Figure 8-12*d* includes representatives of the basaltic minerals, olivines (Fo), pyroxenes (En, Di), and plagioclase (Ab); the line nepheline (Ne)-quartz (Qz) shows the degree of silica-saturation. In a generalized basalt tetrahedron these iron-free minerals would be replaced by the normative minerals of the rock. The planes En-Di-Ab (plane of silica saturation) and Fo-Di-Ab (critical plane of silica undersaturation) divide the tetrahedron into three volumes, each of which encloses a specific group of basaltic types. The tholeiites have hypersthene (Hy) in the norm; these are divided into quartz-normative tholeiites (silica-oversaturated in volume *A*) and olivine-normative tholeiites (silica-saturated in volume *B*). The silica-undersaturated alkali olivine basalts lie in volume *C* on the Ne side of the critical plane Fo-Di-Ab. High-alumina basalts plot closer to the plagioclase composition (Ab) than the normal basalts. Highly alkaline under-saturated rocks (basanites and nephelinites) contain more normative nepheline and occupy the volume near *D*.

The plane Fo-Di-Ab is a thermal divide at low pressures indicating that tholeiites and alkali olivine basalts cannot be related to each other by liquid-crystal relationships. Yoder and Tilley pointed out that at higher pressures, where plagioclase broke down to yield jadeite and other minerals (Figure 8-2), this barrier would be replaced by another as shown by the dashed lines in Figure 8-12*d*. M. J. O'Hara in 1968 reviewed the limitations which such divides place upon petrogenetic theories. In the system CaO-MgO-Al_2O_3-SiO_2 (Figures 8-12*a* and 8-12*c*) he noted

1. The existence of two low-pressure divides, the planes Fo-Di-An and Fo-Qz-An.

2. The plane En-Di-An at low and intermediate pressures between about 5 and 15 kb.

3. The eclogite plane garnet (Py-Gr)-pyroxene(Di-En) at high pressures above 27 kb.

4. Possibly another high pressure divide Fo-Ga-Px above 40 kb.

Melting Relationships of Peridotites and Basalts

Information about the relationships of peridotites to basalts may be obtained by two approaches:

1. By studying the melting relationships of mineral assemblages in portions of the synthetic model systems shown in Figure 8-12.

2. By working with a series of natural rocks, or with glasses of the same compositions.

The first approach is illustrated in Figure 8-13 and the second in Figure 8-14.

Synthetic model systems. Figure 8-13 is a portion of the system CaO-MgO-Al_2O_3-SiO_2 (Fo-Di-Qz-Al_2O_3 in Figure 8-12*b*) showing paths of melting for the model mantle assemblage Fo + Opx + Cpx + Ga. The range of bulk compositions of mantle peridotites is represented by the small volume *P*. The shaded surface *ABCD* separates the volumes for the primary crystallization of Ol and Opx; the field boundary *AC* gives the compositions of liquids where Ol and Opx coexist also with Cpx; the field boundary *AB* gives the compositions of liquids where Ol and Opx

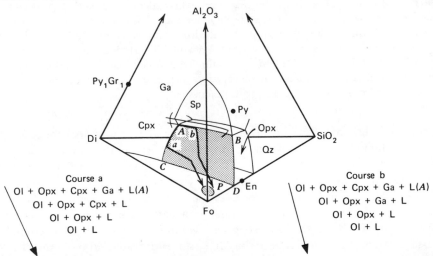

Course a
Ol + Opx + Cpx + Ga + L(*A*)
Ol + Opx + Cpx + L
Ol + Opx + L
Ol + L

Course b
Ol + Opx + Cpx + Ga + L(*A*)
Ol + Opx + Ga + L
Ol + Opx + L
Ol + L

Figure 8-13.   Pattern of melting of a simplified garnet peridotite in the system CaO-MgO-Al_2O_3-SiO_2 (after Kushiro, 1969). Compare Figure 8-12*a*. Peridotite compositions occupy shaded volume P; first liquids form at *A*, and path followed by liquid varies according to the peridotite mineralogy. (From *Tectonophysics*, with permission.)

coexist also with Ga solid solution (note the points for Py and Py_1Gr_1 and see Figure 8-12b); the point A is an isobaric eutectic giving the composition of the liquid formed from the four minerals. The liquid does not change composition or temperature until one of the minerals is dissolved, and then with temperature increasing the liquid will follow either a path such as a or b depending upon the compositions and relative proportions of the Ga and Cpx. The compositions of liquids formed by partial melting of similar mineral assemblages may diverge quite significantly, following either AB or AC because of minor variations in the proportions of Cpx and Ga.

O'Hara and Yoder reported in 1967 the results of melting experiments for mixtures on the join Di-Py and adjacent parts of Wo-En-Al_2O_3 (Figures 8-12b and 8-13) at 30 kb; the results were matched by parallel experiments using similar combinations of natural analyzed minerals separated from eclogite and garnet peridotite nodules from kimberlite. They concluded that the initial liquid produced by melting the assemblage Fo + En + Di + Ga at high pressures (liquid A in Figure 8-13) is silica-undersaturated and probably picritic in character; it is analogous to a hy-normative picrite basalt; orthopyroxene has a reaction relationship with this liquid. These results form the basis for the scheme of generation and fractionation of basaltic magmas in Figure 8-15.

In 1968 I. Kushiro published results of experimental studies in sub-systems Mg_2SiO_4-SiO_2-X, where X represents $CaMgSiO_4$, $CaAl_2O_4$, $MgAl_2O_4$ and $NaAlSiO_4$, in the pressure range 7 to 40 kb. The idea was to determine the effect of increasing pressure on the composition of the univariant liquid A in Figure 8-13 and, in the composition join where $X = NaAlSiO_4$, to determine as well the distribution of alkalis between crystals and liquid as a function of pressure. In each composition join the liquid produced at the beginning of melting coexists with Fo, En solid solution, and one other mineral (Di

solid solution, Sp, Ga, Ab, or jadeitic pyroxene). Three effects were recorded as a function of increasing pressure:

1. The two pyroxene fields expand at the expense of the olivine, so that an increasing proportion of olivine dissolves in the first liquid produced.

2. The first liquid shifts in the direction of decreasing silica content.

3. The first liquid with $X = NaAlSiO_4$ changes from Qz-normative below 10 kb (volume A in Figure 8-12d) to Ol-Hy-normative within the interval 10 to 30 kb (volume B) to Ne-normative at 30 kb (volume C).

4. At a given pressure increasing temperature enriches the liquid in dissolved olivine, producing picritic liquids (paths a and b in Figure 8-13).

From these results Kushiro concluded that quartz tholeiites could be generated by partial fusion of the mantle between about 40 and 100 km, olivine tholeiites between 40 and at least 130 km, and the nepheline-normative basaltic liquids could be produced only at depths greater than 100 km. The results support Kuno's model of magma generation depicted in Figure 8-1d.

Natural peridotite and basalt. The general pattern of the melting interval for peridotites and basalts, both dry and in the presence of water, has been illustrated in Figures 5-15, 6-12, 6-13, 6-18 and 6-19. Following the early work of Yoder and Tilley in 1962 it was 1967 before more experimental details became available.

Figure 8-14a shows results for a garnet peridotite nodule published by K. Ito and G. C. Kennedy in 1967, and Figure 8-14b shows results for a tholeiitic basalt glass published in a companion paper by L. H. Cohen, Ito, and Kennedy. At mantle pressures the Cpx, Sp, and Ga of the peridotite pass into the liquid within 50°C above the solidus producing a wide field for the coexistence of Ol + Opx with liquid. This is

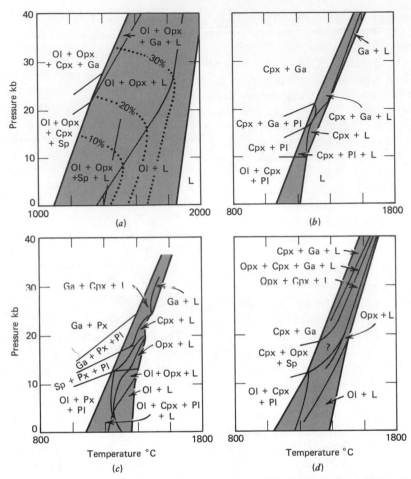

Figure 8-14. Experimentally determined melting relationships for natural peridotite and basalts. (*a*) Peridotite (after Ito and Kennedy, 1967). The dotted lines show percentage of normative olivine in liquid from estimate by O'Hara (1968). (*b*) Tholeiitic basalt (after Cohen et al; 1967). (*c*) Tholeiitic basalt (after Green and Ringwood; 1967; phase boundaries estimated on basis of their data points). (*d*) Olivine-rich tholeiite (after Ito and Kennedy, 1968). Note that there are three different basalts, but the compositions of *b* and *c* do not differ greatly. (With permission from Amer. Jour. Science, and Contr. Mineral. and Petrol. Berlin-Heidelberg-New York: Springer 1967, 1968.)

consistent with results reported by I. Kushiro, Y. Syono, and S. Akimoto in 1968 for a spinel peridotite nodule. Once the first liquid forms at 20 kb there is not much increase in the amount of liquid between 1320 and 1600°C, and presumably little change in composition. In 1968 O'Hara concluded that melting should proceed in step-like form. In contrast Green and Ringwood assume that the percentage of liquid formed will be nearer to a linear function of

temperature above the solidus (Figures 8-17 and 8-19). At 20 and 40 kb the liquid is picritic as shown by the dotted lines. These are estimates by O'Hara for the normative olivine content of the liquid in a partially melted peridotite: with increasing pressure the liquid dissolves an increasing quantity of olivine.

The olivine tholeiite glass of Figure 8-14*b* has 14% normative olivine; subsolidus results have been reviewed in Figures 5-14*b*,

5-15*b*, and 5-17. The solidus of the basalt and eclogite is coincident with that of the peridotite within the limits of experimental measurement, and the melting interval is almost exactly the same as the interval for the solution of Cpx + Ga in the peridotite. The liquidus phase is Ol to about 9 kb, Cpx between 9 and about 24 kb, and Ga at higher pressures. The olivine tholeiite liquid with composition similar to that of the oceanic tholeiites is therefore not in equilibrium with Ol-bearing peridotite at pressures greater than 9 kb or about 30 km and it cannot be an unmodified partial fusion product of peridotite from depths below 30 km. Ito and Kennedy concluded that this cast serious doubt that basalts are primary magmas from the mantle and favored O'Hara's proposal that they are residual liquids of well advanced crystal fractionation processes. They shared O'Hara's view that the common basalts may be derived from deep-seated picritic liquids which undergo extensive olivine fractionation during uprise (Figure 8-1*e*).

In 1967 D. H. Green and A. E. Ringwood published results for a glass of olivine tholeiite composition, with 20% normative olivine; this is equivalent to the tholeiite believed to be the parental magma in Hawaii. The subsolidus transitions for this material were reviewed in Figures 5-15*a* and 5-17. The phase boundaries given in Figure 8-14*c* are my attempt to correlate the run points plotted between 5 and 27 kb by Green and Ringwood. The liquidus phase is Ol to about 12 kb, Opx to about 20 kb, Cpx between 20 and 24 kb, and Ga at higher pressures. Opx is not recorded in runs with liquid present at pressures above 18 kb. There are two significant differences between results for this basalt and that of Figure 8-14*b*; the melting interval is wider at low pressures because of the higher content of normative Ol, and Opx occurs as the liquidus phase instead of Cpx between 12 and 20 kb. These results formed the basis for a comprehensive series of experiments designed to trace paths of fractional crystallization of basaltic mag-

mas at different pressures and the scheme is summarized in Figure 8-16; orthopyroxene plays a dominant role in this fractionation scheme. Tilley and Yoder suggested that the orthopyroxene might be metastable, crystallizing in place of stable clinopyroxene. Green and Ringwood repudiated this suggestion reporting runs which indicated that clinopyroxene rather than orthopyroxene may crystallize metastably in short runs.

In 1968 Ito and Kennedy presented results for an olivine tholeiite and a picrite which, together with the results in Figure 8-14*a* and *b*, provide a picture of the melting relationships on the composition join basalt-peridotite. They extended this toward nepheline basanite compositions with preliminary data on four other compositions. Results for the crystalline olivine tholeiite composition with 26% normative olivine are shown in Figure 8-14*b*. The phase boundaries drawn by Ito and Kennedy must be regarded as schematic; they are based on only 11 run points between 10 and 20 kb, with three more at 40 kb. A discontinuous reaction series, Ol $\rightarrow$ Opx $\rightarrow$ Cpx, occurs during crystallization at high pressures. Although this material has composition very similar to the olivine tholeiite in Figure 8-14*c*, there are significant differences in the two sets of results. The liquidus phase is Ol at 20 kb and Opx is given as the liquidus phase at pressures greater than about 21 kb. Ol is replaced by Opx at 12 kb in Figure 8-14*c*.

Iron capsules were used by Ito and Kennedy and platinum capsules by Green and Ringwood. Ito and Kennedy attributed the different results to errors introduced by the platinum capsules. Green and Ringwood evaluated the loss of iron to the platinum capsule and concluded that the effect is not a major factor in modifying either the major element composition or the crystalline phases of the experimental runs. According to Ito and Kennedy, on the other hand, the platinum capsules probably dissolved appreciable amounts of iron, causing the normative olivine to decrease by "25% or more during the course of most runs, accompanied by an increase in hypersthene."

Ito and Kennedy concluded that their additional results supported their previous conclusion. The discontinuous reaction series gives the prospect of fractional crystallization of the picritic liquid at depth with the formation of SiO_2-poor alkali basalts and other liquids.

Generation and Fractionation of Basaltic Magmas

There have been many schemes developed for basalt petrogenesis and some of these are illustrated in Figure 8-1. The most comprehensive schemes with an experimental basis are those presented by O'Hara and by Green and Ringwood.

Petrogenetic scheme of M. J. O'Hara 1965 and 1968. O'Hara's conclusion that extrusive basalts are residual liquids of advanced crystal fractionation and not primary magmas was discussed in connection with Figure 8-1e, and the experimental basis for his alternative model was outlined in connection with Figures 8-12 and 8-13. His model for a comprehensive scheme for the genesis of basaltic igneous magmas was a stimulating departure from the standard views of the time. A simplified version is shown in Figure 8-15 with an addition on the right from his 1968 paper. O'Hara did not indicate specific pressure intervals; I have added estimates of these to permit comparison with the fractionation scheme of Green and Ringwood in Figure 8-16.

O'Hara distinguished between high, intermediate, and low pressure regimes on the basis of subsolidus mantle mineralogy: garnet, spinel, and plagioclase peridotites. There are three subdivisions, $L1$, $L2$, and $L3$, in the low pressure regime. From available data he estimated the composition of the univariant liquid produced from the four minerals (equivalent to the composition A in Figure 8-13) as a function of pressure. In Figure 8-15 this liquid is shown as hy-normative picrite

in the H regime, Ne-normative picrite in the I regime, and varying from an alkali olivine basalt-like magma, to a high-alumina basalt-like magma, to a tholeiitic magma in the regimes $L1$, $L2$, and $L3$ respectively. This shows the types of primary magmas generated at various depths, and it shows how the liquid composition would change during steady movement upward if it remained just saturated with Ol, Opx, Cpx, and an aluminous mineral. The inevitable eruptive product is a Qz-normative tholeiite. There are many alternative pressure-temperature-time paths by which a deep primary liquid could reach the surface. The effects of interrupting the ascent with fractionation occurring at depth within the specific pressure regimes are illustrated by the horizontal arrows. High-alumina basalt is produced by fractionation in the $L2$ pressure regime, and when this rises to the surface it is split on the low-pressure thermal divide (D) between Ne-normative and Hy-normative basalts. Green and Ringwood disagree with O'Hara's interpretation of the effect of eclogite fractionation in the high pressure regime, H; they conclude that this does not produce silica-poor residual liquids. For partial melting of peridotite in the presence of water in the low pressure regime $L2/L1$ or for a basaltic liquid becoming water-saturated under these conditions O'Hara included the generation of parent andesite liquid but he subsequently concluded that this was not important in the generation of andesite provinces.

In 1968 O'Hara reviewed the available experimental data of the type illustrated in Figures 8-13 and 8-14 with special attention to the compositions of liquids in equilibrium with Ol and Opx at various pressures. He presented various subprojections of the data onto planes within the system CaO-MgO-Al_2O_3-SiO_2 and noted that

"the results obtained by ten workers in three different laboratories are wholly consistent with each other. Moreover these results can be interpreted at each pressure to yield a wholly consistent phase equilibria diagram despite the gross simplification involved in reducing the analyses to a system of four effective components prior to making the subprojections within that system.''

D. H. Green has argued that the use of projections in which Mg and Fe are equated and on which basalts, peridotites, and minerals with very different Fe/Mg ratios are plotted should be used only with great caution for the derivation of liquidus phase fields and cotectics.

On the basis of his review O'Hara revised his model for the evolutionary paths for basalt magma. He rejected the concept that tholeiitic magmas *in general* pass through a nepheline-normative stage at depth and introduced explicitly the concept that olivine fractionation has been a continuous factor in the evolution of all volumetrically important basalt magmas. His revised petrogenetic scheme, making provision for advanced stages of partial melting, is more complex than his 1965 model and that of Green and Ringwood (Figure 8-16). He presented a schematic phase diagram for peridotite, like Figure 8-14*a*, and considered the composition of erupted surface liquids when a magma batch was transported from a specific depth-temperature area at such a speed that it fractionated only olivine during ascent; the source areas of various erupted types are illustrated schematically with respect to the pressure regimes on the right-hand side of Figure 8-15 with temperature above the peridotite solidus indicated by the arrow.

Petrogenetic scheme of D. H. Green and A. E. Ringwood 1967. Green and Ringwood determined the phase relationships for glass with the composition of an olivine tholeiite from Hawaii believed to represent closely the primary magma existing in the depth range 20–60 km (Figure 8-14*c*). Partial analyses of Ol, Cpx, Opx, and Ga coexisting with the high pressure liquids were obtained using a microprobe, and these results were used to estimate the directions of isobaric fractionation of the basaltic magma during crystallization at various pressures. Additional basaltic glasses lying on the calculated fractionation trends were prepared. A few experiments on these compositions permitted them to follow the fractionation path into regions where there would have been very little liquid remaining in the original basalt; thus they examined the later stages of fractionation in detail. The program was very well conceived.

Figure 8-16 shows a simplified version of their fractionation scheme related to the subsolidus mineralogy of mantle peridotite. This incorporates later deductions by Green (1969). Note the absence of spinel peridotite in their model and the greater depth for garnet. Green and Ringwood concluded that there are three distinct trends of fractionation characteristic of the pressure intervals corresponding to:

1. Low pressure or shallow crustal fractionation above 15 km (5 kb) to yield Qz-normative residual liquids.

2. Intermediate pressure fractionation at depths of 15 to 35 km (9 kb experiments) to yield high-Al_2O_3 olivine tholeiites with 3–10% normative Ol.

3. High pressure fractionation at depths of 35–70 km (13.5 and 18 kb experiments) to yield Ol-rich alkali basaltic magmas.

They concluded that garnet does not play a significant role in the genesis of magmas by fractional melting at depths less than 100 km. The olivine tholeiite and tholeiitic picrite starting materials in Figure 8-16 are assumed to represent 20 to 40% partial melting of the mantle peridotite, much more than in O'Hara's scheme. Figures 8-16 and 8-15 merit careful comparison. Figure 8-16 includes two sequences added to the 1967 diagram by Green: the high pressure fractionation sequence and the two trends to olivine nephelinites produced by increasing water content and fractionation of aluminous orthopyroxene.

Note that according to Figures 8-16 and 8-14*c* fractionation at intermediate and high

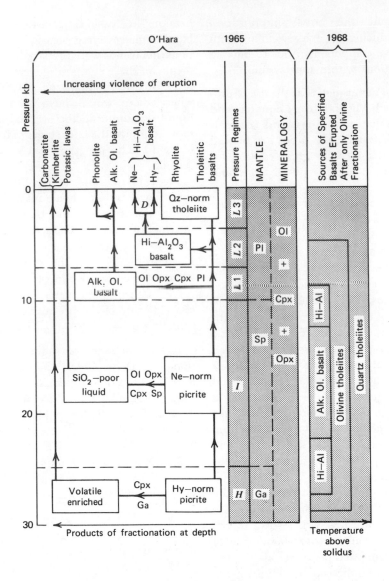

Figure 8-15. Summary of suggested basalt fractionation schemes dependent upon depth of origin and rate of migration to the surface according to O'Hara (1965, 1968). The main portion is based on O'Hara's Table 1 (1965). The main vertical sequence involves steady movement towards the surface through the successive pressure regimes (H = high pressure regime, I = intermediate, L = low, subdivisions 1, 2, and 3. Pressure scale is my estimate to permit comparison with Figure 8-16). Interrupted or delayed ascent at various depths yields the fractionation products shown. Minerals fractionating are listed near arrows. The right hand portion is a revised scheme (1968) showing the character of the erupted liquid when a magma batch formed and fractionated at various levels is taken from a particular pressure and temperature at such a speed that its composition remains near the boundary of, but always just within the olivine primary phase volume; hence it fractionates olivine only during its ascent. This is possible because of the increase in normative olivine content of liquids with pressure, as shown in Figure 8-14a. Olivine nephelinites and melilitites are erupted from fractionated materials at depth, with temperatures below that of the solidus depicted in the figure.

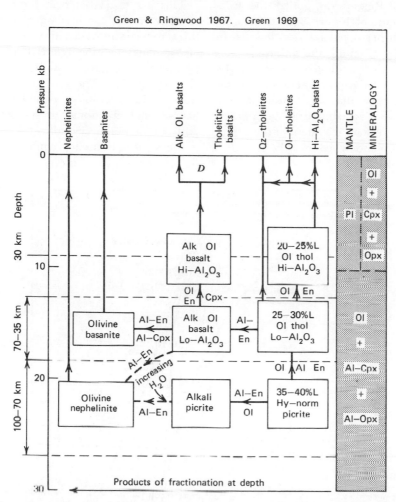

Figure 8-16. Simplified representation of Green and Ringwood's deduced crystal fractionation relationships among various basaltic magmas at moderate to high pressures; "closed system" fractionation. Based on Green and Ringwood (1967) and Green (1969). Contrast the mantle mineralogy between 10 and 30 kb (35 to 100 km depth) in this figure and Figure 8-15. Minerals fractionating are listed near arrows. Compare with Figure 8-17.

pressures is dominated by aluminous orthopyroxenes. Both O'Hara and Kushiro (1969) questioned the validity of this fractionation scheme; the olivine tholeiite treated as a parental basalt cannot be the partial melting product of peridotite between 13 and 18 kb because olivine is not a liquidus phase (Figure 8-14c). Green replied in 1969 that he and Ringwood had discussed the absence of olivine in their original paper. He concluded that careful consideration of the experimental data shows that while olivine would be the liquidus phase of a primary olivine tholeiite magma formed at 18 kb cooling results in precipitation of major orthopyroxene and very minor olivine; if so then the dominant role attributed to orthopyroxene fractionation is justified. Additional experimental data supporting this conclusion was presented by Green in 1971.

Green and Ringwood applied the fractionation scheme of Figure 8-16 to magma generation through partial melting as in Figure 8-17 by assuming the following:

"In one sense, fractional melting may be regarded as the reverse of fractional crystallization, providing that the nature of the crystalline phases are similar in both cases. This relationship is independent of the actual proportions of phases which may be present." (p. 164).

This is a crucial step in the development of their model for magma generation and this statement needs clarification. The process of fractional melting referred to is actually partial melting under equilibrium conditions (Figure 8-17). The reverse of equilibrium partial melting is equilibrium crystallization not fractional crystallization. The path of fractional crystallization of a liquid is the same as that for equilibrium crystallization only if the minerals involved exhibit

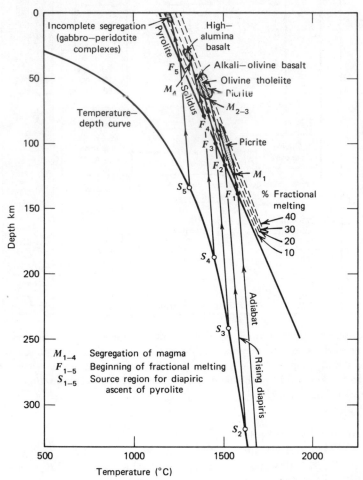

Figure 8-17. A model for magma formation by fractional melting of mantle pyrolite (after Green and Ringwood, 1967). S_1–S_5 represent arbitrary source regions from which there is diapiric ascent of bodies of subsolidus pyrolite to intersect the pyrolite solidus at points F_1–F_5. Partial melting begins at these points and the rising crystal-liquid mushes follow the courses indicated until segregation of magma from residual crystals at M_1–M_4. The nature of the magma is determined by equilibria occurring at M_1–M_4 and not at F_1–F_4. Compare Figures 8-16, and 6-6. (From Contr. Mineral. and Petrol., **15**, 163, Fig. 12. Berlin-Heidelberg-New York: Springer, 1967, with permission.)

no solid solution; that is certainly not true for the systems considered here. Fractional crystallization in a system with olivines, pyroxenes, and plagioclase can produce liquid paths diverging considerably from the paths of equilibrium crystallization with the divergence increasing at lower temperatures; the liquids produced by fractionation in such a system will persist to temperatures which may be considerably lower than the equilibrium solidus temperature.

The liquid fractionation paths depicted in Figure 8-16 cannot be the reverse of the equilibrium melting paths implied in Figure 8-17 unless the minerals involved maintained constant composition during the processes. Green and Ringwood stated that the Fe/Mg ratios of olivines and pyroxenes in the successively studied basaltic liquids are similar to the Fe/Mg ratios of olivines and pyroxenes of mantle-derived ultramafic rocks, and this is the justification for treating the partial melting process as essentially the reverse of the fractionation process.

Figures 3-12 and 6-18 suggest that unusual conditions are required for the temperature to exceed the dry solidus of mantle peridotite. The most reasonable explanation involves the diapiric uprise of mantle material resulting from some gravitational instability as illustrated in Figure 8-17. This process has been depicted schematically in Figure 6-6, and the successive stages can be followed there as well as in Figure 8-17. The process that causes gravitational instability and initiates diapiric uprise is a matter for speculation. One possibility is that upward migration of juvenile water from the deep mantle could operate as shown in Figures 8-22 and 8-23.

The sequence shown in Figure 8-17 is:

1. Solid mantle material rises in diapiric form from a source S under adiabatic conditions.

2. Magma generation occurs when the temperature of the rising mass exceeds the solidus at a point F.

3. The partially melted mass continues to rise adiabatically with the degree of melting increasing as the temperature interval increases above the solidus probably with the liquid remaining in equilibrium with the residual unmelted crystals of the mush. The adiabat with liquid forming has a different slope.

4. Magma segregation occurs at some point M where the extent of partial melting is sufficient, 20 to 40%, and the liquid leaves the residual refractory crystals.

5. Magma rises upward as an independent magma body, cooling, crystallizing, and fractionating.

The critical factor is the composition of the liquid when magma segregation occurs, which is controlled by the depth and the temperature interval above the solidus. Figure 8-17 show the relative positions for the formation of liquids of various basaltic compositions taken from the fractionating scheme in Figure 8-16. The dashed lines above the solidus show the percentage of melting of the mantle.

The linear pattern of melting assumed by Green and Ringwood for dry pyrolite is an idealized situation. In systems where the percentage of melting has actually been measured, the pattern is step-like rather than linear as indicated in Figure 8-19*b* (Robertson and Wyllie, 1971). The dry melting curve here is my estimate based on the phase relationships and the results reported by Ito and Kennedy for a natural peridotite (Figure 8-14*a*). They reported that at 20 kb there was little increase in liquid content between 1320 and 1600°C. The curve is similar to that suggested by O'Hara in 1968. He discussed the partition of elements between crystals and liquid in five stages of fusion without giving temperatures; these stages are shown in Figure 8-19*b*.

Effect of Water

Figure 6-18 shows the general pattern of the melting interval introduced below the

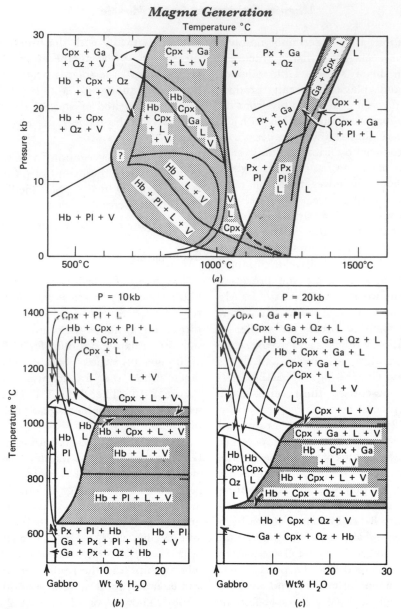

Figure 8-18.   (*a*) Experimentally determined phase relationships in gabbro (Figure 8-14) and gabbro in the presence of excess water (unpublished data by Lambert and Wyllie; see Lambert and Wyllie, 1968, 1970, Hill and Boettcher, 1970). See Figure 6-12. The interval for the coexistence of amphibole (hb) and garnet (ga) is probably too narrow, because of the reluctance of garnet to nucleate; with seeded runs the garnet curve would probably be determined at lower temperatures. Zoisite, epidote, and kyanite or sillimanite are minerals whose presence was suspected but not proved at pressures greater than 10 kb. Abbreviations: ga — garnet, cpx — clinopyroxene, px — pyroxene, pl — plagioclase, hb — amphibole, qz — quartz, L—liquid, V—aqueous vapor phase. (*b*) and (*c*) Schematic isobaric *T-X* sections for 10 kb and 20 kb for gabbro-water, interpolated from the dry and water-excess results in Figure 8-18*a*. (Compare Figures 8-4*b* and 8-11). An important boundary for experimental determination is the vapor-absent liquidus below the field of L; changes in liquidus phases with pressure (Figure 8-14) and with water content may have a strong influence on fractionation products. A congruent reaction is depicted for amphibole in the vapor-absent region; an incongruent reaction is an alternative arrangement.

dry peridotite solidus by the addition of small amounts of water. Experimental results for peridotite-water are limited to reconnaissance experiments near the solidus, but we do have fairly complete data for the melting relationships of a gabbro which illustrate the problem involved. The melting intervals of the rock dry (Type I system) and in the presence of excess water (Type IV system) were compared in Figure 6-12. In Figure 8-18a the detailed phase relationships within the melting intervals are shown. This gabbro contains only 8.6% normative olivine and it becomes a quartz eclogite at high pressures.

Given the results shown in Figure 8-18a, isobaric sections through the water-deficient region can be constructed in the same way that Figure 8-11 was obtained from Figure 8-10 for the system granodiorite-water. The schematic results so obtained for gabbro-water at pressures of 10 and 20 kb are shown in Figures 8-18b and 8-18c.

There are significant differences between the phase relationships at 10 and 20 kb. Notice the distribution of quartz in Figure 8-18c for 20 kb; this ensures that the first liquids produced on partial melting in Types II, III, and IV systems are all silica-rich. For the Type II compositions at 10 kb (vapor-absent subsolidus; Figures 6-13 and 6-19b for an isopleth) the assemblage is transformed from hornblende eclogite to hornblende gabbro with increasing temperature, whereas at 20 kb the assemblage is a quartz-hornblende-pyroxenite. Note the melting pattern of Type II systems; liquid undersaturated with water is produced within the temperature interval where hornblende dissociates (*m'm* in Figure 8-4). In contrast, for the Type III system the first liquid is water-saturated however small the water content.

Information about peridotites and various basaltic compositions that is vital for comprehension of the petrogenesis of basaltic magmas and for evaluation of the effect of water on the physical properties of the mantle includes: (a) the percentage of liquid produced in the vapor-absent region as a function of pressure

(depth), temperature, and water content and (b) the composition of the liquid as a function of the same variables. For a given water content the pattern of phase relationships can change from Type II (as in Figure 8-4b) to Type III (as in Figure 8-4a) where the hornblende breakdown curve passes below the solidus. The composition and water content of the liquids present near the mantle solidus are therefore markedly dependent on the stability of hydrous minerals such as amphiboles and phlogopite.

Amount of liquid. As a guide for experimentation D. H. Green has presented a comprehensive scheme for peridotite-water in the form of a petrogenetic grid which provides an internally consistent working model for mantle source composition, derivative liquids, peridotitic residues, and magmatic cumulates from the liquids at high and low pressures. This is based on reconnaissance experiments in water-deficient regions with peridotites and basaltic compositions. The model is illustrated in Figures 8-19a and 8-20.

Figure 8-19a shows the linear relationship assumed by Green and Ringwood for the percentage of melting above the solidus temperature of mantle peridotite. Compare Figure 8-17. The minerals coexisting with liquid for specific percentage intervals of melting are shown between Figures 8-19a and 8-19b. The other curve in Figure 8-19a is Green's estimate of the percentage of melting produced with 0.1% water at each temperature above the wet solidus for peridotite at a pressure of about 25 kb; the solidus is at 1080°C where amphibole breaks down (Figures 6-18b and 8-20). The position of this curve is based on the measured depression of the liquidus in the vapor-absent region by known amounts of water, but the data and method have not been published as yet. The corresponding contours for percentage of liquid developed as a function of pressure with 0.1% water present are plotted on Figure 8-20.

Green concluded that the near-liquidus role of orthopyroxene in water-bearing

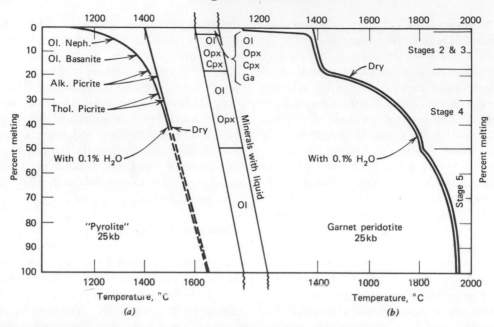

Figure 8-19. Fusion curves for peridotite dry, and in the presence of 0.1% water, showing the percentage liquid produced at successive temperatures above the solidus, and the minerals coexisting with the liquid in different parts of the melting interval (after Wyllie, 1971). (*a*) Dry curve from Green and Ringwood (1967), see Figure 8-17; curve with H_2O and estimated liquid compositions after Green (1970). (*b*) Dry curve estimated from partial experimental data by Ito and Kennedy (1967), see Figure 8-14*a*, with stages according to O'Hara (1968). The curve with H_2O contrasts with curve in *a*, and corresponds closely to Ringwood's (1969) estimate, except for the formation of a finite amount of liquid within a few degrees of the solidus in the *b* model, consequent upon dehydration of amphibole. (From Jour. Geophys. Res., **76**, 1328, 1971, with permission.)

nephelinitic magmas at high pressures may permit fractionation of magmas through olivine-rich basanites to olivine nephelinites at 20–25 kb and through picritic basanites and picritic nephelinites to olivine melilite nephelinites at about 27 kb. From these experimental studies he inferred that the highly alkaline, undersaturated magmas could form by partial melting of mantle peridotite in the presence of water with the general distribution as shown in Figure 8-19*a* for 25 kb and in Figure 8-20 for a range of pressures. The effect of 0.1% water on the magma generation scheme shown in Figure 8-17, as shown in Figure 8-20, is to lower the temperature of formation of the magma types shown in Figure 8-17 and to introduce the wide zones for the generation by fusion of the highly alkaline undersaturated liquids.

Green's curve for melting of peridotite with 0.1% H_2O indicates surprisingly high percentages of liquid at subsolidus temperatures. According to the curve in Figure 8-19*a* at the dry solidus temperature 0.1% of water is sufficient to produce about 17% melting and to dissolve all of the garnet and most of the clinopyroxene; Figure 8-20 indicates 25% of liquid at the dry solidus. Until this curve can be considered as securely based on experimental data I propose as an alternate pattern that shown in Figure 8-19*b*. A finite amount of H_2O-undersaturated liquid is produced within a few degrees of the solidus where hornblende breaks down (Type II composition, vapor-absent subsolidus in Figure 8-18*c*), and very little additional liquid is produced until the temperature approaches closely that of the dry solidus; at higher temperatures it follows the dry curve, rather than approaching it asymptotically as indicated by Green's curve between 30 and 40% melting. I have not attempted to estimate the actual percentages of melting because I do not have adequate data, and the percentages guessed are simply to illustrate a pattern which contrasts with that of Green. Distinction between these two patterns is significant because the amount of interstitial silicate liquid produced by traces of water below the dry melting temperature affects markedly the physical properties of the mantle, and the prospects that the liquid can escape from its host for independent uprise as a magma.

Composition of liquid. The composition of the liquid produced is in dispute. In 1955 A. Poldervaart suggested that magmas more silicic than basalt "may be formed in proportionately smaller amounts than basaltic magma by relatively low temperature partial melting of ultramafic material in the presence of high concentrations of water" (pp. 139–140). A series of experiments by I. Kushiro and associates in increasingly complex synthetic systems proves that the water-saturated liquid developed from Fo + En-bearing assemblages at high pressures is silica-saturated. Extrapolation of these results to the natural rocks leads Kushiro to conclude that equivalent liquids developed in the

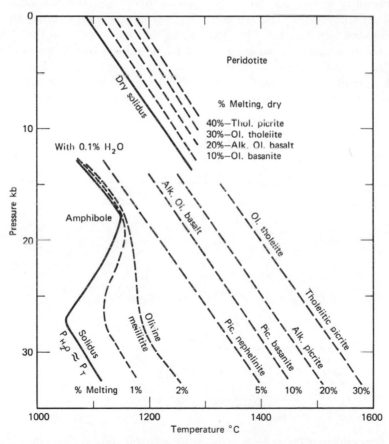

Figure 8-20. Petrogenetic grid for basaltic liquids derived by partial melting of mantle peridotite dry (Figure 8-17), and in the presence of 0.1% water (Figure 8-19*a*), according to Green (1971). (From Phil. Trans. Royal Soc. London, with author's permission.)

mantle are tholeiitic basalts or andesites; O'Hara made the same suggestion in 1965.

This interpretation is in direct conflict with the inferences of Green in Figures 8-19a and 8-20 that these liquids are nephelinitic. Green and Kushiro presented their respective interpretations in 1969. In his 1970 paper Green concluded that both sets of experimental data were probably correct and described preliminary experiments suggesting that at 22.5 kb the conclusions from Kushiro's synthetic systems cannot be extrapolated to a typical basalt. He doubted that a quartz-normative tholeiitic or andesitic liquid could coexist with olivine at water pressure greater than 10 kb.

Figure 8-21 shows that Kushiro's experiments have advanced a long way from the first runs with enstatite-water. The mineral components involved are Fo, En, Qz, Jd, Ne, An, and Ca-tschermaks molecule. The surface *VRBT* is the Fo-Opx liquidus boundary at 20 kb pressure under anhydrous conditions. The point *B* gives the composition

of the first liquid formed by partial fusion of an assemblage including Fo, Opx, Cpx, and an aluminous phase; it is probably in the Ne-normative region of the tetrahedron at 20 kb. The point *A* is the first liquid formed by partial fusion of Fo, Opx, Cpx, and Ga (a synthetic garnet peridotite) in the presence of water. This liquid is silica-saturated and Qz-normative at 20 kb and probably up to 25 kb. Kushiro informed me that he has since analyzed the liquid *A* using a microprobe confirming that its composition is andesitic. Above 30 kb, the first liquid *A* may be silica-undersaturated.

By analogy with our review of Figures 8-4 and 8-8 Figure 8-21 shows that however small the water content of the synthetic garnet peridotite, the first liquid formed is H_2O-saturated, with a composition at *A*. For water-deficient conditions the vapor dissolves in the liquid and with increasing temperature and decreasing P_{H_2O} the liquid composition follows a path close to *AB* but not coincident with it. The liquid composition thus moves

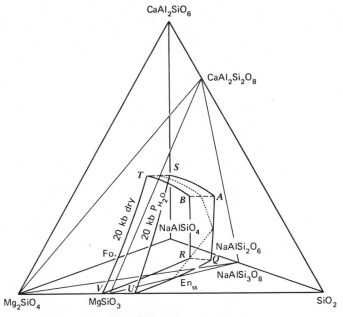

Figure 8-21. The positions of the forsterite-orthopyroxene liquidus boundaries in the system forsterite-nepheline-$CaAl_2SiO_6$-SiO_2-H_2O at 20 kb under water-saturated and dry conditions. The dotted line indicates a trace of the intersection of the plane $MgSiO_3$-$CaAl_2Si_2O_8$-$NaAlSi_3O_8$ with the volume *A-Q-U-S-T-V-R-B* (after Kushiro, 1970. Courtesy of Carnegie Institution).

further away from silica with progressive fusion, and when it crosses the plane En-An-Ab its composition becomes Ol-Hy-normative, still tholeiitic. If the liquid crosses the plane Fo-An-Ab it becomes Ne-normative. How much andesitic liquid can be generated and how far the liquid can move from A toward B depends on the water content.

This model assumes that no hydrous minerals are stable at the solidus: we are dealing with a Type III system (Figure 8-4a). If the pressure is appropriate for the stability of hornblende (or other hydrous mineral) then we have the situation in Figure 8-4b and the system may be either Type II or Type III depending upon the water content. If there is free vapor present as well as the hydrous mineral (Type III) the solidus temperature is lower than for the Type II assemblage with no vapor (Figure 8-4b). The apparent conflict of interpretation between Green and Kushiro may be resolved when all of the variables are considered. The pressure (depth) interval affects the composition of the water-saturated liquid and the stability of hydrous minerals; with hydrous minerals stable the water content defines the system as Type II or III. In different parts of a petrogenetic grid such as that proposed by Green (Figure 8-20) we have the following conditions, each of which may involve a liquid of different composition: (a) Type III system with no hydrous minerals, (b) Type III system with hydrous minerals, and (c) Type II system. The liquid produced in the Type II system is water-undersaturated and it may be significantly different from the water-saturated Type III liquid. The Type III liquid without hydrous minerals will have a composition different from a Type III liquid coexisting with amphibole. Kushiro's results at 20 kb suggest that Type III liquid could be andesitic and Type II liquid could be nephelinitic.

Initiation of diapiric uprise. Figure 8-17 shows a model for magma generation with diapiric uprise of mantle initiated by some unspecified gravitational instability. Figures 8-22 and 8-23 illustrate how juvenile water rising into dry peridotite from deep within the mantle (along the geotherm) would cause incipient melting in a layer ab just above the solidus for peridotite-water. This would lower its density and viscosity relative to the surrounding mantle, and it would tend to rise as shown in Figure 8-23b. It would rise adiabatically as in Figures 8-17 and 8-22 and when the layer ab reached the position cd it would be about 100°C hotter than the surrounding mantle. Continued uprise would carry it through the level near the dry solidus where significant melting occurs (Figure 8-19b), and in the layer at ef magma generation would occur in the normal way and the subsequent events could be described according to the dry model, Figure 8-17 for example.

Figures 6-18b, 6-19b, and 6-20 were used to explain the low-velocity zone in terms of incipient melting due to traces of water. If water rose upward from below the low-velocity zone, as in Figures 8-22 and 8-23a, the amount of interstitial liquid in the base of the zone would increase above the rising water. Increased liquid content would cause decreased density and viscosity, and diapiric uprise could thus be initiated. Magma generation caused by rising diapirs from the top of the low-velocity zone has often been suggested, but if the diapirs begin at deeper levels near the base of the zone, passing through material already partially melted, this could introduce considerable variety into the processes for distribution of trace elements and rare earth elements among the crystals and liquid of the rising diapir and the partially melted mantle in the low-velocity zone.

Basalt Petrogenesis

Current models for the generation and fractionation of basaltic magmas with an experimental basis include the following:

1. The compositions of primary basalts are controlled by the depth of partial fusion of mantle peridotite (Kushiro; Figure 8-1*d*).

2. Tholeiitic basalts are formed by olivine fractionation during uprise of a deep-seated parental picrite liquid with silica-poor alkalic liquids produced by deep-seated fractionation including eclogite fractionation (O'Hara, Yoder, Ito, and Kennedy; Figures 8-1*e* and 8-15).

3. The compositions of basalts are controlled by the degree of partial melting of dry mantle peridotite at the stage where the liquid is segregated, by the depth of segregation and the depth of subsequent fractionation (Green and Ringwood, Figures 8-16 and 8-17).

4. The presence of water in the mantle extends model (3) by introducing zones for small degrees of partial melting below the dry peridotite solidus (Green; Kay, Hubbard, and Gast, 1970; Figures 6-18*b* and 8-20). The low temperature hydrous liquids are silica-saturated according to Kushiro, silica-poor and alkalic according to Green.

These different interpretations emphasize the need for detailed experimentation under controlled conditions in water-deficient peridotite-basalt systems as illustrated in Figures 8-18*b* and 8-18*c*.

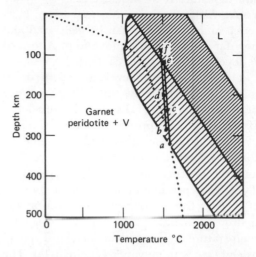

Figure 8-22. Peridotite-water isopleth from Figure 6-18*b*, and dotted oceanic geotherm from Figure 3-12. Shows the diapiric uprise of layer *ab* (Figure 8-23*a*) to successive positions *cd* (8-23*b*) and *ef* (8-23*c*), where basaltic magma is generated. Compare Figure 8-17. (From Wyllie, 1971, Jour. Geophys. Res., **76**, 1328, with permission.)

GENERATION OF BATHOLITHS AND ANDESITES

In 1950 Tilley drew attention to the problem of accounting for the voluminous andesites of the orogens. The plutonic rocks of batholiths approach chemical equivalence with andesites and their associated volcanic rocks and both associations are developed in the same tectonic environment. Although rocks in either association may have an origin independent of the other association, it makes sense to consider their petrogenesis jointly. It is generally held that at least some intrusions in batholiths represent andesitic magmas that failed to reach the surface for eruption.

The bitterness of the Granite Controversy subsided during the 1960's as many petrologists accepted an experimentally-based anatectic model.

We concluded that the normal product of anatexis in the crust is a mush composed of crystals and H_2O-undersaturated granite liquid. The temperatures required for the formation of liquid of intermediate composition, even under the most favorable conditions, are so high (Figures 8-10 and 8-11) that it seems improbable that andesite liquids could be generated in the crust. With this process excluded there remain the following possibilities for the origin of batholithic magmas and andesite magmas:

1. Crustal anatexis yielding a crystal mush of intermediate composition, composed of crystals with water-undersaturated granite liquid.

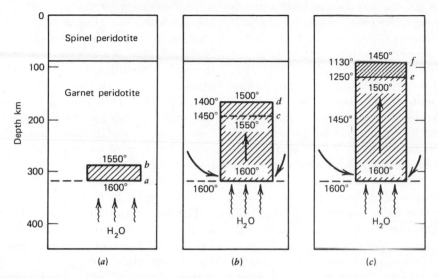

Figure 8-23. Schematic mantle sections showing the diapiric uprise of layer *ab* at the base of the low velocity zone into successive positions *cd* (*b*) and *ef* (*c*) on the adiabatic curves *ace* and *bdf* in Figure 8-22. Water migrating from the deep mantle increases liquid present in layer *ab*, causing diapiric uprise from the base of the low-velocity zone, and basaltic magma generation towards the top of the zone. (From Wyllie, 1971, Jour. Geophys. Res., **76**, 1328, with permission.)

2. Derivation from parent basaltic magma by fractional crystallization, assimilation, hybridism, or some combination of these.

3. Mantle anatexis producing primary andesitic magma

4. Anatexis of crustal material under mantle conditions caused by downward transport of sediments and lithosphere in Benioff zones.

The first process provides a satisfactory explanation for many plutonic rocks of batholiths, but it cannot explain the eruptive andesites. This leads us to mantle conditions for the origin of andesites and possibly to mantle material which suggests in turn that mantle-derived material may be involved in the development of some batholithic intrusions.

The second process is the classic model, and the conditions under which basaltic magmas may yield andesitic derivatives have been discussed by E. F. Osborn. In 1959 he demonstrated that paths of crystallization for liquids in the system $MgO-FeO-Fe_2O_3-SiO_2$

varied according to the oxygen fugacity, and this has since been extended to more complex synthetic systems including plagioclase components. Applying the results to natural magmas leads to the conclusion that there are two reaction series as illustrated in Figure 4-5. In a closed system with oxygen fugacity remaining low the fractionation trend is towards liquid enrichment in Fe/Mg, whereas if the oxygen fugacity is maintained at a high level the residual liquids follow the calc-alkaline trend with enrichment in silica and alkali feldspar. In 1969 Osborn reviewed the problem in some detail and concluded that andesites were produced by fractional crystallization of olivine basalt magma in orogenic regions under conditions of high oxygen fugacity, produced by flow of water from the surrounding geosynclinal rocks into magma and migration away of hydrogen. According to H. P. Taylor in his 1968 comprehensive review of the oxygen isotope geochemistry of igneous rocks, the oxygen isotopes of andesites make it reasonable to assume that these were derived from a parent basaltic magma by some process of magmatic differentiation.

There appears to be little or no continental crust beneath some island arcs (Figures 5-1, 7-4*b* and 7-6*b*) which suggests that, in this environment, the andesites and associated lavas may be derived from the mantle. In 1959 when few petrologists paid much attention to the mantle J. T. Wilson argued that the andesites of island arcs were derived by partial melting of a small fraction of mantle peridotite between 70 and 700 km depth, and that they rose along fractures associated with deep earthquakes. Figures 8-19, 8-20, and 8-21 illustrate the current dispute about the composition of the liquid produced by partial fusion of peridotite in the presence of water; it may be Ne-normative or Qz-normative. Resolution of these different interpretations is important because of their implications with respect to the origin of andesites and the evolution of continents. We saw in Chapter 6 that the upper mantle is composed of peridotite and eclogite. T. H. Green and A. E. Ringwood presented experimental data in 1968 on a series of rocks with compositions between basalts and rhyolites, and they concluded that partial melting of eclogite or hydrous basalts at depth would yield andesitic magmas. They also reviewed hypotheses for the origin of the calc-alkaline suite. In a similar review O'Hara agreed with their conclusion noting that "Although the experi-

mental evidence for this is based upon some exceptionally long linear extrapolations of liquidus temperatures (essentially a nonlinear function) through a minimum of fixed points the basic proposition that the partial melting product is silica-enriched cannot be in doubt." (1968, p. 101.) Thus primary andesite magmas can be generated in the upper mantle from eclogite and possibly from peridotite in the presence of water under some conditions. Origin from mantle material is favored by some trace element data and isotope data as reviewed by S. R. Taylor in 1969.

The fourth process has recently been hailed as the answer to all of the problems (Chapter 14). Partial fusion of sediments and rocks of the oceanic crust and upper mantle at high mantle pressures and varied temperatures occurs because of the downward movement of a slab of lithosphere at the Benioff zones. This produces magmas of basic, intermediate, and acid composition formed under dry or wet conditions, either within the down-going lithosphere or above it, for eruption at the surface or for intrusion into the cores of the young mountain chains developing adjacent to the oceanic trenches. There are plenty of variables to manipulate in order to obtain the solutions needed. What we lack are adequate constraints to facilitate selection among the solutions proposed. The study of trace elements and isotope evolution trends may eventually provide definitive evidence.

9. Geosynclines and the Orogenic Cycle: Classical Views

INTRODUCTION

For the most part this chapter is concerned with the classical treatments of the orogenic cycle, which include no satisfactory mechanism, published before plate tectonics introduced a new conceptual framework for orogenesis. Here we examine the geosynclinal concept as it was developed through a century by standard geological and stratigraphical methods of unravelling the history and paleogeography of a mountain chain. The recent models for mountain building are reviewed in Chapter 14.

In theories for the evolution of mountain chains, or the geotectonic cycle, the concept of geosyncline and orogenic cycle has been a dominant theme since 1859 when J. Hall realized that the Appalachian Mountains in New York State had been formed by elevation of a pile of sediments that were originally deposited in a subsiding trough. In 1873 J. D. Dana gave to such troughs the name geosynclinal which was later changed to geosyncline. This revolutionary concept of a great inversion of relief from depressed zone to folded mountain chain was followed by innumerable geological syntheses coordinating all of the subjects of geology and all of the processes outlined in Chapter 4.

The generalized sequence of the geosynclinal phase, the tectogenic phase, and the orogenic phase as traditionally presented in textbooks includes the following:

1. The accumulation of sediments in a subsiding trough, the geosyncline, along with marginal or submarine eruption of basic and ultrabasic lavas, including spilites and the ophiolite suite.

2. Folding, dislocation, and overthrusting of the rocks in the geosyncline.

3. Regional metamorphism and the emplacement of batholiths.

4. Uplift and the formation of marginal troughs with renewed sedimentation; widening of the geosynclinal zone, and repetition of (2) and (3).

5. Epeirogenic uplift with volcanic eruptions of basalts, andesites and rhyolites, and comagmatic plutonic intrusions.

6. Peneplanation.

The scheme has been modified repeatedly with some geologists extending the idea of a geosyncline to any subsiding region in which a significant thickness of sediments accumulates and others restricting the term only to troughs which give rise to strongly folded alpine-type mountain ranges. It has been generally assumed that orogenic activity implied the former presence of a geosyncline, and that high folded mountain ranges have formed only on the sites of geosynclines. With the revolution of the 1960's, revival of the theory of continental drift and development of the "new global tectonics" (Chapters 11 and 14), however, many geologists claim that the conventional cycle is obsolete and new models for mountain building involve movement and collision of crustal plates. The classical descriptive models are based on much geological data, and students should know this basis before being launched on a collision course.

CLASSIFICATION OF GEOSYNCLINES

Development of the concept of geosynclines is illustrated by various classification schemes. The ideas of Hall and Dana strongly influenced subsequent American thought which has differed in significant aspects from European views. The main differences between them arise because each group developed the concept with a different mountain range as the model. The standard example for American geologists is the Appalachian geosyncline and for European geologists it is the Alpine geosyncline as interpreted by Haug in 1900. Americans have often cited the coastal plain of the Gulf of Mexico as a standard example of a present day geosyncline, but Europeans have considered the Indonesian Archipelago as a standard example.

Both schemes are characterized by a great thickness of sediments, but whereas American geologists have contended that shallow-water sediments characterize subsiding basins European geologists recorded the occurrence of deep-water sediments and concluded that geosynclines were deep, elongated troughs. In the American view a progressively rising geanticline supplied sediment to the neighboring geosyncline in such a way that the rate of sedimentation just kept pace with the rate of subsidence. According to the European view this delicate balance was not maintained and the history and duration of a geosyncline depends on the relative rates of subsidence and sedimentation. Americans have considered geosynclines to occupy sites marginal to continents but Europeans have regarded them as forming either marginal to continents or between continental masses.

Tectonic Elements

The geosynclinal concept as originally developed in both America and Europe embodies a major paleogeographic distinction between one type of mountain chain, that characterized by strong folding, and the rest which can be defined as intracontinental or intracratonic. A geosyncline has a long history and it comprises several tectonic elements each distinguished by specific features and events in space and time. The classification of some of these individual elements as geosynclines has led to such broadening of the original concept that the term geosyncline has been applied to almost any tectonic element that involves subsidence and sedimentation. The main trends are illustrated in Table 9-1 which summarizes and compares various classifications.

Significant contributions were made by H. Stille who recognized cratons and orthogeosynclines as the two major crustal divisions. He subdivided cratons into hochkraton—stable continental crust, and tiefkraton—stable oceanic crust; and he distinguished between eugeosynclines and miogeosynclines. Eugeosynclines, found in the internal zones of the geosynclinal system (internides) most distant from the craton, are characterized by basic lavas and ophiolites; whereas miogeosynclines, found in the external zones (externides) of the system nearer the craton, are free or almost free of igneous activity. He also clarified the tectonic history of geosynclines. Epeirogenesis (undation) occurring over large areas for long periods of time produced geosynclines, geanticlines, and uplift of mountain chains with no

TABLE 9-1 Various Classifications of Geosynclines and Tectonic Elements (Modified Version of Table 1 of Aubouin, 1965)

Stille 1935–40	Kay 1951	Krumbein and Sloss 1963; Badgley, 1965	Sinityzn and Peyve 1950	Aubouin 1965
Orthogeosynclines eugeosyncline miogeosyncline	*Orthogeosynclines* eugeosyncline miogeosyncline	Orthogeosyncline Miogeosynclinal transitional zone	*Primary geosyncline*	*Geosynclines* eu-furrows mio-furrows eu-ridges mio-ridges
	Epieugeosyncline	*Postorogenic basins*	*Secondary geosynclines*	Back-deep Intra-deep
Parageosynclines	*Intracratonal geosynclines* exogeosyncline zeugeosyncline autogeosyncline	*Intracratonic basins* marginal basin yoked basin interior basin	*Residual geosynclines*	Foredeep Intracratonic furrows Basins
	Taphrogeosynclines Paraliageosynclines	Rift valley Coastal geosyncline		Trenches
Hochkraton Tiefkraton	Craton	Craton stable shelf unstable shelf	Platform	

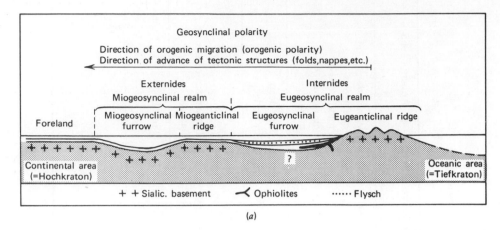

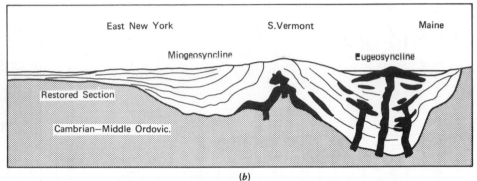

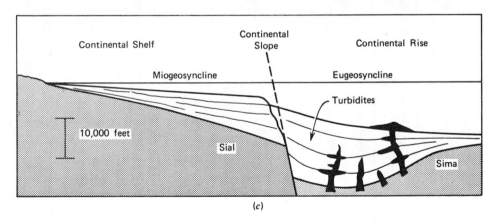

Figure 9-1. Reconstruction and interpretations of eugeosynclines and miogeosynclines. (*a*) Elementary couple (after Aubouin, 1965) showing conditions at the orogenic (terminal) stage of the geosynclinal period, when the emergent eugeanticlinal ridge was supplying detrital flysch material to the eugeo-synclinal furrow (Figure 9-4). (Reproduced by permission, Elsevier Publishing Co.) (*b*) Restored sections of Cambrian and Ordovician in geosynclines from New York to Maine (after Kay, 1951, with permission of The Geol. Soc. Amer.). (*c*) Reconstruction of the section depicted in *b* at the end of Trenton time (after Dietz, 1963). This is analogous to the presumed existing situation off the eastern United States (Figure 9-6). (With permission from Journal of Geology, Copyright 1963 by the University of Chicago.)

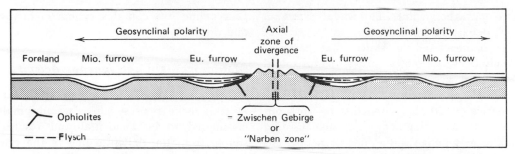

Figure 9-2. Characteristics of a divergent bicouple giving rise to a mountain chain of bilateral symmetry (after Aubouin, 1965). The two couples are depicted as they would appear during the orogenic (terminal) stage of the geosynclinal period (Figure 9-4). It is arbitrarily assumed for convenience that both couples have reached the same stage at the same time. (With permission of Elsevier Publishing Co.)

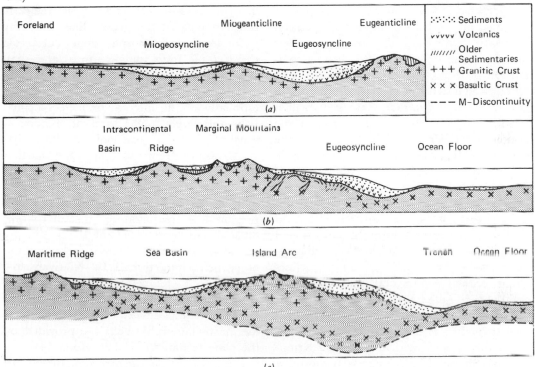

Figure 9-3. Tectonic frameworks (after Matsumoto, 1967). (*a*) Classical examples in Europe adapted from Aubouin (Figure 9-1a). (*b*) Based on Mesozoic examples in East Asia. (*c*) Based on Cenozoic examples on the west side of the Pacific. (From Tectonophysics, with permission.)

structural deformation. Orogenesis (undulation) was very limited in space and time producing intensely folded structures during brief episodes of deformation. In addition he prepared a detailed scheme for the correlation of igneous activity and metamorphism with the stage of evolution of the geosyncline.

The mountain chains rising from orthogeosynclines characterized by tight folds and nappes he termed alpinotype ranges, and other mountains arising from block faulting he termed germanotype ranges. Stille related germanotype mountains to a new class of geosyncline that was not subjected to orogenesis—parageosynclines. This was an extension and departure from the original concept of geosynclines, and he stated that parageosynclines were "second order" features and not "true" geosynclines.

Stille's classification was introduced into America by M. Kay, the first to work out the detailed development of systems consisting of a eugeosyncline and a miogeosyncline. His restored section of the Champlain and Magog belts from New York to New Hampshire is reproduced in Figure 9-1*b* from his 1951 memoir on North American Geosynclines. This memoir also contains Kay's subdivisions of Stille's parageosynclines into several groups classified on the basis of either their position in space relative to the craton or their distribution in time. Table 9-1 shows a comparison of these types with the names used for equivalent tectonic elements by European geologists. Kay's intracratonal geosynclines include exogeosynclines near the edge of the craton (fore-deep), autogeosynclines on the craton without associated highlands (classic basins), and zeugeosynclines which contain sediment from eroded highlands within the craton (introcratonic furrows). Kay also defined three late-cycle geosynclines formed in deformed eugeosynclines (back-deeps and intra-deeps), taphrogeosynclines or sediment-filled rift valleys, and paraliageosynclines along present continental margins (north coast of the Gulf of Mexico). Kay noted that the propriety of considering parageosynclines as geosynclines is debatable, and Table 9-1 shows that other American geologists Krumbein and Sloss and Badgeley prefer not to apply the term geosyncline to these tectonic elements.

According to the original concepts developed by Hall, Dana, and Haug the term geosyncline is restricted to a major paleogeographic feature. It is unfortunate, but not important, that the part of the Appalachian geosyncline originally studied by Hall has since proved to be a secondary feature. Aubouin classified as geosynclines only the major mobile belts, the orthogeosynclines and primary geosynclines of other authors. A geosyncline is a complex system of furrows and ridges.

Metamorphism and Igneous Activity

The development of metamorphic and igneous rocks in geosynclines was originally considered to be incidental and resulting simply from the downwarping of the crust and deep burial of sediments with consequent increase in pressure and temperature. It is now known that regional metamorphism and the associated granitic intrusions are bound up with the orogenic deformation of the geosynclinal sediments and not with the early downwarping of the geosyncline. In many examples the metamorphic zones cut across the tectonic units and are thus later than the major tectonic movements. The late-tectonic regional metamorphism is developed especially in the internal or eugeosynclinal belts and rarely in the external or miogeosynclinal belts. It was only in 1961 that A. Miyashiro drew attention to the fact that many geosynclinal zones, especially in the circum-Pacific region, contain paired metamorphic belts; one belt characterized by high pressure-low temperature metamorphism, often including glaucophane schists, and by abundant basic and ultrabasic rocks; and the other belt characterized by relatively low pressure-high temperature metamorphism with abundant syntectonic and late-tectonic and granitic intrusions. Feasible explanations for the metamorphic belts are provided by the plate tectonics model (Chapter 14).

It is obvious from the preceding discussions that the evolution of a geosyncline is marked at successive stages by specific types of igneous activity. The synthesis and classification proposed by Stille appear to be typical of the eugeosynclinal zones (internides). He recognized three main events which are correlated with Aubouin's scheme for geosynclinal evolution in Figure 9-4. The initial pretectonic magmatic episode produced basic submarine volcanic eruptions yielding the ophiolite suite and spilites; subsequent uplift and erosion may expose contemporaneous but deep-seated ultramafic rocks of the root-zones (Table 6-4). A synorogenic magmatic episode produces syntectonic "concordant" granitic intrusions associated with the regional meta-morphism, and late-tectonic "discordant"

granitic intrusions after the principal orogenic phase. The postmagmatic episode is subdivided into three parts: eruption of andesitic lavas, intrusions of granites and granodiorites, and finally eruption of basaltic lavas.

EVOLUTION OF GEOSYNCLINES

The evolution of a geosyncline and the formation of a mountain chain occupies a long period in geological history, and it is far more complex than indicated by the simple cycle outlined at the beginning of this chapter.

Geosynclinal Couples

Aubouin's book presenting a classical treatment of geosynclines was published in 1965. He reviewed the geology of the Mediter-ranean mountain chains of the Alpine cycle with specific reference to the Hellenides (including much of Albania, Yugoslavian Macedonia, and Greece) as a basis for com-parison, and he concluded that they showed a characteristic pattern of organization and evolution. The fundamental unit is the eugeosynclinal-miogeosynclinal couple which is illustrated schematically in Figure 9-1a. According to Aubouin, the eugeosynclinal domain (internides) comprises a eugeo-synclinal furrow and a eugeanticlinal ridge bordering the ocean. Between the internides and the continental craton (foreland) is the miogeosynclinal domain (externides) com-prising a miogeosynclinal furrow and a miogeanticlinal ridge. The two couples are illustrated as if they had reached the same stage of development, but this is not always so: they can reach different stages at the same time. A similar bicouple is illustrated in Figure 9-1b by Kay's section of the Champlain and Magog belts in New York and New Hampshire.

These fundamental units may be paired, according to Aubouin, forming a divergent

bicouple as illustrated schematically in Figure 9-2. This type of geosyncline produces a mountain chain with bilateral symmetry. The two eugeanticlinal ridges may have an intermediate hinterland between them, they may coalesce to form a single ridge, or they may be absent altogether so that the two eugeosynclinal furrows become one. Aubouin also illustrated a complex system of four eu-miogeosynclinal couples forming two di-vergent bicouples and one convergent bi-couple; the axial zone of convergence corre-sponding to two miogeanticlinal ridges is situated between two divergent bicouples formed simply by repeating Figure 9-2.

Kay demonstrated that the geological development of North America was domi-nated by a series of eugeosynclinal-miogeosyn-clinal couples similar to those depicted in Figures 9-1a and 9-1b. He stated that no examples of these types of geosynclines are being formed today. R. S. Dietz, however, suggested that the Atlantic and Gulf Coast sedimentary prisms may be modern examples —an actualistic interpretation of the normal geosynclinal couple. Figure 9-1c is Dietz's suggested reconstruction of the same geosyn-clinal couple depicted by Kay in Figure 9-1b, drawn by analogy with the existing situation off the eastern United States. The eugeosyn-cline is considered to be the continental rise. This is a prism of sediments deposited on oceanic crust at the base of the continental slope largely by turbidity currents laying down turbidites (greywackes). The sediments lap onto the sialic continental slope. Plutonic and volcanic rocks invade the eugeosynclinal sediments. The continental shelf undergoes

TABLE 9-2 Sedimentary and Igneous Rocks Characteristic of Geosynclines
(Modified After Mitchell and Reading, 1969)

Atlantic Type		Andean Type		Island Arc Type		Japan Sea Type
Miogeosyncline	Eugeosyncline	Mountains	Trench	Islands	Trench	Margin of Restricted Basin
Continental Crust	Oceanic Crust	Continental Crust	Oceanic Crust	Intermediate Crust	Oceanic Crust	Intermediate, Modified Crust
Abundant A and B.	Common C. Rare D. Abundant E.	Rare A and B. Rare to abundant F. Abundant H. Common I.	Abundant C. Rare to common E. Common to rare G.	Locally abundant B. Rare C. Abundant F and G. Common I.	Abundant C. Common G.	Abundant A and E. Locally common B. C present if basin floor oceanic. Tuffs of F. Rare G.

Characteristic rock types:

A. Shallow marine and coastal plain clastic sediments
B. Carbonate sediments
C. Interbedded pelagic sediments, tholeiitic lavas, and ultrabasic rocks
D. Tholeiitic volcanic turbidites
E. Compositionally mature turbidites
F. Calc-alkaline volcanic rocks and minor intrusions
G. Calc-alkaline volcanic turbidites
H. Continent-derived coarse clastic sediments
I. Intermediate or acidic plutonic rocks

isostatic subsidence, and this permits a wedge of epicontinental deposits to build up on the marginal flexure of the continental shelf. These shallow-water sediments make up the miogeosyncline. Comparing Figure 9-1c with 9-1b, we see that there are three elements deleted from the classical couple to transform it into Dietz's actualistic couple. These are the outer half of the miogeosyncline, the island arc (eugeanticlinal ridge), and the tectonic borderland (miogeanticlinal ridge). Also a continental slope is added between the geosynclinal prisms. Dietz suggested that the onset of orogeny occurred when the oceanic sea floor became uncoupled from the continental crust and moved down beneath the continent impelled by sea-floor spreading (Chapters 12 and 14).

Two other types of tectonic profiles for geosynclinal systems have been constructed by T. Matsumoto following a review of the circum-Pacific orogenic system. Figure 9-3a is adapted from Aubouin's cross-section (Figure 9-1a). Figure 9-3b shows the reconstruction of a eugeosyncline at the continental margin of east Asia during the Mesozoic period, and Figure 9-3c is a profile showing sites of sediment accumulation in the region of island arcs and ocean trenches in the western Pacific during the Cenozoic period. Some of these such as the Marianas arc appear to be formed on the ocean floor far from a continent, whereas others, such as the Japanese arc, are superimposed on older, sialic basement structures. The rock assemblages characteristic of the geosynclinal environments illustrated in Figures 9-1 and 9-3 are listed in Table 9-2. Matsumoto concluded that the tectonic framework of the circum-Pacific mobile belt may have changed with time from the classical type of Figure 9-3a during the Paleozoic, through the Andean type in Figure 9-3b during the Mesozoic, to the island arc type in Figure 9-3c during the Cenozoic. Thus the complexity of the orogenic cycle becomes even more apparent when we add the dimension of time to the geosynclinal organization.

Geosynclinal Evolution According to J. Aubouin

Figure 9-4 summarizes Aubouin's scheme for the evolution of a eumiogeosynclinal bicouple corresponding to that of Figures 9-1a and 9-3a. This is based on his reconstruction of the geological history of the Hellenides, but he claims that it depicts a general pattern. J. Debelmas, M. Lemoine and M. Mattauer disputed this claim in 1967. They questioned the fitness of Aubouin's model to explain the Hellenides and concluded that it could only provide a very schematic picture of the tectonic style of Mediterranean Alpine ranges. They praised Aubouin's book as a source of valuable information but felt that the tectonic inventory was insufficient.

Figure 9-4a lists three major periods in the evolution of a geosyncline. The geosynclinal period (I), which is of much longer duration than the late-geosynclinal (II) and post-geosynclinal periods (III), is subdivided into the generative (A), the development (B), and the orogenic or terminal stages (C). The development stage is further subdivided into pre-flysch (i) and flysch periods (ii). Figure 9-4b shows the main events occurring at any time in each part of the geosyncline, and Figure 9-4c illustrates these events in six schematic geological sections corresponding to specific times in 9-4b. Note the time scale in Figures 9-4a and 9-4b.

Figure 9-4 illustrates geosynclinal polarity and in particular orogenic polarity. Different regions of the geosyncline experience similar events but at different times. The orogenic stage of the geosynclinal period (IC) begins first in the internal eugeanticlinal zone of the geosyncline and then migrates outward toward the miogeosynclinal zones. The time of transition from development stage (IB) to orogenic stage (IC) and from the geosynclinal period (I) to the late-geosynclinal period (II) thus depends on geographic position within the geosynclinal system.

For a period of about 120 m.y. after the initial subsidence the emission of ophiolites

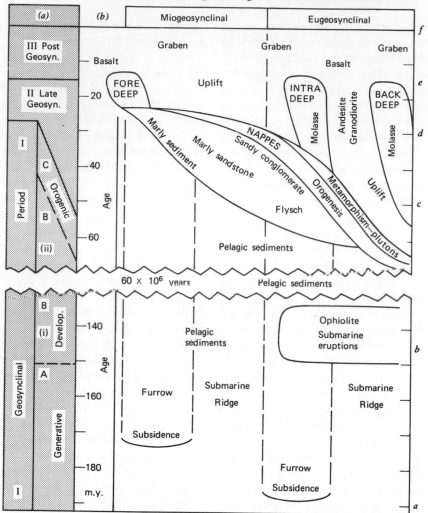

Figure 9-4. (*a*) Evolutionary pattern of a geosyncline, according to Aubouin (1965). Periods:—I. Geosynclinal, II. Late-geosynclinal, III. Post-geosynclinal. Stages: IA, Generative. IB, Development. IC, Orogenic (terminal). The development stage is divisible:—IB(i) pre-flysch period. IB(ii) flysch period. (*b*) Diagrammatic representation of geosynclinal polarity in a bi-couple (Figure 9-1*a*), according to Aubouin (1965). The orogenic stage (IC) begins at the eugeanticlinal ridge and migrates across the system from the interior to the exterior, preceded by the flysch deposits. This is illustrated in Figure *c* by generalized geological cross sections for specific times, *a*, *b*, to *f*. (*c*) Schematic geological cross sections showing the paleogeographical evolution and tecto-orogenic development of a geosynclinal couple as illustrated by the Alpine cycle, based on the Hellenides (after Aubouin, 1965). Sections *a* to *e*, geosynclinal and late-geosynclinal periods. Section *f*, post-geosynclinal period. (With permission of Elsevier Publishing Co.)

is the only activity in the geosyncline apart from the slow accumulation of deep-water, pelagic sediments both in the furrows and on the submarine ridges. Then the orogenic or terminal stage of the geosynclinal period (IC) begins with orogenesis and uplift of the eugeanticlinal ridge. This is followed by rapid erosion, and terrigenous sediments are poured into the eugeosynclinal furrow from the rising highlands. These are poorly sorted sediments of greywacke type collectively termed flysch in the Alpine chains of the

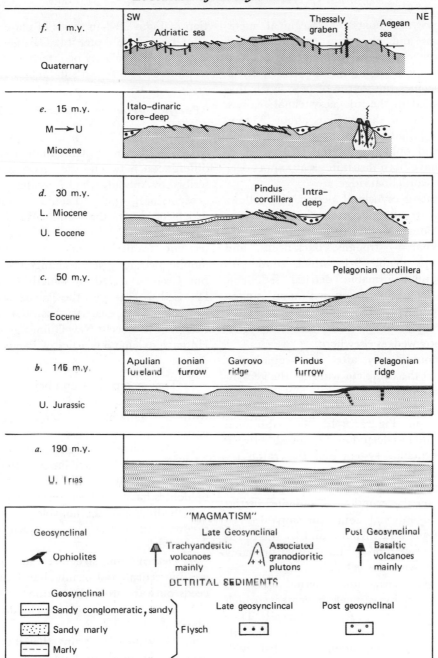

f. 1 m.y.

Quaternary

SW — Adriatic sea — Thessaly graben — Aegean sea — NE

e. 15 m.y.

M →U

Miocene

Italo–dinaric fore–deep

d. 30 m.y.

L. Miocene

U. Eocene

Pindus cordillera — Intra-deep

c. 50 m.y.

Eocene

Pelagonian cordillera

b. 146 m.y.

U. Jurassic

Apulian foreland — Ionian furrow — Gavrovo ridge — Pindus furrow — Pelagonian ridge

a. 190 m.y.

U. Trias

"MAGMATISM"

Geosynclinal

Ophiolites

Late Geosynclinal

Trachyandesitic volcanoes mainly

Associated granodioritic plutons

Post Geosynclinal

Basaltic volcanoes mainly

DETRITAL SEDIMENTS

Geosynclinal

Sandy conglomeratic, sandy

Sandy marly

Marly

Flysch

Late geosynclincal

Post geosynclinal

(c)

Mediterranean region. The deposition of greywackes marks the beginning of the flysch period of the development stage (IBii) which comes before the orogenic stage (IC); but Figure 9-4*b* indicates that in fact the flysch period does not begin until after the orogenic stage has commenced in the inner zones. With continued uplift of the eugeanticlinal ridge and the formation of a cordillera the flysch deposits pile up against the flank of the ridge and migrate across the furrow. As the flysch migrates, with the deposition of successively younger material from the interior to the exterior zones, the zone of orogenesis also migrates. Regional metamorphism and plutonism accompany and follow the orogenesis. Subsequent uplift is followed by erosion and the folded flysch itself becomes the source of new flysch that is deposited in the miogeosynclinal regions. Cross-section *d* shows that about 35 m.y. after the initial onset of orogenesis the sediments and ophiolites of the eugeosynclinal furrow have been thrust in great nappes over the miogeanticlinal ridge.

The late-geosynclinal period (II) follows immediately after the orogenic stage in any zone. Alongside the rising cordillera are formed narrow troughs, back-deeps, and intra-deeps (Table 9-1) which subside to receive unconformable detrital sediments derived from the young mountain ranges. These sediments are termed molasse in the Alpine chains of the Mediterranean region and clastic wedges elsewhere.

By about 50 m.y. after the beginning of orogenesis the complete width of the geosynclinal system is in the late-geosynclinal period of evolution (II) as shown by the cross-section *e* in Figure 9-4c. An additional trough, the foredeep, has developed between the geosynclinal system and the craton, receiving molasse deposits from the highlands rising from the geosyncline. This period is marked by the eruption of andesites in the tectonized internal zones accompanied by some basalts and rhyolites and by the intrusion of granodioritic batholiths probably associated with the volcanic activity.

The post-geosynclinal period (III) is characterized by vertical movements which often cut across the geosynclinal trends. They produce regional arching and tensional rift valleys or graben, often associated with basaltic volcanism. Simultaneously basaltic lavas are erupted on the foreland regions of the craton well beyond the geosynclinal belt.

Aubouin's book was published just before plate tectonics introduced many new concepts related to the significance and evolution of geosynclines. It would be an instructive exercise for students to attempt an interpretation of Figure 9-4 in terms of plate tectonics after reading Chapter 14. Can geosynclinal and orogenic polarity be interpreted in terms of plate collisions and subduction zones?

"Pacific" geosynclines of K. A. W. Crook

In 1969 K. A. W. Crook noted systematic differences between the geosynclines described by Aubouin which he called "Atlantic" geosynclines, and "Pacific" geosynclines which occur in the circum-Pacific region. According to Matsumoto (Figure 9-3) the tectonic framework of the circum-Pacific mobile belt may have changed with time, but Crook referred to something different. He used as examples the Paleozoic Tasman geosyncline of eastern Australia and the Mesozoic to Recent New Guinea geosyncline. He gave as characteristics of Pacific geosynclines:

1. They comprise a number of subparallel volcanic and nonvolcanic troughs and highs with the nonvolcanic elements near the craton.

2. The volcanic troughs contain volcanic-terrigenous flysch-like sediments; the volcanic troughs and highs may lack serpentinites; ophiolites have not been recognized.

3. Sediment in the nonvolcanic troughs is predominantly terrigenous flysch of sialic derivation.

4. Deformation structures are predominantly vertical and terminal tectonism proceeds outward from the craton.

5. Successive pairs of troughs do not display polarity.

6. They do not occupy the sites of older geosynclines and probably develop on a largely simatic floor.

7. They lie between sialic cratons and simatic ocean floors.

Crook concluded that although Atlantic and Pacific geosynclines are superficially similar there are differences between them which are significant and which appear to reflect their different geotectonic environments. Whereas Atlantic geosynclines may develop on the sites of older geosynclines, discordant to their trends, Pacific geosynclines appear to be newly formed on a largely simatic floor, marginal to sialic cratons. The Pacific geosynclines do not exhibit the sedimentary and igneous polarity characteristic of Aubouin's model (Figure 9-4c). The difference may be readily explicable in terms of plate tectonic theory if an Atlantic geosyncline is considered as two Pacific geosynclines which have collided. Figure 14-17 illustrates collision sequences.

EVOLUTION OF CONTINENTS

The hypothesis that continents have grown by a process of marginal accretion was introduced in Chapter 7. The essence of this hypothesis is contained in the concept of geosynclines and the orogenic cycle with the geosynclines developing marginally to continental cratons; the idea is inherent in the early writings of Hall and Dana. New volcanic material is added to the crust from the mantle, dispersed along with pre-existing rocks by sedimentary processes, and the whole complex of rocks is then welded onto the continental plate by metamorphism and mountain building. There are many examples, however, where orogenic belts intersect each other rather than developing in successive, concentric arcs, and one school of geologists maintains that geosynclines form along zones of weakness within continental cratons. Another view is that the continents are not growing at all, but that continental masses are being destroyed by a process of oceanization; continental crust founders and is engulfed in basaltic magma (Chapter 15). In Chapter 14 we examine a fourth view that the oceans open and close with orogenesis and mountain building resulting from collisions of lithospheric plates. Although the size of a continental plate may be increased by marginal accretion resulting from collision at one stage, at another stage a continent may be separated into two plates by rifting and drifting.

One question relevant to all hypotheses of continental evolution is to what extent the continents are increasing in total volume by the addition of new material from the mantle. As noted in Chapter 7 the problem is complicated by the fact that during orogenesis a part of the existing crust is reworked and incorporated into the belts of younger material, and it is difficult to distinguish the regenerated older material from the younger material. We know that magmas from the mantle are added to the crust during the orogenic cycle (Figures 4-1 and 9-4), but we do not know how much of this material is derived from crustal rocks recycled through the mantle (Chapter 14). Strontium and lead isotope evolution trends should eventually solve these problems.

Detailed field mapping, structural studies, and petrological studies have much to contribute, and they provide the essential framework for the interpretation of isotope data. H. R. Wynne-Edwards in 1969 published Figure 9-5 which shows the geological evolution of an area comprising 13,000 square miles of the Grenville province of southwestern Quebec. This is based on systematic mapping and the recognition of structural and textural criteria for subdividing the plutonic and metamorphic rocks. Figure 9-5 merits careful study. The map legend at the right of the diagram representing the rocks as they are now is constructed from a series of steps, like building blocks, added to it at different times. Additions of material represented by the rise in steps up the "lithology" axis are shown for successive orogenies; the Kenoran about 2500 m.y. ago; the Hudsonian about 1750 m.y. ago; the Elsonian about 1400 m.y. ago; and the Grenville about 950 m.y. ago.

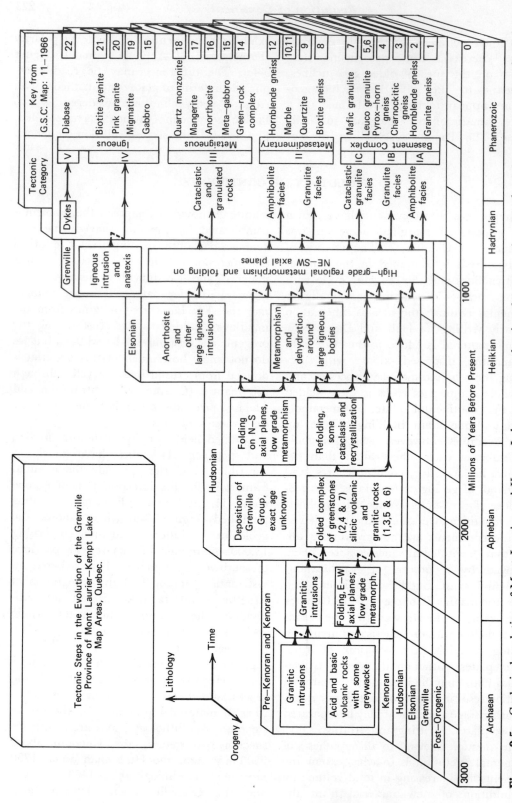

Figure 9-5. Geological evolution of Mont Laurier and Kempt Lake map areas showing the progressive growth of the present geology from the Archaean by a series of tectonic steps in which older rocks were reprocessed and new ones were added (after Wynne-Edwards, 1969, Figure 12, p. 80. With permission of the Geological Association of Canada.)

During each orogeny the older rocks were modified by folding, metamorphism, and sometimes melting, and generally updated by the redistribution of radiogenic products. The recycling process is represented by the steps along the "orogeny" axis. The arrows show the evolutionary paths followed by each block from the time it was added as new material. Rocks in the Grenville province thus show continued tectonic overprinting extending back 2500 m.y. to the Archean.

The Grenville tectonic cycle has been completed from initial deposition to erosion close to base level. The complete cycle includes about 1000 m.y. of sediment deposition and burial with active orogenesis and uplift occupying only 200 to 300 m.y. Most of the rocks of the Grenville province are much older than the Grenville orogeny and therefore they cannot be regarded as new contributions to the continental crust about 950 m.y. ago (Figures 7-13 and 7-14).

CONTEMPORARY GEOSYNCLINES

Criteria for the recognition of contemporary geosynclines include present physiography, whatever geological history can be deciphered from available evidence, and projections into the future. Geosynclines must be of appropriate dimensions and they should exhibit features such as those outlined in Figures 9-1 to 9-4. Mountain systems have arcuate and linear sections, lengths of thousands of kilometers, and widths of a few hundreds of kilometers. Deep troughs are characteristic of the generative and much of the development stage of the geosynclinal period, and thick piles of sediment or severe folding are characteristic of later periods. Our attention is therefore directed to the continental margins: the system of deep ocean trenches and the continental rises (Figure 2-2).

We know that sedimentary and volcanic deposits of geosynclinal dimensions are accumulating on and oceanward of modern continental margins, around island arcs, and in some small ocean basins (Figure 7-8). The characteristics of continental margins were outlined in Chapters 2 and 7. Figure 9-6 shows the distribution of three types according to A. H. Mitchell and H. G. Reading: Atlantic type margins lacking an ocean trench: Andean type margins with a mountain belt bordered by a submarine trench; and Island-arc type margins sepa-

rated by a small ocean basin from the island arc and associated ocean trench.

In their review of continental margins, geosynclines, and sea-floor spreading Mitchell and Reading distinguished five possible types of contemporary geosynclines which are related to the three types of continental margins. An Atlantic type geosyncline is shown in Figure 9-1c, an Andean in Figure 9-3b, and an Island-arc located on and around active island arcs as shown in Figure 9-3c. The small sediment-filled ocean basins on the concave side of many island arcs are the Japan Sea type of geosyncline (Figure 9-3c), and geosynclines of Mediterranean type occur in small oceans between continents. This subdivision of the sites of contemporary sediment accumulation is not a classification of geosynclines equivalent to those in Table 9-1. It is rather an actualistic interpretation of the environments where the classified geosynclines (Table 9-1) may be forming today. Figure 9-6 shows their global distribution. There is no reason to suppose that geosynclinal sequences which appear very similar in the geological record should all have formed in the same kind of environment. On the other hand detailed geological study of geosynclinal sequences should permit paleogeographic reconstruction of the former environment in terms of Atlantic, Andean, or Island-arc types, as previously mentioned

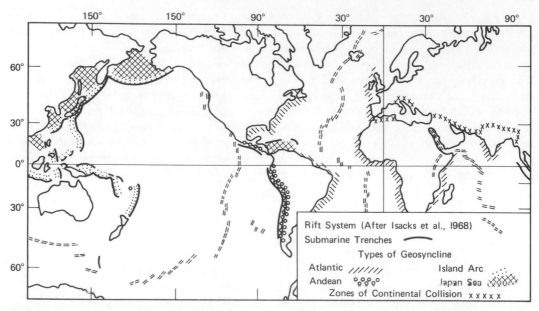

Figure 9-6. Positions of modern geosynclines in relation to world system of rifts and submarine trenches. See also Figures 2-2 and 2-3 (after Mitchell and Reading, 1969, from Journal of Geology with permission. Copyright 1969 by The University of Chicago).

for the circum-Pacific orogenic zone in connection with Figure 9-3. The associations and sequences of sediments and igneous rocks characteristic of each of these environments are summarized in Table 9-2.

Mountain building connected with these geosynclines may be one of three types; Andean, Island arc, or Himalayan. Mitchell and Reading related these to plate tectonic models. An Atlantic type geosyncline may change to a geosyncline of Andean or Island-arc type with the onset of orogeny. Andean and Island-arc type orogenies occur within and are associated with the development of their respective geosynclines. Himalayan type orogeny develops only if the continents drift (Chapter 14), and it then results from the collision of continents; it may affect any type of geosyncline because any one could be trapped between the moving continents.

Finally let us compare some aspects of the geology of a specific modern geosyncline with the evolutionary scheme of Aubouin shown in Figure 9-4. The impressive arcuate

sweep of the Sunda Islands and associated ocean trenches (Figure 9-7) has been linked with the concept of geosynclines since the beginning of this century. The islands are arranged in a double arc. The inner volcanic arc includes the large islands of Sumatra, Java, Bali, Lombok, Sumbawa, and Flores. This arc constitutes a cordillera with Mesozoic sediments partially covering an ancient crystalline basement. Unconformable late-Tertiary sediments also are present, and igneous activity has produced andesite lavas and granodiorite batholiths. The outer, nonvolcanic arc consists of smaller, scattered islands which are merely the points of emergence of a linear shelf. These include the Mentawi Islands, Timor, and Tanimbar Islands. Basement rocks are not exposed and the islands are composed of intensely folded sediments often of deep-water facies. In Timor an ophiolite nappe transported from the north occurs above the recently folded sediments. There is a fairly deep marine trough between these two arcs and the deep Indonesian trench lies outside the outer arc.

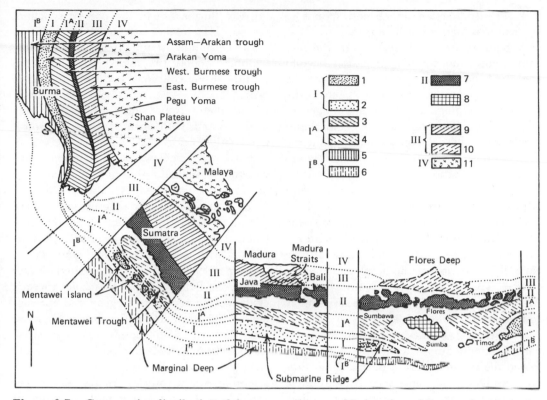

Figure 9-7. Comparative distribution of the structural zones of Indonesia and Burma, demonstrating the transition of structures of the Sunda Islands to the folded chains of Burma, which are quite clearly in the post-geosynclinal stage (Figure 9-4a) (after Aubouin, 1965, from Umbgrove, 1940). 1 = zone of strong Laramide folding; Miocene epochs of compression, weak in Burma, strong in other sectors; zone of strong negative anomalies of gravity. 2 = Submarine continuation of the same zone. 3 = Inner trough, accompanying zone I, filled up with Tertiary sediments in Burma sector. 4 = Submarine inner trough. 5 = Outer trough, accompanying zone I, filled up with Tertiary sediments in Burma sector. 6 = Submarine outer trough, marginal deep. 7 = geanticlines, Tertiary up to Recent volcanism; Miocene diastrophism moderate. 8 = Sumba Island; Upper Miocene folding; Tertiary volcanism; no recent volcanoes; interruption of zones I, I^A, I^B, and of zone of strong negative anomalies of gravity. 9 = Tertiary sedimentation troughs; moderate folding towards the end of the Pliocene. 10 = Submarine continuation of the same troughs. 11 = Regions above sea-level in Tertiary and Pleistocene times. (Reproduced by permission of Elsevier Publishing Co.)

The inner arc shows a positive isostatic gravity anomaly whereas the outer arc and the marine trenches are associated with an extreme negative anomaly.

A cross-section through this island arc shows similarities with the cross-sections *d* and *e* in Figure 9-4c. Aubouin stated that in passing from the inner arc to the outer trench, four features can be correlated directly with the schematic sections:

1. The inner arc with its late-Tertiary back-deep and andesitic and plutonic igneous activity corresponds to the cordillera of the eugeanticlinal ridge.

2. The marine trench between the island arcs corresponds to the intra-deep.

3. The outer arc represents another cordillera with material from the eugeosynclinal furrow being overthrust onto the miogeanticlinal ridge.

4. The Indonesian trench corresponds to the external miogeosynclinal furrow.

Aubouin stated that apart from the fact that there is no continental craton beyond the outer trench the present geology of the Sunda Islands corresponds very closely to the situation existing about 30 m.y. ago in Figure 9-4. The absence of continental craton adjacent to the miogeosynclinal furrow is an unusual arrangement according to other interpretations of geosynclinal couples (Figure 9-1). The inner volcanic arc and adjacent trench are in the late-geosynclinal period (II), the outer arc is in the orogenic stage of the geosynclinal period (IC), and the external trench appears still to be in the early flysch period of the development stage (IBii). The scale and curvature of the Mediterranean-Alpine system are very similar to those of the Sunda Islands.

The structural zones of the Sunda Islands can be extended northward into those of the folded mountain chains of Burma as illustrated in Figure 9-7. This suggests that these zones form part of a single geosyncline. According to Aubouin the alpine-type mountains of Burma are in the post-geosynclinal period (III) further advanced in evolution than the Sunda Islands. According to plate tectonic models the folded Alpine system resulted from collision producing a double system on the Asian continent. The Sunda Islands then represent only the northern side of this system with open ocean to the south and southwest.

This brief review of a contemporary geosyncline indicates that at any given point in time the stages of evolution of a geosyncline not only vary in cross-section but also along its length. Similar conclusions were reached by R. J. Roberts (1969) in connection with the eugeosynclinal zones of the western Cordillera of the United States during the Paleozoic and Mesozoic; there is evidence for repeated orogeny and epeirogeny in different parts of the system for more than 550 m.y. with considerable age variation for the activity both along and across the system. A geosyncline is thus a very complex assemblage of furrows and ridges subjected to similar events at different times in different places.

CAUSES OF SUBSIDENCE AND UPLIFT

The orogenic cycle incorporates most of the geological processes, and these are governed to a large extent by the occurrence and the rate of subsidence and uplift of segments of the Earth's crust. What is less certain is the cause of these major movements although it seems clear that they are second-order effects of major, global processes. The magnitude of the effects, in terms of depth or pressure variations and changes in temperature distribution, can be estimated from study of metamorphic rocks in conjunction with laboratory experiments on metamorphic reactions at known pressures and temperatures.

Subsidence of Geosynclines

Figure 9-8 summarizes in diagrammatic form some of the mechanisms that have been proposed to account for the formation of a geosynclinal trough. It has often been claimed that the development of a geosyncline is a self-generating process as shown in Figure 9-8a. The weight of accumulated sediments is presumed to load the crust sufficiently to produce continued downsinking. Removal of magma from the depths and its eruption at the surface could also contribute to subsidence (Figure 9-8b). The operation of subcrustal currents producing downbuckling of the trough and the formation of a tectogene as depicted in Figure 9-8c is the process that has been invoked most frequently during the last 40 or 50 years. Figure 9-8d and Chapter 14 show a layer of the crust or lithosphere being carried down

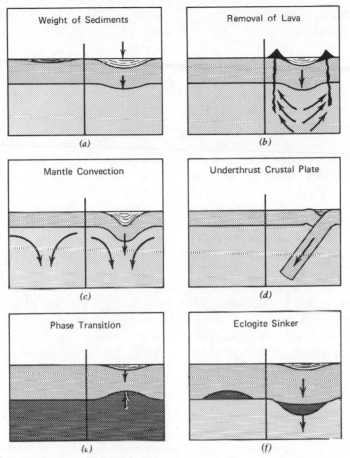

Figure 9-8. Diagrammatic representation of hypotheses for the mechanism of producing subsidence for a geosynclinal trough.

into the mantle as a result of sea-floor spreading (Chapter 12) producing a trough at the surface. According to the phase transition hypothesis cooling at depth causes the conversion of gabbro into eclogite (Chapter 5) with contraction of the deep rocks and the resultant subsidence at the surface. This is illustrated in Figures 9-8e and 12-25 and discussed in more detail in Chapter 10. Figure 9-8f shows a crustal section of gabbro overlying peridotite with a layer of eclogite forming from the gabbro. The denser eclogite would tend to sink dragging down a section of the crust and forming a depression at the surface.

According to Figures 9-8c and d, the vertical movements associated with mountain building are a consequence of horizontal movements occurring beneath the crust, whereas according to Figures 9-8e and f the vertical movements develop independently of lateral movements and the controlling forces are related to changes in heat distribution and local mass readjustments in response to gravity. These represent two viewpoints that are strenuously maintained by different groups of geologists and geophysicists. The intense folding in orogenic belts is developed either as a direct result of compression arising from lateral movement of crustal slabs or as an indirect result of vertical movements with consequent instability and gravity tectonics; compression is then a secondary effect of the vertical movements.

Metamorphic Rocks as Guides to Geosynclinal Conditions

Figure 4-1 shows that subsidence and burial of sediments in the orogenic cycle causes metamorphism, and Figure 4-4 shows the general range of depths (pressures) and temperatures for metamorphic rocks. Figure 9-9 is a more detailed version of the crustal part of Figure 4-4 showing the distribution of the main metamorphic facies in terms of depth and temperature.

The relative positions of the metamorphic facies have been known ever since P. Eskola introduced the concept in 1915. N. L. Bowen described a petrogenetic grid in 1940 suggesting that laboratory measurements at high pressures and temperatures would permit calibration of the metamorphic reactions that form the boundaries of metamorphic facies. Apparatus became available for such experiments in 1949 (Chapter 8) and in 20 years sufficient accurate experimental data has been gathered to provide diagrams such as Figure 9-9. According to F. J. Turner, who presented an experimental appraisal of critical metamorphic reactions in his 1968 book, Bowen's abstract concept has now become an instrument increasingly capable of effective use in calibrating temperature and pressure gradients in metamorphic terranes.

Figure 9-9 shows Turner's preferred estimates of the positions of selected metamorphic facies in terms of depth and temperature based on the measured positions of mineralogical reactions inferred to have occurred in the formation of various metamorphic rocks. Figures 4-4 and 7-11 show estimated continental geotherms for comparison with Figure 9-9.

If sediments are buried in a geosyncline they will be subjected to increases in pressure and temperature as they follow a geotherm. They become metamorphosed according to the position of the geotherm and the changes in temperature distribution caused by the subsidence of cool sedimentary rocks or by the uprise of magma and they subsequently become uplifted and exposed by erosion. The sequence of metamorphosed rocks in the roots of each mountain chain is unique,

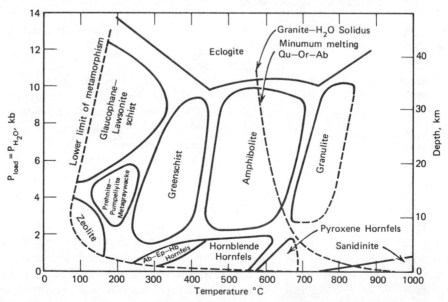

Figure 9-9. Tentative scheme of metamorphic facies in relation to total pressure (assumed equal to P_{H_2O}) and temperature; all boundaries are gradational (after Turner, 1968). See Figure 4-4. (From "Metamorphic Petrology". Copyright McGraw-Hill Book Company, 1968. Used with permission.)

and the mineralogy of the rocks contains the imprint of the range of pressures and temperatures occurring in the sequence during its metamorphism. From study of the mineralogy the positions of the rocks on Figure 9-9 can be located. In some terranes the rock sequence is from greenschist, through amphibolite, to granulite facies, and in others blueschists (glaucophane schists) and eclogites are formed.

The development of granulite facies rocks requires temperatures considerably higher than those indicated by the continental shield geotherm (Figure 3-12) suggesting that regional metamorphism is accompanied by higher than normal heat flow from the mantle. On the other hand the formation of blueschists requires very high pressures or depths of burial with very low temperatures. Any scheme of global tectonics must provide an explanation for the extraordinarily wide range of temperatures to which metamorphic rocks have been subjected at equivalent depths in different geosynclines.

Theories of Global Tectonics

There have been five main global theories: (a) the Earth is rigid and contracting, (b) the Earth is rigid and expanding, (c) the Earth is mobile with the continents drifting over the mantle, (d) the Earth is mobile with the ocean floors spreading apart probably because of convection in the mantle, and (e) the Earth is mobile with rigid plates moving over the asthenosphere. Theories (c), (d) and (e) are not independent of each other (Chapter 14).

In the nineteenth century the contraction theory was accepted as the cause of folding, thrusting, and mountain building. In its classical form the contraction theory holds that the Earth is contracting because it is cooling, and the outer relatively cool and rigid zone is therefore compressed. Lord Kelvin's celebrated model of an Earth cooling from a molten state had to be modified with the discovery of radioactivity and the study of the distribution of radioactive elements within the Earth, but the contraction theory persists and most textbooks have been written with this concept as their foundation. Applications to global and regional tectonics were reviewed in the 1959 book by J. A. Jacobs, R. D. Russell, and J. T. Wilson. Evidence of worldwide rifts in the ocean basins indicative of tension has made this theory less attractive.

One of the pioneers of the expansion hypothesis, O. C. Hilgenberg, suggested a crustal expansion rate of 0.17 cm/year up to the early Mesozoic, followed by a faster rate

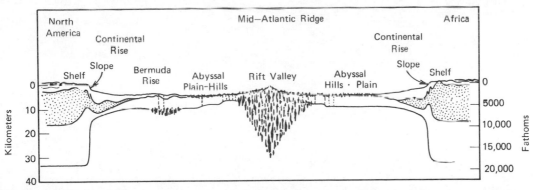

Figure 9-10. Physiographic provinces and trans-Atlantic structure (after Heezen et al; 1959). Based on scattered seismic-refraction measurements in the North Atlantic which have been projected along province boundaries. The topographic profile was pieced together from continuously recorded echo-sounding profiles from New York to Spanish Sahara. (With permission of the Geological Society of America.)

of 1.1 cm/year. Decrease in G, the gravitational constant, would produce expansion. L. Egyed published papers advocating a uniform rate of radius increase of about 0.05 cm/year based on an inferred phase transformation in the Earth's core, and estimates that the areas of the crust covered by shallow seas have decreased with time. The theory was revived in the 1950's by several sets of independent interpretations some of which are illustrated in Figure 12-10. Carey's ideas are outlined in Chapter 11. The worldwide rifts mentioned above were interpreted by B. C. Heezen as rents in the crust resulting from expansion (Figure 9-10). Reviews of the expansion theory are given by Irving (1964, p. 288–292) and Holmes (1965, p. 965–975). Available evidence does not support this process as a cause of global tectonics.

Continental drift is reviewed in Chapter 11, sea-floor spreading in Chapter 12, and plate tectonics in Chapter 14. The occurrence of mantle convection would not preclude either contraction or expansion of the Earth.

The relative significance of vertical and horizontal movements as driving forces in tectonics is still in dispute (Figure 9-8), but advocates of horizontal movement of lithospheric plates have certainly dominated the literature since 1967. Vertical movements are reviewed in Chapter 10 and horizontal movements in Chapters 11-14.

10. *Tectonic Significance of Phase Transitions*

INTRODUCTION

There is no doubt that parts of the outer shell of the Earth have undergone repeated subsidence and uplift, and this is the reason for the operation of most of the geological processes outlined in Chapter 4. Without subsidence there would be no great piles of sedimentary rocks, and without uplift there would be no mountains or exposed metamorphic rocks. The vertical movements may involve large blocks of the crust or elongated belts of the crust. There is some question as to whether these movements are a secondary effect caused by mantle convection and lateral movement of the lithosphere or whether they are caused directly by volume changes at depth within the crust and mantle. Volume changes may be caused by localized heating or cooling, by phase changes such as melting and magma crystallization, or by solid-solid transitions such as those involved in the transformation of gabbro to eclogite.

In the preceding chapter we noted that one possible cause for subsidence in geosynclines was conversion of gabbro to eclogite (Figure 9-8). In Chapter 5 we noted that in tectonically active regions the geophysical evidence concerning the nature of the crust-mantle boundary was consistent with the existence of material involved in this phase transition (Figures 5-4 and 5-18). Furthermore in Chapter 6 we concluded that phase transitions occur at several depths within the mantle and specifically within

the upper 100 km of the mantle. Migration of a phase transition zone at depth must have some effect on the elevation of the surface. Thermal or pressure perturbations at depth may initiate changes of phase, and these changes, involving latent heat, must in turn exert considerable influence on the temperature distribution within the earth.

Successive papers have reviewed various aspects of the problem, and we will examine some of these in historical sequence because this provides a convenient development from simple to increasingly complex models. In all models reviewed the problem was simplified by treating the phase transition as univariant rather than divariant. It appears from Figure 5-17 that the gabbro-eclogite transition may be considered to approximate a pair of univariant reactions. The effects of expansion and contraction at depth resulting from changes in temperature distribution were described by H. H. Hess for the peridotite-serpentinite transition and by J. F. Lovering for the gabbro-eclogite transition. Lovering discussed conversion of gabbro to eclogite or of eclogite to gabbro with resultant contraction or expansion respectively. Contraction at depth would produce subsidence at the surface, and expansion at depth would produce uplift at the surface. G. C. Kennedy considered in addition the effects of isostasy, and he pointed out that crustal thickening caused by downward migration of the Moho would be followed by isostatic uplift, whereas crustal thinning would be followed by isostatic subsidence.

The isostatic effect is illustrated schematically in Figure 10-1. Figure 10-1*a* is a standard representation of blocks of wood (*B*) floating in isostatic equilibrium in a fluid (*E*). The surface level of the fluid is maintained constant in all figures by some method not illustrated. The lower boundaries between the blocks *B* and the fluid *E* represent the Moho. We will change conditions for the middle block with the other two remaining fixed for reference. We can represent downward migration of the Moho by placing an additional block of wood beneath the central block as shown in Figure 10-1*b*. The thickened, light crust is floated up to higher

levels as shown in Figure 10-1*c* in order to restore isostatic equilibrium. Similarly in Figure 10-1*d* migration of the Moho upward can be represented by cutting off a piece of the middle block and removing it. The thinner, light crust therefore loses buoyancy and sinks for restoration of isostatic equilibrium as illustrated in Figure 10-1*e*.

Kennedy also considered the effects of loading or unloading on the surface with consequent pressure perturbations at depth. This led him to the problem of the effects of sediment deposition, which produces first a pressure perturbation followed by a temperature perturbation because of the thermal blanket effect of a column of sediments. It is this problem which has occupied the attention of subsequent workers.

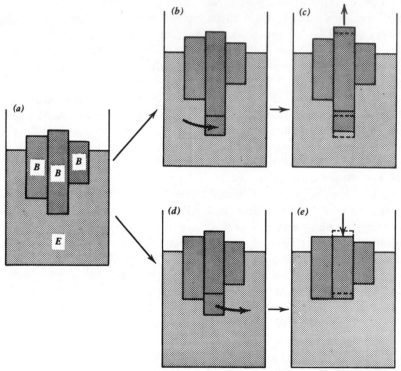

Figure 10-1. Illustration of isostatic adjustment resulting from motion of a phase transition boundary. Wooden blocks, *B*, floating in fluid, *E*, represent crustal basalt in isostatic equilibrium in mantle eclogite. The level of the fluid is maintained constant in all vessels by a device not illustrated. (*a*) Original condition. (*b*) Additional block of *B* is inserted below the center block; this displaces fluid, and is equivalent to conversion of eclogite to basalt and thickening of the crust. (*c*) The center block floats up until it reaches a new equilibrium position. (*d*) A portion of the centre block of (*a*) is cut off and removed; this is replaced by fluid, and is equivalent to conversion of basalt to eclogite and thinning of the crust. (*e*) The center block sinks down until it reaches a new equilibrium position.

EFFECT OF THERMAL PERTURBATIONS

Peridotite-Serpentinite Model of H. H. Hess

Hess drew attention to the tectonic implications of serpentinization of mantle peridotite in 1955. This is a chemical reaction rather than an isochemical phase transition, and it requires the addition of water and possibly the migration of other constituents in solution. The precise chemistry of the serpentinization process remains a matter of debate. Hess assumed that serpentinization of peridotite was accompanied by expansion but this too remains debatable. It was shown in Figure 5-7 that with the present temperature distribution in the Earth the upper mantle in many tectonic environments lies within the stability field of serpentine, so that if water were available the mantle peridotite would become serpentinite. The possible consequences of this are illustrated in Figure 10-2.

Figure 10-2a shows the position of the reaction curve for the dehydration of serpentine, which is taken as the dehydration boundary for serpentinite and the initial

position of the geotherm beneath the oceanic section shown in Figure 10-2b. The point of intersection, A, provides the level in the mantle of Figure 10-2b above which the peridotite lies within the stability field of serpentinite. If water and other volatiles slowly leak from the deep mantle along a favorable zone then peridotite would be converted to serpentinite above the level A as shown in Figure 10-2c. The volume increase so produced would cause the crust to rise with the amount of uplift increasing as the serpentinization worked gradually upward. If the top of the serpentinization zone reached the base of the crust the additional water would then be ejected through the crust with no further tectonic effect. If warming occurred at depth the geotherm in Figure 10-2a would move to higher temperatures and the intersection point A would migrate toward B. This would cause dehydration of serpentinite at the base of the zone as illustrated in Figure 10-2d, and the water would migrate upward and escape through the crust. The decrease in volume so produced would cause subsidence at the surface. Figure 10-2e shows

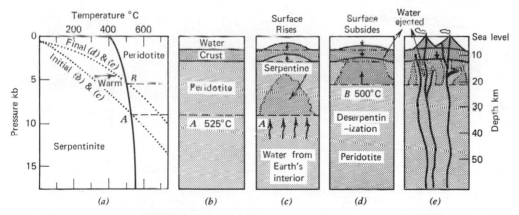

Figure 10-2. Tectonic effect of serpentinization of a peridotite mantle. (*a*) Peridotite-serpen. nite reaction boundary based on Figure 5-7a. Initial geotherm intersects boundary at *A* for the sections (*b*) and (*c*), and warming at depth shifts the intersection point to *B* for (*d*) and (*e*). (*b*) Initial section, showing mantle peridotite at temperature below 525°C that is in the stability field of serpentinite, if water should become available. (*c*) Effect of migration of water upward across boundary *A*. (*d*) Effect of warming at depth, and deserpentinization. (*e*) Effect of volcanic activity in producing localized dehydration. (With permission of The Geological Society of America. (*b*) to (*e*) after Hess, 1955.)

that local volcanic activity could also cause deserpentinization with resultant subsidence at the surface.

Hess discussed applications of this kind of process to oceanic features with specific reference to the mid-Atlantic ridge and the guyots of the Pacific Ocean. The amount of serpentinite formed, and therefore the amount of expansion produced, is dependent on the amount of water available. Hess' later development of the idea of serpentinization of peridotite beneath the midoceanic ridges and its lateral migration forming the crust of the ocean basins is reviewed in connection with sea-floor spreading in Chapter 12. Hess also suggested that the elevation of the Colorado Plateau probably results from an expansion reaction, and that possibly serpentinization of peridotite below the Moho might account for this epeirogenic movement.

Gabbro-Eclogite Model of G. C. Kennedy

Figure 10-3 illustrates Kennedy's explanation for the uplift of plateaus. The initial geotherm intersects the basalt-eclogite phase transition boundary at point A, which produces the crustal section in Figure 10-3b.

Warming at depth changes the position of the geotherm with the point of intersection moving down to B and producing the crustal section in Figure 10-3c. Migration of the Moho from A to B causes conversion of eclogite to basalt with a volume expansion of at least 10%. This expansion causes uplift at the surface. The thickened crust is then uplifted further by isostatic adjustment as illustrated in Figures 10-3d, 10-1b, and c. This causes the Moho to be moved upward as well which means that the point of intersection B in Figure 10-3a also has to change. This means in turn that the temperature distribution at depth must change again. Now we begin to see that the problem involves many factors. One most important factor neglected so far is the latent heat of the gabbro-eclogite phase transition. Heat is absorbed when eclogite is converted to gabbro, so that as soon as migration begins from A toward B the phase change tends to oppose the regional warming trend and to maintain the Moho in its initial position.

The reverse process of cooling at depth would cause the Moho to migrate upward with conversion of basalt to eclogite and resultant contraction and subsidence at the

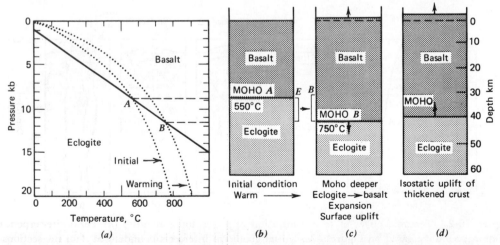

Figure 10-3. G. C. Kennedy's model illustrating qualitatively the uplift of plateaus. (a) The initial geotherm intersects the basalt-eclogite phase transition boundary at A. With warming at depth, the geotherm moves and the point of intersection migrates to B. (b) Crustal model corresponding to initial condition in (a). (c) Crustal model corresponding to later condition in (a). (d) Later isostatic uplift of the thickened crust.

surface. The effect of isostasy would then produce heat, opposing the cooling effect and opposing migration of the Moho. Obviously the rate of heat transfer to and from the phase boundary is an important parameter in this problem.

EFFECT OF PRESSURE PERTURBATIONS

The effect of a change in pressure at depth can be illustrated to a first approximation using Kennedy's example of persistent mountain ranges. Figure 10-4 illustrates a geotherm intersecting the phase transition boundary at point A and producing the crustal section in Figure 10-4b. This section includes a mountain range with average height above sea level of 4 km. Kennedy pointed out that as the mountains are eroded the pressure at the Moho beneath the mountains decreases causing downward migration of the Moho with deepening of the light roots that float the mountains upward again. This is illustrated schematically in Figure 10-4. Suppose that the mountain range could be removed by erosion instantaneously without producing any change in the position of the geotherm. This would provide a pressure decrement at depth of about 1 kb. The depth scale in Figure 10-4 is fixed by the geometry of the Earth, and the pressure scale therefore moves downward after erosion. At each depth the pressure becomes about 1 kb less than it was before erosion. The phase transition is a function of pressure rather than depth, and the position of the phase transition after erosion has moved deeper by about 1 kb as shown in Figure 10-4a. The point of intersection corresponding to the Moho therefore moves to point B giving the crustal section in Figure 10-4c. This requires conversion of eclogite to gabbro, expansion, and uplift. Again we have neglected the opposing effect of the heat absorbed by the phase transition which would tend to transfer the geotherm to lower temperatures and thus move the intersection point B to lower pressures and lower temperatures along the

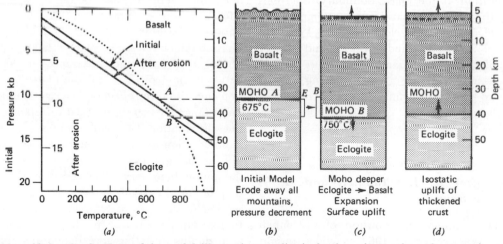

(a)   *(b)*   *(c)*   *(d)*

Figure 10-4. G. C. Kennedy's model illustrating qualitatively the rejuvenation of mountains and the permanence of continents. (*a*) and (*b*) Initial condition, with the Moho intersection at A. The mountains are completely eroded away, with no temperature change at depth, causing the phase transition boundary to move downward in (*a*) to a new intersection point B. The Moho thus moves to new position in (*c*). (*c*) Crustal section after erosion, showing uplift caused by conversion of eclogite to basalt. (*d*) Later isostatic uplift of thickened crust.

new phase transition boundary. The additional uplift resulting from isostatic adjustment is illustrated in Figure 10-4*d*. Thus the effect of erosion is to rejuvenate the mountains.

The deposition of sediments in a subsiding basin provides pressure increments at depth, with results opposite to those illustrated in Figure 10-4 as shown in Figures 10-5*a*, *b*, and *c*. The problem was outlined in a general way by Kennedy. The first effect of pressure loading

by sediments would be the opposite of that of unloading by erosion which is illustrated in Figure 10-4. Initially the pressure at the Moho would increase as sediments displaced water causing upward migration of the Moho, conversion of gabbro to eclogite, and therefore continued subsidence at the surface. The column of sediments behaves as a thermal blanket, however, and the problem involves both pressure and temperature perturbations.

SEDIMENT DEPOSITION: SUBSIDENCE FOLLOWED BY UPLIFT

It is generally considered that the deposition of large thicknesses of sediments such as those occurring in geosynclines cannot be explained by the classical theory of isostasy; but the sequence of subsidence, sediment deposition, and uplift becomes explicable if there is a phase transition involved in the region of the crust-mantle boundary. The problem of determining the changes in depth of a phase transition boundary in response to both temperature and pressure perturbations is difficult because it involves a nonlinear condition at the phase boundary. Solutions to the problem have been attempted by MacDonald and Ness (1960), Wetherill (1961), van de Lindt (1967), and Joyner (1967). O'Connell and Wasserburg (1967) used analytic expressions applying to the initial part and the final part of the motion of a phase boundary.

Dynamics of Motion of a Phase Boundary after R. J. O'Connell and G. J. Wasserburg

The detailed contribution of O'Connell and Wasserburg in 1967 provides insight into the characteristics of the problem of sediment loading at the surface with a phase transition boundary at depth. They presented approximate analytic solutions with both impulsive and continuous loading. The

effects of isostasy, sedimentation, and erosion were considered as modifiers of the essential process of the response of the phase transition to changes in pressure. We should be aware of their conclusions before proceeding with review of the various models developed by others between 1960 and 1967.

They covered first the steady state behavior and then the dynamic problem. They obtained numerical results for various models introducing the effects of thermal blanketing, the time at which this would cause reversal of the motion of the phase boundary, and the effects of isostasy. They concluded that the important parameters affecting motion of the phase boundary were the latent heat of the phase transition, the difference in slope between the geotherm and the phase transition curve, and the effect of isostasy. Of minor importance are convective heat transport and the distribution of heat sources. In the sedimentation model it is the rate of removal of heat that governs the rate of movement of the phase boundary. The redistribution of the latent heat liberated may produce significant changes in temperature distribution within the earth, extending to considerable depths and therefore depending upon deep-seated thermal conditions (Figure 10-5). The long-term motion of the phase boundary depends primarily upon the overall geometry of the model and the boundary condition at depth.

The Models of G. J. F. MacDonald and N. F. Ness and G. W. Wetherill

There are two ways to examine the motion of a phase boundary resulting from sedimentation. One is to consider steady-state or equilibrium configurations, and the more difficult approach is to determine the motion of the phase boundary as a function of the sedimentation rate and the thermal properties of the material involved in the phase transition. MacDonald and Ness used both approaches (1960), and Wetherill (1961) determined the effect of several variables by using steady-state calculations. His results differed from those of MacDonald and Ness for reasons discussed in his paper.

We can use Wetherill's model in Figure 10-5 to illustrate the problem before considering the models in detail. The first effect

of sediments accumulating in a trough would be to increase the pressure at depth with very little change in temperature. The pressure scale and the phase transition curve would then move upward relative to the fixed depth scale as shown in Figure 10-5a. Figure 10-5c shows a hypothetical stage where the original 1 km water in the trough has been completely replaced by 1 km of sediments. This would produce a pressure increment of about 0.14 kb and, neglecting isostasy and heat effects, the phase boundary would therefore migrate upward to the point of intersection B. This would cause conversion of gabbro to eclogite, contraction, and resultant subsidence of the trough. Thus the short-time effect of rapid sedimentation, as in a geosyncline, would be one of sinking. The sediments accumulating in the trough, however, are of low thermal conductivity and

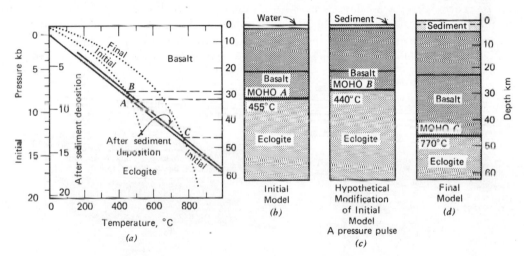

Figure 10-5. G. W. Wetherill's steady state crustal sections showing the effect of deposition of 6 km of sediments in a trough originally 1 km deep. (*a*) Initial condition: assumed phase transition boundary is intersected by the geotherm at *A*, corresponding to crustal section in (*b*). After deposition of 5 km sediment the final equilibrium situation provides intersection *C* with the phase transition boundary (migrated upward after sediment deposition) and crustal section (*d*). (*b*) Initial crustal section. (*c*) Hypothetical modification of initial model: 1 km of water in (*b*) is replaced by 1 km sediment with no change in temperature at depth. Phase transition boundary in (*a*) moves upward (contrast Fig. 10-4) producing new intersection point *B*. This would cause the thinned crust to sink in order to restore isostatic equilibrium; compare Fig. 10-1 (*d*) and (*e*). (*d*) Final equilibrium situation after deposition of 5 km of sediments. The effect of the sediment as a thermal blanket causes heating at depth, with the equilibrium geotherm intersecting the equilibrium phase transition boundary (only approximate in (*a*)) at point *C*, with crustal thickening.

they are richer in radioactive materials than the surrounding rocks. The new sediments act as a thermal blanket with their own heat source, and given sufficient time the temperature at depth will slowly rise causing the phase discontinuity to reverse its motion and migrate downward as illustrated in Figures 10-3 and 10-5. This causes conversion of eclogite to gabbro, expansion, and uplift. The crust thickens and the column of sediments in the geosyncline rises to form a mountain range as shown in Figure 10-5d, the final equilibrium state after deposition of 5 km of sediment.

Figures 10-5b and d show the equilibrium states calculated by Wetherill before and after deposition of 5 km of sediments Figure 10-5a shows the initial position of an assumed phase boundary and the calculated geotherm for the initial crustal model adopted by Wetherill. This is a trough of water 1 km deep above a two-layer crust with total thickness 30 km; the depth to the Moho, the phase change discontinuity, is therefore 31 km (Figure 10-5b). This model is in isostatic equilibrium. After deposition of 5 km of sediments with properties shown in Figure 10-5d, Wetherill calculated that for isostatic and thermal equilibrium the Moho was depressed to a depth of 46.7 km below the surface, and the surface of the sediments was elevated 1.7 km above sea level. The final temperature distribution was as shown in Figure 10-5a. For convenience to prevent overcrowding of the diagram the final position of the phase transition curve in Figure 10-5a is made to coincide with that drawn for the hypothetical intermediate stage.

MacDonald and Ness (1960) studied the motion of the phase change boundary with respect to time for several crustal models. The results of their time-dependent study of "Model C" are shown in Figure 10-6b by the dashed lines. The position and slope of phase transition curve in Model C are very similar to those used by Wetherill (Figure 10-5a). Sedimentation begins initially in a basin 3 km deep, with the phase boundary at a depth of 30 km. The initial effect of

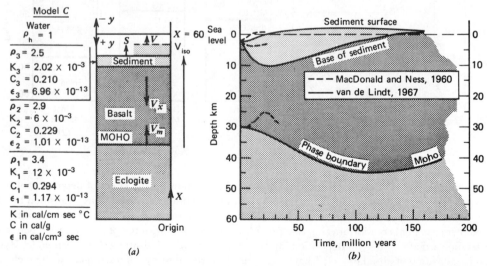

Figure 10-6. Time-dependent solutions for sediment deposition in a trough. (a) Crustal section with physical properties of Model C of MacDonald and Ness. The dynamic model is from van de Lindt, and the meaning of the velocity symbols and distance symbols is explained in the text. (b) The motions of the sediment surface, base of sediments, and phase transition boundary at depth calculated for Model C in (a). The calculations of MacDonald and Ness neglected isostasy whereas van de Lindt included effects of isostatic adjustment. (From Jour. Geophys. Res., **72**, 1289, 1967, with permission.)

sedimentation causes the phase boundary to rise, but the trough becomes filled with sediments in about 8×10^6 years. After 21×10^6 years the phase discontinuity has migrated downward again, and the sediments have become uplifted to a height of 240 m above sea level. MacDonald and Ness drew attention to the three principal time constants involved in the problem: the rate of sedimentation, the rate for obtaining thermal equilibrium, and the rate of establishment of isostatic equilibrium. Rates of sedimentation can be estimated from geological observations. They assumed that the time constant for attainment of isostatic equilibrium is small compared with the time required to attain thermal equilibrium, and they estimated that the time scale for approaching thermal equilibrium would be of the order of 100×10^6 years. The time interval of 21×10^6 years plotted in Figure 10-6b is thus not long enough for equilibrium conditions to be attained. Because of the mathematical difficulties, MacDonald and Ness neglected in their calculations the effects of isostasy and the heat carried by mass transport as the layers moved up or down. Wetherill concluded that both of these factors are likely to be important, and that in order to make further progress in solving the problem it was necessary to obtain time-dependent solutions with these effects considered; it was six years before such solutions were published (Figures 10-6 and 10-7).

Time-Dependent Solution of W. J. van de Lindt

W. J. van de Lindt (1967, 1968) used the crustal model illustrated in Figure 10-6a to calculate the movement of the Moho and the base and surface of the sediments as a function of time. Figure 10-6a provides a fairly simple picture of the various factors, velocities, and relative velocities involved in the problem, so we will examine his treatment of this model in some detail. The numerical results shown in Figure 10-6b are for Model C of MacDonald

and Ness. The origin of the coordinate system is chosen below the Moho and the system is at rest compared to the upper mantle; however, it is moving with respect to a coordinate system fixed in the center of the Earth, because of the isostatic adjustment taking place as a result of the surface loading by sediments. The assumption that this adjustment is instantaneous will not cause large errors as the time necessary to reach isostatic equilibrium is about 10^4 years, whereas the time constants involved in heat conduction are of the order of 40×10^6 years.

The depth of the surface of the Earth beneath the water (constant sea level is assumed throughout) is y, and if the surface rises above the water y is negative. The rate of sedimentation or erosion is taken as proportional to y. The sedimentation speed is s; that is, a layer of thickness s is added to the surface per unit time. The speed with which the surface of the sediment approaches the surface of the water is v. The speed of isostatic adjustment is v_{iso}, and at the depth of isostatic equilibrium in the mantle the pressure will remain unchanged although sedimentation occurs, giving

$$dP/dt = \rho_s s - \rho_h v + \rho_1 v_{iso} = 0 \quad (10\text{-}1)$$

$\rho_s s$ is the weight of the sediment added, $\rho_h v$ is the weight of displaced water, and $\rho_1 v_{iso}$ is the amount of mantle material added to the vertical column because of the isostatic adjustment. If the upward velocity of the Moho with respect to the mantle material is v_m, and v_x is the downward velocity of the material immediately above the Moho, then

$$v_m \rho_1 = v_m \rho_2 + v_x \rho_2 \quad (10\text{-}2)$$

and hence v_x is determined as a function of v_m.

The speed v of the Earth's surface is given by:

$$v = v_{iso} + s - v_x$$

$$= v_{iso} + s - v_m \frac{\rho_1 - \rho_2}{\rho_2} \quad (10\text{-}3)$$

From Equations 10-1 and 10-3 v and v_{iso} can be expressed in terms of s and v_m, which gives

$$v = \frac{(\rho_1 - \rho_s)}{(\rho_1 - \rho_h)} s - \frac{\rho_1(\rho_1 - \rho_2)}{\rho_2(\rho_1 - \rho_h)} v_m$$

$$\text{for} \quad y \geqslant 0 \quad (10\text{-}4)$$

If the Earth's surface is above water there is no longer any displacement of water and the equation becomes

$$v = \frac{(\rho_1 - \rho_s)}{\rho_1} s - \frac{(\rho_1 - \rho_2)}{\rho_2} v_m$$

$$\text{for} \quad y < 0 \quad (10\text{-}5)$$

The velocity of the Moho is a function of rates of heat transfer as indicated in a general way in our discussions of Figures 10-3, 10-4, and 10-5. This problem was tackled by van de Lindt as follows. He assumed that the density, ρ, specific heat, c, heat conductivity, k, and radioactive heat source strength, ϵ, in each layer is constant, although they may vary from layer to layer as shown in Figure 10-6a. In each layer, i, the heat conduction equation is valid

$$\frac{\partial T}{\partial t} = \frac{ki}{\rho_i c_i} \frac{\partial^2 T}{\partial x^2} + \frac{\epsilon_i}{\rho_i c_i} - v_i \frac{\partial T}{\partial x} \quad (10\text{-}6)$$

where T is the temperature and t the time. The first term on the right hand side is the diffusion term, the second is the radioactive heat source term, and the third represents the heat transport due to mass transport. The velocity v_i is zero for the material below the Moho, and it will have the same value for all layers above the Moho. At all interfaces between layers with different physical properties, except the Moho, the interface conditions are

$$T_i = T_{i+1} \quad (10\text{-}7)$$

and

$$k_i \left(\frac{\partial T}{\partial x}\right)_i = k_{i+1} \left(\frac{\partial T}{\partial x}\right)_{i+1} \quad (10\text{-}8)$$

which describes the continuity of heat flux across the boundary. It is assumed that the heat flux from the interior of the Earth is known and constant in time at the origin giving

$$- k \left(\frac{\partial T}{\partial x}\right)_{x=0} = Q \quad (10\text{-}9)$$

At the Moho the boundary conditions are

$$T_1 = T_2$$

and

$$k_1 \left(\frac{\partial T}{\partial x_1}\right) - k_2 \left(\frac{\partial T}{\partial x_2}\right) = \rho_1 r v_m \quad (10\text{-}10)$$

where r is the latent heat of the phase transition. This assumes slow rates of sedimentation and only a small temperature rise at the Moho, otherwise r would have to be replaced by $(r - c\Delta T)$.

When slow sedimentation starts the reaction curve for the phase transition moves from A as illustrated in Figure 10-5a, and the Moho attempts to follow to B while the latent heat released moves the geotherm to higher temperatures tending to maintain the Moho at the deeper level. Diffusion begins to take place, and the Moho eventually moves to its new position of equilibrium. The rate of temperature increase is

$$\left(\frac{dT}{dt}\right)_{\text{Moho}} = g \left(\frac{dP}{dt}\right)_{\text{Moho}} \quad (10\text{-}11)$$

where g is the slope of the reaction curve, (dT/dP). The total rate of change of pressure at the Moho is given by

$$\left(\frac{dP}{dt}\right)_{\text{Moho}} = \rho_s s - \rho_h v - \rho_1 v_m \quad (10\text{-}12)$$

and substitution of (10-4) into (10-12) and the result into (10-11) gives the rate of temperature rise at the Moho as

$$\left(\frac{dT}{dt}\right)_{\text{Moho}} = g \left[\rho_s - \rho_h \frac{\rho_1 - \rho_s}{\rho_1 - \rho_h}\right] s$$

$$- g \left[\rho_1 - \rho_h \frac{\rho_1(\rho_1 - \rho_2)}{\rho_2(\rho_1 - \rho_h)}\right] v_m$$

$$\text{for} \quad y \geqslant 0 \quad (10\text{-}13)$$

If y is negative there is no longer a displacement of water and Equation 10-13 becomes

$$\left(\frac{dT}{dt}\right)_{\text{Moho}} = g(\rho_s s - \rho_1 v_m) \quad \text{for} \quad y < 0$$

$$(10\text{-}14)$$

These are the equations necessary for solution of the problem, and van de Lindt solved it on a digital computer by numerical means. His results for Model C are shown in Figure 10-6 and contrasted with the very different results obtained by MacDonald and Ness for the same model. The principal cause of the difference is the larger amount of sediment deposited, and this was caused in turn by taking into account the isostatic adjustment. The results of van de Lindt are closer to the steady-state results of Wetherill which were also obtained by isostatic considerations (Figure 10-5).

The original trough with a depth of 2.7 km becomes filled to sea level in about 1.8×10^7 years after which the land rises above sea level. The maximum thickness of the sediment is about 10 km. Notice that the initial movement of the Moho in response to the rapid sedimentation is not upward as predicted in the simple explanations (Figure 10-5c) and in the results of MacDonald and Ness (Figure 10-6). Thinning of the crust for about 20×10^6 years, however, does indicate upward movement at the phase boundary relative to the layers above. Similar calculations for other models by Joyner (1967) do indicate initial upward movement of the Moho. The downward movement of the Moho which accompanies thickening of the crust produces upheaval of the sediments, and the maximum depth to the Moho in Figure 10-6 is 45 km (compare Figure 10-5d). The maximum elevation of the sediments is about 3 km. After about 120×10^6 years the Moho starts to rise again aiding the effects of erosion in lowering the elevated sediments and introducing the prospect of oscillatory motions of the Earth's surface about sea level. Using a modified computer program van de Lindt extended the time scale to

400×10^6 years, with the results shown in Figure 10-7a. A second interval of sediment deposition occurred after about 210×10^6 years with elevation of the sediment surface above sea level occurring again about 30×10^6 years later. This was followed by more than 150×10^6 years with little change; the sediment surface remained about 1.5 km above sea level during this erosion cycle.

The effects of varying the values for the parameters listed in Figure 10-6 were calculated and illustrated by van de Lindt, and he concluded that the values of density and thermal conductivity of the sediments are important whereas the influence of the slope of the phase transition curve and the latent heat turn out to be small.

O'Connell (1968) offered a critique of van de Lindt's analysis, and van de Lindt (1968) corrected some errors in his original paper. He stated that none of his computations were affected by the errors. O'Connell also noted that in contrast to the conclusion of van de Lindt he and Wasserburg had shown that the difference in slope between the phase transition curve, the temperature distribution in the Earth, and the latent heat of the phase change are the parameters that primarily determine the initial dynamic response of the model. He suggested that van de Lindt did not detect the importance of these factors because of the specific models investigated. Van de Lindt replied that for his model and within the range of parameters investigated the velocity of the Moho is relatively insensitive to these factors, but he did not claim that this was true for all models. MacDonald and Ness had previously concluded that the amplitude of movements of a phase boundary depended most critically upon the relative slopes of the phase transition reaction curve and the geotherm.

Time-Dependent Solution of W. B. Joyner

All of van de Lindt's models had initial basin depths of 2.7 km. Joyner (1967)

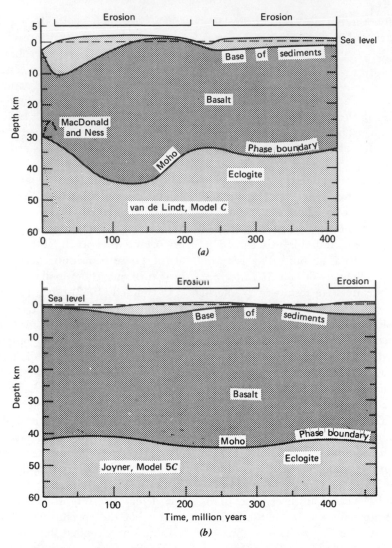

Figure 10-7. Time-dependent solutions for sediment deposition in a trough, illustrating cycles of sedimentation and erosion. Compare Fig. 10-6. (*a*) Solution of van de Lindt in Fig. 10-6(*b*) extended for twice the time. (*b*) Solution of Joyner for deposition in shallow-water. (From Jour. Geophys. Res., **72**, 4977, 1967, with permission. See also Figure 10-6.)

published the results of finite-difference calculations for the same problem using similar basic assumptions but different techniques of implementation and obtained results that he considered to be in general agreement with those of van de Lindt if allowance is made for the differences in parameters. Joyner's models had initial basin depths of 0.5 or 1.5 km and the sedimentation rate was assumed to be constant in most models.

Joyner examined specifically the problem of thick accumulations of sediments deposited in shallow waters of epicontinental seas. Basins with initial depths of 2.5 to 3 km approach the depth characteristics of ocean basins, and they are incompatible with initial crustal thicknesses of 30 km. Such models are useful, however, in establishing the validity of the phase change hypothesis to explain thick sedimentary deposits. Joyner selected a phase transition line based on the diagram of Yoder and Tilley (Figure 5-14a), which has a more gentle slope, (dP/dT), than those used in the models discussed in this section (Figures 10-4 and 10-5). He obtained results for other transition curves indicating that the behavior of the models was quite insensitive to the slope of the transition, which he explained because the effect of changing the slope was compensated by changing ΔS in order to satisfy the Clapeyron equation for the transition. His results indicated that the thickness of sediments deposited depends strongly on the initial water depth and is rather insensitive to variations in deposition rate and the other parameters.

The cyclic results obtained by Joyner for his model 5c are compared in Figure 10-7b with those of van de Lindt. In this model Joyner used a reaction curve with a slope of 66°C/kb which is similar to the curves used in Figures 10-4 and 10-5. The initial crustal thickness is 41 km and the basin, initially 500 meters deep, fills with 3 km of sediments in 118×10^6 years. Contrast this with the deep water model of van de Lindt in Figure 10-7a, where 10 km of sediments accumulate much more rapidly in a basin initially 2.7 km deep. In Joyner's model the crust thins slightly during sediment deposition, and this is followed by slight thickening as the phase change migrates downward after about 100×10^6 years. A period of erosion lasts for 185×10^6 years, during which all of the sediments are eroded away, and then a second period of deposition occurs and the basin receives 3 km of sediments again. The second deposition period lasts for about 100×10^6 years and is followed in turn by a second period of erosion. During the second deposition period the water depth never exceeds 170 meters while 3 km of sediments are deposited. Joyner examined a number of different models, and in all of them deposition was followed by uplift with uplift cycles lasting as long as 200 to 300 million years.

Oscillatory Movements

The results reviewed above confirm that if there is a phase transition in the depth range 30 to 60 km, with characteristics approximating that of the gabbro-eclogite phase transition, then the process of sedimentation will cause migration of the phase transition boundary or zone, with continued subsidence occurring at the surface followed later by uplift of the sediments. Oscillatory motions can occur in this way. Any phase transition at depth will presumably have similar capabilities, the response at the surface varying according to the nature and depth of the phase transition.

Figures 10-5, 10-6, and 10-7 illustrate some quantitative results for various models. In their detailed analysis O'Connell and Wasserburg derived analytic approximations to the actual solution of the problem which permits specific geophysical models to be considered without the need of obtaining numerical solutions for each one. They concluded that for a phase change at a depth of about 40 km the initial upward motion of the phase boundary would be reversed after about 20 million years. It is this time interval which limits the thickness of sediments that can accumulate in a basin and not the initial depth of the water. Therefore when a basin is not filled within 20 million years the sediment thickness depends almost exclusively on the sedimentation rate. For several examples they calculated the maximum thicknesses of sediments deposited in basins with initial depths of 3 km or 5 km and the maximum final elevations. Calculated sediment thicknesses ranged from 8.6 km to 21 km, and the maximum final elevations ranged from 1.0 km to 5.7 km assuming no erosion.

The differences among these various numerical results and the others reported in the papers reviewed are neither unexpected nor discouraging. MacDonald and Ness pointed out in 1960 that a judicious choice of thermal constants could lead to any desired amplitude for the movements. Rather it is encouraging that the various calculations can yield amplitudes and time scales that are in general agreement with geological observations and deductions. Whatever scheme of global tectonics one prefers the effect of phase transitions at depth on near-surface geologic processes must be taken into consideration.

GRAVITY AND PHASE TRANSITIONS: CENTRIFUGE MODELS OF H. RAMBERG

If the material within the Earth undergoes a phase transformation it may become more or less dense than the material around it. As we saw in the preceding section the contraction or expansion at depth will change the surface elevation. In addition superimposed and modifying this effect there may be vertical movement of large bodies of material relative to the surrounding rocks. Under the influence of gravity the material changed through phase transformation will tend either to sink if it has become more dense or to rise if it has become less dense, provided that the mechanical properties of the rock layers will permit the movement. The significance of such processes is illustrated by two examples. If gabbro at the base of the crust (density 3.0) is transformed into eclogite (density 3.5) it is considerably more dense than the rocks around it and the mantle peridotite (density 3.3) below it. Gravity tends to make it sink (Figures 9-6*f* and 12-25*a*). Partial melting of mantle peridotite producing a mush of crystals with interstitial basaltic magma is accompanied by a density decrease. Gravity then drives the crystal mush upward if it can overcome the mechanical strength of the overlying rocks. In Chapter 6 we interpreted the existence of the low-velocity zone in the upper mantle as being due to the presence of interstitial liquid; the low-velocity zone is also of lower density than the mantle above it. The low-density layer is then a potential source of material rising buoyantly under the influence of gravity, as proposed in several models of magma genesis (Figure 6-6, Chapter 8).

In a series of papers since 1963 H. Ramberg has considered the role of gravity in the movement of material within the crust and upper mantle and the tectonic effects produced by such movements. This work, both theoretical and experimental, was brought together in a 1967 monograph. The analytical part of the volume treats rocks as continua in a mechanical sense and uses the theories of fluid dynamics and strength of materials. Unfortunately most rocks are not simple in their mechanical behavior, and the complex geometrical structure of realistic tectonic systems usually defines a rigorous theoretical analysis. Ramberg therefore investigated the effect of gravity on various geometrical structures by constructing dynamic scale models. The Earth's gravity field was imitated at the appropriate scale by the centrifugal force in a large-capacity centrifuge. This permitted the construction of models using more viscous materials than those used in previous model studies.

In previous small-scale model studies conducted in the normal gravity field it was necessary to use mechanically weak and soft materials in order to achieve collapse and flow. These studies have provided valuable information, but they can never correctly imitate gravity-generated features in large structures. In centrifuged models the centrifugal force per unit mass can be made several thousand times stronger than the gravitational force per unit mass, and scale can be maintained when using materials several thousand times stronger and correspondingly more viscous than the materials used in noncentrifuged models of the same size. Materials used in the latter type of experiments include asphalt, heavy oil, soft wax, and wet clay. Materials used by Ramberg include wax, modelling clay, and various putties. The scarcity of material with suitable properties remains a problem if consistent model ratios are to be maintained. Any pair of materials gives a defined model ratio of density, but their strength and viscosity ratios would usually be out of proportion for the defined model ratio. The more complex the structure modelled and the greater the number of rheological properties significant for the tectonics involved the less likely is it that the scale model will be reasonably consistent. Nevertheless the approach is a considerable advance over other model studies, and the structural patterns produced bear striking similarities to some crustal structures. This suggests strongly that gravity is an effective force in deep-seated tectonics with many resultant effects in the superstructure.

The procedure is to prepare a model using layers of different materials with the bottom of the supporting caps in the centrifuge model shaped parallel to the equipotential surface during rotation; the centrifugal force is then directed perpendicular to the layers. Because of the rigidity of the materials used the structures adopted because of deformation during rotation are preserved when rotation ceases, and at the end of a run the model is removed and sliced perpendicular to the original layers for inspection.

Dome Models

It is generally agreed that gravity has played a dominant role in the emplacement of salt domes, batholiths, and mantled gneiss domes. The low-velocity, low-density layer in the upper mantle may be a source for similar gravity-generated dome-like bodies. Figure 10-8a shows a diagrammatic cross-section through an initial model of silicone putty with density 1.12 g/cm³ between layers of painter's putty with density 1.85 g/cm³. There is a small bulge in the buoyant layer which initiated the doming. Figure 10-8b shows a cross-section through the model after a run of four minutes at 700 g in the centrifuge. This shows the typical geometric shape of domal structures in

general with a trunk region (T) rising from the root region and spreading out over the more dense material in the form of a hat (H). The arrows show the marginal sink of the more dense layer. Various stages in the development of such a structure have been examined by centrifuging for a shorter period. At a later stage in the evolution the hat portion may become detached from the trunk forming a surface layer or a sill below a higher level strong layer.

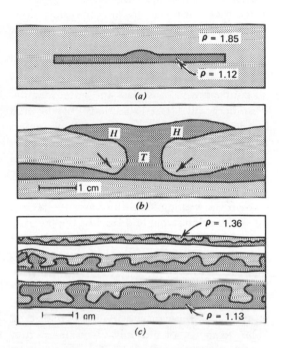

Figure 10-8. Results of centrifuged models illustrating the role of gravity in the formation of tectonic structures (after Ramberg, 1967). (a) Model 026 prior to run in centrifuge. Buoyant layer of silicone with initiating bulge, embedded in painter's putty. (b) Model 026 after run of 4 minutes at 700 g in the centrifuge, showing dome formation. (c) Profiles through three models with inverted density stratification run in centrifuge: for 2.5 min at 800–2000 g, for 1.5 min at 2000 g, and for 5 min at 800 g–2000 g. The dark grey bottom layers have a density of 1.13 g/cm³, the light top layer a density of 1.36 g/cm³. Both materials are silicone (viscosity of the order of 10⁵–10⁶ poises). (With permission of Academic Press.)

According to theory the diameter of domes and the spacing between them should be proportional to the thickness of source layers in models. In the three examples shown in Figure 10-8c the thickness of the domes and the spacing between them is roughly proportional to the thickness of the layers. This was found to be true for many of the models run.

The models illustrated in Figure 10-8 are built of materials representing rocks. It is impossible to include material representing magma in these models because the scaled viscosities turn out to be impossibly low. Even water is much too viscous to represent the silicate liquids. Water at 20°C has a viscosity making it a model material suitable for representing obsidian at 800°C; obsidian of course is a viscous glass and not a magma. Ramberg illustrated sections of layered models which started with small chambers at the bottom filled with a solution of $KMnO_4$. In most models, the solution passed right through the overburden quite rapidly and was extruded at the surface leaving a brown stain along its path of ascent. Passage of the solution strongly disturbed the layering at several levels. The buoyant rise of a solution, with properties representing an approach towards a fluid magma, produces geometric patterns quite unlike the domes developed in models with less contrasted viscosity of buoyant layer and overburden. The geometry of plutonic intrusions is thus clearly related to the physical properties of the intruding material compared with those of the country rock.

Models of Subsiding Sheets

The uprise of a buoyant mass of rock or magma must be balanced by the subsidence of more dense material. Figure 10-9 shows the structures developed when dense horizontal layers sink through less dense material. Figure 10-9a shows a central double layer of putty, $P + D$, suspended in a layer of lighter material, S. This rests on a denser layer and the whole model is covered by a thin over-

burden of modelling clay, M, and silicone putty alternating with layers of modelling clay, L. Figure 10-9b shows a sketch of the cross-section through the model after centrifuging. As the more dense double layer $P + D$ began to sink through the lighter layer, S, it assumed first an anticlinal shape and then the complementary uprise of the lighter material produced a central dome structure which burst through the sinking layers. These were then separated into the two portions shown. The general arrange-

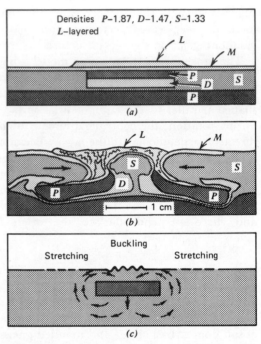

Figure 10-9. Results of centrifuged scale models illustrating the role of gravity in the formation of tectonic structures (after Ramberg, 1967). (a) Sections through a model of unstable density stratification prior to run in centrifuge. Model composed of painter's putty, $\rho = 1.87$ g/cm³; silicone, $\rho = 1.33$ g/cm³; dark silicone, $\rho = 1.47$ g/cm³; silicone putty with sheets of modelling clay; silicone putty, $\rho = 1.35$ g/cm³; modelling clay. (b) Section through model in a after run in centrifuge. (c) Flow lines around a heavy body sinking through a viscous mass. Deformation in competent crust indicated. (Figures 9-8f and 12-25a.) (With permission of Academic Press.)

ment of flow lines around a subsiding body sinking through a viscous mass is depicted in Figure 10-9c. The effect of similar flow lines is evident in the structure shown in Figure 10-9b. Layer D has flowed out sideways from beneath the heavier layer P and then upward and inward over the top of the layer despite the fact that its flow has been through the less dense material S in defiance of gravity. Sinking of the heavy layer P has produced very marked effects on the surface layers, M and L. The layer M was first dragged downward as P subsided, and then when two portions of P were forced apart as the buoyant dome S burst through it the sequence S-M was overturned. The layered overburden, L, was dragged down by the sinking slab, and

severely compressed and folded by the inward flow of layers S above the subsiding layers. These surface strata although much too light to sink by themselves have been pulled down to considerable depth in narrow wedges immediately overlying the heavy layer P. Figure 10-9c shows that the deformation in a competent crust overlying a sinking body would involve compression and buckling immediately above the sinking layer with stretching of the layer on either side.

The geological applications of Figures 10-8, 10-9, and the many other models illustrated in the book were reviewed by Ramberg. We shall have cause to remember these figures when we consider the effects of horizontal movements in Chapter 14.

PHASE TRANSITIONS AND MANTLE CONVECTION: ANALYSIS BY J. VERHOOGEN

We have examined the effect of temperature and pressure perturbations on the migration of a phase boundary at depth, and the effect of gravity in causing vertical movements of bodies with density different from that of their surroundings. Now we must consider what happens if a moving body reaches a level where a phase boundary or phase transition zone exists. The problem has been examined in the context of whether phase transitions in the upper mantle will enhance or hinder mantle convection and opinions are divided. In 1965 J. Verhoogen concluded that phase transitions in the upper mantle probably form no serious obstacle to mantle convection. The examples which follow are based on Verhoogen's treatment.

Figures 10-10a, 10-11a, and 10-12a show a schematic phase diagram for a univariant transition phase 1 ⇌ phase 2; the heavy dashed lines represent the geotherm for the depth interval in the upper mantle that is represented. The diagrams to the right of the phase diagrams represent sections through the same depth interval of the mantle. In the following discussion "phase 1" and "phase 2"

refer to the complete mantle mineral assemblages containing these two phases respectively. The temperature of the mantle at any depth is given by the geotherm, and temperatures are written alongside the mantle sections. The point of intersection of the geotherm with the univariant curve in the phase diagram gives the depth within the mantle column of the boundary between the more dense phase 1 and the overlying phase 2. We are going to consider the movement of bodies of material through this stable, phase-stratified mantle.

The moving bodies are represented by the circles in the mantle sections, and the mineral assemblages present at specific depths are indicated by open circles for phase 2, closed circles for phase 1, and part-filled circles for a body consisting of both phases 1 and 2. The changes in temperature as a function of position at depth of the moving bodies are plotted on the phase diagram as heavy lines with arrows showing the upward or downward direction of movement. At any depth the temperature of a moving body can be read from the phase diagram; these temperatures are written alongside the circles in the mantle sections. For each phase diagram numbers have been assigned to the temperatures specified in the sequence of increasing temperature, and the appropriate points on the phase diagram have been designated by these numbers. This shows directly in the mantle sections whether the moving bodies at any depth are hotter or cooler than the mantle surrounding them. The symbol with the higher subscript is the higher temperature; T_4 is warmer than T_2.

Consider first a small mass of phase 1 at point 5 in Figure 10-10a. This is represented by the closed circle at temperature T_5 just at the phase boundary in the mantle section of Figure 10-10b. Let this mass be displaced upward from point 5 to a level with lower pressure as shown by the rising column of the mantle section in Figure 10-10b. This moves it into mantle composed of phase 2 and therefore phase 1 in the body begins conversion into phase 2. This transition (from more dense to less dense phase) is endothermic, and assuming that the reaction runs adiabat-

ically the temperature of the moving body decreases. As long as phase 1 is undergoing transition to phase 2 the body is constrained to follow the path of the phase boundary. The simultaneous decrease in pressure and temperature of the body consequent upon its upward displacement is therefore given in Figure 10-10a by the path 5-2. If the transition is completed at point 2 then the rising body consists of phase 2 at temperature T_2 within mantle of the same phase at a higher temperature T_4. If displacement were continued upward as indicated in Figure 10-10b, then the path followed would be an adiabat such as 2-1 in Figure 10-10a. At point 1 the body at temperature T_1 is cooler and therefore it remains more dense than the surrounding mantle of the same phase at higher temperature T_3. The system is thus stable with respect to upward movement of phase 1 starting from point 5. The density difference between the rising body and the surrounding mantle tends to reverse any upward motion.

Consider now a small mass of phase 2 at point 5 in Figure 10-10a, represented by the

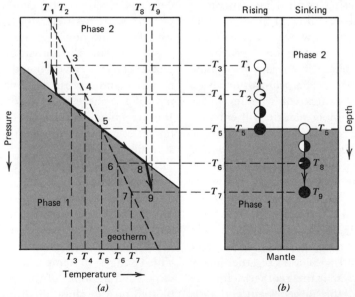

Figure 10-10. Mantle convection and phase transitions; an illustration of treatment by Verhoogen (1965). Shows the depth-temperature paths followed by material moving from position 5. (a) Transition boundary between phase 1 (more dense) and phase 2 (less dense) in mantle material, and a geotherm. (b) Mantle sections along the geotherm in a. Temperature variations and phase transitions of material moving up or down from the position 5 (Figure a) on the transition surface in the mantle.

open circle at temperature T_5 just at the phase boundary in the mantle section of Figure 10-10b. Let this mass be displaced downward from point 5 as shown in the sinking column of the mantle section, and the body then follows the path 5-8-9 in Figure 10-10a, with phase 2 changing to phase 1 as shown in Figure 10-10b. At all points between 5 and 8 the density of the sinking body is less than that of the surrounding mantle and its temperature is higher. The tendency therefore is for reversal of the downward movement.

Therefore the mantle system with a univariant phase transition occurring at the level shown in Figure 10-10b is completely stable with respect to vertical movements beginning at the level of the phase transition boundary.

Figure 10-11 illustrates the situation when a small mass of phase 1 is displaced upward from point 8 on the geotherm at temperature

T_8 well below the level of the phase transition boundary in the mantle column. The temperature of the rising body will fall as shown by the adiabatic line 8-7 until it reaches the transition boundary in the phase diagram (Figure 10-11a) at some distance below the boundary in the mantle (Figure 10-11b). At point 7 the rising body at temperature T_7 will begin to transform from phase 1 to phase 2, surrounded by cooler mantle at temperature T_6 which remains as phase 1. The density of the rising mass thus decreases even further relative to the surrounding mantle and upward movement is enhanced. During the endothermic reaction the path of the rising body is constrained to follow the line 7-5 as shown in Figure 10-11a until all of phase 1 has been converted to phase 2. The body then consists of phase 2 at temperature T_5 surrounded by more dense mantle of phase 1 at a lower temperature T_3; the body therefore continues to rise following an adiabatic line 5-4 with density remaining less than that of the surrounding, cooler mantle.

Provided that the phase transformation is completed before the mass reaches point 2 in Figure 10-11a then the density of the rising body remains always less than that of the mantle and upward motion can continue right through the phase transition level. If the phase transformation were not completed by the time the body reached point 2, then its path would continue along the phase boundary to higher levels beyond 2 producing a situation where the body consisting of phase 1 and phase 2 would be surrounded by mantle of phase 2 only, at a higher temperature. The body would then be of greater density than the mantle and the upward motion would be opposed.

Verhoogen estimated the depth at which upward movement would have to start (point 8 in Figure 10-11a) in order that the transformation be completed by point 2. The results are strongly dependent upon the thermodynamic properties of the phase transition, and upon the relative gradients of

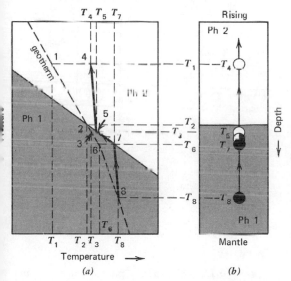

Figure 10-11. Mantle convection and phase transitions; an illustration of treatment by Verhoogen (1965). (a) Transition boundary and geotherm as in Figure 10-10a. Shows the temperature-depth path followed by material moving upwards from position 8. (b) Temperature variations and phase transitions of material moving up through mantle from position 8 in Figure a.

the geotherm and the adiabatic curve. One set of assumptions gave a vertical distance between points 8 and 2 of about 400 km, and another gave a value of about 2000 km. The inhibition of upward movement is small if ΔS of the reaction is small and if the univariant reaction curve is very steep. Similar considerations are applicable to the downward movement of a body starting at point 1 in Figure 10-12a and following initially the adiabatic path 1-2.

Most phase transitions in the mantle are likely to be divariant or multivariant rather than univariant (Chapter 6), and Verhoogen also analyzed the conditions for movement of material through a divariant transition

interval. He showed that for vertical movement of material through such a zone within the mantle the adiabatic gradient need be only slightly greater than the gradient of the initial geotherm.

Verhoogen concluded that although phase transitions act as filters tending to stop small perturbations, they permit large, deep-seated vertical movements to pass through. The analysis assumes that reaction rates are such that the moving bodies are able to transform into the equilibrium phase assemblage stable for the local pressure-temperature conditions. According to Verhoogen transformation rates under upper mantle conditions are not likely to hinder convection.

METASTABLE PHASE TRANSITIONS CAUSE EARTHQUAKES ACCORDING TO J. G. DENNIS AND C. T. WALKER

In their first paper on metastable phase transitions in 1965 Dennis and Walker expressed dissatisfaction with the elastic rebound theory of earthquake source mechanism, and proposed instead that energy was stored chemically and released abruptly in a spontaneous phase transition from metastable material. Their scheme is illustrated in Figure 10-12 using the same approach adopted for Figures 10-10 and 10-11.

Consider the downward movement of a body of mantle material from position 1 in Figure 10-12a. This will follow the adiabatic path 1-2 with temperature becoming progressively less than that of the surrounding mantle so that its increase in density will encourage continued downward movement. According to Verhoogen's evaluation of this problem, when the body reaches point 2 its transformation to phase 1 begins and its path then moves along the line 2-6 (compare path 8-7-5 for the rising body in Figure 10-11a). Dennis and Walker suggested that because of the high activation energy for solid state reactions involving breakdown of the lattice the probability of transition at or near equilibrium, at temperature T_2, was

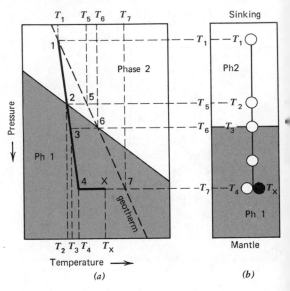

Figure 10-12. Metastable phase transitions in the mantle; an illustration of the scheme proposed by Dennis and Walker (1965) and Walker and Dennis (1966). (a) Transition boundary and geotherm as in Figures 10-10a and 10-11a. Shows path of material moving downwards from position 1, and passing metastably through the phase transition boundary. (b) Mantle section showing temperature variations and phase transitions of material moving down from 1 in Figure a.

infinitesimal and that the body would continue down the adiabatic curve 2-3-4 as metastable phase 2 surrounded by mantle in the form of phase 1 at a higher temperature. They stated that with increasing temperature (as along 2-3-4) the probability of nucleation of phase 1 would increase; at point 4 this occurs and they suggested that the heat liberated during the exothermic reaction would accumulate, accelerating the rate of nucleation and reaction to such an extent that an implosive transition would occur. This is represented in Figure 10-12 by the body of phase 2 at temperature T_4 at point 4 changing at this depth to phase 1 at some higher temperature T_x.

In the preceding section we referred to Verhoogen's consideration of transformation rates. He concluded that the transitions in the mantle would probably run at rates comparable with the velocities of vertical movements. If this is true then the path 1-2 could not continue far in the direction 2-3 in Figure 10-12a without nucleation of phase 1 beginning; how far along the path 2-3-4 the metastable material would have to travel before it had the capability of a spontaneous implosion is not known. Walker and Dennis recognized this problem in a second paper in 1966 after they learned that the olivine-spinel transition in Fe_2SiO_4 composition was completely and readily reversible in the laboratory at temperatures greater than 700°C. Therefore they suggested that downward migration in the upper mantle must begin at levels where the temperature was considerably less than 700°C, and that spontaneous transformation must occur when the temperature of the sinking body reached a critical value in the region of 700°C; this limit corresponds to T_4 in Figure 10-12.

In order to present a limiting scheme that they considered workable they depicted downward migration of mantle from beneath the continental crust at a temperature of about 500°C (T_1) reaching a depth of about 300 km at about 600°C corresponding to point 2 where the normal temperature T_5 was about 1350°C. The phase transition in the mantle (point 6) was passed at a depth of almost 400 km; and the maximum depth of spontaneous transformation of the metastable material was 700 km at a temperature of 700°C (T_4), where the temperature of the surrounding mantle was about 1650°C (T_7). The difference in temperature of about 950°C between sinking body and surrounding mantle at a depth of about 700 km seems rather extreme (difference between T_4 and T_7), and maintaining mantle phase 2 in a metastable condition through a depth of about 400 km (between points 2 and 4) in the temperature range 500–700°C (T_2 to T_4) also seems rather unlikely. For less extreme examples their proposal that a moving body of metastable phase 2 could undergo spontaneous transformation somewhere along the path 2-3-4 above and below the location of the phase transition in the mantle at 6 remains a possibility. Whether such a spontaneous transformation would amount to an implosion remains somewhat more speculative.

There is another problem, and that is the lower density of the moving body than the surrounding mantle along the path 3-4. Here we have metastable phase 2 surrounded by more dense phase 1, and this will oppose continued downward migration. Therefore we cannot expect the limit of point 4 to extend much below point 3. The 300 km depicted as a maximum in their 1966 model is surely excessive. It seems that this proposed mechanism can only be effective if there is some force driving the less dense, metastable material into the more dense mantle. Figure 14-8 illustrates possible mechanisms for sinking lithosphere slabs. It is left as an exercise for students to consider whether or not these mechanisms and the temperature distribution in Figure 14-13 are consistent with the proposal of Walker and Dennis.

11. *Continental Drift: Debate of the Century*

INTRODUCTION

The development of the geological sciences has been shaped by a series of great controversies. These include the Neptunists versus the Plutonists during the period 1775–1825, Catastrophism versus Uniformitarianism between about 1775 and 1835, and the vigorous debate about continental drift during the present century—Mobilists or Drifters versus Stabilists or Fixists.

The idea of a mobile Earth, with the crust floating on a molten interior, was familiar to geologists by 1900, but the concept of continental drift did not gather a large following until 1915 when Alfred Wegener published the first edition of his book *The Origin of Continents and Oceans*. Lateral migration of the continents provided an alternative mechanism to that of contraction for compressing the rocks of mountain ranges.

The evidence presented by Wegener and his followers failed to convince many scientists of the validity of continental drift, and a heated controversy continued between 1910 and 1960. Then the concept of sea-floor spreading was proposed by H. H. Hess with tectonic implications propounded by R. S. Dietz. This was supported by a variety of new evidence from paleomagnetism, geochronology, and marine geology and geophysics. The concept received tremendous impetus when papers published in 1966 and

1967 indicated that measurements of three different features of the Earth all change in the same ratios: these are the ages of magnetic polarity reversals in lava sequences, the depths of reversals of remanent magnetization in deep-sea cores, and the widths of the linear magnetic anomalies parallel to midoceanic ridges. The relationships between a time scale and distances are translated into a velocity of crustal spreading away from the midoceanic ridges. Converts flocked to the Mobilist camp, but some steadfast Fixists such as Harold Jeffreys and the Russian school led by V. V. Beloussov maintained that the theory has no foundation.

The theories of continental drift and sea-floor spreading were combined in schemes of plate tectonics where thin slabs of rigid lithosphere (including continent and ocean crust) moved over the less viscous asthenosphere in the upper mantle. By 1970 with the results of the Joides deep-sea drilling project presented as additional support for sea-floor spreading Earth scientists were coping with a ruling theory. It becomes almost futile to argue against a ruling theory, especially one which appears to solve so many problems. But other problems do remain and in 1970 A. A. Meyerhoff published the first in a series of papers arguing against the drift theory.

The successive discoveries and steps leading to the formulation and modification of the theory of plate tectonics contain many lessons for students; in the next four chapters we shall examine the developments in some detail.

THE DEBATE UNTIL 1950

Continental drift is an old idea formulated initially to explain the striking parallelism of the Atlantic coasts. The geometric fit of the continental margins, however, was not acceptable by itself as critical evidence that the Atlantic Ocean was once closed. Indeed the close fit has been considered as evidence either favorable or unfavorable to continental drift depending on the viewpoint of the scientist. Those against continental drift maintained that if a catastrophic event powerful enough to split a continental mass had occurred, then the newly generated continental margins would be stretched and severely deformed and they could not possibly fit so well.

Alfred Wegener and others compiled an impressive list of evidence supporting continental drift, based on aspects of geography, geodesy, geophysics, geology, paleontology, and paleoclimatology. Counterarguments were erected opposing each line of evidence, and it seemed that the theory could not be proved or disproved. The theory suffered from a lack of quantitative data, and the type of evidence put forward was perhaps psychologically unacceptable to many geologists.

Opponents of drift following the lead of eminent geophysicists such as Harold Jeffreys argued against the theory partly on the grounds that there was no satisfactory explanation as to why the continents had drifted and, furthermore, that the known physical properties of the Earth were such that the proposed lateral migration of the continents was impossible. Proponents of drift, on the other hand, argued that geological facts should not be ignored simply because there was no explanation available for them. Then other experts disputed the geological

"facts." The arguments continued in this indeterminate vein, like some medieval philosophical controversy, until a stalemate was reached in the 1940's. Just about everything that could be argued for and against continental drift had been written not once but many times and the debate faded for lack of additional evidence.

The Theory of Continental Drift

All theories of continental drift require that before the Mesozoic the continents were grouped together into a single block or into two blocks. Wegener proposed the name *Pangaea* for the single continent which he believed existed in the Late Carboniferous as illustrated in Figure 11-1*a*. The shaded areas on the continents represent shallow seas. According to Wegener *Pangaea* began to split up during the Jurassic period, with the southern continents moving either westward or toward the equator, or both. South America and Africa began to drift apart during the Cretaceous a little more than 70 m.y. ago. Figure 11-1*b* shows the distribution of the continents during the Eocene about 45 m.y. ago. Opening of the North Atlantic he thought was mainly accomplished during the Pleistocene, and Figure 11-1*c* shows that Greenland and Norway only started to part company about 1.5 m.y. ago. The Indian peninsula, which was originally long (Figure 11-1*a*), drifted northeastward and was compressed into fold mountains against the Asian continent; and similarly the European Alps and the Atlas Mountains of North Africa were caught between Africa and Europe, and can be interpreted as an extension of the Himalayan chain. Wegener suggested that as the front margins of the

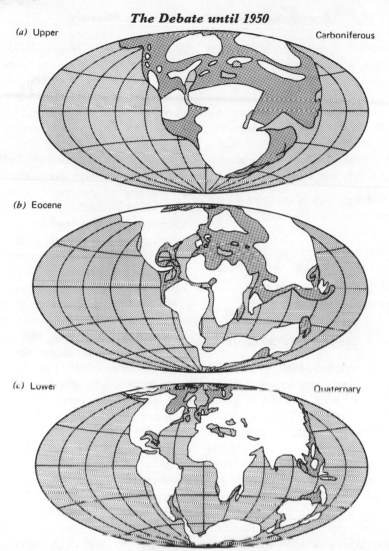

(a) Upper   Carboniferous

(b) Eocene

(c) Lower   Quaternary

Figure 11-1. Stages of continental drift according to Wegener (1915; reprinted version, 1966). Compare Dietz and Holden (1970). (From "The Origin of Continents and Oceans" by A. Wegener, Dover Publications, Inc., New York. Reprinted with permission of the publisher.)

moving continents met the resistance of the ocean floor, they became compressed and folded into mountain ranges. In this way he explained the western cordillera of the Americas and the mountain ranges of New Zealand and New Guinea. His explanation for the tapering ends of Greenland and South America and for the island arcs including the Antilles and those on the east of Asia is that these represent stragglers left behind in the wake of the moving islands. Detached fragments formed the long strings of islands.

The leading proponent of two primordial continents was A. L. du Toit who maintained that from the middle of the Paleozoic era to the beginning of the Tertiary the continent of Laurasia in the northern hemisphere was separated from the continent of Gondwanaland in the southern hemisphere by Tethys, a huge, geosynclinal ocean from which the Alps and Himalayan mountain chains eventually emerged. This grouping of the continents is consistent with the major structural features of the Earth as they exist at present.

Figure 2-3 shows the distribution of Tertiary mountain ranges in the form of two great rings; one surrounding the northern continents of North America and Eurasia (Laurasia) and the other surrounding the southern land masses of South America, Africa, Arabia, India, Australia, and Antarctica (Gondwanaland). The major structural features of the Earth thus consist of:

1. The Laurasian group of continents with the intervening basins of the North Atlantic and Arctic Oceans.

2. The Gondwanaland group of continents with the intervening basins of the South Atlantic and Indian Oceans.

3. The Pacific Ocean basin.

Evidence Cited for Continental Drift

The geological evidence which was debated for half a century has been reviewed many times before and we need consider only the outline of the controversy. There are four main lines of evidence that the continents were once joined together: (a) paleoclimatology, (b) paleontology, (c) the geometrical fit of the continents, and (d) the matching of stratigraphy and truncated structures across oceans.

The anomalous paleolatitudes deduced from paleoclimatology in the 1920's provided strong evidence for continental drift. Climatic zones in the late Paleozoic have been determined from the distribution of glacial deposits, desert sand and deduced wind directions, salt deposits, coal deposits, ancient coral reefs, and other fossil plants and animals indicative of climate. The paleoclimatic zones so deduced suggest either that the continents have moved relative to their present latitudes, or that there have been major climatic fluctuations with time at specific latitudes with stationary continents. (A. A. Meyerhoff reached a different conclusion in 1970; see Chapter 15). The concept of polar wandering was introduced in the nineteenth century to account for variations in climatic and biologi-

cal sequences in different parts of the southern continents. The distribution of Permocarboniferous tillites in Australia, India, South Africa, and South America has been cited as evidence that these continents were in contact, or at least in close proximity, during this period.

Fossil plants and animals believed to be incapable of crossing deep water are distributed in limited geographic regions on both sides of the Atlantic and on continents separated by other oceans. Their distributions suggest that there were former land connections between Europe and North America, and between South America, Antarctica, Australia, India, and Africa. The land connections apparently required in earlier times appear to be nonexistent after the Jurassic. There were vigorous paleontological discussions about the reality and location of hypothetical land-bridges, which were supposed to provide temporary land connections across oceans from one continent to another. The theory of continental drift makes the land bridges unnecessary.

Attempting to fit together the continents marginal to the Atlantic Ocean has been a popular pastime since the last century. Figure 11-1a shows Wegener's reconstruction for the carboniferous period before the onset of drifting. Although most scientists are impressed by the approximate fit of the Atlantic coastlines others remain sceptical.

If the continents were once joined then the structures, ages, and petrology of the rocks must match across the joins. There are remarkable similarities between the stratigraphic and lithological columns for Brazil and southwest Africa from the Silurian through Cretaceous periods and several structural links between groups of rocks in Africa and South America have been proposed. There have been many attempts to reconstruct a single predrift geosyncline from the Caledonian belts of Norway, East Greenland, Scotland, Ireland, and the Appalachian chain. During the great debate the reality of the claimed structural fits across the

Atlantic Ocean was disputed by many geologists.

Mechanisms Postulated for Drifting the Continents

A major problem facing proponents of continental drift was to find a mechanism capable of producing the breakup and drifting apart of the continents. None of the proposed mechanisms was considered convincing by opponents of the theory, and I suspect that those who proposed them were not fully convinced either.

In the nineteenth century it was suggested that the Atlantic Ocean opened up early in the history of the Earth as a direct result of the loss of the moon from the site of the Pacific Ocean. The moon was also involved in the hypothesis of F. B. Taylor, who proposed in 1910 that the moon was captured by the Earth only in Cretaceous times and was at first much nearer to the Earth than it is today. This produced tidal forces which increased the rate of the Earth's rotation, and also caused the continents (Laurasia in the north and Gondwanaland in the south) to slide away from the polar regions towards the equator. The great Tertiary mountain ranges, and mountainous loops and arcs, were pictured as being raised up in front of the sliding continental masses. Ocean basins were produced from the stretched and broken regions left behind or between the sliding continents. A major weakness of the theory is that if mountain building is correlated with capture of the moon it gives no explanation for the orogenic cycles prior to the Tertiary. Also tidal forces strong enough to move the continents on such a scale and to cause the uplift of great mountain ranges, would surely have slowed the Earth's rotation almost to a standstill within a very short time.

Wegener described the drift of the continents as *Polflucht*, the "Flight from the Poles," and he suggested that this resulted from the gravitational attraction between the continents and the Earth's equatorial bulge. He also postulated a general drift to the west reasoning that the differential attractions of the sun and moon would cause the continents to lag behind the rotation of the Earth as a whole. Both of these forces are quite inadequate to overcome the frictional resistance to motion of the continents although Wegener maintained that even such small forces, if maintained for millions of years, would eventually displace the continents.

In his book, *Our Wandering Continents*, dedicated to Wegener in 1937, Du Toit proposed a model whereby a continent slides over the mantle by the action of gravity. Marginal loading of a continental block by a geosynclinal pile causes the continent to become slightly tilted and gravity causes it to slide toward the ocean. The central part of the continent is thus subject to tension which causes rifting, permitting the uprise of magma which contributes to further splitting and arching of the rifted region, which in turn makes the continent slide further outward. This original hypothesis lacked quantitative evidence, and the idea received little attention. Recent interpretations of sea-floor spreading, however, include mechanisms driven by gravity (Figure 14-8c).

In the nineteenth century we already find the essence of the ideas of polar wandering, mantle convection, and sea-floor spreading which have become topics of major concern to geologists and geophysicists since 1950. The study of paleoclimates and paleontological sequences early led to the conclusion that the geographic poles had shifted relative to the positions of the continents, but attempts to trace the locus of polar displacements through geological time always led to contradictions. The concept of polar wandering was generally distrusted, and it has been difficult to place any precise meaning on the term. It is now usually used to convey the idea that while the geographic poles remain fixed relative to the rotating Earth, an outer shell of the Earth becomes decoupled from the mantle and shifts as a whole relative to the poles. Others have understood by polar wandering a turning over of the entire body of the globe relative to its axis of rotation. Another interpretation, coming with the advent of paleomagnetic studies, implied that the magnetic poles wandered relative to the axis of rotation or geographic poles. Polar wandering can occur without continental drift, which implies movement of one continent relative to another, but continental drift cannot occur without polar wandering. If continental drift occurs then different continents will have different paths of polar wandering.

In 1881 Osmond Fisher published a book, *Physics of the Earth's Crust*, which postulated the existence of convection currents in the Earth's fluid interior, with uprise beneath the oceans causing expansion at the surface by the addition of volcanic rocks in the median position and descent of the currents beneath the continents. As the oceans expanded the continents contracted to form fold mountains.

Fisher's ideas were regarded as wild speculations and therefore ignored. The idea of convection in the mantle had been suggested before and it became popular later, but the Earth's interior is now known to be solid.

The first to suggest that continental drift might be explained in terms of convection within a solid mantle appears to have been Arthur Holmes. The model that he described in 1928 (published in 1931) is remarkably close to the concept of sea-floor spreading developed in 1960 by Hess and Dietz; although it could more aptly be termed sea-floor stretching. Figure 11-2 shows Holmes' diagrams for the convective-current mechanism using gravitational energy and thermal energy within the Earth to engineer continental drift and the development of new ocean basins. In Figure 11-2*a*, we have a thick basaltic crust with an overlying sialic layer forming the continents. A current ascending at *A* spreads out laterally beneath

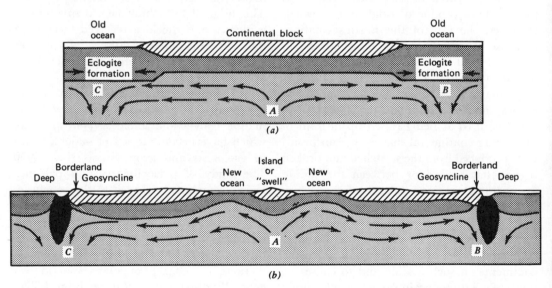

Figure 11-2. Convection current mechanism for causing continental drift (after Holmes, 1931). (*a*) Convection currents in the mantle create a region of tension in the continent, and convert basalt to eclogite where the currents descend. (*b*) The continent is split apart, and the continental masses are carried by the mantle convection currents, leaving a stretched and broken region between them; this becomes new ocean, and the size of the ocean grows with time as basaltic magmas are erupted above the ascending limb. (Compare with model for sea-floor spreading in Figures 12-22, 12-23, and 13-9.) Eclogite formed at *B* and *C* sinks (Figures 9-8(*f*), 10-9(*c*), and 12-25(*a*)), thus making room for the advancing continents.

the crust and extends the continent producing a stretched region or a disruptive basin between two main blocks. These are carried apart on the backs of the currents as depicted in Figure 11-2*b*. The migration is accomplished by removal of the obstruction of the old oceanic crust at *B* and *C* where the currents turn downward. Cooling at *B* and *C* causes conversion of basalt to eclogite. The heavier eclogite sinks thus making room for the advancing continents. The foundered masses of eclogite at *B* and *C* join in the main convective circulation melting at depth to form basaltic magma and rising again in ascending currents as at *A*. The basaltic magma heals gaps in the disrupted continent and contributes to the new ocean floor.

Holmes stated that the formation of a new ocean floor would involve the discharge of a great amount of excess heat. Figure 11-2*b* shows a fragment of sial left behind within the new ocean as an example of how Iceland may have formed, but this Holmes considered as only a local effect; normally, the newly exposed ocean floor consisted of basaltic material only.

Whereas Wegener and others had considered the continents as active elements drifting over and through the substratum, Holmes introduced the concept of passive continents being carried as if on a conveyor belt. This concept, and his use of eclogite as a sinker (Figure 9-8*f*), have become essential parts of recent theories.

DEVELOPMENTS SINCE 1950

The debate about continental drift stagnated in the 1940's, but the controversy was revived in the 1950's by the work of P. M. S. Blackett and S. K. Runcorn on paleomagnetism. Their results indicated that the former positions of the magnetic poles had changed relative to the continents, and continental drift is a process that could explain these changes. This new evidence led many geophysicists to consider the theory of continental drift seriously, while many geologists remained unimpressed by, and suspicious of, this new approach. We shall review the evidence in Chapter 12.

During the same period exploration of the ocean floor by marine geologists and geophysicists such as B. C. Heezen and M. Ewing showed that the midoceanic ridges were more nearly continuous than previously suspected, and that the ocean basins contained far less sediment than previously assumed. The mid-Atlantic ridge was found to be remarkably parallel to the continental borders of the Atlantic Ocean. These observations led geologists to conclude that the ocean basins were relatively young, and that some kind of upwelling occurred beneath the rifted oceanic

ridges. The formulation of these ideas into Hess's concept of sea-floor spreading is reviewed in Chapter 12.

During the 1960's former lines of evidence for continental drift were made more acceptable to scientists trained in physics and chemistry by the use of computers and isotopes (a) to match geometrically the borders of continents on opposite sides of oceans, and (b) to match structural provinces in terms of the isotopic ages of rocks. Examples of these approaches are reviewed below.

Other developments during the 1960's are reviewed in detail in Chapters 13 and 14.

These include:

1. The crucial interpretation of linear magnetic anomalies in terms of dated polarity reversals of the Earth's magnetic field and spreading from the oceanic ridges.

2. The interpretation of seismic data along ridges, island arcs, and transform faults.

3. The interpretation of sediment cores recovered from the deep ocean floor during the Joides program.

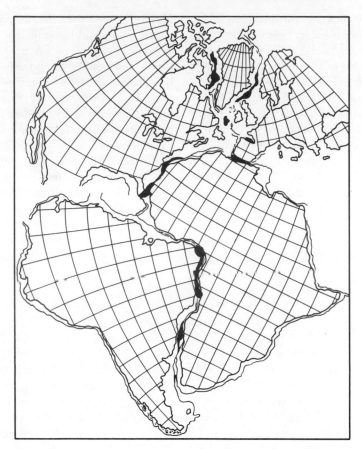

Figure 11-3. Computer fit of the continents around the Atlantic Ocean (after Bullard et al., 1965, with permission). Black areas show overlap of continental shelves. Drake et al. (1968) proposed an alternative fit based on a change in the pattern of magnetic anomalies which occurs on both sides of the Atlantic Ocean (north but not south). They fitted together the boundaries between the marginal "quiet magnetic zones" and the normal, higher amplitude anomaly pattern. Vogt et al. (1970) reviewed this and other hypotheses for the quiet magnetic zones. (Royal Society, London.)

Geometrical Fit of the Continents

Continental fits such as proposed by Wegener (Figure 11-1) are not accepted by all scientists. This is clearly shown by a remark of H. Jeffreys in 1964 according to E. C. Bullard: "I simply deny there is an agreement." In order to place the geometrical fitting of continental masses on a more objective basis Bullard and his associates therefore examined the fit of the continents around the Atlantic Ocean by numerical methods, and they found remarkably good

fits at the 500 fathom (about 915 m) contour which lies on the steep part of the continental slope. A computer was used first to fit South America to Africa with the result shown in Figure 11-3. Overlaps and gaps are indicated. The largest misfit, of 270 km, is at the Niger delta which is a recent addition to the continental edge (since the Tertiary). A similar computer procedure brought together from around the North Atlantic the continental masses of North America, Greenland, and Europe. The fit, shown in Figure 11-3, has omitted Iceland (composed of Tertiary and

Recent igneous rocks). The root-mean-square errors for these fits are 30 to 90 km. The fit of the southern block to the northern block is much poorer, and distortion of the continents to avoid overlaps requires rotation of Spain to close up the Bay of Biscay as shown in Figure 11-3. The reconstruction shows large gaps in the Caribbean and the Mediterranean, and most of Mexico and Central America has been omitted. Adjustments will have to be made to include the Paleozoic rocks known in Oaxaca, Mexico. The root-mean-square misfit is about 130 km. Note the relative rotations of the continental masses required in order to make the fit; this is shown by the lines of latitude and longitude in Figure 11-3.

There have been many estimated fits of Australia to Antarctica, mostly of the sketch-map variety because the problem of fitting the continents around the Indian Ocean is more difficult than that for the Atlantic Ocean. Using a corollary of the technique adopted for construction of Figure 11-3, W. P. Sproll and R. S. Dietz determined a good computerized fit for Australia against Antarctica in 1969, with a root-mean-square misfit of only 51.9 km at the 1000 fm (about 1,830 m) contour on the continental shelf. The fit is geologically permissible. A year later they published a result for the juxtaposition of Africa and Antarctica, seeking the Gondwana supercontinent. Also in 1970 A. G. Smith and A. Hallam published a reconstruction of the complete Gondwanaland, by bringing South America, Africa, Arabia, Australia, Antarctica, India, Madagascar, and New Zealand together by computer fit at the 500 fathom isobath, with the constraint that they investigated only configurations that were geologically reasonable. Their solution for the Africa-to-Antarctica fit differs in some respects from that of Dietz and Sproll, because the continent edges were fitted at different depths. The reconstruction is remarkably close to that published by Du Toit in 1937 when far less data were available to guide the matching of geological structures.

Du Toit's reconstruction is used in Figures 7-15 and 11-5.

In 1970, R. S. Dietz and J. C. Holden used the new geometrical and geological fits to repeat Wegener's reconstruction of *Pangaea* (Figure 11-1) with cartographic precision. They presented maps illustrating the breakup and dispersion of the continents during the past 180 m.y. Absolute geographic coordinates were assigned for the continents as well as for the active ocean rift zones and the oceanic trenches as they migrated to their present positions. They also extrapolated present day plate movements to predict the appearance of the world 50 million years from now.

Matching Age Provinces on Continental Reconstructions

Computer fits of continental margins such as that shown in Figure 11-3 give a less subjective framework than estimated fits for the comparison of structural trends from one continent to another, and recent work involving age measurements is beginning to provide more precise data. A good example is the matching of geological age provinces in West Africa and Northern Brazil, reported by P. M. Hurley and others in 1967 and illustrated in Figure 11-4.

West Africa is divided into two major age provinces, with K-Ar and Rb-Sr age determinations generally in the range 2000 m.y. in Ghana, the Ivory Coast, and regions to the west; and in the range 550 m.y. in the eastern part of Dahomey, Nigeria, and regions to the east. The sharp boundary between these provinces, shown by the dashed line in Figure 11-4, appears to head southwestward from a point near Accra. If Africa and South America had been together according to the fit in Figure 11-3 at the time the boundary was formed, the boundary would have entered Brazil just east of São Luis. Age analyses were therefore made on specimens collected near São Luis to see if this boundary did extend into South America. Both K-Ar and

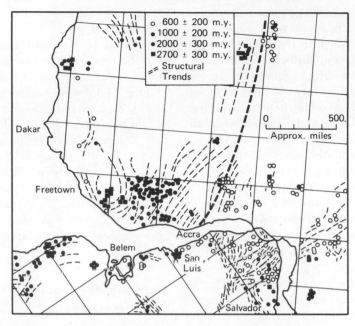

Figure 11-4. West Africa and South America shown fitted together according to the reconstruction of Figure 11-3. In West Africa the 2000-million-year Eburnean age province (solid circles) adjoins the 550-million-year Pan-African age province (open circles); the boundary between them is shown by the heavy dashed line. If Africa and South America were once joined together, this line would have entered Brazil near São Luis. The age measurements for Brazil appear to show the same age provinces as those in West Africa, with the boundary at the predicted location. There may be a similar correlation between West Africa and the east coast of Brazil north of Salvador (after Hurley et al., 1967, Science, 157, 495–500. Copyright 1967 by the American Association for the Advancement of Science. Reproduced by permission).

whole-rock Rb-Sr age measurements were made on the same samples where possible in order to obtain added information on the history of the basement rocks. The results show that the same age boundary does appear at almost exactly the predicted location. For specimens near the boundary, the K-Ar ages are in the range 410 to 640 m.y., and the whole-rock Rb-Sr ages are still in the 2000 m.y. range. Slightly further to the east the whole-rock age also has dropped to 665 m.y. The measurements plotted suggest the possibility of another trans-Atlantic age-boundary correlation further south towards Salvador where a 500 m.y. province gives way to basement exposures of 2000 m.y. age overlain or intruded by rocks of various ages down to 500 m.y. Geological studies have subse-

quently confirmed the existence of a Precambrian tectonic province in this part of Brazil, perpendicular to the coast, which is repeated in Gabon, Africa, in the position expected from the geometric fit of Figure 11-3 (Allard and Hurst, 1969).

Similar links have been noted between Europe and North America. The Lower Paleozoic carbonatites and alkalic rocks associated with the St. Lawrence graben system in Canada have been correlated with the similar rocks in Greenland and Scandinavia. These rocks appear to belong to a single alkalic rock province, defined by a rift system extending at least from central Canada to eastern Sweden on the predrift reconstruction of Figure 11-3. Age determinations indicate that this rift system was active throughout its length about 565 m.y. ago (Doig, 1969).

Age determinations for massif-type anorthosites range from 1100 to 1700 m.y., but they cluster about 1300 ± 200 m.y. The global distribution of the anorthosites takes on a special significance when plotted on predrift reconstructions of the continents such as Figure 11-3 and Du Toit's reconstruction of Gondwanaland (Figure 11-5). The anorthosites then appear to lie within broad belts, at least one connecting America and Europe in Laurasia, and another traversing the continents forming Gondwanaland. N. Herz (1969) thought that the distribution and reasonably consistent ages of the anorthosites suggested that they were the product of a unique cataclysmic event or a thermal event that was normal only at an early time in Earth history. Of interest to us at present, however, is the fact that these anorthosite belts become apparent only if we assume that the continents have drifted apart since the anorthosite emplacement.

Recent compilations of the ages of basement rocks on the continents lends support to the former existence of two ancient continental cratons. Figure 11-5 shows again the predrift reconstruction of the continents, and the shaded areas are the continental blocks having apparent ages greater than 1700 m.y. The ancient cratons are not scattered uniformly throughout the predrift continental crust, and they occupy areas that can be enclosed within two rather smooth ovoid boundaries. One area is centrally located within the supercontinent of Laurasia, and the other similarly within Gondwanaland. These areas are chopped up by younger transcurrent, geologically active belts, and encircled almost entirely by belts of younger continent (Figure 7-15). This grouping of the ancient continental cratons makes it difficult to conceive of a series of drift motions, with splitting of the continents followed by gathering together again, earlier than the break-up that apparently occurred during the last 200 m.y. This evidence thus favors

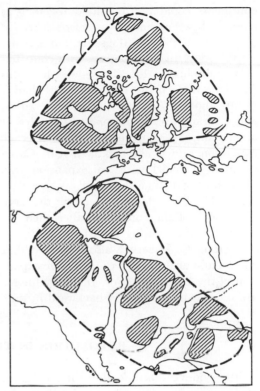

Figure 11-5. A predrift reconstruction in which the continental blocks having apparent ages >1700 million years (hatched areas) appear to be in a coherent grouping within two restricted regions. These blocks are transected and circumscribed by belts of younger rocks. Compare with Figure 11-4. It seems unlikely that during the time between 1700 and 200 million years ago the continents were scattered and drifting, only to be reassembled with this degree of ordering at 200 million years ago. Instead, there appear to have been nonmoving ancient nuclei and continental accretion up to the time of the great drift. (After Hurley and Rand, 1969.) Compare Figure 7-15. See Dietz and Sproll (1970), Smith and Hallam (1970), Schopf (1970), Dietz and Holden (1970). Copyright 1969 by the American Association for the Advancement of Science. Reproduced by permission.

the idea that continental drifting is a unique event which occurred only recently in geological time.

Twenty years detailed study of global tectonics convinced S. W. Carey that the continents could not be fitted together as claimed by Wegener on a Paleozoic Earth (Figure 11-1) if it had the same radius as the Earth today. In a symposium with contributions published in 1958 he showed how this could be done if the Permian Earth had a radius about 0.75 times the present radius, and a surface area about one half its present area. This is equivalent to an average expansion rate of about 0.5 cm/year. Carey adopted a model with an expansion rate increasing with time, however, as shown in Figure 12-11 in order to account for the large dispersions of the continents after their inferred break-up during the Mesozoic Era.

In 1969 R. Meservey also suggested that continental drift might require a large expansion of the Earth's interior. If the continents are moved simultaneously backward in time from their present positions to the relative positions shown in Figure 11-5, the perimeter of the Pacific Ocean must increase in order for it to pass over the Earth's circumference, and then decrease as the continents are reassembled on one side of the Earth. The size of the Pacific Ocean would increase from its present 35% of the Earth's surface to more than 50%. Meservey concluded that the evidence was strong that the combined length of the linkages around the Pacific perimeter decreases as we go back in time at least to 80 million years ago. If this conclusion is valid then there is no topologically possible transformation of the continents on an Earth of the present size that can explain continental drift. The apparent paradox could be resolved if the Earth had expanded during the last 150 million years.

REVOLUTION IN THE EARTH SCIENCES

As the evidence documenting the movement of lithospheric plates accumulated in the literature during 1966, 1967, and 1968 review articles began to appear in which proponents of the Mobilist view point announced a revolution in Earth sciences. J. Tuzo Wilson considers the revolution to be similar to, and as significant as, that which changed the approach to chemistry about 1800, that which occurred in biology about a century ago with the introduction of Darwin's theory of evolution, and that which occurred in physics when classical views were replaced by modern. The revolution, embracing the essential ideas of continental drift that had been debated for half a century, shows a promise of advancing the study of Earth sciences from the stage of data-gathering into the stage of formulation of a precise, comprehensive theory of global geology and use of the theory to make predictions. The new global tectonics challenges all of the past tectonic theories based on fixist or stabilist concepts. In particular the geosynclinal theories reviewed in Chapter 9 require re-evaluation within the new conceptual framework.

The December 1968 issue of *Geotimes* printed (a) an exposition of the revolution by Wilson, (b) a letter from V. V. Beloussov maintaining that the theory of continental drift and the concept of sea-floor spreading should serve only as working hypotheses along with others such as oceanization of the continental crust, and (c) a reply from Wilson restating his contention that the revolution will unite branches of the Earth sciences, formerly fragmented, into a new unified science of the dynamic Earth. I recommend this as required, critical reading for all students.

Whether or not textbooks should be rewritten and education reorganized in accordance with the developing ideas of global tectonics, as Wilson suggests, there is no doubt that the conventional geological subjects have benefited from critical reexamination in the context of global geology. This approach has certainly given geologists a different frame of reference for interpretation of their data, and all indications are that during the 1970's most of the classical and rather descriptive approaches to geology will be tested and modified in terms of the movements and interreactions of continents and oceanic plates. Wilson maintains that because the Earth is a single system the new theory of global tectonics should be learned first, and the traditional subjects can then be studied and more easily understood with the principles of Earth behavior providing the basis.

In the next three chapters we examine in detail the evidence which heralded this revolution.

12. *Paleomagnetism, Polar Wandering, and Spreading Sea Floors*

INTRODUCTION

Interest in continental drift was revived during the 1950's by paleomagnetic studies. Interpretations of fossil magnetism in rocks suggested that through geologic time the position of the Earth's magnetic axis had migrated relative to its rotational axis. The results of paleomagnetism and of marine geology and geophysics set the scene for Hess's formulation of the concept of sea-floor spreading. The subsequent discovery that the polarity of the magnetic field had reversed periodically provided a means for dating the linear magnetic anomalies that figured so prominently in the development of the concept (Chapter 13). The hypothesis of mantle convection as the driving force for horizontal motions at the Earth's surface was revived in various forms.

THE 1950'S: PALEOMAGNETISM AND POLAR WANDERING

All rocks exhibit magnetic properties one of which is a fossil magnetism or natural remanent magnetism (NRM), which was acquired during the formation of the rock and possibly modified afterward. The direction of magnetization was in response to the prevailing magnetic field, and NRM therefore provides the possibility of determining the direction of ancient geomagnetic fields and, in principle, the intensity of the field. The fossil magnetism is due to only a small proportion of minerals in rocks—the accessory iron oxides and sulfides. The intensity of magnetization of the rocks is low for this reason, and because magnetization occurred in a weak magnetic field. The magnetization of rocks has been studied for many years and by 1930 several important points had been clarified. It had been established that the Earth's magnetic field had not changed greatly in the recent geological past. The great majority of observations have been published since 1950, and it was not until then that the potential application of paleomagnetism to paleogeographic problems was put into practice. An interesting account of the historical development of this approach was given in 1967 in a book *Debate About the Earth* by H. Takeuchi, S. Uyeda, and H. Kanamori. E. Irving in 1964 published a book, *Paleomagnetism*, which summarized and reviewed all available paleomagnetic data.

Remanent Magnetism in Rocks

Remanent magnetism, or remanence, is that magnetization remaining in a substance in zero applied field. Thermoremanent magnetization (TRM) is the remanence acquired upon cooling through a certain temperature interval in the presence of a magnetic field. A mineral is ferromagnetic only at temperatures below its Curie point usually about 500°C. The magnetic minerals crystallize from lavas or magmas at temperatures greater than the Curie point, but they do not become magnetized until the temperature falls below 500°C. It has been shown experimentally that igneous rocks acquire a strong thermal remanent magnetism as they pass through the temperature interval 500 to 450°C. The magnetic minerals mainly responsible are within the system FeO-Fe_2O_3-TiO_2. TRM is many times stronger than that which could be induced in a rock at room temperature. It is not only strong but stable. Therefore igneous rocks provide excellent fossil magnets remaining unaffected by most of the later disturbances that may occur.

A small proportion of the detrital fragments in sediments are magnetic, and these have a tendency to align themselves in the direction of the prevailing geomagnetic field during deposition. The rock so formed then acquires detrital remanent magnetization, sometimes called depositional remanent magnetization (DRM). Many factors may disturb this alignment when the particle hits bottom or during compaction and cementation. It is therefore convenient to recognize depositional DRM acquired due to particle alignment during sedimentation, and post-depositional DRM acquired by particle rotation after deposition but before consolidation. An inclination error arises in many sediments; the declination of the magnetized sediment agrees with the magnetic field, but the inclination is often less than the field inclination. The difference may reach 25°. A bedding error is introduced if the surface of deposition is tilted. DRM is about a hundred times weaker than TRM and less stable.

Many rocks contain magnetic minerals which were formed at low temperatures by chemical processes or during metamorphism at temperatures below the Curie point. The growth of magnetic minerals produces chemical remanent magnetization (CRM) in the rocks, and the direction of magnetization then corresponds to the prevailing field at the time of alteration and not at the time of formation of the rock.

The intensity of primary magnetization may decay with time, which is called viscous demagnetization or viscous decay. Also a new magnetization may be acquired at temperatures below the Curie point over long time spans, and this is called viscous remanent magnetization (VRM). Viscous decay and VRM in a prevailing field different from that of the original NRM will tend to obscure the primary magnetization.

If a rock is anisotropic the TRM may be deflected away from the prevailing field towards a direction of "easy" magnetization. Stresses arising from cooling or tectonic causes may also affect the remanence. Sometimes noticeable components of secondary or temporary magnetization are added to a sample between collection and measurement in the laboratory.

Paleomagnetic Measurements

Paleomagnetic surveys are made in rock units of known geological age such as a set of sedimentary beds, lava flows, or intrusive rocks which might span a time interval of 10^3 to 10^6 years. The meaning and accuracy of paleomagnetic interpretations is completely dependent on a knowledge of the geology of the samples. An adequate sampling scheme is essential for satisfactory results. Samples oriented with respect to the geological structure are collected so that their attitude when formed is known. The direction and intensity of NRM is then measured. As discussed in the preceding section NRM is usually complex; the primary or original magnetization may have been modified by secondary magnetizations. These components are resolved by using various techniques of which demagnetization in stages is the most important. The unstable components of NRM are removed by alternating magnetic fields (magnetic cleaning), by heat (thermal cleaning), by chemical treatment, or by combinations of these techniques. The more detailed the tests conducted, the more certain are the paleomagnetic results. All paleomagnetic studies must be evaluated in terms of the stability and reliability of the reported remanent magnetization.

The basic assumptions in paleomagnetic studies are that the Earth's magnetic field has always approximated a dipolar field (Chapter 3) and that the mean direction of permanent magnetization of the rocks at the place of observation represents the mean direction of the ancient geomagnetic field during the time of formation of the rocks. There is good evidence that the first assumption is valid at least as far back as mid-Tertiary and furthermore that the magnetic poles remained close to the present geographic poles in this interval. The second assumption is not necessarily valid because the geomagnetic pole position measured is that corresponding to the time of magnetization, which may be later than the time of formation or the physical age of the rock. Consideration of the fossil magnetism with respect to some geological feature of the rock unit, such as a fold, often permits a conclusion about the acquisition of NRM relative to the times of formation of the rock unit and of the geological feature of the unit.

The final value obtained for each sample is a geographic direction with an angle of dip approximating to that of the ambient geomagnetic field at its time of magnetization. This is plotted on an equal-area or stereographic projection as a point, and points are plotted for every sample measured in the rock unit. Figure 12-1 shows examples of NRM directions measured in four separate rock units. Figure 12-1a shows a close grouping of single determinations from 13 samples of recent sediments. They cluster around direction F of the present dipole field of the Earth. Twenty-one samples from the Miocene sediments of the Arikee Formation in South Dakota show a high dispersion of directions in Figure 12-1b. Figure 12-1c is an

example of a smeared distribution in Triassic sediments of Sidmouth, England, with the directions strung out approximately along the great circle passing through the direction F of the present axial dipole field of the Earth at Sidmouth and the poles $M_S{}^-$ and $M_S{}^+$ of the axis of stable magnetizations determined for Triassic sediments elsewhere in England. Close groupings of directions are often obtained from observations from the same lava flow as shown in Figure 12-1d.

In a paleomagnetic study a question always remains about the different components contributing toward the NRM and their timing relative to the age of the rock. If the measured directions are closely grouped and divergent from the present field as in Figure 12-1d, this indicates the absence of viscous components imposed recently and of CRM from recent weathering. Consistency tests are of greatest significance if applied to results for rock types of different origins and composition. Figure 12-2 shows good agreement between results obtained for three groups of Upper Triassic rocks of New Jersey: red sediments, dolerite intrusions, and basalt lava flows. The mean directions do not differ significantly among the groups. This indicates the absence of inclination error in the sediments and implies that the effect of cooling stresses in the igneous bodies was negligible. The mean direction is considered to be reliable.

The directions recorded for samples in a rock unit represent former magnetic fields as they existed through an appreciable time interval, 10^3 to 10^6 years. In view of the secular variations known to occur in the Earth's magnetic field (Chapter 3) it is not surprising that the directions never agree exactly. Average values have to be calculated from sets of observations such as those shown in Figures 12-1 and 12-2. Statistical tests for the accuracy of calculated mean directions are described in Irving's book. It is necessary to provide some measure of the dispersion of observations about the mean direction and to estimate the accuracy with which the

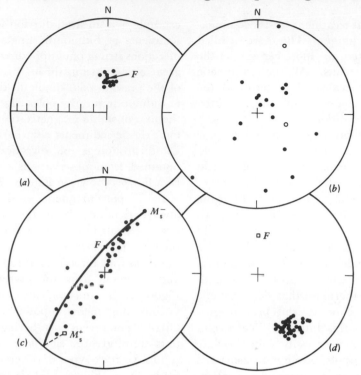

Figure 12-1. Examples of observations (all NRM directions) from a single locality. Directions with positive (negative) inclination are plotted as dots (circles) on a polar equal area net. (*a*) Payette Formation, Glenns Ferry, Idaho, United States. (*b*) Arikee Formation, South Dakota, United States—the scatter in directions is very great ($k = 3$). (*c*) Triassic marls from Sidmouth, England—the great circle through the poles M_s^-, M_s^+ of the axis of stable magnetization and the direction F of the geocentric axial dipole field is marked. (*d*) Cenozoic lava flow, Australia (after Irving, 1964, Figure 4.1, with permission of John Wiley and Sons).

mean is defined. The 95% circle of confidence is used to compare two mean directions. If the circles do not intersect the two mean directions may be judged as differing significantly (Permian and Triassic circles in Figure 12-6*b*). When the circles overlap a further test is required. The mean paleomagnetic pole is defined relative to the present geographic pole for a specific sampling unit covering an interval of time and not for the entire Earth at an instant of time. Thus it is not the ancient equivalent of the present geomagnetic pole. Estimates of the positions of mean paleomagnetic poles are the basic information needed for study of the variations in the Earth's field and for the comparison of results from different rock units on

different continents. No significant shift of paleomagnetic poles relative to the present geographic poles has been detected since mid-Tertiary times, but evidence indicates that they could have shifted considerably earlier. Paleoclimatic studies compared with paleomagnetic results, however, seem to indicate that the ancient magnetic poles have remained approximately coincident with ancient geographic poles as might be anticipated from ideas about the origin of the Earth's magnetic field (Chapter 3).

Interpretation of Paleomagnetic Pole Positions

Mean paleomagnetic directions are ex-

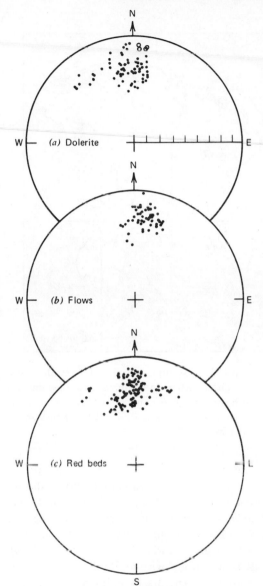

Figure 12-2. Paleomagnetic directions in the Newark Series of New Jersey. The directions in the red sediments are those of NRM. The NRM of the igneous rocks contains viscous components directed along the present field. These have been minimized by magnetic cleaning in 150 oe (peak) alternating magnetic field and the directions after this treatment are given here. Equal-area projection (after Irving, 1964, Figure 5.1, with permission of John Wiley and Sons).

and the angle of dip from the horizontal (compare Figure 3-13*b* for declination and dip of the present magnetic field). The paleomagnetic pole of a rock unit from a given land mass tells us the direction of the land relative to the paleomagnetic pole and the paleolatitude of the rock unit relative to this pole (Figure 3-13*a*). The land mass could be anywhere on a small circle centered on the paleomagnetic pole.

It was in the 1950's that paleomagnetic studies were extended back in time beyond the Tertiary, and results obtained from Triassic red beds in England by P. M. S. Blackett's group were the first to reopen the question of the possibility of continental drift. Clegg, Almond, and Stubbs (1954) found that the geomagnetic field direction in the Triassic diverged about 30° from the present north geographic pole, and it had a dip of about 30° in contrast to the 65° dip at the present time. This proved that the paleomagnetic pole and England had moved relative to each other, but provided no information about the relative movements of England, the paleomagnetic pole, or both relative to the present geographic coordinate system. They assumed that the geomagnetic poles had remained approximately coincident with the present geographic poles, and explained the change in angle of declination as a rotation of England through about 30° since the Triassic. The change in dip was interpreted in terms of a change in latitude, England having migrated northward since the Triassic (note the dips at various latitudes in Figure 3-13*a*). This was startling new evidence for continental drift.

Additional data were gathered for different land masses during different geological periods, and the method of presentation preferred by Blackett's group is shown in Figure 12-3. This shows for various geological periods the ancient latitudes and orientations of Europe, North America, India, and Australia with respect to the paleomagnetic pole, which is assumed to have remained coincident with the present north geographic

pressed in polar coordinates in terms of the angle of declination from the present meridian

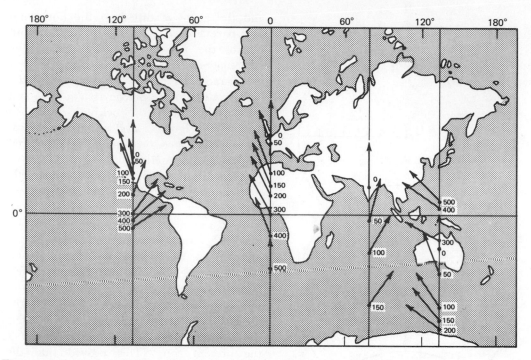

Figure 12-3. Paleolatitudes and orientations of North America, Europe, India, and Australia with respect to the magnetic pole at various geological periods. The numbers indicate the ages of points in millions of years. The paleomagnetic pole is assumed to have remained coincident with the present geographic pole (after Blackett et al., 1960, as redrawn by Takeuchi et al., 1967, with permission of the authors. Reproduced by permission of Freeman, Cooper & Co.).

pole. Results for each continent are given on a line of longitude passing through a town located approximately at the center of the continent, and the ancient latitude and orientation of each continent are given in terms of the reference town. No ancient longitudes are given because they are indeterminable, although this does not mean that there has been no longitudinal movement. Figure 12-3 contains all of the information that can be provided by the measurement of paleomagnetic poles.

S. K. Runcorn and his colleagues examined the data with the initial assumption that the land masses remained fixed in respect to the present geographical coordinates. For England and other European countries, they located the average positions of the paleomagnetic poles for each geological period since the Precambrian and plotted them on a

world map. The older the period the farther removed was the paleomagnetic pole from its present position, as shown in Figure 12-4a. Figure 12-4a can be considered as a diagram presenting in graph form the relative positions of Europe and its paleomagnetic pole at different periods, with no implications for the absolute movements of poles or continents. This is a path of apparent polar wandering, which may reflect polar wandering, continental drift, or some combination of both. The data from a single region are insufficient to illustrate what their relative contributions may have been. The figure shows in a different way the same kind of data illustrated for Europe in Figure 12-3. The study of paleoclimatology, however, had earlier produced a locus of polar wandering for the geographic north pole, which is reasonably similar to the locus of the paleomagnetic poles shown in

Figure 12-4. This suggests that the geographic pole and the paleomagnetic pole had moved together relative to the present geographic coordinates. In 1954 Runcorn and his colleagues interpreted the paleomagnetic data for Europe in terms of true polar wandering without invoking appreciable continental drift.

The locus of apparent polar wandering based on North American samples is shown in Figure 12-4b. It is quite similar to that obtained for Europe but not identical. The paths shown in Figure 12-4 are from Runcorn's 1962 book, and they are based on considerably more specimens than the first paths for Europe and North America which were published in 1954 and 1956 respectively. Yet according to Runcorn they are not different in broad outline from the early paths. The two paths are compared in Figure 12-5. Their approximate parallelism suggests that true polar wandering has occurred. Runcorn and his colleagues concluded that there is a consistent distance of about 30° of longitude between the two curves from Precambrian to Triassic. Therefore a polar wandering hypothesis alone is inadequate, because if only the poles had moved each continent would have an identical locus of polar wandering. These paleomagnetic data are satisfied, however, if continental drift according to Wegener's scheme is added (Figure 11-1). If North America were moved 30° of longitude to the east the two curves would become effectively coincident from the Precambrian to the Triassic and the Atlantic Ocean would be closed. If the continents had moved apart in the Triassic the coincidence of the polar paths from the Triassic to the present is also explained.

Extension of paleomagnetic research to other continents revealed that each continent had a different locus of apparent polar wandering. Whereas it is fairly straightforward to bring together the paths shown in Figure 12-5 by closing the Atlantic Ocean, it becomes much more difficult to obtain a solution for the relationship between polar wandering and continental drift when so many curves are involved. Not only continental drift but continental rotation must be invoked (Figure 12-3). Thus the collection of additional paleomagnetic data raised additional problems of interpretation.

Alternate interpretations have been proposed. Not everyone was convinced about the statistical validity of the distinction between the two paths for Europe and North America (Figure 12-5), and it was suggested by F. H. Hibberd in 1962 that a single path of polar wandering would satisfy most of the data from all continents with all discrepancies between this and previously published loci resulting from secondary magnetization (see also Figure 12-6). Figure 12-1 shows that secondary magnetization can produce rather drastic effects. The assumption of a dipolar field prior to the Tertiary has been challenged. Some consider it more reasonable to assume the existence of a multipolar field than multiple loci of polar wandering.

Despite these reservations the paleomagnetic results by the end of the 1950's were forcing geologists and geophysicists into a reconsideration of continental drift. This led to a symposium organized for the Royal Society of London in 1964 by P. M. S. Blackett, E. Bullard, and S. K. Runcorn which was published as a special volume in 1965. In his introduction Blackett noted that it was advances in two virtually new subjects, the study of magnetism of rocks and the study of the floors of the oceans, that were helping to overcome the former widespread objections to the concept of continental drift. In the final discussion, after 320 pages of text, M. G. Rutten stated that remembering the violently opposed continental drift discussions during the 1920's, the

"papers of this symposium have shown that, apart from paleomagnetic data, nothing much has been changed . . . It still depends on which part of the geological data one finds most strongly heuristic, if one is a 'drifter' or a 'fixist' . . . The only way to remain fixist

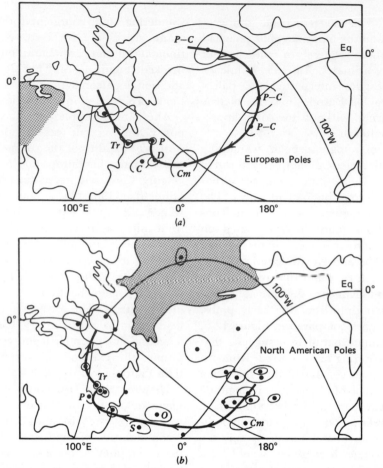

Figure 12-4. Polar wandering curves based on measurements on rocks in (*a*) Europe, and (*b*) North America. Key: *P-C*, Precambrian; *Cm*, Cambrian; *O*, Ordovician; *S*, Silurian; *D*, Devonian; *C*, Carboniferous; *P*, Permian; *Tr*, Triassic (based on Figure 19 of Runcorn, 1962, by permission of Academic Press, Inc.).

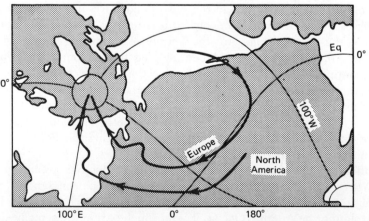

Figure 12-5. Comparison of polar wandering curves for Europe and North America, redrawn from Figure 12-4 (after Runcorn, 1962, by permission of Academic Press, Inc.).

now, is to disbelieve paleomagnetism, a position which becomes more and more awkward as its methods tend to become better substantiated. In future we shall have to base all of our geological theoremata on the data supplied by paleomagnetism."

PALEOMAGNETISM SINCE 1960: MAGNETIC REVERSALS

Polar Wandering and Continental Drift

Figure 12-5 summarizes the paleomagnetic results of the 1950's which suggested that true polar wandering and continental drift had both occurred. Improvement of techniques and the accumulation of more data permitted additional tests for consistency (Figure 12-2), with revision and reevaluation of the significance of the apparent polar wandering curves. Several compilations and reviews were published. The review by A. Cox and R. R. Doell in 1960 represents the state of the art at the end of an exciting decade. Irving's book of 1964 includes more results and shows how rapidly data accumulated.

The accumulation of data did not solve all problems and lead to general acceptance of the thesis portrayed in Figure 12-5. This is demonstrated in striking fashion by three reviews published almost simultaneously. In a symposium volume edited by Runcorn in 1967 J. Hospers and K. M. Creer reviewed essentially the same data for North America and Europe, and Creer extended his synthesis worldwide to include all paleomagnetic data available. Hospers concluded that the data suggest true polar wandering at least since the Cambrian and cast serious doubt on the reality of continental drift. Creer prepared apparent polar wandering curves for seven continents, superimposed the Upper Paleozoic portions for five of them, and concluded that the circumstantial evidence for drift now appears to be overwhelming. In 1968 I. A. Rezanov examined the data in Irving's book, together with all available data for the U.S.S.R., and concluded that paleomagnetic data are still so unreliable that they cannot be used as evidence either for or

against the relative drift of continents. We cannot spare the pages to attempt resolution of such conflicting interpretations; all we can do is to expand slightly this outline of their conclusions.

Irving reexamined the apparent polar wandering curves of Figure 12-5, using all pole estimates available by late 1963 that fulfilled his minimum reliability criteria. Instead of the previous simple pattern, Figure 12-6a shows what Irving referred to as rather curious changes in the relative longitudes for the pole paths of North America and Europe-northern Asia. The North American path begins to the west of the Eurasian path in the Cambrian. They cross around Silurian times and again between Triassic and Cretaceous times. Irving concluded that the differences between the Lower Paleozoic curves may or may not be fortuitous, and that many more results are needed before this feature can be adequately discussed. He emphasized that it would be many decades before an adequate coverage would be available. On the other hand there is little doubt, on the basis of the 1963 evidence, that the longitude difference in the Permian poles is significant and incidentally the most marked in the diagram; 95% circles of confidence for several of the mean poles are shown in Figure 12-6b.

Figure 12-6a shows in addition to Irving's plotted poles and paths six poles determined by Hospers using paleomagnetic triangulation. The procedure used is to take widely separated sampling sites from the same continent using rocks of the same age. The paleomagnetic pole for the continent at this time lies at the point of intersection of the paleomeridians drawn through each site. If everything is as it should be the pole positions

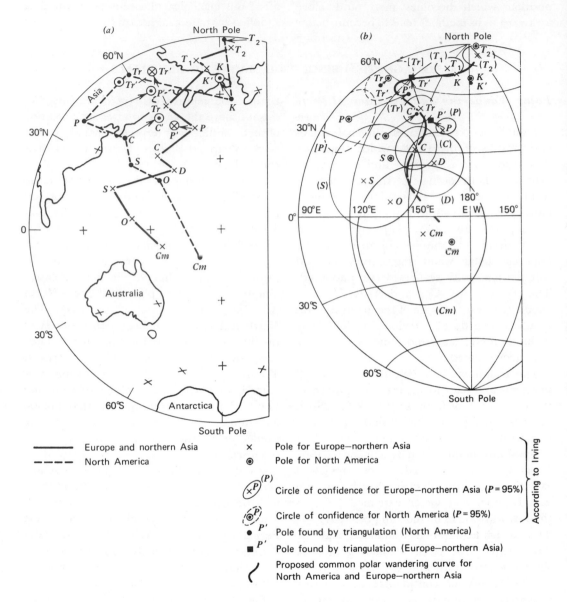

Figure 12-6. Polar wandering curves after Hospers (1967). (a) The Cambrian and younger pole positions for Europe and North America are shown, according to data evaluated by Irving in 1964. Separate paths for each continent are shown, and these paths intersect each other twice. Contrast Figure 12-5. The poles that Hospers located by triangulation are distinguished from the others by symbols with a prime (e.g. *Tr'*). The difference between Irving's poles and the triangulated poles is shown by arrows. Key: T_2, Upper Tertiary; T_1, Lower Tertiary; *K*, Cretaceous; *J*, Jurassic; *Tr*, Triassic; *P*, Permian; *C*, Carboniferous; *D*, Devonian; *S*, Silurian; *O*, Ordovician; *Cm*, Cambrian. (b) Hospers' preferred single polar wandering curve common to Europe-northern Asia and North America, based on all previous data and the new triangulated data. (Reproduced by permission of John Wiley and Sons.)

on the meridians calculated in the usual way from the inclination (dip) should coincide with the intersection point determined by triangulation. In order to test the reliability of Irving's poles Hospers compared them with the six triangulation poles. For North America he found that results from Carboniferous and Permian samples (sediments) show a statistically significant difference, whereas Triassic and Cretaceous data (including igneous samples) do not. For the Eurasian poles he found no statistical difference in Permian samples but a significant difference for the Triassic results. Hospers reviewed the many sources of systematic error in paleomagnetic inclinations measured in sediments, and concluded that these errors are likely to decrease the inclination and therefore provide a calculated pole position too far from the sampling site (Figure 3-13a). He therefore considers the paleomagnetic intersections to be more reliable. Figure 12-6a shows that the new pole positions found by triangulation bring the separate polar wandering paths closer. Although Irving had concluded that the longitude difference between Permian poles is significant, Hospers's largest revision closes the gap between the two Permian poles.

Hospers presented a single polar wandering curve for North America and Europe-northern Asia using his six preferred triangulation points and the other poles from Irving. Figure 12-6b shows the curve passing among the pole positions transferred from Figure 12-6a, together with some of Irving's 95% circles of confidence. The curve is nowhere outside the circles of confidence for the poles used and is within a few degrees of the triangulated poles. Hospers concluded that on this common polar wandering curve pole positions for the Upper Tertiary, Cretaceous, and Triassic times may be fixed with considerable confidence and for Permian and Carboniferous times with less accuracy. The Silurian and Cambrian poles are uncertain, and any comparison of great circle distances between Eurasian and North American

poles is meaningless for Cambrian and Silurian times. The rate of polar wandering indicated is about 90° in 600 m.y. or an average of about 1 or 2 cm per year. Creer felt that Hospers's conclusions were not justified. He stated that Hospers's preference for triangulated poles over Irving's calculated mean poles only makes sense if it can be shown that the inclinations of the remanent magnetism of the rock formations studied are likely to be systematically low, and this is not so.

Creer's first step in the synthesis of worldwide paleomagnetic data was to construct apparent polar wandering paths for the seven continents: South America, Africa, Australia, Europe-Russia, North America, India, and Australia. These were based for the most part on Irving's data compilation with inclusion of all other reliable data. For many of the poles data were sparse. No Cambrian poles and only one Ordovician pole was used. The principal paleomagnetic argument supporting the drift hypothesis is the divergence of apparent polar wandering curves from the geographic poles when drawn on the present globe, and Creer found marked divergence of the curve for Europe from those for Australia and India.

More data are available for Europe-Russia than for any other land mass, and one might assume therefore that the pole path was reasonably established. Nevertheless Creer modified the curve for Eurasia quite significantly by introducing a remagnetization hypothesis to explain some peculiarities about NRM of Devonian red-beds. He presented an up-to-date version of the original polar wandering curve for Eurasia (Figure 12-4a), which he based on essentially the same data as the curve in Figure 12-6a. This is shown in Figure 12-7a for the south poles with standard error circles; the curve is not reconcilable with curves for South America and Africa, and Creer resolved this problem by means of evidence that two significantly different paleomagnetic poles could be derived from studies of Devonian

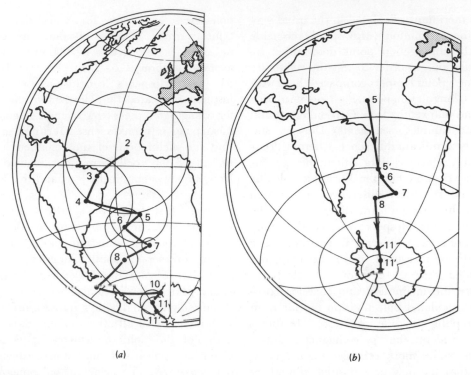

(a)   (b)

Figure 12-7. Polar wandering curve revised by Creer (1967). (a) Polar wandering curve (south pole) for Europe from Cambrian onward based on Irving's (1964) data. (Compare 12-4(a) and 12-6(a).) Standard error circles are plotted. (b) Polar wandering curve (south pole) for Europe and Russia; this is Creer's revised version of the curve in Figure (a). Key: 2, Cambrian; 3, Ordovician; 4, Silurian; 5, Devonian; 5′, Devonian, presumed Carboniferous magnetic age; 6, Carboniferous; 7, Permian; 8, Triassic; 9, Jurassic; 10, Cretaceous; 11, Tertiary; 11′, Upper Tertiary. (Reproduced by permission of John Wiley and Sons.)

red-beds. According to the remagnetization hypothesis one of these represents the true Devonian geomagnetic field, and the other the Carboniferous or permo-Carboniferous field. The revised polar wandering curve thus derived is shown in Figure 12-7b.

Creer then proceeded with paleogeographic reconstructions. The apparent polar wandering curves for five continents—South America, Africa, North America, Eurasia, and Australia—indicated similar polar movements of about 50° during the Upper Paleozoic. By taking a set of spherical shells, on each of which is drawn a particular continent and its polar wandering curve, Creer was able to superimpose the Upper Paleozoic portions of the curves with the Carboniferous poles

coincident. This brought the continents into close proximity as shown for four of them in Figure 12-8. He therefore concluded that during the Upper Paleozoic the paleomagnetic pole had moved about 50° relative to the *fixed* distribution of continents shown in Figure 12-8. This reconstruction differs from Wegener's (Figure 11-1a) and the computer fit (Figure 11-3) mainly because there is a gap between Laurasia and Gondwanaland of about 1500 km at its narrowest, between Africa and the eastern United States. Thus according to the paleomagnetic method of continental reconstruction a broad Tethys Sea completely separated the supercontinents of Laurasia and Gondwanaland during the Upper Paleozoic. Study of the divergence of the Mesozoic curves for the individual continents away from the superimposed master Paleozoic curve of Figure 12-8 shows, in principle, the sequence and direction of breakup of the continents. According to the scheme in Figure 12-8 the initial movements apparently occurred between the Carboniferous and the Permian when North America, Europe, and Australia were displaced from South America and Africa.

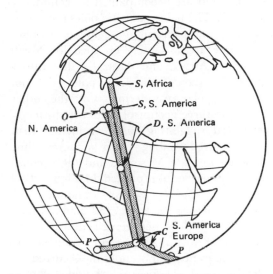

Figure 12-8. Sketch of photograph of polar wandering curves (south poles) for Europe, North America, Africa, and South America. The curves were drawn on spherical shells, and shifted over a globe until the wandering curves coincided (after Creer, 1967). Superposition of the Upper Paleozoic portions of the curves brings the continents into close proximity. Compare Figures 11-1(*a*) and 11-3. Key: *S*, Silurian; *O*, Ordovician; *D*, Devonian; *C*, Carboniferous; *P*, Permian. See paper by McElhinny and Luck (1970) for reconstruction of Gondwanaland from a common polar-wandering path for Lower Paleozoic data from southern continents. (Reproduced by permission of John Wiley and Sons.)

These stimulating experiments in paleogeography offer tantalizing glimpses of what paleomagnetic methods may provide in the future when more data are available and their interpretation is better understood.

Irving stated in 1964 that many decades of effort were yet required, and Rezanov would certainly maintain that Creer had over-extended the data. Creer himself emphasized two of the problems which limit interpretation of paleomagnetic data: the fact that the magnetic age of many sediments is younger than the fossiliferous age, and the difficulty in establishing equality of geological age for rock formations such as unfossiliferous red-beds in widely separated continents. A few quotations from Rezanov's translated paper will serve to illustrate his contention that even the most recent paleomagnetic data are too unreliable to be of any use in connection with the hypothesis of continental drift.

"We assembled only the most reliable determinations of the poles.

"So far as we could see, availabilities of large numbers of paleomagnetic measurements lead almost invariably to a prodigious scattering of paleomagnetic poles, and it becomes evident that no opinion at all may be formed on the continental drift on the basis of one single determination of the pole. But this is precisely the case with South America, in which the Silurian, Devonian, Carboniferous, and Permian paleomagnetic poles were found, each on the basis of one single determination. For Africa, two Carboniferous and two Permian poles are available, both for the same district; it develops that they are 5000 and 7000 km apart, respectively.

"The example with the 'migration' of the poles of the Nizhnyaya Tunguska area is a clear illustration of some kind of major error in the paleomagnetic determinations of the pole. Three regions are indicated, 6000–9000 km from each other, and the pole 'jumps' therein from region to region.

"It should be essential to investigate the causes responsible for inconsistencies of the paleomagnetic data for one and the same continent prior to drawing such responsible conclusions (tentative as they may be) in regard to such tremendous horizontal shifts and rotations within the continents.

"The width of the Atlantic Ocean is smaller than the range of reproducibility of the paleomagnetic measurements.

"But even this accuracy is unattainable for geological periods older than the Carboniferous. The scattering is as broad as 10,000 kms, i.e. the maximum possible discrepancy when magnetic axes of the earth are perpendicular to each other, in the Rhipean, Cambrian, Ordovician, and Silurian."

Many of the uncertainties arise from measurements of sedimentary rocks. A 1967 paper by C. S. Grommé, R. T. Merrill, and J. Verhoogen is an encouraging sign that we can expect better interpretations as more measurements are obtained for igneous rocks dated radiometrically. These authors determined the directions of NRM for two igneous intrusions in California with potassium-argon apparent ages of 136 m.y. and 142 to 129 m.y., and they obtained a mean paleomagnetic pole to which they assigned an averaged age of 138 m.y. In Figure 12-9 the new determination was compared with the five other Cretaceous paleomagnetic poles available for radiometrically dated igneous rocks in North America. The 95% confidence circles or ovals overlap and none of the six pole positions is significantly different from any other. This indicates that from about 138 to 84 m.y. ago either polar wandering relative to North America did not occur or its rate was too slow to be detected. Apparent polar wandering did occur earlier than 138 m.y. according to the only two other Mesozoic paleomagnetic poles available from radiometrically dated igneous rocks. These are shown in Figure 12-9 as *WM*, the White Mountain series dated at 180 m.y., and as *NG*, the Newark Group dated at about 202 m.y. The directions for the Newark Group are shown in Figure 12-2.

These data also provide new evidence that the Earth's field was a geocentric dipole farther back in time than the Cenozoic Period. The Cretaceous paleomagnetic poles from widely spaced localities in North America are coincident, and the paleomeridians have a common intersection.

Figure 12-9 compares the North American poles with 20 reliable late Mesozoic poles from other continents. The authors took the seemingly drastic step of omitting sedimentary rocks, but they pointed out that of the total of about 60 independent Jurassic and Cretaceous poles listed by Irving in 1964 only three from sedimentary rocks are reliable and are accompanied by confidence intervals. The paleomagnetic poles representing large and overlapping segments of Mesozoic time are so closely grouped for North America, Africa, and Australia that differences between individual poles do not indicate polar wandering. On the other hand the three groups of poles are distinctly separated suggesting that continental drift has occurred among them at some time between the late Mesozoic and the present.

The dated results plotted in Figure 12-9 show no apparent polar wandering for North America between 138 and 84 m.y. ago, nor for Africa between 209 and 109 m.y. ago, nor for Australia between 178 and 93 m.y. ago. Apparent polar wandering occurred for North America between 202 and 138 m.y. ago. These results suggest that effectively no true polar wandering occurred during the Jurassic and much of the Cretaceous between about 200 and 100 m.y. ago. The apparent polar wandering for North America between 202 and 138 m.y. ago the authors therefore interpret as continental drift.

They extended these conclusions by considering other published evidence that Australia showed very little apparent polar wandering from late Carboniferous to very early Tertiary (about 300 to 60 m.y. ago), and that considerable apparent polar wandering for Africa occurred between early Permian and late Triassic times (about 275 to 200 m.y. ago) and again since about the middle of the Cretaceous (about 100 m.y. ago). Thus the periods of apparent polar wandering for North America and Africa did not coincide in time within the Mesozoic

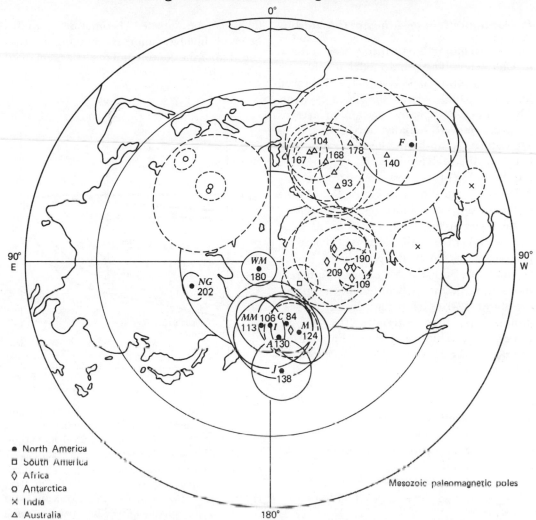

Figure 12-9. Equal-area map of northern hemisphere showing selected Mesozoic paleomagnetic pole positions for six continents (after Grommé et al., 1967); 95% confidence ovals or circles are shown as solid lines for North America poles and as dashed lines for others. Radiometric ages (in millions of years) are given where available. *J* is new determination by authors. (From Jour. Geophys. Res., **72**, 5661, 1967, with permission.)

Era. From the Australian evidence they concluded that true polar wandering during the Mesozoic was either absent or greatly subordinate. Therefore the apparent polar wandering for Africa and North America they interpreted in terms of continental drift.

The results shown in Figure 12-9 for North America suggest a period of continental drift followed by a long period of essentially no drift. Considering also the other results referred to above there is apparently a pattern for continents to experience long periods of essentially no drift preceded and followed by shorter periods of relatively rapid drift. The periods of drift for different continents do not necessarily coincide. Apparent polar wandering, and hence by implication continental drift, is an intermittent process.

Paleomagnetism and the Earth's Radius

The expanding Earth theory was outlined in Chapter 9. The usual model involves an originally continuous layer of sial broken into continental slabs as the Earth expands. The continents remain constant in area as they are moved radially outward, and as expanding ocean-basin floor is produced between them. This was B. C. Heezen's interpretation of the cross-section through the Atlantic Ocean shown in Figure 9-9. As the radius of the Earth increases the geocentric angle between two fixed points on a stable continental mass will decrease. Therefore paleomagnetic methods potentially permit calculation of the Earth's radius at different times in the past if the field remained dipolar.

Figure 12-10 following Irving (1964) summarizes three estimates (by Egyed, Carey, and Hilgenberg) of the rate of increase of the Earth's paleoradius expressed as fractions of the present radius which is set at 1.0. These rates were inferred independently of paleomagnetism.

Several methods have been used to estimate the paleoradius of the Earth using paleomagnetic methods. The simplest uses the equation

$$r_p = d/[\cot^{-1}(\tfrac{1}{2}\tan I_1) - \cot^{-1}(\tfrac{1}{2}\tan I_2)] \tag{12-1}$$

where I_1 and I_2 are paleomagnetic inclinations for two sites of the same age on the same stable continental mass and d is the distance measured on the surface between the two geomagnetic latitude circles. Cox and Doell averaged 80 values calculated for the Permian and obtained the value 6310 km, with the standard deviation of the mean being 230 km. This result is shown on Figure 12-10. The method is subject to uncertainties involving the average direction of magnetization of the sample and its time of magnetization compared to its time of formation and to uncertainties about matching the presumed or apparent ages of widely spaced rocks. It is doubtful that radius changes of less than 20% can be detected. Estimation of radius changes involves another assumption, that the area of the continental mass has remained unchanged by tectonic activity.

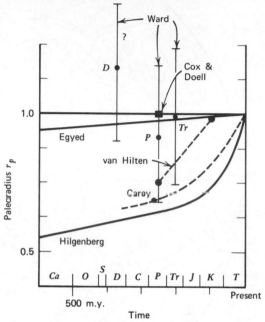

Figure 12-10. Changes in the Earth's radius since the beginning of the Cambrian (modified after Irving, 1964, Figure 10.31). Geological time is set out horizontally, and the paleoradius vertically; $r_p = 1.0$ is the present radius. The hypotheses of expansion are: Egyed 0.05 cm/year; Hilgenberg (1962); Carey (1958) $r_p = 0.75$ for late Paleozoic and the dashed line being surmise from his text. The values determined paleomagnetically by Ward (1963) are D, Devonian; P, Permian; Tr, Triassic, the limits being estimates of the standard deviations of the means. Points P, K connected by dashed line are from van Hilten (1968); Cox and Doell (1961) gave value for Permian of 6310 km with standard deviation 230 km. (Reproduced by permission of John Wiley and Sons.)

More generalized procedures include that of M. A. Ward who used the criterion that the most probable paleoradius is that for which the dispersion of paleomagnetic poles for a continent is a minimum. Three results determined for the Devonian, Permian, and Triassic have standard deviations of 20%;

there is no systematic trend of the mean values with time. D. van Hilten modified this method to take into account possible deformations of the continent and found it suitable only for the paleomagnetic data available in the Permian and Cretaceous Periods. The results are shown in Figure 12-10 without error bars and connected tentatively by the dashed line.

The paleomagnetic results are inconsistent with the hypotheses of Carey and Hilgenberg, but not sufficiently accurate to test the slow expansion rates proposed by Egyed. It appears therefore that the expanding Earth theory alone is inadequate to explain continental drift.

Magnetic Reversals in Igneous Rocks

The discovery by B. Brunhes in 1906 that some rocks are magnetized in a direction opposite to that of the Earth's present magnetic field received very little attention for many years. In the 1950's there developed a controversy over the origin of the normal (negative) polarities and the reversed (positive) polarities. It was established that the total numbers of reversed and normal rocks are about equal since the Precambrian. Many reversals of polarity were found in stratigraphically dated sequences of Mesozoic and Tertiary rocks, apparently occurring at intervals of a few thousand years. This suggested that the change of polarity was caused by a reversal of the Earth's magnetic field but the rocks were not accurately dated, and it could not be proved that the reversals occurred simultaneously in different locations. The only alternative solution appeared to be that certain rocks were capable of selfreversal due to some chemical or mineralogical peculiarity. Evidence against this is that recent lavas always have normal polarity. It is known, however, that self-reversal can occur. T. Nagata reported in 1952 that when he cooled a dacite lava from a temperature above its Curie point it became magnetized in a direction opposite

to the ambient field. At about the same time it was shown theoretically that there are several possible mechanisms by which a rock could acquire reverse magnetization especially in rocks containing two magnetic minerals. The complex relationships among the titanomagnetites and ilmenite-hematite series of minerals in igneous rocks suggested that selfreversals might have occurred. The concept that the Earth's magnetic field has reversed at intervals was a difficult one for many to accept.

Many laboratory and geological tests were devised to settle the controversy, and one of these involved study of the magnetization of igneous rocks and their associated thermally metamorphosed rocks. This is a particularly useful test because the country rock and igneous rock are usually of very different materials. According to the hypothesis of field reversals the metamorphic rock on cooling should acquire the same polarity as the igneous rock. If the hypothesis of self-reversals is to account for the observed frequency of normal and reversed polarities, then in roughly half of the cases studied the polarities of the metamorphosed rocks should be reversed compared to that of the igneous body. Results obtained by 1964 from this test and others were reviewed in Irving's book. A comparison of polarities observed in igneous rocks and their baked contacts from all continents and all geological periods except the lower Paleozoic produced 85 examples where the polarities of igneous rock and contact agreed, and only two where they disagreed.

An example illustrating the effects on several different rock types in the vicinity of an extensive igneous intrusive complex is illustrated in Figure 12-11. A 5000 ft. section exposed in the dry valleys of South Victoria Land, Antarctica, consists of a basement complex intruded by dikes overlain by Beacon sandstone. The whole sequence is intruded by the Jurassic Ferrar dolerite with three thick sheets making up almost one-half the total section. The schematic section shown in Figure 12-11 is through the horizontal component of magnetization. The arrows show the directions of magnetization observed in each unit based on detailed collections from over 100 sites. The remanence directions after magnetic cleaning are uniform throughout. Petrological estimates suggest that the temperature of the sediments exceeded 160°C when the dolerites were emplaced, and it appears that the sandstones and basement rocks were remagnetized at this time. In none of the country rocks was reversed polarity discovered.

By the time Irving published his book in 1964 the evidence was very strong that most reversed polarities were caused by reversals of the Earth's magnetic field, but the con-

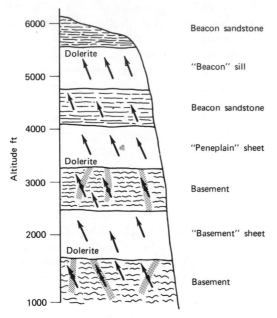

Figure 12-11. Schematic geological section in South Victoria Land (after Irving, 1964, Figure 7.26). Basement dikes, black, Admiralty Granites, wavy lines. Vertical section through horizontal component of magnetization which is approximately N 250°E. The arrows give, in a schematic fashion, the directions observed in each unit. The result is based on collections from over a hundred collecting sites by Bull, Irving, and Willis (1962). (Reproduced by permission of John Wiley and Sons.)

clusive test that polarities should be constant in all rock units of the same age on all continents had not been made and occasional selfreversal for individual rocks was not precluded. If reversed polarities were produced by selfreversals then the contrasting polarities should be randomly distributed in space and time. Irving illustrated the two available examples of reversals dated radiometrically by independent groups of workers, and his figures are combined in Figure 12-12. This shows encouraging agreement; a pattern of alternating polarities with periodicity varying from 0.2 to 1.5 m.y. Irving summarized the status of results for late 1963 and included the following statements:

1. Recent and late Pleistocene rocks are magnetized in the same sense as the present field (negative polarity).

2. In early Pleistocene and older rocks frequent reversals of polarity occur in a manner unrelated to the rock type with a time scale of 10^6 years.

3. All late Carboniferous and Permian rocks (except the Upper Tartarian) are positively polarized indicating a time scale of about 50×10^6 years.

4. In rock formations with both polarities only a few percent of samples show intermediate directions suggesting that the transition period is short compared to the time for which constant polarity is maintained.

These first two studies of radiometrically dated lavas appeared to confirm previous conclusions from stratigraphic studies that polarity intervals were of about the same duration, and these were termed polarity epochs. Extension of lava studies with carefully chosen material and precisely determined polarities and ages, however, revealed the existence of polarity intervals with shorter durations of about 10^5 years. These were termed polarity events. Figure 12-13 shows how the successive discovery of polarity events changed the apparent distribution and duration of polarity intervals. By 1969 there were available 150 radiometric

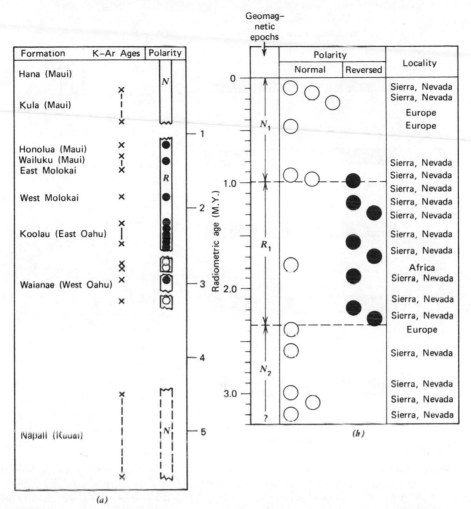

Figure 12-12. Polarity epochs dated radiometrically in the Hawaiian Islands, California, and elsewhere (from comparison by Irving, 1964, Figures 7.3 and 7.4). (*a*) Hawaiian Islands. The levels studied paleomagnetically and radiometrically are marked by dots (positive) and circles (negative) in the third column and crosses in the second column. (*b*) California and elsewhere. (Reproduced by permission of John Wiley and Sons.)

ages and polarity determinations meeting reasonable standards of reliability and precision in rocks less than 4.5×10^6 years old. In order to resolve the fine structure of reversals indicated by Figure 12-13 at least 450 determinations would be required from lavas whose ages are uniformly spaced at intervals no greater than 10^4 years. One difficulty here is in the nature of lava flows. The interval between successive flows may be

a few months or 10^5 years, and the task of locating volcanic formations with the required ages to give suitable coverage is a formidable one.

Figure 12-14 confirms that the polarity epochs and events are synchronous in widely spaced parts of the Earth, and no reasonable doubt remains that these events are produced by rather rapid switching of the Earth's magnetic dipole. The figure also

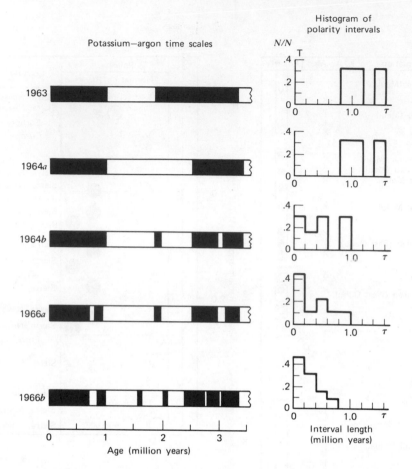

Figure 12-13. Successive versions of the radiometric time scale for reversals, showing how the discovery of polarity events changed the apparent distribution of polarity intervals. In the corresponding histograms, N_T is the total number of polarity intervals and N is the number in each class interval of the histogram. Compare Figures 12-14 and 12-21. (With permission; from A. Cox, Science, **163**, 237–245, 1969. Copyright 1969 by the American Association for the Advancement of Science.)

illustrates the problem of resolving the events caused by the uncertainties of K-Ar age determinations. The polarities may be intermixed for about 10^5 years on either side of a boundary between normal and reversed magnetization; the epochs are clearly demarcated but the events with durations of only 1 or 2×10^5 years are not. The transitions that bound epochs and events are even briefer time intervals as indicated by Figure 12-19.

Additional events have been discovered since the compilation of data in Figure 12-14,

from deep-sea cores as well as from lavas, and the geomagnetic reversal time scale as far back as 4.5×10^6 years including all reliable data published to the end of 1968 is shown in Figure 12-21. Attempts to extend the time scale further back in time have been made, but errors in K-Ar age determinations become too large for accurate work. Statistical analysis of the magnetic polarity of rocks as a function of their K-Ar ages has shown that the dating precision of rocks about 2.5 m.y. old is 3.6%, which agrees with independent estimates of precision

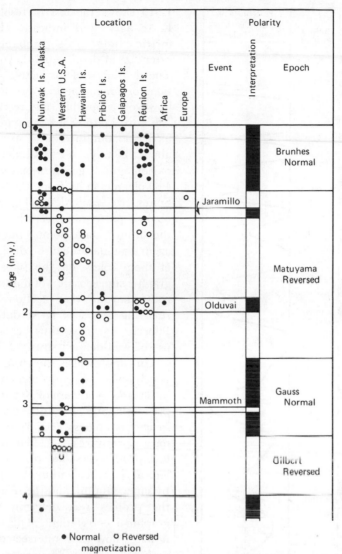

Figure 12-14. Ages and direction of magnetization of lavas from different continents, with corresponding polarity epochs and events. Compare Figures 12-13 and 12-21. (Adapted after Opdyke and Foster, 1967, by Bullard, 1968. Reproduced with permission.)

based on known sources of analytical error. For older rocks the dating error becomes large relative to the lengths of polarity epochs and much larger than the average event. For rocks 10 m.y. old dating precision of 3% gives a dating error of 3×10^5 years.

The difficulty in working with older rocks was illustrated by G. B. Dalrymple *et al.* in

Figure 12-15. This shows the known time scale for geomagnetic polarity, extended in a hypothetical pattern from 3.5 to 12 m.y. to provide the basis for developing a probability model. Using the probability model they calculated the curve for the percentage of rock samples that would have a normal polarity as a function of their K-Ar ages assuming a dating precision of 3%. Figure

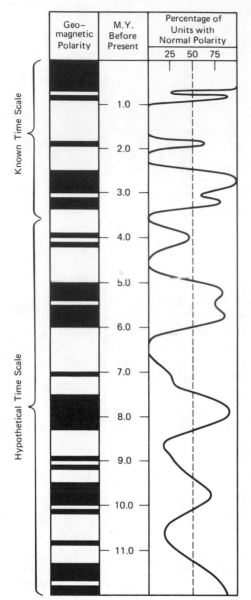

Figure 12-15. Percentage of samples that would have normal polarity as a function of their K-Ar ages, assuming a dating precision of ± 3.0% (standard deviation) and an infinite sample density. The left-hand column shows the assumed polarity of the geomagnetic field. The assumed polarities prior to 3.6 m.y. are entirely hypothetical and are intended only to show the loss of resolving power of K-Ar dating for the earlier part of the reversal time scale (after Dalrymple et al., 1967, by permission of North-Holland Publishing Company, Amsterdam).

12-15 shows the loss of resolving power of K-Ar ages, and indicates that the radiometric time scale probably cannot be extended in detail much beyond 5 or 6 m.y. The definition of distinctive polarity transitions or longer periods of uniform polarity, however, is possible (Figure 12-16).

Figure 12-16 shows the type of results that have been obtained in the study of lavas up to about 20 m.y. old. Results of combined paleomagnetic-geochronometric studies of 44 volcanic units from the western United States are plotted in Figure 12-16*a*. No obvious pattern of reversals is apparent from the limited data points, but four epoch or event boundaries can be distinguished, three of them solely on the basis of the stratigraphic superposition of normal and reversed rocks. Partial compensation for the lack of resolving power of K-Ar dating in rocks of Pliocene age may thus be obtained by the use of well-defined stratigraphic successions.

This approach is shown in Figure 12-16*b*, which gives the results of a cooperative effort in Iceland by a team of 10 authors studying a predominantly basaltic succession ranging in age from oldest Tertiary to young Quaternary. Cores were collected from 21 overlapping lava profiles, designated *A* to *V*. The chronological sequence within each profile is known by superposition, and the relationship among the profiles was determined from stratigraphical correlations. The number of lava flows or flow units intersected was 1140 but, with allowances for overlap of one profile with another, the total succession comprises 900 separate lava flows or flow units totalling 8.8 km in thickness. Measurements were made on 2200 oriented samples from 1070 flows. They recorded 551 normal flows, 406 reversed flows, 73 anomalous flows, and they rejected 40. Most anomalous flows occurred between normal and reversed flows, and these are considered to give information about the intermediate field during the transition interval. Icelandic basalts are poor material for K-Ar dating, but 10 samples from six flows have been dated indicating a maximum age

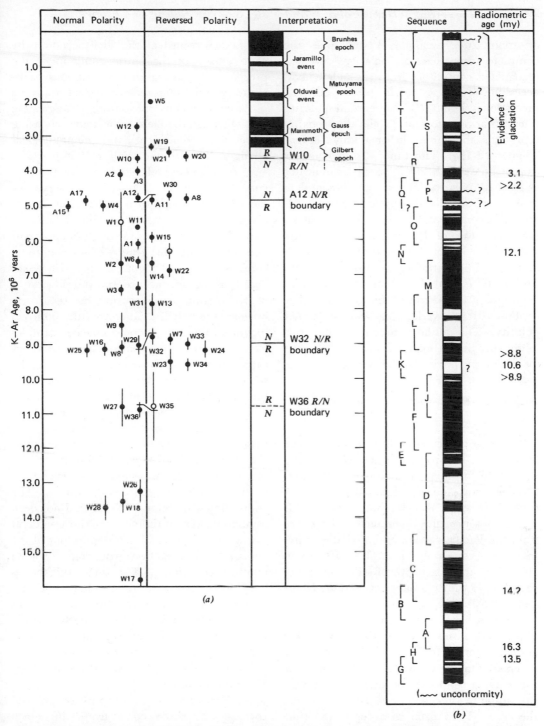

Figure 12-16. Magnetic polarity and K-Ar age determinations of volcanic rocks. (*a*) Upper Cenozoic rocks from the western United States (after Dalrymple et al., 1967). Shaded circles = primary data, open circles = secondary data, *x* = previously published data for the Gilbert epoch. The vertical bars are the estimated precision of the K-Ar ages at the 68% confidence level. The horizontal bars indicate the superposition relationships between flows of opposite polarity. The previously determined polarity time scale for the interval from 0 to 3.6 m.y. ago is shown (compare Figures 12-13 and 12-14). (*b*) Icelandic lavas. Generalized succession of polarity zones in overlapping lava profiles, designated *A*, *B*, etc. (after Dagley and nine co-authors, 1967). (Reproduced by permission of North-Holland Publishing Company, Amsterdam, and of *Nature*).

of about 20 m.y. Figure 12-16*b* gives the complete sequence of polarity reversals recognized continuous from the oldest lavas accessible with a small break between successions *O* and *P*. The normal polarity interval at the top of profile *V* probably represents the Jamarillo or Olduvai event (Figure 12-16*a*). The times of the reversals cannot be defined, but the analysis indicates at least 61 polarity intervals or 60 complete changes of polarity in this succession which gives an average rate of at least 3.0 inversions/ 10^6 years. Two of the normal polarity zones are noteworthy for their stratigraphic length and by implication for their duration. Profiles *E*, *F*, and *J* contain about 76 nonoverlapping lavas, and profiles *L*, *M*, and *N* contain 101 nonoverlapping lavas of normal polarity. The authors suggested that this might provide a pair of marker horizons for dating purposes near 10^7 years.

Magnetic Reversals in Deep-Sea Sediments

The paleomagnetic study of sediment cores from the oceans has proceeded sporadically since 1938, but it was not until 1964 that reversely magnetized sediments were discovered. C. G. A. Harrison and B. M. Funnel observed five reversals in a short core from the Pacific Ocean. Sedimentation rates in deep oceans range from 1 to 10 mm in 10^3 years, and the depth of sediment corresponding to the boundary between the Gauss and the Gilbert epochs would thus be somewhere between 3 and 35 m. The continuous nature of deep-sea sedimentation may provide a more complete record of geomagnetic reversals than that available from volcanic rocks. Data from the volcanic rocks provide the absolute time scale needed to calibrate the cores. Then from core depth and time scale rates of sedimentation can be estimated. Because of its close connection with the evolution of man the Pleistocene has been a battleground for scientists since it was named more than 100 years ago, and the

incomplete record of the Pleistocene on the continents kept the battle going. Now the prospect of dating the continuous sedimentary and faunal records in deep-sea cores provides a means for dating stratigraphic levels through the Pleistocene, and correlating these with biological and climatological changes. (Ericson and Wollin, 1968.)

The paleomagnetic study of long piston cores at Lamont Geological Observatory picked up momentum in 1966 when it was observed that cores from high latitudes around Antarctica were strongly magnetized and stable. The stratigraphy of seven of these cores is shown in Figure 12-17, with zones based on the appearance and disappearance of radiolarian species and the record of normal (negative) and reversed (positive) magnetization of closely spaced samples from the cores. The paleomagnetic stratigraphy and the radiolarian zones have the same time dependence throughout the area indicated on the location map. The sequence of normally and reversely magnetized sections in the left-hand core duplicates exactly the magnetic stratigraphy for lava flows shown in Figure 12-14. The 1966 version of the radiometric time scale for geomagnetic reversals is correlated in Figure 12-17 with the cores. The Jaramillo event (Figure 12-14) had not been established at that time but the short segment of negative polarity is recorded in four of the cores. The correlation from one core to another is striking. The zone thicknesses vary from one core to another reflecting different sedimentation rates in different locations, but the relative lengths of the magnetic zones in each core are effectively the same as the relative durations of the polarity epochs and events determined by K-Ar dating of lavas.

In Figure 12-18 the depths of polarity reversals in three cores from the north Pacific have been plotted against the geomagnetic reversal time scale. The lines through these points give the average sedimentation rates. For two of the cores rates of 0.75 cm and 1.3 cm/10^3 years are obtained,

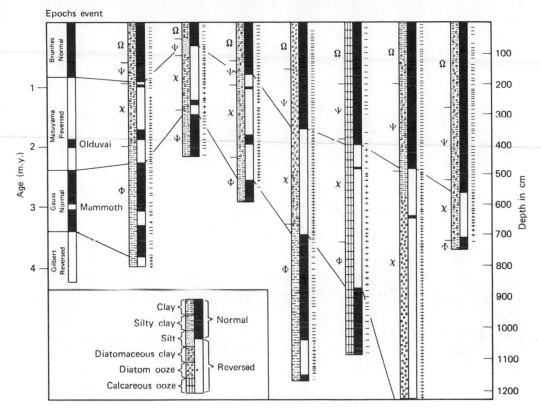

Figure 12-17. Magnetic stratigraphy in deep-sea cores from the Antarctic, with stratigraphic correlation and radiolarian faunal zones (greek letters). Normally and reversely magnetized levels are correlated across the Antarctic and with the dated polarity epochs (after Opdyke et al., 1966). See Foster and Opdyke (1970) for extension of the magnetic stratigraphy and its correlation with the sea-floor spreading anomaly sequence to 9 m.y. or more (Figures 13-13 and 13-15). (Reproduced with permission from Science, **154**, 349–357, 1966. Copyright 1966 by the American Association for the Advancement of Science.)

and the third core gives an average rate of 0.36 in the upper part and 0.80 in the lower. A well marked volcanic ash layer occurs in all three cores at different depths, and the correlation is confirmed because for all three cores Figure 12-18 gives an age of about 1.2 m.y. for this layer. These results indicate that average sedimentation rates in deep-sea environments remain constant through long periods although there are probably short-term fluctuations that average out. Assuming a constant rate of sedimentation and using diagrams such as Figure 12-18 it becomes possible to estimate the durations of the polarity events which are too short to be

measured accurately by dating lavas. The Jaramillo event appears to have lasted from 0.95 to 0.89 m.y. ago, and the Olduvai event extended from 1.95 to 1.79 m.y. ago. Given an average rate of sedimentation for a core it is now possible to estimate the date of anything that is recorded in the core, provided that the cores are amenable to paleomagnetic study. Unfortunately not all of them are.

These deep-sea cores with a geomagnetic time scale not only provide information about stratigraphy, paleontology, and sedimentation rates but also about the history of the Earth's magnetic field. Figure 12-19 shows the results of a detailed study of a portion of

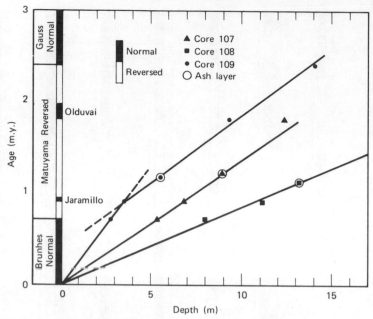

Figure 12-18. Rate of sedimentation for three cores from the North Pacific, given by plot of depths of polarity reversals against the geomagnetic polarity time scale (after Ninkovich et al., 1966). (Reproduced by permission of North-Holland Publishing Co., Amsterdam.)

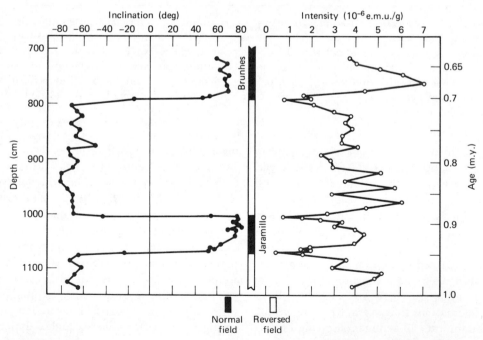

Figure 12-19. Magnetic inclination and intensity of magnetization of a portion of a core from the North Pacific (after Ninkovich et al., 1966). (Reproduced by permission of North-Holland Publishing Co., Amsterdam.)

a core from the north Pacific. The core depth scale is shown on the left and the polarity epochs and events in the center. The corresponding ages are given on the right. In an attempt to examine the nature of the Earth's magnetic field during a transition interval the cores were closely sampled across the reversals. The measured values of inclination and intensity of magnetization for samples taken at 1 cm intervals are shown in the figure. In a uniform sediment the intensity of magnetization provides a measure of the magnetizing field. The results indicate that the intensity of the Earth's field decreases through about 10,000 years by 60 to 80% before there is any change in dip, and then the field reverses during an interval of 1000 or 2000 years, and the field intensity then builds up again for another 10,000 years. This behavior is compatible with a reversal of the Earth's dipole field accompanied by the usual fluctuating nondipole field.

Measurements of the Earth's dipole moment show that the dipole field decreased between 1835 and 1965 at a uniform rate from 8.5 to 8.0×10^{25} gauss cm^3. If the rate remains constant the moment will pass through a zero point about 2000 years from now and then reverse its polarity. There is evidence, however, that this change is a part of a well-defined dipole fluctuation with cycle duration of about 10^4 years, and the probability that a geomagnetic reversal will result from the decrease in dipole moment currently in progress is 5% (Cox, 1969).

The continued paleomagnetic study of sedimentary cores has provided valuable information about the duration of longer events and has helped to extend the time scale back into the Gilbert reversed epoch, but it has not resolved the frequency of very short events. There are a number of inconsistencies and discrepancies in the detail of these events. The known Olduvai and Jaramillo events, for example, are not recorded in all cores as can be seen in the results of the first detailed study in Figure 12-17. The Gilsa

event (Figure 12-21) is represented in some cores and not in others. Some of the discrepancies may be due to disturbances in the coring process which appears to be something of an art; soft bands may be squeezed out or disturbed sediment sucked in. Slumping or turbidity currents could cause removal or addition of material at some sites on the deep-ocean floor and this becomes more likely approaching the continental margins.

The problem of short period polarity events in deep-sea sedimentary cores was reviewed by N. D. Watkins in 1968. He selected seven cores from the Southern Ocean and sampled them at 10 cm intervals reduced to 2 cm in the regions of polarity changes. All results are shown in Figure 12-20a. The normal and reversed polarities are correlated with the standard polarity time scale on the right. Figure 12-20b is a log of the inclination of the remanent magnetization for the core at the right of Figure 12-20a. Repeated measurements and resampling at 2 cm intervals prove that there is no question about the reality of the short polarity events shown in Figure 12-20. Whether or not they really reflect the ancient geomagnetic polarity behavior is regarded by Watkins as "much more problematical." He is convinced that there must exist a natural source or sources creating inconsistencies in the results for the upper Matuyama paleomagnetic stratigraphy, and his review led him to conclude that the main difficulties result from pene-contemporaneous organic redeposition of sediment. Watkins suggested that benthonic faunal activity may create a zone of redeposition up to 50 cm deep, and he discussed the possible effects of such activity. These include modification of the true position of a polarity boundary, reduction of the thickness of a given polarity zone, creation of new boundaries or "split" events, and the complete loss of short events. Reliable delineation of the fine structure of the geomagnetic history of deep-sea cores can only be expected for cores from areas of high sedimentation rates and no significant biological redeposition.

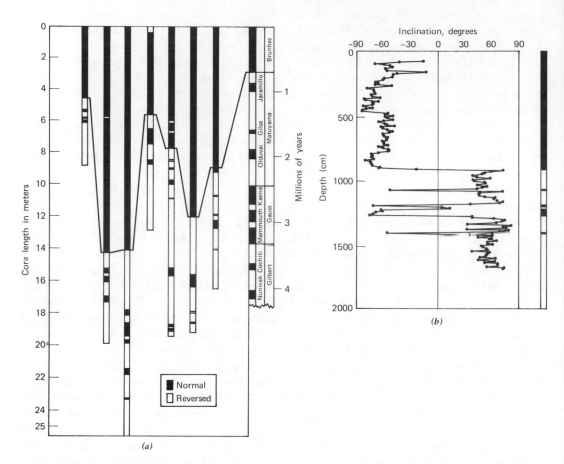

Figure 12-20. Short period polarity events in deep-sea sediments (after Watkins, 1968). (*a*) Polarity of remanent magnetism in seven selected deep-sea sedimentary cores from the Southern Ocean. Polarity time scale at right due to Cox and Dalrymple (1967). First column shows polarity, either normal (black) or reversed (clear); names of geomagnetic events in second column; names of geomagnetic epochs in third column. (*b*) Inclination of remanent magnetism in specimens from core on right of Figure (*a*), following demagnetization at 150 oersteds. Inclination is negative when normal polarity. Polarity log at right: black is normal, clear is reversed. (Reproduced by permission of North-Holland Publishing Co., Amsterdam.)

Figure 12-21 published in January 1969, compared with Figure 12-14 published in 1967, shows the discovery of several short events in only two years, and it is predicted and expected that others remain to be discovered. The time scale for Figure 12-21 is based on K-Ar dating, and each short horizontal line shows the measured age and the magnetic polarity of one volcanic cooling unit. The duration of events is based in part on paleomagnetic data from sediments and in part on interpretation of the profiles through the linear magnetic anomalies over mid-oceanic ridges. Interpretation of these anomalies depends upon the theory of sea-floor spreading. Therefore let us turn next to this theory which has proved to be the central theme for the revolution of the 1960's.

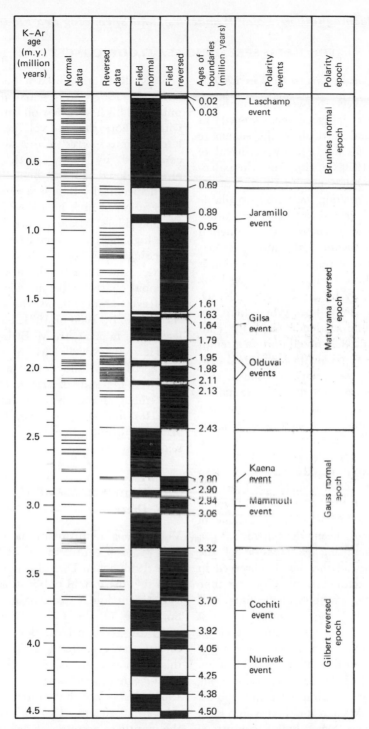

Figure 12-21. Time scale for geomagnetic reversals (after Cox, 1969). Each short horizontal line shows the age as determined by potassium-argon dating and the magnetic polarity (normal or reversed) of one volcanic cooling unit. Included are all published data which meet reasonable standards of reliability and precision. Normal-polarity intervals are shown by the solid portions of the "field normal" column and reversed-polarity intervals, by the solid portions of the "field reversed" column. The duration of events is based in part on paleomagnetic data from sediments and magnetic profiles. Compare with Figure 12-13. (From A. Cox, Science, **163**, 237–245, with permission. Copyright 1969 by the American Association for the Advancement of Science.)

THE 1960'S: SPREADING SEA-FLOOR CONCEPT

The Bandwagon Began to Roll in 1960

In the controversy between the mobilists and the fixists, alternatively known as horizontalists and verticalists, the stalemate in the 1940's was followed by a revival of interest in the 1950's arising from the paleomagnetic interpretations and the increasing amount of information becoming available about the physiography and physical properties of the ocean basins and continental margins. The reluctance of many geologists and geophysicists to reopen the discussion is indicated by the contents of the 1959 book *Physics and Geology* by J. A. Jacobs, R. D. Russell, and J. T. Wilson. They were not impressed by the evidence cited for continental drift, and they noted that no attempt had been made to reconcile continental drift with modern observations of the ocean. Although they concluded that the paleomagnetic evidence for polar wandering was rather convincing, they added that polar wandering could hardly be considered a major cause of orogenesis; it was possibly a by-product. They noted that convection current theories were widely held at this time, but they felt that arguments in favor of convection were inconclusive, and none of the theories had been developed into a specific or convincing explanation of the formation of any of the details observed in mountains or continents. By 1960 the time was ripe for new syntheses of the data, and it is notable that the attention of most investigators was directed toward the ocean basins rather than toward possibly drifting continents. A short time after the publication of this volume J. T. Wilson had become one of the most active contributors to the mobilist thesis and a spokesman for the revolution in the earth sciences. After half a century of indecisive arguments the theory of continental drift supported by the concept of sea-floor spreading generated such enthusiasm in the early 1960's that it took on the character of a bandwagon.

S. K. Runcorn edited a book, *Continental Drift*, in 1962, fifty years after Wegener published his first review on continental drift, with the hope that the volume would stimulate a serious interest in this subject formerly considered by many earth scientists as already closed. In this book B. C. Heezen noted specifically the ideas generated by the observation of high heat flow from the mid-oceanic ridges. Many scientists realized that such high heat flow could be accounted for by rising convection cells in the Earth's mantle, and these could also explain the tensional character of the ridges. According to Heezen among those who had expressed such views by 1960 or 1961, either in public lectures or in print, are M. Ewing, E. Bullard, H. Hess, R. Revelle, M. Menard, and R. Dietz. Of these and doubtless many others who were in a state of ferment over the heady excitement of rapidly accumulating data it was Hess and Dietz who developed the convection current scheme into the concept of sea-floor spreading.

The Contributions of H. H. Hess and R. S. Dietz

Voluminous literature on sea-floor spreading as a cause of continental drift has appeared since 1961, when Dietz introduced the term with the first formal publication on the topic "Continent and Ocean Basin Evolution by Spreading of the Ocean Floor." Hess is generally given priority for originating the theory, however (and Dietz acknowledges this), although his paper on the "History of the Ocean Basins" was not published until 1962. Hess wrote that although Holmes and others had suggested convection currents in the mantle to account for deformation of the Earth's crust, mantle convection was still considered a radical hypothesis not widely accepted by geologists and geophysicists. By 1960, however, scientists were beginning to realize that acceptance of mantle convection would permit the construction of a reasonable story to describe the evolution of ocean basins with whole realms of previously unrelated facts falling into a regular pattern. Hess's theory proposed that the sea floor is essentially the out-cropping of the peridotite mantle partly hydrated to form serpentinite. This is covered by a thin veneer of sediments and volcanic rocks. The major structures of the sea floor are direct expressions of the convection process, with the midoceanic ridges marking the sites of rising limbs of mantle-convection cells and the oceanic trenches being associated with convergences or descending limbs of convection cells. The continents are carried passively on the convecting mantle, and they do not plow through the oceanic crust as proposed by earlier hypotheses of continental drift. The leading edges of continents are strongly deformed where they impinge upon the downward moving limbs of convecting mantle, and the cover of oceanic sediments and volcanic rocks may also ride down into what Hess called the "jaw crusher" of the descending limb to be metamorphosed and eventually welded onto the continents.

Hess's concept of the rising convection cell at the midocean ridges is illustrated in Figure 12-22. Hess was impressed by the uniform thickness, 4.7 ± 0.7 km, of the main crustal layer (layer 3) beneath the oceans. He argued that basalt flows could not conceivably be so uniform in thickness over such a large area, and that it was more likely that the bottom of this layer represented a present or past isotherm, the temperature at which peridotite was converted into serpentinite in the presence of water. Figure 10-2 shows the intersection of a geotherm with the serpentinization reaction. This occurs at a temperature near 500°C, and in Figure 12-22a Hess has adopted the 500°C isotherm as the level at which serpentinization occurs. Water migrating upward with the rising limb of a mantle convection cell coexists with unaltered peridotite until the temperature falls to 500°C at the isotherm, and here the water

reacts to produce serpentinite. Divergence of the rising limb at the surface carries the layer of serpentinized mantle across the ocean floor, and the 500°C isotherm migrates to deeper levels as shown in Figure 12-22a. But in the absence of water the boundary between unaltered and partly serpentinized mantle remains at the depth corresponding to the isotherm beneath the midocean ridge. This explanation for the uniform thickness of the main crustal layer requires that the temperature distribution beneath the midocean ridge remains uniform for periods of time of the order of 10^8 years, which presents some problems of its own.

For most of the ocean basin the Moho is defined as the boundary between crustal material with P-wave velocity of 6.7 km/sec (range from 6.0 to 6.9) and mantle material with P-wave velocity of 8.1 km/sec, as shown in Figure 12-22b. This boundary is not found beneath the ocean ridge crest where the seismic wave velocities are considerably lower. Hess suggested that the anomalous seismic velocities could be explained by the higher temperature of the rising material, together with fracturing where the convective flow changes direction from vertical to horizontal. The fractures are healed as the partly serpentinized peridotite and the underlying peridotite move together to the flanks of the ridge. A new crustal layer is thus generated from mantle material at the midoceanic ridges, and this layer spreads across the ocean basin floor to a descending limb of the convection cell. Here the process is reversed and the descending material is dehydrated when its temperature reaches 500°C releasing water upward to the sea. It has been shown experimentally that the strength of serpentinite drops appreciably near its dehydration temperature, and this may have significant tectonic implications if it occurs beneath ocean trenches assumed by Hess to be the sites of descending limbs of the convection cells.

According to the theory of sea-floor spreading if the rate of movement for convection is

1 or 2 cm per year, the floors of the ocean basins must be completely renewed every 200 or 300 million years. The sea floors and the shapes of the ocean basins are thus comparatively young features compared to the ancient continental blocks. This could account for the relatively small thickness of sediments on the ocean floor and for the apparent absence of sediments older than the Jurassic period. Figure 12-22*b* shows how younger sediments should show progressive overlap on a midocean ridge if the mantle does move laterally away from the ridge crest. This prediction has since been con-

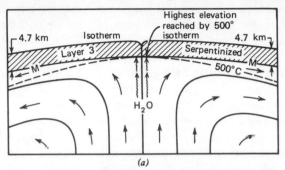

(a)

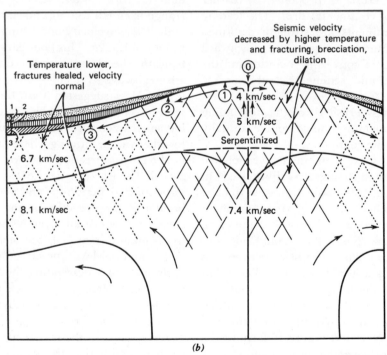

(b)

Figure 12-22. Representation of sea-floor spreading by Hess (1962). (*a*) Diagram to portray highest elevation that 500°C isotherm can reach over the rising limb of a mantle convection cell, and expulsion of water from mantle which produces serpentinization above the isotherm. (*b*) Diagram to represent (1) apparent progressive overlap of ocean sediments on a mid-ocean ridge which would actually be the effect of the mantle moving laterally away from ridge crest, and (2) the postulated fracturing where convective flow changes direction from vertical to horizontal. Fracturing and higher temperature could account for the lower seismic velocities on ridge crests, and cooling and healing of the fractures with time, the return to normal velocities on the flanks. (Reproduced by permission of The Geological Society of America.)

firmed in striking fashion by the results of JOIDES, the deep sea drilling program, in the Atlantic (Figure 14-7).

If new sea floor is generated at the mid-ocean ridges the older, displaced sea floor must be carried downward at the sites of converging convection cells. Continental masses are initially moved along in conveyor-belt fashion until they attain a position of dynamic balance overlying a convergence. There the continents come to rest, but the mantle continues to shear under and descend beneath them possibly forming an ocean trench. If a new region of divergence develops beneath a continental mass the mantle convection currents will tend to rift the continent apart. Presumably the Atlantic Ocean marks an ancient rift which separated North and South America from Europe and Africa. The Indian Ocean Rise may extend through the Red Sea into the African Rift Valleys, tending to fragment the continent. Downward movement of convection cells beneath continents or at continental margins places the continents under compression, which accounts for orogenesis and alpine folding. If the continental block is being carried along passively by the convecting mantle the margin is tectonically stable, but if the downward convection limb is un-coupled from the continent marginal mountain ranges tend to be formed. It is anticipated that the sites of divergence and convergence may change with time causing new tectonic patterns.

Holmes's early model of mantle convection and continental drift is remarkably similar to the model of sea-floor spreading developed by Hess and Dietz, but there is a significant difference which sets the scheme of Hess and Dietz apart not only from Holmes's model, but also from the many similar convection models that were published in the early 1960's. In 1931 it was generally believed that sialic patches, and even a thin layer of sial, existed above the thick basaltic layer forming the oceanic crust. Holmes therefore depicted his convection cells operating beneath the

oceanic crust which was stretched, thinned, and in part disrupted as the continental blocks were carried apart (Figure 11-2). He did refer to the addition of new basaltic material from the mantle to the crust. Hess and Dietz on the other hand proposed that there is no crust in the usual sense beneath the oceans. The thin veneer distinguished by seismic measurements is actually the exposed mantle surface, the upper part of an active convection cell, whose properties have been modified by serpentinization where the cell approaches the surface at a midoceanic ridge (Figure 12-22*a*).

Some Other Convection Models

In a detailed review, "Sea Floor Relief and Mantle Convection," H. W. Menard in 1965 concluded that neither the facts of marine geophysics nor the hypothesis of some kind of convection in the mantle required acceptance of the bolder hypothesis of sea-floor spreading as developed by Hess and Dietz. Menard was not convinced that the sea floor is periodically swept clean by convection, nor that the oceanic crust is easily created or destroyed. He noted that because the oceanic crust does exist the hypothesis of sea-floor spreading is meaningless unless this crust has the same composition as the mantle. A phase change from basalt to eclogite is not acceptable (Chapter 10), and Menard argued against Hess's serpentinization hypothesis of Figure 12-22. He showed that the 500°C isotherm was never likely to have been at the right depth to produce a uniform layer of serpentinized peridotite with the thickness of the present crust, and that the 500°C isotherm beneath the crust of the midocean ridges could rise to within 3 km of the ocean floor. Even if the 500°C isotherm did not rise above the 5 km depth at the ridge crest (Figure 12-22) Menard doubted that in such a thermally unstable environment the proposed limitation of crustal thickness by the isotherm could produce the uniform thickness required. He also considered it implausible that the crust could be transported from the ridge crests for several thousand kilometers without change in thickness.

Menard did agree, however, that the characteristic features of ocean rises and ridges are readily explained by transient convection cells in the mantle beneath the crust as illustrated in Figure 12-23. The sea-floor crust is stretched and thinned and crustal blocks are moved about differentially along faults. Menard's examination of the distribution of ocean rises on polar projection maps led him to conclude that 87% of the rises lie on circles centered on continental shields with most of the circles having about the same radius. About one-half the length of the rises, including of course the midoceanic ridges, is centered relative to the ocean basins. The circular pattern of ocean rises and ridges suggested to Menard that perhaps the primary control of convection is the upward separation of light from dense fractions of the mantle under continents, with the denser material sinking and spreading out

from below the continents. At some distance from the center, the material moves upward to complete the cell and to form the oceanic rises. The evidence from marine geophysics suggests that material rises and spreads laterally under the crust, but it does not indicate anything about the depth of or even the existence of a deep return flow in an organized convection cell.

According to Hess and Dietz mantle convection cells reach the ocean floor (Figure 12-22), but Menard pictures them moving beneath a thin ocean crust (Figure 12-23). J. T. Wilson discussed the same problem in 1963, and placed the convection cells deeper within the mantle beneath a lithosphere about 100 km thick (Figure 12-24a). The lithosphere is a tectonic unit including the crust and a rigid layer of the upper mantle. Horizontal flow is restricted to the weaker asthenosphere layer which is equated with the seismic low-velocity zone, and both upward and downward flow are similarly restricted to narrow zones within the mantle.

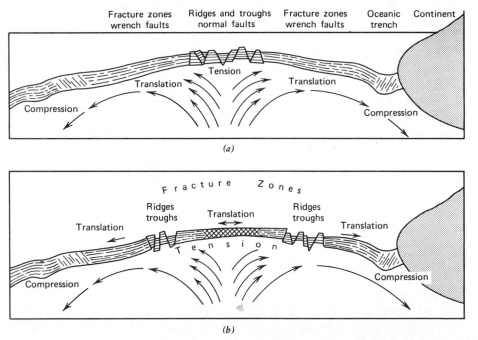

Figure 12-23. Convection current hypothesis; topographic and tectonic effects resulting from upwelling beneath the oceanic rise system. (From "Marine Geology of the Pacific" by H. W. Menard. Copyright 1964, McGraw-Hill Book Company. Used with permission.)

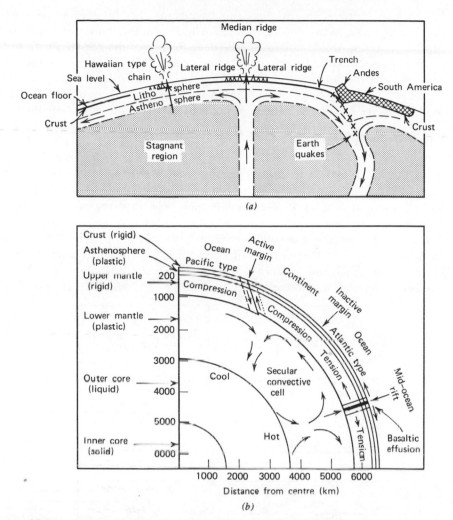

Figure 12-24. Alternative mantle convection models. (*a*) Diagrammatic section showing a type of convection which might explain the origin of pairs of lateral ridges, median ridges, Hawaiian-type chains and mountain building (after Wilson, 1963). (*b*) Convection currents confined to depths below the Benioff seismic zones (Figure 2-6) (after Bernal, 1961). Note that only shallow-focus earthquakes are associated with the mid-oceanic ridge system (Figures 2-5 and 14-3). (Reproduced by permission, from *Nature*.)

Downward flow beneath continental margins such as the Andes occurs along the inclined zones of deep-focus earthquakes (Figure 2-6). The convection current carries the lithosphere over the asthenosphere and drags it downward beneath the oceanic trench and marginal mountains. See Figure 14-5.

Figure 12-24*a* also illustrates a possible explanation for linear chains of volcanoes in the ocean basins. Active volcanoes are formed near the crest of the ridge and the horizontal currents carry the volcanic piles successively off the ridge, detaching them from their source. This process could explain the aseismic lateral ridges extending from median ridges. Similarly a fixed source of lava rising from stagnant mantle beneath the asthenosphere could produce a Hawaiian-type chain of volcanoes. If the lava source were in the asthenosphere then a single active volcano would be formed, and this would migrate as the asthenosphere source migrated carrying the lithosphere and super-incumbent volcano with it.

J. D. Bernal inferred that the upper mantle was rigid to a depth of 900 km because of the existence of deep earthquake zones (Figure 2-6), and he proposed that convection cells are restricted to the lower mantle as illustrated in Figure 12-24*b*. This model includes the asthenosphere shown in Figure 12-24*a*, but it plays no role in the convective scheme. The convection cells place the upper mantle and crust in tension beneath the midoceanic ridges, drag the ocean floor and western hemisphere continents to the west, and create

the Pacific trenches by overthrusting along the inclined plane defined by intermediate and deep-focus earthquakes. He also indicated the existence of a rift zone extending to 900 km in depth beneath the midoceanic ridges, but this is unlikely because all earthquakes in these regions are of shallow-focus types.

S. K. Runcorn noted in 1962 that the scale of continental drift required mantle-wide convection cells. He suggested that gradual growth of the core would cause the convection cells to decrease in size and increase in number in a series of rather abrupt changes in convection patterns. Each change in convection pattern could produce drifting and repositioning of the continents.

In 1968 J. C. Maxwell suggested that

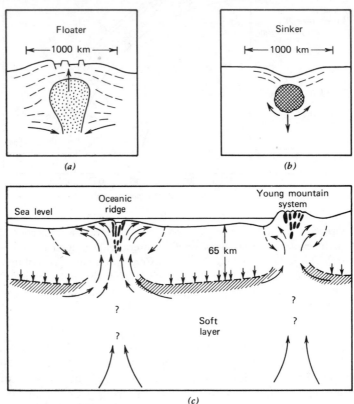

Figure 12-25. Vertical movements as alternative to closed convection cells (after **Maxwell**, 1968). (*a*) Sinker (Figures 9-8(*f*) and 10-9). (*b*) Floater (Figure 10-8). (*c*) Postulated association of oceanic ridges and young mountain systems with diapiric mantle material. Black blobs represent mantle rocks brought up into the crust (Figure 6-6). (Reprinted by permission, *American Scientist*, journal of The Society of the Sigma Xi.)

closed convection cells were not required to explain the features of ocean basins and continental margins. Figures 12-25a and b show the initiation of convection or advection by the uprise or sinking of a portion of the mantle that has become slightly less dense or more dense respectively than its surroundings. These two types, the floater and the sinker, may accomplish the transport of large volumes of material but they may occur independently of each other, and therefore they are not necessarily components of a convecting system. Maxwell proposed the scheme illustrated in Figure 12-25c. Both midoceanic ridges and young mountain systems are loci of long-continued high thermal energy and hot mantle material probably rising from the low-velocity zone. Diapiric uprise of this material is driven by the sinking of cool areas of normal crust making up the continents and ocean basins. Maxwell accepts sea-floor spreading because he states that the rising hot mantle material must spread laterally from the oceanic ridges if the process is to continue, but this system seems to be incapable of bringing about continental drift.

The enthusiasm with which various convection models were developed in the early 1960's to explain the observations of marine geophysics and to support continental drift met with opposition. For example G. J. F. MacDonald published a series of papers using evidence from gravity and heat flow measurements to show that continents have deep structures, and that chemical differences between oceanic and continental regions must extend downward to several hundred kilometers (Chapter 7). This implies that if convection cells exist within the Earth, they cannot extend upward to near-crustal levels. This would only permit schemes of the type shown in Figure 12-24b. MacDonald's comparison, however, of the figure of the Earth derived from satellite gravity data with the ideal ellipsoid of revolution assuming hydrostatic equilibrium, demonstrates that stress differences exist in the mantle, and that an average viscosity of 10^{26} poises is required. This value is incompatible with a convecting mantle. Despite these arguments, evaluation of possible mantle convection cells has continued, and we shall return to this topic in Chapter 14.

13. *Magnetic Anomalies in the Ocean Basin*

INTRODUCTION

Neither paleomagnetism nor magnetic anomaly is listed in the index of the symposium on the *Crust of the Earth* edited by A. Poldervaart, and published in 1955. In their contribution reviewing the geophysical contrasts between continents and ocean basins M. Ewing and F. Press stated that:

"Magnetic measurements are as yet meager, but those available indicate that over large oceanic areas the magnetic field is unusually smooth. As data accumulate, valuable conclusions may be drawn from edge effects of continents and from correlation with submarine topography."

The increasing value of magnetic data was confirmed by the fact that in a 1966 symposium on "The History of the Earth's Crust" at the Goddard Institute for Space Studies in New York, edited in 1968 by R. A. Phinney, no fewer than six of 16 contributed papers were concerned with geomagnetism or magnetic anomalies of the ocean basins, and revolutionary conclusions were indeed being drawn. The interpretations and correlations were not related merely to local features of submarine topography, but to worldwide oceanic processes.

During the interval 1955 to 1966 there had been three major developments related to geomagnetism. Two of these have been reviewed in Chapter 12: the location of paleomagnetic pole positions with respect to continental masses at specific times during geological history and the recognition and dating of reversals of the Earth's magnetic dipole. The third was the discovery of linear magnetic anomalies over the ocean basins and their correlation with the polarity reversal time scale.

Several papers suggesting some relationship between the magnetic anomalies and sea-floor spreading had appeared by 1966, but the parts of the theory did not fall into place until the Goddard conference. The editor, Phinney, introduced the symposium volume with a historical sketch of recent developments. He wrote that

"In July 1966, the results of Vine and Pitman and Heirtzler were known to only a few colleagues. Word got around during the fall, by preprint, and at the Goddard conference, slightly preceding formal publication in *Science*. By January 1967 the impact was such that nearly 70 papers on sea-floor spreading had been submitted for the April meeting of the American Geophysical Union."

The linear anomalies in what Ewing and Press described as an unusually smooth magnetic field thus deserve our careful attention.

Oceanographic cruises from the Lamont Geological Observatory and the Scripps Institution of Oceanography initiated the intensive collection of magnetic data. In

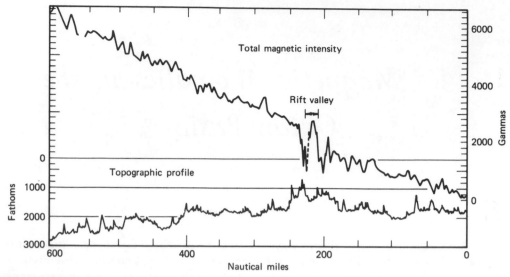

Figure 13-1. Profile of total magnetic intensity and topography, mid-Atlantic ridge (after Heezen et al., 1959). Magnetic values in gammas relative to an arbitrary zero. (Reproduced by permission of The Geological Society of America.)

their special paper on "The Floors of the Oceans. I, North Atlantic" B. C. Heezen, M. Tharp, and M. Ewing of Lamont reported that the first magnetic data for the Atlantic Ocean from a ship-towed magnetometer were obtained in 1948, and that by 1957 nearly 20 crossings had been made of the crest of the mid-Atlantic ridge. These revealed a characteristic pattern of a large positive anomaly of more than 500 γ over the rift valley with negative anomalies of

300 to 500 γ over the adjoining rift mountains. This is illustrated by Figure 13-1 which gives a topographic profile of the mid-Atlantic ridge and the profile of total magnetic intensity. They also reported rough fields with 5- to 15-mile wide 100 to 200 γ anomalies over oceanic ridges and rises. It is now known that a large part of the ocean basins is covered by a magnetic pattern in the form of stripes, but these were not reported until 1958.

1958–1968: LINEAR MAGNETIC ANOMALIES

1958: Discovery in the Pacific by R. G. Mason

Research ships from the Scripps Institution had been towing total field magnetometers behind them since 1952, but the isolated, long profiles provided no basis for quantitative interpretation. In 1955 the U.S. Coast and Geodetic Survey Ship appropriately named *Pioneer* began surveys with closely spaced lines, and an extensive area of the Pacific was

mapped with the detail and accuracy associated with airborne magnetometer surveys on land. The first magnetic results were reported by R. G. Mason in 1958. The survey was extended to an area 250–300 miles wide off the foot of the continental slope between latitudes 32 and 52°N as shown in Figures 13-2 and 13-3. The detail for a portion of this survey is related to the geomorphic provinces and principal fault zones in Figure 13-3. Further cruises were undertaken by Scripps ships in 1958, 1959, and 1960 in order to extend the survey to the west on both sides of the Pioneer and Mendocino faults.

The magnetic maps were constructed by first plotting the measured field value along the ship's track and contouring the result with an interval of 50γ. The standard deviation of a single observation is about 15γ, but the relative error between any two points on the same traverse is negligible. It is unlikely that normal diurnal variations, whose effects were not removed, have any significant effect on the positions and magnitudes of the more important anomalies. The regional magnetic field was determined by a smoothing process for mean observed fields in 20 mile squares, and this was subtracted from the observed field to obtain the magnetic anomaly maps shown in Figures 13-2 and 13-3. Figure 13-2 shows the anomaly pattern in terms of positive (black) and negative (white), and Figure 13-3 shows contour intervals of 250γ.

The magnetic survey revealed a narrow pattern of remarkably straight anomalies of about 400γ magnitude and about 30 km width, trending north-south for about 1000 km. The largest anomaly is less than 2% of the Earth's field. The more prominent anomalies are remarkably regular throughout their lengths, and the pattern is clearly offset by the Murray, Pioneer, and Mendocino fracture zones. The offset suggests for the Murray fault zone a right-lateral displacement of about 155 km in the eastern section, and more than 640 km on the western section; a left-lateral slip of about 265 km across the Pioneer fault; and about 1185 km across the Mendocino fault zone. These displacements are of the same magnitude as those required by continental drift, but the fact that the magnetic pattern is preserved in the upper part of the oceanic crust indicates that this too is rigid and not a sea of sima through which the sialic continents can sail.

The remarkable regularity of the magnetic pattern points to a simple cause but

no satisfactory explanation was forthcoming at this stage. It was clearly established that the pattern showed little correlation with topography over most of the area. The anomalies could be produced by flat, slablike structures approximately underlying the areas of positive anomaly, and there are two factors that restrict the possible depths and thicknesses of such slabs. Because of the sharpness of the anomalies their sources must lie at shallow depths beneath the ocean bottom. The anomalies require magnetization contrasts corresponding to susceptibilities in the range 0.005 to 0.015, which is appropriate for the contrast between magnetic basalts and relatively nonmagnetic sediments. The upper limit to the polarization contrast that can be assumed gives a minimum thickness for the production of anomalies with the observed magnitude.

Figure 13-4 shows the profile of a positive magnetic anomaly along the line A-A just above the Mendocino fault in Figure 13-3. Also shown are three schematic crustal sections through infinite north-south structures with the shapes of the slabs adjusted by trial and error so that the computed theoretical anomaly for each structure (not shown in the figure) fits the observed anomalies. The first example represents an isolated body of magnetic basalt within the second layer of the crust, the second an elevated fault block of the main basaltic crustal layer, and the third a vertical slab of the main crustal layer plus the second layer with magnetic contrasts compared with the adjacent crust possibly caused by intrusion of highly magnetic material from the mantle. Any one of these structures would explain the observed anomalies in two-dimensional profiles at least, but no topographic or seismic expression of any of these possible structures has been established in the surveyed area.

The various explanations proposed to account for the anomalies include the following:

1. The excess magnetic material represents the topography of the upper surface

of the more magnetic crust, which has been covered for the most part by a blanket of sediments.

2. Lava flows have spread out over the floors of the ocean filling pre-existing troughs or depressions due to block faulting or of unknown origin.

3. More strongly magnetic material has been injected into old lines of weakness in the ocean floor possibly connected with the

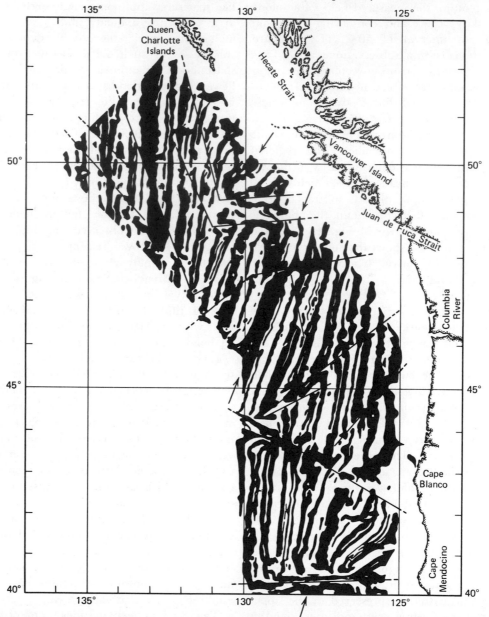

Figure 13-2. Summary diagram of total magnetic-field anomalies southwest of Vancouver Island. Areas of positive anomaly are shown in black. Straight lines indicate faults offsetting the anomaly pattern; arrows, the axes of the three short ridge lengths within this area—from north to south, Explorer, Juan de Fuca, and Gorda ridges (after Raff and Mason, 1961). (Reproduced by permission of The Geological Society of America.)

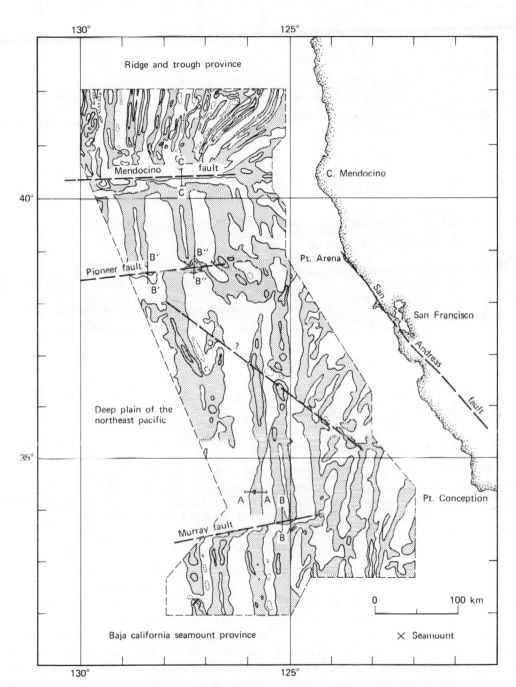

Figure 13-3. Skeleton magnetic map, showing geomorphic provinces, principal known faults, and location of profiles. The contour interval is 250 gammas. Areas of positive anomaly are stippled. Negative contour lines are broken (after Mason and Raff, 1961). (Reproduced by permission of The Geological Society of America.)

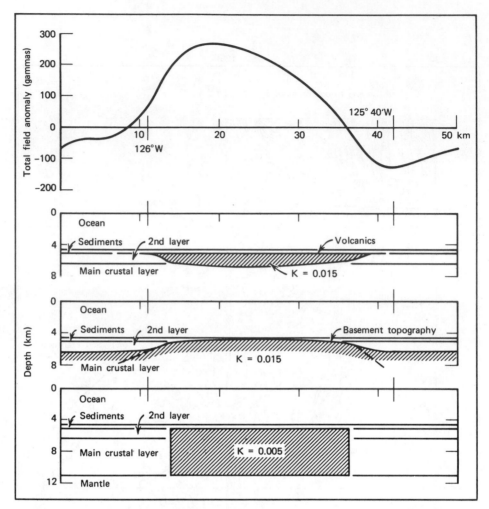

Figure 13-4. Possible interpretations of magnetic profile *a-a* in Figure 13-3 (after Mason and Raff, 1961). The theoretical anomaly computed for each of the three infinite north-south striking structures fits the observed anomaly almost exactly. (Reproduced by permission of The Geological Society of America.)

Earth's spin or mechanisms involving Earth tides.

4. Among other suggestions A. D. Raff in 1961 noted that the magnetic anomalies in the area of the *Pioneer* survey and in some other parts of the Pacific run generally parallel to the ridges there, and that "it looks as though the two are related."

5. The lineation may be the record of a regenerative process of the oceanic crust along oceanic rises, either by global expan-

sion or by the rise of mantle material along oceanic ridges, and bands of lineations should then run parallel to the ridges.

6. The occurrence of linear anomalies in bands parallel to the oceanic rises could be explained by the hydration of olivine of the peridotite mantle into serpentine and magnetite according to the sea-floor spreading hypothesis (Figure 12-22).

7. F. J. Vine, D. H. Matthews, and L. W. Morley proposed a specific model in 1963

correlating the anomalies with sea-floor spreading and reversals of the Earth's magnetic field, and this has subsequently been generally accepted.

1963: Explanation by F. J. Vine, D. H. Matthews, and L. W. Morley

By 1963 many profiles showing bathymetry and the associated total magnetic field, similar to that given in Figure 13-1, had established the existence of a central magnetic anomaly over the oceanic ridges in the North Atlantic, the Antarctic, and the Indian Oceans, and a consistent pattern of anomalies on the ridge flanks. Largely on the basis of measurements made from *H.M.S. Owen* in 1962 during the International Indian Ocean Expedition Vine and Matthews suggested in *Nature* in September, 1963, that the pattern of linear magnetic anomalies discovered in the Pacific (Figure 13-2) was due to strips of the sea floor being magnetized in opposite directions. They explained the existence of such strips by proposing that as new oceanic crust is formed over a convective upcurrent in the mantle beneath an oceanic ridge, according to the idea of sea-floor spreading, it becomes magnetized in the direction of the Earth's magnetic field when it cools below the Curie temperature. If the Earth's field reverses periodically as sea floor spreading occurs, then successive strips of oceanic crust paralleling the crest of the ridge will be alternately normally and reversely magnetized. The successive strips will then cause an increase or decrease in the total intensity of the ambient magnetic field, thus producing the series of linear anomalies.

In an article published in 1964 Morley and Larochelle referred to the presentation of essentially the same hypothesis by L. W. Morley at the Annual Meeting of the Royal Society of Canada in Quebec City in June 1963. Apparently similar conclusions were reached independently. Morley suggested that a nearly unbroken record of the reversals of the Earth's magnetic field may exist in the permanent magnetization of the rocks on the ocean floors, and he suggested that determination of the rate of mantle convection might give the history of the reversals of the Earth's field.

The idea was developed in detail by Vine and Matthews. They found that the magnetic effects of some individual topographic features of igneous origin on the crest of the Carlsberg Ridge in the Indian Ocean could be distinguished. Two volcano-like features were studied. One had negative anomalies as expected for normal magnetization in this low magnetic latitude, but the other had a pronounced positive anomaly suggesting that it was reversely magnetized. The estimated effective susceptibilities of the material forming the two features was ± 0.0133. Computed magnetic profiles across the Ridge assuming normal magnetization were found to bear little resemblance to the observed profiles. This suggested to Vine and Matthews that whole blocks of the survey area might be reversely magnetized. Computed profiles for models with blocks of the crust alternately normally and reversely magnetized did fit reasonably well with the observed profiles, and this led to formulation of the hypothesis outlined above.

The models shown in Figure 13-5 were developed to show how the reversed magnetization hypothesis could account for the shapes of the anomaly profiles encountered in the Pacific (Figures 13-2, 13-3, and 13-4), and the pronounced central anomaly which occurs over the oceanic ridges. The crust is divided into strips 20 km wide which are alternately normally and reversely magnetized. The thickness of the magnetic crust is limited by the depth to the Curie-point isotherm, which was assumed to be 20 km below sea level for the deep ocean, and 11 km beneath the center of the ridges where heat flow is higher. The Curie-point isotherm is probably not far below the 500°C isotherm, which Hess estimated to be at somewhat higher levels (compare Figure 12-22a). The effective susceptibility adopted for the magnetic crust

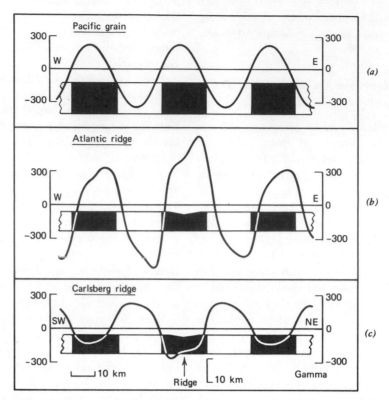

Figure 13-5. Magnetic profiles computed for various crustal models (after Vine and Matthews, 1963). Crustal blocks marked as normally or reversely magnetized. Effective susceptibility of blocks, 0.0027, except for the block under the median valley in profiles *b* and *c*, 0.0053. (*a*) Pacific grain. Total field strength, $T = 0.5$ oersted; inclination, $I = 60°$; magnetic bearing of profile, $\theta = 073°$. (*b*) Mid-Atlantic Ridge. $T = 0.48$ oersted; $I = 65°$; $\theta = 120°$. (*c*) Carlsberg Ridge, $T = 0.376$ oersted; $I = -6°$; $\theta = 044°$. (By permission of *Nature*.)

is 0.0027, considerably less than that derived for the isolated features described above. For the block beneath the median valley in the two Ridge profiles, the effective suscepti-bility was doubled, to 0.0053, because this most recent block is the only one which has a uniformly directed magnetic vector. The effective magnetization of adjacent blocks is reduced by the intrusion and extrusion of new material in a wide zone extending beyond the limits of the central zone. The computed profiles shown in the figure for each crustal section illustrate the essential features of the observed profiles, and confirm that these can be produced by alternating directions of

magnetization, without recourse to major inhomogeneities of rock type within the main crustal layer, or to unusually strongly mag-netized rocks.

Let us note that at this time, not one of the three basic assumptions of the hypothesis was generally accepted: (a) reversal of the Earth's magnetic field, (b) the contribution of remanent magnetism to oceanic magnetic anomalies, and (c) sea-floor spreading. In addition three observations and inferences were not explained by the hypothesis:

1. The best known examples of linear anomalies were those in the northeast Pacific (Figure 4-33), but these did not

appear to be parallel to any existing or pre-existing oceanic ridges.

2. In surveys of known ridges elsewhere, the only anomaly that could be correlated from one profile to another was the central anomaly, and the existence of linear anomalies paralleling the central anomaly was not established.

3. The hypothesis did not explain the observed variations in the amplitude and wavelength of anomalies on either side of the ridge.

The hypothesis was at first received with some scepticism, but rather suddenly between 1965 and 1967 the bandwagon which had been rolling since 1960 gathered such momentum that it carried most scientists along with it. The developments during this interval have had, and will continue to have, such far-reaching implications for the geophysical sciences that we must examine in some detail the successive stages. I think that there are many lessons for students to learn in following the development of the ideas feeding this revolution; for it *has* to be accepted as a revolution. With evidence and interpretations being exchanged at meetings, with so many papers being published within a few months of each other, and with different journals having publication delays varying from 2 to 12 months or more after receipt of the manuscript, it is not easy to decipher the succession. As examples of the ways in which scientific theories develop I have selected two trends which appear quite clear. Magnetic survey work proceeded apace over many oceans, and Vine tested all new data against the sea-floor spreading hypothesis claiming proof by 1966. J. R. Heirtzler and his associates at Lamont, who were actively engaged in gathering new data, felt that some other interpretation was required, but they were unable to produce one. Then some time during 1966 they were persuaded that sea-floor spreading did provide the solution, and by 1968 they had out-spread Vine, one of the original spreaders.

1965-1966: Confirmation by F. J. Vine and J. T. Wilson

By 1965 two of the basic assumptions of the Vine-Matthews hypothesis, field reversals and the contribution of remanence to the magnetic anomalies, had become accepted, and the periodicity of the field reversals was being defined (Figure 12-13). Then J. T. Wilson showed that the apparent absence of oceanic ridges in the northeast Pacific is due to the complications caused by the large horizontal fault zones (Figures 2-3, 13-2, and 13-3) whose effect was not considered in the original hypothesis. In his 1964 book, *Marine Geology of the Pacific*, H. W. Menard had previously identified a belt of ridges and troughs extending north of the Mendocino fracture zone and vanishing, apparently at another fault, at about 50°N.

In order to explain two puzzling features about these lines of large horizontal shear Wilson developed the concept of transform faults. The two features are that many of the dislocations terminate abruptly, and recent seismic activity is often confined to short parts of the fracture zones. The horizontal movements had conventionally been interpreted as due to transcurrent faults, but this tacitly assumed that the faulted medium is continuous and conserved. If new crust is being generated on ridges, then other kinds of faults have to be envisaged. Figures 13-6c and *d* show how a sinistral transcurrent fault would offset a ridge. Figure 13-6a shows a dextral ridge-ridge type transform fault connecting two expanding ridges. In 13-6b the fault is shown after a period of ridge expansion, the arrows with solid heads indicating sea-floor spreading from the ridges. Motion has not changed the apparent offset of the ridge, and the relative motion is limited to the fracture between the ridge crests; note that the motion along the shear between the ridge crests is in the reverse direction to that expected if the fault were regarded as transcurrent. D. C. Krause (1966, p. 425) independently developed a

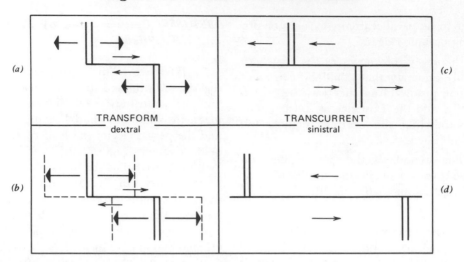

Figure 13-6. Motion on transform faults compared with transcurrent faults (after Wilson, 1965). (*a*) Dextral ridge-ridge type transform fault connecting two expanding ridges. (*b*) Fault shown in (*a*) after a period of movement. Note that motion has not changed the apparent offset. (*c*) Sinistral transcurrent fault offsetting a ridge, with offset in the same sense, but motion in the opposite sense to the transform fault in (*a*). (*d*) Fault shown in (*c*) after a period of motion. Note that the offset has increased. Open-headed arrows indicate components of shearing motion. Solid-headed arrows indicate ocean floor spreading from the ridge axis. (From Science, **150**, 482–485, 1965, with permission. Copyright 1965 by the American Association for the Advancement of Science.)

similar fault concept for an expanding ocean basin.

Application of the concept of transform faults by Wilson to the northeast Pacific in 1965 suggested that the Juan de Fuca Ridge, southwest of Vancouver Island, was linked to the East Pacific Rise by the San Andreas transform fault. Figure 13-2 includes the region of the Juan de Fuca Ridge and any interpretation of the Ridge should obviously be compatible with the magnetic observations. Wilson identified a crustal block with intense parallel anomalies striking N20°E, instead of the usual north-south, and located an axis about which the anomalies are symmetrically arranged. This axis coincides with the physiographic expression of the Ridge crest. This confirmed that at least some of the linear magnetic anomalies in the Pacific should be related to oceanic ridges, and also that these were parallel to a central anomaly over the crest of a ridge.

In a companion paper to Wilson's contribution Vine and Wilson reexamined the Vine-Matthews hypothesis with a polarity reversal time scale to guide them. The three models shown in Figure 13-7 are based directly on the ridge models illustrated in Figures 13-5*b* and 13-5*c*, with the magnetized blocks of crust extending from 3 to 11 km below sea level. Each model in Figure 13-7 represents a 200 km section symmetrically disposed about a ridge crest at point *O*. According to the sea-floor spreading hypothesis the age of the crust at any distance from the ridge crest is a function of the spreading rate, and for the assumed spreading rates of 1 and 2 cm per year per limb in Figures 13-7*a* and 13-7*b*, the age of the crust at any distance from the ridge crest *O* is readily calculated. The polarity reversal time scale shown as 1964*b* in Figure 12-13 has major reversals at 1, 2.5, and 3.4 m.y., and short events at about 1.9 and possibly 3 m.y. These ages are shown below the crustal sections at appropriate distances from the ridge center, and the corresponding alternately magnetized blocks of the crust are fitted

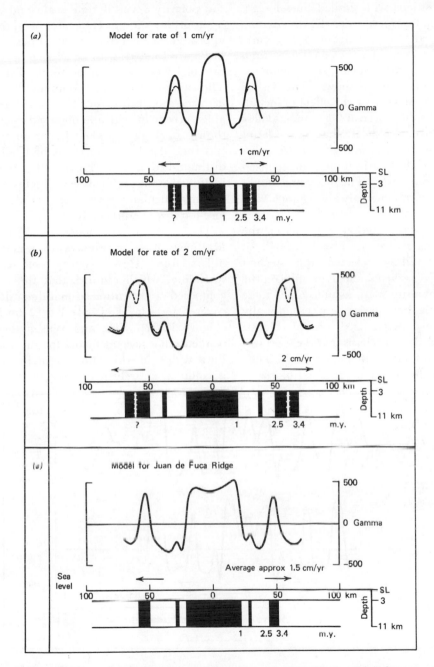

Figure 13-7. Models and calculated total field magnetic anomalies resulting from a combination of polarity reversals for the Earth's magnetic field and ocean-floor spreading (after Vine and Wilson, 1965). Normally and reversely magnetized blocks of crust are distinguished . (*a*) and (*b*) assume uniform rates of spreading. (*c*) Deduced from the gradients on the map of observed anomalies. The dashed parts of the computed profiles show the effect of including what was in 1965 considered to be a possible reversal at 3 m.y. (compare polarity sequence with Figure 12-13). (From Science, **150**, 485–489, 1965, with permission. Copyright 1965 by the American Association for the Advancement of Science.)

to these positions with dashed lines for the possible 3 m.y. reversal. For the distribution of magnetized blocks so deduced in Figures 13-7a and 13-7b and for the conditions stipulated in the figure legend, the computed magnetic profiles are as shown. The dashed parts of the profiles show the effect of including the possible reversal at 3 m.y. These computed profiles indicate that with a faster spreading rate, the profile gives more information about the distribution of magnetized blocks within the crust. The shape and amplitude of the central anomaly also change with spreading rate.

Figure 13-7c is a computed model for the Juan de Fuca Ridge (see also Figure 13-9). Vine and Wilson selected the steepest gradients of observed anomaly profiles on the magnetic intensity map, assumed that these delineated the boundaries between normally and reversely magnetized crustal blocks, and concluded on this basis that the crust was magnetized as shown in Figure 13-7c. They related the alternately magnetized blocks to

the polarity reversal time scale and added appropriate dates. The rate of spreading appeared to be rather erratic (dashed line in Figure 13-10a), with an average spreading rate of 1.5 cm per year per limb of the cell. The anomaly profile computed for this deduced model in Figure 13-7c shows reasonable agreement with the observed anomalies shown in Figure 13-8a. Unfortunately at this time the Jaramillo event had not been distinguished and in Figure 13-7 this was identified as the Olduvai event (Figure 12-14), and the calculated spreading rate was therefore too low (Figure 13-10).

The generalized model of Vine and Matthews used in Figures 13-5 and 13-7 differs from Hess's concept of sea-floor spreading, which requires that the crust is composed of serpentinized mantle peridotite, covered by a layer of only 1 or 2 km basalt (Figure 12-22). Vine and Wilson therefore worked out a second model for the Juan de Fuca Ridge, in which the magnetic material is confined entirely to layer 2, between 3.3

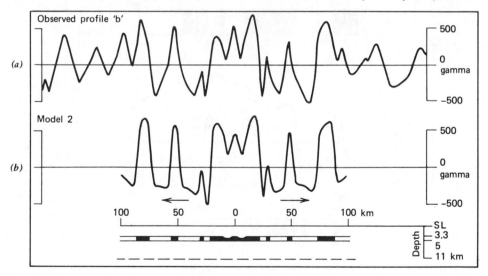

Figure 13-8. Interpretation of magnetic anomalies associated with the Juan de Fuca Ridge (after Vine and Wilson, 1965). (a) Observed profile 'b' across the Juan de Fuca Ridge; note its symmetry. (b) Model of magnetized crust and calculated anomaly assuming a strongly magnetized basalt layer only. Normal or reverse magnetization is with respect to an axial dipole vector; axial dipole dip taken as +65°. Effective susceptibility taken as ±0.01, except for the central block, +0.02. (From Science, **150**, 485–489, 1965, with permission. Copyright 1965 by the American Association for the Advancement of Science.)

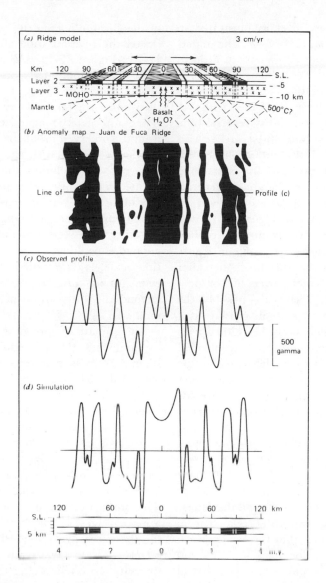

Figure 13-9. The tape-recorder, conveyor belt scheme for sea floor spreading summarized by Vine (1968). (*a*) A schematic representation of the crustal model, applied to the Juan de Fuca Ridge, southwest of Vancouver Island (Figures 13-2 and 13-8). Shaded material in layer 2, normally magnetized; unshaded, reversely magnetized. (*b*) Part of the summary map of magnetic anomalies recorded over the Juan de Fuca Ridge (Figure 13-2). Black, areas of positive anomalies; white, areas of negative anomalies. (*c*) A total-field magnetic anomaly profile along the line indicated in *b*. (*d*) A computed profile assuming the model and reversal time scale. Intensity and dip of the Earth's magnetic field taken as 54,000 gamma and +66°; magnetic bearing of profile 087°. (1 gamma = 10^{-5} oersted.) (S.L. = sea level.) Note: Throughout, observed and computed profiles have been drawn in the same proportion: 10 km horizontally is equivalent to 100 gamma vertically. Normal or reverse magnetization is with respect to an axial dipole vector, and the effective susceptibility assumed is ±0.01 except for the central block at a ridge crest (+0.02). (From F. J. Vine in "The History of the Earth's Crust", ed. R. A. Phinney, Figure 1, p. 75. Copyright 1968, Princeton University Press. Used with permission of the author.)

and 5 km below sea level as illustrated in Figures 13-8b, 13-9a, and 13-9d. The underlying serpentinite would contribute little to the observed anomalies. Figure 13-9 is a schematic representation of this crustal model, published later by Vine, with yet another revision of the polarity reversal time scale. Figures 13-9c and 13-9d are revised versions of Figures 13-8a and 13-8b. In this model the effective susceptibility is assumed to be considerably higher than in the generalized model. Figure 13-8b was derived in precisely the same way as Figure 13-7c, and comparison of the two computed profiles with the observed magnetic profile in Figure 13-8a suggests that the specific model after Hess is an improvement over the original generalized model. These are possibly the two extremes because a main crustal layer of serpentinite would surely be riddled with basic intrusions. Vine and Wilson concluded that the success of both models confirms the essential feature of the Vine-Matthews hypothesis that steep magnetic gradients are due to boundaries between normally and reversely magnetized crust.

At the Goddard conference in November 1966 (published 1968) and in a *Science* article of December 1966, Vine reviewed new evidence from several magnetic surveys which provided convincing confirmation that linear magnetic anomalies can be correlated from one profile to another, that these do parallel ridge crests, and that for many latitudes and orientations, the anomalies are symmetrical about the axis of the ridge (Figures 13-8a, 13-12b). This evidence he regarded as virtual proof of sea-floor spreading and its implications.

The successive revision of the polarity reversal time scale shown in Figure 12-13 has complicated the interpretation and dating of the anomaly profiles, as shown by Figures 13-8, 13-9, and 13-10 for the Juan de Fuca Ridge. These figures used the polarity time scales labelled in Figure 12-13 respectively as 1964b, 1966a, and 1966b. Since the publication of Figures 13-7 and 13-8, recognition

of the Jaramillo event in igneous rocks produced the reversal scale of 1966a in Figure 12-13 (Figure 12-14). Vine noted that he and Wilson could have used the magnetic profile shown in Figure 13-8 to predict the then-unknown Jaramillo event; the distribution of alternately magnetized blocks inferred from the observed profile in Figure 13-8 does not agree closely with the reversal scale of Figure 12-14 (excluding the short Mammoth event). This raises the possibility that if spreading rates remain reasonably constant, which might be anticipated for inertial reasons, then the spacing of the linear anomalies could be used to check the reversal time scale, to date more accurately the short events, and to extrapolate the time scale beyond the 4 m.y. limit of the K-Ar dating method.

An estimate of spreading rate is obtained by following the procedure described for Figure 13-7c. Normal-reverse crustal boundaries inferred from the magnetic profiles are correlated with the polarity epochs and events. In Figure 13-10a the distances of the inferred boundaries from the ridge crests are plotted against the 1966a time scale for the reversal sequence (Figures 12-13 and 12-14). Each point shows the distance from a ridge crest of a specific normal-reverse boundary. The horizontal time scale thus gives the apparent ages, or the duration of spreading, for the successive boundaries in the crustal sections. There are small departures from linearity for each of the three ridges illustrated, and the best straight line through each set of points gives the average spreading rate listed. The dashed line shows the earlier interpretation of Vine and Wilson for the Juan de Fuca Ridge (Figures 13-7 and 13-8), which was based on identification of the then-unknown Jaramillo event as the Olduvai event; this implied a more erratic spreading and slower average rate.

The deviations from linearity for the East Pacific Rise and the Juan de Fuca Ridge are exactly analogous, except that the magnetic profile for the South Pacific provides evidence that the Mammoth event near 3 m.y. is multiple (and it is so plotted in Figure 13-11). Figure 13-10*b* gives straight lines for the average spreading rates of these two ridges, and the distances of the inferred normal-reverse boundaries have been re-plotted on these lines; this causes slight changes in the apparent ages of the boundaries. Equivalent normal-reverse boundaries on each ridge remain very similar in apparent age, as shown by the dashed lines, and Vine suggested that the average ages of these equivalent pairs might be considered, tentatively, as revised values for the polarity reversal time scale. The slightly revised scale is shown in Figure 13-10*b*, and compared with the K-Ar scale at the top of the diagram which has been transferred from Figure 13-10*a*.

The polarity reversal time scales in Figure 13-10 are correlated with anomaly profiles extending out to 150 km from the crest of the East Pacific Rise, but anomalies have been recorded for much greater distances. If we assume that spreading from the East Pacific Rise proceeded at a constant rate of 4.4 cm per year for a period of 11.5 m.y., we can plot the positions of normal-reverse boundaries inferred from the anomalies measured at distances out to 500 km from the Rise crest as shown in Figure 13-11. These points, located on the spreading rate line in terms of distance, provide the sequence of normal and reversed polarities shown on the horizontal time scale. The time scale out to 5.5 m.y. is based on the plot for the East Pacific Rise together with the similar plot for the Juan de Fuca Rise using the average apparent ages for equivalent boundaries as in Figure 13-10*b*; but beyond 5.5 m.y. the time scale is based only on the East Pacific boundaries. The extrapolation depends upon the assumption of constant spreading rate, and this is therefore a reversal time scale relative to the East Pacific Rise. This pattern of longer and shorter events can be correlated with scales obtained from other ridge systems, and with scales obtained by other techniques,

such as the uncalibrated sequence of polarity zones in Figure 12-16.

Comparison of the anomaly profiles for the East Pacific Rise and the Juan de Fuca Ridge indicates that changes in spreading rate do occur. Figure 13-10*a* shows that the rate of spreading from the Juan de Fuca Ridge remained fairly constant at 2.9 cm per year for a period of 4 m.y. However, the inferred boundaries for distances greater than about 150 km from the Juan de Fuca Ridge crest depart from linearity when plotted against the time scale relative to the East Pacific Rise as shown in Figure 13-11. One interpretation of Figure 13-11 is that spreading from the Juan de Fuca Ridge decelerated about 5.5 m.y. ago from a rate of about 4 cm per year to its present rate of 2.9 cm per year. Alternatively, if we assumed that spreading from the Juan de Fuca Ridge had remained constant, we would obtain a different reversal time scale, and the points for the East Pacific Rise when plotted against this scale would have departed from the straight line. This would imply an acceleration of spreading from the East Pacific Rise about 5.5 m.y. ago.

From his review of the available data on magnetic anomalies in the Pacific Ocean Vine concluded that the north-south anomalies of the northeast Pacific (Figure 13-2) are related to a former crest of the East Pacific Rise, presumably overridden by the North American continent and possibly now located beneath the area of the Colorado Plateau uplift. He therefore constructed a profile just north of the Mendocino fracture zone, extending from the Gorda Ridge at $127\frac{1}{2}°$W out to the boundary of the north-south anomalies at 168°W (Figures 13-3 and 13-16 for the geography). If a constant spreading rate of 4.5 cm per year is assumed for this 3500 km section of the Pacific crust, the profile can be calibrated beyond 11 m.y. by using the same procedure described for Figure 13-11. This extends the time scale for reversals out to more than 80 m.y. which indicates that spreading was initiated from the

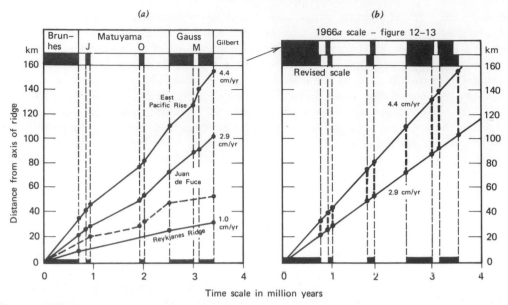

Figure 13-10. Magnetic anomalies and polarity reversals (after Vine, 1966). (*a*) Inferred normal-reverse boundaries within the crust plotted against the reversal time scale (Figure 12-13, 1966*a* scale). The dashed line represents a similar plot for the Juan de Fuca Ridge if one assumes an earlier time scale, as did Vine and Wilson in Figures 13-7 and 13-8. Note the similar deviations from linearity for the East Pacific Rise and Juan de Fuca Ridge. The average spreading rates are given for each ridge, and two of these are illustrated in Figure *b*. (*b*) Distances of polarity reversal boundaries from ridges plotted on the straight lines giving average spreading rates from Figure *a*. This suggests that the reversal time scale requires slight revision (assuming that there was a constant spreading rate for the past 4 m.y.). The revised sequence of reversals on the time scale is based on the dashed lines through the plotted points, and compared with the previous reversal time sequence from *a* at the top of the figure. (From *Science*, **154**, 1405–1415, 1966, with permission. Copyright 1966 by the American Association for the Advancement of Science.)

East Pacific Rise in the late Cretaceous. Vine emphasized the speculative nature of this time scale and stressed the necessity for using it to predict and correlate anomalies in other oceanic areas. This procedure has since been used extensively (Figure 13-13).

1965-1966: Scepticism and Conversion of J. R. Heirtzler and Lamont Associates

The marine magnetics program at the Lamont Geological Observatory was initiated by M. Ewing. The early magnetic measurements in the Atlantic Ocean (Figure 13-1) were subsequently extended to the Pacific, Antarctic, and Indian Oceans. At a symposium on "The World Rift System" in

1965, M. Talwani, X. Le Pichon, and J. R. Heirtzler reviewed the patterns of magnetic anomalies over the Midocean Ridge Rise system, dividing them into "axial anomalies" and "flank anomalies." Figure 13-12*b* shows axial anomalies lying within the lines *BB*, and the rather abrupt change to more irregular, longer-wavelength anomalies on the flanks; in some locations, the amplitudes become higher too. Their review ended with the following comments:

"We are not sure of the ultimate origin of the ridge magnetic anomalies. However we do believe that the symmetry of the magnetic pattern and its parallelism to the strike of the ridge requires that the magnetic anomalies owe their existence to the formation of the

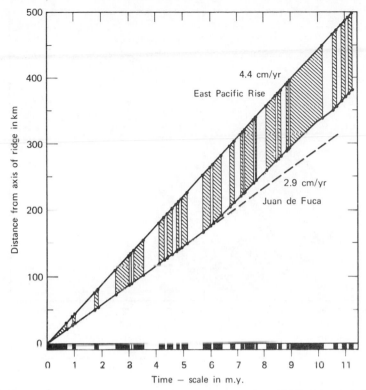

Figure 13-11. Extrapolation of the reversal time scale (after Vine, 1966). Magnetic boundaries across the East Pacific Rise, deduced out to 500 km from the crest on the basis of magnetic anomalies, and plotted on a line representing a constant spreading rate of 4.4 cm/year (extrapolation of line in Figure 13-10*b*). Similar boundaries for the Juan de Fuca Ridge are plotted out to 150 km assuming a constant spreading rate of 2.9 km/year. The reversal time scale out to 5.5 m.y. is based on both plots. Beyond that time, the time scale is based on the East Pacific Rise boundaries, and the distances of the boundaries for the Juan de Fuca Ridge then depart from the line of constant spreading rate; beyond 5.5 m.y., the Juan de Fuca points trace a line indicating faster spreading rates (assuming constant East Pacific rates). (From Science, **154**, 1045–1415, 1966, with permission. Copyright 1966 by the American Association for the Advancement of Science.)

ridge, which in turn is caused by a change of density in the upper mantle . . . However we feel that the variation in amplitude of the axial anomalies as well as the completely different character of the flank anomalies argues against the Vine and Matthews hypothesis." (Irvine, 1966, p. 346.)

The Reykjanes Ridge southwest of Iceland was selected in 1961 for magnetic survey because the crest was known to have a large anomaly, and there were indications that it had a linear character. Comparison with the long, regular anomalies of the northeast

Pacific appeared worthwhile. A proposal was submitted in 1962 and the survey was carried out during October and November 1963 as a cooperative project by Lamont Geological Observatory and the U.S. Naval Oceanographic Office. As shown in Figure 13-12*a*, an area about 350 km square was surveyed, located about 350 km southwest of Keflavik in Iceland. Aeromagnetic survey aircraft operating from Keflavik flew 58 lines approximately perpendicular to the ridge axis, at 0.46 km altitude, with track spacing of 5–10 km. This survey had more accurate

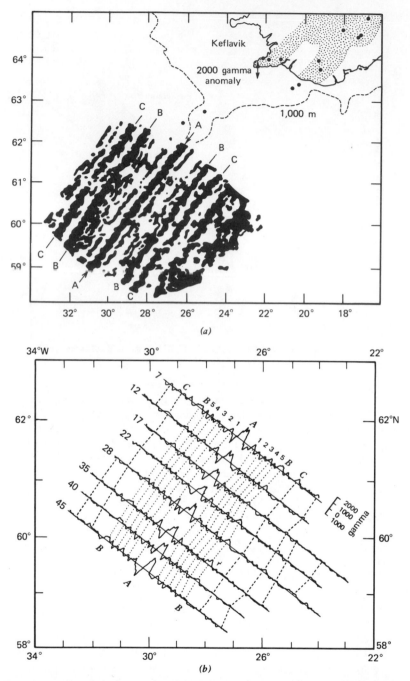

Figure 13-12. Magnetic anomalies over Reykjanes Ridge south of Iceland (after Heirtzler et al., 1966). (*a*) Black striped area indicates positive magnetic anomalies; the central anomaly over the ridge is marked by the arrows. (*b*) Eight of the 58 profiles of total magnetic field across the Ridge, projected on lines at right angles to the Ridge. Note the continuity of the central anomalies along the ridge axis, and the symmetry about this axis of the profiles. (From *Deep-sea Research*, by permission of Pergamon Press.)

navigation than other extensive marine surveys completed by that time. The anomalies were determined from the total geomagnetic field intensity measurements by subtracting values for a regional field which was a least squares fit to a simple linear function of latitudes and longitudes. The result is shown in the anomaly map of Figure 13-12*a*, which was published in 1966. This was the first study to reveal in detail the magnetic pattern over a large area of the midoceanic ridge in a place where its identification as a ridge was beyond doubt.

Clear details of the anomalies are given in Figure 13-12*b* by the profiles projected on lines perpendicular to the ridge. The axial anomaly *A* over the ridge of the crest is an outstanding feature, with a width of 40 km and an amplitude of 3000 γ, peak-to-peak. The axial anomalies consist of six pairs of symmetrically disposed anomalies each about 15 km wide with 100 to 300 γ amplitudes. The larger wave-length anomalies of the flank provinces can be correlated from one profile to another through long distances (dotted lines in Figure 13-12*b*). In the axial area between the lines *BB*, the regional field is depressed by several hundred gammas (this does not show on the anomaly map and profiles). The authors suggested that this might be caused by elevation of the Curie isotherm (500°C) because of high heat flow in this axial region, with the result that the amount of magnetized material in the crust decreases progressively towards the axis. They concluded that except for the body causing the axial anomaly, the source of the magnetic anomalies in the axial zone lay entirely within the 2–4 km thick upper layer.

In their discussion of the origin of the anomalies Heirtzler *et al.* (p. 440) concluded that the Vine-Matthews hypothesis "in its present form does not explain the characteristic change in magnetic pattern from the axial zone to the flanks and the difference between the axial anomaly and the adjacent ones."

The change from axial anomalies to flank anomalies was previously listed as the third

difficulty facing the Vine-Matthews hypothesis, and by 1966 it appeared to be the main argument of the Lamont group against the hypothesis. In 1966 Vine suggested that the change might reflect an increase in frequency of reversals of the Earth's field together possibly with a decrease in its intensity. Because this type of change would be worldwide, the boundary should occur at different distances from ridge axes, according to the spreading rate in a region. Vine reported from a preliminary examination of many ridge profiles that such a change may have occurred approximately 25 m.y. ago. Before he published this tentative explanation, however, the Lamont group had been converted to the Vine-Matthews hypothesis, following Hess's model for sea-floor spreading (Figure 13-9 and not Figures 13-5 and 13-7).

Acceptance of the hypothesis must have come between the end of 1965 and the fall of 1966. The paper on the Reykjanes Ridge was submitted in November, 1965, and published in June, 1966. W. C. Pitman and J. R. Heirtzler submitted a paper to *Science* on September 12, 1966, which was published on December 2 just two weeks before Vine's paper in the same journal. The Lamont contribution described the magnetic anomalies over the Pacific-Antarctic Ridge, which were as linear and symmetrically disposed about the ridge as those illustrated in Figure 13-12 for the Reykjanes Ridge.

Impressed by this symmetry the authors applied the Vine-Matthews hypothesis to the ridge, using normally and reversely magnetized blocks with upper surface set by the average bathymetry and a thickness of 2 km. The crust sloped upward toward the ridge crest in this model, reflecting the sea-floor surface, rather than remaining horizontal as in Figure 13-9. The anomalies were correlated with the K-Ar reversal time scale back to the Gilbert epoch as in Figure 13-10*a*, which gave an average spreading rate of 4.5 cm per year. Normally and reversely magnetized blocks were then placed beneath the remaining observed anomalies to the edge of the profile. Assuming a constant spreading rate, Pitman and Heirtzler extended the polarity reversal scale back to 10 m,y., just as Vine did in Figure 13-11. This reversal time scale relative to the Pacific-Antarctic Ridge was then applied to the Reykjanes Ridge, using Vine's spreading rate of 1 cm per year (Figure 13-10*a*). The scale was contracted by a factor of 4.5 to allow for the difference in rates, and a magnetic profile for the Reykjanes Ridge was computed from the crustal model so obtained. The similarity of computed and observed anomaly profiles was considered good enough to indicate constant spreading rates from these two ridges at least for the past 10 m.y. This led the Lamont investigators to conclude that:

"We feel that these results strongly support the essential features of the Vine and Matthews hypothesis and of ocean-floor spreading as postulated by Dietz and Hess . . . permits one (using a constant spreading rate) to date reversals of the geomagnetic field back to 10 million years ago."

At the Goddard symposium on "The History of the Earth's Crust" held at essentially the same time, November 1966, Heirtzler stated that: "several studies of magnetic anomalies nearing completion at Lamont not only support axial spreading but indicate that this process extends completely from the ridge axis to the continental shelf in many areas." (Phinney, 1968, p. 90.) He noted that the following criteria must be satisfied before a series of magnetic anomalies could be associated with ocean-floor spreading: (a) the anomalies must be linear in a direction parallel or nearly parallel to ridge axis; (b) the anomalies must be in accord with the history of field reversals found in other oceans, modified by the presence of fracture zones. Since 1966 the efforts of many scientists at the Lamont Geological Observatory have been directed toward extrapolation and testing of the hypothesis in diverse ways.

1968: Extrapolation by J. R. Heirtzler, G. O. Dickson, E. M. Herron, W. C. Pitman, III, and X. Le Pichon

A series of four contiguous papers in the March 1968 issue of the *Journal of Geophysical Research* showed that a magnetic anomaly pattern, parallel to and bilaterally symmetrical about the midoceanic ridge system, exists over extensive regions of the North Pacific, South Pacific, South Atlantic, and Indian Oceans. The selected results compared in Figure 13-13 demonstrate further that the pattern is the same in each of these oceanic areas, and that the pattern may be simulated in each region by the same sequence of crustal blocks. The blocks comprise a series of normally and reversely magnetized material 2 km thick. The observed anomaly pattern and the general configuration of the crustal model conform to the prediction of Vine and Matthews (Figure 13-9 modification of Figure 13-5) and this is regarded by the Lamont investigators as strong support for the concept of sea-floor spreading. The authors did note, however, that if the Vine-Matthews theory is basically in error, then the conclusions of their analysis do not apply.

Figure 13-13 shows observed magnetic profiles for each of the oceans; the ship tracks are given in the original papers. Each profile has been projected perpendicular to the ridge axis. Some of the key anomalies have been numbered for reference purposes, and the dashed lines show the correlation of these anomalies from one profile to another. The crustal models that could explain the observed anomalies according to the Vine-Matthews theory are shown in the usual way by the strips of alternately black (normally magnetized) and white (reversely magnetized) units. Simulated profiles computed for the crustal models are shown between the observed profiles and the models. The ridge crests of each profile have been plotted directly above each other, and their positions indicated by 0 km. A distance scale is given

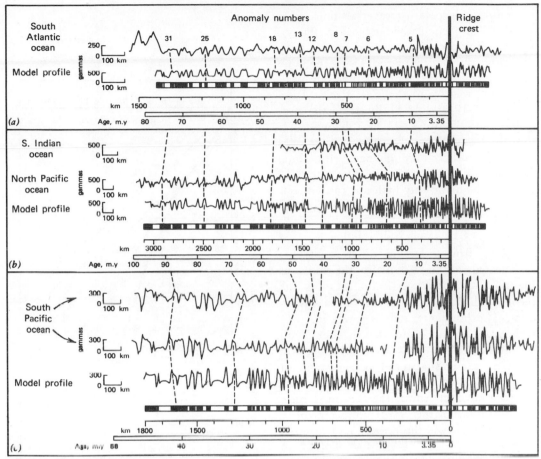

Figure 13-13. Sample magnetic profiles from various oceans (after Heirtzler et al., 1968; reviewing data in three companion papers by Dickson et al., 1968, Le Pichon and Heirtzler, 1968, and Pitman et al., 1968). Beneath each observed profile is a theoretical profile for comparison, calculated from the assumed sequence of crustal blocks normally magnetized (black) and reversely magnetized (white). Each block is taken to be 2 km thick. With each model is a time scale constructed arbitrarily by assuming an age of 3.35 m.y. for the end of the Gilbert reversed epoch (see Figures 12-14 and 12-21), and assuming a constant spreading rate for 50 to 100 million years beyond that time. Dashed lines relate similarly shaped anomalies identified by the numbers at the top of Figure (*a*). Note that specific anomalies in different profiles do not have the same "ages" on the three time scales. (From Jour. Geophys. Res., **73**, 2119–2136, 1968, with permission.)

below each crustal model showing the distances of anomalies from the ridge crests. Notice that there are three different distance scales in the figure. An arbitrary age scale is given with each model, based on the date of 3.35 m.y. for the beginning of the Gauss normal polarity epoch (Figures 12-14 and 12-21), and assuming a constant rate of spreading. These are not true age scales, as shown by the fact that specific anomalies in different profiles do not have the same "age" on the scales. The main objective of the fourth paper in the series was to derive a time scale for the sequence of geomagnetic field reversals predicted by the crustal models.

We have already discussed the procedure adopted by Vine, Pitman, and Heirtzler for extending the polarity reversal time scale relative to a particular ridge back to 10 m.y., by assuming a constant spreading rate from that ridge during this period. Using the same procedure Vine extrapolated the scale back to 80 m.y. relative to the East Pacific Rise. Figures 13-10 and 13-11 illustrate the procedure, but Figure 13-11 also demonstrates that the rates of spreading of one ridge relative to another may change with time. Assumptions of constant spreading rate must therefore be made only with due caution and after comparison with relative reversal time scales for other ridges.

Figure 13-14 shows the relative rates of spreading from ridges in different oceans; its construction is essentially the same as Figure 13-11. The vertical axis gives the recorded distances of anomalies from a ridge crest. The horizontal axis is a relative time scale based on the assumption that the South Atlantic ocean floor has been spreading at a constant rate of 1.9 cm per year for the past 80 m.y. This rate was calculated from the distance covered within the 3.35 m.y. since the beginning of the Gauss normal polarity epoch. The ages of the normal and reversed magnetic intervals listed for the South Atlantic crustal model in Figure 13-13 were obtained by dividing the distances to the model magnetized bodies by the spreading rate of 1.9 cm per year. Both distances and ages are shown on the axis of Figure 13-14, and these may be compared with the crustal model polarity sequence in Figure 13-13. The ages of the specific anomaly numbers at the bottom of the diagram are located with respect to this time scale by their distance from the crest of the South Atlantic Ridge.

The time scale extrapolated from the South Atlantic profile was selected as a standard because:

1. The anomaly pattern for the South Indian Ocean is not sufficiently long;
2. The North Pacific profile is distorted in the ridge axis;
3. The spreading in the South Pacific appears to have changed relative to both the

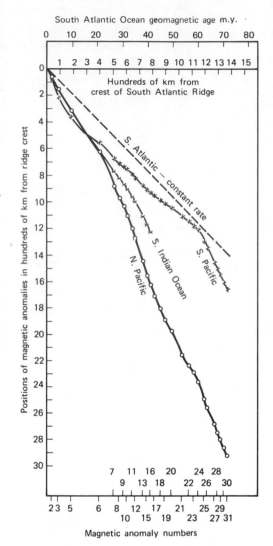

Figure 13-14. Relative rates of spreading from ridges in different oceans (after Heirtzler et al., 1968). The distance of a given anomaly in the South Atlantic from the ridge crest, plotted against the distance of the same anomaly in the South Indian, North Pacific, and South Pacific Oceans (data from Figure 13-13). Anomaly numbers are given, facilitating comparison with Figure 13-13. The South Atlantic anomaly distance scale is used to provide a relative time scale, assuming constant spreading rate (compare Figure 13-11). (From Jour. Geophys. Res., **73**, 2119–2136, 1968, with permission.)

North Pacific and the South Atlantic in the vicinity of anomalies 5 and 24, suggesting that these are real changes in rate;

4. There is evidence that the spreading rate from both the North and South Atlantic Ridge has been constant for the past 10 m.y.

The authors emphasized the possible error inherent in the extrapolation and this warning should be heeded. The numbers listed as "ages" in Figures 13-13 and 13-14 do not represent the passage of the stated number of years. All "ages" are relative to an assumed constant rate of spreading. In Figure 13-13 each age scale is derived independently of the others for each specific ridge. In Figure 13-14 all ages are relative to the assumed constant spreading rate for the South Atlantic. Each anomaly has the same age in Figure 13-14 but not according to the "age scales" in Figure 13-13. The age of 80 m.y. just beyond anomaly 32 compares with a 73-m.y. date given by Vine for the same point on a magnetic profile.

In Figure 13-14 the distances to model magnetized bodies in the other oceans have been plotted against the South Atlantic relative time scale. The relative spreading rate curves illustrated are continuous, but nonlinear. The straight dashed line shows where the spreading rate curve for the South Atlantic would plot. Time scales for the North Pacific and the South Pacific crustal models relative to the South Atlantic standard could be taken directly from the horizontal axis, and adjusted to fit the model blocks for the crustal sections in Figure 13-13. The nonlinear age scales so determined would differ from the scales given in Figure 13-13; each specific anomaly would then be located at the same age on the three scales.

Figure 13-15 shows a geomagnetic reversal time scale, relative to the South Atlantic spreading, derived from Figures 13-13 and 13-14. The anomaly numbers are plotted against their ages. It is not legitimate to compare this hypothetical polarity time scale with the fine detail of marine core paleo-magnetic data (Figures 12-17 and 12-20a), but comparison with polarity scales such as those in Figure 12-16 may prove to be useful. It now appears to be generally agreed that the linear magnetic anomalies of the ocean basins do provide a record of the geomagnetic polarity reversals. Even events of quite short duration have been inferred from profiles in regions of rapid spreading, but it is unlikely that all short events have been distinguished. The greatest uncertainty is in the assignment of ages. Despite the assumptions and extrapolations involved in the provisional age scale of Figure 13-15, scientists directly involved in this research consider it unlikely that the ages can be wrong by as much as a factor of two.

Figure 13-14 demonstrates considerable variations in spreading rate, but there are no systematic gaps in the anomaly patterns, which suggests that any cessation of spreading occurred simultaneously in all oceans. Some variations in spreading rate were apparently rather abrupt. For example the ratio between spreading rates in the North and South Pacific was constant (at about one) between anomalies 31 and 24, increasing between anomalies 24 and 18 (to more than three), and after anomaly 5 it decreased again to about 0.7. There have been several suggestions that sea-floor spreading has been episodic with worldwide discontinuities. This is consistent with paleomagnetic evidence that drifting of continents has been intermittent, and other evidence comes from the distribution of sediments.

In 1962 ships from the Lamont Observatory began a reconnaissance survey of sediment thickness in the Pacific Ocean using a continuous seismic profiler. Results published in 1968 are shown in Figure 13-16. The upper layer of sediments (acoustically transparent) ranges in thickness from 20 m to 1 km. Beneath the transparent layer over most of the western North Pacific, but not in the eastern Pacific, is an acoustically opaque layer with thickness usually between 20 and 300 m. Evidence from cores indicates that the

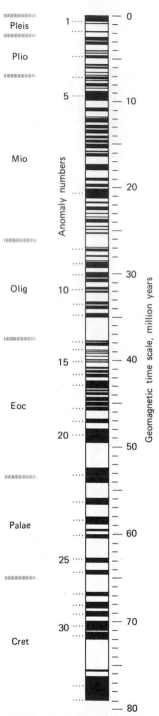

transparent layer consists of Cenozoic sediments, and that the pre-Cenozoic sediments represented by the opaque layer are limited to the western part of the Pacific basin. The limited sediment core data are consistent with the concept of a spreading ocean floor. The present pattern of biological productivity produces a high sedimentation rate in the equatorial belt, caused by upwelling along the equator. Figure 13-16 indicates that this pattern has continued throughout the Cenozoic with no indication of a change in the position of the equator during this time. The uniform thickness of the transparent layer along the axis of the equatorial belt is sharply reduced across the East Pacific Rise. It shows a marked decrease in thickness from 99° to 107°W, and less pronounced thinning between 150° and 168°W.

Similar patterns have been detected in many traverses across the midoceanic ridge system. J. Ewing and M. Ewing reported in 1967 that at the axes of ridges sediments are too thin to be resolved in the profiler records. At a distance of 100 to 400 km from the axis there is an abrupt increase in thickness of the sediments. Beyond the discontinuity the flank sediments are of almost constant thickness. Ages for the discontinuities, estimated by dividing the distance from the ridge crest by the spreading rate calculated for the region from the magnetic anomaly pattern, are about 10 m.y. in all regions. If uniform spreading rates are assumed it is very difficult to explain the discontinuity. It requires unlikely changes in the rates of sediment accumulation on a worldwide basis. Intermittent sea-floor spreading is the preferred interpretation.

These studies of sediment distribution led the authors to propose three main episodes of sea-floor spreading:

Figure 13-15 continued

Figure 13-15. The geomagnetic time scale (after Heirtzler et al., 1968), relative to the South Atlantic spreading (Figure 13-14). From left to right: Phanerozoic time scale for geologic eras, numbers assigned to bodies and magnetic anomalies, geomagnetic field polarity with normal polarity periods black (Figure 4-8). (From Jour. Geophys. Res., **73**, 2119–2136, 1968, with permission.)

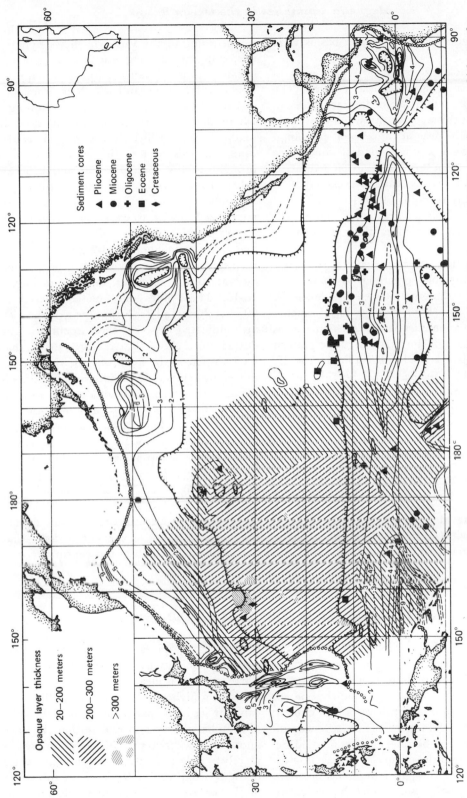

Figure 13-16. Sediment thickness and pre-Pleistocene core samples of the north Pacific (after Ewing et al., 1968). Contours show thickness of upper transparent layer in tenths of seconds of reflection time; each contour represents approximately 100 meters. Thicknesses of opaque and lower transparent layer are represented by hatching. Ages of cores generally correspond to a depth of about 10 meters in the sample. Open circles show axes of trenches. (J. Ewing *et al.*, Geophysical Monograph, **12**, 147, 1968, with permission of Amer. Geophys. Union.)

1. Mesozoic when the ocean basins were formed.

2. Early Cenozoic during which most of the midocean ridge area was created. The pre-Cenozoic sediments in the Pacific Ocean suggest that this cycle began when the region now at 145°W was at the axis of spreading. Then followed a long period of quiescence during which most of the Cenozoic sediments were deposited; this may have exceeded 15 m.y., possibly reaching 30 or 40 m.y.

3. Latest Cenozoic beginning 10 m.y. ago during which the crestal regions were generated.

Identification of the two latest major readjustments of spreading is based on changes in sediment thickness at crest-flank and ridge-basin boundaries, which coincide closely with anomalies 5 and 32. Anomaly 5 is the outer limit of the axial magnetic pattern which is shown as the line *BB* in Figure 13-12*b*. Anomaly 32 coincides with the ridge-basin boundaries in the Atlantic and the opaque layer boundary in the Pacific (Figure 13-16), which is dated as Mesozoic. Therefore Le Pichon (1968) suggested that the provisional scale in Figure 13-15 should be modified by dating anomaly 32 at 60 m.y. (early Paleocene) instead of 77 m.y., and by placing an interruption of spreading about 10 m.y. long at anomaly 5, preceded by a general slowing down of the movement. This adjusted time scale is consistent with the available data.

According to the Vine-Matthews hypothesis, mapping the magnetic anomalies of the ocean basins is tantamount to mapping isochrons for the ocean basin floor. The locus of a specific anomaly is a line of constant age, where the age corresponds to the time of magnetization, which is interpreted as the time that this part of the ocean floor was brought to the surface at an active ocean ridge. The areas in which the sequence of anomalies shown in Figure 13-15 had been recognized by 1967 are summarized in Figure 13-17. Instead of showing magnetic anomalies as in Figures 13-2 and 13-12, the map shows lines corresponding to provisional ages at 10 m.y. intervals, taken from Figure 13-15. These lines may be considered as 10 m.y. "growth lines" from the ridges which are indicated by the zero isochron. The dotted lines show the fracture zones that have displaced the isochrons. In view of the preceding remarks about intermittent spreading and the uncertainty of the age scale in Figure 13-15 the dates assigned to the isochrons must be regarded as provisional. However, the ability to contour the ocean basins with isochrons, even if they are of uncertain ages, provides us with the prospect of unravelling the history of the ocean basins, and the resultant movement of the continents, with precision of detail inconceivable during the debate about continental drift that occupied the first half of this century.

PROBLEMS AND INTERPRETATIONS SINCE 1968

At the end of his comprehensive review on "Reversals of the Earth's Magnetic Field" in 1968 Sir Edward Bullard wrote:

"The lecture on which this paper is based was given in June 1967; it was a well chosen time, the threefold story of the reversals of the field had just become clear and could be easily and elegantly set out. In the few months needed to write the paper there has been an avalanche of new results which has revealed many discrepancies and many matters needing elucidation; the usual chaos of the Earth sciences is clearly about to be reestablished at a higher level of understanding. We can now see many particular facts in the light of a global theory and realize that they are anomalous, whereas previously they appeared merely as isolated facts. The worldwide nature of the reversals can tie together the spreading of the ocean floor and the volcanic and

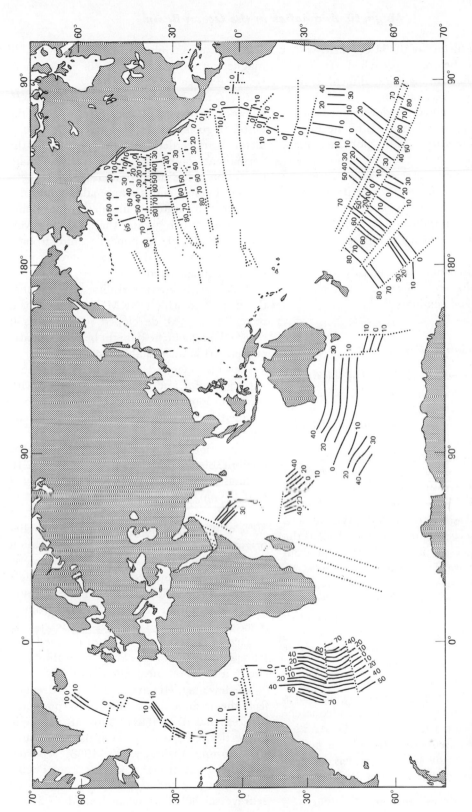

Figure 13-17. Isochron map of the ocean floor according to the magnetic anomaly pattern using time scale in Figure 13-15. Numbers on isochron lines represent age in millions of years. Dotted lines represent fracture zones (after Heirtzler et al., 1968, from Jour. Geophys. Res., **73**, 2119–2136, with permission).

sedimentary history of the last 50 m.y. with a detail never before approached. It is not surprising that ambiguities, difficulties and contradictions are emerging."

We have examined the development of the threefold story of the reversals of the Earth's field and its implications for global tectonics, as they appeared in 1968. We should now review the discrepancies referred to by Bullard as well as more recent evidence. Unfortunately, if this book is to cover the range of material indicated by its title and yet remain within the length limits of economic feasibility, there is simply no space for an adequate discussion. I can only introduce selected topics and suggest a few references for reading and seminars.

Magnetization of Basalt

Bullard reviewed evidence that the titano-magnetite minerals in reversely magnetized lavas are more highly oxidized than in normally magnetized lavas and stated that the undoubted statistical correlation of oxidation and magnetic polarity must be considered one of the major unsolved problems of Earth science. In a detailed study of over 550 specimens from 14 lavas and seven dikes in Iceland N. D. Watkins and S. E. Haggerty in 1968 reported that the oxidation spectra in each body, and from baked sub-basaltic sediments, exclude the existence of self-reversals of magnetic polarity. The results show a strong correlation between the percentage of reversed magnetic polarity specimens and higher oxidation in the lavas but no correlation in the dikes. There is at present no satisfactory explanation.

The state of magnetization of submarine basalts, and the causes of magnetic anomalies in oceanic crustal material, are not adequately understood. In a 1968 review N. D. Watkins listed the assumed properties of the mobile basaltic blocks involved in models for sea-floor spreading such as Figure 4-40. These are:

1. A high intensity of magnetization, J.

2. An effectively uniform intensity of magnetization, in terms of instrument response at sea level.

3. A Q factor greater than unity. $Q = J/XH$, where X is the susceptibility and H is the ambient geomagnetic field intensity. The Q factor is thus defined as the ratio of the permanent (or remanent) intensity of magnetization to the induced intensity of magnetization. Typical oceanic anomalies could not be caused by induced magnetization alone. If Q is not much greater than 1.0, however, then the induced component would strongly diminish any negative anomaly due to a reversely magnetized body.

4. A high magnetic stability, S. S describes the degree to which NRM is retained in a rock. The NRM developed in an original reversed magnetic field might be subsequently obliterated in an unstable rock.

Watkins reviewed results indicating that variations of J and S in lavas correlate with the oxidation state, as revealed by the titano-magnetite minerals. The Q factor in many basaltic bodies shows considerable variation with position in the body, which is probably attributable to initial cooling conditions. In Icelandic basaltic rocks Watkins reported that Q values of less than 1.0 were found in over 20% of lava specimens, and more than 30% of dike specimens, showing that bodies originally possessing reverse magnetization may not always produce a negative magnetic anomaly. Watkins concluded that single bodies of basaltic material may develop sufficient variation in $J, X, Q,$ and S to produce a series of anomalies parallel to the cooling surface of the body. A series of inclined and truncated bodies of this kind could thus give rise to a series of linear anomalies.

The few magnetic studies of basalts dredged from the ocean floor that have been completed suggest that they have unusual magnetic properties, compared to continental basalts. For example Ozima *et al.* (1968) reported that three out of eight samples from Pacific seamounts and the deep ocean crust showed selfreversal of TRM when heated to 300°C in air. When heated, all eight samples showed a much greater increase in saturation magnetization, I_s, than do continental basalts. The authors suggested that a "certain physical-chemical condition prevailing in the deep ocean bottom is responsible for their peculiar magnetic properties." They also pointed out that hot emanations or lavas rising along fractures would produce an increase in I_s, and might conceivably cause linear magnetic anomalies. A process of this type, however, is not likely to produce magnetic anomalies with widths of the order of tens of kilometers.

Direct measurement of magnetic polarity in basalts dredged from the Reykjanes Ridge at approximately 60°N (Figure 13-12) was reported by J. de Boer, J. Schilling, and D. C. Krause in 1969. The original orientations of the basalts on the ocean floor, the positions in which they were magnetized, were deduced from standard volcanic criteria such as the shape of pillows, the location of pillow necks and stems, and vesicle distribution. Cores were drilled from the oriented specimens and the average directions of magnetization for each specimen gave the directions of remanent magnetization. The measurements indicated that basalts from the axial zone (zone *A* in Figure 13-12) were magnetized in a field with normal polarity along a direction roughly parallel to that of the present field. A specimen dredged from an adjacent area with negative anomaly (between *A* and 1 in Figure 13-12) was magnetized with inclination $-20° \pm 5°$, in opposition to the $+74°$ inclination of the present field at the recovery site. This suggests that the basalt was magnetized in a reversed magnetic field supporting the Vine-Matthews theory.

Selected Interpretations

In order to examine anomalies the trend of the regional geomagnetic field must be removed from the observed magnetic field. Bullard (1967) deplored the proliferation of methods employed to remove the trend, and the problems thus introduced when results from different surveys were fitted together. The problems should be eased in the future if an October, 1968, recommendation is followed. The International Association of Geomagnetism and Aeronomy—International Union of Geodesy and Geophysics recommended that the International Geomagnetic Reference Field be used as the common survey datum. This field is described by a spherical harmonic analysis of degree and order 8 and it includes coefficients of annual change.

Before 1968 many surveys had employed methods where the trend was estimated from the survey data itself. This procedure invariably produces a series of anomalies of alternating sign, even if all of the variations in total field intensity are positive (or negative) with respect to the smoothed geomagnetic field derived from global observations. According to Watkins and Richardson (1968) it has yet to be satisfactorily established that all the linear anomalies of the ocean basins are truly of variable sign. This is germane to interpretation of the anomalies in view of Watkins's conclusions that linear variations in the regional field can be produced by variations in the magnetization of a single, cooling basaltic body.

The estimated rate of spreading from a ridge has been correlated with a number of physical features. For faster spreading there appear to be more earthquake epicenters on the transform faults compared to the ridges. The East Pacific Rise, with a fast spreading rate, exhibits smoother topography with no central rift valley than the rugged and more slowly spreading mid-Atlantic ridge. According to Menard (1967) the topographic relief is proportional to the thickness of layer 2 and inversely proportional to the spreading rate. He calculated the discharge of lava from the crests of ridges necessary to form layer 2 by multiplying spreading rate times the thickness times a unit length. The estimated discharge on each flank is 25 to 45 km³/

10^6 yr/km, which gives 5 to 6 km³/year for the whole ridge system. A rate of 1.8 km³/year for the age of the Earth would be sufficient to produce the total volume of the continents.

"Some Remaining Problems in Sea-Floor Spreading" were discussed at the 1966 Goddard symposium by Menard (Phinney, 1968), and he drew attention to features of the ridges and transverse fracture zones that were more complicated than the simple spreading theory implies. He suggested that the East Pacific Rise was once straight, and that portions of the crest have migrated with spreading from each short segment now being caused by a separate convection cell. Menard showed in Figure 13-18 how one-sided convection, with the ridge crest migrating at the same rate that the sea floor appears to

spread, could produce a series of symmetrical anomalies indistinguishable from the model with uniform spreading from a fixed crest (Figure 13-9).

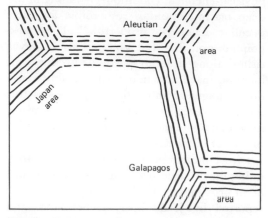

Figure 13-19. Schematic map of magnetic anomalies as they would appear if they were a paleomagnetic record produced by very large polygonal sections of the earth's crust receding from each other. Only parts of this scheme are confirmed by data (after Raff, 1968, Jour. Geophys. Res., **73**, 3699–3705, with permission).

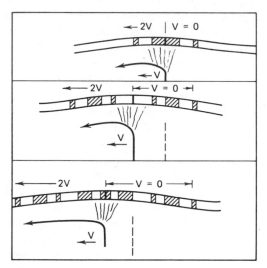

Figure 13-18. A means of producing symmetrical magnetic anomalies without moving one flank of a rise. The underlying convection cell migrates at the apparent spreading rate. One side spreads at twice the apparent rate and the other is fixed. This is one extreme possibility, the other extreme is that the convection is fixed and the two flanks spread at the same rate. (From H. W. Menard in "The History of the Earth's Crust", ed. R. A. Phinney, Figure 3 on p. 113. Copyright 1968 by Princeton University Press. Reprinted by permission of Princeton University Press.)

A. D. Raff in 1968 related the schematic map of anomalies in Figure 13-19 with the approximate configuration of three radial rifting lines existing in the Galapagos area, and the sharply bending anomalies in the Aleutian area. He suggested that this pattern could be produced by upwelling of mantle material at a center with radial spreading giving the effect of large polygonal plates receding from each other. The development of the "magnetic bight" near the Aleutians has been the subject of much discussion.

Detailed analyses of the magnetic anomaly pattern and fracture zones of the northeast Pacific (Figures 13-2 and 13-3) have led to conclusions that the process of sea-floor spreading was episodic (see also discussion of Figure 13-16). Peter *et al.* (1970) suggested two changes in the direction and rate of spreading; one between anomalies 22 and 24 and the second between anomalies 5 and 7 (see also Bassinger *et al.*, 1969). Menard and

Atwater (1968) assumed five events of reorganization. Vine and Hess (1970) reviewed evidence that the geometry of spreading changed within the past 5 m.y. in the Pacific and also in the North Atlantic, and they suggested that seismically inactive structures may well have been active at some time during the Tertiary. It can be anticipated that if there is a change in the spreading pattern, the change from one crustal geometry to another will be rapidly accomplished (Menard and Atwater, 1968).

It will be some time before agreement is reached about interpretation of magnetic anomaly patterns in terms of changing rates of movement, and of regional or global periods of cessation of spreading. In December 1969 D. A. Emilia and D. F. Heinrichs reviewed 15 published magnetic profiles from the Atlantic, Pacific, and Indian Oceans, using diagrams similar to Figures 13-10 and 13-11. The diagrams could be interpreted in two ways:

1. If the postulated relations between magnetic anomalies and the paleomagnetic time scale (as in Figures 13-13 and 13-15) are assumed correct, then the results indicate variable worldwide increases and decreases in the rates of spreading. This differs from episodic spreading, where all movement stops completely for a period.

2. If the spreading rates have remained nearly constant for the past 3.32 m.y., some of the accepted relations between the anomalies and the paleomagnetic time scale are incorrect.

They preferred the second interpretation, because it does not require complication of the basic spreading hypothesis. They concluded that the anomaly previously correlated with the Olduvai event is probably caused by the Gilsa event (Figure 12-21), and that the Olduvai event is represented by a minor anomaly previously unassigned to the Earth's polarity history. No particular magnetic anomaly had previously been assigned to the Gilsa event. J. D. Phillips and B. P. Luyendyk concluded in 1970 that the deep-sea drilling results

(Chapter 14) and oceanic magnetic anomalies together suggested that the rate of motion between the North American and African plates (Figure 14-3) had been relatively constant through the Cenozoic. D. P. McKenzie and W. J. Morgan showed in 1969 that a complex series of tectonic events can be produced by the evolution of triple junctions (Figure 13-9) without changes in the direction or magnitude of the relative motion between plates (Chapter 14).

During the next few years we can anticipate that interpretations of linear anomalies in terms of crustal structure, faulting, and relative movements will undergo successive modifications. For example in 1969 G. Peter and R. Lattimore reinterpreted the magnetic structural map of the Juan de Fuca and Gorda Ridges which we reviewed in Figures 13-2, 13-3, and 13-8 to 13-11. The 1970 review of magnetic data bearing on sea-floor spreading north of Iceland by P. R. Vogt, N. A. Ostenso, and G. L. Johnson illustrates well some of the problems and the choices that have to be made between alternative interpretations.

The theoretical model for interpretation of the linear anomalies is most elegant, but the complications introduced by real rocks and real geological structures inevitably lead to some speculative correlations and applications. Recent results suggest that the layer of oceanic rocks responsible for the magnetic anomalies may be much thinner than indicated in the models reported in this chapter. As Bullard noted, the acquisition of new data is likely to reestablish the usual chaos in the Earth sciences, but we do now have a universal model as a point of departure.

A Note of Caution from N. D. Watkins and A. Richardson

Watkins stated in 1968 that we know so little about the magnetic properties of submarine basalts and other oceanic crustal material that this permits a considerable degree of circular reasoning in interpretations of linear anomalies. Any model of blocks based on the geomagnetic polarity reversal scale can be fitted to the observed anomalies by suitable variations of block depths, widths (spreading rate variations), and intensity of magnetization; and Watkins added that magnetic details may optionally be interpreted as spurious, minor or multiple events, or possibly inherent magnetic features of finite bodies. Watkins and Richardson (1968) therefore concluded that to invoke variable spreading rates seems to be forcing data to fit what is still a hypothesis and should be discouraged until the geomagnetic polarity time scale beyond 4.5 m.y. is established.

The significance of highly symmetrical anomalies across some ridges and the repeat of a standard pattern fitting the radiometric polarity time scale in different oceans were not denied by Watkins and Richardson; but they made the point that if areas of active crustal spreading are to be delineated accurately, it is necessary to consider alternative causes of linear magnetic anomalies when the classic pattern is not clear. They emphasized that when geological information about the structure and rock types was available, this can not be ignored in the interpretation of the anomalies.

Most of our knowledge of the midoceanic ridges is based on individual traverses, and there have been few detailed studies of specific areas. Data collected on cruises from Woods Hole Oceanographic Institution in 1964 and from the Scripps Institution in 1965 permitted van Andel and Bowin to present a detailed geological and geophysical study of the crest and upper western flank of the mid-Atlantic ridge between 22° and 23°N. They prepared a schematic structural diagram for a 25 km cross section including the median valley and, on the basis of this and the known bathymetry, Watkins and Richardson presented the hypothetical geological cross section shown in Figure 13-20b. The picture is one of basalts overlying their low grade metamorphic equivalents, greenstones and greenschists. Following metamorphism,

uplift of up to 3 km occurred along faults dipping 30 to 45° toward the median valley; the overburden was displaced by sliding away from the crest. Eruption of basalts into the valleys followed the rifting. The vertical relief of up to 2 km completes the schematic section shown in the figure.

Figure 13-20d is the observed total magnetic profile at 22.5°N, and model a shows the conventional reversed polarity block model used to interpret the profile d. Figure 13-20a shows also the bathymetry of the sea floor. Notice the volume of rock above the magnetized strip which is ignored with respect to the development of anomalies.

Watkins and Richardson stated that "it is virtually impossible to see how the essential assumption that vertically-sided finite blocks of alternating polarity and high J and S values can be applied to the bathymetry depicted in" Figure 13-20b. Also the vertical movement required by the regional tectonic analysis argues against the idealized uniform crustal spreading blocks depicted in Figure 13-20a. They therefore sought an alternative interpretation based on the geological features.

Figure 13-20c is a version of model b showing the assumed distribution of basalts with normal (black) and reversed (white) polarity. Magnetic properties derived from local dredged rocks were used. The greenstone and greenschist have negligible magnetization. The lower depth of 5.7 km below sea level for model c is a standard value (model a). The computed profile for model c shown in Figure 13-20e demonstrates that the observed profile in d can be explained by a series of faulted geological features with the basic stratigraphy illustrated by van Andel and Bowin (model b). An exact fit could be obtained by suitable variation of the subjectively selected parameters, and the amplitude of computed profile e would be reduced to the same as observed profile d by halving the J and Q values of the basalt.

The magnetic anomalies *can* be correlated with the standard reversed polarity block

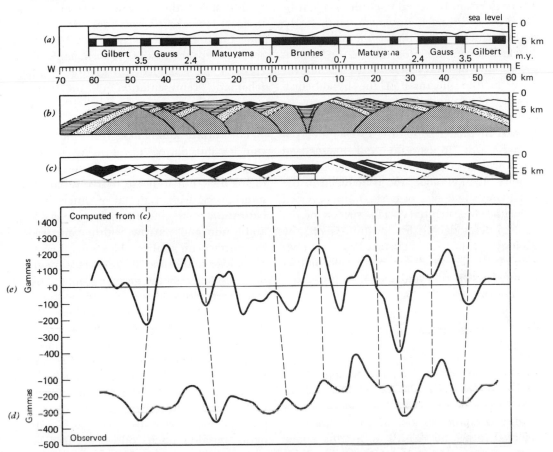

Figure 13-20. Interpretation by Watkins and Richardson (1968). Observed (*d*) and computed (*e*) west-east total field magnetic sea-level profiles across the mid-Atlantic ridge at latitude 22.5° North. Horizontal and vertical scales for models *a*, *b* and *c* are the same, given in km at the bottom and right hand side of the diagram. Model *a* shows the conventional crustal spreading finite block model used to interpret the magnetic profile when averaged with other profiles (Figure 13-13). The additional horizontal scale in model *a* is in millions of years. Note the difference between the bathymetry and the limits of the blocks. Model *b* employs the known bathymetry, and is a geological model which is hypothetical, although the central segment between 5 km west and 15 km east of the central part of the median valley (0 km) is essentially from Figure 7 of van Andel and Bowin (1968). Geological symbols used in model *b* are: fine dots = greenschist; coarser dots = greenstone; horizontal shading = basalt; black = young basalt. Model *c* shows the geometry and polarity of the blocks employed for the computation of curve *e*: black = normal polarity; white = reversed polarity (except for the greenschist and greenstone which is magnetically negligible and therefore without polarity). Curve *e* employs magnetic parameter averages obtained from local dredged material as follows: 'young basalt' has $J = 9.7 \times 10^{-3}$ emu/cm³, χ = (susceptibility) = 0.7×10^{-3} c.g.s. and $Q = 28.4$; 'basalt' has $J = 4.3 \times 10^{-3}$ emu/cm, $\chi = 2.4 \times 10^{-3}$ c.g.s., $Q = 3.6$; both the greenschist and greenstone have effectively J, χ, and $Q = 0$. Vertical scales for *d* and *e* in gammas. (Reproduced by permission of North-Holland Publishing Company, Amsterdam.)

model shown in Figure 13-20a, but this neglects the known relief and volume of rocks above the magnetized blocks. A better approximation to actual conditions might be obtained by using the average depth to the ocean floor as the top of the magnetized blocks following the procedure of the Lamont group, but the total relief of 2 km is about the same as the thickness of the hypothetical blocks. The magnetic anomalies cannot be correlated with the topography alone. The geological model in Figure 13-20c, incorporating both topography and magnetized structural layers, does provide a good correlation; but this model too is artificial in the sense that the choice of normal and reversed polarities was arbitrary. The success of the geological model in this location where the geology can be reconstructed, however, does emphasize the need for more information about the structure and magnetic properties of the oceanic crust.

Near-Bottom Magnetic Results from the "Fish"

Since 1967 ships from the Scripps Institution of Oceanography have taken a closer look at the properties of the ocean floor by towing a submersible instrument package referred to as the "Fish" at heights above bottom between 35 and 180 m. The Fish records its height above the sea floor and the magnetic field intensity, while the towing ship makes conventional bathymetric and magnetic measurements at the surface. Figure 13-21 shows results from two locations. Figure 13-21a is from a traverse across the East Pacific Rise just south of the Gulf of California. Figure 13-21b is a traverse across anomaly number 10 in the northeast Pacific, at latitude 32°25′N, just east of longitude 126°W (Figures 13-15 and 13-3).

Each diagram shows a profile of the bottom topography, the path of the Fish relative to the bottom, the total magnetic field intensity as recorded along the Fish path, and the total magnetic intensity as measured at the

surface. The magnetic intensity measured at the surface is correlated in the usual way with uniformly magnetized blocks of crust. Figure 13-21a shows the Brunhes normal polarity interval extending across the Rise crest and the Jaramillo event in the Matuyama reversed epoch on the western flank of the crest. I have added to Figure 13-21b a schematic representation of the younger normal event of anomaly 10 (Figure 13-15).

The surface anomalies give a picture that can be interpreted in terms of uniformly magnetized blocks of crust, but the measurements at depth indicate that the blocks are interspersed with material of different magnetization producing anomalies of very large amplitude and narrow width compared to the surface anomalies. The smoother trace and lower amplitude of the surface magnetic field results from the greater distance of the instrument from the magnetic bodies.

The deep anomalies in Figure 13-21a show the same general character right across the portion of the East Pacific Rise traversed. Superimposed on the profile are three large offsets of about 2000γ each, occurring at the transitions between normally and reversely magnetized crustal material as inferred from the surface magnetic profile. Each offset occurs within a horizontal interval of about 140 m. From the estimated spreading rate of 3 cm per year this indicates that it takes no more than 4700 years for the Earth's field to reverse, and it implies a maximum limit of about 280 m for the width of the intrusion center at the ridge crest. This is surprisingly narrow: other estimates range up to 10 km for the intrusion center.

In both locations the topography is lineated parallel to the surface magnetic anomalies. Larson and Spiess concluded that the topography of the 200 m abyssal hills forming the relief on the East Pacific Rise is not responsible for the deep anomalies. They suggested that the anomalies are compatible with the hypothesis that the magnetic field has under-

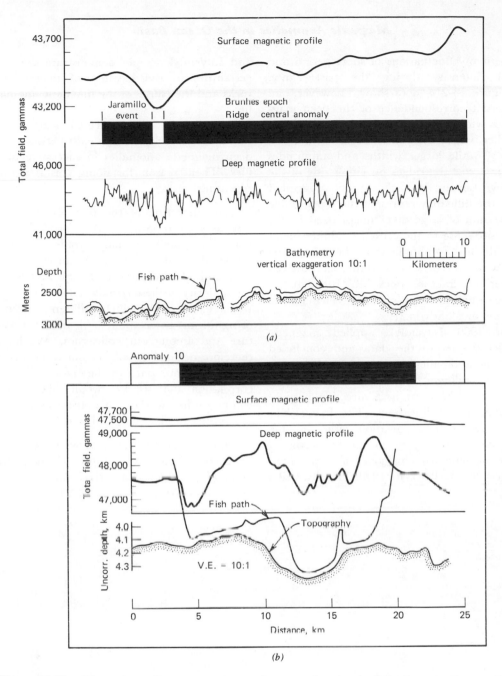

Figure 13-21. Magnetic profiles measured near the ocean floor by the "Fish". Each diagram shows bottom topography, total magnetic field intensity recorded along the Fish path, and measured at the surface. The polarity reversal sequences associated with the surface anomalies are shown. (a) Profile across the East Pacific Rise crest from west (at left) to east (after Larson and Spiess, 1969). Surface profiles are vertically exaggerated ten times with respect to the deep magnetic profiles. The trace above the bathymetry is the track of the instrument above the ocean floor. The sediment cover is illustrated as true thickness to the basement reflector. (b) Profiles across the younger normal event of anomaly 10 (Figure 13-15) in the north-east Pacific, latitude 32°25′ N, just east of longitude 126° W (Figures 13-15 and 13-13) (after Luyendyk et al., 1968). (Reproduced by permission. (a) Science, **163**, 68–71, 1969. Copyright 1969 by the American Association for the Advancement of Science. (b) Jour. Geophys. Res., **73**, 5951, 1968.)

gone many fluctuations of short period and small intensity during the past 2 m.y.

Figure 13-21*b* does show, however, an apparent correspondence of the deep magnetics with the topography. Low amplitude smaller width anomalies occur within the valley, while larger widths and amplitudes occur across the ridges on either side of the valley. Luyendyk *et al.* in 1968 suggested that the hills may represent the topographic expression of large dikes magnetized in the opposite sense to the crust containing them, and the low amplitude anomalies within the valley might represent hydrothermally altered material in fracture zones. Most other studies indicate no correspondence between magnetic profiles and bathymetry.

In 1969 Luyendyk applied magnetic model studies to the data and concluded that the anomalies are caused by slow, continuous changes in magnetization within layer 2 of the crust in a direction perpendicular to the lineations. The fluctuations are due to time variation either in the paleofield intensity or in the properties of injected magmas. This conclusion was questioned by G. Peter in 1970. His discussion

and Luyendyk's reply demonstrate the uncertainties in the location of magnetized rocks and the causes of the magnetic anomalies.

Luyendyk *et al.* confirmed statistically that a true correlation does exist between the deep magnetic anomalies in successive profiles. Their reason for doing this is worth repeating:

"The literature of the past few years is replete with reports on the correlation of marine magnetic anomaly profiles by their supposed similarity over distances up to thousands of kilometers. Although we believe the majority of these correlations to be valid, we have noticed the widely different concepts of significant correlation, both in the literature and among our colleagues. We have therefore decided to subject our assertion of lineation of our data to the rigors of cross-correlation analysis. We would like to encourage other workers in marine magnetics to conduct similar statistical tests."

Figure 13-21 confirms that we have much to learn about the causes of magnetic anomalies in the ocean basins.

14. *Plate Tectonics*

INTRODUCTION

Whatever the detailed explanations may be for the development of the linear magnetic anomalies their correlation with reversals of the Earth's magnetic field imprinted upon new oceanic crust generated at the mid-oceanic ridges gave considerable momentum to the spreading floor theories of Hess and Dietz (Chapter 12). Earth scientists were quick to realize the significance of this correlation, and in 1967 and 1968 four major papers appeared introducing what has become known as plate tectonics. The increasing support for many aspects of the sea-floor spreading hypothesis led Vine and Hess to reformulate the concept in 1968 in terms of the new evidence which had become available. Their review was published in 1970.

Figure 13-17 shows a pattern of recent motion of the ocean floors away from three active ridges in the Atlantic, Pacific, and Indian Oceans. A fourth active ridge traverses the Arctic Ocean. According to the concept of sea-floor spreading the uprise of mantle material and its spreading from the oceanic ridges causes the ocean floor and coupled continental blocks to move away from the ridges. The ocean crust goes down into the mantle at the oceanic trenches, which form sinks for the system. Figure 14-1 is a provisional attempt by Vine and Hess to show the relationships between continental and oceanic crust. The ridge crests, fracture zones, and isochrons have been transferred from

Figure 13-17 and a 65 m.y. growth line has been added in the Pacific. Although there are many gaps in the isochron map there is no major gap in geographical coverage. This permits delineation of the shaded areas in Figure 14-1 as oceanic crust created within the past 65 m.y. during Cenozoic time.

Figure 14-1 shows that although there are oceanic trenches available for large segments of the spreading Pacific Ocean floor, other plates can not be accommodated in trenches. The general picture indicates that the continents on either side of the Atlantic are coupled to the blocks spreading from the mid-Atlantic ridge, and that the North American continent has over-ridden the northern part of the East Pacific Rise. The location of sinks between the ridges in the Atlantic and Indian Oceans is not obvious. Spreading rates are slower in the Atlantic than in the Pacific, and possibly this could be related to the inference that spreading from the mid-Atlantic ridge is driving large continental plates westward over the faster-spreading Pacific Ocean floor.

The concept of plate tectonics has developed rapidly since 1967, and it has major implications for all aspects of geology. Already the topic merits a whole book in itself. In the first section I attempt to show the stages of formulation of the concept by seven scientists, but after that I can only consider briefly a few selected fragments of the total picture.

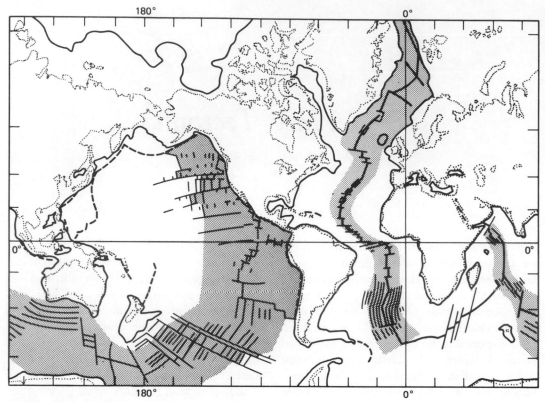

Figure 14-1. Relationships between continental and oceanic crust (after Vine and Hess, 1970). The shaded area is oceanic crust created within the past 65 m.y. (Cenozoic time); ridge crests, fracture zones, and isochrons transferred from Figure 13-17; dashed lines are trenches, see Figure 2-2. (With permission of John Wiley and Sons.)

THE CONCEPT OF TECTONICS ON A SPHERE

The concept of plate tectonics combines the satisfactory parts of the hypotheses of continental drift and sea-floor spreading. The Earth's surface is considered to be made up of a few rigid crustal plates or blocks which are in motion relative to each other. In the early 1960's there were many convection cell models published with ocean ridges representing sites of upwelling and ocean trenches the sites of convergence and descending limbs of cells (Chapter 12). Since 1967 more attention has been paid to deciphering the relative surface motions than to the energy source for the motion. The boundaries of the rigid plates are not necessarily related to convection cells in the mantle,

and the lack of a satisfactory driving mechanism remains a major problem.

Wilson's transform fault concept was extended to a spherical surface, as illustrated in Figure 14-2, by D. P. McKenzie and R. L. Parker in 1967 and by J. W. Morgan in 1968. Euler's theorem states that a layer on a sphere can be moved to any other conceivable orientation by a single rotation about a properly chosen axis through the center of the sphere. The relative motion of two layers can therefore be described by a rotation about some axis. This is the theorem used by Bullard and associates to fit together the continents on either side of the Atlantic (Figure 11-3).

Paving Stones of the North Pacific by D. P. McKenzie and R. L. Parker

In 1967 McKenzie and Parker outlined the key elements of plate tectonics, which they termed a paving stone theory, defining ridges and trenches respectively as lines along which crust is produced and destroyed. They tested the theory in the North Pacific with particular reference to focal mechanisms of earthquakes and to volcanism.

They determined the pole of relative rotation between the Pacific and the North American continent from the slip directions of two widely spaced faults, the San Andreas fault and the site of the 1964 Alaskan earthquake in the Kodiak Island region. Construction on a sphere gave a pole position of 50°N, 85°W. According to the paving stone theory, all slip vectors of the North Pacific earthquakes should lie on small circles around this pole. They found that about 80% of 80 published fault plane solutions for shallow earthquakes during and after 1957 had slip vectors in agreement with the prediction. Figure 14-4 shows the general consistency of the slip directions around the Pacific. The area that they examined amounts to about a quarter of the Earth's surface, and they expected it to be equally applicable to the other three-quarters. Subsequent global analyses confirmed their predictions.

McKenzie and Parker also noted that the distribution of trenches, active andesitic volcanoes, and intermediate- and deep-focus earthquakes is controlled by the trend of the faults which in turn influences the fault plane behavior for the shallow earthquakes. The shallow earthquakes associated with these features have overthrust fault solutions; they are not caused by faults of strike-slip transform nature.

Aseismic Crustal Blocks of the World by W. J. Morgan

When Morgan formalized the concept of plate tectonics in 1968, he divided the surface of the Earth into about twenty plates in each of which there is no distortion of any kind. The boundaries between plates are of three types which are shown in Figure 14-1. At the extensive ridge type new surface is generated and crustal surface is destroyed at the compressive type including trenches. The third type is represented by the great faults at which crustal surface is neither created nor destroyed. Figures 2-5 and 14-3 show that present earthquake activity is largely concentrated in narrow belts which are essentially continuous, and these belts form the boundaries to a series of plates which are essentially aseismic. Shallow-focus earthquake activity delineates the boundaries of the plates that are in relative motion across the surface of the Earth. Intermediate- and deep-focus earthquakes appear to be restricted to compressive boundaries such as trenches and young fold mountain chains as shown in Figure 14-3. Notice that although there is a circum-Pacific earthquake belt (Figure 2-5) the intermediate- and deep-focus earthquakes are not continuous around the Pacific.

Figure 14-2 shows Wilson's transform fault concept on the Earth's surface. Blocks 1 and 2 are crustal plates, the short double lines represent offset midoceanic ridges, and the series of curved lines crossing the ridges represent transform faults. If spreading occurs from the oceanic ridges it is convenient to assume that block 1 remains stationary, and to consider the movement of block 2 relative to block 1. According to Euler's theorem the relative motion can be described by rotation of block 2 about a specific axis with pole at A. This is defined by two parameters to locate the pole and a third to specify the angular velocity. For a given angular velocity the relative velocity between the blocks increases as the sine of the angular distance from the pole A. For relative movement to continue about pole A, the transform faults where blocks 1 and 2 are in contact must lie on small circles centered on A. This gives us two methods for calculating the positions of poles of relative motion, or poles of opening, from measurable features at the Earth's surface: (a) the first method involves location of the pole that fits the strikes of fracture zones, and (b) the second involves location of the pole best fitting the variation in spreading rate.

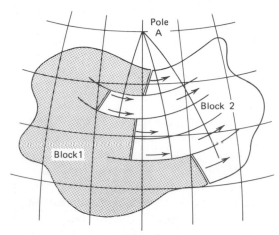

Figure 14-2. Transform faults on a spherical surface. On a sphere the motion of block 2 relative to block 1 must be a rotation about some pole. All faults on the boundary between 1 and 2 must be small circles concentric about the pole A. (From W. J. Morgan, Jour. Geophys. Res., **73**, 1959, 1968, with permission.)

Morgan examined the traces of the fault zones offsetting the mid-Atlantic ridge and the Pacific-Antarctic ridge, and the old and new fracture zones in the North Pacific Ocean. Each group of fracture zones satisfied the conditions required by Figure 14-2, and Morgan estimated the positions of poles of relative motion by drawing great circles perpendicular to the strikes of offsets of the ridges, or of fault segments, and determining the circles of intersection. The spreading velocities estimated in different parts of the mid-Atlantic ridge from the magnetic anomalies agreed roughly with the velocity pattern predicted for the pole located from the fracture zones.

Global Patterns of Surface Motion by X. Le Pichon

Le Pichon had access to more complete data than Morgan on the pattern of spreading (Chapter 12) and the locations and extents of some large fracture zones, and in 1968 he redetermined the three poles of rotation located by Morgan, with good agreement. He also located poles for the fracture zones in the Indian Ocean and the Arctic Ocean. The circles numbered 1 to 5 in Figure 14-3 are the positions of the five poles for the principal lines of opening in the Atlantic (1), North Pacific (2), South Pacific (3), Arctic (4), and Indian Oceans (5). Pole number 6 is a preliminary result for the S.W. Indian Ocean representing motion of Antarctica relative to Africa, which was deduced by assuming closure of Gondwanaland and summing angular velocities; Le Pichon's result differs significantly from that reported by Morgan. Angular velocities were derived for each pole from the available spreading rate data. The general picture is one of simplicity with the two major openings in the Atlantic and the Pacific occurring about approximately the same axis, not greatly inclined from the Earth's rotational axis (poles 1, 2, and 3), and with these openings being linked by two oblique openings, one in the Indian and one in the Arctic Ocean.

Le Pichon concluded that his results were detrimental to the hypothesis of an expanding Earth. Expansion of the Earth would be approximately the same along all radii, because the Earth has remained spherical. If the opening were due to expansion, however, the calculated rates of opening along great circles indicate differential expansion by as much as 500 km along different radii during the past 10 m.y.

Given a spherical Earth of constant radius and the derived poles and rotational vectors of the five major openings, Le Pichon attempted the first worldwide analysis for the movement of the major blocks relative to each other, predicting rates of crustal exten-

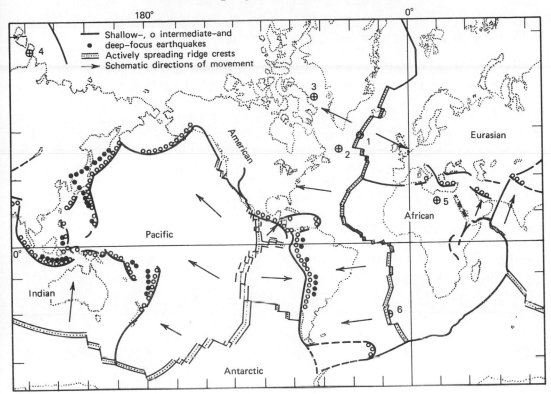

Figure 14-3. Summary of seismicity of the Earth (Figure 2-5) and aseismic plates (after Vine and Hess, 1970). Plates assumed by Le Pichon (1968) are named, and instantaneous centers of rotation deduced for plate pairs are plotted as numbered circles with internal crosses. Centers are: 1, for South Atlantic (America–Africa); 2, for North Pacific (America–Pacific); 3, for South Pacific (Antarctica–Pacific); 4, for Arctic Ocean (America–Eurasia); 5, for N.W. Indian Ocean (Africa–India); 6, for S.W. Indian Ocean (Antarctica–Africa). Spreading rates at ridge crests are indicated schematically and vary from 1 cm/year near Iceland to 6 cm/year in the equatorial Pacific. (With permission of John Wiley and Sons.)

sion, and rates of compression at the plate boundaries. Instead of the 20 crustal blocks considered by Morgan, Le Pichon chose six major blocks in such a way that the problem could be solved. Smaller plates were incorporated into the chosen six which are shown in Figure 14-3: these are the Pacific, American, African, Eurasian, Indian, and Antarctic blocks. Boundaries to the blocks are shown by:

1. Actively spreading ridge crests whose spreading rates are indicated by the separation of the pair of lines; they vary from 1 cm per year near Iceland to 6 cm per year in the equatorial Pacific.

2. Transform faults.

3. Oceanic trenches or Tertiary mountain belt systems.

All boundaries of the plates have shallow-focus earthquakes associated with them, and the compressive boundaries (3) also have associated intermediate- or deep-focus earthquakes. The directions of relative movement of these major blocks are shown by the arrows. Le Pichon concluded that this picture was in reasonable agreement with physiographic, seismic, and geological data and then proceeded to study and reconstruct surface movements since the Mesozoic. If the Antarctic block is considered as fixed relative to the Earth's rotational axis then all other plates and plate boundaries will move across the mantle.

J. D. Phillips and B. P. Luyendyk in 1970 considered both the deep-sea drilling results and magnetic anomalies. They located the relative motion pole for the North Atlantic somewhat further south and east than that given in Figure 14-3, and their angular rate of opening of the central North Atlantic differs markedly from rates offered by Morgan and Le Pichon. Their data show only one relative rotation pole during the late Cenozoic contrary to conclusions published by others in 1969. They favored the idea that the central North Atlantic and the South Atlantic represent different plate systems.

The New Global Tectonics by B. Isacks, J. Oliver, and L. R. Sykes

The third major paper of 1968, by Isacks, Oliver, and Sykes, dealt specifically with the strong support given by the observations of seismology to the concepts of sea-floor spreading and plate tectonics. The correlation of earthquakes with the boundaries of plates, and of intermediate and deep earthquakes with compressive boundaries, has already been noted. Figure 14-4 shows again the ridges, transform faults, and trenches bounding the plates given in Figure 14-3, and also the slip vectors determined from studies of shallow-focus earthquakes. Each arrow depicts the motion of the plate on which it is drawn relative to the

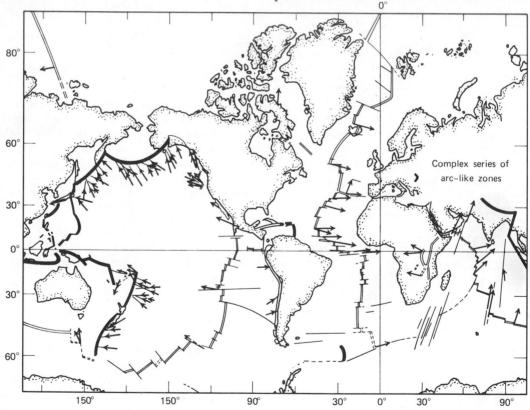

Figure 14-4. Summary map of slip vectors derived from earthquake mechanism studies (after Isacks et al., 1968). Arrows indicate horizontal component of direction of relative motion of block on which arrow is drawn to adjoining block. Crests of world rift system are denoted by double lines; island arcs, and arc-like features, by bold single lines; major transform faults, by thin single lines. (From Jour. Geophys. Res., **73**, 5855, 1968, with permission.)

plate on the other side of the tectonic feature. These directions correspond remarkably well with the movement directions of Figure 14-3 derived by Le Pichon from different evidence. In particular many of the vectors of differential movement calculated by Le Pichon for arcs, shown in his Figure 6, are very similar in direction to the slip vectors of Figure 14-4. McKenzie and Parker had previously reviewed the slip vectors around the North Pacific.

Figure 14-5 is a block diagram illustrating schematically the main features of sea-floor spreading, plate tectonics, or the "New Global Tectonics." The rigid lithosphere, including the crust and part of the upper mantle, is on the order of 100 km thick. The asthenosphere, a layer of no strength on the appropriate time scale, corresponds to the low-velocity zone extending down beneath the lithosphere to several hundred kilometers. The mesosphere makes up the greater part of the mantle and is relatively passive with strength. The lithosphere is continuous except at the boundaries of the moving plates. The diagram shows plates moving away from midoceanic ridges and down into the asthenosphere at island arcs with relative surface movement between plates occurring across transform faults. A return flow compensating for the movement of the lithosphere is shown in the asthenosphere with uprise at the opening ridge crests.

Seismology also provides evidence about movements at the ridges and trenches which supports this model. In connection with Figure 13-6 we noted that the relative motion across a transform fault is the reverse of that expected from a transcurrent fault, and that relative motion was restricted to the portion of the fault between the ridge crests. Shallow earthquakes along the oceanic ridges occur only along the ridge crest where they result from normal faulting (tension) and along the fracture zone between ridge crests. They do not extend beyond them as they would for transcurrent faults. Furthermore focal mechanism solutions indicate motions appropriate for transform faults.

Island arcs and ocean trenches have long been interpreted as sites of compression and downward movement of material, but

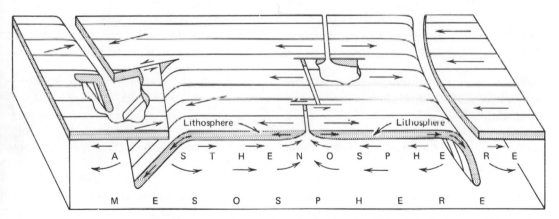

Figure 14-5. Relative movements of plates (after Isacks et al., 1968). Block diagram illustrating schematically the configurations and roles of the lithosphere, asthenosphere, and mesosphere in a version of the new global tectonics in which the lithosphere, a layer of strength, plays a key role. Arrows on lithosphere indicate relative movements of adjoining blocks. Arrows in asthenosphere represent possible compensating flow in response to downward movement of segments of lithosphere. One arc-to-arc transform fault appears at left between oppositely facing zones of convergence (island arcs), two ridge-to-ridge transform faults along ocean ridge at center, simple arc structure at right. See Richter (1969) for criticism and reply by Isacks et al. (1969). (From Jour. Geophys. Res., **73**, 5855, 1968, with permission.)

analyses of gravity data and focal mechanisms for earthquakes indicate tension at trenches. The seismic evidence reviewed by Isacks et al. gives strong support for the picture of a slab of relatively cool lithosphere being transported deep into the mantle as illustrated in Figure 14-5. Remarkable evidence is the discovery of anomalous zones about 100 km thick, with low attenuation of seismic waves and relatively high seismic velocities, which occupy the positions of the down-going lithosphere in Figure 14-5. The deep seismic zones beneath island arcs (Figure 2-6) are situated on the upper sides of the anomalous zones. The changing pattern of seismic activity with depth beneath the island arcs is also consistent with the model in Figure 14-5. Infrequent shallow earthquakes beneath the trench axis indicate extension and normal faulting probably caused by downward bending of the lithosphere (Figure 14-9). A thin zone of active seismicity beneath the inner margin of the trench is characterized by dip-slip mechanisms indicating underthrusting of the oceanic plate. This zone is continuous with the deep zone (Figure 2-6), for which the 1968 evidence favors maximum compressive stress parallel to the dip of the seismic zone which is the presumed direction of the slab. The maximum depths of earthquakes divided by estimated spreading rates indicate that the underthrusting of the lithosphere is about equivalent to the amount of spreading during the past 10 m.y. More recent work shows that for Benioff zones with no known deep earthquakes, or for those with a gap in seismicity between 300 and 500 km depth, the intermediate focus earthquakes indicate tension in the downdip direction. The deep-focus earthquakes, if they occur, have nearly vertical axes of compression. This information according to T. J. Fitch and P. Molnar in 1970 is consistent with models in which gravitational body forces are important factors in the dynamics of the lithosphere (Figure 14-8).

Figure 14-6 shows the relationships of continents and oceans to the lithosphere plates of Figure 14-5, as depicted schematically by J. F. Dewey and J. M. Bird in 1970 for some specific cross sections. They also presented a more detailed version incorporating more geological information (Figure 14-16). Figure 14-6 shows the continents as superficial passengers on rigid plates up to 150 km thick. Oceanic plates are being consumed in marginal trenches, against a continent in *a* and against an island arc in *b*. Small ocean basins are trapped within the complex trench-transform fault systems associated with island arcs in the Western Pacific as in *c* and *d*. Continents are too buoyant to be carried down into the mantle, and *g* represents the Indian plate partly underthrust beneath the Asian plate with the Himalayan uplift resulting.

The seismic evidence is not everywhere as clearly favorable for plate tectonics. I. S. Sacks in 1969 reported that preliminary results from earthquakes in the Peru-Brazil region contradict the pattern expected for the simple model of downgoing lithosphere shown in Figure 14-6b. There is a low-Q zone beneath the Andes, but the region has high-Q pathways through it and the difference in Q between them is probably more than an order of magnitude. The holes are not distributed in any systematic manner relative to the trench-mountain system.

C. F. Richter commented on the new global tectonics paper in 1969 congratulating Isacks, Oliver, and Sykes for their service in coordinating the expanding subject and noting some points which they had cautiously avoided or not fully considered. The most significant questions that he raised were in connection with the tectonics of Japan and the Alpide belt. In particular he stated that the tectonic complexity of Japan is too often obscured in global discussions by mapping on a very small scale, and he presented the major features which have to be explained by any general global scheme. In their reply the authors agreed that, although they believe the overall fit of seismological data with Le Pichon's plate model is highly significant,

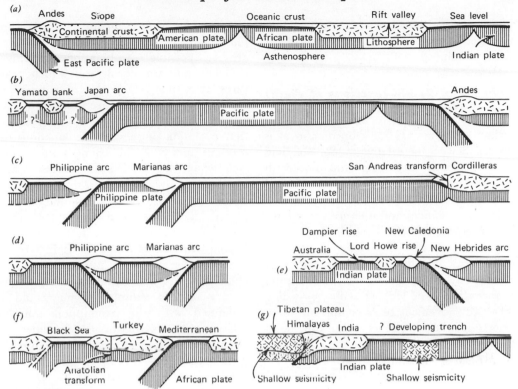

Figure 14-6. Schematic sections showing plate, ocean, continent, and island arc relationship (after J. F. Dewey and J. M. Bird, Jour. Geophys. Res., **75**, 2625, 1970, with permission).

"so significant that the hypothesis is effectively proved," the real Earth was certainly more complex on a finer scale and more detailed studies of selected regions should tend to increase the complexity of the model.

Small Crustal Plates by P. Molnar and L. R. Sykes

Morgan considered global tectonics in terms of 20 major plates, and Le Pichon managed with only six as shown in Figure 14-3. He chose six major blocks in such a way that he could solve specific problems. Figure 14-3 shows that in the region of the Caribbean and Middle America there are areas bounded by linear tectonic features which were treated as parts of the larger American and Antarctic plates. The Caribbean plate underlies the Caribbean Sea and is bounded by the Middle America arc,

the Cayman trough, the West Indies arc, and the seismic zone through northern South America. The Cocos plate is bordered by the East Pacific rise, the Galapagos rift zone, the north-trending Panama fracture zone near 82°W, and the Middle America arc.

From their study of the seismicity of the region Molnar and Sykes in 1969 concluded that the tectonics is dominated by the interaction of these two small plates which are moving relatively to the larger plates around them. The plates are nearly aseismic. Molnar and Sykes determined the focal mechanisms of 70 earthquakes and worked out the relative movements of the plates, the rates of motion, and the locations of underthrusting. This is one of the few parts of the world where an active oceanic ridge approaches an island arc, and elucidation of the tectonics in such regions is a necessary test for the global theories. Figure 14-6 indicates the existence of small plates associated with the island arc systems of the Pacific Ocean.

Evolution of Triple Junctions of Plates by D. P. McKenzie and W. J. Morgan

Plate tectonics as developed above is concerned with the relative motions of rigid plates presently occurring and not with changes that occurred through geological time. We saw in Chapter 13 that patterns of magnetic anomalies and fracture zones have been interpreted in terms of changes in the spreading pattern. The changes proposed include (a) changes in the directions of spreading, (b) changes in the relative rates of spreading, and (c) episodic spreading where all movement stops completely for a period. If plate tectonics is to provide a complete theory of global tectonics it must include geological time as a variable.

In 1969 McKenzie and Morgan discussed and classified the geometry and stability of all possible triple junctions, where three plates meet in a point (Figure 13-9). There are 16 possible combinations of ridges, trenches, and transform faults which, taken three at a time, can form triple junctions. All except two of these are stable in certain conditions. McKenzie and Morgan showed that evolution of these triple junctions can produce many of the changes in tectonic style which would otherwise appear to have been caused by a change in the direction or magnitude of the relative motion between plates. They reexamined the magnetic lineations and fracture zones in the northeast Pacific (Figures 13-2 and 13-3) in terms of the evolution of triple junctions and concluded that the main features of the geological history of this complex region could be explained with the assumption that the relative velocities of the major plates had remained unchanged during the Tertiary. This and other examples show how a complex series of events can be produced simply by the geometry of plates.

THE RESULTS OF DEEP SEA DRILLING

Hess's 1962 model for sea-floor spreading away from the midocean ridge is shown in Figure 12-22b. It predicts progressive overlap of ocean sediments onto the ridge; the greater the distance from the ridge the older the deepest sediments should be. The results from deep-sea drilling in the South Atlantic, on leg 3 of the *Glomar Challenger* cruise, have confirmed this prediction. This cruise, under the aegis of the Joint Oceanographic Institutes for Deep Earth Sampling (JOIDES) has as its primary aim the recovery of complete sedimentary sections of the ocean floor. Results were presented by A. E. Maxwell and seven coauthors in 1970.

Figure 14-7a shows eight sites from which long cores were obtained on a traverse across the mid-Atlantic ridge near 30°S. Basement composed of basalt was reached at seven sites. Each site was located within the well-defined magnetic anomaly pattern for this latitude (Figure 13-13) which gives a predicted age for the basaltic basement rocks if a constant rate of sea-floor spreading is assumed (Figures 13-14, 13-15, and 13-17). Abundant calcareous microfossils were used to assign ages to the sediments in the core. Figure 14-7a gives the magnetic age of the basement and the paleontological age of the deepest sediment just above the basement. The agreement between them is reasonably good.

The ridge has numerous offsets and changes of strike (Figure 14-4), and it is not always easy to locate a point from which a given site should have originated according to sea-floor spreading. The distance of each site from the ridge crest was estimated in two ways: (a) by measuring the linear distance to the nearest ridge axis, and (b) by rotating the site back

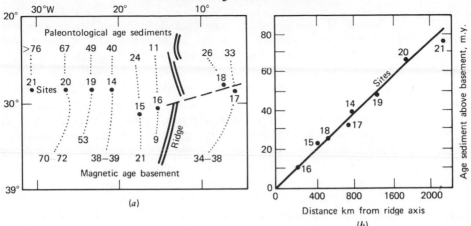

Figure 14-7. Results obtained from deep drilling in the South Atlantic Ocean. (*a*) Trend of the axis of the Mid-Atlantic Ridge in the South Atlantic as determined from geophysical evidence plus earthquake epicenter data. The drilling sites are shown relative to the ridge axis. For each site, the diagram shows the paleontological age of the deepest sediment just above basement, and the magnetic age of the basement according to the magnetic anomaly and the geomagnetic time scale (Figures 13-13 and 13-15). (*b*) The age of the sediment from *a* plotted as a function of distance from the ridge axis. The points lie close to a straight line, indicating a constant spreading half rate of 2 cm/year (compare Figure 13-14, where constant rate was assumed for the South Atlantic). After Maxwell *et al.*, Science, **168**, 1047–1059, 1970. Copyright 1970 by the American Association for the Advancement of Science.

to the ridge along a small circle about Morgan's rotation axis with pole at 62°N, 36°W (compare pole 1 in Figure 14-3). These distances are compared in Figure 14-7*b* with the paleontological age of the sediment immediately above the basalt basement. The relationship is clearly linear, corresponding to a spreading half-rate of 2 cm/year. The studies of magnetic anomalies had previously given the same rate assuming constant spreading in the South Atlantic (Figures 13-13 and 13-14). The agreement between the paleontological and magnetic estimates of sea-floor spreading is regarded as strong support for the hypothesis of sea-floor spreading and the concept of magnetic stratigraphy. It has yet to be established, however, that the basalt reached is definitely basement rock. Deeper drilling may possibly reveal more sediments below basaltic layers.

Similar deep-drilling results in the Pacific Ocean, reported in 1970 by A. G. Fischer and nine coauthors, have yielded a wealth of useful data, but they have not yet proved as directly relevant to interpretation of sea-floor spreading. It has been tentatively inferred that the ocean crust becomes progressively older from the East Pacific Rise to Hawaii and further west, but chert in the older sediments prevented penetration into the basement below the sediments for a definitive test. The results of drilling suggest that the "opaque layer" of Figure 13-16 is an alternating sequence of pelagic ooze and chert.

MECHANISM OF PLATE TECTONICS

Plate tectonic theory has had considerable geometrical success in accounting for complex tectonic phenomena, but there remains the major problem of the energy source and the mechanism which maintains the motion. Whatever drives the motions provides the energy for earthquakes and volcanism, and the radioactive decay of U, Th, and K is probably the only source of energy large enough. Some kind of thermal convection is required to convert this into heat. The problem relates directly to the geochemistry of the mantle and crust as well as to the mode of convection within the mantle.

Figure 14-8a shows a standard Rayleigh-Benard mantle convection model, with flow in the asthenosphere exerting viscous drag to the strongly coupled lithosphere and forcing it into the mantle where the cell turns downward (see also Figures 12-23 and 12-24). This produces tension in the surface slab and compression in the down-going slab.

Figure 14-5 depicts an alternate form of convective motion in which the lithosphere forms the upper part of a convection cell, as originally proposed by Hess (Figure 12-22). In connection with Figure 14-5 Isacks et al. suggested that the pattern of flow in the mantle may be controlled largely by the motions of the cold plates of lithosphere. In Figure 14-8b it is assumed that the cold slab of lithosphere is denser than the asthenosphere, and therefore it sinks, pulling behind it the surface litho-

sphere, under tension. This process would be aided by the conversion of crustal gabbro to eclogite in the sinking slab. D. P. McKenzie reviewed the consequences and causes of plate motions in 1969, and he refers to this mechanism as an extreme form of convection which takes place if viscosity is a rapidly varying function of temperature. He presented a rudimentary analysis of the process and concluded that, despite its appeal, this mechanism probably cannot maintain the motion of the large plates. In Figure 14-5 the arrows in the asthenosphere represent possible compensating flow in response to the downward motion of the lithosphere plates. McKenzie emphasized that the mantle and lithosphere are in thermal and mechanical contact, and that motions cannot occur in either without the other being affected. Thus in Figure 14-5 sinking of the cold lithosphere causes motion in the mantle, and in Figure 14-8a motion in the mantle moves the lithosphere. Probably all of the mechanisms illustrated schematically in Figure 14-8 are in operation, and the relative contributions of each will not be

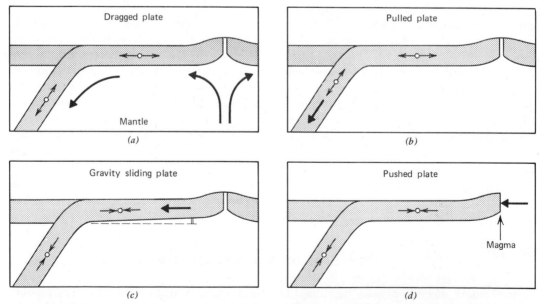

Figure 14-8. Diagrammatic representation of mechanisms proposed for moving plates of lithosphere (compare Figure 14-5). States of compression and tension in the plates are indicated. See text for discussion.

known until a detailed three-dimensional analysis has been completed on the problem. Apparently complete solution is beyond the capability of the present generation of computers.

In 1969 A. L. Hales suggested that gravitational sliding of the lithosphere on a partly melted asthenosphere (Figure 6-20) could play an important role in sea-floor spreading as illustrated in Figure 14-8c. The ocean floor is assumed to be horizontal, with the surface of the low velocity zone dipping gently away from the ridge. He calculated that a slope of 1/3000 for the surface of the low velocity layer would be sufficient for sliding at a rate of 4 cm/year, and that if a block 2000 km long and 100 km thick is tending to slide toward a continent in this way, then the stresses at the interface with the continent would be of the order 2×10^8 dyne/cm^2. The stresses are more than enough to cause failure. The potential energy per year is of order 10^{24} ergs over a front of 2000 km, and this is comparable with the energy released in earthquakes. Hales presented these numbers only to show that stresses and energies of the right order are possible. Similar effects might be produced if the surface of a lithosphere slab dips away from the ridge possibly because of distortion arising from convection as in Figure 14-8a.

L. Lliboutry proposed a fourth process in 1969 as shown in Figure 14-8d. He suggested that the lithosphere plates are pushed apart (upslope) by compressed magma which fills a deep crack connected with magma at the top of the asthenosphere. As the plates diverge the magma solidifies on the walls of the crack and erupts at the surface inter-mittently; but once initiated the crack never disappears. Neither currents in the astheno-sphere nor sinking of the plates at one end are excluded as contributory forces to plate motion.

Lliboutry also discussed the way in which the horizontal lithosphere beneath the ocean floor could change its configuration so abruptly to a near-planar slab dipping steeply into the

mantle. He pointed out that the usual picture of a plastic extension of the upper lithosphere and a plastic compression of the lower lithosphere would produce the curvature shown in Figure 14-9a and that this would be retained in the softer asthenosphere. As an alternative he described the mechanism illustrated in Figure 14-9b. The oceanic lithosphere impinging upon the continental lithosphere initially bends a little as the continental lithosphere exerts a resisting force dipping at about 45°. When the shear stress in a vertical plane reaches the shear strength value, a vertical fault is produced originating an earthquake. The oceanic lithosphere is thus sheared into vertical fragments, but the sequence of fault blocks

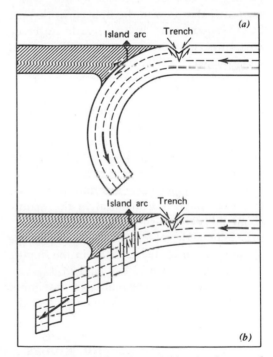

Figure 14-9. Lithosphere sinking in the mantle beneath island arcs. Mechanism at a sink. (a) Plastic bending of the lithosphere. (b) Discontinuous shearing by faulting along vertical planes. Only this process agrees with the plane form of the sunk lithosphere, and with the focal mechanism of shallow earthquakes. (From Lliboutry, L., Jour. Geophys. Res., **74**, 6525, 1969, with permission.)

as a whole takes on the shape of an inclined plate. In 1970 A. Malahoff reviewed the mechanisms for gravity and thrust faults under the oceanic trenches and presented a picture that seems to be consistent with Figure 14-9b. He suggested that between 0 and 10 km below the trench failure was largely by gravity faults; between 10 and 40 km depth there were largely low dipping (20°) thrust faults; and between 40 and 600 km there were possibly step faults allowing the plate to move downward as shown in Figure 14-9b. He suggested that the gravity faulting beneath the trench explained why their sedimentary fill was usually undistorted.

In 1970 N. L. Carter and H. G. Ave'Lallemant described deformation experiments on dunite and peridotite which suggested that flow in the upper mantle may be governed primarily by a non-linear creep law. In a companion paper (Ave'Lallemant and Carter, 1970) they proposed that syntectonic recrystallization is an important or dominant mode of flow in the upper mantle. They examined their results with reference to the motions of spreading lithosphere plates, and presented schematic "flow-fields" corresponding to the hypotheses where:

1. The displacements are due to drag by creep of convecting material in the mantle (Figure 14-8a).

2. Where the motion of the lithosphere is opposed by drag due to creep in the mantle below (Figures 14-8b, 14-8c, and 14-8d).

They illustrated the planes of maximum shearing stress for these two models and the fabrics that would develop by syntectonic recrystallization during flow. The preferred crystal orientations influence the seismic wave velocities, and they compared the effect of the proposed tectonite fabrics with recorded seismic anistropy in the upper mantle. The problem remains inconclusive but the approach is promising.

H. R. Shaw suggested in 1970 that the dissipation of tidal energy in the solid Earth may play a crucial role in magma generation and in the mechanism of sea-floor spreading. Only about 3% of the heat flow from the surface results from tidal energy, but this is enough to produce about 30 km^3 of magma a year. The tidal energy is strongly focused by dissipation mechanisms and can locally dominate other energy sources. Shaw concluded that dissipation is maximized in a partly molten region with a liquid fraction of 10 to 30%. Magmatic injection in the vicinities of oceanic ridges (Figure 14-11) transfers heat both from the stored heat in the mantle and from tidal sources. The possible tidal magma production alone can account for much of the average spreading rate. A compensating mass transfer from deeper regions is required to maintain a quasi-steady state. Thus a tidal-magmatic mechanism can act as a trigger to convective circulation. He also discussed the heat flux from continents and oceans and concluded that globally constant heat flux is consistent with equality of heat fluxes due to mass transfer and radioactive heat production in continental and oceanic sections explained in terms of the tidal-magmatic model.

MIDOCEANIC RIDGES AND MANTLE CONVECTION

Interest in sea-floor spreading was originally concerned mainly with the formation of new oceanic crust at the ridge crests, and the magnetic lineations have apparently left a record of the spreading history. The topography and crustal structure of the ridges is known in some detail as well as the gravity and heat flow profiles transverse to them (Figures 3-17a, 7-4, and 7-5). There have been several attempts to fit the observed heat flow to theoretical models of temperature distribution and convection.

The lithosphere acts as a thermal boundary layer supporting large temperature and density gradients because of its rigidity. There are two models for sea-floor spreading which relate to the ridge and its anomalies. Spreading is associated with (a) uprise of hot mantle material intruded in a narrow zone, or (b) uprise of hot mantle material in broad convective zone beneath the ridge. According to the first model the heat flow is not the surface expression of temperature anomalies in the upper mantle but a consequence of sea-floor spreading. The heat flow then cannot be used to infer mantle temperature distributions. This model was explored by D. P. McKenzie in 1967. According to the second model, as reviewed by E. R. Oxburgh and D. L. Turcotte in 1968, the geological features and geophysical anomalies associated with the ridges are surface expressions of the mantle conditions.

Thermal Structure

McKenzie presented a model for the temperature distribution in a lithosphere plate 50 km thick moving with constant velocity away from a vertical boundary at constant temperature. As the plates move apart hot material wells up from the mantle and forms a linear ridge along the boundary. He found good agreement with the observed width and shape of the heat flow anomalies and the free-air gravity anomalies transverse to the ridges. He concluded that both gravity and oceanic heat flow are probably controlled by the strength and thermal properties of the lithosphere and that it is not necessary for the upper mantle to be hotter beneath the ridges than it is elsewhere. In 1969 N. H. Sleep concluded from mathematical models that gravity, heat flow, and ridge topography cannot be expected to distinguish between McKenzie's preferred model and that of convective upwelling in the mantle.

A boundary-layer theory for the structure of two-dimensional convection cells was applied to mantle convection by Oxburgh and Turcotte, and the results beneath ocean ridges are shown in Figure 14-10. It is not

known whether the mantle behaves as a plastic or a viscous body. Oxburgh and Turcotte adopted a model with a constant, Newtonian viscosity independent of temperature, and they neglected the volume heat release due to radioactivity within the cell. K. E. Torrance and D. L. Turcotte have since obtained a numerical solution for mantle convection using a temperature and depth dependent viscosity appropriate for diffusion creep. They gave the resulting flow, temperature, and viscosity distributions. Oxburgh and Turcotte concluded that although their assumed viscosity relations were certainly wrong, the provisional agreement between the model and natural phenomena suggests that the simplifying assumptions had not too great an effect.

Figure 14-10 shows the steady-state distribution of isotherms beneath an ocean ridge predicted by the Oxburgh-Turcotte model. The temperature gradients are restricted to thin horizontal boundary layers and to thin vertical plumes on the boundaries between cells. The cores of cells are abiabatic. The mushrooming isotherm distribution for the upper ascending limb is not likely to be realized in detail because heat will be transferred rapidly upward by magmas from this region. The lines with arrows show the streamlines derived from velocity vectors for particle motion.

Petrological Structure

The general pattern of the petrology of the upper mantle beneath the ridges and oceanic lithosphere can be determined from this model for temperature distribution, and the schematic phase diagrams presented in Figures 6-18 and 6-19 for peridotite and gabbro, dry and in the presence of traces of water. Although modification of the temperature distribution and experimental determination of the phase diagrams will change the details, the general pattern may not be changed significantly. Each point in Figure 14-10 is fixed by a specific pressure (depth)

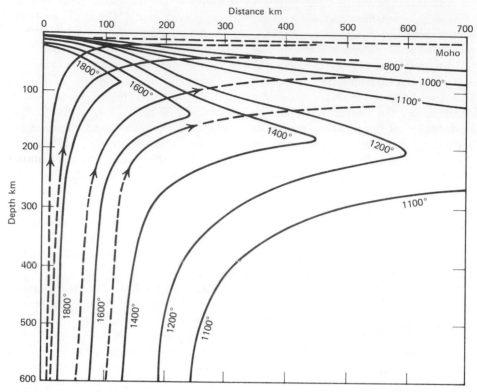

Figure 14-10. Model for convection cells rising beneath a midoceanic ridge, with steady-state thermal structure computed for material with viscosity independent of temperature. Lines with arrows show the streamlines for particle motion (after E. R. Oxburgh and D. L. Turcotte, Jour. Geophys. Res., **73**, 2643, 1968, with permission).

and temperature, and the phase assemblage for a given material at each point can be determined from the appropriate figure, 6-18 or 6-19. Figure 14-11*a* shows the distribution of phase assemblages for a mantle composed of peridotite with a trace of water, and Figure 14-11*b* shows the phase assemblages produced in a hypothetical mantle of gabbroic composition with a trace of water. Figures 14-11*a* and 14-11*b* together provide a picture of the petrology of the suboceanic mantle and specifically the mantle beneath oceanic ridges assuming that the mantle is composed of peridotite together with material of gabbroic composition. The layered sequence for mantle some distance from the ridge is similar to that summarized in Figure 6-20. The

presence of a trace of water produces a zone of incipient melting which is equated with the low-velocity zone. Material of gabbroic composition exists as eclogite. The low-velocity zone increases in thickness considerably near the ridge crest extending downward to the level of the olivine-spinel transition (Chapter 6).

There is a large zone of partial melting where the temperature exceeds the dry solidus of peridotite, and gabbro in most of the corresponding zone in Figure 14-11*b* is completely melted. The zone is more than 400 km wide and 200 km thick. Oxburgh and Turcotte noted that only mantle material ascending along streamlines within 100 km of the plume center would pass through this fusion zone, and that this material then forms

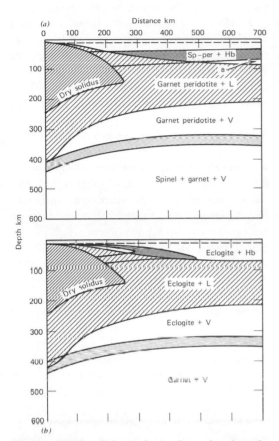

Figure 14-11. Schematic sections showing the petrology of peridotite (*a*) and eclogite (*b*) with traces of water in the suboceanic mantle, extending from the crest of a midoceanic ridge. These sections are derived from the temperature-depth (pressure) distribution in Figure 14-10, and the schematic isopleths for these materials shown in Figures 6-18(*b*) and 6-19(*b*). Compare Figure 6-20 for standard layered mantle sequences. Note the large zone in Figure 14-11(*b*) for partial melting above the dry solidus temperatures. The low-density layer of gabbro and garnet granulite in (*b*) exhibits a general pattern similar to that required by models in Figure 7-5 for the structure of midoceanic ridges. Compare Figure 5-18(*b*). (From P. J. Wyllie, Jour. Geophys. Res., **76**, 1328, 1971, with permission.)

the upper 130 km of the horizontal limb. If all of the basaltic magma generated within the fusion zone escapes to the surface at or near the ridge, then this 130 km layer would be composed of residual peridotite. If some of the interstitial magma crystallizes as it is transported laterally through the upper boundary of the fusion zone then we have to consider phase assemblages in the upper mantle for material of basaltic composition.

Figure 14-11*b* shows that basaltic magma emerging from the fusion zone above 80 km depth crystallizes as gabbro, and that below 100 km it crystallizes as eclogite. Lateral transportation of the lithosphere causes cooling of the gabbro as shown by the streamlines in Figure 14-10, and if equilibrium is maintained the gabbro is transformed into eclogite through a wide zone of garnet granulite (Figure 5-15). This illustrates the dynamic model for the suboceanic mantle proposed by F. Press in 1969 (Figure 3-10). The low density layer of gabbro and garnet granulite exhibits a general pattern similar to that required by the models in Figure 7-5 for the structure of midoceanic ridges.

In 1967 F. Aumento presented the results of petrological study of a variety of extrusive rocks dredged from the first area of the mid-Atlantic ridge, at 45°N, to have been sampled and surveyed in detail. He reported a complete and continuous sequence from olivine tholeiites to alkali basalts together with high-alumina equivalents. Using available experimental data as a guide (Chapter 8) he developed a scheme of multiple cycles of partial melting beneath the ridge axis. A continuously variable decreasing degree of partial melting at 35–60 km depth could produce a continuous range in liquid compositions. With Figure 14-10 available Oxburgh and Turcotte considered similar models for magma generation, and P. W. Gast in 1968 placed the petrogenetic scheme on a more quantitative basis with special reference to the origin of tholeiitic and alkaline basalts and to trace element fractionation.

LITHOSPHERE CONSUMPTION BENEATH ISLAND ARCS

The concept of plate tectonics requires that the lithosphere extends down into the mantle beneath island arcs (Figures 14-5 and 14-6). The intimate connection between surface structures and deep structures indicated by seismology is persuasive evidence. Whether the rigid slabs of lithosphere move down into the asthenosphere as a result of impelling forces or as a result of their own negative buoyancy (Figure 14-8), it is clear that the lithosphere constitutes and controls some kind of convective motion within the mantle. The descending lithosphere is a strong heat sink, yet the island arc areas above the cold slabs exhibit high heat flow anomalies and volcanic activity. The main features associated with sinking lithosphere which need explanation include:

1. Intermediate- and deep-focus earthquakes.
2. Ocean trenches 3 to 4 km deeper than the neighboring ocean basin floor.
3. Large negative gravity anomaly over the trench with smaller positive anomaly over the arc.
4. Abrupt increases in heat flow profiles along the line of the island arcs (Figures 3-17*b* and 3-17*c*).
5. Andesite volcanism in belts usually less than 300 km wide along island arcs, occupying the ocean side of the zone of high heat flow.

These topics were reviewed in 1969 by D. P. McKenzie and in 1970 by E. R. Oxburgh and D. L. Turcotte.

Thermal Structure

The factors influencing the temperature distribution in a downgoing slab of lithosphere include: (a) the descent rate, (b) adiabatic compression of the lithosphere as it moves into the mantle, (c) radioactive heating, (d) energy generated by phase changes in peridotite and gabbro of the lithosphere as the pressure increases, and (e) stress or frictional heating at the boundaries between lithosphere and asthenosphere.

McKenzie showed that there is a close relation between the temperature structure in a descending slab and that in the lithosphere near spreading ridges. He calculated the temperature distribution in a 50 km-thick slab with the geometry shown in Figure 14-12*a*, assuming that the spreading rate had been constant during the past 10 m.y. The thermal conductivity of the cold lithosphere is so small that the inner part of the slab is heated only very slowly (Figure 14-13). McKenzie gave temperatures on isotherms in dimensionless units, because of the uncertainty of the parameters used in the calculations, including the temperature of the adiabatic mantle. He concluded that earthquakes are restricted to those regions of the mantle which are colder than a certain, unspecified temperature. The earthquakes should thus occur in a thin layer within the lithospheric slab. The calculated temperature distribution could not explain the heat flow anomaly, and probably would not account for the volcanic activity either. McKenzie and other authors have appealed to frictional or stress heating at contacts between lithosphere and asthenosphere to explain these phenomena.

Movement of a large rigid slab into the asthenosphere would set up stresses and govern the flow of mantle material in the absence of other forces. Figure 14-12*a* shows McKenzie's calculated distribution of stream lines for flow in the mantle produced in this way. Shearing stresses are exerted on the lithosphere boundaries, and calculated contours are shown in Figure 14-12*b*. The stresses behind the arc are much greater than those in front. Viscous dissipation within the mantle produces heat from the mechanical energy which moves the plates against the viscous forces. The heat generated by friction between the mantle and the surface of the sinking slab would take 300 m.y. to be conducted to the surface, and this therefore cannot account for the surface heat flow anomaly. Others have appealed to diapiric uprise of hot material or magmas as factors contributing to the heat flow, but the magnitude of these effects appears to be small compared with the gross regional heat flux.

McKenzie cited three effects which could account for the heat flow anomaly:

1. The calculated contours for stress heating in Figure 14-12c show considerable heat generation at shallow depths, and this could diffuse to the surface in geologically reasonable times.

2. The calculated flow lines behind the island arc show hot mantle being carried closer to the base of the lithosphere.

3. Thinning of the lithosphere by this flowing mantle could possibly lead to even higher levels for the hotter material of the asthenosphere.

Oxburgh and Turcotte presented a possible distribution of isotherms under an island arc system, with the constraint that temperatures in the zone of movement between sinking lithosphere and mantle reached the melting temperatures of the oceanic sediments and crustal rocks, and were buffered by these temperatures. This assumption allowed the temperature along the fault zone to be prescribed. McKenzie concluded that the earthquake zone is not the physical boundary between lithosphere and asthenosphere, but a cooler layer within the slab, and this is consistent with the seismological data. The frictional zone of heating and magma generation then does not coincide with the earthquake zone, as Oxburgh and Turcotte and many others (Figure 8-1d) have assumed.

Minear and Toksöz used a quasi-dynamic scheme and a finite difference solution of the conservation of energy equation to determine the effects of the several factors listed above

on the temperatures in a downgoing slab. Thermal regimes were calculated for a 160 km-thick slab downwarping at 45° to the horizontal, with the complexity of the physical model being increased in a series of steps in order to demonstrate the effects of the various heat generating processes and the spreading rate. Figure 14-13a shows the distribution of isotherms taken from their Figure 9 which assumes a 1 cm/year spreading velocity, shear-strain heating of 1.6×10^{-4} ergs/cm^3 sec along the top edge of the slab and 1.0×10^{-5} ergs/cm^3 sec along the ends and bottom, and no contribution from phase changes or adiabatic compression. The dominant pattern is one of isotherms depressed deeply into the mantle. The zone of shear-strain heating with temperatures greater than those in normal mantle is less than 30 km wide above 200 km and the greatest temperature increases occur at depths less than 100 km (compare Figure 14-12c). Minear and Toksöz concluded that the temperatures were just high enough to produce melting along the upper edge of the slab, but this conclusion depends upon the results used for estimating the solidus temperature of the material.

Magma Generation

Using the phase diagram for peridotite in the presence of a trace of water given in Figure 6-18b, and assigning the appropriate phase assemblage to each pressure (depth)-temperature point in Figure 14-13a, we obtain Figure 14-13b. No melting occurs in peridotite (or in gabbro or eclogite) anywhere in this mantle section in the absence of water. Given a trace of water, incipient melting occurs through the shaded zone. If this layer of incipient melting does correspond to the low-velocity zone, as suggested in Figure 6-20, Figure 14-13b shows that the low-velocity zone is not continuous beneath island arcs. The upward migration of the olivine-spinel transition zone within the slab is shown.

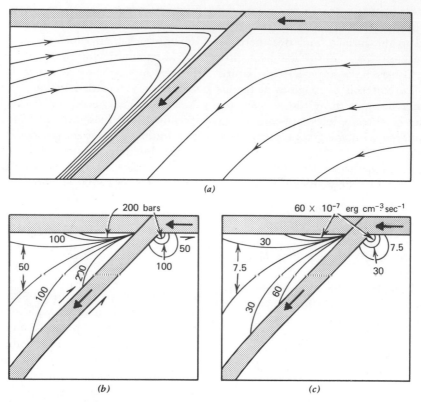

Figure 14-12. Stress heating caused by flow within the mantle associated with a downgoing slab of lithosphere (after McKenzie, 1969). (*a*) Stream lines for flow within the mantle. Motion is with respect to the 50 km thick plate behind the island arc, and is driven by the motion of the other plate and of the sinking slab. Thermal convection outside the slab is neglected. (*b*) Shear stresses in bars caused by the flow in Figure (*a*). The half arrows show the direction of the forces exerted by the fluid on the plates and slab. The stresses on the plate behind the island arc exceed those on the plate in front. (*c*) Stress heating caused by viscous dissipation within the flow in Figure (*b*), in units of 10^{-7} erg cm^{-3} sec $^{-1}$. The heating within the mantle is more intense behind than in front of the arc. (From Geophysical Journal, with permission.)

Magma may be generated beneath an island arc system in three ways:

1. The oceanic crust at the surface of the lithosphere, including siliceous and water-bearing sediments, could be partially fused by either thermal conduction from. the mantle or frictional heating. A small increment of temperature added to Figure 14-13*a* would cause partial melting of eclogite or peridotite, dry, in a zone near the surface of the slab at depths between about 150 and 300 km. Oxburgh and Turcotte discussed the mechanism for the formation of the main

members of the calc-alkaline suite by a process of continuous variation in temperature, pressure, composition of material being fused, and water content as melting occurs down the frictional heating zone and inward into the sinking lithosphere. T. Hatherton and W. R. Dickinson reviewed the relationship between andesitic volcanism and seismicity beneath island arcs in 1969 and concluded that the andesites are formed by fusion along the seismic zone, and that their potassium content increases as a function of depth of origin.

2. Dehydration of the lithosphere and upward migration of water into hotter mantle overlying the cold slab could produce partial melting of the mantle, as proposed, for example, by W. Hamilton and by A. R. McBirney in 1969. This process has been postulated as a source of intermediate and acid magmas for andesites and batholiths.

3. The hotter boundary zone comprising both mantle and upper lithosphere is potentially unstable, and diapiric uprise of solid material under adiabatic conditions

may be initiated from this zone leading to magma generation according to the Green and Ringwood model in Figure 8-17. Figure 14-13b shows that if water is present along this high temperature boundary the peridotite of mantle and lithosphere will contain interstitial liquid, and influx of more water from the downgoing slab into any part of this layer would increase the percentage of liquid present, reduce the viscosity, and facilitate diapiric uprise according to the scheme shown in Figures 8-21 and 8-22. This process

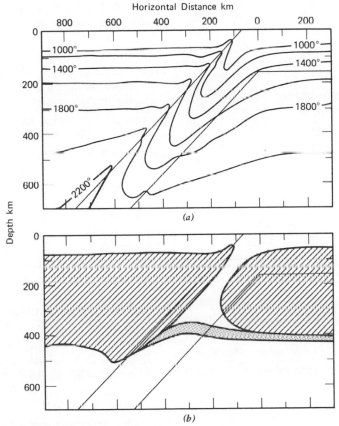

Figure 14-13. Schematic sections through the mantle and slab of lithosphere in mantle beneath an island arc. The lithosphere, 160 km thick, dips at 45° from the axis at 0 km. (a) Isotherms computed by Minear and Toksöz (1970), according to their Figure 9, which takes into account shear-strain heating along the edges of the lithosphere slab. (b) Petrology of the mantle section, assuming peridotite with traces of water (as in Figure 6-18b) and the temperature distribution in (a). The shaded zone shows incipient melting, interrupted by the slab. Uprise of the olivine-spinel transition (Figures 6-17 and 6-18a) within the cold slab is shown. The corresponding diagram for gabbroic material is similar. Note that normal fusion temperatures for the generation of basaltic magmas are not reached anywhere, according to this temperature distribution and the extrapolated dry peridotite solidus used (Figure 6-18(a)). (From P. J. Wyllie, Jour. Geophys. Res., **76**, 1328, 1971, with permission.)

would produce basaltic magmas from peridotite or andesites from eclogite at depths considerably above the Benioff zones (contrast Figure 8-1*d*). If water is abundant liquids of intermediate composition may be generated in similar fashion from peridotite. The compositions of the liquids produced depend upon a number of variables (Chapter 8).

Future correlation of the observed products of volcanoes with the compositions of liquids from various crystalline materials under known experimental conditions should permit us to locate the sources of various magmas beneath island arcs and beneath ocean ridges. For example it may prove possible to relate the variation in potash content across island arcs to the breakdown of hornblende, muscovite, and phlogopite in a partially hydrated downgoing lithosphere slab, as suggested by several investigators. If the sources of specific magmas can be located, this would provide invaluable fixed points in terms of depths (pressures) and temperatures for many geophysical calculations and models such as the temperature distribution shown in Figures 14-10 and 14-13*a*. This in turn would place limits on assumed physical properties of mantle materials. The way in which petrology, experimental petrology, and geophysics are coming to be mutually dependent is very heartening. Until more precise data are available, we will have to be satisfied with the general patterns such as Figures 14-11 and 14-13*b* based on extrapolated experimental data (Figures 6-18 and 6-19) and possible temperature distributions such as Figures 14-10 and 14-13*a*.

Structures Associated With Trenches

If slabs of lithosphere move down into the mantle beneath the island arcs we would expect to see evidence of compression. The seismic data show that the lithosphere beneath the trenches is in tension, so that gravity faults and vertical movements characterize this region, and that this is replaced

beneath the inner margin of the trench by underthrusting (Figure 14-9). The details are not yet clear, and it is not known how or where the overthrust fault approaches the surface of the Earth.

Figure 14-14*a* is a summary by J. F. Dewey and J. M. Bird of the main geological structures associated with trenches. This shows the sites of blueschist metamorphism (glaucophane schist facies in Figure 9-10) and the injection of serpentinite from the descending slab. This is a mechanism suggested by R. S. Dietz in 1963 for the origin of ultramafic rocks of orogenic belts (Table 6-4). The sediments in the trench are undeformed, but they may show extreme deformation at the foot of the inner trench wall. It is generally assumed that some of the sediment is carried into the mantle with the oceanic plate, and that some of it is mechanically plastered by folding on to the continental margin or island arc.

There has been considerable discussion about the behavior of sediments in oceanic trenches, reviewed in 1970 by D. W. Scholl and associates. The problem is very complex, because it depends upon erosion rates from adjacent land masses, which depend in turn rather sensitively upon the climate and the paleogeography. Although some trenches exhibit clear evidence of compressional deformation as indicated in Figure 14-14*a*, the sediments at the base of the continental margin in the Peru-Chile trench and others around the Pacific basin appear to exhibit only extensional features. This observation is incompatible with the generally accepted model of spreading. C. K. Seyfert suggested in 1969 that the downward movement of sediments occurs landward of the axes of oceanic trenches and that the sediments are folded, faulted, and metamorphosed as they are carried beneath the continent or island arc. There is a sharp seismic contact between the undeformed sediments and the material underlying the lower part of the continental slope, which might be explained if the transition from undeformed to deformed and

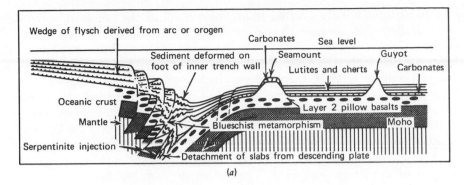

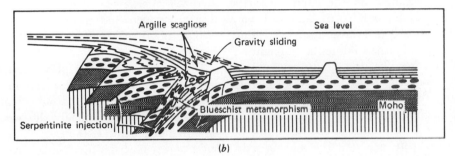

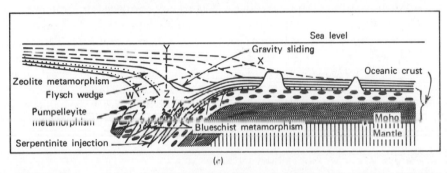

Figure 14-14. Structures associated with oceanic trenches (after Dewey and Bird, 1970). (*a*) Main geological structures. (*b*) and (*c*) Possible sequence of events in the growth of an island arc, as described in text. (From Jour. Geophys. Res., **75**, 2625, 1970, with permission.)

metamorphosed rocks is of tectonic form. This is a plausible explanation, but it is not altogether satisfactory to explain the apparently general lack of compressional features of sediments in trenches.

In their 1970 review Scholl and coauthors estimated the volume of undeformed trench sediments off Chile. If plate convergence had occurred at the 5 to 10 cm/year rates implied by the magnetic data during late Cenozoic time, and if those sediments deformed or consumed by the compression were undetect-

able because of the nature of their deformation as proposed by Seyfert, then the volume of undeformed sediments remaining should be considerably less than the volume of material supplied to the trenches during this period. Their estimates of the volume of terrigenous sediments contributed by continental erosion in late Cenozoic time are, in fact, quite close to the estimated volume in the trenches. They speculated that sinking of the oceanic lithosphere may take place seaward of the trench, with the trench fill and bedrock floor remaining partially or totally isolated from much of the differential movement between the plates.

During the Mesozoic and Cenozoic periods, according to J. Gilluly in 1969, the volumes of sediment derived from the North American continent should have been about the same in the Atlantic and Pacific Oceans. In fact it is estimated that the volume of terrigenous Atlantic sediment is at least six times as great as that of the Pacific sediment. The paucity of sediment off the Pacific coast is readily accounted for if North America began to drift westward in Triassic time overriding the Pacific and large volumes of sediment on a Benioff zone like that now presumed to underly western South America. Gilluly suggested that the onset of the tremendous plutonism of the Mesozoic of western America might have been related to the beginning of continental drift, and that the missing sediments off the Pacific coast were dragged down into the mantle along a sinking plate of lithosphere, fused, and then intruded as batholiths. This theme was also presented by W. Hamilton in 1969 and W. G. Ernst in 1970.

Figures 14-14*b* and 14-14*c* illustrate the possible sequence of events in the growth of a single island arc according to Dewey and Bird. With the beginning of plate descent a trench forms by the intricate thrusting of wedges of oceanic crust and mantle to form a small ridge. Submarine gravity slides of sediment carry blocks of basic and ultrabasic rocks from the thrust wedges into the trench, and these are carried down where they undergo strong deformation and blueschist metamorphism. When the descending plate reaches a depth of over 100 km, the former oceanic crust fuses yielding calcalkaline magmas, with basaltic magmas being generated at higher levels. A volcanic island arc is thus developed and its erosion produces a wedge of flysch sediments as shown in Figure 14-14*c*.

During periods of fast plate consumption blueschist metamorphism is active producing chaotic mélanges, or "argille scagliose," which may rise isostatically during periods of slow movement. If the rate of plate consumption is slow the flysch may build across the trench onto oceanic sediments, producing enormous apparent thicknesses (measured along *WX* instead of *YZ*). Trenches and their associated sedimentary and metamorphic rocks are likely to be extremely complicated. The chaotic mélange terrains of many mountain belts are now being examined as possible positions of earlier trenches and Benioff zones.

GEOSYNCLINES, MOUNTAIN BUILDING, AND SEA-FLOOR SPREADING

The concept of plate tectonics is forcing geologists to reexamine nearly everything that they thought they knew about the origin of rocks and mountain ranges. The classical concepts of eugeosynclinal and miogeosynclinal sequences outlined in Chapter 9 appear to be too simple to derive maximum benefit from the plate tectonics model. A Penrose Conference of The Geological Society of America was held in December 1969 at Asilomar, Pacific Grove, California, to discuss "The meaning of the new global tectonics for magmatism, sedimentation, and metamorphism in orogenic belts." There is no printed record of the proceedings of these informal conferences, but reports of the meetings by W. R. Dickinson were published in *Geotimes* and *Science* in April and June respectively 1970. Dickinson noted the impact felt from the ideas of young research workers, citing specifically the rapt attention paid to the authoritative discussion of time and motion along the San Andreas Fault system by graduate student Tanya Atwater. More experienced workers were anxious to explore the implications of plate tectonic theory for their own interpretations. Marshall Kay, for example, inquired how his concepts of geosynclines (Chapter 9) should be translated to mesh with the fresh concepts of orogeny.

The term "subduction zone" was found useful by most of the 95 participants at the conference. This was applied to any elongate region along which crustal rocks are led to descend, by folding, faulting, or a combination of both, relative to masses in an adjacent block. The usage implied the presence of a plate juncture of convergence.

Eugeosynclines may include sedimentary sequences that have formed as (a) trench complexes, (b) island arc complexes, or (c) continental rise deposits. Miogeosynclines may include sedimentary sequences that have formed as (a) continental shelf and slope deposits, (b) deposits trapped in elongate traps between island arcs and trenches, or (c) clastic wedges in foredeep complexes (Kay's exogeosynclines, Table 9-1). If oceans can be shown to open and close, then in time any sediment wedge associated with a continental margin will eventually become involved with a subduction zone and orogeny. Dickinson concluded that, although the classic concept that geosynclinal downbowing exerts direct causative control on the position and extent of orogeny is no longer tenable, the geosynclinal theory remains valid if the causative function is replaced by the notion of coincidence, or consequence; the sediment piles relate to orogeny in the sense that "to exist is to be deformed." Each large orogenic belt should offer a different sequential history, although certain sequences are more apt to be duplicated.

The topics that received particular attention included: (a) andesite chains, (b) batholith belts, (c) ophiolite complexes, (d) glaucophane schist or blueschist metamorphism, (c) relationship of continental structures to subduction zones, (f) nomenclature of tectonic elements and stratigraphic facies in orogenic belts, and (g) the meaning of geosynclinal theory in the new conceptual framework.

Andesites and batholith belts were taken as extrusive and intrusive phases of magma-

tism related to inclined seismic zones in the mantle above intermediate- and deep-focus earthquakes (Figure 14-15). Ophiolite complexes and some other ultramafic rocks of the orogenic belts were interpreted as oceanic crust formed at midocean ridges at high temperatures, and incorporated tectonically at low temperatures into orogenic belts at subduction zones (Figures 14-16e and 14-16f). The formation of blueschists was attributed to rapid descent of cold lithosphere to great depths along subduction zones in trench complexes (Figures 14-14 and 14-15a).

In Chapter 9 we reviewed the classification of contemporary geosynclines proposed by Mitchell and Reading in 1969 as they are related to three types of continental margins. We can now relate them also to plate tectonics as in Figures 14-6, 14-15, 14-16, and 14-17. Atlantic-type geosynclines develop where there is no differential movement between continent and ocean; Andean-type geosynclines develop when there is differential movement; and island arc-type geosynclines develop where a lithosphere plate descends some distance from a continental margin, with Japan Sea-type geosynclines developing behind some island arcs. If spreading ceases or is renewed, or if the position of descending lithosphere changes, the type of geosyncline also will change and theoretically there are 12 possible changes. Mountain building associated with geosynclinal development may be Andean, island arc, or Himalayan. Himalayan orogeny is produced when two continents collide (Figure 14-17c). A comprehensive review was published by Dewey and Bird in 1970.

Dewey and Bird presented Figure 14-15 as a model for the evolution of a cordilleran-type mountain belt (Andean) formed where a trench originates near to a continental margin of Atlantic-type. This is a development and refinement of schemes presented early in the sea-floor spreading game by R. S. Dietz (1963). The general features of Figure 14-14 are incorporated in the trench-formation stage. The sequence is self-explanatory. It is

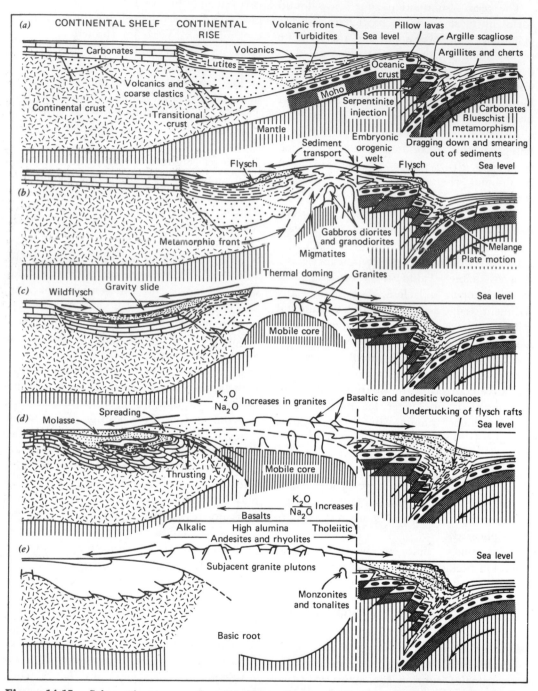

Figure 14-15. Schematic sequence of sections illustrating a model for the evolution of a cordilleran-type mountain belt developed by the underthrusting of a continent by an oceanic plate (after Dewey and Bird, 1970, Jour. Geophys. Res., **75**, 2625, with permission).

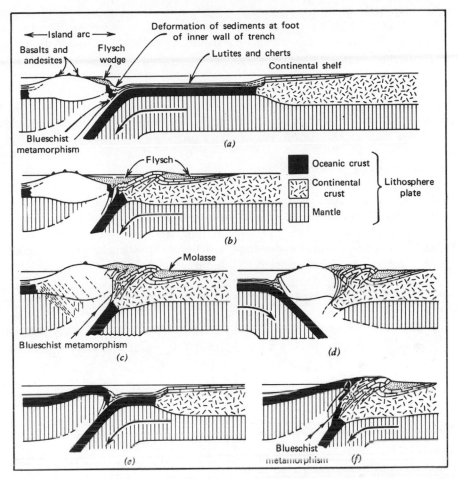

Figure 14-16. (a)–(d) Schematic sequence of sections illustrating the collision of a continental margin of Atlantic type with an island arc, followed by change in the direction of plate descent. (e)–(f) Proposed mechanism for thrusting oceanic crust and mantle onto continental crust (after Dewey and Bird, 1970, Jour. Geophys. Res., **75**, 2625, with permission).

based on an analysis of the north-western parts of the Appalachians for Ordovician time (Bird and Dewey, 1970) and on the general structure of the Mesozoic Cordilleran system of western North America. This figure should be compared with Figures 9-1, 9-3, and 9-4. The continental shelf edge in Figure 14-15a marks the boundary between the miogeosynclinal and the eugeosynclinal assemblages in Figure 9-1.

According to plate tectonics the continents are carried passively as part of the rigid lithosphere. Oceanic lithosphere moves down into the asthenosphere along lines marked by ocean trenches. Figures 14-1 and 14-6 show that continued spreading eventually brings continental masses to ocean trenches, and Figure 14-16 illustrates the consequence of the collision of a continental margin of Atlantic type with an island arc according to Dewey and Bird. The low density continental mass cannot be carried into the mantle because buoyancy forces will oppose the motion (Chapter 10). Continents once formed are therefore very difficult to destroy. McKenzie pointed out in 1969 that if an island arc attempts to consume a continent (Figures 14-16*b* and 14-16*c*) large stresses will develop and the island arc is likely to flip so that it consumes oceanic crust originally behind the arc, as in Figure 14-16*d*. Convection, as represented by the descending lithosphere plate, is now changed. McKenzie suggested that the whole pattern of mantle convection must depend on the motion of the continents.

Mitchell and Reading presented a simple model in which the development of continental margins and geosynclines was related to the oscillation of continents between ocean rises. This is illustrated in Figure 14-17. Each diagram shows two continents moving between ocean ridges considered as fixed in

position relative to one another, in both plan view and cross section. The geosynclines are distinguished as active if the lithosphere is descending, or passive if there is no differential movement between margin and ocean floor.

In Figure 14-17*a* continent I moves westward, approaches the ocean ridge, and in Figure 14-17*b* its motion has been reversed and it is moving east. The Andean-type geosyncline on the now passive western margin will be eroded and an Atlantic-type geosyncline will develop. An island arc-type geosyncline is formed near the eastern margin, and the former Atlantic-type geosyncline is now enclosed in a small ocean basin or geosyncline of Japan Sea-type.

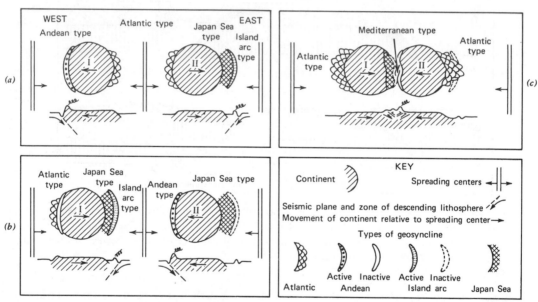

Figure 14-17. Simplified model to show how continents may oscillate between ocean rises as a result of changes in pattern of ocean-floor spreading (after Mitchell and Reading, 1969). Atlantic-type geosynclines develop on passive continental margins where the continent is moving with the lithospheric plate of the adjacent ocean floor. Andean-type geosynclines develop where spreading lithosphere descends beside a continent. Island arc-type geosynclines develop where lithosphere descends some distance from a continent. Japan Sea-type geosynclines occur between continents and islands arcs. The collision of two continents may produce Mediterranean-type geosynclines. Compare Figures 14-14, 14-15, and 14-16. See Figure 9-5 (*a*), (*b*), and (*c*). Each diagram shows both plan view and cross section of two continents moving between ocean ridges considered as fixed in position relative to one another. The geosynclines are distinguished as active if the lithosphere is descending, or passive if there is no differential movement between margin and sea floor. For discussion of spreading episodes and succession of one type of geosyncline by another, see text. (From Jour. Geology with permission. Copyright 1969, University of Chicago Press.)

Continent II in Figure 14-17a moves eastward with geosynclines equivalent to those near Continent I in Figure 14-17b. In Figure 14-17b it has reversed direction, and an active Andean-type geosyncline forms across the earlier Atlantic-type geosynclinal deposits. The now inactive island arc on the east is eroded to sea level and may subside. The Japan Sea-type geosynclinal basin persists.

Continued convergence of the two continents results in their collision in Figure 14-17c and the development of a Himalayan-type orogeny. The original arcuate shape of the Andean- and island arc-type geosynclines of Figure 14-17b may be destroyed. Small ocean basins of Mediterranean-type may remain where the two continents are not in contact.

Mitchell and Reading continued their examples one stage further, with the combined continent in Figure 14-17c separating again as a result of resumption of spreading along the line of the Himalayan-type orogenic belt. Figure 14-17 is sufficient to indicate the complexity of superimposed geosynclinal types that can be expected according to this model of oscillating continents, which is a necessary development of plate tectonics.

We have already referred to the application of plate tectonic theory to western North America by Hamilton, Gilluly, and Ernst. There are many other papers reevaluating geological interpretations. Two notable contributions of 1970 applied the theory to the geology on either side of the Atlantic. The paper by Bird and Dewey on the evolution of the Appalachian Caledonian orogenic belt is an outstanding demonstration of interpretation of complex geology in terms of plate tectonic theory. They examined in detail the sequences of sedimentation-deformation-metamorphism patterns of the New England and Newfoundland segments of the Appalachian orogen and its relationship to Atlantic Ocean plate tectonics. It is instructive to compare this paper with the 1963 contribution of Dietz on collapsing

continental rises. The second paper is by A. M. Ziegler, who reviewed the geosynclinal development of the British Isles during the Silurian period. He presented a series of detailed paleogeographic maps, which gave insight into the positions of landmasses and shelf areas, and the areas of turbidite deposition and graptolitic shale deposition. He concluded that each side of the British geosynclinal complex might qualify as a geosyncline in its own right, and that the whole complex should be regarded as the marginal effects of two continents, unrelated until they collided. His conclusions are consistent with Wilson's 1966 suggestion that the Atlantic Ocean was open during the Lower Paleozoic, closed during the Upper Paleozoic and Lower Mesozoic, and opened again along a slightly different line during the present period of spreading from the mid-Atlantic ridge.

After examining the history of western America in terms of plate tectonics, Hamilton in 1970 turned his attention to Asia and the Ural Mountains. He reviewed the Soviet literature and attempted to explain the Uralides, the entire complex of late Precambrian and Paleozoic foldbelts between the Russian and Siberian platforms, as the result of collision of the two platforms. This was suggested as long ago as 1924 but most Soviet geologists have rejected long-distance horizontal transport; they postulate vertically moving elements, such as anticlinoria and synclinoria, assumed to have evolved through long periods. The paleomagnetic data for the Siberian and Russian platforms suggest that these land masses were widely separated in the Cambrian, converged during middle Paleozoic, and collided in the Permian or Triassic. Hamilton concluded that the geology of the Uralides accords with the collision concept with an intervening oceanic plate sliding down a subduction zone. From the igneous and metamorphic rocks of the Uralides, and the structure, he deduced a history of the continental margins before and during collision.

Dewey and Bird concluded that mountain building occurs in two ways:

1. The island arc/cordilleran (Figure 14-15) which is thermally driven and develops on the leading plate edges above a descending plate.

2. The result of collisions, continent/continent (Figure 14-17c) or continent/island arc (Figure 14-16) which are for the most part mechanically driven. They listed six main differences between these types of mountain belts, and emphasized the fundamental importance of the ophiolite suite (Table 6-4) in evaluating the stages of development of mountain belts. It is interesting to note that only a few years ago the ophiolite suite was largely ignored in petrogenetic discussion of ultramafic rocks by geologists not primarily concerned with Alpine geology [Wyllie, 1967 (p. 412), 1969, 1970]. Now these rocks are widely considered to be samples of the oceanic crust or upper mantle, and many occurrences of ultramafic rocks in orogenic belts are being closely re-examined to see if they can be identified as representatives from an ophiolite suite. These rocks are destined to play a key role in the interpretation of plate collisions. Some restraint is necessary, however, because not all ultramafic rocks represent oceanic crust or upper mantle (Chapter 6).

Ophiolites received particular attention from J. C. Maxwell in his 1966 review of Mediterranean geology (published in 1970). He concluded that ophiolite complexes represent ultramafic mantle material, partially melted and differentiated during diapiric uprise into and above lighter crustal rocks (Figure 6-6). He noted the similarity of larger ophiolite masses to the structure of the oceanic crust (Figure 7-8) and suggested that some parts of "oceanic areas" deter-

mined seismically may contain ophiolites overlying sedimentary rocks; false depths to the Moho may thus have been reported. Maxwell summarized his own work and that of European geologists active in the Mediterranean region, and concluded that there is convincing evidence for a continuous sialic basement between Europe and Africa in the later Carboniferous. The marine sediments of the Alpine system appear to have been deposited everywhere on continental crust. Maxwell interpreted the Mediterranean as a strip of continental crust in which massive injections of mantle material accompanied the generation of a dynamic ridge-and-basin movement. This process of oceanization by ophiolites is discussed further in the next chapter. The significance of ophiolites in this interpretation differs from that attributed to them in plate tectonics.

According to continental drift and plate tectonics, there was once a large oceanic area between Africa and Eurasia, which began to close in Permian or Cretaceous time. According to Maxwell the similarity and continuity of the geological histories of Africa and Europe is so striking that if the continents were not now close together, we would certainly postulate that they had once been together and had since drifted apart. The Mediterranean itself developed with widespread deposition of shallow-water Permo-Triassic sediments everywhere disconformable on older sialic rocks. The JOIDES drilling program in the Mediterranean should provide critical evidence for determination of the geological history of the region and for evaluation of Maxwell's conclusion that no lateral displacement of oceanic dimensions can reasonably be postulated between Africa and Eurasia, at least since Mid-Paleozoic.

15. *Global Geology in the 1970's*

INTRODUCTION

The revolution of the 1960's burst into the 1970's with a wave of enthusiasm for re-evaluation and reinterpretation of geology in terms of the new global tectonics. In Chapters 11, 12, 13, and 14 we reviewed the evidence that purports to prove the reality of continental drift in its new guise of plate tectonics. In October of 1970, using this evidence, R. S. Dietz and J. C. Holden presented a new version of Wegmann's original breakup of *Pangaea* (Figure 11-1), together with a prediction of what the world would look like 50 m.y. from now. Los Angeles is due to pass San Francisco in about 10 m.y., and to slide into the Aleutian trench in about 60 m.y.

What started as hypothesis is now hailed as theory. Articles in popular magazines state that the original geopoetry of H. H. Hess (Figure 12-22) is now geofact. Many items of evidence published since about 1966 were accompanied by confident claims that this was the virtual or definitive proof. Finally, in 1970, the results of the JOIDES drilling program brought evidence from the South Atlantic (Figure 14-7) that the overlapping of sediments and the ages of sediments were just as predicted (Figure 12-22). A single proof is sufficient for a theory to become fact. This model suffers from an overdose if one counts all the claims for virtual proof that have been made for it. The new global tectonics has enjoyed phenomenal success as a working hypothesis; so much success, in fact, that it is fast becoming a ruling theory.

There has been continuing opposition to the concept of continental drift. A. A. Meyerhoff has kept track of the pro-drift and anti-drift literature, and he informed me that of the papers published between about 1956 and 1970, antidrift papers comprise at least 21%. In the few years since the formulation of the plate tectonics model the papers with anti-drift arguments are rarely cited. Some of the problems were outlined in Chapter 14; these include the apparently undeformed sediments in many oceanic trenches and the properties of the deep seismic zones beneath the Andes. There are problems related to the geology of the Mediterranean. Other problems are presented by some paleontologic data and the paleoclimatic data reviewed in the following pages. Protagonists of plate tectonics tend to neglect the data and arguments not explained by the ruling theory, arguing that these can probably be ascribed to our lack of understanding rather than to inadequacies of the model, and that time and improved understanding will bring forth explanations.

Every revolution has its counterrevolution. The 1968 debate in *Geotimes* between J. T. Wilson, proclaiming the revolution, and V. V. Beloussov, presenting opposition, was mentioned in Chapter 11. Beloussov has consistently maintained that global tectonics are explained by vertical movements and oceanization rather than by horizontal movements and compression, and he appears to have a large following among Soviet geologists. A. A. Meyerhoff is a crusader, recently joined by C. Teichert and soon to be joined by 13 more coauthors, who is marshalling the rather scattered anti-drift arguments in counterrevolution. He maintains that unless the distribution of paleoclimate indicators and his interpretation of them can be ac-counted for by the new global tectonics, the mobilist concepts must be relegated to the status of speculations.

This is yet another chapter that merits a whole book, but again I must satisfy myself with a brief outline.

PROGNOSIS FROM THE 1969 PENROSE CONFERENCE AND THE 1970 GEODYNAMICS COMMISSION

The topic of the Penrose Conference at Asilomar in December 1969 was "The meaning of the new global tectonics for magmatism, sedimentation, and metamorphism in orogenic belts." The reports of W. R. Dickinson were reviewed in Chapter 14. The classical views of orogenesis outlined in Chapter 9 appear very incomplete compared with the new concepts explored at the Penrose Conference. Stratigraphic approaches to the study of orogenic belts must involve evaluation of lateral movements and their effects, as well as the vertical movements. The topic that received particular attention at the conference was the relationship of continental structures to subduction zones, and the meaning of ophiolite complexes, stratigraphic facies, glaucophane schist or blueschist metamorphism, andesite chains, and batholith belts, with respect to this relationship. These are the rocks that will be studied diligently during the 1970's for clues in unravelling sequences of plate collisions.

The international Upper Mantle Project provided much of the stimulus for the development of plate tectonic theory. During the terminal phase of this Project, 1968-1970, an international Commission was appointed to plan a Geodynamics Project to facilitate extension of long-term research efforts initiated under the aegis of the Upper Mantle Project. Objectives were outlined at a meeting in Flagstaff, Arizona, in June 1970 and published in the November 1970 issue of the *Transactions of the American Geophysical Union*. The Geodynamics Project is an international, interdisciplinary program of research "on the dynamics and dynamic history of the Earth with emphasis on deep-seated foundations of geological phenomena." The Commission of Geodynamics considered the program in four parts:

1. The movement of lithospheric blocks relative to each other, and the concentration of strong tectonic activity in a few relatively narrow mobile belts between blocks. Studies should be concentrated in active belts where movements and deformations can be measured.

2. Studies directed to characterization of the movements of the lithospheric blocks, and the driving forces within the Earth responsible for the movements.

3. Study of the primarily vertical movements that occur within lithosphere blocks, apparently independent of their horizontal movements. Sedimentary basins and elevated plateaus provide evidence for enormous vertical movements, often accompanied by block faulting and volcanism.

4. Systematic studies of past orogenic activity as reflected in the geological record, in order to determine past lateral movements of the lithosphere blocks. Igneous, metamorphic and tectonic relationships in ancient orogenic belts should be compared with similar relationships in recent active belts.

Initiation of the Geodynamics Project ensures that the international and inter-disciplinary cooperation that flourished with the Upper Mantle Project will continue to grow. Whatever happens in the revolution and counterrevolution, there can be no doubt that our understanding of the Earth will advance rapidly.

V. V. BELOUSSOV AND R. W. VAN BEMMELEN PREFER "OCEANIZATION"

V. V. Beloussov struck a note of caution in his 1969 review of the results of the Upper Mantle Project, but this does not appear to have dampened the ardor of enthusiasts for the new global tectonics. Beloussov noted that, while much attention has been paid to data from the oceans in the formulation of the new global tectonics, our knowledge of the structure and development of the oceans is still much more sketchy than data from the continents. He considers that insufficient attention is being paid to continental geology, and historical perspective is thus being lost. The results of two centuries of data gathering from the continents should not be oversimplified in order to bring them down to the level of the schematic data available for the oceans. He suggested that the new oceanic results "have cast a hypnotic spell and thrown a shadow over much that is old and familiar."

Beloussov argued in favor of vertical movements initiated in the mantle as the major cause of folding, magmatism, and metamorphism. There are regions of slow uplift and subsidence on old continental platforms that have been maintained for hundreds of millions of years, and these are difficult to explain by theories requiring large horizontal displacements of the crust. He reviewed the contradictions involved in the idea of large transcurrent or transform fault displacements. The absence of a satisfactory scheme of worldwide convection to explain the apparent movements implied by paleomagnetic data leads him to question the assumption that the geomagnetic field has always had the same structure as it has now (Chapter 12).

The hypothesis of "oceanization" is an alternative to the hypotheses involving continental rift and ocean spreading, and Beloussov maintains that it meets most of the objections raised against the mobilist theories. The tectonosphere is divided into vertical blocks developing more or less independently. Continental crust is destroyed as sinking crustal blocks are replaced by basaltic lava, forming ocean floors. This explains the equality of continental and oceanic heat flow and avoids the conflict between oceanic and continental geology. The deep faults surrounding the Pacific separate areas of continental differentiation and of oceanic homogenization. If sediments older than the Cretaceous are distributed across the ocean floors, they must be contained in layer 2 or even layer 3. Beloussov suggested that intensified subsurface radioactive heating in oceanic regions at the end of the Paleozoic or beginning of the Mesozoic led to violent volcanic processes and oceanization which destroyed the older oceanic crust.

Beloussov emphasized the inadequacy of our knowledge of the Earth and advocated the balanced use of oceanic and continental data of geology, geophysics, and geochemistry. Apparent contradictions should not be artificially avoided by arbitrary assignment of more weight to one aspect or the other; all evidence must be studied and evaluated. No scientist would challenge this approach out of context, but in the context of plate tectonics there is at least a tendency among many geologists to neglect the method of multiple working hypotheses.

Beloussov and his followers are not alone in advocating a process of oceanization as a major factor in shaping the present Earth's surface. In 1968 R. W. van Bemmelen wrote a review paper on the origin and evolution of magmas and the Earth's crust. He argued against the hypothesis of continental growth by addition of magmas from the mantle (Chapters 7 and 8), and described a process of oceanization. The sialic crust was formed early in the Earth's history by the accretion of satellitic material. This crust has been transformed into oceanic crust by the emplacement of basic magmas, either locally within the crust or in regions of mega-shearing.

His views differ from those of Beloussov, because he believes that continental drift and sea-floor spreading do occur. According to his undation theory of the Earth's evolution endogenic energy flows outward from the core and is transformed into various types of energy according to the physico-chemical conditions of the structural layers of the Earth through which the energy passes. Near surface phenomena such as volcanism, seismic activity, and heat flow are the effects of the deeper-seated processes caused by the outward flow of endogenic energy. Undations are accompanied by the flow of material within the Earth (mass-circuits) which deform the geoid: the greatest deformations are called "mega-undations." During the development of a mega-undation the upper mantle is arched and the overlying layers spread under gravity. This can lead to removal of the continental shields from the crest of a mega-undation, with new ocean basin floored by upper mantle material being formed in the wake of the shields. Accumulating bodies of basaltic magma produce geo-undations such as the midoceanic ridges, the tops of which are subjected to gravitational spreading with tension and rifting occurring at its crest. According to van Bemmelen sea-floor spreading is caused by the spreading under gravity of the geo-undatory midocean rise (Figure 14-8c).

A. A. MEYERHOFF MAINTAINS THAT THE ATLANTIC OCEAN HAS BEEN OPEN FOR 800 MILLION YEARS

Not all geologists consider the new global tectonics to be a panacea. A. A. Meyerhoff published two major papers in 1970 marshalling geological and paleoclimatological evidence that appears to be inconsistent with the conclusion that the Atlantic Ocean has closed and opened again during the past 800 to 1000 m.y. In 1971 he was joined in a third paper by C. Teichert.

Meyerhoff reviewed the distribution of paleoclimate indicators. He plotted the distribution of marine evaporites, coal deposits, tillites, and proved desert eolian deposits on a series of world maps, one for each age from the Proterozoic through the Miocene, and interpreted them in terms of modern meteorology, physical oceanography, and climatology. This was emphasized because adherence to outdated climatological concepts has led to "myths widely believed by earth scientists," which Meyerhoff discussed and evaluated.

The maps show that:

1. Coal and evaporite belts are axisymmetric about the present rotation axis.

2. The Earth's two horse-latitude belts have remained coincident through time, as indicated by the desert eolian deposits and present desert belts.

3. Of the world's evaporite deposits, 95% by volume and area occur in regions now receiving less than 100 cm of annual rainfall; these regions reflect very closely the present planetary wind-circulation pattern, and the coincidence demonstrates that this pattern has not changed since middle Proterozoic time, except under local and temporary influences.

These and related facts show that the planetary wind-current and ocean-current pattern has been essentially the same for 800-1000 m.y. The only known explanation for this coincidence is that the rotational pole, the continents, and the ocean basins have been in the same positions since middle Proterozoic time.

Since the Devonian major coal deposits occupy two belts extending to high latitudes. Equatorward from the coal belts are the evaporite belts. Their widths fluctuate through time, indicating that there have been episodic changes in the world temperature. The Earth must have been very warm in "evaporite maxima," when the evaporite belts spread through up to 125° latitude. Glaciation is most common in periods of "glacial maxima," when the coal belts are broadest and the evaporites are restricted to about 40–60° latitude width. Meyerhoff presented a curve showing quantitatively the general distribution since the middle Proterozoic time of evaporite and glacial maxima, showing the fluctuation of world climate with corresponding fluctuations in the widths of climatic zones.

Ocean currents and wind currents may modify the positions of paleoclimate indicators anticipated from their latitude. Their main effect is a northward offset. In his second paper Meyerhoff described the northward deviation from axisymmetry of the Northern Hemisphere evaporite zone in the circum-Arctic and circum-North Atlantic Oceans which characterized several evaporite-maxima periods from later Proterozoic through Early Permian times. He explained these in terms of world-climate fluctuations and the development of major transoceanic sills within the framework of existing continents, ocean basins, and ocean-current systems.

Meyerhoff and Teichert reviewed the formation of coal and glacial deposits during the glacial maximum periods. They pointed out that coal is not a tropical deposit; coal requires cool winters in middle latitudes, abundant moisture, and large swampy areas. Every large coal basin is on one of two high-latitude coal zones or on the eastern sides of continents. Glaciation can occur only if there are elevated regions (regardless of latitude), and sufficient moisture for glaciers to build and ice caps to expand. This requires the presence of a nearby ocean and its associated moisture-laden air. During glacial maximum periods precipitation near ice-cap peripheries can be 45° or more from the poles, and mountain glaciation can occur at any latitude.

Meyerhoff and Teichert pointed out that glaciers cannot form deep in the interiors of continents where there is no water supply. They gave reasons for believing that shallow epeirogenic seas cannot account for the ice-cap and coal distributions. A Gondwanaland reconstruction (Figures 11-1 and 11-5) requires that the Permo-Carboniferous glaciated areas be too far inland for moist ocean-air currents to reach them. "Thus the 'Gondwanaland' hypothesis, in this respect, defeats itself by making an impossibility of one phenomenon it purports to explain." This conclusion has been stated several times during the past 45 years, but the message has been lost.

The distribution of glacial and coal deposits during Carboniferous and Permian times requires certain moisture precipitation and ocean current patterns, and these show that the Atlantic and Indian Oceans have existed since Late Carboniferous or earlier time.

These conclusions, based on "factual, observable data, in sharp contrast to speculations based on recent geophysical-oceanographic studies," convince Meyerhoff and Teichert that, until advocates of the new global tectonics find alternative explanations for the distribution of paleoclimate indicators, the mobilist concepts will have to be regarded as speculations supported by only a fraction of the known geological, paleontological, and paleoclimatological data. We have already noted Beloussov's concern that structural aspects of continental geology have been neglected for the sake of the ocean-based plate theory.

EPILOGUE

In the final sentence of their 1968 paper Isacks, Oliver, and Sykes wrote:

"Even if it is destined for discard at some time in the future, the new global tectonics is certain to have a healthy, stimulating, and unifying effect on all the Earth sciences."

We have seen the stimulation and the unification of effort, and we have seen also the first major dissent. The new global tectonics has enjoyed such phenomenal success as a working hypothesis that it is becoming a ruling theory. The dangers of this situation are shown by a quotation from a distinguished paper by C. E. Wegmann in 1963 (p. 5):

"because commonly *the notions, concepts and hypotheses control the selection of facts recorded by the observers*. They are nets retaining some features as useful, letting pass others as of no immediate interest. The history of geology shows that a conceptual development in one sector is generally followed by a harvest of observations, since many geologists can only see what they are asked to record by their conceptual outfit."

This is not the first time in geological history that a theory has been acclaimed as virtually proven. It has been said that Abraham Werner's promulgation of the Neptunian theory elevated geology to the rank of a real science. For years his theory appeared to be unassailable. He maintained that all geological formations, and rocks of all types except for those actually observed to emerge from volcanoes as lavas, had originated as successive deposits or precipitates from a primeval ocean. Werner's thesis was challenged by the school of Plutonists and subsequently abandoned; this was one of the most celebrated and bitter controversies in science. James Hutton and the Plutonists presented evidence satisfying most geologists that many rocks were formed by the cooling and crystallization of hot material that had risen in a fused condition from subterranean regions. Neptunism, which dominated geological thought for many years, simply disappeared.

Another controversy of long duration concerned the question of whether or not fossils were of organic origin. When they were eventually recognized as the remains of living things, champions of the theological cause maintained that fossils had been carried to their present positions above sea level by the Noachian deluge. Indeed the existence of fossils in mountains was cited as scientific evidence proving that the biblical deluge had occurred. Baron Cuvier discovered that certain fossils were confined to specific rock formations, and he concluded that a series of widespread catastrophes had caused the disappearance of faunas characterizing certain formations. The doctrine of Catastrophism was dominant for nearly three centuries, with Cuvier's major contributions being published in 1811 and 1812. Toward the end of the eighteenth century, Hutton's theory of the Earth emphasized that in the geological record there is "no sign of a beginning—no prospect of an end." The idea that geological history should be explainable in terms of events occurring at present led to the controversy between Catastrophism and Uniformitarianism, as the competing principle was called by Sir Charles Lyell. In continental Europe the principle became known as Actualism. The publication in 1833 of Lyell's book *Principles of Geology* marked the end of Catastrophism, and Uniformitarianism became the new creed. It is now recognized, however, that the uniform flow of geological history has been interrupted by local intermittent catastrophes.

The conceptual development of plate tectonics is indeed gathering a harvest of data, observations, and interpretations. The pieces are fitting together very well, in a most persuasive fashion. But what about those paleoclimate indicators of Meyerhoff? Other dominant theories have been toppled: surely this could not happen to the new global tectonics —or could it?

References

Aubouin, J., 1965, *Geosynclines*, Elsevier, Amsterdam.

Ahrens, T. J., and Y. Syono, 1967, Calculated mineral reactions in the earth's mantle. *J. Geophys. Res.*, **72**, 4181–4188.

Akimoto, S., and H. Fujisawa, 1968, Olivine-spinel solid solution equilibria in the system Mg_2SiO_4-Fe_2SiO_4. *J. Geophys. Res.*, **73**, 1467–1479.

Akimoto, S., and E. Komada, 1967, Effect of pressure on the melting of olivine and spinel polymorph of Fe_2SiO_4. *J. Geophys. Res.*, **72**, 679–686.

Allard, G. O., and V. J. Hurst, 1969, Brazil-Gabon link supports continental drift. *Science*, **163**, 528–532.

Anders, E., 1968, "Chemical processes in the early solar system, as inferred from meteorites," in *Accounts of Chemical Research*, October 1968, American Chemical Society.

Anderson, D. L., 1967, "Latest information from seismic observations," in *The Earth's Mantle*, ed. T. F. Gaskell. Academic, New York.

———, 1967, Phase changes in the upper mantle. *Science*, **157**, 1165–1173.

———, 1968, Chemical inhomogeneity of the mantle. *Earth Planet. Sci. Letters*, **5**, 89–94.

———, 1970, "Petrology of the mantle," in *Fiftieth Anniversary Symposia*, ed. B. A. Morgan. Mineralogical Society of America, Special Paper No. 3.

———, and C. Sammis, 1969, The low velocity zone. *Geofisica Internacional*, **9**, 3–19.

Archambeau, C. B., E. A. Flinn, and D. G. Lambert, 1969, Fine structure of the upper mantle. *J. Geophys. Res.*, **74**, 5825–5865.

Aumento, F., 1967, Magmatic evolution on the mid-Atlantic ridge. *Earth Planet. Sci. Letters*, **2**, 225–230.

Ave'Lallemant, H. G., and N. L. Carter, 1970, Syntectonic recrystallization of olivine and modes of flow in the upper mantle. *Bull. Geol. Soc. Am.*, **81**, 2203–2220.

Badgley, P. C., 1965, *Structural and Tectonic Principles*, Harper and Row, New York.

Barazangi, M., and J. Dorman, 1969, World seismicity map of E.S.S.A., Coast and Geodetic Survey epicenter data for 1961-1967. *Bull. Seismol. Soc. Am.*, **59**, 369–380.

Barth, T. F. W., 1962, *Theoretical Petrology*, 2nd ed., Wiley, New York.

Bassinger, B. G., O. E. DeWald, and G. Peter, 1969, Interpretation of the magnetic anomalies off Central California. *J. Geophys. Res.*, **74**, 1484–1487.

Beloussov, V. V., 1968, An open letter to J. Tuzo Wilson. *Geotimes*, **13**(10), 17–19.

———, 1969, Earth's tectonosphere (results and further problems for investigation). *Intern. Geol. Rev.*, **11** (No. 12 December), 1368–1381.

Benioff, H., 1955, "Seismic evidence for crustal structure and tectonic activity," in *Crust of the Earth*, ed. A. Poldervaart. Geological Society of America Special Paper 62.

Bernal, J. D., 1936, Discussion. *Observatory*, **59**, 267–268.

———, 1961, Continental and oceanic differentiation. *Nature*, **192**, 123–125.

Birch, F., 1952, Elasticity and constitution of the earth's interior. *J. Geophys. Res.*, **57**, 227–286.

———, 1961, Composition of the earth's mantle. *Geophys. J.*, **4**, 295–311.

———, and P. LeComte, 1960, Temperature-pressure plane for albite composition, *Am. J. Sci.*, **258**, 209–217.

Bird, J. M., and J. F. Dewey, Lithosphere plate-continental margin tectonics and the evolution of the Appalachian orogen. *Geol. Soc. Am. Bull.*, **81**, 1031–1060.

Blackett, P. M. S., J. A. Clegg, and P. H. S. Stubbs, 1960, An analysis of rock magnetic data. *Proc. Roy. Soc. (London)*, **A-256**, 291–322.

Blackett, P. M. S., E. Bullard, and S. K. Runcorn, 1965, A symposium on continental drift. *Phil. Trans. Roy. Soc. London*, **1088**.

Boer, J. de, J. G. Schilling, and D. C. Krause, 1969, Magnetic polarity of pillow basalts from Reykjanes Ridge. *Science*, **166**, 996–998.

Boettcher, A. L., 1970, The system CaO-Al_2O_3-SiO_2-H_2O at high pressures and temperatures. *J. Petrol.*, **11**, 337–379.

———, and P. J. Wyllie, 1967, Hydrothermal melting curves in silicate-water systems at pressures greater than 10 kilobars. *Nature*, **216**, 572–573.

———, 1968, Melting of granite with excess water to 30 kilobars pressure. *J. Geol.*, **76**, 235–244.

———, 1968, The quartz-coesite transition measured in the presence of a silicate liquid and calibration of piston-cylinder apparatus. *Contr. Mineral. Petrol.*, **17**, 224–232.

———, 1968, Jadeite stability measured in the presence of silicate liquids in the system $NaAlSiO_4$-SiO_2-H_2O. *Geochim. Cosmochim. Acta.*, **32**, 999–1012.

———, 1969, Phase relationships in the system $NaAlSiO_4$-SiO_2-H_2O to 35 kilobars pressure. *Am. J. Sci.*, **267**, 875–909.

Bowen, N. L., 1928, *The Evolution of the Igneous Rocks*. Princeton University Press, Princeton.

———, 1940, Progressive metamorphism of siliceous limestone and dolomite. *J. Geol.*, **48**, 225–274.

———, and O. F. Tuttle, 1950, The system $NaAlSi_3O_8$-$KAlSi_3O_8$-H_2O. *J. Geol.*, **58**, 489–511.

Boyd, F. R., and J. L. England, 1960, Apparatus for phase-equilibrium measurements at pressures up to 50 kilobars and temperatures up to 1750°C. *J. Geophys. Res.*, **65**, 741–748.

———, 1962, Pyrope. *Carnegie Inst. Wash. Yearbook*, **61**, 109–112.

Brown, H., and C. Patterson, 1947, The composition of meteoritic matter II. The composition of iron meteorites and of the metal phase of stony meteorites. *J. Geol.*, **55**, 508–510.

Bruhnes, B., 1906, Recherches sur le direction d'aimantation des roches volcaniques. *J. Phys. Radium Paris*, **5**, 705–724.

Brune, J. N., 1969, "Surface waves and crustal structure," in *The Earth's Crust and Upper Mantle*, ed. P. J. Hart. Geophysical Monograph 13, American Geophysical Union, Washington, D.C.

Bull, C. E., E. Irving, and I. Willis, 1962, Further paleomagnetic results from South Victoria Land, Antarctica. *Geophys. J.*, **6**, 320–336.

Bullard, E. C., 1967, The removal of trend from magnetic surveys. *Earth Planet. Sci. Letters*, **2**, 293–300.

———, 1968, Reversals of the earth's magnetic field. *Phil. Trans. Roy. Soc. London*, **A-263**, 481–524.

———, J. E. Everett, and A. G. Smith, 1965. The fit of the continents around the Atlantic, *Phil. Trans. Roy. Soc. London*, **A-258**, 41–51.

Bullard, E. C., C. Freedman, H. Gellman, and J. Nixon, 1950, The westward drift of the Earth's magnetic field. *Phil. Trans. Roy. Soc. London*, **A-243**, 67–92.

Bullen, K. E., 1936, The variation of density and the ellipticities of strata of equal density within the earth. *Monthly Notices Roy. Astron. Soc. Geophys. Supplement*, **3**, 395–401.

———, 1967, "Basic evidence for earth divisions," in *The Earth's Mantle*, ed. T. F. Gaskell. Academic, New York.

Burk, C. A., 1968, Buried ridges within continental margins. *Trans. N.Y. Acad. Sci.*, **2-30**, 397–409.

Burnham, C. W., 1967, "Hydrothermal fluids at the magmatic stage," in *Geochemistry of Hydrothermal Ore Deposits*, ed. H. L. Barnes. Holt, Rinehart, and Winston, New York.

Cameron, A. G. W., 1968, A new table of abundances of the elements in the solar system, in *Origin and Distribution of the Elements*, ed. L. H. Ahrens. Pergamon, Oxford.

Carey, S. W., 1958, *Continental Drift, a Symposium*. Geology Department, University of Tasmania, Hobart.

Carmichael, I. S. E., 1963, The crystallization of feldspar in volcanic acid liquids. *Geol. Soc. London Quart. J.*, **119**, 85–131.

Carswell, D. A., 1968, Picritic magma—residual dunite relationships in garnet peridotite at Kalskaret near Tafjord, South Norway. *Contr. Mineral. Petrol.*, **19**, 97–124.

———, 1968, Possible primary upper mantle peridotite in Norwegian basal gneiss. *Lithos*, **1**, 322–355.

Carter, J. L., 1966, Chemical composition of primitive upper mantle. Geol. Soc. Amer. Annual Meeting, San Francisco, Program 35–36.

Carter, N. L., and H. G. Ave'Lallemant, 1970, High temperature flow of dunite and peridotite. *Bull. Geol. Soc. Am.*, **81**, 2181–2202.

Chayes, F., 1966, Alkaline and subalkaline basalts. *Amer. J. Sci.*, **264**, 128–145.

Cohen, L. H., and W. Klement, 1967, High-low quartz inversion: determination to 35 kbars. *J. Geophys. Res.*, **72**, 4245–4251.

Cohen, L. H., K. Ito, and G. C. Kennedy, 1967, Melting and phase relations in an anhydrous basalt to 40 kilobars. *Am. J. Sci.*, **265**, 475–518.

Cox, A., 1969, Geomagnetic reversals. *Science*, **163**, 237–245.

———, and R. R. Doell, 1960, Review of paleomagnetism. *Bull. Geol. Soc. Amer.*, **71**, 645–768.

———, 1961, Paleomagnetic evidence relevant

to a change in the Earth's radius. *Nature*, **189**, 45–47.

Clark, S. P., and A. E. Ringwood, 1964, Density distribution and constitution of the mantle. *Rev. Geophys.*, **2**, 35–88.

Clegg, J. A., M. Almond, and P. H. S. Stubbs, 1954, The remanent magnetism of sedimentary rocks in Britain. *Phil. Mag.*, **45**, 583–598.

Creer, K. M., 1967, "A synthesis of world-wide paleomagnetic data," in *Mantles of the Earth and Terrestrial Planets*, ed. S. K. Runcorn. Wiley-Interscience, New York.

Crook, K. A. W., 1969, Contrasts between Atlantic and Pacific geosynclines. *Earth Planet. Sci. Letters*, **5**, 429–438.

Dachille, F., and R. Roy, 1956, System Mg_2SiO_4-Mg_2GeO_4 at 10,000, 60,000, and about 300,000 psi. *Bull. Geol. Soc. Am.* **67**, 1682–1683.

——, 1960, High pressure studies of the system Mg_2GeO_4-Mg_2SiO_4 with special reference to the olivine-spinel transition. *Am. J. Sci.*, **258**, 225–246.

Dagley, P., R. L. Wilson, J. M. Ade-Hall, G. P. L. Walker, S. E. Haggerty, T. Sigurgeirsson, N. D. Watkins, P. J. Smith, J. Edwards, and R. L. Grasty, 1967, Geomagnetic polarity zones for Icelandic lavas. *Nature*. **216**, 25–29.

Dalrymple, G. B., A. Cox, R. R. Doell, and C. S. Grommé, 1967, Pliocene geomagnetic polarity epochs. *Earth Planet. Sci. Letters*, **2**, 163–173.

Dana, J. D., 1873, On some results of the earth's contraction from cooling, including a discussion of the origin of mountains and the nature of the earth's interior. *Am. J. Sci.*, **5**, 423–443; **6**, 6–14, 104–115, 161–171.

Dennis, J. G., and C. T. Walker, 1965, Earthquakes resulting from metastable phase transitions. *Tectonophysics*, **2**, 401–407.

Dewey, J. F., and J. M. Bird, 1970, Mountain belts and the new global tectonics. *J. Geophys. Res.*, **75**, 2625–2647.

Dickcy, J. S., 1970, "Partial fusion products in alpine-type peridotites: Serrania de la Ronda and other examples," in *Fiftieth Anniversary Symposia*, ed. B. A. Morgan. Mineralogical Society of America, Special Paper No. 3.

Dickinson, W. R., 1970, Global tectonics. *Science*, **168**, 1250–1259.

Dickson, G. O., W. C. Pitman, and J. R. Heirtzler, 1968, Magnetic anomalies in the South Atlantic and ocean floor spreading. *J. Geophys. Res.*, **73**, 2087–2100.

Dietz, R. S., 1961, Continent and ocean basin evolution by spreading of the sea floor. *Nature*, **190**, 854–857.

——, 1963, Alpine serpentines as oceanic rind fragments. *Geol. Soc. Am. Bull.*, **74**, 947–952.

——, 1963, Collapsing continental rises: an actualistic concept of geosynclines and mountain building. *J. Geol.*, **71**, 314–333.

——, and J. C. Holden, 1970, Reconstruction of Pangaea: breakup and dispersion of continents, Permian to present. *J. Geophys. Res.*, **75**, 4939–4956.

——, 1970, The breakup of Pangaea. *Sci. Am.*, **223**, 30–41.

Dietz, R. S., and W. P. Sproll, 1970, Fit between Africa and Antarctica: a continental drift reconstruction. *Science*, **167**, 1612–1614.

Doig, R., 1969, An alkaline rock province linking Europe and North America. Programme and abstracts for joint Annual Meeting Geological Society of Canada and Mineralogical Association of Canada, Montreal, June 1969.

Dorman, J., 1969, "Seismic surface-wave data on the upper mantle", in *The Earth's Crust and Upper Mantle*, ed. P. J. Hart. Geophysical Monograph 13, American Geophysical Union, Washington, D.C.

Drake, C. L., and J. E. Nafe, 1968, "The transition from ocean to continent from seismic refraction data", in *The Crust and Upper Mantle of the Pacific Area*, eds. L. Knopoff, C. L. Drake, and P. J. Hart. Geophysical Monograph 12, American Geophysical Union, Washington, D.C.

Drake, C. L., J. I. Ewing, and H. P. Stockard, 1968, The continental margin of the eastern United States. *Can. J. Earth Sci.*, **5**, 993–1010.

Drever, H. I., and R. Johnston, 1967, Picritic minor intrusions, 71-82, in *Ultramafic and Related Rocks*, ed. P. J. Wyllie, John Wiley and Sons, New York.

DuToit, A., 1937, *Our Wandering Continents*, Oliver and Boyd, Edinburgh.

Egyed, L., 1957, A new dynamic conception of the internal constitution of the Earth. *Geol. Rundschau*, **46**, 101–121.

Emilia, D. A., and D. F. Heinrichs, 1969, Ocean floor spreading: Olduvai and Gilsa events in the Matuyama epoch. *Science*, **166**, 1267–1269.

Engel, A. E. J., 1963, Geologic evolution of North America. *Science*, **140**, 143–152.

——, C. G. Engel, and R. G. Havens, 1965, Chemical characteristics of oceanic basalts and the upper mantle. *Geol. Soc. Am. Bull.*, **76**, 719–734.

Ericson, D. B., and G. Wollin, 1968, Pleistocene climates and chronology in deep-sea sediments. *Science*, **162**, 1227–1234.

Ernst, W. G., 1970, Tectonic contact between the Franciscan mélange and the Great Valley sequence—crustal expression of a late Mesozoic Benioff zone. *J. Geophys. Res.*, **75**, 886–901.

Eskola, P., 1915, On the relations between the chemical and mineralogical composition in the metamorphic rocks of the Orijarvi region. *Bull. Comm. Geol. Finlande*, **44**, 109–145.

Evans, B. W., 1965, Application of a reaction rate method to the breakdown equilibria of muscovite and muscovite plus quartz. *Am. J. Sci.*, **263**, 647–667.

Ewing, J., and M. Ewing, 1967, Sediment distribution on the mid-ocean ridges with respect to spreading of the sea floor. *Science*, **156**, 1590–1592.

——, T. Aitken, and W. J. Ludwig, 1968, "North Pacific sediment layers measured by seismic profiling," in *The Crust and Upper Mantle of the Pacific Area*, eds. L. Knopoff, C. L. Drake, and P. J. Hart. Geophysical Monograph 12, American Geophysical Union, Washington, D.C.

Ewing, M., and F. Press, 1955, "Geophysical contrasts between continents and ocean basins," in *Crust of the Earth*, ed. A. Poldervaart. The Geological Society of America, Special Paper 62.

Fermor, L. L., 1913, Preliminary note on garnet as a geological barometer and on an infraplutonic zone in the earth's crust. *Geol. Survey India*, **43**, Part 1.

Fischer, A. G., B. C. Heezen, R. E. Boyce, D. Bukry, R. G. Douglas, R. E. Garrison, S. A. Kling, V. Krasheninnikov, and A. C. Pimm, 1970, Geological history of the Western North Pacific. *Science*, **6**, 1210–1214.

Fisher, D. E., O. Joensuu, and K. Boström, 1969, Elemental abundances in ultramafic rocks and their relation to the upper mantle. *J. Geophys. Res.*, **74**, 3865–3873.

Fisher, O., 1889, *Physics of the Earth's Crust*, 2nd ed., Macmillan, London.

Fitch, T. J., and P. Molnar, 1970, Focal mechanisms along inclined earthquake zones in the Indonesia-Philippine region. *J. Geophys. Res.*, **75**, 1431–1444.

Foster, J. H., and N. D. Opdyke, 1970, Upper Miocene to Recent magnetic stratigraphy in deep-sea sediments. *J. Geophys. Res.*, **75**, 4465–4473.

Fujisawa, H., 1968, Temperature and discontinuities in the transition layer within the Earth's mantle: geophysical application of the olivine-spinel transition in the Mg_2SiO_4-Fe_2SiO_4 system. *J. Geophys. Res.*, **73**, 3281–3294.

Fyfe, W. S., 1960, Hydrothermal synthesis and determination of equilibrium between minerals in the subsolidus region. *J. Geol.*, **68**, 553–566.

Gast, P. W., 1968, Trace element fractionation and the origin of tholeiitic and alkaline magma types. *Geochim. Cosmochim. Acta*, **42**, 1057–1068.

Gilluly, J., 1969, Oceanic sediment volumes and continental drift. *Science*, **166**, 992–993.

Girdler, R. W., 1967, "A review of terrestrial heat flow," in *Mantles of the Earth and Terrestrial Planets*, ed. S. K. Runcorn. Wiley-Interscience, New York.

Goles, G. G., 1969, "Cosmic abundances," in *Handbook of Geochemistry*, Vol. 1., ed. K. H. Wedepohl. Springer-Verlag, Heidelberg.

Green, D. H., 1964, The petrogenesis of the high temperature peridotite intrusion in the Lizard area, Cornwall. *J. Petrol.*, **5**, 134–188.

——, 1967, "Effects of high pressure on basaltic rock," in *Basalts: The Poldervaart Treatise on Rocks of Basaltic Composition*, Vol. 1, eds. H. H. Hess and A. Poldervaart. Wiley-Interscience, New York.

——, 1969, The origin of basaltic and nephelinitic magmas in the earth's mantle. *Tectonophysics*, **7**, 409–422.

——, 1971, Compositions of basaltic magmas as indicators of conditions of origin: application to oceanic volcanism. *Phil. Trans. Roy. Soc. London*, Ser. A, **268**, 707–725.

——, and A. E. Ringwood, 1963, Mineral assemblages in a model mantle composition. *J. Geophys. Res.*, **68**, 937–945.

——, 1964, Fractionation of basalt magmas at high pressures. *Nature*, **201**, 1276–1279.

——, 1967, The stability fields of aluminous pyroxene peridotite and garnet peridotite and their relevance in upper mantle structure. *Earth Planet. Sci. Letters*, **3**, 151–160.

——, 1967, The genesis of basaltic magmas. *Contr. Mineral. Petrology*, **15**, 103–190.

————, 1967, An experimental investigation of the gabbro to eclogite transformation and its petrological applications. *Geochim. Cosmochim. Acta*, **31**, 767–833.

————, 1968, Genesis of the calc-alkaline igneous rock suite. *Contr. Mineral. Petrol.*, **18**, 105–162.

Grommé, C. S., R. T. Merrill, and J. Verhoogen, 1967, Paleomagnetism of the Sierra Nevada, California, and its significance for polar wandering and continental drift. *J. Geophys. Res.*, **72**, 5661–5684.

Guier, W. H., and R. R. Newton, 1965, The earth's gravity field as deduced from the Doppler tracking of five satellites. *J. Geophys. Res.*, **70**, 4613–4626.

Gutenberg, B., 1926, Untersuchungen zur Frage, bis zu welcher tiefe die erde Kristallin ist. *Zeit. Geophysik.*, **2**, 24–29.

Hales, A. L., 1969, Gravitational sliding and continental drift. *Earth Planet. Sci. Letters*, **6**, 31–34.

Hall, J., 1859, Description and figures of the organic remains of the lower Helderberg Group and the Oriskany Sandstone. Natural History of New York; paleontology. *Geol. Surv.*, **3**, Albany, New York.

Hamilton, W., 1969, Mesozoic California and the underflow of Pacific mantle. *Geol. Soc. Am. Bull.*, **80**, 2409–2430.

————, 1970, The Uralides and the motion of the Russian and Siberian platforms. *Bull. Geol. Soc. Am.*, **81**, 2553–2576.

Harris, P. G., and J. A. Rowell, 1960, Some geochemical aspects of the Mohorovicic Discontinuity. *J. Geophys. Res.*, **65**, 2443–2459.

Harris, P. G., A. Reay, and I. G. White, 1967, Chemical composition of the upper mantle. *J. Geophys. Res.*, **72**, 6359–6369.

Harrison, C. G. A., and B. M. Funnell, 1964, Relationship of paleomagnetic reversals and micropaleontology in two late Caenozoic cores from the Pacific Ocean. *Nature*, **204**, 566.

Hatherton, T., and W. R. Dickinson, 1969, The relationship between andesitic volcanism and seismicity in Indonesia, the Lesser Antilles, and other island arcs. *J. Geophys. Res.*, **74**, 5301–5310.

Haug, E., 1900, Les géosynclinause et les aires continentales. Contribution à l'étude des régressions et des trangressions marines. *Bull. Soc. Geol. France*, **28**, 617–711.

Heezen, B. C., M. Tharp, and M. Ewing, 1959, *The Floors of the Oceans. I. The North Atlantic.* The Geological Society of America, Special Paper 65.

Heirtzler, J. R., X. Le Pichon, and J. G. Baron, 1966, Magnetic anomalies over the Reykjanes Ridge. *Deep-Sea Res.*, **13**, 427–443.

Heirtzler, J. R., G. O. Dickson, E. J. Herron, W. C. Pitman, and X. Le Pichon, 1968, Marine magnetic anomalies, geomagnetic field reversals, and motions of the ocean floor and continents. *J. Geophys. Res.*, **73**, 2119–2136.

Herrin, E., 1969, "Regional variation of P-wave velocity in the upper mantle beneath North America," in *The Earth's Crust and Upper Mantle*, ed. P. J. Hart. Geophysical Monograph 13, American Geophysical Union, Washington, D.C.

Herz, N., 1969, Anorthosite belts, continental drift, and the anorthosite event. *Science*, **164**, 944–947.

Hess, H. H., 1955, "Serpentines, Orogeny, and Epeirogeny," in *Crust of the Earth*, Geological Society of America, Special Paper 62, 391–408.

————, 1962, "History of ocean basins," in *Petrologic Studies: A volume to Honor A.F. Buddington*, eds. A. E. J. Engel, H. L. James, and B. F. Leonard. Geological Society of America, New York.

————, 1964, "The oceanic crust, the upper mantle and the Mayaguez serpentinized peridotite," in *A Study of the Serpentinite*, ed. C. A. Burk. National Academy of Science-National Research Council Publication 1188.

————, 1964, The oceanic crust, the upper mantle and the Mayaguez serpentinized peridotite, in *A Study of the Serpentinite*, ed. C. A. Burk. National Academy of Science-National Research Council Publication 1188.

————, and G. Otalora, 1964, "Mineralogical and chemical composition of the Mayaguez serpentinite cores," in *A Study of Serpentinite*, ed. C. A. Burk. National Academy of Science-National Research Council Publication 1188.

Hibberd, F. H., 1962, An analysis of the positions of the Earth's magnetic pole in the geological past. *Geophys. J.*, **6**, 221–244.

Hilgenberg, O. C., 1962, Rock magnetism and the Earth's paleopoles. *Geofis. Pura. Appl.*, **53**, 52–54.

Hill, R. E. T., and A. L. Boettcher, 1970, Water in the earth's mantle: melting curves of basalt-water and basalt-water-carbon dioxide. *Science*, **167**, 980–981.

Holloway, J. R., and C. W. Burnham, 1969, Phase relations and compositions in basalt-H_2O-CO_2 under the Ni-NiO buffer at high temperatures and pressures. Abstracts with Programs for 1969, Part 7, The Geological Society of America, 104–105.

Holmes, A., 1931, Radioactivity and earth movements. *Trans. Geol. Soc. Glasgow*, **18**, 559–606.

———, 1965, *Principles of Physical Geology*, 2nd ed., Ronald, New York.

Hospers, J., 1967, "Review of paleomagnetic evidence for the displacement of continents, with particular reference to North America and Europe-northern Asia," in *Mantles of the Earth and Terrestrial Planets*, ed. S. K. Runcorn. Wiley-Interscience, New York.

Hsu, L. C., 1968, Selected phase relationships in the system Al-Mn-Fe-Si-O-H: a model for garnet equilibria. *J. Petrol.*, **9**, 40–83.

Hurley, P. M., and J. R. Rand, 1969, Pre-drift continental nuclei. *Science*, **164**, 1229–1242.

Hurley, P. M., F. F. M. de Almeida, G. C. Melcher, V. G. Cordani, J. R. Rand, K. Kawashita, P. Vandoros, W. H. Pinson, and H. W. Fairbairn, 1967, Test of continental drift by comparison of radiometric ages. *Science*, **157**, 495–500.

Hyndman, R. D., I. B. Lambert, K. S. Heier, J. C. Jaeger, and A. E. Ringwood, 1968, Heat flow and surface radioactivity measurements in the Precambrian shield of Western Australia. *Phys. Earth Planet. Interiors*, **1**, 129–135.

Irvine, T. N., 1966, ed. *The World Rift System*, Geological Survey Canada Paper 66-14, Queen's Printer, Ottawa.

Irving, E., 1964, *Paleomagnetism*, Wiley, New York.

Isacks, B., J. Oliver, and L. R. Sykes, 1968, Seismology and the new global tectonics. *J. Geophys. Res.*, **73**, 5855–5899.

———, 1969, *J. Geophys. Res.*, **74**, 2789–2790.

Ito, K., and G. C. Kennedy, 1967, Melting and phase relations in a natural peridotite to 40 kilobars. *Am. Jour. Sci.*, **265**, 519–538.

———, 1968, Melting and phase relations in the plane tholeiite-lherzolite-nepheline basanite to 40 kilobars with geological implications. *Contr. Mineral. Petrol.*, **19**, 177–211.

———, 1970, "The fine structure of the basalt-eclogite transition," in *Fiftieth Anniversary Symposia*, ed. B. A. Morgan. Mineralogical Society of America, Special Paper 3.

Jackson, E. D., 1969, Discussion on the paper "The origin of ultramafic and ultrabasic rocks" by P. J. Wyllie. *Tectonophysics*, **7**, 517–518.

Jacobs, J. A., R. D. Russell, and J. T. Wilson, 1959, *Physics and Geology*, McGraw-Hill, New York.

Jeffreys, H., 1936, The structure of the earth down to the 20° discontinuity. *Monthly Notices Roy. Astron. Soc. Geophys. Supplement*, **3**, 401–422.

Johannes, W., 1968, Experimental investigation of the reaction forsterite + H_2O = serpentine + brucite. *Contr. Mineral. Petrol.*, **19**, 309–315.

Joyner, W. B., 1967, Basalt-eclogite transition as a cause for subsidence and uplift. *J. Geophys. Res.*, **72**, 4977–4998.

Karig, D. E., 1970, Ridges and basins of the Tonga-Kermadec island arc system. *J. Geophys. Res.*, **75**, 239–254.

Kaula, W. M., 1968, *An Introduction to Planetary Physics*. Wiley, New York.

Kay, M., 1951, North American geosynclines. *Geol. Soc. Am. Memoir* **48**.

Kay, R., N. J. Hubbard, and P. W. Gast, 1970, Chemical characteristics and origin of oceanic ridge volcanic rocks. *J. Geophys. Res.*, **75**, 1585–1613.

Keays, R. R., R. Ganapathy, J. C. Laul, E. Anders, G. F. Herzog, and P. M. Jeffery, 1970, Trace elements and radioactivity in lunar rocks: implications for meteorite infall, solar-wind flux, and formation conditions of moon. *Science*, **167**, 490–493.

Kennedy, G. C., 1959, The origin of continents, mountain ranges, and ocean basins. *Am. Sci.*, **47**, 491–504.

———, G. J. Wasserburg, H. C. Heard, and R. C. Newton, 1962, The upper three-phase region in the system SiO_2-H_2O. *Am. J. Sci.*, **260**, 501–521.

Kennedy, W. Q., 1933, Trends of differentiation in basaltic magmas. *Am. J. Sci.*, **25**, 239–256.

———, and E. M. Anderson, 1938, Crustal layers and the origin of magmas. *Bull. Volcanol.*, **2-3**, 23–82.

Kitahara, S., S. Takenouchi, and G. C. Kennedy, 1966, Phase relations in the system MgO-SiO_2-H_2O at high temperatures and pressures. *Am. J. Sci.*, **264**, 223–233.

Knopoff, L., 1967, Thermal convection in the earth's mantle, in *The Earth's Mantle*, ed. T. F. Gaskell. Academic, New York.

Kornprobst, J., 1969, Le massif ultrabasique des Beni Bouchera (Rif Interne, Maroc): Etude des péridotites de haute température et de haute pression, et des pyroxénolites, à grenat ou sans grenat, qui leur sont associées. *Contr. Mineral. Petrol.*, **23**, 283–322.

Kosminskaya, I. P., and S. M. Zverev, 1968, "Deep seismic soundings in the transition zones from continents to oceans," in *The Crust and Upper Mantle of the Pacific Area*, eds. L. Knopoff, C. L. Drake, and P. J. Hart. Geophysical Monograph 12, American Geophysical Union, Washington, D.C.

Kosminskaya, I. P., N. A. Belyaevsky, and I. S. Volvovsky, 1969, "Explosion seismology in the USSR," in *The Earth's Crust and Upper Mantle*, ed. P. J. Hart. Geophysical Monograph 13, American Geophysical Union, Washington, D.C.

Krause, D. C., 1966, "Equatorial shear zone," in *The World Rift System*, ed. T. N. Irvine. Geology Survey of Canada, Paper 66-14, Queen's Printer, Ottawa.

Krumbein, W. C., and L. L. Sloss, 1963, *Stratigraphy and Sedimentation*, 2nd ed. Freeman, San Francisco.

Kuno, H., 1959, Origin of Cenozoic petrographic provinces of Japan and surrounding areas. *Bull. Volcanol.* **2-20**, 37–76.

Kushiro, I., 1968, Compositions of magmas formed by partial zone melting of the Earth's upper mantle. *J. Geophys. Res.*, **73**, 619–634.

———, 1969, "Clinopyroxene Solid Solutions Formed by Reactions Between Diopside and Plagioclase at High Pressures" in Mineralogical Society of America, Special Paper No. 2.

———, 1969, The system forsterite-diopside-silica with and without water at high pressures. *Amer. J. Sci.*, **267A**, 269–294.

———, 1969, Discussion of the paper "The origin of basaltic and nephelinitic magmas in the earth's mantle" by D. H. Green. *Tectonophysics*, **7**, 427–436.

———, 1970, Systems bearing on melting of the upper mantle under hydrous conditions. Carnegie Inst. Wash. Yearbook, **68**, 240–245.

———, 1970, Stability of amphibole and phlogopite in the upper mantle. Carnegie Inst. Wash. Yearbook, **68**, 245–247.

———, and H. S. Yoder, 1966, Anorthite-forsterite and anorthite-enstatite reactions and their bearing on the basalt-eclogite transformation. *J. Petrology*, **7**, 337–362.

Kushiro, I., Y. Syono, and S. Akimoto, 1968, Melting of a peridotite nodule at high pressures and high water pressures. *J. Geophys. Res.*, **73**, 6023–6029.

Lachenbruch, A. H., 1970, Crustal temperature and heat production: implications of the linear heat-flow relation. *J. Geophys. Res.*, **75**, 3291–3300.

Lambert, I. B., and K. S. Heier, 1967, The vertical distribution of uranium, thorium, and potassium in the continental crust. *Geochim. Cosmochim. Acta*, **31**, 377–390.

Lambert, I. B., and P. J. Wyllie, 1968, Stability of hornblende and a model for the low velocity zone. *Nature*, **219**, 1240–1241.

———, 1970, Melting in the deep crust and upper mantle and the nature of the low velocity layer. *Phys. Earth Planet. Interiors*, **3**, 316–322.

———, 1970, Low-velocity zone of the earth's mantle: incipient melting caused by water. *Science*, **169**, 764–766.

Lambert, I. B., J. K. Robertson, and P. J. Wyllie, 1969, Melting reactions in the system $KAlSi_3O_8$-SiO_2-H_2O to 18.5 kilobars. *Am. J. Sci.*, **267**, 609–626.

Larimer, J. W., and E. Anders, 1967, Chemical fractionations in meteorites—II. Abundance patterns and their interpretation. *Geochim. Cosmochim. Acta*, **31**, 1239–1270.

Larson, R. L., and F. N. Spiess, 1969, East Pacific Rise crest: a near-bottom geophysical profile. *Science*, **163**, 68–71.

Lee, W. H. K., and S. Uyeda, 1965, "Review of heat flow data," in *Terrestrial Heat Flow*, ed. W. H. K. Lee. Geophysical Monograph 8, American Geophysical Union, Washington D.C.

Le Pichon, X., 1968, Sea-floor spreading and continental drift. *J. Geophys. Res.*, **73**, 3611–3697.

———, and J. R. Heirtzler, 1968, Magnetic anomalies in the Indian Ocean and sea-floor spreading. *J. Geophys. Res.*, **73**, 2101–2117.

Lliboutry, L., 1969, Sea-floor spreading, continental drift and lithosphere sinking with an asthenosphere at melting point. *J. Geophys. Res.*, **74**, 6525–6540.

Lovering, J. F., 1958, The nature of the Mohorovicic Discontinuity. *Trans. Am. Geophys. Un.*, **39**, 947–955.

———, 1962, "The evolution of the meteorites—evidence for the coexistence of chondritic, achondritic, and iron meteorites in a typical parent meteorite body," in *Researches on Meteorites*, ed. C. B. Moore, 179–197.

Lubimova, E. A.,1967,"Theory of thermal state of the earth's mantle," in *The Earth's Mantle*, ed. T. F. Gaskell. Academic, New York.

Luyendyk, B. P., 1969, Origin of short-wavelength magnetic lineations observed near the ocean bottom. *J. Geophys. Res.*, **74**, 4869–4881.

———, 1970, Reply, *J. Geophys. Res.*, **75**, 6721–6722.

———, J. D. Mudie, and C. G. A. Harrison, 1968, Lineations of magnetic anomalies in the northeast Pacific observed near the ocean floor. *J. Geophys. Res.*, **73**, 5951–5957.

Lyell, C., 1833, *Principles of Geology*, Murray, London.

McBirney, A. R., 1969,"Compositional variations in Cenozoic calc-alkaline suites of Central America," in *Oregon Dep. Geol. Mineral Ind. Bull.*, **65**, 185–189.

McElhinny, M. W., and G. R. Luck, 1970, Paleomagnetism and Gondwanaland. *Science*, **168**, 830–832.

McKenzie, D. P., 1967, Some remarks on heat flow and gravity anomalies. *J. Geophys. Res.*, **72**, 6261–6273.

———, 1969, Speculations on the consequences and causes of plate motions. *Geophys. J.*, **18**, 1–32.

———, and W. J. Morgan, 1969, Evolution of triple junctions. *Nature*, **224**, 125–133.

McKenzie, D. P. and R. L. Parker, 1967, The North Pacific: an example of tectonics on a sphere. *Nature*, **216**, 1276–1280.

MacDonald, G. J. F., 1963, The deep structure of continents. *Rev. Geophys.*, **1**, 587–665.

———, and N. F. Ness, 1960, Stability of phase transitions within the earth. *J. Geophys. Res.*, **65**, 2173–2190.

MacGregor, I. D., 1964, The reaction 4 Enstatite + spinel = forsterite + pyrope. *Carnegie Inst. Wash. Yearbook*, **63**, 157.

Malahoff, A., 1970, Some possible mechanisms for gravity and thrust faults under oceanic trenches. *J. Geophys. Res.*, **75**, 1992–2001.

Markhinin, E. K., 1968, "Volcanism as an agent of formation of the Earth's crust," in *The Crust and Upper Mantle of the Pacific Area*, eds. L. Knopoff, C. L. Drake, and P. J. Hart. Geophysical Monograph 12, American Geophysical Union, Washington, D.C.

Mason, B., 1965, The chemical composition of olivine-bronzite and olivine-hypersthene chondrites. *Am. Museum Novitates*, 2223, 1–38.

———, 1966, Composition of the earth. *Nature*, **211**, 616–618.

Mason, R. G., 1958, A magnetic survey off the west coast of the United States between latitudes 32° and 36°N, longitudes 121° and 128°W. *Geophys. J.*, **1**, 320–329.

———, and A. D. Raff, 1961, Magnetic survey off the west coast of North America, 32°N latitude to 42°N latitude. *Bull. Geol. Soc. Am.*, **72**, 1259–1266.

Matsumoto, T., 1967, Fundamental problems in the circum-Pacific orogenesis. *Tectonophysics*, **4**, 595–613.

Maxwell, A. E., R. P. Von Herzen, K. J. Hsu, J. E. Andrews, T. Saito, S. F. Percival, E. D. Milow, and R. E. Boyce, 1970, Deep-sea drilling in the South Atlantic. *Science*, **168**, 1047–1059.

Maxwell, J. C., 1968, Continental drift and a dynamic earth. *Am. Sci.*, **56**, 35–51.

———, 1970, "The Mediterranean, ophiolites, and continental drift," in *What's New on Earth*. Rutgers University Press, New Jersey.

Maynard, G. L., 1970, Crustal layer of seismic velocity 6.9 to 7.6 kilometers per second under the deep oceans. *Science*, **168**, 120–121.

Melson, W. G., E. Jarosewich, V. T. Bowen, and G. Thompson, 1967, St. Peter and St. Paul Rocks: a high-temperature, mantle-derived intrusion. *Science*, **155**, 1532–1535.

Menard, H. W., 1964, *Marine Geology of the Pacific*, McGraw-Hill, New York.

———, 1967, Sea-floor spreading, topography, and the second layer. *Science*, **157**, 923–924.

———, 1967, Transitional types of crust under small ocean basins. *J. Geophys. Res.*, **72**, 3061–3073.

———, 1968, "Some remaining problems in sea-floor spreading," in *The History of the Earth's Crust*, ed. R. A. Phinney. Princeton University Press, New Jersey.

———, and T. Atwater, 1968, Changes in direction of sea-floor spreading. *Nature*, **219**, 463–467.

Menard, H. W., and S. M. Smith, 1966, Hypsometry of ocean basin provinces. *J. Geophys. Res.*, **71**, 4305–4325.

Merrill, R. B., J. K. Robertson, and P. J. Wyllie, 1970, Melting reactions in the system $NaAlSi_3O_8$-$KAlSi_3O_8$-SiO_2-H_2O to 20 kilobars compared with results for other feldspar-quartz-H_2O and rock-H_2O systems. *J. Geol.*, **78**, 558–569.

Meservey, R., 1969, Topological inconsistency of continental drift on the present-sized earth. *Science*, **166**, 609–611.

Meyerhoff, A. A., 1970, Continental drift: implications of paleomagnetic studies, meteorology, physical oceanography and climatology. *J. Geol.*, **78**, 1–51.

———, 1970, Continental drift, II: high-latitude evaporite deposits and geologic history of Arctic and North Atlantic Oceans. *J. Geol.*, **78**, 406–444.

———, and C. Teichert, 1971, Continental drift, III: late Paleozoic glacial centers, and Devonian-Eocene coal distribution. *J. Geol.*, **79**, in press.

Millhollen, G. L., 1971, Melting of nepheline syenite with H_2O and $H_2O + CO_2$, and the effect of dilution of the aqueous phase on the beginning of melting. *Am. J. Sci.*, **269**, in press.

Minear, J. W., and M. N. Toksöz, 1970, Thermal regime of a downgoing slab and new global tectonics. *J. Geophys. Res.*, **75**, 1397–1419.

Mitchell, A. H., and H. G. Reading, 1969, Continental margins, geosynclines, and ocean floor spreading. *J. Geol.*, **77**, 629–646.

Miyashiro, A., 1961, Evolution of metamorphic belts. *J. Petrol.*, **2**, 277–311.

———, F. Shido, and M. Ewing, 1969, Composition and origin of serpentinites from the mid-Atlantic ridge near 24° and 30° North latitude. *Contr. Mineral. Petrol.*, **23**, 117–127.

Molnar, P., and J. Oliver, 1969, Lateral variations of attenuation in the upper mantle and discontinuities in the lithosphere. *J. Geophys. Res.*, **74**, 2648–2682.

Molnar, P., and L. R. Sykes, 1969, Tectonics of the Caribbean and Middle America regions from focal mechanisms and seismicity. *Geol. Soc. Am. Bull.*, **80**, 1639–1684.

Morgan, W. J., 1968, Rises, trenches, great faults, and crustal blocks. *J. Geophys. Res.*, **73**, 1959–1982.

Morley, L. W., and A. Larochelle, 1964, "Paleomagnetism as a means of dating geological events," in *Geochronology in Canada*, ed. F. F. Osborne. The Royal Society of Canada Special Publications, 8. University of Toronto Press, Toronto.

Muehlberger, W. R., R. E. Denison, and E. G. Lidiak, 1967, Basement rocks in continental interior of United States. *Amer. Assoc. Petroleum Geol. Bull.*, **51**, 2351–2380.

Nagata, T.. 1952, Reverse thermo remanent magnetism. *Nature*, **169**, 704–705.

Newton, R. C., 1966, Some calc-silicate equilibrium relations. *Am. J. Sci.*, **264**, 204–222.

Nicholls, G. D., 1967, "Geochemical studies in the ocean as evidence for the composition of the mantle," in *Mantles of the Earth and Terrestrial Planets*, ed. S. K. Runcorn. Wiley-Interscience, New York.

Ninkovich, D., N. Opdyke, B. C. Heezen, and J. H. Foster, 1966, Paleomagnetic stratigraphy, rates of deposition and tephrochronology in North Pacific deep-sea sediments. *Earth Planet. Sci. Letters*, **1**, 476–492.

O'Connell, R. J., 1968, Critique of a paper by W. J. van de Lindt, "Movement of the Mohorovicic discontinuity under isostatic conditions." *J. Geophys. Res.*, **73**, 6604–6066.

———, and G. J. Wasserburg, 1967, Dynamics of the motion of a phase change boundary to changes in pressure. *Rev. Geophys.*, **5**, 329–410.

O'Hara, M. J., 1965, Primary magmas and the origin of basalts. *Scottish J. Geol.*, **1**, 19–40.

———, 1968, The bearing of phase equilibria studies in synthetic and natural systems on the origin and evolution of basic and ultrabasic rocks. *Earth Sci. Rev.*, **4**, 69–133.

———, and H. S. Yoder, 1967, Formation and fractionation of basic magmas at high pressures. *Scottish J. Geol.*, **3**, 67–117.

Opdyke, N. D., B. Glass, J. D. Hayes, and J. Foster, 1966, Paleomagnetic study of Antarctic deep-sea cores. *Science*, **154**, 349–357.

Osborn, E. F., 1959, Role of oxygen pressure in the crystallization and differentiation of basaltic magma. *Am. J. Sci.*, **257**, 609–647.

———, 1969, The complementariness of orogenic andesite and alpine peridotite. *Geochim. Cosmochim. Acta*, **33**, 307–324.

Oxburgh, E. R., and D. L. Turcotte, 1968, Mid-ocean ridges and geotherm distribution during mantle convection. *J. Geophys. Res.*, **73**, 2643–2661.

———, 1970, Thermal structure of island arcs. *Geol. Soc. Am. Bull.*, **81**, 1665–1688.

Ozima, M., M. Ozima, and I. Kaneoka, 1968, Potassium-argon ages and magnetic properties of some dredged submarine basalts and their geophysical implications. *J. Geophys. Res.*, **73**, 711–723.

Pakiser, L. C., 1965, The basalt-eclogite transformation and crustal structure in the western United States. *U.S. Geol. Survey Prof. Paper*, **525-B**, B1–B8.

———, and R. Robinson, 1966, "Composition of the continental crust as estimated from seismic observations," in *The Earth Beneath the Continents*, eds. J. S. Steinhart and T. J. Smith. Geophysical Monograph 10, American Geophysical Union, Washington, D.C.

Pakiser, L. C., and I. Zietz, 1965, Transcontinental crustal and upper-mantle structure. *Rev. Geophys.*, **3**, 505–520.

Peter, G., 1970, Discussion of paper by B. P. Luyendyk, *J. Geophys. Res.*, **75**, 6717–6720.

———, and R. Lattimore, 1969, Magnetic structure of the Juan de Fuca-Gorda Ridge area. *J. Geophys. Res.*, **74**, 586–593.

Peter, G., B. H. Erickson, and P. J. Grim, 1970, "Magnetic structure of the Aleutian trench and northeast Pacific basin," in *The Sea*, Vol. 4, eds. A. E. Maxwell, E. C. Bullard, E. Goldberg, and J. L. Worzel, Wiley-Interscience, New York.

Phillips, J. D., and B. P. Luyendyk, 1970, Central North Atlantic plate motions over the last 40 million years. *Science*, **170**, 727–729.

Phinney, R. A., 1968, ed., *The History of the Earth's Crust*. Princeton University Press, New Jersey.

Pitman, W. C., and J. R. Heirtzler, 1966, Magnetic anomalies over the Pacific-Antarctic ridge. *Science*, **154**, 1164–1171.

Pitman, W. C., E. M. Herron, and J. R. Heirtzler, 1968, Magnetic anomalies and sea-floor spreading. *J. Geophys. Res.*, **73**, 2069–2085.

Piwinskii, A. J., 1968, Experimental studies of igneous rock series: central Sierra Nevada Batholith, California. *J. Geol.*, **76**, 548–570.

———, and P. J. Wyllie, 1968, Experimental studies of igneous rock series: a zoned pluton in the Wallowa Batholith, Oregon. *J. Geol.*, **76**, 205–234.

Poldervaart, A., 1955, "Chemistry of the Earth's Crust," in *Crust of the Earth*, ed. A. Poldervaart, Geological Society of America, Special Paper 62.

Press, F., 1968, Density distribution in the earth. *Science*, **160**, 1218–1221.

———, 1968, Earth models obtained by Monte Carlo inversion. *J. Geophys. Res.*, **73**, 5223–5234.

———, 1969, The suboceanic mantle. *Science*, **165**, 174–176.

Raff, A. D., 1968, Sea-floor spreading—another rift. *J. Geophys. Res.*, **73**, 3699–3705.

———, and R. G. Mason, 1961, Magnetic survey off the west coast of North America, 40°N latitude to 52°N latitude. *Bull. Geol. Soc. Am.*, **72**, 1267–1270.

Ramberg, H., 1967, *Gravity, Deformation and the Earth's Crust*. Academic, London.

Rezanov, I. A., 1968, Paleomagnetism and continental drift. *Intern. Geol. Rev.*, **10**, 765–776.

Richter, C. F., 1969, Comments on paper by B. Isacks, J. Oliver, and L. R. Sykes, "Seismology and the new global tectonics," *J. Geophys. Res.*, **74**, 2786–2788.

Ringwood, A. E., 1958, The constitution of the mantle—I. Thermodynamics of the olivine-spinel transition. *Geochim. Cosmochim. Acta*, **13**, 303–321.

———, 1958, The constitution of the mantle—II. Further data on the olivine-spinel transition. *Geochim. Cosmochim. Acta*, **15**, 19–28.

———, 1958, The constitution of the mantle—III. Consequences of the olivine-spinel transition. *Geochim. Cosmochim. Acta*, **15**, 195–212.

———, 1961, Chemical and genetic relationships among meteorites. *Geochim. Cosmochim. Acta*, **24**, 159–197.

———, 1962, A model for the upper mantle. *J. Geophys. Res.*, **67**, 857–868.

———, 1966, Genesis of chondritic meteorites. *Rev. Geophys.*, **4**, 113–175.

———, 1966, "The chemical composition and origin of the earth," in *Advances in Earth Sciences*, ed. P. M. Hurley. M.I.T. Press, Cambridge, Mass.

———, 1966, "Mineralogy of the mantle," in *Advances in Earth Sciences*, ed. P. M. Hurley. M.I.T. Press, Cambridge, Mass.

———, 1969, "Composition and evolution of the upper mantle," in *The Earth's Crust and Upper Mantle*, ed. P. J. Hart. Geophysics Monograph 13, American Geophysical Union, Washington, D.C.

———, 1969, Phase transformations in the mantle. *Earth Planet. Sci. Letters*, **5**, 401–412.

———, 1970, Phase transformations and the constitution of the mantle. *Phys. Earth Planet. Interiors*, **3**, 109–155.

———, and D. H. Green, 1964, Experimental investigations bearing on the nature of the

Mohorovicic discontinuity. *Nature*, **201**, 566–567.

———, 1966, An experimental investigation of the gabbro-eclogite transformation and some geophysical implications. *Tectonophysics*, **3**, 383–427.

———, 1966, "Petrological nature of the stable continental crust," in *The Earth Beneath the Continents*, eds. J. S. Steinhart and T. J. Smith. Geophysics Monograph 10, American Geophysical Union, Washington, D.C.

Ringwood, A. E., and A. Major, 1970, The system Mg_2SiO_4-Fe_2SiO_4 at high pressures and temperatures. *Phys. Earth Planet. Interiors*, **3**, 89–108.

Ringwood, A. E., and M. Seabrook, 1962, Olivine-spinel equilibria at high pressure in the system Ni_2GeO_4-Mg_2SiO_4. *J. Geophys. Res.*, **67**, 1975 1985.

Rittmann, A., 1962, *Volcanoes and Their Activity*. Wiley, New York.

Roberts, R. J., 1969, The cordilleran continental margin—continental collisions vs. geotectonic cycles. Abstracts with programs for 1969, Part 7, 286 288. The Geological Society of America.

Robertson, E. C., F. Birch, and G. J. F. MacDonald, 1957, Experimental determination of jadeite stability relations to 25,000 bars. *Am. J. Sci.*, **255**, 115–137.

Robertson, J. K., and P. J. Wyllie, 1971, Rock-water systems, with special reference to the water-deficient region. *Am. J. Sci.*, in press.

Ronov, A. B., and A. A. Yaroshevsky, 1969, "Chemical composition of the Earth's crust," in *The Earth's Crust and Upper Mantle*, ed. P. J. Hart, Geophysical Monograph 13, American Geophysical Union, Washington, D.C.

Rouse, G. E., and R. E. Bisque, 1968, Global tectonics and the Earth's core. *Mines Mag.*, 58, 2–8.

Roy, D. M. and R. Roy, 1954, An experimental study of the formation and properties of synthetic serpentines and related layer silicate minerals. *Am. Miner.*, **39**, 957–975.

Roy, R. F., D. D. Blackwell, and F. Birch, 1968, Heat generation of plutonic rocks and continental heat flow provinces. *Earth Planet. Sci. Letters*, **5**, 1–12.

Runcorn, S. K., 1962, editor, *Continental Drift*. Academic Press, New York.

Sacks, I. S., 1969, Distribution of absorption of shear waves in South America and its tectonic significance. *Carnegie Inst. Wash. Yearbook*, **67**, 339–344.

Scarfe, C. M., and P. J. Wyllie, 1967, Serpentine dehydration curves and their bearing on serpentinite deformation in orogenesis. *Nature*, **215**, 945–946.

Scheidegger, A. E., 1963, *Principles of Geodynamics*. Academic Press, New York.

Schmucker, V., 1969, "Geophysical aspects of structure and composition of the earth," in *Handbook of Geochemistry*, Vol. 1, ed. K. H. Wedepohl. Springer-Verlag, Heidelberg.

Scholl, D. W., M. N. Christensen, R. von Huene, and M. S. Marlow, 1970, Peru-Chile and trench sediments and sea-floor spreading. *Geol. Soc. Am. Bull.*, **81**, 1339–1360.

Schopf, J. M., 1970, Gondwana paleobotany. *Antarctic J. U.S.*, **5**, 62–66.

Seyfert, C. K., 1969, Undeformed sediments in oceanic trenches with sea-floor spreading. *Nature*, **222**, 70.

Shaw, H. R., 1970, Earth tides, global heat flow, and tectonics. *Science*, **168**, 1084–1087.

Sheridan, R. E., R. E. Houtz, C. L. Drake, and M. Ewing, 1969, Structure of continental margin off Sierra Leone, West Africa. *J. Geophys. Res.*, **74**, 2512–2530.

Sleep, N. H., 1969, Sensitivity of heat flow and gravity to the mechanism of sea-floor spreading. *J. Geophys. Res.*, **74**, 542–549.

Smith, A. G., and A. Hallam, 1970, The fit of the southern continents. *Nature*, **223**, 139–144.

Sollogub, V. B., 1969, "Seismic crustal studies in southeastern Europe," in *The Earth's Crust and Upper Mantle*, ed. P. J. Hart. Geophysical Monograph 13, American Geophysical Union, Washington D.C.

Sproll, W. P., and R. S. Dietz, 1969, Morphological continental drift fit of Australia and Antarctica. *Nature*, **222**, 345–348.

Stewart, A. D., 1968, Geology in British universities. *Nature*, **217**, 987–988.

Stewart, D. B., 1967, Four-phase curve in the system $CaAl_2Si_2O_8$-SiO_2-H_2O between 1 and 10 kilobars. *Schweiz. Miner. Petrogr. Mitt.*, **47**, 35–59.

Suess, H. E., and H. C. Urey, 1956, Abundances of the elements. *Rev. Mod. Phys.*, **28**, 53–74.

Sykes, L. R., 1966, The seismicity and deep structure of island arcs. *J. Geophys. Res.*, **71**, 2981–3006.

Takeuchi, H., S. Uyeda, and H. Kanamori, 1967, *Debate About the Earth*. Freeman, Cooper and Co. San Francisco.

Talwani, M., X. Le Pichon, and M. Ewing, 1965, Crustal structure of the midocean ridges, 2. *J. Geophys. Res.*, **70**, 341–352.

Tarakanov, R. Z., and N. V. Leviy, 1968, "A model for the upper mantle with several channels of low velocity and strength," in *The Crust and Upper Mantle of the Pacific Area*, eds. L. Knopoff, C. L. Drake, and P. J. Hart. Geophysical Monograph 12, American Geophysical Union, Washington, D.C.

Tatsch, J. H., 1964, Distribution of active volcanoes; summary of preliminary results of three-dimensional least-squares analysis. *Geol. Soc. Am. Bull.*, **75**, 751–752.

Taylor, F. B., 1910, Bearing of the Tertiary mountain belt on the origin of the earth's plan. *Geol. Soc. Amer. Bull.*, **21**, 179–226.

Taylor, H. P., 1968, The oxygen isotope geochemistry of igneous rocks. *Contr. Mineral. and Petrol*, **19**, 1–71.

Taylor, S. R., 1968, "Geochemistry of Andesites," in *Origin and Distribution of the Elements*, ed. L. H. Ahrens. Pergamon, Oxford.

———, 1969, "Trace element chemistry of andesites and associated calc-alkaline rocks," in *Proceedings of the Andesite Conference*, ed. A. R. McBirney. Oregon Dep. Geol. Mineral Ind. Bull., **65**.

Tilley, C. E., 1950, Some aspects of magmatic evolution. *Geol. Soc. London Quart. J.* **106**, 37–61.

———, and H. S. Yoder, 1964, Pyroxene fractionation in mafic magma at high pressures and its bearing on basalt genesis. Carnegie Inst. Wash. Yearbook, **63**, 114–121.

Tilling, R. I., D. Gottfried, and F. C. W. Dodge, 1970, Radiogenic heat production of contrasting magma series: bearing on interpretation of heat flow. *Geol. Soc. Am. Bull.*, **81**, 1447–1462.

Toksöz, M. N., J. Arkani-Hamed, and C. A. Knight, 1969, Geophysical data and long-wave heterogeneities of the earth's mantle. *J. Geophys. Res.*, **74**, 3751–3770.

Torrance, K. E., and D. L. Turcotte, 1971, Structure of convection cells in the mantle. *J. Geophys. Res.*, **76**, 1154–1161.

Tozer, D. C., 1967, "Towards a theory of thermal convection," in *The Earth's Mantle*, ed. T. F. Gaskell. Academic, New York.

Turcotte, D. L., and E. R. Oxburgh, 1969, Convection in a mantle with variable physical properties. *J. Geophys. Res.*, **74**, 1458–1474.

Turner, F. J., 1968, *Metamorphic Petrology*. McGraw-Hill, New York.

Tuthill, R. L., 1969, Effect of varying f_{O_2} on the hydrothermal melting and phase relations of basalt. *Trans. Am. Geophys. Un.*, **50**, 355.

Tuttle, O. F., and N. L. Bowen, 1958, Origin of Granite in the Light of Experimental Studies in the System $NaAlSi_3O_8$-$KAlSi_3O_8$-SiO_2-H_2O. *Geol. Soc. Am. Memoir*, **74**.

Umbgrove, J. H. F., 1947, *The Pulse of the Earth*. Martinus Nijhoff, The Hague.

Vacquier, V., S. Uyeda, M. Yasui, J. Sclater, C. Corry, and T. Watanabe, 1966, Heat flow measurements in the northwestern Pacific. *Bull. Earthquake Res. Inst. Tokyo Univ.*, **44**, 1519–1535.

Van Andel, Tj. H., and C. O. Bowin, 1968, Mid-Atlantic ridge between 22° and 23° North latitude and the tectonics of mid-ocean rises. *J. Geophys. Res.*, **73**, 1279–1298.

Van Bemmelen, R. W., 1968, On the origin and evolution of the earth's crust and magmas. *Sonderdruck Geolog. Rundschau*, **57**, 657–705.

Van de Lindt, W. J., 1967, Movement of the Mohorovicic discontinuity under isostatic conditions. *J. Geophys. Res.*, **72**, 1289–1297.

———, 1968, Reply. *J. Geophys. Res.*, **73**, 6607.

Van Hilten, D., 1968, Global expansion and paleomagnetic data. *Tectonophysics*, **5**, 191–210.

Vestine, E., I. Lange, L. Laporte, and W. Scott, 1947, The geomagnetic field, its description and analysis. *Carnegie Inst. Wash. Publ.*, **580**.

Verhoogen, J., 1965, "Phase changes and convection in the earth's mantle," in *A symposium on continental drift*, eds. P. M. S. Blackett, E. Bullard, and S. K. Runcorn. *Phil. Trans. Roy. Soc. London*, 1088.

Vine, F. J., 1966, Spreading of the ocean floor: new evidence. *Science*, **154**, 1405–1415.

———, 1968, "Magnetic anomalies associated with mid-ocean ridges," in *The History of the Earth's Crust*, ed. R. A. Phinney. Princeton University Press, New Jersey.

———, and H. H. Hess, 1970, "Sea-floor spreading", in *The Sea*, Vol. 4, eds. A. E. Maxwell,

E. C. Bullard, E. Goldberg, and J. L. Worzel. Wiley-Interscience, New York.

Vine, F. J., and D. H. Matthews, 1963, Magnetic anomalies over oceanic ridges. *Nature,* **199,** 947–949.

Vine, F. J., and J. T. Wilson, 1965, Magnetic anomalies over a young oceanic ridge off Vancouver Island. *Science,* **150,** 485–489.

Vogt, P. R., C. N. Anderson, D. R. Bracey, and E. D. Schneider, 1970, North Atlantic magnetic smooth zones. *J. Geophys. Res.,* **75,** 3955–3968.

Vogt, P. R., N. A. Ostenso, and G. L. Johnson, 1970, Magnetic and bathymetric data bearing on sea-floor spreading north of Iceland. *J. Geophys. Res.,* **75,** 903–920.

Von Herzen, R. P., and W. H. K. Lee, 1969, "Heat flow in oceanic regions," in *The Earth's Crust and Upper Mantle,* ed. P. J. Hart. Geophysical Monograph 13, American Geophysical Union, Washington, D.C.

Walker, C. T., and J. G. Dennis, 1966, Explosive phase transitions in the earth's mantle. *Nature,* **209,** 182–183.

Ward, M. A., 1963, On detecting changes in the Earth's radius. *Geophys. J.,* **8,** 217–225.

Washington, H. S., 1925, The chemical composition of the Earth. *Am. J. Sci.,* **9,** 351–378.

Watkins, N. D., 1968, Short period geomagnetic polarity events in deep-sea sedimentary cores. *Earth Planet. Sci. Letters,* **4,** 341–349.

———, 1968, Comments on the interpretation of linear magnetic anomalies. *Pure Appl. Geophys.,* **69,** 179–192.

———, and S. E. Haggerty, 1968, Oxidation and magnetic polarity in single Icelandic lavas and dikes. *Geophys. J. Roy. Astronom. Soc.,* **15,** 305–315.

Watkins, N. D., and A. Richardson, 1968, Comments on the relationship between magnetic anomalies, crustal spreading and continental drift. *Earth Planet. Sci. Letters,* **4,** 257–264.

Wegener, A., 1966, *The Origin of Continents and Oceans.* Translated by John Biram from the Fourth (1929) revised German edition. Dover, New York.

Wegmann, C. E., 1963, Tectonic patterns at different levels. *Geol. Soc. South Africa,* annexure to **66,** 1–78.

Wetherill, G. W., 1961, Steady-state calculations bearing on geological implications of a phase-transition Mohorovicic Discontinuity. *J. Geophys. Res.* **66,** 2983–2993.

White, I. G., 1967, Ultrabasic rocks and the composition of the upper mantle. *Earth Planet. Sci. Letters,* **3,** 11–18.

Wiik, H. B., 1956, The chemical composition of some stony meteorites. *Geochim. Cosmochim. Acta,* **9,** 279–289.

Wilson, J. T., 1959, Geophysics and continental growth. *Am. Sci.,* **47,** 1–24.

———, 1963, Hypothesis of Earth's behavior. *Nature,* **198,** 925–929.

———, 1965, Transform faults, oceanic ridges, and magnetic anomalies southwest of Vancouver Island. *Science,* **150,** 482–485.

———, 1966, Did the Atlantic close and then re-open? *Nature,* **211,** 676–681.

———, 1968, A revolution in earth science. *Geotimes,* **13**(10), 10–16.

———, 1968, A reply to V. V. Beloussov. *Geotimes,* **13**(10), 20–22.

Winkler, H. G. F., 1967, *Petrogenesis of Metamorphic Rocks.* 2nd ed. Springer-Verlag, New York.

Woollard, G. P., 1968, "The interrelationship of the crust, the upper mantle, and isostatic gravity anomalies in the United States," in *The Crust and Upper Mantle of the Pacific Area,* eds. L. Knopoff, C. L. Drake, and P. J. Hart. Geophysical Monograph 12, American Geophysical Union, Washington, D.C.

Wyllie, P. J., 1967, ed., *Ultramafic and Related Rocks.* Wiley, New York.

———, 1969, The origin of ultramafic and ultrabasic rocks. *Tectonophysics,* **7,** 437–455.

———, 1970, "Ultramafic rocks and the upper mantle," in *Fiftieth Anniversary Symposia,* ed. B. A. Morgan. Mineralogical Society of America Special Paper No. 3.

———, 1971, The role of water in magma generation and initiation of diapiric uprise in the mantle. *J. Geophys. Res.,* **76,** 1328–1338.

Wynne-Edwards, 1969, "Tectonic overprinting in the Grenville Province, southwestern Quebec." Geology Association of Canada Special Paper No. 5.

Yoder, H. S., 1955, "Role of water in metamorphism," in *The Crust of the Earth,* ed. A. Poldervaart, Geological Society of America Special Paper 62.

———, and G. A. Chinner, 1960, Almandite-pyrope-water system at 10,000 bars. *Carnegie Institut. Wash. Yearbook,* **59,** 81–84.

Yoder, H. S., and I. Kushiro, 1969, Melting of a hydrous phase: phlogopite. *Am. J. Sci.*, **267-A**, 558–582.

Yoder, H. S., and C. E. Tilley, 1962, Origin of basalt magmas: an experimental study of natural and synthetic rock systems. *J. Petrol.*, **3**, 342–532.

Ziegler, A. M., 1970, Geosynclinal development of the British Isles during the Silurian Period. *J. Geol.*, **78**, 445–479.

Author Index

Ade-Hall, J. M., 381
Ahrens, L. H., 380, 390
Ahrens, T. J., 126, 379
Aitken, T., 382
Akimoto, S., 123, 125, 126, 194, 379, 385
Allard, G. O., 264, 379
Almond, M., 273, 381
Anders, E., 96, 99, 100, 101, 102, 103, 379, 384, 385
Anderson, C. N., 391
Anderson, D. L., 27, 28, 126, 129, 133, 136, 379
Anderson, E. M., 384
Andrews, J. E., 386
Archambeau, C. B., 22, 379
Arkani-Hamed, J., 33, 390
Atwater, T., 337, 366, 386
Aubouin, J., 213, 214, 215, 216, 217, 219, 220, 222, 223, 226, 227, 228, 379
Aumento, F., 359, 379
Ave'Lallemant, H. G., 26, 33, 356, 379, 380

Badgley, P. C., 213, 216, 379
Bailey, E. B., 170
Barazangi, M., 14, 379
Barnes, H. L., 380
Baron, J. G., 383
Barth, T. F. W., 169, 379
Bassinger, B. G., 336, 379
Beloussov, V. V., 255, 266, 373, 375, 377, 379, 391
Belyaevsky, N. A., 385
Benioff, H., 13, 15, 379
Bernal, J. D., 123, 303, 304, 379
Birch, F., 28, 77, 104, 123, 157, 158, 159, 379, 389
Bird, J. M., 350, 351, 364, 365, 366, 367, 368, 369, 371, 372, 379, 381
Bisque, R. E., 15, 16, 389
Blackett, P. M. S., 261, 273, 274, 275, 379, 390
Blackwell, D. D., 157, 158, 159, 389
Boer, J. de, 335, 379
Boettcher, A. L., 52, 77, 173, 174, 175, 180, 202, 379, 380, 383
Bostrom, K., 115, 382
Bowen, N. L., 111, 113, 170, 171, 173, 174, 181, 182, 230, 380, 390
Bowen, V. T., 386
Bowin, C. O., 338, 339, 390

Boyce, R. E., 382, 386
Boyd, F. R., 52, 77, 171, 380
Bracey, D. R., 391
Brown, H., 97, 380
Bruhnes, B., 285, 380
Brune, J. N., 141, 380
Bukry, D., 382
Bull, C. E., 286, 380
Bullard, E. C., 35, 40, 262, 275, 289, 298, 332, 334, 335, 337, 344, 379, 380, 388, 390, 391
Bullen, K. E., 20, 21, 27, 28, 380
Burk, C. A., 147, 380, 383
Burnham, C. W., 173, 175, 380, 384

Carey, S. W., 232, 265, 266, 284, 285, 380
Carmichael, I. S. E., 182, 380
Carswell, D. A., 112, 117, 380
Carter, J. L., 120, 380
Carter, N. L., 26, 33, 356, 379, 380
Chamberlin, T. C., 59, 60
Chayes, F., 170, 380
Chinner, G. A., 78, 391
Christensen, M. N., 389
Clark, S. P., 28, 29, 32, 127, 171, 381
Clegg, J. A., 273, 379, 381
Coes, L., 171
Cohen, L. H., 52, 81, 82, 83, 86, 87, 89, 175, 193, 194, 380
Cordani, V. G., 384
Corry, C., 390
Cox, A., 277, 284, 288, 295, 296, 297, 380, 381
Craig, H., 98
Creer, K. M., 277, 279, 280, 281, 381
Crook, K. A. W., 222, 381
Cuvier, Baron, 378

Dachille, F., 123, 124, 125, 126, 381
Dagley, P., 291, 381
Dalrymple, G. B., 289, 290, 291, 296, 381
Daly, R. A., 168, 169
Dana, J. D., 211, 212, 216, 223, 381
De Almeida, F. F. M., 384
Debelmas, J., 219
Denison, R. E., 163, 387
Dennis, J. G., 252, 253, 381, 391

DeWald, O. E., 379
Dewey, J. F., 350, 351, 364, 365, 366, 367, 368, 369, 371, 372, 379, 381
Dickey, J. S., 117, 381
Dickinson, W. R., 362, 366, 367, 374, 381, 383
Dickson, G. O., 326, 327, 381, 383
Dietz, R. S., 214, 217, 219, 255, 257, 260, 263, 265, 298, 301, 326, 343, 364, 367, 371, 373, 381, 389
Dodge, F. C. W., 390
Doell, R. R., 277, 284, 380, 381
Doig, R., 265, 381
Dorman, J., 14, 21, 379, 381
Douglas, R. G., 382
Drake, C. L., 65, 66, 67, 262, 381, 382, 385, 386, 389, 390, 391
Drever, H. I., 117, 381
Du Toit, A., 257, 259, 263, 265, 381

Edwards, J., 381
Egyed, L., 284, 285, 381
Emilia, D. A., 337, 381
Engel, A. E. J., 161, 162, 163, 165, 171, 172, 381, 382, 383
Engel, C. G., 382
England, J. L., 52, 77, 171, 380
Erickson, B. H., 388
Ericson, D. B., 292, 382
Ernst, W. G., 366, 371, 382
Eskola, P., 230, 382
Evans, B. W., 52, 382
Everett, J. E., 380
Ewing, J., 330, 331, 381, 382
Ewing, M., 113, 261, 298, 307, 308, 322, 330, 382, 383, 387, 389, 390

Fairbairn, H. W., 384
Fermor, L. L., 72, 382
Fischer, A. G., 353, 382
Fisher, D. E., 115, 382
Fisher, O., 260, 382
Fitch, T. J., 350, 382
Flinn, E. A., 379
Foster, J. H., 289, 293, 382, 387
Freedman, C., 380
Fujisawa, H., 120, 125, 126, 379, 382
Funnell, B. M., 292, 383
Fyfe, W. S., 83, 382

Ganapathy, R., 384
Garrison, R. E., 382
Gaskell, T. F., 380, 384, 386, 390, 379
Gast, P. W., 208, 359, 382, 384
Gellman, H., 380
Gilluly, J., 366, 371, 382
Girdler, R. W., 40, 41, 42, 45, 46, 382
Glass, B., 387
Goldberg, E., 388, 391

Goldschmidt, V. M., 72, 123
Goldsmith, J. R., 52
Goles, G. G., 94, 95, 114, 382
Gottfried, D., 390
Grasty, R. L., 381
Green, D. H., 81, 82, 83, 84, 85, 86, 87, 89, 90, 91, 111, 112, 113, 114, 119, 152, 154, 175, 194, 195, 196, 197, 199, 200, 201, 203, 204, 205, 206, 207, 208, 363, 382, 383, 385, 388, 389
Green, T. H., 210, 383
Greenwood, H. G., 178
Grim, P. J., 388
Gromme, C. S., 282, 283, 381, 383
Guier, W. H., 38, 45, 383
Gutenberg, B., 20, 21, 23, 32, 383

Haggerty, S. E., 334, 381, 391
Hales, A. L., 355, 383
Hall, J., 211, 212, 216, 223, 383
Hallam, A., 263, 265, 389
Hamilton, W., 363, 366, 371, 383
Harris, P. G., 74, 89, 112, 113, 114, 120, 383
Harrison, C. G. A., 292, 383, 386
Hart, P. J., 380, 381, 382, 383, 385, 386, 388, 389, 390, 391
Hatherton, T., 362, 383
Haug, E., 212, 216, 383
Havens, R. G., 382
Hays, J. D., 387
Heard, H. C., 52, 384
Heezen, B. C., 231, 232, 261, 284, 298, 308, 382, 383, 387
Heier, K. S., 156, 157, 158, 384, 385
Heinrichs, D. F., 337, 381
Heirtzler, J. R., 307, 315, 322, 324, 325, 326, 327, 328, 330, 333, 381, 383, 385, 388
Herrin, E., 383
Herron, E. J., 383
Herron, E. M., 326, 388
Herz, N., 265, 383
Herzog, G. F., 384
Hess, H. H., 71, 72, 87, 109, 111, 112, 113, 233, 235, 236, 255, 260, 261, 269, 298, 299, 300, 301, 313, 318, 320, 325, 326, 337, 343, 344, 347, 352, 354, 373, 382, 383, 390
Hibberd, F. H., 275, 383
Hilgenberg, O. C., 231, 284, 285, 383
Hill, R. E. T., 175, 180, 202, 383
Holden, J. C., 257, 263, 265, 373, 381
Holloway, J. R., 175, 384
Holmes, A., 63, 64, 72, 232, 260, 261, 298, 301, 384
Hospers, J., 277, 278, 279, 384
Houtz, R. E., 389
Howell, B. F., 2, 30
Hsu, K. J., 386
Hsu, L. C., 77, 384
Hubbard, N. J., 208, 384
Hurley, P. M., 164, 165, 263, 264, 265, 384, 388

Hurst, V. J., 264, 379
Hutton, J., 378
Hyndman, R. D., 158, 384

Irvine, T. N., 323, 384, 385
Irving, E., 232, 269, 271, 272, 273, 277, 278, 279,
    280, 281, 282, 284, 285, 286, 287, 380, 384
Isacks, B., 348, 349, 350, 354, 377, 384, 388
Ito, K., 81, 82, 83, 85, 86, 87, 88, 89, 90, 91, 121,
    175, 193, 194, 195, 196, 204, 208, 380, 384

Jackson, E. D., 110, 384
Jacobs, J. A., 30, 231, 298, 384
Jaeger, J. C., 384
James, H. L., 383
Jarosewich, E., 113, 386
Jeffrey, P. M., 384
Jeffreys, H., 20, 21, 123, 255, 256, 262, 384
Joensuu, O., 115, 382
Johannes, W., 52, 384
Johnson, G. L., 337, 391
Johnston, R., 117, 381
Joyner, W. B., 238, 243, 244, 245, 384

Kanamori, H., 269, 390
Kaneoka, I., 387
Karig, D. E., 144, 384
Kaula, W. M., 98, 384
Kawashita, K., 384
Kay, M., 208, 213, 214, 216, 217, 366, 367, 384
Kay, R., 384
Keays, R. R., 99, 384
Kelvin, Lord, 231
Kennedy, G. C., 68, 72, 74, 75, 76, 80, 81, 82, 83, 85,
    86, 87, 88, 89, 90, 91, 121, 171, 174, 175, 193,
    194, 195, 196, 201, 204, 208, 233, 234, 236,
    237, 238, 380, 384
Kennedy, W. Q., 170, 384
Kitahara, S., 71, 121, 384
Klement, W., 52, 380
Kling, S. A., 382
Knight, C. A., 33, 390
Knopoff, L., 22, 381, 382, 384, 385, 386, 390, 391
Komada, E., 123, 379
Kornprobst, J., 117, 385
Korzhinskii, D. S., 180
Kosminskaya, I. P., 65, 140, 146, 148, 385
Kossina, E., 7
Krasheninnikov, V., 382
Krause, D. C., 315, 335, 379, 385
Krumbein, W. C., 59, 60, 61, 213, 216, 385
Kuno, H., 172, 193, 385
Kushiro, I., 78, 79, 121, 135, 174, 175, 192, 193, 194,
    199, 205, 206, 207, 208, 385, 392

Lachenbruch, A. H., 157, 385
Lambert, D. G., 379
Lambert, I. B., 52, 122, 135, 136, 156, 157, 158, 174,
    175, 181, 202, 384, 385

Lange, I., 390
Laporte, L., 390
Larimer, J. W., 100, 102, 385
Larochelle, A., 313, 387
Larson, R. L., 304, 341, 385
Lattimore, R., 337, 388
Laul, J. C., 384
LeComte, P., 77, 379
Lee, W. H. K., 40, 41, 42, 43, 44, 45, 385, 391
Lemoine, M., 219
Leonard, B. F., 383
Le Pichon, X., 322, 326, 327, 332, 346, 347, 348,
    349, 350, 351, 383, 385, 390
Leviy, N. V., 22, 23, 390
Lidiak, E. G., 163, 387
Lliboutry, L., 355, 385
Lovering, J. F., 72, 73, 74, 75, 76, 86, 233, 385
Lubimova, E. A., 31, 32, 386
Luck, G. R., 281, 386
Ludwig, W. J., 382
Luyendyk, B. P., 337, 341, 342, 348, 386, 388
Lyell, C., 59, 60, 378, 386

MacDonald, G. J. F., 77, 159, 160, 238, 239, 240, 241,
    243, 245, 305, 384, 389
MacGregor, I. D., 79, 386
Major, A., 123, 125, 126, 127, 389
Malahoff A., 356, 386
Markhinin, E. K., 155, 386
Marlow, M. S., 389
Mason, B., 97, 98, 103, 104, 105, 114, 116, 386
Mason, R. G., 308, 310, 311, 312, 386, 388
Matsumoto, T., 215, 219, 222, 386
Mattauer, M., 219
Matthews, D. H., 312, 313, 314, 318, 320, 323, 325,
    326, 332, 335, 391
Maxwell, A. E., 352, 353, 386, 388, 390
Maxwell, J. C., 304, 305, 372, 386
Maynard, G. L., 147, 386
McBirney, A. R., 155, 363, 386, 390
McElhinny, M. W., 281, 386
McKenzie, D. P., 337, 344, 345, 349, 352, 354, 357,
    360, 361, 362, 369, 370, 386
Melcher, G. C., 384
Melson, W. G., 113, 386
Menard, H. W., 6, 7, 10, 11, 148, 301, 302, 315, 335,
    336, 337, 386
Menard, M., 298
Merrill, R. B., 174, 175, 386
Merrill, R. T., 282, 383
Meservey, R., 266, 387
Meyerhoff, A. A., 255, 258, 373, 376, 377, 378, 387
Millhollen, G. L., 179, 180, 387
Milow, E. D., 386
Minear, J. W., 361, 363, 387
Mitchell, A. H., 218, 225, 226, 367, 370, 371, 387
Miyashiro, A., 113, 216, 387
Molnar, P., 25, 350, 351, 382, 387

Moore, C. B., 386
Morgan, B. A., 379, 381, 384, 391
Morgan, W. J., 337, 344, 345, 346, 347, 348, 351,
  352, 353, 386, 387
Morley, L. W., 312, 313, 387
Mudie, J. D., 386
Muehlberger, W. R., 163, 387
Murthy, V. R., 114

Nafe, J. E., 65, 66, 67, 381
Nagata, T., 285, 387
Ness, N. F., 238, 239, 240, 241, 243, 245, 386
Newton, R. C., 52, 77, 384, 387
Newton, R. R., 38, 45, 383
Nicholls, G. D., 112, 113, 114, 117, 119, 387
Ninkovich, D., 294, 387
Nixon, J., 380
Nockolds, S. R., 112

O'Connell, R. J., 238, 243, 245, 387
O'Hara, M. J., 116, 172, 192, 193, 194, 195, 196, 197,
  198, 199, 201, 204, 206, 208, 210, 387
Oliver, J., 25, 348, 350, 377, 384, 387, 388
Opdyke, N. D., 289, 293, 382, 387
Osborn, E. F., 209, 387
Osborne, F. F., 387
Ostenso, N. A., 337, 391
Otalora, G., 112, 383
Oxburgh, E. R., 32, 33, 357, 358, 359, 360, 361, 362,
  387, 390
Ozima, M., 334, 387

Pakiser, L. C., 24, 87, 88, 142, 149, 154, 388
Parker, R. L., 344, 345, 349, 386
Patterson, C., 97, 380
Percival, S. F., 386
Peter, G., 336, 337, 342, 379, 388
Peyve, 213
Phillips, J. D., 337, 348, 388
Phinney, R. A., 307, 319, 326, 336, 386, 388, 390
Pimm, A. C., 382
Pinson, W. H., 384
Pitman, W. C., 307, 325, 326, 327, 381, 383, 388
Piwinskii, A. J., 175, 187, 388
Poldervaart, A., 148, 149, 150, 153, 154, 155, 205,
  307, 379, 382, 388, 391
Press, F., 27, 28, 29, 90, 131, 136, 307, 359, 382, 388

Raff, A. D., 310, 311, 312, 336, 386, 388
Ramberg, H., 246, 247, 248, 249, 388
Rand, J. R., 164, 165, 265, 384
Reading, H. G., 218, 225, 226, 367, 370, 371, 387
Reay, A., 113 383
Revelle, R., 298
Rezanov, I. A., 277, 281, 388
Richardson, A., 335, 337, 338, 339, 391
Richter, C. F., 349, 350, 388

Ringwood, A. E., 28, 29, 32, 81, 82, 83, 84, 85, 86,
  87, 89, 90, 91, 97, 101, 103, 104, 105, 111,
  112, 113, 114, 116, 117, 119, 120, 123, 124,
  125, 126, 127, 128, 129, 130, 131, 134, 136,
  152, 154, 175, 194, 195, 196, 197, 199, 200,
  201, 203, 204, 208, 210, 363, 381, 382, 383,
  384, 388, 389
Rittmann, A., 169, 389
Roberts, R. J., 228, 389
Robertson, E. C., 77, 389
Robertson, J. K., 176, 187, 188, 201, 385, 386, 389
Robinson, R., 149, 154, 388
Ronov, A. B. 10, 148, 149, 150, 151, 153, 154,
  155, 389
Rouse, G. E., 15, 16, 389
Rowell, J. A., 74, 89, 383
Roy, D. M., 123, 389
Roy, R., 123, 124, 125, 126, 381, 389
Roy, R. F., 157, 158, 159, 389
Runcorn, S. K., 261, 274, 275, 276, 277, 298, 304,
  379, 381, 382, 384, 387, 389, 390
Russell, R. D., 231, 298, 384
Rutten, M. G., 275

Sacks, I. S., 350, 389
Saito, T., 386
Salisbury, R. D., 59, 60
Sammis, C., 135, 379
Scarfe, C. M., 52, 71, 121, 389
Scheidegger, A. E., 7, 389
Schilling, J-G., 335, 379
Schmucker, V., 38, 389
Schneider, E. D., 391
Scholl, D. W., 364, 365, 389
Schopf, J. M., 265, 389
Sclater, J., 390
Scott, W., 390
Seabrook, M., 125, 126, 389
Sederholm, J. J., 171
Seyfert, C. K., 364, 365, 389
Shaw, H. R., 356, 389
Sheridan, R. E., 146, 389
Shido, F., 113, 387
Sigurgeirsson, T., 381
Sinityzn, 213
Sleep, N. H., 357, 389
Sloss, L. L., 59, 60, 61, 213, 216, 385
Smith, A. G., 263, 265, 380, 389
Smith, P. J., 381
Smith, S. M., 6, 7, 10, 11, 386
Smith, T. J., 388, 389
Sollogub, V. B., 143, 389
Speiss, F. N., 340, 341, 385
Sproll, W. P., 263, 265, 381, 389
Steinhart, J. S., 388, 389
Stewart, A. D., 2, 389
Stewart, D. B., 174, 389
Stille, H., 212, 213, 215, 216

Stockard, H. P., 381
Stubbs, P. H. S., 273, 379, 381
Stueber, A. M., 114
Suess, H. E., 94, 95, 389
Sykes, L. R., 15, 348, 350, 351, 377, 384, 387, 388, 390
Syono, Y., 126, 194, 379, 385

Takenouchi, S., 384
Takeuchi, H., 269, 274, 390
Talwani, M., 145, 322, 390
Tarakanov, R. Z., 22, 23, 390
Tatsch, J. H., 13, 390
Taylor, F. B., 259, 390
Taylor, H. P., 209, 390
Taylor, S. R., 160, 161, 210, 390
Teichert, C., 374, 376, 377, 387
Tharp, M., 308, 383
Thompson, G., 113, 386
Tilley, C. E., 80, 81, 83, 86, 87, 89, 170, 171, 172, 175, 191, 192, 193, 195, 208, 245, 390, 392
Tilling, R. I., 157, 390
Toksöz, M. N., 33, 45, 46, 361, 363, 387, 390
Torrance, K. E., 357, 390
Tozer, D. C., 32, 390
Turcotte, D. L., 32, 33, 357, 358, 359, 360, 361, 362, 387, 390
Turner, F. J., 52, 230, 390
Tuthill, R. L., 175, 390
Tuttle, O. F., 171, 173, 174, 182, 380, 390

Umbgrove, J. H. F., 227, 390
Urey, H. C., 94, 95, 103, 389
Uyeda, S., 40, 41, 42, 43, 44, 45, 269, 385, 390

Vacquier, V., 44, 390
Van Andel, Tj. H., 338, 339, 390
Van Bemmelen, R. W., 375, 376, 390
Van de Lindt, W. J., 238, 240, 241, 243, 245, 387, 390
Vandoros, P., 384
Van Hilten, D., 284, 285, 390
Verhoogen, J., 249, 250, 251, 252, 253, 282, 383, 390
Vestine, E., 35, 390
Vine, F. J., 307, 312, 313, 314, 315, 316, 317, 318, 319, 320, 321, 322, 323, 325, 326, 327, 328, 332, 335, 337, 343, 344, 347, 390, 391
Vogt, P. R., 337, 391

Volvovsky, I. S., 385
Von Herzen, R. P., 40, 41, 386, 391
Von Huene, R., 389

Walker, C. T., 252, 253, 381, 391
Walker, G. P. L., 381
Ward, M. A., 284, 391
Washington, H. S., 103, 104, 391
Wasserburg, G. J., 238, 243, 245, 384, 387
Watanabe, T., 390
Watkins, N. D., 295, 296, 334, 335, 337, 338, 339, 381, 391
Wedepohl, K. H., 382, 389
Wegener, A., 255, 256, 257, 258, 259, 261, 262, 266, 275, 298, 391
Wegmann, C. E., 373, 378, 391
Werner, A., 378
Wetherill, G. W., 238, 239, 240, 241, 243, 391
White, I. G., 113, 114, 383, 391
Wiik, H. B., 97, 391
Willis, I., 286, 380
Wilson, J. T., 3, 210, 231, 266, 267, 298, 302, 303, 315, 316, 317, 318, 320, 322, 344, 345, 371, 373, 384, 391
Wilson, R. L., 381
Winkler, H. G. F., 171, 391
Wollin, G., 292, 382
Woollard, G. P., 142, 391
Worzel, J. L., 388, 391
Wyllie, P. J., 52, 71, 77, 107, 108, 115, 116, 118, 121, 122, 132, 133, 135, 136, 173, 174, 175, 176, 187, 188, 201, 202, 204, 208, 209, 359, 363, 372, 379, 380, 381, 384, 385, 386, 388, 389, 391
Wynne-Edwards, H. R., 223, 224, 391

Yaroshevsky, A. A., 10, 148, 149, 150, 151, 153, 154, 155, 389
Yasui, M., 390
Yoder, H. S., 77, 78, 79, 80, 81, 83, 86, 87, 89, 121, 171, 172, 173, 175, 191, 192, 193, 195, 208, 245, 385, 387, 390, 391, 392

Ziegler, A. M., 371, 392
Zietz, I., 24, 142, 388
Zverev, S. M., 65, 146, 385

Subject Index

Acmite, 78
Activation energy, 252
Actualism, 378
Adiabat, 250, 252
Adiabatic mantle, 357
Adiabatic uprise, 200-201, 207, 209
Advection, 305
Acromagnetic survey, 139
Africa, 5, 6, 12, 258, 262-263, 301
  age provinces, 263-264
  continental drift, 256-259, 261-266
  heat flow, 41, 44-45
  land bridges, 258
  lithology, 258
  paleomagnetic poles, 282-283
  polar wandering path, 279-283
  rift valleys, 13, 301
  stratigraphy, 258
  tillites, 258
African plate, 337
Age determinations, isotopic, 59-62
Age provinces, continental data, 161-165
  continental reconstructions, 263-266
  Eburnean province, 264
  Pan-African province, 264
Alaska, 10
Alaskan earthquake, 345
Albite, 52-53, 76-78, 173-174, 179, 191
Aleutians, 336
Alkali basalt, 112-113, 167-168, 170, 172, 359
Alkali olivine basalt, 84-86, 106, 110, 191-192, 196
Alkalic rocks, 264
Alkalic rock complexes, 106-108, 110
Alkalic ultrabasic lavas, 107-108
Alluvial fans, 11
Almandine, 76-77
Alpine crustal type, 141
Alpine geosyncline, 212, 220-221
Alpine tectonics, 350
Alps, 10, 256
Aluminosilicate minerals, 51
Amphibole, 51, 53, 106, 118, 120-121, 134-135, 151, 202, 204
Amphibolite, 53, 67-68, 150-152, 176

Analytical techniques, 168
  activation, 114
  spectrochemical, 115
Anatexis, 49, 189-190, 171, 173, 208-209
Andalusite, 151
Andean-type geosynclines, 218, 225, 367, 370-371
Andes, 10, 63, 303, 350, 373
Andesite, 150, 155-156, 160, 167-168, 170, 175, 209, 211, 217, 222, 226, 345, 360, 363-364, 367, 374
  composition, 155
  continental margins, 155
  island arcs, 155
  origin, 208-210
  Oxygen fugacity, 209
  potassium content, 362, 364
Anelasticity of the mantle, 33
Angular velocity between plates, 345-346, 348
Anorthite, 52, 76-77, 79-80, 82, 173-174, 182
Anorthosites, massif-type, 265
Antarctica, 6, 10, 258, 263, 285
Antarctic Ocean, 6
  magnetic anomalies, 322-332
Anticlinoria, 371
Antilles, 257
Apatite, 151
Apollo 11, 98
Appalachian geosyncline, 212, 216
Appalachian mountain belt, 142-143, 211, 258, 369, 371
Appalachian Province, age, 162
Apparatus, high pressure, 123, 230
  piston-cylinder, 171, 173
Arabia, 258, 263
Aragonite, 151
Arctic Ocean, 6, 258, 343, 346
  heat flow, 41
Arkose, 51
Asia, 5-6, 215, 219, 228, 256-257
  continental margins, 147-148
  heat flow, 41
  seismic wave velocity, 23
Asphalt, 246
Assimilation, 170, 209

Asteroidal bodies, 97, 101-102
Asthenosphere, 63, 231, 255, 302-304, 349, 354-355, 360, 369
Atlantic coasts, geometric fit, 256, 261-263
Atlantic crust, structure, 144-147, 231-232
Atlantic-type geosynclines, 218, 225, 367, 370-371
Atlantic geosynclines, 222-223
Atlantic Ocean, 6, 258-259, 263, 275, 301
    crustal structure, 144-147, 231-232
    heat flow, 41-42, 44
    magnetic anomalies, 322-332, 337
    persistence, 376-377
    seismic refraction profile, 143-144
    seismic wave velocities, 65-66
    spreading rates, 328-329
    ultramafic rocks, 109
Atlantic Ocean sediment, volume, 366
Atlas Mountains, 256
Atmosphere, 1-3, 48-49, 93, 105, 113, 119
Augite, 52
Australia, 6, 222, 258, 263, 272
    heat flow, 41, 156-158
    land bridges, 258
    paleolatitudes, 273-274
    paleomagnetic poles, 282-283
    paleo-orientations, 273-274
    polar wandering paths, 279-283
    rock compositions, 156
    tillites, 258
Autogeosynclines, 213, 216
Axial dipole vector, 318
Axis of rotation, 344

Back-deep, 213, 216, 227
Bali, 226-227
Basalt (*see also* Alkali basalt, tholeiite, high-alumina
        basalt), 68, 71-86, 104, 150-151, 155, 160, 167,
        171, 175, 193-196, 210-211, 217, 222, 236-237,
        260-261
    classification, 191-192
    composition, 115-116
    magnetization, 334-335, 337
    metamorphism, 152
    mineralogy, 190-192
    ocean floor, 334-335, 337-339
    oxidation state, 334-335
Basalt-eclogite phase transition (*see* Gabbro-eclogite
        phase transition)
Basalt petrogenesis, 196-208
Basalt tetrahedron, 171, 191-192
Basaltic crustal layer, 170
Basaltic glassy substratum, 168-169
Basanites, 192, 204
Basaltic magma, 68, 105, 109-111, 131, 134, 223,
        261, 359, 364, 366
    differentiation, 109
    fractionation, 196-208
    generation, 196-208

Basaltic magma (*continued*)
    high pressure, 117
    liquidus minerals, 195
    low pressure, 117
Basement rocks, world ages, 265
Basin and Range Province, Bouguer anomaly, 141
    crustal structure, 88, 141-142
    heat flow, 141, 157-158
    seismic wave velocities, 22, 141
Basins, 8, 216, 244-245
    classified, 213
Batholiths, 167, 170-171, 208-211, 222, 226, 247,
        363, 366-367, 374
Bay of Biscay, 263
Beacon sandstone, 285-286
Beni Bouchera, Morocco, 117
Benioff seismic zone, 13-16, 172, 209-210, 350, 366
Benthonic faunal activity, 295
Beta-phase, 125-128, 134
Biological productivity, 330
Biological redeposition, 295
Biosphere, 49
Biostratigraphic units, 61
Biostratigraphical correlation, 59
Biotite, 51, 186-188
Blueschists, 216, 314
Blueschist metamorphism, 364-367, 374
Bouguer gravity anomalies, 38-39, 141, 143-145
Bouguer gravity correction, 38
Boundary-layer theory, 32, 357
Brazil, 143, 258, 263-264
Brecciation, 108
Brucite, 53, 71
Brunhes epoch, 340
Burma, 227-228

Calc-alkaline magmas and rocks, 174, 181, 185-187,
        189, 209-210, 362, 366
Calcite, 50, 53, 151
Calcium ferrite structure, 130
Caledonian mountain belts, 258
California, 13, 143, 149, 282
Canada, 264-265
Canberra, 123, 175
Carbon dioxide, 52, 173, 180, 189
    in crust, 153-156
    in mantle, 136
    in olivine, 119
Carbonates, 51, 150-152
Carbonatites, 264
Caribbean region, 263, 351
Carlsberg Ridge, 313
Catastrophism, 59, 255, 378
Cayman Trough, 351
Centrifugal force, 246-247
Centrifuge models, tectonic structures, 246-249
    domes, 247-248
    subsiding sheets, 248-249

Chaotic structures, 109
Chemical potential gradients, 55
Chemical variation diagrams, 115-118
Chert, 353
Chesapeake Bay, 143
Chicago, 175
Chlorite, 51, 151
Chondrules, 97-99, 103
Chronostratigraphic units, 61
Circum-Pacific orogenic zone, 215, 219, 222, 226
Clastic wedges (*see* Molasse)
Clays, 150-152, 246
Clay minerals, 50-51, 151
Climate, 1, 364
Climatic zones, 258
Clinohumite, 51, 134
Closed system, 55
Close-packed oxides, 123, 129
Coal deposits, 258, 376-377
Coastal geosynclines, 213
Cocos plate, 351
Coesite, 51-52, 179
Colorado Plateau, 236, 321
  low-velocity zone, 22
  Moho depth, 63
  seismic wave velocities, 22
Columbia Plateau, 149
Computers, 14, 21, 243, 261, 354
Conrad discontinuity, 140
Conservation of energy equation, 361
Continental accretion, 265
Continental collision, 226, 228, 350, 367, 369-372
Continental crust (*see* Crust, continental)
Continental drift, 4, 13, 36, 164, 212, 232, 255-256,
      269, 273-275, 277-283, 298, 305, 309, 329,
      344, 366, 373, 375
  arguments against, 373-377
  debate, 255-261, 332
  evidence, 258-259, 261-266
  mechanisms, 259-261, 353-356
  restrictions, 159
  theory, 256-258
Continental geology, 375
Continental growth, 155-156, 160-165, 375
Continental margins, 8, 11-13, 225, 303, 367
  Andean type, 12, 225
  Atlantic type, 12, 225
  buried ridges, 147
  crust, 139
  crustal structures, 146-148
  crustal thickness, 143-144, 146
  deep-focus earthquakes, 159
  Island arc type, 12, 147, 225
  Sea of Okhotsk, 146
  seismic wave velocities, 65-67, 146
  Sierra Leone, 146
  South Kurils island arc, 146
Continental nuclei, 164

Continental platform, 7-10, 12, 375
Continental rise, 11-12, 217, 225
  area, 10
  deposits, 367
Continental rotations, 263, 273-275, 281
Continental shelf, 8-9, 11, 217-218, 262-263
  area, 10
  deposits, 367
  petrology, 150
  volume, 150
Continental shields, 9, 376
  area, 10
Continental slope, 8-9, 217, 219, 262, 308
  area, 10
  deposits, 367
  petrology, 150
  volume, 150
Continents, 1, 5-6, 9-10
  age data, 161-165
  age provinces, 263-266
  areas, 6, 40
  deep structure, 159-161, 305
  distribution, 5-6, 39
  geometrical fit, 256, 258, 262-266, 280, 344
  heat flow, 40-42
  Moho depth, 63-64
  origin and evolution, 4, 32, 47, 105, 159, 210,
      223-225
  oscillation between ocean ridges, 370-372
  permanence, 47, 68, 237-238
  re-assembled, 164
  rifting, 259
Contracting earth theory, 231, 255
Contraction at depth, 6, 233, 246
Convection (*see* Mantle convection)
Convection cell, migrating, 356
Convective heat transport, 238
Conveyor belt analogy, 261, 301, 319
Coral reefs, 258
Cordierite, 51, 77
Cordillera, 221, 227-228, 257
Cordilleran-type mountain belt, 367-369
Core, 3, 15-17, 20, 30, 232, 376
  composition, 29, 94, 103-105
  density, 28-29, 105
  elastic properties, 20
  electric currents, 36
  fluid motions, 16
  formation and growth, 137, 304
  gravity, 30
  mass, 103, 105
  pressure, 30
  radius, 29
  temperature, 30
Core-mantle boundary, 16, 19, 28, 30
Coriolis force, 1
Corundum, 53, 174
Cosmic abundances of elements, 94-95, 99

Counter-revolution, 4, 373-374
Cratons, 212-213, 216, 222, 265
Creep, diffusion, 357
Creep law, 356
Critical end-point, 174
Crust, 2, 3, 139-165
    andesitic, 160-161
    composition, 54, 89, 139, 148-149, 152-158
    defined, 64
    density distribution, 139
    hydrated mantle, 72
    low-velocity layers, 140
    mineralogy, 150-152, 181-182
    oceanic layers, 66, 71-72
    origin, 48, 155-156, 375-376
    petrology, 54, 150-152
    structure, 139-149
    thickness, 17, 63-67, 141, 244
Crust, basaltic layer, 140, 149, 151, 156
    composition, 153
    petrology, 150-152
    volume, 148
Crust, continental, 140-143, 148, 218, 344
    composite seismic section, 64
    composition, 153
    North America, 142
    Russian, 143
    volume, 148
Crust, continental margins, 146-149
    composition, 154
    rock types, 162
    volume, 148
Crust, granitic layer, 140, 149, 151, 156
    composition, 153
    petrology, 150-152
    volume, 148
Crust, oceanic, 113, 139, 143-148, 218, 236, 344
    area, 10
    Atlantic cross section, 143-144
    Bouguer gravity anomaly, 141
    composite seismic sections, 64
    composition, 153-154
    heat flow, 40-43, 141
    high-velocity crustal layer, 146-147
    layered structure, 143, 148
    magnetization, 309-342
    magnetized basalts, 334-335, 342
    Moho depth, 63-64, 352-353
    seismic velocities, 23, 25, 65-67, 141, 143, 146-147
    subduction, 360-372
    thickness, 141
    volume, 148
Crust, sedimentary layer, 148-149, 156
    composition, 153
    petrology, 150-152
    volume, 148
Crustal blocks, aseismic, 345-346
Crustal compression, rates of, 346-347

Crustal extension, rates of, 346-347
Crustal rocks, magma generation, 181-190
    mineral variation, 181-182
Crustal sections and structures, 64, 139-149, 247
    centrifuge models, 246-249
    continental margins, 140-148
    continents, 140-143
    oceans, 143-147
Crust-mantle boundary, 63-76, 86-91, 233-245
Crystal cumulates, 107, 109-110, 168
Crystal fractionation, 196
Crystal mush, 107, 109, 200, 208, 246
Curie point, 143, 269, 285, 313
Curie point isotherm, 313, 325
Czechoslovakia, 112

Darwin's theory of evolution, 266
Declination, angle of, 273
Deep-sea cores, 288
    intensity of magnetization, 293-295
    magnetic inclination, 293-296
Deep-sea drilling, 255, 301, 337, 348, 352-353
Deformation, 55, 247-248, 366
Deformation experiments, 33, 356
Degree of freedom, 178
Dehydration reactions, 49, 71, 120, 235
    effect on rock strength, 299
Demagnetization, viscous, 170
Density, 207, 242-243
    Earth, 19, 26
    lithosphere, 356-357
    mantle, 27-29
Density ratio, 247
Density stratification, 247-248
Denver, 140
Deposition, sediment, 57
Deserpentinization, 235
Desert eolian deposits, 258, 376
Diapiric intrusion, 109
Diapiric uprise, 107, 200-201, 207, 209, 304-305, 361, 363, 372
Diastrophism, 59
Diatremes, 108
Differentiation, 108, 173, 375
Differentiation indices, 115-117
Diffusion creep, 33
Diffusion term, 242
Diopside, 52-53, 78
    aluminous, 79-80
Diorite, 51, 151, 167, 170, 182
Dip, angle of, 273
Dissociation, congruent, 188, 202
    incongruent, 188, 202
Dissociation reaction, 120
Dnieper graben, 143
Dolomite, 53, 109, 151
Domes, centrifuge models, 247-248
Drifters, 255, 275

Dunite, 67, 111, 113, 114-117, 119, 120, 151, 285, 356
  experimentally deformed, 33, 356
Dynamo theory, 1, 36

Earth, active zones, 12-16
  accretion, 103, 117-118
  accretion, gravitational energy, 31
  cooling, 31
  cross-section, 18
  density, 18, 26-29
  differentiation, 102-103, 137
  elastic properties, 26, 29-30
  electrical conductivity, 26, 30-31
  electric currents, 36
  figure of, 17, 32, 37, 305
  free oscillations, 19, 21, 28-29
  heat production, 2, 99
  mass, 17-18, 29
  molten, 231
  moment of inertia, 18, 26, 29
  origin and history, 5, 30, 49, 102-103
  paleoradius, 284-285
  physical properties, 33-46, 256
  physiographic features, 5-16, 39
  radius, paleomagnetic, 284-285
  rift system, 12-13
  rotational axis, 33-36, 269
  size and shape, 17
  structure, 1, 17
  surface relief, 6-8
  surface, tectonic classification, 140-141
  thermal budget, 17, 26, 30-31, 39-46
  tides, 31, 356
Earthquakes, Alaska, 345
  circum-Pacific belt, 345
  depth of focus, 17, 23
    deep-focus, 13-15, 159-160, 303-304, 345, 347-348, 350, 360, 367
    shallow-focus, 13-15, 43, 303-304, 345, 347-349, 350, 355
    intermediate-focus, 13-15, 303, 345, 347-348, 350, 360, 367
  distribution, 13-16, 347
  energy, 39, 353-355
  epicenters, 13-14, 18, 335
  focal mechanisms, 345, 348-352, 355
  maximum intensity, 23
  metastable phase transitions, 252-253
  North Pacific, 345
  slip directions, 345, 348-349
  theory, 252
  waves (*see* Seismic waves)
Earth's composition, 17, 73, 93
  chondritic model, 99, 118, 161
  evidence from rocks, 105-111
  extraterrestrial evidence, 94-105
  metallic fraction, 103-105

Earth's composition *(continued)*
  silicate fraction, 103-105
Earth's core (*see* Core)
Earth's crust (*see* Crust)
Earth's gravity field, 17, 25-26, 29-30, 33, 37-39, 139, 159, 229, 259, 305
  anomalies, 33, 37-39, 45, 139-140, 142-143, 145, 147
  corrections, 37-39
  equipotential surface, 37
  geoid, 37, 39, 45-46, 376
  potential, 39, 103
Earth's interior, 17-46
Earth's properties, 26-33
Earth's structure, 20-26
Earth's magnetic field (*see also* Magnetic anomalies, Magnetic polarity reversals, Paleomagnetism), 1, 16-17, 33-36, 261, 271, 269-270, 307, 375
  declination, angle of, 34, 36
  direction, 33, 36
  dipoles, 33-36, 271, 275, 282, 284
  dynamo theory, 36
  geomagnetic latitude, 36
  geomagnetic poles, 271-273
  inclination, angle of, 34, 36
  intensity, 33, 36, 334
    horizontal component, 34
    vertical component, 34-35
  magnetic axis, 34, 36, 269
  magnetic dip poles, 34, 36
  magnetic elements, 34
  magnetic equator, 34
  magnetic maps, 34-35, 309-313, 325, 333, 336
  magnetic meridian, 34
  magnetic moment, 36, 295
  magnetic storms, 36
  multipole, 275
  non-dipole, 34, 36
  polarity, 36
  polarity reversal, 285
  regional trend, 325, 335
  spherical harmonic analysis, 35
  variations and fluctuations, 36, 271
Earth's mantle (*see* Mantle)
East Pacific Rise, 12, 343, 351, 353
  crust thickness, 63
  crust-mantle structure, 146
  "Fish" traverse, 340-342
  magnetic anomalies, 320-323, 328
  sediment thickness, 330
Eclogite, 28-29, 51-52, 54-55, 67-69, 72-86, 111, 119-136, 151-152, 170-171, 192, 195, 203, 210, 229, 233, 236-237, 239, 246, 260-261, 354, 358-359, 361-362, 364
  olivine, 121
  quartz, 121, 203
  picritic, 115-117
Ecostratigraphic units, 61

Eh versus pH diagrams, 55
Elements, vertical differentiation (*see also* Radioactive
    elements, Trace elements), 159-161
Element abundances, 94-95, 99
  chondrites, 99-101
  cosmic, 94-95, 99
  solar, 94-95
  stellar, 93-94
Elsonian orogeny, 223-224
Endogenic energy, 376
Endothermic transition, 250
Energy sources, 48, 353
  internal, 48
  solar radiation, 48
England, 271-272
Enstatite, 79, 80, 96
  aluminous, 52, 77, 79-80
Epeirogenesis, 212, 228
Epeirogenic seas, 377
Epidote, 51
Epieugeosynclines, 213
Equatorial bulge, 160, 259
Equipotential surface, 247
Erosion, 1, 48-49, 59, 156, 216, 222, 230, 237-238,
    365
  cycle, 243-244
  rates, 364
Eruption, explosive, 108
Etang de Lers, France, 117
Eugeosynclines, 109, 212-222, 227-228, 366-367, 369
Euler's theorem, 344-346
Eurasia, 5-6, 258
  polar wandering paths, 274-281
Europe, 5-6, 41, 215, 262, 273-274, 301
  heat flow, 41
  land bridges, 258
  polar wandering paths, 174-181
  paleomagnetic poles, 274-281
Evaporite, 51, 150, 376-377
Evaporite maxima, 376-377
Evolution, Darwin's theory of, 266
Exciton, 31
Exogeosynclines, 213, 216, 367
Expansion at depth, 233, 236-237, 240, 246
Expanding earth theory, 231-232, 265-266, 284-285,
    312, 346
Experimental petrology, 364
Experimental studies at high pressures, 230
  crustal compositions, 171-190, 208-210
  effect of platinum capsules, 195-196
  extrapolation, 89, 131-134
  gabbro-eclogite transition, 76-86
  mantle compositions, 119-130, 171-173, 190-210
  metastable crystallization, 83
  olivine-spinel transition, 123-130
  seeding, 86
  shock waves, 93
  uncertainties, 127

Externides, 212
Extraterrestrial bodies,
  chemistry, 93
  cooling rate, 97

Faults, continental margin, 13, 15
  major, 12-13
  Mendocino, 309, 315, 321
  Murray, 309
  oceanic, 13, 15, 308, 311, 315, 332-333
  normal, 349
  Pioneer, 309
  strike-slip, 341
  thrust, 211, 231, 345, 356, 364, 366
  transcurrent, 12-13, 315-316, 349, 375
  transform, 12-13, 261, 315-316, 335, 344-345,
    347-349, 375
Faunas, 59, 292
Fayalite, 77, 120, 123-124, 126, 253
Feldspar, 50-51, 152, 174, 182
Fennoscandia uplift, 32
Ferrar dolerite, 285-286
Ferromagnetic mineral, 269
Field mapping, 223
"Fish", 340-342
Fission tracks, 62
Fixists, 255, 266, 275, 298
Floater, 304-305
Flores, 226-227
Flow concentrates, 110
Flow differentiation, 107
Flow layering, 109
Fluid dynamics, 246
Fluidized system, 108
Flysch, 214, 220, 222, 366
Folding, 55, 211-212, 225, 229, 231, 249, 375
Fore-deep, 213, 216, 222
Fore-deep complexes, 367
Fossils, 258, 378
Fractional crystallization (*see also* Magma fractiona-
    tion), 56, 131, 170, 182
  oxygen fugacity, 56, 209
  iron enrichment, 209
Fracture zones, 25, 109, 143, 343-344, 346
Free-air gravity anomaly, 37, 39
Free-air gravity correction, 37
Frictional heating, 360-363
Forsterite, 53, 71, 79-80, 82, 120
Furrows, geosynclinal, 213-214, 216-217, 228
Fusion paths, 183-187, 192-193
Fusion zone, 358-359

Gabbro, 51-54, 67, 69, 72-86, 90, 106, 109, 111-112,
    119, 121, 133-136, 150-152, 173-176, 182, 202,
    229, 233, 236-237, 239, 246, 354, 357-359, 361
Gabbro-eclogite phase transition, 72-91, 121-122,
    133-134, 236-246, 301
  density variation, 85-86, 88, 91

Gabbro-eclogite phase transition *(continued)*
  effect of rock composition, 84-85
  experimental studies, 76-86
  metastable crystallization, 83
  Moho models, 72-76, 86-91, 236-246
  seismic velocity, 90
  tectonic significance, 236-246
Galapagos rift zone, 336, 351
Galaxy, 94, 102
Garnet, 51-52, 76-81, 84-87, 106, 118, 120, 130, 134,
    151, 190-192, 197, 202, 204
Garnet granulite, 82-83, 85-86, 88-90, 359
Gases, exsolved from magma, 108, 119
Gauss epoch, 292, 327-328
Geanticline, 212
Geochemical cycle, 48
Geochronology, 160, 255, 263-266
Geochemistry, 2, 31, 168
Geodesy, 256
Geodynamics Commission, 374
Geographic poles, 259-260, 271-273, 275
Geography, 256
Gehlenite, 77-78
Geoid *(see* Earth's gravity field)
Geological controversies, 255
Geological curricula, 2-3, 47, 266
Geological cycle, 48-49
Geological periods, 60-61
Geological processes, 3, 33, 47-59, 211, 228, 233
Geological record, 374
Geological time scale, 1, 47, 59-62, 297, 321-332
Geology, 3, 256, 378
Geomagnetic time scale, 61-62, 288-297, 307,
    320-332, 337-338, 353
Geophysical processes, 33, 39
Geophysics, 2, 168, 256
Geosynclinal couples, 214-215, 217-219, 227-228
Geosynclinal polarity, 219, 222-223
Geosynclinal rocks, igneous, 218, 226
  metamorphic, 230-231
  sedimentary, 218, 226
Geosynclines, 3, 68, 211-232, 233, 239-240, 259,
    266, 366-372
  active, 370
  classified, 212-218, 222-223, 225, 367, 370-371
  contemporary, 225-228
  evolution, 217-231
  passive, 370
Geotherm, 54, 70-72, 89, 120, 127-128, 131, 136,
    158, 230, 235, 237
Geothermal gradient, 26, 31-32
Germanate model systems, 93, 124-126, 130-131
Ghana, 263
Gilbert epoch, 292, 325, 327
Gilsa event, 295, 337
Glacial clays, 152
Glacial deposits, 258
Glacial maxima, 376-377

Glaciations, 40, 377
Glass, basaltic, 80-83
Glaucophane schists *(see* Blueschists)
Global geology, 3-4, 33, 266-267, 332, 373-378
Global processes, 228
Global tectonics, 212, 231-232, 265-266, 334
Gneiss, 51, 112, 150-151, 156, 171, 186
Gondwanaland, 257-259, 263, 265, 280, 346, 377
Granite, 51, 54, 104, 150-151, 156-157, 170, 174,
    182, 216-217
  controversy, 170-171, 208
  late-tectonic discordant, 217
  syntectonic concordant, 217
  system, 182-185
Granite-forming cycles, 161-165
Granitophile elements, 156
Granodiorite, 68, 74, 150-151, 167, 170, 173, 177,
    182, 186, 217, 222, 226
Granulite facies, high pressure, 67
Graptolitic shale deposition, 371
Gravitational instability, 201, 207
Gravitational sliding of lithosphere plates, 355
Gravitational spreading, 376
Gravitation constant, universal, 26, 232
Gravity faults, 356, 364
Gravity layering, 109
Gravity settling, 97, 107
Gravity sliding, 109, 366
Gravity tectonics, 229, 246-249
Great Plains, 88
Greenland, 256-258, 262
Greenschists, 338-339
Greenstone, 152, 156, 338-339
Grenville orogeny, 223-225
Grenville province, 162, 223, 225
Greywackes, 217, 220, 221
Grossularite, 52-53, 76-78
Groundmass, 182
Gulf of Mexico, 11, 212, 216
Gutenberg-Wiechert discontinuity, 17
Guyots, 236

Harmonic coefficients, 33, 35-36, 38
Hawaiia, 113, 147, 195, 197, 353
Hawaiian type volcanic chains, 303-304
Heat conduction equation, 32, 242
Heat flow, 30, 33, 39-46, 68, 73, 87, 89, 91, 93,
    140-141, 143-144, 159, 242, 298, 305, 325, 356
  Australia, 156
  United States, 157-158
Heat flow equality, oceans and continents, 68, 158,
    159-160, 375
Heat flow profiles, island arc, 44, 360
  mid-Atlantic ridge, 44
Heat flow provinces, 157-158
Heat flow values, arithmetic means, 41-42
  continents, 41-42, 158
  geometric means, 41-42

Heat flow values *(continued)*
  global analysis, 43-46
  logarithmic, 41-42, 45-46
  measurement, 39-40
  modes, 41
  normal, 41
  oceans, 40-42
  physiographic provinces, 39-40, 43
  review of data, 40-46
  world, 45
Heat generation, 33, 156-157, 159, 238, 331
Heat productivity of rocks, 157-159
Heat transfer, 31-33, 46
  conduction, 31, 131
  convection, 31, 131
  exciton transfer, 31
  magmatic, 357
  photon transfer, 31
  radiation, 31, 131
  rates, 238, 242
Hellenides, 219-21
Hercynite, 77
High-alumina basalt, 170, 172, 191, 196, 359
High pressure experiments, 168, 171-173
High-Q zone, 350
Himalayas, 10, 256, 350
Historical geology, 3
Hochkraton, 212-213
Horizontalists, 298
Horizontal movements, 229, 232, 249, 269, 281,
  373-375
Hornblende, 50-51, 120-122, 134, 136, 187-188,
  203-204, 364
Hudsonian orogeny, 223-224
Hybridism, 209
Hydrogen, 209
Hydrologic cycle, 1, 47-50
Hydrosphere, 1-3, 48-49, 57, 105, 113, 119
Hydrothermal solutions, 31, 49
Hypersthene, 170, 191, 195-196
Hypsographic curve, 7-8
Hypsometry, 7, 11

Iceland, 261-262, 290-291, 323-326, 334, 337
Ice Sheet, 10, 32
Idaho, 272
Igneous activity, geosynclines, 215-217
Igneous processes, 48, 55-57
Igneous rock associations, 105-106, 167-168
  plutonic, 170
  volcanic, 170
Igneous rocks, 48, 50-51, 94
  average composition, 152
  geosynclines, 218
  mineralogy, 55, 181-185, 190-192
  per cent crust, 151
  reversed magnetization, 285-92
  texture, 55

Ilmenite-hematite, 285
Ilmenite structure, 130
Incompressibility, 19, 26
India, 6, 258, 263
  continental drift, 256
  land bridges, 258
  paleolatitudes, 273-274
  paleo-orientations, 273-274
  polar wandering path, 279-281
  tillites, 258
Indian Ocean, 6, 258, 263
  fracture zones, 346
  heat flow, 41-42
  magnetic anomalies, 322-332, 337
  seismic wave velocities, 65-66
  spreading rate, 328-329
  ultramafic rocks, 109
Indian Ocean rise, 301
Indonesia, 10, 13, 212, 227-228
International Geomagnetic Reference Field, 335
International Gravity Formula, 37 39
Internides, 212, 216
Intracratonal geosynclines, 213, 216
Intra-deep, 213, 216
Iodine isotopes, meteorites, 102
Ionic size, 161
Ionosphere, 34, 36
Ireland, 258
Iron, solar abundance, 103
Iron capsules, 195-196
Iron-nickel alloy, 94, 96-99, 104
Iron oxide, 269
Iron sulfide, 269
Island arcs, 9-11, 13-16, 219, 225-227, 257, 349, 367
  Benioff zones, 13-16, 350, 361
  Bouguer anomaly, 141
  crust, 139, 210
  crustal composition, 155-156
  crustal section, 63-64, 141, 144
  geological features, 141
  growth of, 365-366
  heat flow, 43-44, 141, 360-361
  lithosphere consumption, 360-366
  magma generation, 361-364
  phase transition, 89
  rock types, 148, 162, 167, 172, 210
  seismic data, 261
  seismic wave velocities, 23, 25, 64, 141
  thermal structure, 360-361, 363
  ultramafic rocks, 106, 109
Island arc-type geosynclines, 218, 225, 367, 370-371
Isochrons map, ocean floor, 343-344
Isoclinic maps, 34-35
Isodynamic maps, 34
Isogonic maps, 34
Isomagnetic maps, 34, 36
Isoporic maps, 36
Isostasy, 4, 6-9, 25, 233, 240-241, 243

Isostatic compensation, Airy's theory, 38, 63
Pratt's theory, 38
Isostatic gravity anomaly, 38-39, 227
Isostatic uplift, 366
Isotherms, 357-358, 363
Isotope analysis, 168
Isotope evolution trends, 210
Isotopes, age determinations, 59-62
Isotope tracer studies, 160
Ivory Coast, 263

Jadeite, 52, 76-81, 83-85, 116, 121, 130, 134,
173-174, 179-180, 191
Japan, 10-11, 219
crustal structure, 147
heat flow, 41, 43
seismic wave velocities, 23, 155
tectonics, 350
Japan Sea-type geosynclines, 218, 225, 367, 370-371
Jaramillo event, 292-293, 295, 318, 320, 340
Java, 226-227
JOIDES, 255, 261, 301, 352-353, 372-373
Juan de Fuca ridge, 310, 316, 318-323, 337

Kalskaret, Norway, 117
Kamacite, 96
Kaolin, 51
Kermadec-Tonga arc system, 144, 148
Kenoran orogeny, 223-224
Kimberlites, 106-108, 110, 112-113, 134, 193
Kinetics of reactions, 252-253
Kuril Islands, crustal structure, 146-147
seismic wave velocity, 23
volcanic rock composition, 155
volume of eruptions, 155
Kyanite, 52-53, 77, 151, 173-174, 202

Land-bridges, 258
Land hemisphere, 6
Latent heat, 233, 236, 238, 242-243
Laurasia, 257-259, 265, 280
Lava flows, ultrabasic, 106
Layered intrusions, 106-107, 110
Lead isotope evolution trends, 223, 265
Leucite, 174
Lherzolite, 119
Limestone, 51
Liquidus surface, vapor-saturated, 183, 185
Lithification, 57
Lithophile elements, 102
Lithosphere, 25, 29, 63, 209-210, 228, 232-233, 255,
302-303, 349, 354
Lithosphere plate consumption, 253, 350, 360-366
adiabatic compression, 360-361
anomalous seismic zones, 350
attenuated seismic waves, 350
dehydration, 363
descent rate, 360

Lithosphere plate consumption *(continued)*
frictional heating, 360-361
phase transitions, 360-361
radiogenic heating, 360
shear strain heating, 361-363
Lithosphere plates, 266, 345, 347, 349
dynamics, 350
schematic cross-sections, 351
temperature distribution, 357, 360-361
world map, 347
Lithostratigraphic units, 61
Lizard peridotite, 111-112
Lombok, 226-227
Los Angeles, 175, 373
Low-density layer, mantle, 246-247
Low-Q zone, 350
Low-velocity layers, crust, 140
Low-velocity zone, mantle, 21, 22, 25, 29, 31, 33, 63,
131, 134-136, 207-208, 246-247, 302, 305, 349,
355, 358, 361
Low-viscosity layer, mantle, 33
Lunar rocks, 93
Lunar soil, 98

Magma fractionation, 54-55, 167, 171, 190, 195,
196-208
continuous, 172
dependence on pressure, 171, 197-199
eclogite, 196, 208
isobaric, 197
olivine, 193, 197-198, 208
orthopyroxene, 195, 197, 199, 203-204
Magma generation, 116, 134, 167-210, 356, 359
amount of melting, 203-205
composition of liquid, 205-207
crust, 54, 181-190
dependence on depth, 172, 193, 208
effects of water, 201-203
island arcs, 361-364
mantle, 54-55, 190-210
models, 187, 190-193
Magmas, 31, 48-49, 53-55, 59, 107, 167-168, 170,
200-201, 208-209, 230, 233, 248, 259
andesitic, 173, 189-190, 196, 206, 208-209
basaltic, 173, 190, 196-208
granitic, 173, 187, 189
picritic, 190, 193, 195-196, 204, 208
primary, 116, 168-171, 196-197
Magmatic processes, 54-57, 105
Magmatism, 374-375
Magmatist, 171
Magnesite, 53
Magnesium silicate, hydrated, 126
Magnetic anomalies, 1, 4, 5, 32, 34-36, 61, 255,
261-262, 269, 296, 307-342, 343, 348, 352-353,
356
amplitude of linear, 315, 323, 325
as isochrons for ocean floor, 332-333

Magnetic anomalies *(continued)*
  axial (of ocean rise) 322, 324-325, 332
  correlation, 321-332, 340-342
  criteria for sea-floor spreading, 326
  flank (of ocean rise), 322, 324-325, 332
  in single basaltic bodies, 334-335
  interpretations, 309-315, 337-342
  maps, 309-313, 325, 333, 336
  measured near ocean floor, 340-342
  numbering system, 326-330
  sequence, 293
  symmetry (about ocean rise) 322, 324, 326
  wavelength of linear, 315, 325
Magnetic anomaly profiles, 313, 318-321, 324-325, 327,
      336, 338-339, 342
  computed, 309, 312-314, 317-319, 325-326,
      338-339
Magnetic bight, 336
Magnetic cleaning, 270, 273, 286
Magnetic polarity reversals *(see also* Magnetic
      anomalies, paleomagnetism), 36, 61, 255, 261,
      269, 277-297, 313-317, 320, 327, 343
  duration of transition period, 286, 294-295
  history of discovery, 285-289
  igneous rocks, 285-292
  oceanic sediments, 292-297
  probability model, 289-290
Magnetic prospecting, 34
Magnetic surveys, air-borne, 139, 308, 315, 323-326
  ship-towed, 308, 315
Magnetite, 151, 312
Magnetization, correlation with higher oxidation, 334
  intensity of, 338
  primary, 270
  reversed, 285-297
  self-reversed, 285-286, 334-335
Magneto-hydrodynamic phenomena, 36
Magnetosphere, 1
Maine, 214
Mammoth event, 320-321
Mantle, 1-3, 16-17, 20-33, 48, 50-51, 58, 65-76,
      86-91, 93-137, 159-160, 190-210, 233, 234-247,
      249-252, 299-301, 323, 353-366
  heterogeneity, 23-26, 33, 63, 110-113, 116, 131, 159
  low-velocity zone, 21-23, 25, 29, 31, 33, 131-136,
      207-208, 246-247, 302, 305, 349, 355, 358, 361
  magma generation, 190-210
  phase transitions, 119-137, 235-246
  physical properties, 20-33, 36, 39, 44, 63, 89-90, 93,
      103, 119, 123, 130-131, 203, 205, 364
  transition zone, 29, 123-137
Mantle composition, 29, 68, 93-137, 159-160
  Fe/Mg ratio, 104, 130, 136-137
  tabulated, 104, 114
Mantle convection, 25-26, 31-32, 42, 46, 89, 137,
      159-160, 167, 232-233, 249-252, 259-261, 269,
      298-305, 313, 336, 344, 354, 356-360
Mantle-derived rocks, 26, 93, 105-119, 200

Mantled gneiss domes, 247
Mantle mineralogy, 50-51, 93-137, 152, 190-192,
      196, 199-200
  norms, 120
  olivine composition, 120, 123, 127
  tabulated, 120, 130
Mantle petrology, 28-29, 54, 131-137, 357-364
Marble, 51, 150-151
Marginal accretion, continents, 223, 160-165
Marianas arc, 219
Marianas trench, 7, 8
Marine geology, 255, 269
Marine geophysics, 3, 40, 255, 269
Marshall Islands, 147
Mass transport, 242
Matuyama epoch, 295, 340
Mayaguez, Puerto Rico, 112-113
Mediterranean Alpine range, 219-220, 228
Mediterranean geology, 263, 372-373
Mediterranean seas, heat flow, 41
Mediterranean-type geosynclines, 225-370
Mélanges, 366
Melilitites, 198
Melting of minerals and rocks, 50-55, 168, 173-176,
      182-190, 192-196, 201-205
Mentawi Islands, 226-227
Mesosphere, 349
Metamorphic belts, paired, 216
Metamorphic facies, 54, 230
Metamorphic processes, 48, 58-59, 156
Metamorphic reactions, 230
Metamorphic rocks, 48, 50-51, 53, 55, 110, 150, 157
  Australia, 156
  average composition, 152
  geosynclines, 228, 230-231, 364-368, 371
  magnetized, 285
  per cent Earth's crust, 151
  textures, 26, 59
  trenches, 366
Metamorphism, 49, 54, 58, 99, 106, 171, 180, 211,
      215-217, 222-223, 225, 270, 299, 338, 374-375
  paired belts, 216
Metasomatism, 49, 55, 106, 171
Metastable reactions, 55, 83, 252-253
Metastable pyroxenes, 195
Meteorites, 94-105
  accretion temperature, 101
  achondrites, 72-73, 94, 96-98, 102-104, 118
  apparent age, 102
  carbonaceous chondrites, 94-105, 160
  chondrites, 73, 94, 96, 98-103, 118
  classification, 94-101
  composition, 94-102
  earth bombardment, 1
  element abundances, 94-95, 99-101
  genetic relationships, 101-102
  iron, 94, 96-97, 102-103
  metal composition, 97-98, 102

Meteorites *(continued)*
  organic compounds, 98
  parent bodies, 72-74, 100-101, 118
  theories of origin, 101-102
Mexico, 263
Mica, 50, 151
Mica peridotite, 106
Microfossils, calcareous, 352-353
Mid-Atlantic ridge, 12, 143-147, 236, 261, 343
  anomalous mantle, 143-144
  Bouguer gravity anomaly, 143-145
  crustal structure, 144-146
  density variations, 144, 146
  drilling sites, 353
  fracture zones, 346
  heat flow, 44
  magnetic anomalies, 308, 314, 346
  metamorphosed basalt, 152
  Moho depth, 63
  rift valleys, 308, 335, 338-339
  spreading rates, 346
  topography, 308
Middle America arc, 351
Mid-oceanic ridges, 236, 255, 261, 299-300, 304, 343
  earthquakes, 13-15, 303, 343-344
  gravity profiles, 356-357
  heat flow, 43-46, 298, 356-357
  magnetic anomalies, 307-342
  mantle convection, 298-305, 356-359
  overlap of sediments, 300-301, 352-353, 373
  petrological structure, 357-359
  sea-floor spreading, 298-305
  seismic data, 261, 300
  thermal structure, 357-358
  world distribution, 8-9, 12-13, 344, 348
Migmatites, 49, 171
Mineral facies, peridotite, 118, 131
Mineralogy, crustal rocks, 2, 50-51, 151-152, 182-187
  mantle, 50-51, 120, 93-137, 152, 190-192,
    196, 199-200
Mineral reactions, 50-55
Miogeosyncline, 212 222, 227, 366-367, 369
Miscibility gap, 183
Mobilists, 255, 266, 298, 374-375, 377
Model ratios, 247
Mohorovicic discontinuity, 17, 63-76, 86-91, 142-143,
    170, 299-302, 233-246
  as chemical discontinuity, 68-69, 72
  as gabbro-eclogite transition zone, 67-76, 86-91,
    236-246
  as peridotite-serpentinite boundary, 67, 71-72, 87,
    235-236, 299-302
  defined, 64
  depth, 63-64, 66-67, 87-89, 372
  seismic velocities, 20, 23-26, 63-67
  tectonic significance, 233-246
  temperature, 73, 158
  thickness, 64, 87

Molasse, 222, 367
Monte Carlo method, 27-29, 131, 136
Moon, 259
  rocks, 93
  soil, 98
Mountain building, 3-4, 211-232, 366-372
  Andean type, 226, 367-368
  causes, 231-232, 353-356, 366-372
  continent/continent collision, 370, 372
  continent/island arc collision, 369, 372
  cordilleran, 368
  cycles, 161-165
  Himalayan type, 226, 367, 371
  island arc/cordilleran, 368, 372
  island arc type, 226, 367
Mountain ranges, 8-10, 47, 68, 211-212, 215
  persistent, 237-238
  rejuvenation, 237-238
  Tertiary, 9, 258-260, 347
Mount Everest, 7, 8
Multiple working hypotheses, method of, 375
Muscovite, 51-53, 180-181, 186, 364

Nappes, 215, 222
Nebraska, 88
Nepheline, 76-77, 191, 197
Nepheline syenite, 179, 182
Nephelinites, 192, 197-198, 204, 206
Neptunian theory, 255, 378
Nevada, 88
New global tectonics, 4, 15, 212, 266, 348-351, 373,
    375, 377-378
New Guinea, 10, 222, 257
New Hampshire, 216
New Jersey, 271, 273
New York, 214, 216, 231
New Zealand, 6, 10, 144, 257, 263
Niger delta, 262
Nigeria, 263
Nizhnyaya Tunguska, 281
Noachian deluge, 378
Noble gases in meteorites, 102
Nodules, ultramafic, 26
  average composition, 112-113
  eclogite, 106, 109-110
  modal mineralogy, 120
  peridotite, 106, 109-116, 193-194
Non-orogenic areas, heat flow, 43
  seismic velocities, 65-67
North America, 258, 262, 301
  age, 161-165
  crustal composition, 154-155
  crustal sections, 140-143
  heat flow, 41, 157-158
  land bridges, 258
  orogenic trend lines, 163-164
  polar wandering path, 274-283
  paleomagnetic poles, 274-283

North American cordillera, 369
North American plate, 337
North American sediments, volume relations, 366
Norway, 256, 258
Nuclear explosions, underground, 17, 19, 23, 24
Nucleation, 81-83, 86, 202, 253
Nucleosynthesis, 102

Obsidian, 248
Ocean basins (*see also* Crust, oceanic), 1, 5, 7-8,
    10-12, 244, 259-261, 298-305
  isochrons, 332-333, 343-344
  magnetic anomalies, 307-342
Ocean basins, small, 12, 225, 350
Ocean currents, 1
Oceanic crust (*see* Crust, oceanic)
Oceanic plate, underthrusting, 350
Oceanic ridges and rises, 8-11, 15, 144-146, 298-305,
    356-359, 348-351
  Bouguer gravity anomaly, 141, 143-145
  crustal thickness, 141
  Explorer, 310
  geological features, 141, 337-340
  Gorda, 310, 321, 337
  heat flow, 32, 43, 141
  Juan de Fuca, 89, 310, 316, 320-323, 337
  magnetic anomalies, 36, 313-331, 335-340
    central, 313, 318
    flank, 313
  mantle composition, 112-113
  migrating with convection, 336
  peridotite, 106, 109
  petrology, 357-359
  phase transitions, 89-90
  sediment thickness, 330
  seismic wave velocities, 23, 25, 64, 141
  serpentinite, 106, 109, 112-113
  temperature distribution, 32, 357-358
Oceanic trenches, 8-13, 213, 219, 225-228, 299, 303,
    343, 345, 349, 367
  area, 10
  earthquake focal mechanisms, 350
  faulting, 355-356
  gravity data, 350
  heat flow, 43
  Moho depth, 63-64
  peridotite, 106, 109
  sediments, undistorted, 356, 364-365, 373
  sediment volume relations, 364-366
  serpentinites, 106, 109
  structures, 364-366
Oceanization, 160, 223, 266, 372-373, 375-376
Oceans, areas, 6, 40
  distribution, 5-6, 39
  mean depths, 6
Ocean sediments, paleontological age, 352-353
Oil, 59, 246
Olduvai event, 292-293, 295, 318, 320, 337

Olivine, 26, 51, 76, 85, 96, 98-99, 106, 113, 117, 119,
    123-130, 134, 151, 170, 190-192, 199, 201, 312
Olivine-spinel transition, 123-130, 358, 361, 363
Omphacite, 52, 76, 78, 85
Ophiolite suite, 106, 109, 167-168, 211, 212, 216,
    219, 222, 367, 372, 374
Ore deposits, 34, 49
Organic compounds, 98
Orogenic belts, 5, 10, 12-16, 25, 59, 156, 161-165,
    225-228, 230-232, 366-372, 374
  area, 10
  crustal composition, 154
  crustal thickness, 64-67
  heat flow, 43, 156
  metamorphism, 58, 216-217, 230-231, 366-372
  phase transitions, 89
  ultramafic rocks, 106, 108-110, 112
Orogenic cycle, 49, 89, 211-232, 259, 366-372
Orogenic polarity, 219, 222
Orthoclase, 53, 151, 173-174, 181, 187
Orthogeosynclines, 212-213, 215
Overthrusting, 211, 231
Oxidation, correlated with reversed magnetization, 334
Oxygen fugacity, 190, 209
Oxygen isotopes, 209-210
  in meteorites, 102

Pacific-Antarctic ridge, fracture zones, 346
  magnetic anomalies, 325-326
Pacific geosynclines, 222-223
Pacific Ocean, 6, 12-13, 215, 219, 222, 236, 259, 266,
    292, 343
  continental margins, 147-148
  crustal section, 144
  fracture zones, 346
  geomorphic provinces, 311
  heat flow, 41-42, 44, 87
  high seismic velocity layer, 146-147
  magnetic anomalies, 308-314, 322-332, 337
  sediment thickness, 329-332
  sediment volume, 366
  seismic wave velocities, 23, 65-66, 146-147
  spreading rates, 328-329
  ultramafic rocks, 109
Paleoclimatology, 256, 258-259, 274, 373-374,
    376, 378
Paleogeography, 211, 225, 269, 280-281, 364, 371
Paleolatitude, 258, 273
Paleomagnetic inclinations, errors in sediments, 279
Paleomagnetic pole positions, 260-261, 271-277, 307
Paleomagnetic stratigraphy, deep-sea cores, 292-293
Paleomagnetic triangulation, 277-279
Paleomagnetism (*see also* Earth's magnetic field,
    Magnetic anomalies, Magnetic polarity reversals,
    Remanent magnetism), 1, 3, 5, 33, 36, 255,
    260-261, 269-297, 307, 371, 375
  basic assumptions, 271
  global synthesis, 279-281

Paleomeridians, 277-279
Paleontology, 2, 256, 258, 293, 373
Panama fracture zone, 351
Pangaea, 256, 263, 373
Parageosynclines, 213, 216
Paraliageosynclines, 213, 216
Paving stone theory, 345
Pelagic ooze, 353
Peneplanation, 211
Penrose conference, 374
Periclase, 52-53
Peridotite, 51, 54-55, 67-72, 87, 90, 103, 109, 111,
    114, 150-151, 170, 172-174, 190, 193-196, 205,
    207, 210, 229, 246, 299, 312, 318, 356-359,
    361-364
  experimentally deformed, 33, 356
  field associations, 105-106
  garnet-, 112-113, 117, 119, 129, 192-193, 206
  hypothetical mantle, 29, 111, 113-117, 136
  mineral facies, 118-120
  mineralogy, 105-106, 120
  mylonites, 113
  phase diagrams, 118, 120, 132
Peridotite nodules (*see* Nodules)
Peridotite-serpentinite phase transition, 71-72, 233,
    235-237, 298-303
Periodic table, 99
Perovskite structure, 130
Peru-Brazil region, 350
Peru-Chile trench, 364-365
Petrogenesis, 167-173, 192, 196-201, 207-208, 359
Petrogenetic grid, basalts, 203, 205
  metamorphic rocks, 230
Petroleum industry, 59
Petrology, 2, 31, 55-59, 105-110, 131-137, 150-152,
    168-172, 216-217, 223, 357 360, 361-366
Phase relationships, determination of, 182
Phase transition boundary,
  dynamics of motion, 238-246
  effect of pressure perturbation, 233-234, 237-238,
    249
  effect of thermal perturbation, 233 234, 235-237,
    249
  effect on mantle convection, 249-252
  steady-state configurations, 238-241
  time-dependent solutions, 241-246
Phase transitions, 50-55, 67-91, 119-137, 233-253
  gabbro-eclogite, 72-91, 233, 236-237, 246, 260-261
  latent heat, 233, 236, 238, 242-243
  olivine-spinel, 123-130
  peridotite-serpentinite, 71-72, 87, 233, 235-236,
    298-302
  reaction rates, 70, 250, 252-253
  tectonic significance, 63, 86-91, 229, 233-253,
    360-361
Phase transitions in mantle, 33, 47, 119-137, 360-361
  as cause of earthquakes, 252-253
  eclogite, 122, 133

Phase transitions in mantle *(continued)*
  implosive, 253
  in lithosphere slab, 360-361
  metastable, 252-253
  olivine-spinel, 123-130, 253
  peridotite, 119-122, 132
  post-spinel, 129-130
Phenocrysts, 182
Philippines, 10
Phlogopite, 51, 106, 120-121, 134, 203, 364
Phonolite, 168
Photon, 31
Physical sciences, 2
Physiographic features, 2, 8-13
Picritic liquid magmas, 111, 117, 131
Piedmont plains, 11
Plagioclase, 52-53, 76, 80, 82-85, 99, 106, 116,
    118-121, 131, 151, 181, 185, 187, 190-192,
    201-202, 209
Planetary ocean-current patterns, 376-377
Planetary wind circulation, 376-377
Plateaus, 8, 68, 236
Plate boundaries, 344-345, 347, 351
  triple junctions, 336-337, 352
Plate collision, 212, 222, 223, 374
Plates, small crustal, 347, 351-352
Plate tectonics, 4, 5, 109, 159, 211, 216, 222-223,
    226, 228, 231-232, 255-256, 343-373
  angular velocity of plates, 345-346, 348
  mechanism, 353-356
  ruling theory, 373, 378
  working hypothesis, 373, 378
  world-wide analysis, 346-348
Platinum capsules, 195-196
Plutonists, 255, 378
Pneumatolytic gas, 49
Polarity epochs, (*see also* Magnetic polarity reversals),
    286-297
Polarity events (*see also* Magnetic polarity reversals),
    286-297
  resolution of, 288
Polar wandering, 4, 36, 258-260, 272-283
Pole of relative rotation, 345-348, 353
Polflucht, 259
Polymorphs, Al_2SiO_5, 83
Pore fluids, 49, 53, 55, 119, 173, 178, 180-181,
    189-190
Potassium-argon dating, 165, 263-264, 282, 287-291,
    296-297, 320-321, 325
Potassium/rubidium, 98
  mantle, 118
  meteorites, 98
Potassium/thorium, Australian rocks, 156
Potassium/uranium, 98
  Australian rocks, 156
  mantle, 118
  meteorites, 98
Power-creep equation, 33

Pressure perturbations, 233-234, 237-238, 249
Pressure vessel, Tuttle, 171-173
  internally heated, 173
Pressure within the Earth, 26, 29-30
Primary geosynclines, 213, 216
Primary magmas, 116, 168-172, 195-196, 208-210
Protostars, 102
Protosun, 102
Putties, 246-248
P-wave velocities (*see* Seismic wave velocities)
Pyrolite, 28-29, 111-115, 200, 204
Pyrope, 50, 52, 76-79, 82, 84-85, 119
Pyroxene, 50-51, 76, 85-86, 98-99, 106, 118-120,
  129-130, 134, 151, 190-192, 201, 204
Pyroxene granulite, 82, 156
Pyroxenite, 117, 203

Quartz, 50-53, 76-77, 79-80, 151-152, 173-174,
  179-182, 185, 187-188, 190, 202-203
Quiet magnetic zones, 262

Radiative heat transfer, 32
Radioactive elements, 1, 31, 59, 61-62, 225, 231, 240,
  353-354, 356
  Australia, 156-157
  distribution, 42, 68, 73, 99, 156-161
  short-lived, 103
  United States, 157-158
Radioactive heat source term, 242
Radiolarian species, 292
Radiometric ages, 161-165, 261, 165, 286
Rare earth elements, 105, 118, 152, 161, 207
Rayleigh-Benard convection model, 354
Reaction, adiabatic, 250
Reaction rates, 70, 252-253
Reaction relationship, 170, 193
Reaction series, 54-55
Recrystallization, 55, 59
Red beds, 279-280
Red Sea rift, 301
Remagnetization hypothesis, 279-280
Remanent magnetization, 255, 269-271, 334
  chemical, 270
  depositional, 270
  detrital, 270
  inclination error, 270-271
  oceanic anomalies, 314-315
  post-depositional, 270
  secondary, 270, 275
  thermoremanent, 269-270, 334-335
  viscous, 270, 273
Residual geosynclines, 213
Residual liquids, 108, 181, 195-196, 209
Revolution in earth sciences (*see* Counter-revolution),
  3-4, 266-267, 298, 315, 373-374
Reykjanes ridge, magnetic anomalies, 323-326
  magnetized basalts, 335
Rheological properties, 247

Rhyolite, 150, 155, 168, 175, 210-211, 222
Ridges, geosynclinal, 213-214, 217, 226-228
Rift systems, 9, 12-16, 89, 213, 216, 222, 301
Ring complexes, 106
Ring of fire, 13
Rock cycle, 47, 49-51, 55, 156
Rocks, crustal abundances, 151-152
  experimental seismic velocities, 66-67
  magnetized, 269-270
  physical properties, 139
  seismic velocities, 139, 152
Rock units, classification, 61
  succession, 59
Rock-water systems, 173-181
Rocky Mountains, 10, 142, 149
  low-velocity zone, 22
  seismic wave velocities, 22-25
Rubidium/strontium, 98
  mantle, 118
  meteorites, 98
Rubidium-strontium dating, 165, 263 264
Russian platform, 143, 371

St. Lawrence graben system, 264-265
St. Peter and St. Paul Rocks, 113, 119
Salt deposits, 258
Salt domes, 247
Salts, 51
San Andreas fault, 13, 316, 345, 366
Sands, 150-151
Sandstone, 51
San Francisco, 140, 373
Sapphirine, 52, 77
Satellites, artificial, 159-160
  geoid maps, 38-39
  gravitational potential, 39
  gravity surveys, 37, 39, 305
Satellitic material, accretion of, 375
Saturation boundary, 176-177, 188-189
Saturation magnetization, 335
Scandinavia, 264
Schist, 51, 150
Scientific revolutions, 266
Scotland, 170, 258
Sea-floor spreading, 4-5, 72, 109, 152, 159, 219, 229,
  231-232, 255, 259-260, 266, 293, 296, 298-305,
  307, 312-314, 316-317, 319-320, 326, 329-332,
  334, 337, 344, 348, 352-353, 356, 366-372, 375
Sea level, 10
Sea of Okhotsk, crustal structure, 146-148
Seas, epicontinental, 244
Secondary geosynclines, 213
Sedimentary basins, 10
Sedimentary processes, 48-49, 57-58, 223, 244, 374
Sedimentary rocks, 48-49, 51, 55
  Australia, 156
  average composition, 152

Sedimentary rocks *(continued)*
  geosynclines, 218, 230
  per cent crust, 151
  red, 271, 273
  trenches, 366
Sedimentation rate, 239-241, 242-245
  ocean, 292-294
  Pacific Ocean, 330
Sediments, 10-12, 48, 50, 53, 58, 109, 152, 156, 270
  deposition, 55, 234, 238-246
  thermal blanket effect, 234, 238-240
  transportation, 12, 55, 57, 59
Sediments, oceanic, 40, 63, 65-66, 72, 148, 150-151
      210-211, 222, 228
  Atlantic coast prisms, 217
  calcareous, 150, 218
  clastic, 218
  deep-sea cores, 261, 352-353
  deep-water, 212, 226
  Gulf coast prisms, 217
  magnetized, 292-297, 329
  pelagic, 218, 220
  red clays, 150
  shallow-water, 212, 218-219
  siliceous, 150
  terrigenous, 220, 222, 365
  thickness, Pacific, 329-332
  turbidites, 217-218, 371
Seismic waves, 17-26
  attenuation, 25
  attenuation in lithosphere slab, 350
  body waves, 18-26, 29
  reflection, 18
  refraction, 18
  surface waves, 18-19, 28, 147, 159-160
  travel time anomalies, 39
  travel times, 19-20, 29
Seismic wave velocities, 17-26, 63-67, 139-149
  anisotropy, 356
  body waves, 19, 20-26, 28-29, 63-67
  continental margins, 146, 148
  crust, 139, 144, 148
  effect of crystal orientation, 356
  experimental measurements in rocks, 66-67
  low-velocity zone, 134-135
  mantle transition zone, 129-130
  Moho, 63
  oceanic sediments, 65-66, 148
  profiles through the Earth, 20-26
Seismology, 5, 17-26, 30, 63, 261, 315, 348-351,
      360, 364
Serpentine, 51-53, 71, 98, 106, 120-121, 235,
      312, 318
Serpentinite, 68-72, 87, 106, 109, 111-114, 152, 222,
      299, 320, 364
Serpentinite-peridotite associations, 106
Serpentinization, 235-236
Shadow zone, 20

Shale, 51, 150, 151
Shearing stresses, 360
Shields, continental, 10, 141
  Australia, 156
  Bouguer gravity anomalies, 141
  crustal composition, 154
  crustal thickness, 141
  heat flow, 43, 45, 141, 156
  Moho thickness, 64
  North America, 161-163
  seismic wave velocities, 25, 141
  world-wide ages, 164-165
Shock wave experiments, 93, 130, 136
Shore line, 9
Sierra Leone, 143, 146
Sierra Nevada, 142-143
  heat flow, 157
Sierrania de la Ronda, Spain, 117
Siberian platform, 371
Silicate liquid, immiscible, 97
Silicon, in core, 105
Silicone, 247-248
Sillimanite, 52, 77, 151, 181, 202
Sills, trans-oceanic, 377
Sinker, 304-305
  eclogite, 261
Sinking lithosphere slabs, 249, 253
Skye, Scotland, 117
Solar abundances of elements, 93-95, 99-100
  of iron, 103
Solar nebula, 98, 100-103
Solar radiation, 1-2, 36, 48
Solar system, 2, 93, 102
Solidus surface, divariant, 178-181
Solution chemistry, 55
South America, 256-258, 262-263, 301
  age provinces, 263-264
  continental drift, 256
  land bridges, 258
  lithology, 258
  polar wandering path, 279-281
  stratigraphy, 258
  tillites, 258
South Dakota, 271-272
Spain, 6, 263
Spanish Sahara, 231
Spherical harmonic analysis,
  gravity field, 33, 39
  heat flow, 33, 43-45
  magnetic field, 33, 35, 335
  surface topography, 33
  tectonic features, 39
  travel times, seismic waves, 33
Spilites, 211, 216
Spinel, 51, 79-80, 106, 118-120, 123-130,
      134, 190-192
Spreading directions, 352
Spreading geometry, 336-337

Spreading rate, 318, 320-321, 323, 325, 327, 336, 340, 346-347, 352-353, 360
  comparison for major oceans, 328-329
  correlation with physical features, 335-336
  episodic, 329-332, 336-337, 352
  variations, 329, 338
Stabilists, 255, 266
Stars, 94
Staurolite, 51
Steady state deposition, 33
Step faults, 356
Stereosphere (*see* Lithosphere)
Strain rate, 33
Stratiform intrusions (*see* Layered intrusions)
Stratigraphic classification, 59-61
Stratigraphic facies, 367, 374
Stratigraphy, 2, 47, 59, 258, 293
Strength of materials, 246
Strength ratio, 247
Strontium isotope evolution trend, 165, 223
Strontium isotope ratios, 98, 165
  mantle, 118
  meteorites, 98
Strontium plumbate structure, 130
Structural provinces, 258, 261
Structural studies, 223
Subcrustal currents, 228-229
Subduction zones, 222, 367, 371, 374
Submarine canyons, 12
Submersible instrument package (see "Fish")
Subsidence, 4, 6, 49, 58, 67, 233, 238-246, 248-249
  causes, 228-229
  in centrifuge models, 248-249
  isostatic, 219
  rates, 212
Sulfur, in core, 105
Sumatra, 226-227
Sumbawa, 226-227
Sun, composition of, 94, 96
Sunda Islands, 226-228
Sunspot activity, 36
Surface films, 55
Sweden, 264
Switzerland, 112
Syenite, 150-151, 175, 179, 182
Synclinoria, 371
Syntectonic recrystallization, 356
Systems, phase equilibria, CaO-Al_2O_3-alkalis, 116
  CaO-MgO-Al_2O_3-SiO_2, 79, 116, 119, 190-192, 196
  feldspar-quartz-H_2O, 173-174
  FeO-Fe_2O_3-TiO_2, 270
  FeO-MgO-SiO_2, 116
  Fe_2SiO_4-Mg_2SiO_4, 123-126
  gabbro-H_2O, 54, 121, 133-134, 175, 178, 188, 200-203
  gabbro-H_2O-CO_2, 180
  granite-H_2O, 54, 175
  granodiorite-H_2O, 186-188, 203

Systems, phase equilibria (*continued*)
  $KAlSi_3O_8$-$NaAlSi_3O_8$-H_2O, 171, 173, 183
  $KAlSi_3O_8$-$NaAlSi_3O_8$-SiO_2-H_2O, 171, 182-185
  Mg_2GeO_4-Mg_2SiO_4, 123
  MgO-FeO-Fe_2O_3-SiO_2, 209
  $MgSiO_3$-H_2O, 206
  Mg_2SiO_4-$NaAlSiO_4$-$CaAl_2SiO_6$-SiO_2-H_2O, 206-207
  Mg_2SiO_4-SiO_2-X, 193
  muscovite-SiO_2-H_2O, 180
  $NaAlSiO_4$-$KAlSiO_4$-SiO_2, 181
  Ni_2GeO_4-Mg_2SiO_4, 125-126
  peridotite-H_2O, 54, 121, 132, 134-135, 175-176, 178, 203, 208
  residua system, 181-182, 186
  rock-H_2O, 173-175
  silicate-H_2O, 176, 178
  SiO_2, 51
  syenite-H_2O, 175
  tonalite-H_2O, 175
Taenite, 96
Talc, 71
Tanimbar Islands, 226-227
Tape-recorder analogy, 319
Taphrogeosynclines, 213, 216
Tasman geosyncline, 222
Tectogene, 228-229
Tectonic classification of Earth's surface, 140-141
Tectonic cycle, 47, 49
Tectonic elements, 212-216
Tectonic features, world map, 9
Tectonic overprinting, 225
Tectonic significance of phase transitions, 63, 86-91, 229, 233-253, 360-361
Tectonics on a sphere, 344-352
Tectonic structures, centrifuge models, 246-249
Tectonite fabrics, 356
Temperature distribution in earth, 26, 30-32, 40, 70, 73, 158, 167, 170, 228-229, 233, 235, 238, 253
Tethys Sea, 257, 280
Theological cause, 378
Thermal areas, 41, 43
Thermal blanket effect of sediments, 234, 238-240
Thermal cleaning, 270-271
Thermal conductivity, 31, 40, 46, 158, 239, 243, 360, 362
Thermal constants, 245
Thermal divides, 172, 182, 192, 196
Thermal energy, 47, 260, 304-305
Thermal perturbations, 233-234, 235-237, 249
Thermodynamic equilibrium, 50, 55
Tholeiite, 112-113, 167-168, 170, 172, 191-196, 208, 218
  oceanic, 150-151, 167, 171
  olivine, 80, 83, 86, 113, 191, 195, 197, 199, 359
  quartz, 81-84, 191
Tidal energy, 259, 356
  dissipation mechanisms, 356
Tidal-magmatic mechanism, 356

Tiefkraton, 212-213
Tillites, 258, 376-377
Time units, classification, 61
Timor, 226-227
Titanomagnetite, 285, 334
Tonalite, 174-175
Topographic gravity correction, 38
Trace elements, 168, 207, 210
  crust, 105, 156-157
  fractionation, 359
  grain boundaries, 118
  mantle, 118-119
  microinclusions, 118
  peridotite, 111
Trachyte, 168, 182
Transcurrent faults, 12, 13, 315-316, 349, 375
Transform faults, 12, 13, 261, 315-316, 335, 344-345,
    347-349, 375
Transformist, 171
Tremolite, 52
Troilite, 94, 96, 99, 103
Tschermak's molecule, 85, 190-192
Turbidites, 217-218, 371
Turbidity currents, 11, 217, 295
Tuttle pressure vessel, 171, 173

Ubekendt Ejland, west Greenland, 117
Ugelvik, Norway, 112
Ultrabasic lavas, 107-108, 211
Ultrabasic magmas, 107-108
Ultramafic rocks (*see also* Nodules, peridotites),
    105-112, 114-118
  alkali contents, 115-117
  associations, 105-106
  compositions, 111-117
  element abundances, 114-115
  extrapolation to mantle, 110-111
  kimberlites, 106-108, 110, 112-113, 134, 193
  layered intrusions, 106-107, 168
  oceanic types, 109-113
  ophiolites, 106, 109, 167-168, 211-212, 216, 219,
    222, 367, 372, 374
  orogenic types, 26, 106, 108-111, 117, 168, 217,
    364, 372
  petrogenesis, 106-110, 167-168
Unconformity, 49, 59
Uplift, 49, 58-59, 66-68, 211-212, 216, 221-222,
    228-229, 233-234, 236-246
Upper Mantle Committee, 40
Upper Mantle Project, 374
Undation (*see* Epeirogenesis)
Undation theory, 376
Uniformitarianism, 255, 378
United States, 214, 217, 228, 291
  crustal composition, 154
  crustal structure, 87-89, 140-143
  crustal thickness, 24-25
  eastern superprovince, 25, 142-143, 149, 154

United States (*continued*)
  heat flow, 143, 157-158
  isostatic anomalies, 142-143
  seismic wave velocities, 22-25, 64, 142
  tectonic provinces, 22
  western superprovince, 25, 142-143, 149, 154
Universe, origin of, 2
Ural Mountains, 371

Valency, 161
Vapor-absent assemblage, 120-121, 132-136, 176, 178,
    188-189, 204
Vapor-excess reactions, 120-121
Verticalists, 298
Vertical movements, 222, 229, 232, 246-249, 251,
    364, 373-375
Vesicles in basalts, 335
Viscosity, 26, 32-33, 207, 354, 357-358, 363
  magma, 169
  mantle, 32-33, 159, 305
Viscosity ratio, 247
Viscous decay, magnetization, 270
Viscous dissipation, 360, 362
Viscous drag, 354
Volatiles in magmas, 108
Volcanic activity, 235-236, 345
Volcanic-sedimentary cycles, 161-165
Volcanoes, 31, 47, 49, 113, 115, 119, 303-304
  Cenozoic, 9-10
  distribution, 5, 13
  energy, 39, 353-354
  heat flow, 43
  oceanic, 13

Wasatch mountain front, 142-143
Washington, D. C., 140
Water, in crust, 54, 153-156
  in high pressure experiments, 170-181, 201-207
  in magma generation, 55-56, 203-207
  in mantle, 69, 71-72, 113-114, 118-122, 132-136,
    235, 299, 357-359, 361-364
  juvenile, 49, 119, 172-173, 201, 207, 299
  meteoritic, 49, 119
  underground, 40, 48-50
Water-absent systems, 173-176, 187
Water-deficient systems, 176-181, 184, 188, 203,
    206-207, 208
Water-excess systems, 173-176, 187
Water hemisphere, 6
Water pressure, 177-181, 183-186, 230
Water-saturated liquid, 173-176, 179-180, 183-187,
    189, 196, 204
Water-undersaturated liquid, 177-178, 184, 189, 208
Wax, 246
Weathering, 1, 48-49, 57-58, 152, 271
West Indies arc, 351
Williamson-Adams equation, 26
Wind direction, 258

Wollastonite, 77-78
World climate fluctuations, 377
World rift system, 8-9, 12-13, 231, 348

Xenocrysts, 108

Xenoliths, 108
Xenon in meteorites, 102

Zeugeosynclines, 213, 216
Zoisite, 173-174, 202

MENTAL HEALTH MATTERS

This Reader forms part of The Open University course *Mental Health and Distress: Perspectives and Practice* (K257), a second-level undergraduate course and an optional course for the Diploma in Health and Social Welfare. The selection of items is therefore related to other material available to students. A companion book for the course is *Speaking our Minds: an anthology of personal experiences of mental distress and its consequences.* Both books are designed to evoke the critical understanding of students. Opinions expressed in them are not necessarily those of the course team or of The Open University. If you are interested in studying the course or working towards a degree or diploma, please write to the Information Officer, School of Health and Social Welfare, The Open University, Walton Hall, Milton Keynes MK7 6AA, UK.

MENTAL HEALTH MATTERS:
A READER

Edited by

Tom Heller, Jill Reynolds, Roger Gomm,
Rosemary Muston and Stephen Pattison

MACMILLAN in association with

The Open
University

Editorial material, selection, commissioned articles
© The Open University 1996

First published 1996 by
MACMILLAN PRESS LTD
Houndmills, Basingstoke, Hampshire RG21 6XS
and London
Companies and representatives
throughout the world

ISBN 0–333–67847–8 hardcover
ISBN 0–333–67848–6 paperback

A catalogue record for this book is available
from the British Library.

This book is printed on paper suitable for recycling and
made from fully managed and sustained forest sources.

10   9   8   7   6   5   4   3   2   1
05   04   03   02   01   00   99   98   97   96

Typeset by Footnote Graphics, Warminster, Wilts

Printed and bound in Great Britain by
T. J. Press (Padstow) Ltd

Contents

Acknowledgements   ix
Notes on the Contributors   xi
General Introduction   xxi

I Understanding Mental Health and Distress: Debates and Theories

1 Introduction   3

2 Psychological approaches to mental health and distress
   Rudi Dallos   6

3 The nature of psychiatric disorders
   R. E. Kendell   17

4 'Schizophrenia' re-evaluated
   Mary Boyle   27

5 Life events, loss and depressive disorders
   George W. Brown   36

6 Women and madness: the mental asylum
   Phyllis Chesler   46

7 The cultural context of mental distress
   Richard Warner   54

8 Labelling mental illness
   Thomas J. Scheff   64

9 On being sane in insane places   70
   David L. Rosenhan

10   Reversing deviance   79
      Roger Gomm

11   Some problematic aspects of dementia   87
      Tom Kitwood

12   Families and the experience of mental distress   97
      David W. Jones

II   **Mental Health Policy, Social Inequality and Civil Rights**

13   Introduction   107

14   Mental health and inequality   110
      Roger Gomm

15   George III and changing views of madness   121
      Linda Jones

16   Professionals, the state and the development of mental
      health policy   134
      Joan Busfield

17   Scare in the community: Britain in moral panic   143
      Matt Muijen

18   Beyond the asylum   157
      Julian Leff

19   Media images of mental distress   163
      Greg Philo, Jenny Secker, Steve Platt, Lesley Henderson,
      Greg McLaughlin and Jocelyn Burnside

20   Towards understanding suicide   171
      Steve Taylor and Alan Gilmour

21   Two notions of risk in mental health debates   181
      David Pilgrim and Anne Rogers

22   The history of tranquilliser use   186
      Jonathan Gabe

23   The black experience of mental health law   196
      Deryck Browne

24   The need to change mental health law   205
     Nigel Eastman

III Involving Users in Mental Health Services

25   Introduction   215

26   The history of the user movement in the United Kindom   218
     Peter Campbell

27   The lives of 'users'   226
     Peter Barham and Robert Hayward

28   Structuring effective user involvement   238
     Guillermo Garcia Maza

29   Mental health services that empower women   242
     Jennie Williams and Gilli Watson

30   Asian women speak out   252
     Steve Fenton and Azra Sadiq

31   Focusing on health: focus groups for consulting about health
     needs   260
     Tang My and Christopher Cuninghame

32   Developing crisis services   266
     Liz Sayce, Yvonne Christie, Mike Slade and Alison Cobb

IV Examining Practice

33   Introduction   277

34   Using research to change practice   282
     Glenys Parry

35   Reviewing advances in psychiatry   290
     Rosalind Ramsay and Tom Fahy

36   Normalising professional skills   297
     David Brandon

37   Developing a bridge to women's social action   304
     Sue Holland

38   Working psychotherapeutically with adult survivors of child
     sexual abuse   309
     Fiona Gardner

39   Treating anorexia nervosa   319
     Janice Russell

40   Rehabilitating voice-hearers   326
     Marius Romme

41   Communicating as if your life depended upon it: life history
     work with people with dementia   333
     John Killick

42   Practising cultural psychiatry: the doctor's dilemma   339
     Sourangshu Acharyya

43   Maintaining an emergency service   346
     Peter Tyrer

44   Working with refugees and torture survivors: help for the
     helpers   355
     Jane Shackman and Jill Reynolds

45   Doing being human: reflective practice in mental health work   364
     Tom Heller

Index   371

Acknowledgements

The authors and publishers wish to thank the following for permission to use copyright material: American Association for the Advancement of Science, for material from David Rosenhan, 'On Being Sane in Insane Places', *Science*, 1791 (1973) Copyright © 1973 American Association for the Advancement of Science; The American Psychiatric Press Inc and the author, for adapted text from George W. Brown, 'Loss and Depressive Disorders', Paul Hoch lecture (March 1994), reproduced in B. P. Dohrenwend (ed.), *Adversity, Stress and Psychpathology* (1966); Artesian Books, for material from Fiona Gardner, 'Psychotherapy with Adult Survivors of Child Sexual Abuse', *British Journal of Psychotherapy*, 6(3) (1990); The Athlone Press Ltd, for Tom Kitwood, 'Some Social Aspects of the History of Social Dementia', in German Berrios and Roy Porter (eds), *The History of Clinical Psychiatry: The Origin and History of Psychiatric Disorders* (1995); Blackwell Science Ltd, for material from Sourangshu Acharyya, 'The Doctor's Dilemma: the Practice of Cultural Psychiatry in Multicultural Britain', in J. Kareem and R. Littlewood (eds), *Intercultural Therapy: Themes, Interpretations and Practice* (1992); BMJ Publishing Group, for material from Nigel Eastman, 'Mental Health Law – Civil Liberties and the Principle of Reciprocity', *British Medical Journal*, 308 (1994), Rosalind Ramsay and Tom Fahy, 'Recent Advances on Psychiatry', *British Medical Journal*, 311 (1995), and Janice Russell, 'Treating Anorexia Nervosa: Clinical Concerns, Personal Views', *British Medical Journal*, 311 (1995); The British Psychological Society, for material from Mary Boyle, 'Form and Content, Function and Meaning in the Analysis of "Schizophrenic" Behaviour', *Clinical Psychological Forum* (September 1992); Cambridge University Press, for material from Peter Tyrer, 'Maintaining an Emergency Service', and Liz Sayce, Yvonne Christie, Mike Slade and Alison Cobb, 'Users' Perspective on Emergency Needs' in M. Phelan, G. Strathdee and G. Thornicroft (eds), *Emergency Mental Health Services in the Community* (1995); Churchill Livingstone, for material from Robert Kendell, 'The Nature of Psychiatric Disorders', in R. E. Kendell and A. K. Zealley (eds), *Companion to Psychiatric Studies*, 3rd edn, (1993); Commission for Racial Equality, for material from Steve Fenton and Azra Sadiq, The Sorrow in My Heart', in *The Sorrow in My Heart: Asian Women Speak Out About Depression* (1993); Dementia Services Development Centre, for John Killock, 'Please Give Me Back My Personality!' in *Writing and Dementia* (1994); the Controller of Her Majesty's Stationery Office, for Crown copyright material; Health Education Authority, for material from G. Philo, J. Secker, S. Platt, G. McLaughlin and J. Burnside, 'The Impact of the Mass Media on Public Images of Mental Illness: Media Content and Audience Belief', *Health Education Journal*, 53 (1994); Macmillan Magazines Ltd, for material from Julian Leff, 'Beyond the Asylum', *Health Service Journal* (22 June

1995); and Tang My and Christopher Cunninghame, 'Ways of Saying', *Health Service Journal* (15 September 1994); Macmillan Press Ltd, for material from David Brandon, 'The Implications of Normalisation Work for Professional Skills' in S. Ramon, *Beyond Community Care: Normalisation and Integration Work*, MIND/ Macmillan (1991); Reed Business Publishing, for material from Matt Muijen, 'Scare in the Community: Britain in Moral Panic', *Community Care* (7–13 September 1995); Routledge, for material from Deryck Browne, 'Sectioning: the Black Experience', and Sue Holland, 'Interaction in Women's Mental Health and Neighbourhood Development' in S. Fernando (ed.), *Mental Health in a Multi-ethnic Society: A Multi-disciplinary Handbook* (1995), Peter Barham and Robert Hayward, 'Community Mental Patients?', in Peter Barham and Robert Hayward (eds), *Relocating Madness: Frm the Mental Patient to the Person* (1995), Richard Warner, 'Cultural Explorations of Psychotic Behaviour; in *Recovery from Schizophrenia: Psychiatry and Political Economy*, Routledge & Kegan Paul (1985), and Joan Busfield, 'Professionals, the State and Development of Mental Health Policy' in *Managing Madness: Changing Ideas and Practice*, Unwin Hyman (1986); Marius Romme and Sandra Escher, for 'Rehabilitation' in Marius Romme and Sandra Escher (eds), *Accepting Voices*, MIND (1993); Sage Publications Ltd, for material from Phyllis Chesler, 'Women and Madness: the Mental Asylum', *Feminism and Psychology*, 1(2) (1994); *Schizophrenia Bulletin*, for material from Thomas Sheff, 'Labelling Mental Illness' from 'Schizophrenia as Ideology', *Schizophrenia Bulletin*, 1, Fall (1970). Every effort has been made to trace the copyright-holders, but if any have been inadvertently overlooked the publishers will be pleased to make the necessary arrangement at the first opportunity.

Notes on the Contributors

Sourangshu Acharyya co-founded the first psychotherapy centre providing psychotherapy and counselling to ethnic and cultural minorities in the UK in 1983. He has worked as the Medical Director for Southend Community Care Services NHS Trust. Last December he resigned his post as Medical Director to resume his full-time clinical work as a consultant psychiatrist. He has published a number of papers and is a co-author of the seminal textbook *Intercultural Therapy*. His main areas of treatment are affective disorders, stress-related disorders and trans-cultural psychiatry. He is currently writing a short primer entitled *Psychiatry in the Surgery*.

Peter Barham is a psychologist who is at present a research associate at the Wellcome Institute for the History of Medicine, and honorary research fellow in the history of medicine at University College London. He is also founder and adviser to the Hamlet Trust. His books include *Schizophrenia and Human Value* (1993) and *Closing the Asylum* (1992).

Mary Boyle is head of the doctoral programme in clinical psychology at the University of East London. She has written widely on the concept of schizophrenia and on alternatives to psychiatric diagnosis as well as on gender issues in clinical psychology.

David Brandon is professor in community care at Anglia Polytechnic University and also a Buddhist monk. He has direct experience of mental illness and is editor of the international mental health journal *Breakthrough*. He is author of *Zen in the Art of Helping* and *Advocacy − Power to People with Disabilities* among others.

George Brown graduated in anthropology from University College London, and joined the Medical Research Council Social Psychiatry Unit in 1956, where during research with schizophrenic patients he helped to develop the widely used Camberwell Family Interview and the

Expressed Emotion measures. In 1968 he joined Bedford College and started a programme of research, largely with Medical Research Council support, on the aetiology and course of depressive disorders. He was elected a Fellow of the British Academy in 1986, Honorary Fellow of the Royal College of Psychiatry in 1987 and member of the Academia Europaea in 1991.

Deryck Browne is Policy Development Officer with the NACRO Mental Health Unit. His publications on issues concerning race and mental health include 'Black People, Mental Health and the Courts: Race Issues in Research on Psychiatry and Criminology' (in *Racism and Criminology*, eds D. Cooke and B. Hudson, London: Sage, 1995).

Joan Busfield is a professor of sociology and pro-vice-chancellor at the University of Essex. She initially trained as a clinical psychologist at the Tavistock Clinic and now researches in the field of mental disorder. Her most recent book is *Men, Women and Madness: Understanding Gender and Mental Disorder* (1996).

Peter Campbell is a mental health system survivor. In the last ten years he has been involved in action by mental health service users locally and nationally. He was a founding member of Survivors Speak Out and Survivors' Poetry and was active for many years with Camden Mental Health Consortium in North London. He was trained to work with pre-school children and now works as a writer, freelance trainer and performing poet. He is particularly interested in mental health nursing and the positive aspects of madness.

Phyllis Chesler is the author of eight books, including *Women and Madness* and *Mothers on Trial: The Battle for Children and Custody*, which have been translated into many foreign languages. Dr Chesler is a founder of the Association for Women in Psychology, and the National Women's Health Network. She is a professor of psychology and women's studies at the College of Staten Island, City University of New York (CUNY). She resides in Brooklyn with her son.

Yvonne Christie is a black woman from the African Diaspora who is presently working as an independent consultant and trainer in the field of mental health. Yvonne's special interests lie in two main areas: (1) that of developing mental health service approaches which people want to use and which are responsive to individuals, and (2) developing initiatives which address the race, gender and cultural backgrounds of their users. She is a mother of two children and a member of a large extended family, and lives in south London. Her other writings include *African Women in the Diaspora*.

Alison Cobb is a policy officer at the mental health charity MIND. Recent areas of work include drug treatments and ECT, crisis services and discrimination.

Christopher Cuninghame is community health adviser for Save the Children, working with the Fund's projects in the UK. Previously, he worked overseas and in community child health, public health and health promotion in the NHS. He is particularly interested in the links between community health issues in the UK and in development settings in other countries.

Rudi Dallos is a senior lecturer in psychology at The Open University and also practises as a clinical psychologist for the NHS with a special interest in family therapy. His research and clinical interests are in the development and treatment of problems in relation to family belief systems, power and ideologies of mental health. He teaches on several family therapy courses in the southwest of England, has published and edited a range of books in these areas and has contributed to Open University courses and materials on aspects of clinical psychology, relationships and family life.

Nigel Eastman is head of the Academic Section of Forensic Psychiatry, St George's Hospital Medical School, London. He is doubly qualified as a psychiatrist and barrister and has a particular interest in the relationship between law and psychiatry. His other academic interests are in forensic psychiatry, health services research and in the assessment of need for services in a multi-agency context. He is also a practising consultant forensic psychiatrist. He is secretary of the Mental Health Law Sub-Committee of the Royal College of Psychiatrists and a member of the Mental Health and Disability Committee of the Law Society. He is adviser in forensic psychiatry to the South Thames Regional Health Authority.

Tom Fahy is a consultant psychiatrist at the Bethlem and Maudsley NHS Trust, and senior lecturer in psychological medicine at the Institute of Psychiatry. He is an assistant editor of the *British Journal of Psychiatry* and the *Psychiatric Bulletin*. His clinical and research interests are in the field of community psychiatry, including case management, and psychiatric services for ethnic minorities.

Steve Fenton has carried out research in the area of health and ethnicity over the last decade. He has specialised in mental health and cultural similarities and dissimilarities in concepts of mental distress. In 1991–3 he was engaged in the Bristol Black and Ethnic Minority Health Survey; the report was published in 1993 and followed by further publications. He is a senior lecturer in sociology.

Jonathan Gabe is senior research fellow in the Department of Social Policy and Social Science at Royal Holloway, University of London. He is currently a co-editor of the journal *Sociology of Health and Illness*. His research interests are in the areas of mental health, health policy and health care professions. Recent publications include (as joint editor with D. Kelleher and G. Williams) *Challenging Medicine* (1994) and *Medicine, Health and Risk* (as editor, 1995).

Fiona Gardner is a psychoanalytic psychotherapist currently working with young people for the Bath Mental Health Care Trust, and in private practice. She was a founder member of the Avon Counselling and Psychotherapy Service in Bristol, and previously worked in child guidance and social services. Her research degree was on gender differences in child psychiatry. She has published on gender and mental health, child abuse, and psychotherapy training. Her current research is on aspects of professional development and education.

Alan Gilmour is project manager for the Centre for Practice Development with the Mental Health Services of Salford NHS Trust. He has recently co-ordinated the publication of *The Spirit of the Act*, an open learning manual for mental health workers, and continues his interest in suicidology and risk assessment.

Glasgow Media Group is a collective of academics from a number of different universities who write and produce work on issues in media. The contact is Greg Philo, research director of the Glasgow University Media Unit.

Roger Gomm is a lecturer in the School of Health and Social Welfare at the Open University; he trained as an anthropologist and has spent most of his life teaching sociology and social administration to social workers and health professionals.

Robert Hayward is a psychologist who has held appointments at Bradford University and the Institute of Psychiatry. For six years he was director of operations and training for the Hamlet Trust, a mental health organisation that develops and supports self-help initiatives in mental health in eastern and central Europe. He is currently working as a mental health training consultant to NGOs. He is joint author with Peter Barham of *Relocating Madness: From the Mental Patient to the Person* (1991).

Tom Heller is a general practitioner in a multiply deprived area of Sheffield. Since 1985 he has been Senior Lecturer at the School of Health and Social Welfare at The Open University and has written educational material on drug use and abuse, coronary heart disease, mental health in older people and various aspects of cancer prevention. He is an honorary lecturer at the University of Sheffield.

Sue Holland is a consultant clinical psychologist with South Buckinghamshire NHS Trust, working specifically with black and Asian minorities. She has pioneered 'social action psychotherapy' services in working-class multi-racial London neighbourhoods for twenty years. She has been employed in the voluntary sector, the NHS and social services. Recently she was given the newly established award for challenging inequality of opportunity by the British Psychological Society.

David Jones is a research fellow with the School of Health and Social Welfare, the Open University. The research on families and mental illness was completed for his PhD studies at the London School of Economics, where he also teaches on the Social Work Diploma. He is also honorary therapist with Haringey Healthcare Trust.

Linda Jones is a senior lecturer and dean of the School of Health and Social Welfare at The Open University. Since joining The Open University in 1990 she has contributed to social health studies courses, including K258 Health and Wellbeing and K301 Promoting Health. Her research interests focus on the social history of public health and on contemporary health promotion, in particular policies for children's health.

Robert Kendell is now president of the Royal College of Psychiatrists, but from 1974 to 1991 he was professor of psychiatry at the University of Edinburgh. His research was focused mainly on issues of diagnosis and classification in psychiatry and on the epidemiology of schizophrenia and puerperal psychoses. He was also dean of the Edinburgh Medical School in 1986–90 and Chief Medical Officer in the Scottish Office in 1991–95.

John Killick is a freelance writer. In April 1996 his poetry collection *Wind Horse* was published. He has worked as writer in residence for Westminster Health Care from 1992 and between 1994 and 1996 this work was financially supported by the Arts Council of England. He concentrates on working with people with dementia, and the Dementia Services Development Centre of the University of Stirling published his account of his methods *Please Give Me Back My Personality* in 1994.

Tom Kitwood is senior lecturer in psychology at the University of Bradford, and leader of Bradford Dementia Group. His present research is

concerned with the social psychology surrounding the dementing process, and his work involves him in constant contact with persons who
have dementia and their carers. Recent books include *Concern for Others*
(1990), *Person-to-Person: A Guide to the Care of Those with Failing Mental
Powers* (with Kathleen Bredin, 1991), and *The New Culture of Dementia
Care* (as editor and contributor, 1995).

Julian Leff is professor of social and cultural psychiatry at the Institute
of Psychiatry, London. He was doing research and clinical work at Friern
Hospital when the decision was made to close it and Claybury Hospital.
He helped establish the Team for the Assessment of Psychiatric Services
in 1985 to evaluate the consequences of this decision, and has been
honorary director of the Team since then. He also conducts research on
families with a schizophrenic member, and in transcultural psychiatry.

Guillermo Garcia Maza is assistant manager for Chesterfield Support Network, which supports self-help groups and people with enduring mental
health problems. He is at the moment carrying out a research project to
evaluate the effectiveness of user participation structures in the Network.

Matt Muijen is director of the Sainsbury Centre for Mental Health. He
is a psychiatrist by background. His main interest is the development
of mental health services and the Sainsbury Centre undertakes research, service evaluation, training and service development, towards
this end.

Rosemary Muston is a course manager at The Open University of Health
and Social Welfare. Prior to this she worked for many years in the
National Health Service.

Glenys Parry is professor associate in the department of psychology and
the Centre for Health and Related Research at the University of Sheffield,
and is director of research and development for Community Health
Sheffield NHS Trust. She is a clinical psychologist and cognitive analytic
psychotherapist, supervising and training in this method. Her interests
include psychological processes in social support and stressful life events,
and links between psychotherapy research, policy, audit and service
evaluation. She has recently led the NHS Executive review of strategic
policy on NHS psychotherapy services in England.

Stephen Pattison is a Senior Lecturer in Health and Social Welfare at
The Open University. He used to be a mental health care chaplain.

David Pilgrim is a Consultant Clinical Psychologist in the Communicare
NHS Trust, Blackburn, and Honorary Senior Research Fellow in Clinical

Psychology at the University of Liverpool. His books include *Clinical Psychology Observed* (with Andy Treacher, 1992), *A Sociology of Mental Health and Illness* (1993) and *Mental Health Policy in Britain* (1996), both with Anne Rogers.

Rosalind Ramsay is an honorary lecturer at the Institute of Psychiatry. She is an Assistant Editor of the *British Journal of Psychiatry* and the *Psychiatric Bulletin*. Her research and clinical interests are in the provision of mental services and women's mental health.

Jill Reynolds is a lecturer in the School of Health and Social Welfare at the Open University. She has worked previously as a social worker, as a trainer with voluntary organisations working with refugees and as a lecturer on social work qualifying courses. She has written on social work and refugees, on working with interpreters and on gender issues in social work education.

Anne Rogers is a reader in sociology in the National Primary Care Research Centre based at the University of Salford. She has co-authored *A Sociology of Mental Health and Illness* (1993) and *Mental Health Policy In Britain* (1996) (both with David Pilgrim as well as *Experiencing Psychiatry* (1992) (with David Pilgrim and Ron Lacey).

Marius Romme has been professor in social psychiatry at the University of Maastricht since 1974. Before that he was the director of a community mental health centre in Amsterdam. Since 1987 he has been involved with Sandra Escher in the study of hearing voices. This concerns research but also bringing together networks to promote change in the attitude towards the hearing voices phenomenon.

David Rosenhan was professor of psychology at Stanford University at the time of writing the article reproduced in this volume. His 'pseudo patient experiment' was an important element of the challenge to orthodox psychiatric ideas in the 1970s.

Janice Russell is a senior lecturer in the Department of Psychological Medicine, University of Sydney, Australia. Originally trained as a physician, she also includes in her research interests psycho-endocrinology and biological aspects of the eating disorders. She is the director of the Eating Disorders Units of the Northside Clinic, Greenwich, and of the Concord Hospital in Sydney.

Azra Sadiq works at the Centre for the Study of Minorities and Social Change in the Department of Sociology at Bristol University.

Liz Sayce is policy director of MIND, the National Association for Mental Health. MIND works throughout England and Wales to promote mental health and to improve the quality of life of people with mental health problems. Liz Sayce has a background in social work and mental health research and has published work on a number of issues in British mental health policy, including community mental health centres, psychiatric case registers and user involvement.

Thomas Scheff was professor of sociology at the University of California, Santa Barbara at the time of writing the article reproduced in this volume. His writings on the sociology of mental illness are regarded as classic statements in the sociology of deviance.

Jane Shackman is training officer and Social worker at the Medical Foundation for the Care of Victims of Torture. She has worked for many years with people in exile in the UK. She has also developed teaching materials for use in employing, working with and training community interpreters. She does some consultancy work with NGOs that work in areas of conflict.

Mike Slade is a trainee clinical psychologist at the University of Surrey, an honorary lecturer in the Department of Psychiatry at the Institute of Psychiatry, and an honorary mental health policy adviser with the Department of Health. He has previously been a computer scientist and a registered mental nurse. Since gaining his psychology degree with the Open University, he has been involved in research into how to identify people with severe mental health problems, and methods of assessing their needs and quality of life.

Tang My came to Britain in 1980 as a refugee from Vietnam. Since 1981 she has been working as an interpreter/translator and as a community and social worker. She was qualified as a social worker in 1987 and is now working for Deptford Vietnamese Project – a health and social services funded scheme to help the Vietnamese refugees in Lewisham to settle and access local services.

Steve Taylor teaches medical sociology and medical law at the Universities of London and Coventry. The study of suicide and deliberate self-harm is one of his research areas and he has written a number of articles and produced two books on this topic: *Durkheim and the Study of Suicide* (1982) and *Suicide* (1988).

Peter Tyrer is professor of community psychiatry at Imperial College School of Medicine at St Mary's Hospital. He has been involved in developing front-line mental health services since 1973 and has carried out extensive work into different models of community psychiatric care and their effectiveness and cost implications.

Richard Warner graduated from Medical School in London, trained at social psychiatry centres in Oxford and the borders of Scotland and earned a graduate degree in anthropology from the University of Colorado. He has practised community psychiatry in Colorado since 1971 and has been the medical director of the Mental Health Centre of Boulder County, Boulder, Colorado, since 1976. He is a Professor in the departments of anthropology and psychiatry of the University of Colorado. He is the author of *Recovery from Schizophrenia: Psychiatry and Political Economy* (second edition, 1994) and the editor of *Alternatives to the Hospital for Acute Psychiatric Treatment* (1995).

Gilli Watson is a clinical and community psychologist. She has a long-standing commitment to developing mental health services that address the impact of social inequalities on women's health. She has published in this area and contributes to clinical training and practice.

Jennie Williams is a clinical psychologist who works as senior lecturer in mental health at the Tizard Centre, University of Kent. Since the early 1970s her main concern has been the effects of social inequalities on mental health and mental health services.

General Introduction

This collection of writings aims to demonstrate that mental health really does matter. It matters to individuals. It matters to groups of people in families, networks and communities. It matters to societies, states and cultures. Promoting and maintaining positive mental health is one of the most important challenges facing Western societies. This goes far beyond a concern with the minority of people who experience severe mental distress. Even so, critically understanding the ways that people who experience mental distress have been viewed and treated by others provides insights that can contribute to better promotion of mental health in the future.

To study mental health and distress is to embark upon a fascinating, interdisciplinary journey into understanding. The editors have chosen diverse examples of writing from many disciplines to illuminate the meanings and practices related to mental health and mental distress. Some of the articles in this collection are reprinted from other sources, while others have been newly commissioned for this volume.

Much of the chosen material offers a sharp challenge to standard psychiatric practice and current social policy. Articles have been chosen that illustrate disputes between currently dominant ideologies and alternative interpretations of mental health and mental distress. The aim is to explore the edges between competing explanations and different practices. This can enhance debate about necessary changes in the mental health sphere.

Mental Health Matters is not concerned with theoretical approaches alone. Many articles describe or analyse practical approaches to mental health and distress. The four parts have been designed to move from theoretical and contextual debates (Parts 1 and 2) to practical and practice-based examples of mental health work (Parts 3 and 4). However, readers will find that they get as much from dipping into the book according to their needs and interests as from reading it from end to end in the order laid down.

The book is in four parts:

Part I   Understanding Mental Health and Distress: Debates and Theories

In this part a variety of perspectives is described and explained, giving an insight into some of the theoretical concepts concerning mental health and mental distress. Some of the more dominant discourses from psychiatry and psychology are challenged by articles which present different sociological, gender-related and anthropological insights.

Part II   Mental Health Policy, Social Inequality and Civil Rights

Mental distress is closely related to social and material deprivation. This implies that the policies appropriate to promoting mental health should primarily be economic and social rather than medical and curative. Such policies would aim to improve the conditions of large numbers of people. In fact, there never has been a comprehensive mental health policy in the United Kingdom. Instead, there have been a variety of policies for dealing with a minority of individuals who are clearly designated as 'mentally ill'. This part looks at the historical evolution of these mental illness policies, relating them to their social context and effects. Particular attention is paid to their potentially coercive nature.

Part III   Involving Users in Mental Health Services

This part focuses on the way that users of the mental health services can be involved in the development of their own 'care', and also in policy formulation regarding the development of services. Although service providers are required to seek 'users' views', in practice people do not always feel that they have been heard. These difficulties, and how to overcome them, especially in relation to particularly marginalised or disadvantaged groups, are the subject of this part.

Part IV   Examining Practice

This part illustrates humane and innovative ways in which the practice of mental health work can be carried out. Drawing upon research of various kinds, many of the articles suggest ways of changing and improving practice. A range of approaches is described and its theoretical bases explained in order to enrich critical, reflective, mental-health-care practice.

This collection of readings forms part of the material of study for students on an Open University Course, **K257 Mental Health and Distress: Perspectives and Practice.** The course uses structured teaching texts and audio tapes as well as the material in this volume to guide students through the debates and discourses that are outlined here. Priority is given to the views and experiences of people who are users of the mental health services, or who consider themselves to be survivors of that system. A companion volume to this, edited by Jim Read and Jill Reynolds, *Speaking Our Minds: an Anthology of Personal Experience of Mental Distress and its Consequences* (Macmillan, 1996), has been published simultaneously. The short, personal accounts contained therein about experiencing mental distress, views of different forms of treatment and venues for 'care', as well as positive proposals for change within the mental health system, complement the more formal articles in this book. Together, the books provide a unique and comprehensive overview of contemporary perspectives in mental health.

The material in this reader has been chosen to be challenging and thought-provoking. Discussion of mental health and distress can bring about intense feelings. We suggest that you be mindful of the potential impact of material in this book upon yourself.

Selecting what to reprint in this volume from the mountains of already published material, and who to commission to write new material for it, has been an exciting process for the five editors. Although we are all white and European, with professional backgrounds, we have come from different backgrounds and disciplines, including medicine, social work, psychology, anthropology and theology. The collection we have chosen reflects some of the tensions that exist between us personally and as members of different disciplines. We hope it presents a multi-faceted, multi-professional perspective on the complex, contested area of mental health. If readers find here material that challenges current perceptions and stimulates them to think critically about the way they feel and think about ideas and practices in the mental health sphere, then we will have added some force to our conviction that mental health matters.

Part I

UNDERSTANDING MENTAL HEALTH AND DISTRESS: DEBATES AND THEORIES

1

Introduction

This part gives a broad introduction to some of the theories and debates that inform mental health work. No collection of articles and extracts could include the entire range of understanding in this complex area or encompass everything that might be relevant. The articles provide insight into the tangled and often conflicting understandings of mental health and distress.

Rudi Dallos's article (Article 2) outlines psychological frameworks that have been influential in mental health theory and practice. Discussion focuses on five frameworks (Biological and Medical, Behavioural, Psychodynamic, Humanistic and Systemic) which propose distinct explanations for the causes of mental health problems and imply different methods of treatment. Dallos identifies three levels of analysis, the societal, interpersonal or individual, and discusses these in turn.

Robert Kendell, President of the Royal College of Psychiatry (Article 3), draws on biological explanations which view psychological problems as having much to do with neurological and metabolic factors. While recognising that social circumstances and psychological experiences may play an enormous part in engendering mental distress, he argues that 'disorders' such as schizophrenia and depression may be linked to particular defects in the brain, for instance to deficits in the neuro-transmitters which carry signals between brain cells. This implies that mental illness has similar causes to physical illnesses and should be treated in much the same way. Thus, mental illnesses can usefully be identified, classified and diagnosed in terms of definable syndromes or clusters of symptoms.

Mary Boyle (Article 4) makes a plea for the re-evaluation of 'schizophrenic' behaviour. She suggests, for example, that close analysis of the content of hallucinations, or the possible meaning behind behaviour that could be considered 'psychotic', is an important part of developing an understanding of the phenomena currently labelled 'schizophrenia'. Boyle argues that the form and content of the 'madness' sometimes gives clues that might help to make sense of otherwise apparently bizarre thoughts or behaviour.

In 'Life Events, Loss and Depressive Disorders' George Brown (Article 5) summarises his research work over many years. Through detailed longitudinal research he has demonstrated some of the factors that lead to the increased chance of the development of depression in women. The combination of low self-esteem, difficult environmental situations, nasty life events and the possible effects of biological vulnerability have a cumulative effect, the magnitude of which he describes.

Phyllis Chesler (Article 6) outlines the classical feminist argument that insists that there is a double standard of mental health and humanity: one for women and one for men. She produces evidence from within mental institutions that indicates that the way female mental patients are treated mirrors the female experience within the family. Although this article was first published in 1972, and in the USA, there is much that continues to resonate concerning the conditions of women inside mental institutions and in society at large. One area of concern to Chesler which has changed, at least in England and Wales, is that women no longer outnumber men in admissions to psychiatric hospitals. Particularly amongst young adults the trend is now reversed, but the reasons for this are not yet clear and it is likely that psychiatric treatment continues to be sex-differentiated (Payne, 1995).

The cultural context of mental distress is explored by Richard Warner (Article 7). Warner takes us on an anthropological journey, not just into different geographical regions, but through a brief description of some of the ways that the people living in other cultures may understand and contextualise mental distress. He concludes that mental states similar to the Western symptom-complex labelled 'schizophrenia' are common throughout the world. Certain features of the way that these manifestations are coped with in other societies bring about better rates of recovery than in industrialised societies. In particular, this form of behaviour may not be stigmatised or linked to violence and disruption.

Thomas Scheff (Article 8) describes the social process through which mental illness is 'created' by labelling people as mentally ill. Two important aspects of this process are the assumptions that the social majority makes about normality and the tolerance levels that members of society have for people who appear to be very different from themselves. This theme concerning the labelling of social deviants is extended in 'On Being Sane in Insane Places', the report of a famous experiment by David Rosenhan (Article 9). Eight people (pseudopatients) gained admission to psychiatric hospitals through the simple device of saying that they had been hearing voices. Admission seems to have been straightforward and thereafter they ceased simulating any symptoms of abnormality. In these circumstances their very normality was interpreted as psychiatric symptomatology; the persistence and power of the labelling process is thus vividly revealed.

Roger Gomm's chapter 'Reversing Deviance' (Article 10), details attempts to counteract the labelling approach. The leading role in changing images of mental illness has to come from the people so labelled. Certain policies and approaches can also be used within the mental health world to reverse some of the adverse effects of the labelling and stigmatisation process. The concept of normalising professionals and services so that they are non-stigmatising to users instead of normalising people who have mental health problems can help to reverse assumptions about who and what needs to change in order to bring about a more mentally healthy world. The implications for practice are examined in Part 4 (see Brandon, Article 36).

The received wisdom that Alzheimer's Disease is a progressive, organically-based, neurological disorder that inevitably leads to problems of memory and behaviour is challenged by Tom Kitwood (Article 11). He puts forward a strong case for considering dementia to be determined as much by psychological and social considerations as by changes in the brain. He poses the question as to whether stress or particular social conditions can *cause* the biochemical and immunological changes that then go on to affect brain functioning.

Finally, David Jones (Article 12) turns to the experience of people in the families of those who are mentally distressed. Jones argues that the experience and perspective of families is often unhelpfully neglected in focusing upon distressed individuals. The article considers the difficult process that families often have to go through in coming to terms with the reality of a severely distressed person.

References

Payne, S. (1995) 'The rationing of psychiatric beds: changing trends in sex-ratios in admission to psychiatric hospital', in *Health and Social Care in the Community*, 3 (5), 289–300.

2

Psychological approaches to mental health and distress

RUDI DALLOS

The purpose of this article is to outline the dominant psychological frameworks that have been influential in mental health theory and practice. The discussion will focus on five frameworks: Biological and Medical, Behavioural, Psychodynamic, Humanistic and Systemic. These overlap to some extent, but do propose significantly distinct explanations for the causes of mental health problems and in turn imply different methods of treatment.

Levels of analysis

The frameworks can be seen as located at one of three levels of analysis: societal, inter-personal or individual. The individual level has traditionally been the most dominant, with its suggestion that problems of mental health have causes which can be traced to individual factors, for example, biological abnormalities or deficits (biological/medical), faulty learning experiences (cognitive/behavioural) or emotional traumas (psychodynamic). Systemic/Interactional explanations (for example, family therapy) suggest that mental health problems arise from relationship difficulties, such as conflicts and stresses in families.

Biological and medical frameworks

Frameworks which draw on biological explanations view psychological problems as resulting predominantly from physical causes, such as defects in the functioning of the brain and other organs, hereditary biological factors and the results of damage or accidents. Some suggest that 'disorders' such as schizophrenia and depression may be linked to particular defects in the brain, for instance to deficits in the neuro-transmitters which carry signals between brain cells (McKenna, 1987). In addition medical frameworks embody the idea that mental health problems are like a form of physical illness (Kendell, 1993).

This view carries with it the proposition that mental illnesses can be identified and classified – diagnosed in terms of definable syndromes or clusters of symptoms. These are seen to offer a clear picture of development of the illness and prognosis. Such classification is also seen as a useful framework to guide research and to guide treatment through the characterisation of the syndromes comprising the illnesses.

Drug treatment

The implication for treatment here is that problems are largely to be treated at the level of the individual and are in the main determined by the person's biology. Treatment is mainly by means of psychotropic (mood-altering) drugs which may help to alleviate symptoms such as distress, confusion or excessive agitation. There have been extensive debates about the value of drug-based treatments, and especially about the indiscriminate prescription of tranquillisers and antidepressant drugs. The positive aspects are that many seriously disturbed patients no longer need to be physically restrained and can live more normal lives. However, long-term use of medication can lead to addiction, damaging physical side-effects, psychological dependence and possibly a reduced sense of autonomy and control over one's own life.

Psycho-education

Psycho-educational approaches, whilst fundamentally having a biological view of causation, also recognise that, as in physical illnesses, a variety of environmental and social factors interact with the illness. Education and advice on ways of recognising and coping with stress can be helpful in alleviating it and this broadly constitutes a form of psycho-education. However, the term psycho-education is also employed more specifically in an approach which proposes that mental illnesses are susceptible to various forms of stress and that the severity of the illness is mediated by the level of stress in the person's social environment, in particular their family and other intimate relationships. As an example, assisting families through an educational programme to maintain the emotional atmosphere within acceptable limits has been found to reduce the incidence of relapse and the need for re-hospitalisation (Brown *et al.*, 1962; Leff and Vaughn, 1981).

Behavioural frameworks

Behavioural approaches assume that some psychological problems are acquired through learning experiences, and that they are subsequently maintained by the patterns of events (rewards and punishments) in the environment. The classic experiments on conditioning have shown that negative states such as fears or phobias can be created by repeatedly pairing a neutral object with an aversive or frightening stimulus, for example electric shock or a loud noise. The focus of treatment is predominantly on reversing the negative or inappropriate learning experiences which have caused and may continue to maintain or determine the problems.

Different techniques of treatment have been developed from behavioural principles, including systematic desensitisation, behavioural modification and cognitive behavioural approaches.

Systematic desensitisation

This technique (Wolpe, 1958) is typically used to assist people with phobias, such as an excessive fear of spiders, public speaking, going out, heights and so on. It usually consists of the following stages:

(1) The person suffering from the phobia is taught relaxation techniques. This helps the person to become calm and pleasantly relaxed at will.
(2) Following this, a reduced, tolerable exposure to the fearful object or situation is presented so that the person remains relaxed.
(3) The severity of the stimulus is gradually increased in stages. For example, someone frightened of heights may practice ascending to the first floor of a building, then gradually to the second and eventually to the top.

Behaviour modification

Behaviour modification procedures concentrate on building or replacing 'undesirable' behaviours with more 'desirable' ones (Sheldon, 1972; Cullen *et al.*, 1981). Problems, like all behaviours, are seen to be maintained by the situation that the person is in. Systematic assessment is conducted in order to identify the factors in a person's present environment which are maintaining the problematic behaviours. For example, sympathy, attention, time off from school and so on may function as rewards for a young child's anxiety which inadvertently maintains his or her problem of anxiety. Behavioural approaches often involve the client in detailed record-keeping prior to and during treatment in order to identify such inadvertent rewards. This self-monitoring has repeatedly been found to produce significant behavioural change in itself. Possibly the active involvement in this recording indirectly generates some sense of control and autonomy and may also produce some insights which help to promote change.

Cognitive-behavioural approaches

Learned Helplessness

Seligman's (1975) research shows that if animals are allowed to learn ways of avoiding negative experiences, such as electric shocks, and then prevented from using the avoidance strategies they have learned, that they become apathetic or 'helpless'. In one experiment people were engaged in solving a range of simple problems which were replaced (without their knowledge) by impossible ones. Following frustrating attempts to solve these, even when then presented with more simple (and solvable) problems, people continued to fail or dismissed any successes as due to luck. Consistent with this model, deprived, disadvantaged and powerless environments are seen to be conducive of depressive and other mental disorders.

Beck's model of depression

A related approach is Beck's (1967) cognitive model of depression, which suggests that depression is produced and maintained by a person's characteristically gloomy and pessimistic style of thinking. These thoughts (cognitions) are seen to be maintained by a consistent perceptual bias towards the negative aspects of experience, and by systematic, logical errors of interpretation. A person may make excessive generalisations based upon very limited information. Also, an over-selective use of information may mean that only negative experiences are noted and all positive reactions are ignored. For instance, based upon a few isolated incidents of rejection, a person may conclude that he or she is totally worthless and liked by nobody. Beck adds that depressive thinking is rooted in a triad of dysfunctional assumptions which lead to self-perpetuating and automatic negative cognitions:

- Negative view of self: *I am worthless.*
- Negative view of circumstances: *Everything is bleak and I cannot manage the demands that people put upon me.*
- Negative view of the future: *Things will only get worse and there is nothing I can do to change them.*

Beck proposes that these gloomy thoughts may be triggered by real-life experiences, such as loss, failure or rejection. Subsequently, though, these depressive cognitions filter and restrict a person's thinking so that he or she is virtually unable to recognise any positive experience.

Beck's therapy aims to teach people to modify their immediate cognitive responses to potentially upsetting situations, and then to restructure the fundamental beliefs on which their depressive responses are based.

Cognitive approaches are also used to cope with a range of anxiety-related disorders, particularly those associated with anxieties about controlling spontaneous functions, for example, sexual problems (Kaplan, 1974).

Psychodynamic approaches

In essence psychodynamic approaches argue that problems are determined by the history of a person's prior emotional experiences, especially childhood ones. By and large these are regarded to be outside the person's conscious awareness. Treatment therefore focuses on the individual and attempts, by making the memories of these experiences conscious, to enable the person to be more free and autonomous:

(1) One major contribution that psychodynamic theory offers is the hypothesis that clinical problems are largely rooted in negative childhood emotional experiences which serve to disrupt the normal path of development through the psychosexual stages. Resulting problems may be manifest in childhood, adolescence or in various guises in adult life.

(2) Difficulties are seen to be located in the unconscious. Therefore the memories of these events, including the thoughts and emotions surrounding them, are not readily accessible but need to be teased out using a variety of techniques, such as dream analysis, free association and so on. This contrasts with the cognitive approaches which tend to assume that cognitions are consciously available for negotiation between the psychologist and the client.

(3) Freud, and other psychodynamic theorists in differing degrees, proposed that sexuality was the cornerstone of emotional problems. The conflict between society's prohibitions regarding sexuality and a person's sexual desires was seen as resulting, for many people, in profound emotional problems which may need therapy.

(4) In particular, Freud suggested that problems are related to inabilities to resolve sexual feelings within a family. The Oedipus and Electra triangle was a central concept suggesting that a young child is caught between ambivalent feelings of desire, guilt and anger as a result of antagonism towards the same-sex parent, and sexual feelings for his or her opposite-sex parent.

An important contribution to the practice of therapy is the emphasis on the therapeutic relationship and transference. Freud argued that his patients inevitably transferred their feelings (carried over from other earlier relationships, especially with their parents) on to him in various ways. The therapist could be seen as a punishing father about whom the client was ambivalent: on the one hand the client would hate and want to attack him and, on the other hand, would harbour unacceptable sexual feelings for him. Treatment involved the client working through these feelings with the therapist and thereby becoming more aware and in control of them (strengthening the ego). Later Freudians, such as Klein and Fairbairn, emphasised that it was also inevitable that the therapist could in turn project his or her feelings on to the client (counter-transference). Rather than attempting to eliminate such feelings, the dynamics of the therapeutic relationship were seen as a main point of focus allowing the person to use the relationship with the therapist to express and find new, more constructive ways of dealing with the feelings.

Freud himself pioneered the technique of psychoanalysis, a one-to-one confrontation between analyst and client, often more than once a week over many years (see, for example, Freud, 1962). This is obviously a time-consuming and expensive treatment. Attempts have been made to develop brief psychodynamic techniques which are more specifically focused on relieving specific problems. Many of these do not focus on sexuality and transference but take a broad range of emotional factors into account.

Group therapies are another way in which psychodynamic therapies can be used to reach more people. These employ various techniques to bring out and work through unresolved emotional problems with other members of the group. One therapist can work with up to a dozen patients at a time, and therapeutically utilize various psychodynamic processes, such as transference, denial and projection between group members (Bion, 1961).

More recent developments explore the effects of early relationships and attachments on current relationships (Scharff, 1982). In family-focused therapy, for example, the aim is to resolve the unconscious tensions and emotions in the whole family as well as in the individual displaying the problems. The aim is to ensure that the family environment is subsequently more supportive and relapse is less likely to occur (Boszromenyi-Nagy and Sparks, 1973). Theories of psycho-social transition focus on the ways in which people respond to loss and how they can be helped to reconstruct their lives when faced with catastrophic change.

Humanistic frameworks

Humanistic psychology is also concerned with unconscious processes, but views people as potentially creative and motivated by a need to grow and develop (Vondarcek and Corneal, 1994). In contrast to psychodynamic approaches, the unconscious is seen as a positive entity which can be fruitfully explored and developed in various ways through art, music, writing, drama, conversation and self-reflection. In general it is suggested that a person's conscious and unconscious states can be integrated, and the aim is to encourage a sense of autonomy, control and freedom. A holistic approach attempts to integrate aspects of the person: behaviour, emotions, cognitions, sensations, dreams and fantasies. It is argued that through negative aspects of relationships with others, people may have come to feel bad, disgusted or ashamed about aspects of themselves. Humanistic therapies, such as Carl Rogers' client-centred counselling (1965), are concerned to understand how people experience themselves and how this relates to their problems. Treatment aims to provide integration by providing a context of 'non-evaluative warmth' or validation – a supportive environment in which people are able to experiment with new roles and are assisted to regain their sense of self-agency and purpose. There is an emphasis on helping people to integrate their feelings and their conscious awareness. This involves encouraging people to 'get in touch' with and accept their positive and negative feelings and to communicate more clearly and freely about these.

In Rogers' non-directive counselling approach, the therapist tries to avoid making interpretations and refrains from giving advice because this confuses the therapist's own needs with the client's needs. Initating and advising can make the therapist feel good but may actually cause the client to feel stupid (Kelly, 1955).

In reality a position of total non-directiveness is almost impossible to achieve; a film of Rogers working with clients reveals that he does in fact make interpretations and even offers advice. Rogers in fact states that the therapist should not appear to be some kind of unresponsive robot but a warm, caring human being who, at least to some extent, reveals something of his or her own thoughts and feelings.

However, the emphasis is on encouraging the client to 'do the work' of solving his or her own problems with some assistance from a therapist in order to encourage the growth of self-confidence and respect. Some varieties of humanistic techniques go even further and dispense with the therapist role altogether. In co-counselling or leaderless groups, people teach each other basic skills and then mutually assist one another in applying them to solving their own problems. Humanistic ideas have had a considerable impact in encouraging forms of self-help group in the community, such as alcoholics' groups, family support groups and women's groups.

Systemic frameworks

Systemic frameworks view problems as residing not simply or predominantly in an individual but in the current patterns of actions and in the communications within relationships.

Systemic family therapy

It is suggested that though symptoms may appear to emerge predominantly in one member of a family, this is essentially a facet of their disturbed relationships (Bateson, 1956; Jackson, 1968). The approach was initially given impetus by the discovery that children's symptoms were frequently a distressed response to being drawn into conflicts between their parents (Rutter, 1975; Haley, 1976; Minuchin *et al.*, 1978). Such observations led to the suggestion that many symptoms could be seen as functioning to avoid, or distract attention from, other areas of conflict in family relationships. To label one member of the family as 'ill', it is suggested, might delay or prevent necessary changes in the disturbed relationships in the family as a whole.

Systemic family therapy approaches focus on patterns of actions and communications in relationships. They use feedback to focus on how people's actions are continually influenced by information about the effects of previous actions. Of particular interest are vicious or pernicious cycles. An example would be a young man's response to his mother's interference, which was to retreat into his shell, sleep a lot and cease to take care of himself. This made his mother interfere even more in his life in an angry and critical way, accusing him of being lazy and telling him to get a job. The young man would, eventually, react to this heightened conflict and tension at home by acting is some bizarre and disturbed way which could result in another episode of hospitalisation.

Systemic approaches have increasingly come to recognise that the repetitive relationships patterns that can be observed in families in part arise from their beliefs and understandings. Furthermore, these family beliefs are in turn seen to be related to wider societally shared assumptions, for instance about relationships, gender, mental health and family life (Dallos, 1991).

Common threads and contrasting aspects

There are clear differences between the various frameworks but also there are important features that they share, especially in the 'messy' business of clinical practice where many practitioners use an eclectic and pragmatic approach. The frameworks differ in terms of the level of analysis at which they are predominantly focused but also on the important dimension of agency and determinism.

People can be seen as having agency, as being active, autonomous and capable of reflecting on their experiences and making choices about their lives. Alternatively, deterministic explanations assume that people are more or less formed by their biological, historical or social circumstances, and have little control over the effects that these have on them.

The frameworks however share many important features:

Empathy

Most approaches recognise that successful treatment involves the need to establish a positive, trusting relationship with the client, his or her family, and others (Rogers, 1965, Yates, 1983). If for no other reason, this is necessary in order to gain the cooperation of the client, especially in the sometimes-tedious record-keeping that is required in behavioural programmes.

The therapeutic relationship

There is general recognition of the need to consider the relationship between the client and the therapist. Both client and therapist bring a history of previous experiences to the therapeutic encounter. The therapist has to be able to modify his or her relationship with the client, couple or family in order to overcome anxieties associated with change which are a frequent feature of psychological problems.

Communication

All forms of therapy require clear communication at different levels between the client and the therapist. For many patients, the inability to communicate about their problems is a critical feature, and arguably symptoms often emerge as a form of indirect communications, expressing what people want to say, but feel they are not allowed to say. Family therapists often encourage family members to demonstrate how they communicate at home. Even therapists who work individually may try to explore communication with others as directly as possible. For example, Perls' (1969) gestalt therapy encouraged people to imagine some real aspects of their lives, such as a parent sitting in an empty chair, and to express their feelings to the imagined person in the chair, rather than simply attempt to talk about the feelings and cognitions. This may help to show the therapist more directly how they communicate with others and their underlying feelings when they do so.

Pernicious Cycles

A number of orientations include the idea that problems can escalate through a cycle of interaction between the behaviour and its consequences (feedback). The classic example is that of the person who starts drinking to gain some relief from worries and problems. Unfortunately, the problems become worse as a result of the alcoholic's inability to deal with them effectively because of the effects of drink (Heather and Robertson, 1985). This in turn can aggravate the problem. Family therapists likewise talk about 'attempted solutions' to problems, such as a father nagging a wayward adolescent child to behave more responsibly, which can further anger the child, leading to more of the same behaviour.

References

Bateson, G. (1972) *Steps to an Ecology of Mind*, New York, Basic Books.

Bateson, G., Jackson, D., Haley, J. and Weaklands, J. (1956) 'Towards a theory of schizophrenia', *Behavioural Science*, vol. 1, 251–264.

Beck, A. T. (1967) *Depression: clinical, experimental and theoretical aspects*, New York, Harper & Row.

Bion, W. R. (1961) *Experiences in Groups*, London, Tavistock.

Boszromenyi-Nagy, I. and Sparks, G. (1973) *Invisible Loyalties*, New York, Harper & Row.

Boyle, M. (1990) *Schizophrenia: A Scientific Delusion?*, London, Routledge.

Brown, G. W., Monck, E. M., Carstairs, G. M. and Wing, J. K. (1962) 'The influence of family life on the course of schizophrenic illness', *British Journal of Preventative Social Medicine*, vol. 16, 55–68.

Cullen, C., Hattersley, J. and Tennant, L. (1981) 'Establishing behaviour: the constructional approach', in Davey, G. (ed.), *Applications of Conditioning Theory*, London, Methuen.

Dallos, R. (1991) *Family Belief Systems, Therapy and Change*, Milton Keynes, Open University Press.

Freud, S. (1962) *Two Short Accounts of Psychoanalysis*, Harmondsworth, Pelican.

Haley, J. (1976) *Problem Solving Therapy*, San Francisco, Jossey-Bass.

Heather, N. and Robertson, I. (1985) *Problem Drinking: The New Approach*, Harmondsworth, Pelican Books.

Jackson, D. (1957) 'The question of family homeostasis', *Psychology Quarterly Supplement*, vol. 31, 79–90.

Jackson, D. (ed.) (1968) *Therapy, Communication and Change*, Palo Alto, CA, Science and Behavior Books Inc.

Kaplan, H. A. (1974) *The New Sex Therapy*, New York, Brunner Mazel.

Kelly, G. (1955) *The Psychology of Personal Constructs*, New York and London, Norton.

Kendell, R. E. (1993) 'The Nature of Psychiatric Disorders', in Kendell, R. E. and Zealley, A. (eds), *Companion to Psychiatric Studies*, 5th edn, Edinburgh, Churchill Livingstone.

Leff, J. and Vaughn, C. (1981) 'The role of maintenance therapy and relatives' expressed emotion in relapse of schizophrenia: a two-year follow-up', *British Journal of Psychiatry*, vol. 139, 102–10.

McKenna, P. J. (1987) 'Pathology, phenomenology and the dopamine hypothesis of schizophrenia', *British Journal of Psychiatry*, vol. 151, 288–301.

Minuchin, S., Rosman, B. and Baxter, L. (1978) *Psychosomatic Families: Anorexia Nervosa in Context*, Cambridge, MA, Harvard University Press.

Perls, F. S. (1969) *Gestalt Therapy Verbatim*, Moab, Real People Press.

Rogers, C. (1965) *Client Centred Therapy*, New York, Houghton Mifflin.

Rutter, M. (1975) *Helping Troubled Children*, Harmondsworth, Penguin.

Scharff, D. E. (1982) *The Sexual Relationship: an object relations view of sex and the family*, Boston, Routledge & Kegan Paul.

Seligman, M. (1975) *Helplessness*, San Francisco, Freeman.

Sheldon, B. (1982) *Behaviour Modification*, London, Tavistock.

Vondarcek, F. W. and Corneal, S. (1994) *Strategies for Resolving Individual and Family Problems*, Pacific Grove, Brook Cole.

Wolpe, J. (1958) *Psychotherapy and Reciprocal Inhibition*, Stanford, CA, Stanford University Press.

Yates, A. J. (1983) 'Behaviour therapy and psychodynamic theory: basic conflict or reconciliation or integration?', *British Journal of Clinical Psychology*, vol. 22(2), 107–127.

3

The nature of psychiatric disorders

R. E. KENDELL

And little by little I can look upon madness as a disease like any other.
Vincent van Gogh

Historical assumptions about madness

Madness has a long history. Most cultures have recognised people who were either temporarily or permanently 'deranged' and most languages have a word for the phenomenon. The cause of this derangement was even more puzzling two thousand or two hundred years ago than it is today and many different assumptions have been made about its origins. At one time or another madness has been attributed to divine intervention, evil spirits, fevers, heredity, unbridled passions, strong liquor, the influence of the moon and blows to the head. Madmen have been regarded with revulsion, pity, hilarity, reverence and indifference and their disordered proclamations treated with derision, mockery, horror, or, occasionally, with grave respect. But throughout this kaleidoscope jumble of attitudes and opinions the dominant assumption has generally been that madness was a disease of some kind. Certainly, attempts to alleviate madness were usually made by the same people, and by the same means, as attempts to cure illnesses of other kinds. Whether he was priest, shaman, physician or apothecary, the appointed authority used much the same ceremonies, spells, potions or medicaments to treat madness as he did to treat a wide variety of other more obviously medical disorders.

In the fourth century BC Plato discussed madness in his *Phaedo* and distinguished 'madness given us by divine gift' from natural madness. The

Edited and revised from Kendell, R. E. and Zealley, A. K. (1993) *Companion to Psychiatric Studies*, Edinburgh, Churchill Livingstone, 1–7.

former was inspired by Apollo, Dionysus, Aphrodite or the Muses; the latter originated in physical disease. His contemporary Hippocrates, or the corpus of writings attributed to him, described at least five forms of madness – phrenitis (acute mental disturbance with fever), mania (acute mental disturbance without fever), melancholia (chronic mental disturbance), epilepsy (convulsions) and hysteria (paroxysmal conditions seen in women) and regarded all of them as medical conditions requiring treatment (Menninger, 1963). Both aetiological hypotheses, the supernatural and the medical, persisted for the next two thousand years. For Richard Burton in the seventeenth century and John Wesley in the eighteenth century, religious melancholy was, quite literally, ensnarement by the Devil, but physicians – Greek, Roman, Arab and European – almost invariably accepted Hippocrates' assumptions. Celsus, Areteus and Galen, writing in the first and second centuries AD, all regarded melancholia and other forms of madness as diseases, and Galen's four humours provided an explanatory framework which was at least as plausible for malancholia (attributed, as the name implies, to an excess of black bile if the patient was sad and fearful, or of yellow bile if he was angry and agitated) as for illnesses of other kinds. In the eleventh century AD the great Persian philosopher physician Ayicenna incorporated Galen's teachings about madness and the four humours into his vastly influential *Canon of Medicine*. As a result, a coherent set of assumptions about the nature and causes of madness were shared by the Christian and Moslem worlds for several hundred years. Although by the eighteenth century respect for ancient authorities, and for Galen's humoral pathology, had finally crumbled, the central assumptions of Hippocrates, Galen and Ayicenna about the nature of madness still held sway. When Lady Mary Wottley Montagu commented in a letter that 'madness is as much a corporeal distemper as the gout or asthma' she was simply voicing the commonplace views of her contemporaries, both educated and uneducated (Porter, 1987).

Towards the end of that century, however, an important change took place. It slowly became clear that the armamentarium of eighteenth-century medicine – the special diets, bleeding, purging, emetics and blistering – however valuable they might be for fashionable 'nervous disorders' like hypochondriasis and hysteria, had little effect on madness itself, a state of affairs emphasised in England by the impotence of George III's physicians to alleviate his recurrent bouts of insanity. At the same time the waning of religious fervour that accompanied this 'Age of Reason' meant that the theological assumptions of Burton and Wesley were no longer acceptable. An explanatory and therapeutic vacuum was thus developing, and the managers of the new private madhouses moved into it. Francis Willis, the clergyman who was eventually called in to treat the mad king, was typical of this new breed. Willis was a charismatic healer with extensive personal experience of handling lunatics by virtue of having lived for years with several in his house. His therapeutic powers depended partly on this experience and partly on his personality, on what he would have called moral authority and we might regard as psychotherapy. The success of the York Retreat, which had been opened by the Quaker William Tuke in 1796, reinforced Willis' influence, for at least in its early days The Retreat was demonstrably more effective than other madhouses at calming and curing the insane, despite the fact that it used no medicaments and employed no physician.

The success of the novel regime of William Tuke's Retreat, together with a widespread revulsion against the brutal repressions of the eighteenth century, symbolised by Pinel's historic decision to unchain his patients in the Bicêtre, led in the first half of the nineteenth century to a widespread commitment to 'moral treatment' of the insane. Indeed, this provided the basic therapeutic regime of the new lunatic asylums which were fast replacing the private madhouses. Moral treatment assumed that kindness, an ordered regime with regular occupation, regular religious observance, wholesome food and an avoidance of passion and excess would lead to the recovery of most forms of insanity and that medications, and indeed physicians, were generally unnecessary. At the same time autopsies were making it increasingly clear that insanity was not accompanied by the visible pathologies that characterised so many other diseases.

Madness becomes mental disease

In the early years of the nineteenth century a new term, mental disease, began to replace the old terms madness and lunacy, and there was much debate about whether a mind could be diseased in the absence of any disease of the brain, and whether philosophers might not be better fitted to treat diseased minds than physicians. Although it was eventually conceded that mental disease still lay within the province of medicine, and members of the emerging psychiatric profession like Benjamin Rush insisted that the fundamental pathology of diseases of the mind was somatic (Rush himself believed it lay 'primarily in the blood vessels of the brain'), the new term had come to stay (Hunter and Macalpine, 1963). Eminent nineteenth-century alienists like Griesinger repeatedly emphasised their conviction that mental illnesses were diseases of the brain ('Psychische Krankheiten sind Erkrankungen des Gehirns') but the doubts remained, even within their own profession; and it remained the case throughout that century and the first third of the next that mental diseases were largely uninfluenced by the physician's pharmacopoeia and, apart from general paralysis, could not be shown to be accompanied by brain pathology, either macroscopic or microscopic.

The territory of psychiatry is still formally described as mental illness or mental disorder, even though these terms now embrace a far broader range of conditions than they did when they were first introduced. (The American Psychiatric Association uses 'mental disorders' and the World Health Organisation 'mental and behavioural disorders' as their main generic terms.) The implication still persists, therefore, that psychiatric disorders are disorders of the mind, and fundamentally different from disorders of the body. Although the philosophical problem of the relationship between mind and body remains unsolved, and may indeed be insoluble, we are far better informed than the early nineteenth-century physicians who coined the term 'mental disease'. We are aware of a wide range of situations in which alterations to the structure or functioning of the brain result in predictable changes in the subjective experience of the subject, or in his behaviour or cognitive performance. We can also demonstrate that mental activity is accompanied by increased metabolic activity in discrete areas of the brain.

The cerebral substrate of mental phenomena

Stimulating electrodes deep within the brains of conscious subjects can evoke sensory experiences, emotions or vivid memories which have the subjective characteristics of familiar mental phenomena and come and go as the stimulating electrode is switched on and off. Pharmacological agents of various kinds will induce mood states, improve or impair memory, alter perceptions or induce unconsciousness. Lesions in specific areas of the brain will produce predictable changes in the subject's ability to speak, to understand the speech of others, to remember past events, to recognise familiar people or objects, or to hear, see, taste or smell.

In short, we are increasingly able to control, restrict or enhance mental activity by manipulating the functioning or integrity of the brain, and to observe the biochemical events accompanying conscious mentation. We are not in a position to say, and perhaps never will be, that all mental activity has a neurophysiological substrate but we can observe, and influence, an increasingly wide range of close spatial and temporal relationships between mental activity and the structure or functioning of the brain.

It is therefore reasonable for us to assume that all mental activity, normal or abnormal, is accompanied by, and could not take place without, transmission of nervous impulses within and energy consumption by particular neural networks in the brain.

If this philosophical position is accepted, as it is in one form or another by most contemporary neuroscientists, it follows that there is, strictly speaking, no such thing as disease of the mind or mental disorder and that Griesinger was right – mental illnesses are diseases of the brain, or at least involve disordered brain function – because all mental events are accompanied by and dependent on events in the brain. (Thomas Szasz was also right: mental illness is a myth, though not for the reasons he believed.) Acceptance of this argument does not, however, commit psychiatrists to a crude somaticism. It does not imply that we should ignore the role of social and psychological factors in the genesis of psychiatric disorders, still less that we should abandon psychological forms of treatment, or discard concepts like grief, hostility and loneliness in favour of catecholamine and neuropeptide assays.

The functions of the cerebral hemispheres

Despite the impressive advances of the last two decades we know far less about the structure and functioning of the human brain than we do about any other organ in the body. That is why we understand so little about the aetiology of most psychiatric disorders. We know quite enough, though, to realise that the functions and physiology of the brain are far more complex than those of any other organ, probably by several orders of magnitude. So we should not be surprised if psychiatric disorders have characteristics which are not shared with other diseases. One of the most important distinguishing characteristics of psychiatric disorders is the contribution which the patient's previous experience and current psychological and social predicaments – his childhood upbringing, recent life events, the fact that he always feels unwanted, or is lonely or demoralised – make to their aetiology. This is hardly surprising, however, if one reflects that it is one of the brain's most important functions to keep a detailed record of past experience. Sensory perceptions can only be recognised and assessed in the light of a memory of the implications and sequelae of similar perceptions in the past; and appropriate decisions about future behaviour can only be made after reviewing the consequences of previous responses in similar situations. Moreover, because man is pre-eminently a social species with a uniquely long period of dependent immaturity, appraisals are almost bound to concern his relationships with other people, particularly his own family.

Stressful past experiences and the patient's appraisal of his current social environment play a part in the genesis of many illnesses. Asthma, migraine and acute myocardial infarction are obvious examples. They play a much greater role in the genesis of psychiatric disorders, however, and for good reasons. To remember what has happened in the past, to appraise current situations in the light of that memory, and to create moods and action plans appropriate to these appraisals are among the brain's most important functions. It should not surprise us, therefore, that psychiatric illnesses characteristically involve disorders of perception, memory, cognition, mood and volition. Memory is of central importance because it is involved in most of these activities. Information processing is the brain's most basic function and the means by which meaning, which is derived from the interrelationships between different items of information, is attributed to events and symbols. The brain's memory stores are as crucial to this role as the memory banks of a computer are to its functioning. It is no coincidence, therefore, that memories and meanings play a key role in most psychological theories of the aetiology of psychiatric disorders.

Mental and physical illness

If psychiatric disorders generally involve disordered brain function, and there is, strictly speaking, no such thing as disease of the mind or mental disorder then a further question obviously arises: what is the difference between psychiatric disorders like schizophrenia and obsessional disorder, and diseases of the brain like encephalitis and Parkinson's disease? Or indeed between psychiatric disorders and bodily disease in general?

The answer is that the difference is no more fundamental than the difference between, say, gastrointestinal and cardiovascular disorders. Psychiatric disorders tend to involve the patient's whole personality, his social behaviour and his ability to make rational responses to both incoming sensory information and internal cognitive assessments because they involve dysfunctions of the cerebral mechanisms responsible for perception, memory, cognition and mood. Their effects are therefore more global, pervasive and subtle than those of disorders of, for example, the gall bladder or the hip joint. But this is a difference of degree, not a qualitative difference. In fact the classification of individual disorders as 'mental' or psychiatric is in large measure determined by the fact that, for historical reasons, they or related conditions have generally been treated by psychiatrists. If anorexia nervosa were still treated mainly by gynaecologists and endocrinologists it would be regarded, and classified, as a gynaecological or endocrine disorder. If alcoholism and other forms of drug dependence and abuse were usually treated by specialist 'alcohologists' with no psychiatric training they would almost certainly be classified separately from mental or psychiatric disorders. The fact that multi-infarct dementia is regarded and classified as a psychiatric rather than as a vascular disorder well illustrates the utilitarian basis of disease classifications. What matters is not aetiology or pathology so much as which medical speciality has traditionally treated the disorder and possesses the requisite diagnostic and therapeutic expertise. The distinction between 'neurological' disorders of the brain like Parkinson's disease and psychiatric disorders like schizophrenia is particularly artificial and can only be understood in the light of the different historical origins of psychiatry and neurology, and the unfortunate nineteenth century dichotomy between mind and brain. Eventually the distinction is likely to be abandoned. For the time being, however, the skills required to diagnose and treat Parkinson's disease and multiple sclerosis on the one hand and schizophrenia and obsessional disorders on the other are probably sufficiently different to justify two different specialities.

The distinction between mental and physical illness, and the mind/body distinction from which it is derived, has encouraged patients and many doctors to believe that the two are fundamentally different. Both are apt to assume that mental illness is evidence of a certain lack of moral fibre, and that, if they really tried, psychiatric patients ought to be able to control their anxieties, their despondency and their strange preoccupations and 'snap out of it'. It is true, of course, that we all believe in 'free will'; we believe that we ourselves and other people can exercise a certain amount of control over our feelings and behaviour. But why should we expect people suffering from phobic or depressive illnesses to be able to exercise more control over their symptoms than those suffering from, say, myxoedema or arthritis? In a similar vein, patients complaining of intense fatigue or abdominal pain are often dismayed to be told that they have a depressive illness, and interpret such a diagnosis as meaning that their doctor does not believe that they are exhausted by the slightest exertion, or are really in pain, and is dismissing their complaints as 'all in the mind'. All too often doctors make similar assumptions and perceive it as their job to decide whether such patients are ill, or 'just depressed'. They may even attempt to reassure them by saying 'There's nothing wrong, all the tests are negative, you're just depressed,' or words to that effect.

In reality, depressive and other psychiatric illnesses do not differ in any material respect from so called bodily illnesses. It has already been demonstrated that they are almost bound to involve cerebral dysfunction of some kind; and the evidence that there is an important genetic component to the aetiology of depressions and panic disorder implies that there must be biological differences, qualitative or quantitative, between people who are and are not liable to depressive illnesses and anxiety states. The symptoms of depression and phobic anxiety are just as 'real' and painful as those of other illnesses and can no more be overcome by an effort of will than the symptoms of myxoedema. They can, however, be treated effectively with antidepressant drugs which have no effect on mood in other people, and the panic attacks of phobic subjects can be precipitated by intravenous lactate infusions which are similarly without effect in other people. Again, these differences in response imply the existence of underlying biological differences.

The implications of the terms 'mental illness' and 'mental disorder' are therefore seriously misleading, and have had a baneful influence on medical and lay attitudes for nearly two hundred years. In the words of those appointed by the American Psychiatric Association to draft DSM-IV (the diagnostic manual used in the USA), mental disorder 'could not be a more unfortunate term, preserving as it does an outdated mind–body duality. There is much that is physical in the so-called mental disorders, and much mental in the so-called physical disorders. The term organic disorder is an equal abomination' (Frances *et al.*, 1991).

The role of cerebral pathology

Insisting that psychiatric illnesses are not mental illnesses does not imply that they are illnesses of the brain, or bodily illnesses. Neither minds, nor brains nor bodies become ill in isolation. Only people, or in a wider context organisms, become ill or develop diseases. The most characteristic symptom of so-called bodily illness is pain, a purely subjective or mental phenomenon. The first symptom of most systemic illnesses, from the common cold to typhoid fever, is vague generalised malaise, another purely subjective experience. On the other hand, we know, or at least are justified in assuming, that psychiatric conditions like paranoid psychoses and obsessional illnesses whose manifestations may be entirely subjective must involve cerebral dysfunctions.

Many of these cerebral dysfunctions are genetic in origin. We already know, for example, that genetic factors contribute to the aetiology of schizophrenic and affective psychoses, Alzheimer's disease, most depressive illnesses, panic disorder, obsessional disorders, alcohol dependence, and many kinds of mental handicap. It is likely, too, that some life experiences result not merely in the registration of a memory but in enduring changes in brain morphology. It is well established that the richness and complexity of synaptic pathways in the developing brain and the thickness of the cerebral cortex are permanently influenced by the complexity and significance of incoming sensory information (Diamond, 1988). A young animal kept in the dark throughout the early weeks of life, for example, does not develop normal visual association areas in its occipital cortex. The consistency and chronicity of the complex of symptoms developing after overwhelming terrifying experiences (post-traumatic stress disorder) strongly suggests that enduring changes in cerebral functioning have occurred, that the revealing phrase 'scarred for life' is more than a metaphor.

It must not be assumed, however, that all psychiatric disorders are necessarily based on cerebral dysfunctions of some kind. In the first place, what we regard as illness or disorder appears to shade insensibly into normality. Many studies of the distribution of psychiatric symptoms in the general population have failed to reveal any boundary or discontinuity between depressive or anxiety disorders and the temporary emotional disturbances of everyday life. More fundamentally, illness is itself a socially defined concept and involves value judgements which are liable to change from time to time and place to place. Someone who develops depression of mood, insomnia, anhedonia, impaired concentration and weight loss after losing his job or his reputation is regarded as having a psychiatric illness, namely a depressive disorder. Someone who develops identical symptoms after losing his parent or spouse, on the other hand, is not, probably because grief after bereavement is both expected and esteemed, and we are loath to label something we esteem as illness. It is unlikely, though, that there is any difference in the underlying mechanisms. Similarly, paedophilia is regarded as a psychiatric disorder but homosexuality is not, not because of any assumed difference in the underlying causes of the two phenomena, but because our culture strongly disapproves of the former but not of the latter. Because the concept of illness or disorder involves a value judgement, the phenomena we choose to label as psychiatric disorder and those which eventually prove to be rooted in a cerebral dysfunction of some kind will only be identical if we deliberately define the terms psychiatric disorder and cerebral dysfunction to achieve this end. All we can or need say at present is that we have strong presumptive evidence that most severely handicapping psychiatric disorders are probably rooted in cerebral dysfunctions, and that these may be either inborn or acquired, or a combination of the two.

The future

It must be emphasised once more that the twin propositions that psychiatric illnesses do not differ in any fundamental way from other illnesses, and that all psychiatric disorders in which there is a substantial impairment of function almost certainly involve a cerebral malfunction of some kind, do not imply that psychological and social factors do not play a major aetiological role in many psychiatric disorders, or that psychological and social therapies are inappropriate or ineffective. To take just three examples, the evidence that recent life events play a major role in the genesis of depressive illnesses, that the emotional atmosphere within the family is a major determinant of early relapse in schizophrenia and that psychological treatments can be effective in depressive, phobic and obsessional disorders is beyond challenge, and in each case perfectly compatible with an assumption of cerebral dysfunction.

If we are convinced, however, that even neurotic illnesses and other so-called minor disorders probably involve a malfunctioning neuronal substrate we should be prepared to devote the main thrust of our research to exploring this neuronal substrate, and to elucidating the mechanisms by which stress of various kinds causes it to malfunction. Psychiatry already possesses a variety of highly effective pharmacological therapies. The most important of these – the neuroleptics, the tricyclic antidepressants and lithium – were all discovered by chance, without any prior understanding of the cerebral mechanisms they influenced. How much more effective then can we expect the new generation of neuropharmacological agents to be that are developed by design to correct identified neuronal dysfunctions. Yet even though the most important future developments in treatment and prevention are likely to be derived from the biological sciences, there is no reason in principle why psychological and social interventions should not continue to play a major and even an expanding role. The various disorders resulting from substance abuse illustrate this very well. We can look forward with confidence to a time when the means by which alcohol and other psychoactive substances produce intoxication, dependence and tissue damage are understood at a cellular and molecular level. Almost certainly this knowledge will bring with it new means of detecting, reducing and perhaps reversing intoxication, dependence and tissue damage, and of identifying high-risk individuals in advance. This does not mean, though, that it will no longer be important to understand the social forces and psychological pressures that lead people into careers of abuse and dependence, or that political, social and economic measures may not continue to be the most effective means of combating drug misuse on a national scale. However detailed our understanding of underlying neuronal mechanisms eventually becomes, and however potent the resulting pharmacological and other physical therapies may be, the understanding and treatment of psychiatric disorders is always going to require a rounded appreciation of the social setting in which they develop, and an empathic understanding of the patient's inner feelings.

References

Diamond, M. C. (1988) *Enriching Heredity: the impact of the environment on the anatomy of the brain*, New York, Free Press.

Frances, A. J., First, M. B., Widiger, T. A. *et al.* (1991) 'An A to Z guide to DSM-IV conundrums', *Journal of Abnormal Psychology* vol. 100, 407–412.

Hunter, R. and Macalpine, I. (1963) *Three Hundred Years of Psychiatry 1535–1860*, London, Oxford University Press.

Menninger, K. (1963) *The Vital Balance: the life process in mental health and illness*, New York, Viking Press.

Porter, R. (1987) *Mind-Forg'd Manacles: a history of madness in England from the Restoration to the Regency*, London, Athlone Press.

4

'Schizophrenia' re-evaluated

MARY BOYLE

If we look at the mainstream literature on schizophrenia, we can see that, with a few exceptions (e.g., Benjamin, 1989; Strauss, 1989) relatively little attention has been devoted to the individual and specific *content* of hallucinations and delusions. Instead, researchers and clinicians have mainly directed their attention to what we might call structural aspects, to general features said to characterise psychotic hallucinations and delusions. These include the spontaneous occurrence of hallucinations and a belief in the reality of the experience (Al-Issa, 1977) and, for delusions, the conviction with which the belief is held, the extent to which it is preoccupying, its absurdity – as seen by others – and the involvement of personal reference (Oltmanns, 1988).

For both hallucinations and delusions, considerable emphasis has also been placed on the lack of known cultural or stimulus determinants. Although, as we shall see, one or two of these features do refer to content, they do so only in a very general way so that when it's been decided that someone is deluded or has hallucinations, it is rare to find in mainstream literature the suggestion that *what* they believe of *what* voices say to them might be significant in understanding the development of their problems or in helping them overcome them.

Ignoring content

Before I discuss how we might remedy this neglect, I want to consider what I think is an important prior issue, and that is *why* the study of content has been so neglected. We're not so naïve as to suppose that the kinds of questions which researchers ask, and the ways in which they

Originally 'Form and Content, Function and Meaning in the Analysis of "Schizophrenic" Behaviour', *Clinical Psychology Forum* (September 1992), 10–15.

conceptualise human experience are merely a function of the importance of those particular questions, of their potential usefulness in pushing forward the boundaries of science. Rather, we know that professional, social, political and psychological factors can determine what is seen as worthy of study. With this in mind, I'd like to suggest four major reasons why the study of the individual content of hallucinations and delusions has played such a small part in the traditional research literature.

The first is that individual content has not been seen as important in terms of the medical model within which the experiences of hallucinations and delusions have often interpreted. Once it has been decided that someone is deluded, has hallucinations, is suffering from schizophrenia, then, it's implied, you might as well study the delirious speech of a patient with an infectious fever or the precise pitch of a tubercular cough in the hope of finding the causes and cures of these illnesses, as study *what* the 'schizophrenic' belives or *what* their voices say. Within this model, hallucinations and delusions are symptoms which the person suffers from, rather than meaningful experiences which they might, at least in part, actively construct.

A second, and closely related reason for the neglect of content, is that the study of *form* – of general features common to a group – is seen as more scientific. This view has been strongly put in relation to schizophrenia by the American psychologist Ogden Lindsley, but it has also been put forward in relation to psychological processes in general. Lindsley has put it like this:

> [A definition of psychotic symptoms] in terms of their frequency and intensity ... frees the investigator from concern with hallucinatory content and bizarreness. It brings the hallucinatory symptom into the body of natural science (Lindsley, 1963, 296).

Valentine has expressed it even more strongly:

> Psychological science can never be concerned with the content of behaviour (because this must necessarily vary) but only with the principles of adaptation. This puts it squarely in the realm of biological science (Valentine, 1982, 5).

This kind of belief has its roots in the idea that the goal of science, including psychological science, is to discover universal laws applicable across people, time and place. Within this belief system, as Lindsley's quote shows, the individual content of hallucinations and delusions is seen not only as idiosyncratic, as not amenable to systematic study, but almost as getting in the way of scientific progress. By contrast, the study of form, of general characteristics, is seen as linking psychology to the natural sciences.

I want to come back to this point later, but I'll go on to a third, and related, reason why I believe content has been underemphasised. It's one which arises from the great concern of psychiatry and psychology to distinguish the normal from the abnormal, the mentally ill from the mentally healthy. Not only is it believed that to do this scientifically, the distinction must be based on shared, stable features of 'mental illness' rather than apparently idiosyncratic content, it's recognised that to pay too much attention to content might be professionally damaging. As Malcolm Lader puts it:

> To rest the diagnosis of a delusion solely on the content of firmly held beliefs exposes psychiatrists to the charge that they are acting as society's suppressors of heterodox ideas. When these beliefs are political or religious, the danger increases (Lader, 1977, 151).

In other words, if psychiatrists pay too much attention to content it might come to look as if they are acting as social agents in the suppression of certain belief systems, rather than as disinterested scientists applying general rules to distinguish the normal from the abnormal, and to classify different types of abnormality.

The fourth and final reason for the neglect of content I want to consider is perhaps the most complex and is, paradoxically, related to the fact that content, at least of belief systems, *is* taken into account in making a diagnosis of schizophrenia.

Included in almost all lists of defining criteria for delusions is the idea of a bizarre belief being not culturally shared. What this actually means is that neither those around the 'deluded' person nor the examining psychiatrists can understand how the belief came to be formed or make any sense of it within the person's cultural background. The content of the belief is bizarre, irrational and unintelligible with reference to any theories held by observers. But having made the judgement that the content of a belief is meaningless, then it seems to be assumed that we can safely ignore it or, at least, see it as an epiphenomenon, an idiosyncratic consequence of a biological illness.

The idea that the beliefs of people diagnosed as schizophrenic are meaningless and unintelligible has been seriously challenged in the last few years but particularly in relation to people from ethnic minority cultures living in this country. Two of the most prominent writers in this area have been Roland Littlewood and Maurice Lipsedge. The starting point of their discussion was, they say, their realisation that diagnoses of schizophrenia, when made about groups from ethnic minorities, often conveyed the doctors' lack of understanding rather than the presence of the 'key symptoms' by which schizophrenia is conventionally recognised by British psychiatrists. Patients appeared to be regarded as unintelligible because of their cultural background. Littlewood and Lipsedge go on to claim that, 'with a sympathetic knowledge of another's culture and of their personal experience it is possible to understand much of what otherwise appears as inexplicable irrationality', and that, 'there is always an interrelation between personal experience and cultural preoccupations which is not haphazard ... and which can be understood historically' (Littlewood and Lipsedge, 1989, 216–17).

Littlewood and Lipsedge have provided a detailed and important account of the relationship between what seem to be bizarre and meaningless beliefs and experiences, and the cultural background and personal experience of people from ethnic minorities who might be diagnosed as schizophrenic. This account has paid close attention to both form and content in a way which is unusual in both psychiatry and psychology. I don't wish to find fault with this work because I believe it to be an important step in the right direction. But what seems to me to be highly significant about these ideas, which, after all, give a central role to the content, to the intelligibility, of bizarre beliefs and experiences, is that the ideas were developed mainly in relation to people with non-Western cultural backgrounds. In other words, this analysis hasn't amounted to any real change in the way we think about delusions and hallucinations in Western groups.

If we ask why this kind of analysis, this attempt to make the unintelligible intelligible, has not been taken up enthusiastically in relation to Western culture, then one answer may lie in two important functions served by our seeing 'psychotic' behaviour as culturally and personally unintelligible. The first function is apparent in our society's tendency to deny, or, at least to de-emphasise, our capacity for bizarre beliefs and experiences. It is as if, for example, religious groups, spiritualists and so-called fringe movements have become the safe repositories of this capacity. In these contexts, and when we cannot find others to validate our experiences, then it is as if such beliefs and experiences are so threatening to our rational image of ourselves and our society that we deal with them by denying that their occurrence and their content can be understood within rational, scientific and technologically sophisticated Western culture. So such experiences can only be made intelligible by seeing them as a product of brain dysfunction.

The second, professional, function may lie in the centrality of the idea of irrationality to definitions of madness (Szasz, 1987). In the absence of reliable evidence that everyone diagnosed as schizophrenic behaves as they do because of brain dysfunction, and in the absence of any biological criteria in official diagnostic systems, then the notion of irrationality, of a lack of social and cultural determinants of behaviour, becomes crucial, however inappropriately, as indirect evidence of organic disorder. Thus to search for meaning in 'psychotic' behaviour, in other words to make the behaviour intelligible in its social context, comes close to suggesting that we do not need to seek biological explanations or, at least, not within a simplistic reductionist model.

These, I think, are some of the reasons why the content of psychotic behaviour has received relatively little attention in the traditional psychiatric literature. Together, they constitute a powerful barrier to attempts to understand psychotic phenomena.

Before I turn to a brief discussion of how we might approach the study of content, I'd like to return to one of the reasons I suggested for its neglect – that the study of form is more scientific because it can result in the development of general laws of behaviour. The idea that we can develop such laws, independent of time, place and person, has been strongly criticised over the last decade, particularly within the social constructionist movement in psychology (Gergen, 1985). These critiques have been closely associated with the development of theoretical frameworks for relating discourse and personal accounts to a cultural setting (Potter and Wetherell, 1987; Hollway, 1989). So we have at least the beginnings of a theoretical rationale and a set of research methods to justify and make possible the systematic study of content (see also Gains, 1988).

Attending to content

I want now to suggest some possible ways in which we might approach the study of the content of psychotic experiences. I must emphasize that these are not independent and that any separation of them is rather artificial.

The first is to develop general categories of content which may be used for the study of groups or individuals. Such methods have long been used by anthropologists (e.g. Wallace, 1959). The categories could include, for example, whether the message is from someone living or dead, whether it is positive or negative, and whether it commands action or merely comments on behaviour. Categorisations like these are, of course, already in use to some extent as, for example, when we distinguish persecutory delusions from delusions of grandeur. This in itself should alert us to the fact that merely categorising experiences in this way may be of little theoretical or practical consequence unless we attempt to go beyond description and cataloguing and try to make sense of the experience within a cultural framework.

A second approach to content is to see it as communication using unusual symbols or using metaphor as if it were literal. This idea has a long history but it has never been seriously pursued in the traditional literature. Bleuler himself provided examples of such communication and claimed that:

> When one patient declares that she is Switzerland, or when another wants to take a bunch of flowers to bed with her so that she will not awaken any more

these utterances seem to be quite incomprehensible at first glance. But we obtain a key to the explanations by virtue of the knowledge that these patients readily substitute similarities for identities ... they employ symbols without any regard for their appropriateness in the given situation (Beuler, 1950, 428).

Lucy Johnstone has given an example of this mechanism within a family where the over-protected daughter was diagnosed as schizophrenic (Johnstone, 1989). Littlewood and Lipsedge have also emphasized the role of metaphor and symbol in psychotic behaviour. Indeed, their task can be seen as one of decoding such speech and making sense of it within the person's cultural and social framework. But, like Beuler, they want to retain the idea of *inappropriate or inaccessible* metaphors or symbols and to label such phenomena as manifestations of illness, as if we could scientifically adjudicate on how people *ought* to use metaphor.

A third approach to the analysis of the content of 'psychotic' behaviour, but which can also incorporate the study of form, is to look at the possible functions served by what are called delusions and hallucinations. It is very difficult indeed to integrate the idea of function into a medical model of 'schizophrenia'. Function implies, after all, that the behaviour may be purposeful and the content meaningful, in contrast to the medical model's depiction of it as a response to disordered brain chemistry, even if mediated by environmental events. But like any behaviour, what we think of as hallucinations and delusions with a particular content, can fulfil a number of functions including the following:

- *Confer special status, or give access to special knowledge or skills.* Clearly the status will vary across cultures, and the opportunities for special status within our society are much more limited than in others. Nevertheless, we would be naïve to assume that this function is not important. We all remember Doris Stokes and, at a less exalted level, I'm reminded of a woman who was understandably reluctant to relinquish the idea that she was telepathic and accept instead that she might be mentally ill. Even to believe that you're being persecuted in some way, that you are the object of malicious intentions, is perhaps better than acknowledging that you are of no particular interest to anyone.
- *Provide comfort.* In their study of hallucinations in the non-psychiatric population, Posey and Losch (1983) highlighted the extent to which certain hallucinated voices were seen as comforting. One of my students was once told by a woman whose hallucinations she was trying to 'extinguish' that the woman was grateful for her trouble, but that she preferred to keep her voices as they were company for her. Fonagy and Slade (1982) in a study which also aimed to 'extinguish' hallucinations, described a woman who stopped telling the experimenters about, but apparently didn't stop experiencing, the voice of her boss saying he loved her, and continued to report only voices

which told her to steal. Clearly, the woman's behaviour in this study cannot be understood without reference to the content of her hallucinations.

- *Make decisions or provide guidance.* Julian Jaynes (1976) has developed a fascinating theory of consciousness which is based on the idea that prior to about 3000 BC much of our action was unconscious and that decisions were taken and guidance provided by hallucinated voices, interpreted as the voices of gods. As part of his evidence, Jaynes uses the *Iliad*, in which men are consistently told what to do by the voices of gods. Jaynes claims that this 'bicameral' mind broke down as social complexity increased and with the development of writing. It's very difficult to know if this account is accurate, but what is clear is that accounts of hallucinated voices providing guidance and directing behaviour are both ancient and plentiful (see also Nydegger, 1972). Closely related to this is the idea of hallucinated voices fulfilling a fourth function of
- *Removing responsibility for negatively judged behaviour.* It's difficult to believe that the fact that Peter Sutcliffe, the 'Yorkshire Ripper', claimed to have heard the voice of God telling him to kill prostitutes rather than, say, to love his neighbour is of no theoretical or practical significance. Yet the fact that he and the woman mentioned earlier who heard the voice of her boss saying he loved her, were given the same diagnostic label, suggests that the similarities between them outweight the differences.

To these major functions can be added that delusions and hallucinations may provide a fantasy escape from an intolerable reality; may give access to the sick role and may provide an acceptable explanation for puzzling or aversive events.

Finally, we should remember that what the 'deluded' person tells us might be true. Maher (1988) has reminded us of the work of Southward early in this century. He found that a number of supposedly deluded patients showed, at post mortem, undetected bodily pathology which made intelligible their delusional beliefs about their bodies.

What do we know?

Having suggested that we should try to render the form and content of psychotic behaviour intelligible by looking at it in relation to the person's context and culture, I want to sound a note of extreme caution. It would be easy to imagine that we already 'know' about our own culture and about people's experience in it – indeed John Wing has claimed that a psychiatric examination must be grounded in a thorough familiarity with the subjective experience of human beings, as if such familiarity were already on offer from training courses (Wing, 1978).

Wing is apparently suggesting that our task is to take this 'knowledge' and compare it with the form and content of the psychiatric patient's experience to see if they 'match'. If they don't, then we may judge the patient's experience to be irrational and the patient possibly to be mentally ill. I find this approach at best naïve and at worst dangerous. I don't believe that at the moment we have an extensive knowledge of the subjective experience of human beings and I certainly don't believe that what knowledge we might have is widely shared amongst psychologists and psychiatrists. Had such knowledge existed and been used, feminists and anti-racists would not have been able to show how the subjective experiences of some groups, particularly women and people from ethnic minorities have often been rendered invisible or invalid by psychological theory and method.

Making better sense of psychotic experiences is an important goal, but I doubt that we can achieve it until we make more realistic sense of ourselves.

References

Al-Issa, I. (1977) 'Social and cultural aspects of hallucinations', *Psychological Bulletin*, vol. 84, 570–587.

Benjamin, L. (1989) 'Is chronicity a function of the relationship between the person and the auditory hallucination?', *Schizophrenia Bulletin*, vol. 15, 291–310.

Bleuler, E. (1950) *Dementia Praecox or The Group of Schizophrenias*, trans J. Zitkin, New York, International Universities Press.

Fonagy, P. and Slade, P. D. (1982) 'Punishment versus negative reinforcement in the aversive conditioning of auditory hallucinations', *Behaviour Research and Therapy*, vol. 20, 483–492.

Gains, A. D. (1988) 'Delusions: culture, psychosis and the problem of meaning', in Oltmanns, T. F. and Maher, B. A. (eds), *Delusional Beliefs*, New York, Wiley.

Gergen, K. J. (1985) 'The social constructionist movement in psychology', *American Psychologist*, vol. 40, 266–275.

Hollway, W. (1989) *Subjectivity and Method in Psychology*, London, Sage.

Jaynes, J. (1976) *The Origins of Consciousness in the Breakdown of the Bicameral Mind*. London, Allen Lane.

Johnstone, L. (1989) *Users and Abusers of Psychiatry*, London, Routledge.

Lader, M. (1977) *Psychiatry on Trial*, Harmondsworth, Penguin.

Lindsley, O. (1963) 'Direct measurement and functional definition of vocal hallucinatory symptoms', *Journal of Nervous and Mental Diseases*, vol. 136, 293–297.

Littlewood, R., and Lipsedge, M. (1989) *Aliens and Alienists: ethnic minorities and psychiatry*, 2nd edn, London, Unwin Hyman.

Maher, B. A. (1988) 'Anomalous experience and delusional thinking: the logic of explanations', in Oltmanns, T. F. and Maher, B. A. (eds), *Delusional Beliefs*, New York, Wiley.

Nydegger, R. V. (1972) 'The elimination of hallucinatory and delusional behaviour by verbal conditioning and assertive training: a case study', *Journal of Behaviour Therapy and Experimental Psychiatry*, vol. 3, 225–227.

Oltmanns, T. F. (1988) 'Approaches to the definition and study of delusions', in Oltmanns, T. F. and Maher, B. A. (eds), *Delusional Beliefs*, New York, Wiley.

Posey, T. B. and Losch, M. (1983) 'Auditory and hallucinations of hearing voices in 375 normal subjects', *Imagery, Cognition and Personality*, vol. 3, 99–113.

Potter, J. and Wetherell M. (1987) *Discourse and Social Psychology: beyond attitudes and behaviour*, London, Sage.

Strauss, J. S. (1989) 'Subjective experiences of schizophrenia: towards a new dynamic psychiatry', *Schizophrenia Bulletin*, vol. 15, 179–187.

Szasz, T. (1987) *Insanity: the idea and its consequences*, New York, Wiley.

Valentine, E. (1982) *Conceptual Issues in Psychology*, London, Allen & Unwin.

Wallace, A. F. C. (1959) 'Cultural determinants of response to hallucinatory experience. *Archives of General Psychiatry*, vol. 7, 198–205.

Wing, J. (1978) *Reasoning about Madness*, Oxford, Oxford University Press.

5

Life events, loss and depressive disorders

GEORGE W. BROWN

This paper explores how far life events that provoke depressive disorders do so because of loss. It deals with clinically relevant depression at a 'caseness' level of the order defined by DSM-III major depression (Finlay-Jones *et al.*, 1980), and only with studies of women. A convincing aetiological model for men that is emerging is unlikely to differ a great deal from that for women (Bolton and Oatley, 1987; Eales, 1985).

It is difficult to conceive of an alternative to a broadly social explanation for the large differences in rates of depression that can occur in populations. For example, two recent studies that have revealed a one-year prevalence of major depression among women in a Basque-speaking rural community of 3 per cent (Gaminde *et al.*, 1993) and one of 30 per cent in a black urban setting in Zimbabwe (Broadhead and Abas, 1994a), a tenfold difference. The differences were exactly paralleled by the frequency of adverse life events. Both used the same research instruments. Studies of Western urban populations have rates falling roughly between these extremes (e.g. Brown and Harris, 1978). There are, of course, bound to be significant biological risk factors: these can be regarded as contributing to variability in risk within populations, with differences across populations largely driven by psychosocial factors.

Current research suggests that the bulk of depressive disorder is the result of a failure to meet goals derived from evolutionary-based needs. These include being admired, forming friendships, having a core adult attachment figure, having children and so on. These goals are almost entirely social in nature. Thus, rates of depression are likely to be largely the result of psychosocial processes. However, the gaps between goals

Edited and adapted from the Paul Hoch Lecture, 'Loss and Depressive Disorders', delivered in New York in March 1994. Published in full in B. P. Dohrenwend (ed.) (1996) *Adversity, Stress and Psychopathology*, Washington, DC: The American Psychiatric Press, Inc.

and their fulfilment must generally have particular qualities for clinical depression to emerge.

The role of life events in the onset of clinical depression

The study of life events has been critical in developing a radical social perspective. The instrument used in the research described here, the Life Events and Difficulties Schedule (LEDS), has been good enough to establish that certain kinds of nasty event do occur quite soon before onset in the majority of depressive disorders and that they are of aetiological significance (Brown and Harris, 1986, 1989).

The LEDS has now been used in some twenty studies, covering depressed patients and non-patients. It requires skilled interviewing focused on the broad context of the event and its timing *vis à vis* other events and any onset. Findings concerning the onset of depression have been broadly comparable. In adult samples between two-thirds and 90 per cent of episodes have at least one severely threatening life event occurring not long before onset. It must be emphasised that long-term threat is involved; such events must have long-term threatening implications at a point in time some 10 to 14 days after their occurrence. Events with short-term threat, however nasty (e.g., a child admitted to hospital, apparently dangerously ill, but recovering within a few days) do not provoke depression (Brown and Harris, 1978).

Table 4.1 summarises the findings of one study in terms of onsets in a 12-month period. This longitudinal study was carried out with 400 women with a child living at home in Islington, an inner-city area in North London (Brown *et al.*, 1987). The women were largely working class. A quarter were single mothers. They were followed up at 12-month intervals. The table deals with the 303 women at risk of an onset – at the time of our first contact they did not have depression at a clinical level. The findings are fairly typical.

Two results are relevant. First, 29 of the 32 women developing a

TABLE 4.1 Onset of depression among 303 women in terms of provoking agent status

Provoking agent status	No. of onsets	Onset rate (%)
A. No provoking agent	2/153	1
B. Provoking agent:		
i. Major difficulty only	1/20	5
ii. Severe event	29/130	22
Total onset rate	32/303	11

depressive disorder had a severe event in the six months before onset, most within a matter of weeks. Secondly, only about a fifth of those experiencing a severe event in the year went on to develop depression. This latter result raises the issue of vulnerability, a matter which will be considered after the nature of provoking life events has been examined.

The question of endogenous depression

There can be little doubt of the existence of genuine 'endogenous' depressive disorders. But one of the puzzling findings of life-event research is that events appear to play a significant role in all forms of depression (e.g., Brown and Harris, 1978; Paykel *et al.*, 1971). Recent studies of patient populations have been variable, but they have consistently failed to show that endogenous and non-endogenous conditions, when defined in clinical terms, are all that different (e.g., Bebbington and McGuffin, 1989; Katschnig *et al.*, 1986).

A recent study of depressed patients seen by psychiatrists in North London may possibly throw light on this puzzling conclusion. It distinguished melancholic/psychotic and non melancholic/psychotic patients on the basis of the presence of 'endogenous' symptoms such as delusions, early waking and retardation (Brown *et al.*, 1994). Those with a melancholic/psychotic condition formed only some third of the sample as a whole. All patients tended to experience a severe event before onset except for the melancholic/psychotic with a prior onset. The result may go some way to explain the inconsistency of findings in the literature, since the proportion of such patients may be greater in tertiary treatment centres that have sometimes been studied. It also suggests some biological process such as 'scarring' is at work in the melancholic/psychotic group that increases the chance of a spontaneous episode after the first (Post *et al.*, 1986). A key point for understanding depression as a whole is that melancholic/psychotic conditions are unlikely to form more than a tenth of the total range of clinically relevant depression to be found when whole populations are considered.

The role of loss

In considering the nature of provoking severe events, the starting point is the experience of loss. The LEDS relies on investigator-based judgements along the lines of the notion of the likely response of a particular individual to a specific event once her biography and immediate circumstances have been taken into account (Brown, 1989). The ratings take account of an event's likely impact on central goals, plans and concerns.

This contextual approach is used to delineate the severely threatening events mentioned earlier. Thus, a young university student, finding out she was pregnant eight weeks after her boyfriend, who had planned to marry her, had suddenly left her, would be judged to have had a severe event, irrespective of anything she might say about her actual reaction. This is a typical example of a severe event. Usually, there is not much doubt about the nastiness of such events.

The evidence also suggests that a core identity is likely to be involved in events provoking depression. In addition, each severe event has been considered in terms of the presence of loss, extending the concept to cover not only loss of a person, but that of a role or cherished idea about oneself or someone close. Such loss has been contrasted with danger, that is, the threat of future loss. A particular event may reflect both danger and loss. Several studies have shown that at least three-quarters of severe events leading to depression involve a loss (Finlay-Jones and Brown, 1981; Finlay-Jones, 1989). Despite this, there are reasons for questioning its centrality for clinical depression.

Loss is by no means always involved. One woman had had a serious problem for two years with the hyperactivity of her eight-year old son. The event which led to onset was a teacher complaining about the child in front of many other mothers. It must be doubtful if anything had been lost at this particular point in time. She had long been aware of her son's difficulties and had received complaints from the school well over a year before, albeit not in public. This kind of example points to the need for an alternative perspective.

The role of humiliation/entrapment

A further perspective underlines situations leading to powerlessness, defeat and the lack of any way forward. Responses to the experience of defeat have been viewed as critical for human depression. They have been thought to originate in an evolutionary sense from either the activity of defending territory, or submission following being 'out-ranked' in a group-living species (e.g. Price *et al.*, 1994). With this perspective in mind, Gilbert (1992) has outlined a number of depression-provoking situations which closely follow those reached by LEDS life-event research:

1 direct attacks on a person's self-esteem that force them into a subordinate position;
2 events that undermine a person's sense of rank, attractiveness and value, particularly via the consequences of the event for core roles; and
3 events where escape is blocked.

Ideas like these go beyond the concept of loss. In discussing human despair Unger (1984), for example, also underlines the key importance of what he calls the experience of imprisonment. This can occur in the 'blocked escape' described by Gilbert when we are unable to free ourselves from an unrewarding setting, or in grief when despair arises from a disbelief in the ability to reaffirm an identity in the absence of the relationship.

To reflect these ideas we have used material collected about severe events, or related series of severe events, in the Islington general population sample with a two-year period being taken, rather than the one-year used earlier (see Brown *et al.*, 1994). A *hierarchical* scheme based on contextual ratings of likely response to the events was used (see Table 5.2). The overall risk of onset from a severe event was approximately 1 in 6 in the 6 months following the event.

There were large differences in risk of depression by event category (see Table 5.2). The relatively low risk associated with a loss-alone event, except the category of 'death', and the fact that a third of the humiliating and entrapping events did not involve loss, suggests that something more than loss is usually necessary to bring about a depressive onset.

The likely importance of humiliation and devaluation of self and entrapment is seen in another result. Separations associated with humiliation (category A (i) in Table 5.2) were further divided into whether the woman took *some* initiative after learning of an infidelity or marked violence. When the category of woman who clearly took the initiative with regard to separation is taken into account, a clear gradient emerges. Hammen (1988) has made the point that the one finding that has clearly emerged from the extensive research on the reformulated learned helplessness model involves the importance of lack of control.

TABLE 5.2   Onset by type of severe event over 2 year period in Islington community series.

Hierarchical event classification	No. of onsets	Onset rate (%)
A. All 'humiliation' events:	31/102	30
i. Humiliation: separation	12/34	35
ii. Humiliation: other's delinquency	7/36	19
iii. Humiliation: put down	12/32	38
B. All 'trapped' events (and not A)	10/29	34
C. All 'loss' events; (and not A or B)	14/157	9
i. Death	7/24	29
ii. Separation subject initiated	2/18	11
iii. Other key loss	4/58	7
iv. Lesser loss	1/57	2
D. All 'danger' events (and not A, B or C)	3/89	3
E. All severe events	58/377	15

To summarise: severe events provoking onset of depression will tend to have many, but not all, of the following characteristics:-

High commitment in the role area involved in the event
Loss – defined in broad sense
Devaluation in one's own or other's eyes
Experience of defeat
Entrapment
Lack of a sense of control.

It also might be added that almost all events involved some kind of inter-personal crisis.

Psychosocial vulnerability

The general conclusion concerning type of depression-provoking life events fits well into current findings concerning vulnerability to depression seen in psychosocial terms. Two background factors measured at the time of first interview in Islington proved highly predictive of onset of depression in the following 12 months:

- *negative psychological factors* (negative evaluation of self or chronic sub-clinical symptoms);
- *negative environmental factors* (negative interaction with a partner or child in the home or lack of a close confidante in the case of single mothers).

The order of prediction can be judged by the fact that, while only 23 per cent of the 303 women at risk for developing a depressive disorder had both risk factors, three-quarters of all onsets taking place in the following 12-month period occurred among this group (Brown, Bifulco and Andrews, 1990).

There is a conceptual similarity between humiliation and entrapment events and these two risk factors. This is clear with low self-esteem, which is a key component of the negative psychological factor. However, low self-esteem, is not essential for the development of depression. Some depressed women in Islington had high self-esteem when seen during the episode (Brown, Andrews, Bifulco and Veiel, 1990). It would appear quite possible following, say, an entrapment event, to retain high self-esteem and in no way to blame oneself for the event, yet still to develop depression (Gilbert, 1992). On present evidence it is, however, likely that low self-esteem prior to the experience of a severe event will raise the chances of experiencing it in terms of defeat and general hopelessness.

Figure 5.1 shows the results of putting together the type of provoking

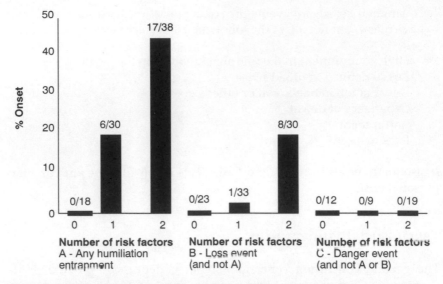

FIGURE 5.1   Rates of onset of depression in follow-up year by severe event type and background risk among 130 Islington women

event and vulnerability. For the one-year follow-up period a severe event, however nasty, was typically not enough to produce depression; neither was vulnerability on its own. Both were usually required. The size of the interactive effects are impressive. For example, for loss without humiliation/entrapment, onset was almost absent without the presence of both background risk factors. One implication is that if risk of onset is considered in terms of traditional event-type categories such as loss of job, move of house, illness and divorce, only a very small proportion of most types are likely to lead to depression.

Recent research also suggests that psychosocial factors are equally involved in determining the *course* of a depressive disorder (Brown *et al.*, 1992). Studies concerning determinants of chronicity underline the importance of 'negative' predictors such as an interpersonal difficulty (but no other kind) at point of onset and the experience of neglect and abuse in childhood (Brown and Moran; Brown *et al.*, 1994).

Some final comments

Many questions remain. First, why should a life event usually be necessary to provoke depression? There seems to be a tendency for human beings to adapt to adversity and deprivation, not in the sense of ruling out suffering, but apparently enough to ward off any onset of depression at a level of clinical severity. It may reasonably be suspected that life events are usually necessary to provoke onset, even when levels of background deprivation and nastiness are great. It should also be remembered here that we do not always recognise the degree of our dependence on a particular social arrangement until that arrangement goes wrong.

Secondly, why, typically, do psychosocial vulnerability factors also have to be present for clinical depression to emerge? It may be that for depression to occur, the sense of defeat and hopelessness brought about by the event must be matched by an inability to contemplate dealing with its implications.

Two broad processes are likely to prove critical:

- A *cognitive-emotional* one in which interpretations of hopelessness and helplessness are brought to the situation. The Islington research has made clear the predictive importance of low self-esteem (defined on the basis of negative comments) as a vulnerability factor, and how such negative evaluation of self is highly related to difficulties in the women's current environment (Brown, Andrews, Bifulco and Veiel, 1990; Brown, Bifulco, Veiel and Andrews, 1990; Brown, Bifulco and Andrews, 1990). While complex transactional processes involving personal characteristics and environment are undoubtedly involved, on current evidence a person's doubts about her ability to deal effectively with the consequences and implications of an event probably typically have a realistic basis.
- *Biological vulnerability* stemming from ongoing stressors. Some kind of depletion in the brain of a 5HT transmitter is one possibility. This may, indeed, be one way in which self-esteem is lowered (Deakin, 1990).

Conclusion

Some representative findings concerning life events have been considered. The LEDS measure upon which the findings are based is certainly crude. However, it has been good enough to get research underway and has provided a considerable number of replicable findings.

Life events are the very phenomena that our brains have evolved to deal with. What happens in the outside world can have profound biological implications. Furthermore, the study of life events, at least when collected by semi-structured instruments such as the LEDS, inevitably leads to a concern with needs, plans and goals which are easily overlooked in the current emphasis on cognitive and neurophysiological processes. The human experience is centrally about wanting, about goals, about meaning – in ways not necessarily easy to articulate. If social science research is not grounded in this experience, its contributions to psychiatry will often lack illuminative power.

References

American Psychiatric Association (1987) *Diagnostic and Statistical Manual of Mental Disorders* (3rd edn, revised), Washington, DC, American Psychiatric Association.

Bebbington, P. and McGuffin, P. (1989) 'Interactive models of depression', in Paykel, E. and Herbst, K. (eds), *Depression, an integrative approach*, London, Heinemann Medical.

Bolton, W. and Oatley, K. (1987) 'A longitudinal study of social support and depression in unemployed men', *Psychological Medicine* vol. 17, 453–460.

Broadhead, J. and Abas, M. (1994a) 'Depression and anxiety among women in an urban setting in Africa' (manuscript).

Broadhead, J. and Abas, M. (1994b) 'Life events and depression and anxiety amongst women in an urban African setting' (manuscript).

Brown, G. W. (1989) 'Life Events and Measurement', in Brown, G. W. and Harris T. O. (eds), *Life Events & Illness*, New York, Guildford Press; London, Unwin & Hyman, 49–94.

Brown, G. W. (1992) 'Social Support: an investigator-based approach', in Veiel H. O. F. and Baumann U. (eds), *The Meaning and Measurement of Social Support*, Washington, DC, Hemisphere Publishing Corporation, 235–257.

Brown, G. W., Andrews, B., Bifulco, A. and Veiel, H. (1990) 'Self-esteem and depression: 1. Measurement issues and prediction of onset', *Social Psychiatry & Psychiatric Epidemiology*, vol. 25, 200–209.

Brown, G. W., Bifulco, A. and Harris T. O. (1987) 'Life events, vulnerability and onset of depression: some refinements', *British Journal of Psychiatry*, vol. 150: 30–42.

Brown, G. W., Bifulco, A. and Andrews, B. (1990) 'Self-esteem and depression: 3. Aetiological issues', *Social Psychiatry & Psychiatric Epidemiology*, vol. 25, 235–243.

Brown, G. W. and Harris T. O. (1978) *Social Origins of Depression: a study of psychiatric disorder in women*, London, Tavistock Publications; New York, Free Press.

Brown, G. W. and Harris T. O. (1986) 'Establishing Causal Links: the Bedford College studies of depression', in Katschnig, H. (ed.), *Life Events and Psychiatric Disorders'*, Cambridge, Cambridge University Press.

Brown, G. W. and Harris T. O. (1989a) *Life Events & Illness*, New York, Guildford Press; London, Unwin & Hyman.

Brown, G. W. and Harris T. O. (1989b) 'Depression', in Brown, G. W. and Harris T. O. (eds), *Life Events & Illness*, New York, Guildford Press; London, Unwin & Hyman.

Brown, G. W., Harris, T. O. and Eales, M. J. (1993) 'Aetiology of anxiety and depressive disorders in an inner-city population. 2. Comorbidity and adversity', *Psychological Medicine*, vol. 23, 155–165.

Brown, G. W., Lemyre, L. and Bifulco, A. (1992) 'Social factors and recovery from anxiety and depressive disorders: a test of the specificity', *British Journal of Psychiatry*, vol. 161, 44–54.

Brown, G. W. and Moran, P. (1997) 'Clinical and psychosocial origins of chronic depressive episodes: 1. A community survey'. *British Journal of Psychiatry* forthcoming.

Brown, G. W., Harris T. O., Hepworth, C. and Robinson, R. (1994) 'Clinical and psychosocial origins of chronic depression. 2. A patient enquiry', *British Journal of Psychiatry*, vol. 165, 457–465.

Deakin, J. W. (1990) 'Serotonin subtypes and affective disorder' in Idzidowski, C. and Cowen, P. (eds), *Serotonin – Sleep and Mental Disorder*, Oxford, Blackwell Scientific.

Eales, M. J. (1985) 'Social factors in the occurrence of depression, and allied disorders, in unemployed men', PhD thesis, Royal Holloway & Bedford New College, University of London.

Finlay-Jones, R. (1989) 'Anxiety' in Brown, G. W. and Harris T. O. (eds), *Life Events & Illness*, New York, Guilford Press; London, Unwin & Hyman, 95–112.

Finlay-Jones, R., Brown, G. W., Duncan-Jones, P., Harris T. O., Murphy, E. and Prudo, R. (1980) 'Depression and anxiety in the community', *Psychological Medicine*, vol. 10, 445–454.

Finlay-Jones, R. and Brown, G. W. (1981) 'Types of stressful life event and the onset of anxiety and depressive disorders', *Psychological Medicine*, vol. 11, 803–815.

Gaminde, I., Uria, M., Padro, D., Quarejeta, I. and Ozamiz, A. (1993) 'Depression in three populations in the Basque country – a comparison with Britain', *Social Psychiatry and Psychiatric Epidemiology*, vol. 28, 243–251.

Gilbert, P. (1992) *Depression: The Evolution of Powerlessness*, Hove (UK) and Hillsdale (USA), Lawrence Erlbaum Associates.

Hammen, C. (1988) 'Depression and cognition about personal stressful life events', in Alloy, L. B. (ed.), *Cognitive Processes in Depression*, New York, Guilford Press.

Katschnig, H., Pakesch, G. and Egger-Zeidner, E. (1986) 'Life stress and depressive subtypes: a review of present diagnostic criteria and recent research results', in Katschnig, H. (ed.), *Life Events and Psychiatric Disorders: controversial issues*, Cambridge; Cambridge University Press.

Paykel, E. S., Prusoff, B. A. and Klerman, G. L. (1971) 'The endogenous–neurotic continuum in depression', *Journal of Psychiatric Research*, vol. 8, 73–90.

Post, R. M., Rubinow, D. R. and Ballenger, J. C. (1986) 'Conditioning and sensitisation in the longitudinal course of affective illness', *British Journal of Psychiatry*, vol. 149, 191–201.

Price, J., Sloman, L, Gardner, R. Jr, Gilbert, P. and Rohde, P. (1994) 'The social competition hypothesis of depression', *British Journal of Psychiatry*, vol. 164, 309–315.

Unger, R. M. (1984) *Passion: An Essay on Personality*, New York, Free Press.

6

Women and madness: the mental asylum

PHYLLIS CHESLER

As early as the sixteenth century women were 'shut up' in madhouses (as well as in royal towers) by their husbands (Dershowitz, 1969). By the seventeenth century, special wards were reserved for prostitutes, pregnant women, poor women, and young girls in France's first mental asylum, the Salpêtrière (Foucault, 1961; Szasz, 1970; Rosen, 1968). By the end of the nineteenth and throughout the twentieth century, the portraits of madness, executed by both psychiatrists and novelists, were primarily of women.

Today, more women are seeking psychiatric help and being hospitalized than at any other time in history. Significantly more women are being 'helped' than their existence in the population at large would allow us to predict.

This increase may be understood, not only in the context of the 'help-seeking' nature of the female role, or the objective oppression of women, but in the context of at least three recent social trends. Traditionally, most women performed both the rites of madness and childbirth more invisibly – at home – where, despite their tears and hostility, they were still needed. While women live longer than ever before, and longer than men, there is less and less use, and literally no place, for them in the only place they 'belong' – within the family. Many newly useless women are emerging more publicly into insanity. Their visibility is also greater, due to our society's relatively successful segregation of violence (into ghettos and jails) and 'madness' (into hospitals and doctors' offices). Finally, the prevalence of female mental patients is related to our society's atheism, or worship of the tangible: no longer are women sacrificed as voluntary or involuntary witches. They are, instead, taught to sacrifice themselves for newly named heresies (Szasz, 1970).

The patriarchal nature of psychiatric hospitals has been documented

Edited from Chesler, P. (1972) *Women and Madness*, New York, Avon, chapter 2.

by M. Foucault (1961), T. Szasz (1970), E. Goffman (1970) and T. Scheff (1966). Journalists, social scientists, and novelists have described, deplored, and philosophized about the prevalence of overcrowding, understaffing, and brutality in America's public mental asylums, jails, and medical hospitals. It is obvious that *state* mental asylums are the 'Indian reservations' for America's non-criminally labeled poor, old, black, Latin, and female populations. It is also obvious that the state hospital, much like the poor or workhouse of old, functions as a warning specter, particularly to those women involved in earlier or more part-time phases of their 'careers' as psychiatric patients.

Mental asylums rarely offer asylum. Both their calculated and their haphazard brutality mirrors the brutality of 'outside' society. The 'scandals' about them that periodically surface in the media are like all atrocities – only everyday events, writ large. Madness – as a label or reality – is not conceived of as divine, prophetic, or useful. It is perceived as (and often further shaped into) a shameful and menacing disease, from whose spiteful and exhausting eloquence society must be protected. At their best, mental asylums are special hotels or college-like dormitories for white and wealthy Americans, where the temporary descent into 'unreality' (or sobriety) is accorded the dignity of optimism, short internments, and a relatively earnest bedside manner. At their worst, mental asylums are families bureaucratized: the degradation and disenfranchisement of self, experienced by the biologically owned child (patient, woman), takes place in the anonymous and therefore guiltless embrace of strange fathers and mothers. In general psychiatric wards and state hospitals, 'therapy', privacy, and self-determination are all either minimal or forbidden. Experimental or traditional medication, surgery, shock, and insulin coma treatment, isolation, physical and sexual violence, medical neglect, and slave labor are routinely enforced. Mental patients are somehow less 'human' than either medical patients or criminals. They are, after all, 'crazy'; they have been abandoned by (or have abandoned dialogue with) their 'own' families. As such, they have no way – and no one – to 'tell' what is happening to them.

The mental asylum closely approximates the female rather than the male experience within the family. This is probably one of the reasons why Erving Goffman, in *Asylums*, considered psychiatric hospitalization more destructive of self than criminal incarceration. Like most people, he is primarily thinking of the debilitating effect – *on men* – of being treated like a woman (as helpless, dependent, sexless, unreasonable – as 'crazy'). But what about the effect of being treated like a woman when you *are* a woman? And perhaps a woman who is already ambivalent or angry about just such treatment?

Perhaps one of the reasons women embark and re-embark on 'psychiatric careers' more than men do is because they feel, quite horribly, at 'home' within them. Also, to the extent to which *all* women have been poorly nurtured as female children, and are refused 'mothering' by men as female adults, they might be eager for, or at least willing to settle for, periodic bouts of ersatz 'mothering', which they receive as 'patients'. Those women who are more ambivalent about or rejecting of the female role are often eager to be punished for such dangerous boldness – in order to be saved from its ultimate consequences. Many mental asylum procedures *do* threaten, punish, or misunderstand such women into a real or wily submission. Some of these women react to such punishment (or to a dependency-producing environment) with increased and higher levels of anger and sex-role alienation. If such anger or aggressiveness persists, the women are isolated, strait-jacketed, sedated, and given shock therapy. One study published by four male professionals in 1969 in the *Journal of Nervous and Mental Diseases* (Ludwig *et al.*, 1969) describes how they attempted to reduce the aggressive behavior of a thirty-one-year-old 'schizophrenic' woman by shocking her with a cattle prod whenever she 'made accusations of being persecuted and abused; made verbal threats; or committed aggressive acts'. They labeled their treatment a 'punishment program' and note that the 'procedure was administered against the expressed will of the patient'.

Celibacy is the official order of the asylum day. Patients are made to inhabit an eternal American adolescence, where sexuality and aggression are as feared, mocked, and punished as they are within the Family. Traditionally, mental hospital wards are sex-segregated; homosexuality, lesbianism, and masturbation are discouraged or forcibly interrupted. The heterosexual dances sponsored in the late 1950s and 1960s were like high school proms, replete with chaperones, curfews, and frustration.

The female-'dominated' atmosphere of hospitals means a (shameful) return to childhood, for both men and women. However, the effect of sexual repression, for example, is probably different for female than for male patients. Women have already been bitterly and totally repressed sexually; many may be reacting to or trying to escape from just such repression, and the powerlessness that it signifies, by 'going mad'. Many male patients may be escaping the demands of a compulsive and aggressive heterosexuality by 'going mad'. Its absence is *perhaps* not as psychologically or physiologically devastating as it is in the case of women.

Female patients, like female children, are closely supervised by other women (nurses, attendants) who, like mothers, are relatively powerless in terms of the hospital hierarchy and who, like mothers, don't really 'like' their (wayward) daughters. Such supervision, however, doesn't protect the female as patient-child from rape, prostitution, pregnancy, and the blame for all three – any more than similar 'motherly' supervision protects the female as female child in the 'real' world, either within or outside the family. Over the years, there have been numerous newspaper accounts of the prostitution, rape, and impregnation of female mental patients by the professional and non-professional staff, and by male inmates.

Both sexual repression and sexual abuse still haunt women in their 'asylums'. They do not escape enforced maternity either. A study published in the *American Journal of Orthopsychiatry* (Wignall and Meredith, 1968) documented a 366 per cent increase in the *recorded* pregnancy and delivery rate among psychiatrically hospitalized women in Michigan state hospitals from 1936 to 1964. Sixty per cent of the responding mental asylums in this study did not and/or would not prescribe contraceptives or perform abortions.

The female social role and psychiatric symptons: depression

Why are women psychiatrically 'disturbed' and hospitalized? Why do they seek private therapy? What is 'schizophrenia' or madness like, or about, in contemporary women?

Two researchers recently stated that men are really as 'psychologically disturbed' as women are:

> There is no greater magnitude of social stress impinging on one or the other sex. Rather [each sex] tends to learn a different style with which it reacts to whatever fact has produced the psychological disorder (Dohrenwend and Dohrenwend, 1969).[1]

I would not so much disagree with this statement as qualify it in several important ways. Many men *are* severely 'disturbed' – but the form their 'disturbance' takes is either not seen as 'neurotic' or is not treated by psychiatric incarceration. Theoretically all men, but especially white, wealthy, and older men, can act out many 'disturbed' (and non-'disturbed') drives more easily than women can. The greater social tolerance for female 'help-seeking' behavior, or displays of emotional distress, does not mean that such conditioned behavior is either valued or treated with kindness. On the contrary. Both husbands and clinicians experience and judge such female behavior as annoying, inconvenient, stubborn, childish, and tyrannical. Beyond a certain point, such behavior is 'managed', rather than rewarded: it is treated with disbelief and pity, emotional distance, physical brutality, economic and sexual deprivation, drugs, shock therapy, and long-term psychiatric confinements.

Given the custodial nature of asylums and the anti-female biases of most clinicians, women who seek 'help', women who have 'symptoms', are actually being punished for their conditioned and socially approved self-destructive behavior. Typically 'female' and 'male' symptomatology appear early in life. Studies of childhood behavior problems have indicated that boys are most often referred to child guidance clinics for aggressive, destructive (anti-social), and competitive behavior; girls are referred (if they are referred at all), for personality problems, such as excessive fears and worries, shyness, timidity, lack of self-confidence, and feelings of inferiority (Macfarlane *et al.*, 1954; Philips, 1957; Peterson, 1961; Terman and Tyler, 1970). Similar, sex-typed symptoms exist in adults also:

> the symptoms of men are also much more likely to reflect a destructive hostility toward others, as well as a pathological self-indulgence ... Women's symptoms, on the other hand, express a harsh, self-critical, self-depriving and often self-destructive set of attitudes (Philips, 1957).

A study by E. Zigler and L. Phillips, comparing the symptoms of male and female mental hospital patients, found male patients significantly more assaultive than females and more prone to indulge their impulses in socially deviant ways like 'robbery, rape, drinking, and homosexuality' (Zigler and Philips, 1960). Female patients were often found to be 'self-deprecatory, depressed, perplexed, suffering from suicidal thoughts, or making actual suicidal attempts'.

Most women display 'female psychiatric symptoms such as depression, frigidity, paranoia, psychoneurosis, suicide attempts, and anxiety. Men display 'male' diseases such as alcoholism, drug addiction, personality disorders, and brain diseases.

Typically female symptoms all share a 'dread of happiness' – a phrase coined by Thomas Szasz to describe the 'indirect forms of communication' that characterize 'slave psychology'. The analogy between 'slave' and 'woman' is by no means a perfect one. However, there is some theoretical justification for viewing women, or the sex-caste system, as the prototype for all subsequent class and race slavery (Engels, 1942). Women were probably the first group of human beings to be enslaved by another group. In a sense, 'woman's work', or woman's psychological identity, consists in exhibiting the signs and 'symptoms' of slavery – as well as, or instead of, working around the clock in the kitchen, the nursery, the bedroom, and the factory (Lorenz, 1970).[2]

Depression

Women become 'depressed' long before menopausal chemistry becomes the standard explanation for the disease. National statistics and research studies all document a much higher female to male ratio of depression or manic-depression at all ages. Perhaps more women *do* get 'depressed' as they grow older – when their already limited opportunities for sexual, emotional, and intellectual growth decrease even further. Dr Pauline Bart (1971) studied depression in middle-aged women and found that such women had completely accepted their 'feminine' role – and were 'depressed' because that role was no longer possible or needed.

Traditionally, depression has been conceived of as the response to – or expression of – loss, either of an ambivalently loved other, of the 'ideal' self, or of 'meaning' in one's life. The hostility that should or could be directed outward in response to loss is turned inwards toward the self. 'Depression' rather than 'aggression' is the female response to disappointment or loss. The research and clinical evidence for any or all of these views is controversial. We may note that most women have 'lost' – or have never really 'had' – their mothers; nor is the maternal object replaced for them by husbands or lovers. Few women ever develop strong socially approved 'ideal' selves. Few women are allowed, no less encouraged, to concern themselves with life's 'meaning'. (Women lose their jobs as 'women', rather than any existential hold on life's meaning.) In a sense, women can't 'lose' what they've never had.

Women are in a continual state of mourning – for what they never had – or had too briefly, and for what they can't have in the present, be it Prince Charming or direct worldly power. It is not very easy for most women to temper, idle, or philosophize away their mourning with sexual, physical, or intellectual exercises. When female depression swells to clinical proportions, it unfortunately doesn't function as a role-release or respite. For example, according to a journal study, published by Dr Alfred Friedman (1970), 'depressed' women are even *less* verbally 'hostile' and 'aggressive' than non-depressed women; their 'depression' may serve as a way of keeping a deadly faith with their 'feminine' role. 'Depressed' patients were actually less verbally hostile than 'normal' control patients – and their verbal hostility and 'resentment' decreased even further as they 'improved' – i.e., became less 'depressed' according to clinical and self-rating. (Depressed male patients were more verbally hostile than their female counterparts.) Dr Friedman's interpretation of this finding is as follows: he hypothesizes that the 'depressive' *usually* expresses very little verbal (or other forms of) hostility, and becomes 'depressed' only when the 'usual defenses break down'.

It is important to note that 'depressed' women are (like women in general) only *verbally* hostile; unlike most men, they do not express their hostility physically – either directly, to the 'significant others' in their lives, or indirectly, through physical and athletic prowess. It is safer for women to become 'depressed' than physically violent. Physically violent women usually lose physical battles with male intimates; are abandoned by them as 'crazy' as well as 'unfeminine'; are frequently psychiatrically or (less frequently) criminally incarcerated. Further, physically strong and/or potentially assaultive women would gain fewer secondary rewards than 'depressed' women; their families would fear, hate, and abandon them, rather than pity, sympathize, or 'protect' them. Psychiatrists and asylums would behave similarly: hostile or potentially violent women (and men), who are oppressed and powerless, are, understandably, hardly ever treated ethically or legally – or kindly – by others.

Notes

1. Olle Hagnell quotes another theory of conditioned female 'patient' behaviour made by H. Holter in *Holter* (1966).
2. Konrad Lorenz, a noted writer on animal behaviour, has recently been quoted as saying, 'There's only one kind of people at a social disadvantage nowadays – a whole class of people who are treated as slaves and exploited shamelessly – and that's the young wives. They are educated as well as men and the moment they give birth to a baby, they are slaves ... they have a 22-hour workday, no holidays and they can't even be ill.' Interview, *New York Times*, 5 July 1970.

References

Bart, Pauline (1971) 'Portnoy's mother's complaint', in Gornick, V. and Moran, B. (eds), *Woman in Sexist Society: Studies in Power and Powerlessness*, New York, Basic Books.

Dershowitz, Allan M. (1969) 'Preventive detention and the prediction of dangerousness. some fictions about predictions', *Journal of Legal Education*, vol. 23.

Dohrenwend, Bruce and Dohrenwend, Barbara (1969) *Social Status and Psychological Disorders*, Chichester, Interscience.

Engels, Friedrich (1942) *The Origins of the Family, Private Property and the State*, New York, International Publishers.

Foucault, Michel (1967) *Madness and Civilization: a history of insanity in the age of reason* (1961), trans. Richard Howard, London, Tavistock, 1967.

Friedman, Alfred L. (1970) 'Hostility factors and clinical improvement in depressed patients', *Archives of General Psychiatry*, vol. 23.

Goffman, Erving (1970) *Asylums*, Harmondsworth, Penguin Books.

Holter, H. (1966) *A Prospective Study of the Incidence of Mental Disorders: the Lundby project*, Stockholm, Svenska Bokforlaget.

Ludwig, Arnold, Marx, Arnold J., Hill, Phillip A. and Browning, Robert M. (1969) 'The control of violent behavior through faradic shock: a case study', *Journal of Nervous and Mental Diseases*, Vol. 148, 1969.

MacFarlane, Jean *et al.* (1954) *A Development Study of the Behavior Problems of Normal Children Between Twenty-one Months and Thirteen Years*, Berkeley, University of California Press.

Peterson, D. R. (1961) 'Behavior problems of middle childhood', *Journal of Consulting Psychology*, vol. 25.

Philips, L. (1957) 'Cultural *vs* intra-psychic factors in childhood behavior problem referrals', *Journal of Clinical Psychology*, vol. 13.

Phillips, Leslie (1969) 'A social view of psychopathology', in London, Perry and Rosenhan, David (eds), *Abnormal Psychology*, New York, Holt, Rinehart & Winston.

Rosen, George (1968) *Madness in Society: Chapters in the Historical Sociology of Mental Illness*, New York, Harper & Row.

Scheff, T. J. (1966) *Being Mentally Ill: a sociological theory*, Chicago, Aldine Press.

Szasz, Thomas S. (1970) *The Manufacture of Madness*, New York, Harper & Row.

Szasz, Thomas S. (1961) *The Myth of Mental Illness: Foundations of a Theory of Personal Conduct*, London, Harper & Row.

Terman, L. M. and Tyler, L. E.(1970) 'Psychological sex differences', in Carmichael, L. (ed.) *Manual of Child Psychology*, New York, John Wiley.

Wignall, C. M. and Meredith, C. F. (1968) 'Illegitimate pregnancies in state institutions', *Archives of General Psychiatry*, vol. 18.

Zigler, E. and Phillips, L. (1960) 'Social effectiveness and symptomatic behaviors', *Journal of Abnormal and Social Psychology*, vol. 61.

7

The cultural context of mental distress

RICHARD WARNER

Maria, a young Indian woman living in a village on Lake Atitlan in Guatemala, alienates her close relatives and the people of the community by her irresponsible behavior before finally suffering a full-blown psychotic episode. She hallucinates, believing that spirits are surrounding her to take her to the realm of the dead, and she walks about the house arguing with ghosts. A local shaman perceives that she is *loca* (crazy) and diagnoses her as suffering the effect of supernatural forces unleashed by the improper behavior of certain relatives. He prescribes a healing ritual which calls for the active participation of most of her extended family. Her condition requires her to move back to her father's house, where she recovers within a week. Benjamin Paul, the anthropologist who describes Maria's case, points out several features of interest. Maria is never blamed for her psychotic behavior or stigmatized by her illness, because her hallucinations of ghosts are credible supernatural events and she is innocently suffering the magical consequences of the wrong-doing of others. The communal healing activities lead to a dramatic reversal of Maria's course of alienation from family and community. In the West, a psychotic episode is likely to lead to increased alienation. In the case of Maria, conflict resolution and social reintegration are central to her recovery and result from the folk diagnosis and treatment of her symptoms (Paul, 1967).

The Folk Diagnosis of Psychosis

Throughout the non-industrial world, the features of psychosis are likely to be given a supernatural explanation. The Shona of Zimbabwe, for

Edited and adapted from Warner, R. (1994) *Recovery from Schizophrenia: Psychiatry and Political Economy*, 2nd edn, London and New York, Routledge & Kegan Paul, 162–70.

example, believe visual and auditory hallucinations to be real and sent by spirits (Gelfand, 1964). In Dakar, Senegal:

> one can have hallucinations without being thought to be sick. A magical explanation is usually resorted to and native specialists are consulted. There is no rejection or alienation by society. The patient remains integrated within his group. As a result, the level of anxiety is low (Collomb, 1966).

The psychiatrist who gives this report claims that 90 per cent of the acute psychoses in Dakar are cured because the patient's delusions and hallucinations have an obvious culturally relevant content, and he or she is not rejected by the group.

Similarly, in the slums of San Juan, Puerto Rico:

> If an individual reports hallucinations, it clearly indicates to the believer in spiritualism that he is being visited by spirits who manifest themselves visually and audibly. If he has delusions ... his thoughts are being distorted by interfering bad spirits, or through development of his psychic faculties spirits have informed him of the true enemies in his environment. Incoherent ramblings, and cryptic verbalizations indicate that he is undergoing a test, an experiment engineered by the spirits. If he wanders aimlessly through the neighbourhood, he is being pursued by ambulatory spirits who are tormenting him unmercifully (Rogler and Hollingshead, 1965).

In many cases where a supernatural explanation for psychotic features is used, the label 'crazy' or 'insane' may never be applied. I once remarked to a Sioux mental health worker from the Pine Ridge Reservation in South Dakota that most Americans who heard voices would be diagnosed as psychotic. Her response was simple. 'That's terrible.'

Nigerian Attitudes to Mental Illness

Urban and rural Yoruba with no formal education, from the area of Abeokuta in southwestern Nigeria, were asked their opinions about descriptions of typical mentally ill people. Only 40 per cent of those questioned thought that the paranoid schizophrenic described was mentally ill (Erinosho and Ayonrinde, 1981). (Some 90 to 100 per cent of Americans label the subject of this vignette as mentally ill, D'Arcy and Brockman, 1976.) Only 21 per cent of the uneducated Yoruba considered the description of the simple schizophrenic to be a mentally ill person. (Some 70 to 80 per cent of American respondents call this hypothetical case mentally ill, D'Arcy and Brockman, 1976.)

What is perhaps even more impressive than the details about labeling psychosis in this Nigerian study is the very high level of tolerance revealed. More than 30 per cent of the uneducated Yoruba would have been willing to marry the paranoid schizophrenic person described and

55 per cent would have married the simple schizophrenic. In contrast, when skilled workers from the area of Benin in midwestern Nigeria were asked their opinions about someone specifically labeled a 'nervous or mad person,' 16 per cent thought that all such people should be shot and 31 per cent believed that they should be expelled from the country. These educated Nigerians conceived of mad people as 'senseless, unkempt, aggressive and irresponsible (Binitie, 1970).

Malaya

In Nigeria it appears that the label 'mad,' 'crazy' or 'mentally ill' is only applied to highly disruptive individuals and brings with it harsher treatment. The same pattern has been observed in a Malay village in Pahang state. Here the term for madness, *gila*, is only applied to violent people. 'Madmen' are always handed over to authorities outside the village for permanent banishment. Within the community of over 400 people, however, are many probable psychotics who have never been labelled mad – twelve who are 'eccentric,' including senile elderly people and marginally functional hermits; and one 'person with less than healthy brains' who spends a good deal of time praying and reading in solitude. Five people exhibiting *latah* – a so-called culture-bound psychosis – were also identified (Colson, 1971): but this condition may not be a psychosis in the proper sense of the term (Leff, 1981).

Laos

Although Dr Westermeyer in some of his publications disputes that psychotics often escape being labeled *baa* (insane) in Laos, his own observations are very close to the findings in Nigeria and Malaya. Lao villagers are apparently slow to apply the term *baa*, and a person so labeled tends to have a chronic illness, usually of several years' duration, and to be highly disruptive, assaultive or bizarre. Hallucinations are never mentioned by the villagers as a feature of insanity. Unless there are local conditions restricting the development of brief psychoses so common elsewhere in the Third World, then one must assume that the reason there are so few acute cases in Dr Westermeyer's Lao sample is that they are not considered by the villagers to be *baa*. Interestingly, the severely psychotic *baa* individuals in Laos are not exiled or assassinated but continue to receive food, shelter, clothing and humane care and are restrained and incarcerated only as long as their violent behavior requires (Westermeyer and Wintrob, 1978). It is apparent that labeling is an important issue only insofar as it affects management. As we shall see in the next example, it is the concept of illness which lies behind the label which is also critical in determining care and treatment.

Four East African Societies

Anthropologist Robert Edgerton, describing attitudes to psychosis among tribesmen of four East African pastoral and farming societies, confirms that violence and destructiveness are emphasized in descriptions of psychosis (*kichaa*) and hallucinations are virtually never mentioned Edgerton (1971). Most commonly reported features of psychosis include murder, assault, arson, abuse, stealing and nakedness. The pastoralists whose homesteads are more widely dispersed and who are more free to move away from disagreeable circumstances are less concerned than the farmers about the social disruption of psychotics (Edgerton, 1971). The intriguing conclusion of Edgerton's survey is that the tribal view of the cause of psychosis determines not only the manner of treatment but also the level of optimism about recovery (Edgerton, 1971). The Pokot of northwest Kenya and the Sebei of southeast Uganda have a naturalistic conception of the cause of psychosis. They implicate a worm in the frontal portion of the brain and are very pessimistic about the possibilities of cure. The Kamba of south-central Kenya and the Hehe of southwest Tanzania, on the other hand, attribute the cause of psychosis to witchcraft or stress and are optimistic about curing such disorders. The two tribes which are most unsure about their respective theories of causation, the Pokot and the Hehe, also tend to be more ambivalent about the curability of the condition. The Kamba and the Hehe, holding a supernatural theory for the cause of psychosis, favor the use of tranquilizing herbs and ritual in treatment. The pessimistic Sebei and Pokot, with the naturalistic belief system, are much more inclined to treat psychotics harshly, as illustrated by the remarks of a Pokot shaman: 'I am able to cure mads. I order the patient tied and placed upon the ground. I then take a large rock and pound the patient on the head for a long time. This calms them and they are better' (Edgerton, 1971). The Pokot and Sebei recommend that psychotics should be tied up forever, allowed to starve, driven away to die or killed outright.

Stigma

Life for some psychotics in the Third World, according to a few of these reports, is not a bed of roses. But we should not allow reports of harsh treatment of the most severely disturbed psychotics in some areas to obscure the central facts. Many people who would be considered psychotic in the West are not so labeled in the Third World, especially if their condition is brief or not disruptive. Many more, though labeled 'crazy' like Maria the Guatemalan Indian woman, are treated vigorously and optimistically with every effort to reintegrate them rather than reject them.

Psychiatrists working in the Third World have repeatedly noted the low level of stigma which attaches to mental disorder. Among the Formosan tribesmen studied by Rin and Lin, mental illness is free of stigma (Rin and Lin, 1962). Sinhalese families freely refer to their psychotic family members as *pissu* (crazy) and show no shame about it. Tuberculosis in Sri Lanka is more stigmatizing than mental illness (Waxler, 1977). The authors of the WHO follow-up study suggest that one of the factors contributing to the good outcome for schizophrenics in Cali, Colombia, is the 'high level of tolerance of relatives and friends for symptoms of mental disorder' – a factor which can help the 'readjustment to family life and work after discharge' (WHO, 1977).

The possibility that the stigma attached to an illness may influence its course is illustrated by research on Navajo epileptics conducted by anthropologist Jerrold Levy in cooperation with the Indian Health Service. Sibling incest is regarded as the cause of generalized seizures, or Moth Sickness, in Navajo society, and those who suffer from the condition are highly stigmatized for supposed transgressions of a major taboo. It is interesting to learn that these individuals are often found to lead chaotic lives characterized by alcoholism, promiscuity, incest, rape, violence and early death. Levy and his co-workers attribute the career of the Navajo epileptic to the disdain and lack of social support which he or she is offered by the community (Levy *et al.*, 1979). To what extent, we may wonder, can features of schizophrenia in the West be attributed to similar treatment?

High Status in Psychosis

It seems strange in retrospect that tuberculosis should have been such a romantic and genteel illness to eighteenth- and nineteenth-century society that people of fashion chose to copy the consumptive appearance (Sontag, 1979). Equally curious, the features of psychosis in the Third World can, at times, lead to considerable elevation in social status. In non-industrial cultures throughout the world, the hallucinations and altered states of consciousness produced by psychosis, fasting, sleep deprivation, social isolation and contemplation, and hallucinogenic drug use are often a prerequisite for gaining shamanic power (Eliade, 1972; Black Elk, 1971). The psychotic features are interpreted as an initiatory experience. For example, whereas poor Puerto Ricans who go to a psychiatric clinic or insane asylum are likely to be highly stigmatized as *locos* (madmen), schizophrenics who consult a spiritualist may rise in status. Sociologists Lloyd Rogler and August Hollingshead report: 'The spiritualist may announce to the sick person, his family, and friends that the afflicted person is endowed with *facultades* (psychic faculties), a matter of prestige at this level of the social structure' (Rogler and Hollingshead, 1965).

The study indicates that Puerto Rican schizophrenics who consult spiritualists may not only lose their symptoms, they may also achieve the status of mediums themselves. So successful is the social reintegration of the male Puerto Rican schizophrenics studied that, after some readjustment of family roles, their wives found them *more* acceptable as husbands than did the wives of normal men.

Similar folk beliefs exist in Turkey. Dr Orhan Ozturk, a psychiatrist in Ankara, writes:

> A person may be hallucinated or delusional, but as long as he is not destructive or very unstable he may not be considered insane ... Such a person may sometimes be considered to have a supernatural capacity for communication with the spirit world and may therefore be regarded with reverence and awe (Ozturk, 1964).

Ruth Benedict tells us that Siberian shamans who dominate the life of their communities:

> are individuals who by submission to the will of the spirits have been cured of a grievous illness ... Some, during the period of the call, are violently insane for several years; others irresponsible to the point where they have to be constantly watched lest they wander off in the snow and freeze to death ... It is the shamanistic practice which constitutes their cure (Benedict, 1934).

Several other writers have suggested that indigenous healers who have suffered psychotic episodes may find their elevated status and well-defined curing role to be a valuable defense against relapse (Ackernecht, 1943; Silverman, 1967). Psychiatrist Fuller Torrey argues, however, that few shamans can be psychotic. The role is too responsible and demanding, he claims, for a schizophrenic to manage (Torrey, 1972, 1980). While, no doubt, many healers are not psychotic, Dr Torrey underestimates the importance of features of psychosis as an initiatory experience. He is neglecting on the one hand, the heightened possibility of complete remission for Third World psychotics and, on the other hand, the capacity of schizophrenic individuals to be completely functional in some areas of their lives despite islands of illogical thinking. One well-known North American Indian medicine man with whom I am familiar would doubtless be diagnosed schizophrenic by a Western psychiatrist by virtue of his extremely tangential and symbolic speech, which is often incomprehensible, his inappropriate emotional responses and his hallucinations. This man, however, is highly respected by his community and often travels the country on speaking engagements. The psychotic may be able to function well as a shaman, argues anthropologist Julian Silverman of the US National Institute of Mental Health, because 'the emotional supports ... available to the shaman greatly alleviate the strain of an otherwise excruciatingly painful [schizophrenic] existence. Such supports are all too often completely unavailable to the schizophrenic in our culture' (Silverman, 1967).

Healing Ceremonies

Being thought of as a spiritualist or healer is not the only way Third World psychotics may gain status. Curing rituals for those with mental disorders may also enable the individual to increase his or her social status and redefine his or her social role. Anthropologist Ralph Linton observes that low-status individuals among the Tanala of Madagascar, such as second sons and childless wives, may rise in status as a result of the elaborate healing rite for mental illness (Linton, 1956). Patients who participate in the curing possession cults in Trinidad (Mischel and Mischel, 1958) among the Yoruba of Nigeria (Prince, 1964) and in the Zar cult of northern Ethiopia (Messing, 1967) have all been observed to achieve an elevation of social status as a consequence of their membership.

Initiation into these cults also provides new friends, ongoing group support and the opportunity for social involvement, and similar benefits appear to result from other healing rites. Robin Fox, a British anthropologist, gives a detailed account of a clan cure for a 40-year-old woman with a chronic mental disorder in the Pueblo Indian community of Cochiti in New Mexico. The woman is a member of the Oak clan by birth, but by undergoing a healing ritual which entails adoption also into the Water clan, she acquires additional supportive relatives, a new social role and a new home. She subsequently shows complete recovery (Fox, 1967).

Social Consensus

There is some anthropological evidence that broad group participation in healing not only aids the reintegration of the patient but is also a necessary and powerfully effective element in the treatment of emotional illness. The French anthropologist Claude Levi-Strauss, for example, analyzes the effectiveness of a highly respected Kwakiutl shaman from British Columbia who is skeptical of his own healing powers. Levi-Strauss concludes that the shaman is effective despite his cynicism because 'the attitude of the group' endorses his treatment. The social consensus is more important than the attitude of the healer or even of the patient (Lévi-Strauss, 1972).

A related example of the importance of social consensus in the outcome of mental illness is provided by anthropologist Lloyd Warner's discussion of the role played in the voodoo death of an Australian aborigine by his own social group after he has been 'boned' by an enemy. First the victim's kin withdraw their support and he becomes an isolated and taboo person. Then the community conducts a mourning ritual to protect the group from the soul of the 'half dead' man. Unless the group attitude is reversed by the performance of a counter-ritual, the victim shortly dies (Warner, 1937). These examples illustrate, on the one hand, the powerful effect of social rejection and stigma on the course of emotional illness and the importance of social acceptance and reintegration: on the other hand, they suggest that any form of treatment which does not receive full community endorsement (and much of institutional psychiatry in the West falls into this category) has a limited chance of success. This analysis, for example, would predict that the Kamba and the Hehe of East Africa who are optimistic about the treatment of mental illness would have better recovery rates from psychosis than the Pokot and Sebei who have no confidence in the ability of their doctors to effect a cure. Edgerton's study presents no evidence, unfortunately, to indicate whether or not this is the case.

Understanding the potential of social consensus to affect outcome allows us to explain why even those individuals who are treated in modern Western-style hospitals and clinics in the developing world rather than by indigenous therapists may experience a higher recovery rate from psychosis. It is not the specific treatment technique which is critical (as long as it is not too regressive) but the social expectations that are generated around the episode of illness. The treatment approaches of the psychiatric clinic may well be supplemented by community diagnosis, rediagnosis and indigenous healing ceremonies which facilitate social reintegration of the sick person. Even among relatively Westernized city-dwellers, according to a report from Senegal, traditional cultural beliefs persist which help to alleviate psychological distress and mental disorder (Beiser and Collomb, 1981). The existence of a social consensus for recovery and the willingness and capacity of the community to re-integrate the psychotic person are, no doubt, strongly influenced by whether he or she can serve a useful social role. The benefits of traditional community life for the psychotic are less likely to persist in the face of changing patterns of labor use which increase the risks of unemployment and dependency.

Summary

- The folk diagnosis of insanity stresses violence and disruption, and many psychotics from the developing world escape this label.
- Many psychotics in the Third World are not stigmatized and some may even rise in status.
- Although some psychotics in non-industrial societies may be brutally

treated, in the majority of cases vigorous and optimistic efforts are made to achieve a cure.

- Curing rituals encourage broad community involvement and aid the social reintegration of psychotics.
- The optimistic social consensus mobilized by the curing ceremony may aid recovery from emotional disorders.

References

Ackernecht, E. H. (1943) 'Psychopathology, primitive medicine and primitive culture', *Bulletin of the History of Medicine*, vol. 14, 30–67.

Beiser, M. and Collomb, H. (1981) 'Mastering change: epidemiological and case studies in Senegal, West Africa', *American Journal of Psychiatry*, vol. 138: 455–459.

Benedict, R. (1934) *Patterns of Culture*, Boston, Houghton-Mifflin, 267–268.

Binitie, A. O. (1970) 'Attitude of educated Nigerians to psychiatric illness', *Acta Psychiatrica Scandinavica*, vol. 46, 391–398.

Black Elk (1971) *The Sacred Pipe*, Baltimore, Penguin.

Collomb, H. (1966) 'Bouffées délirantes en psychiatrie Africaine,' *Transcultural Psychiatric Research*, vol. 3: 29–34.

Colson, A. C. (1971) 'The perception of abnormality in a Malay village', in Wagner, N. N. and Tan, E. (eds), *Psychological Problems and Treatment in Malaysia*, Kuala Lumpur, University of Malaya Press.

D'Arcy, C. and Brockman, J. (1976) 'Changing public recognition of psychiatric symptoms? Blackfoot revisited', *Journal of Health and Social Behavior*, vol. 17, 302–310.

Edgerton, R. B. (1966) 'Conceptions of psychosis in four East African societies', *American Anthropologist*, vol. 68, 408–425.

Edgerton, R. B. (1971) *The Individual in Cultural Adaption*, Berkeley, University of California Press, 188.

Eliade, M. (1972) *Shamanism: Archaic Techniques of Ecstasy*, Princeton, Princeton University Press/Bollinger Paperback.

Erinosho, O. A. and Ayonrinde, A. (1981) 'Educational background and attitude to mental illness among the Yoruba in Nigeria', *Human Relations*, vol. 34: 1–12.

Fox, J. R. (1967) 'Witchcraft and clanship in Cochiti therapy,' in Middleton, J. (ed.), *Magic, Witchcraft and Curing*, Garden City, New York, 255–284.

Gelfand, M. (1964) 'Psychiatric disorders as recognized by the Shona', in Kiev, A. (ed), *Magic, Faith and Healing*, New York, Free Press, 156–73.

Leff, J. (1981) *Psychiatry Around the Globe: a transcultural view*, New York, Marcel Dekker, 19.

Levi-Strauss, C. (1972) *Structural Anthropology*, Harmondsworth, Penguin, 180.

Levy, J. E., Neutra, R. and Parker, D. (1979) 'Life careers of Navajo epileptics and convulsive hysterics', *Social Science and Medicine*, vol. 13: 53–66.

Linton, R. (1956) *Culture and Mental Disorders*, Springfield, Ill, Charles C. Thomas.

Messing, S. D. (1967) 'Group therapy and social status in the Zar cult of Ethiopia', in Middleton, J. (ed.), *Magic, Witchcraft and Curing*, Garden City, New York, Natural History Press, 285–293.

Mischel, W. and Mischel, F. (1958) 'Psychological aspects of spirit possession', *American Anthropologist*, vol. 60, 249–260.

Ozturk, O. M. (1964) 'Folk treatment of mental illness in Turkey,' in Kiev, *Magic, Faith and Healing*.

Paul, B. D. (1967) 'Mental disorder and self-regulating process in culture: A Guatemalan illustration', in Hunt, R. (ed.), *Personalities and Cultures: readings in psychological anthropology*, Garden City, New York, Natural History Press.

Prince, R. (1964) 'Indigenous Yoruba psychiatry', in Kiev, *Magic, Faith and Healing*, 84–120.

Rin, S. and Lin, T. (1962) 'Mental illness among Formosan aborigines as compared with the Chinese in Taiwan', *Journal of Mental Science*, vol. 108: 134–138.

Rogler, L. H. and Hollingshead, A. B. (1965) *Trapped: families and schizophrenia*, New York, Wiley, 254.

Silverman, J. (1967) 'Shamans and acute schizophrenia', *American Anthropologist*, vol. 69, 21–31.

Sontag, S. (1979) *Illness as Metaphor*, New York, Vintage Books.

Torrey, E. F. (1972) *The Mind Game: witchdoctors and psychiatrists*, New York, Emerson Hall.

Torrey, E. F. (1980) *Schizophrenia and Civilization*, New York, Jason Aronson.

Warner, W. L. (1937) *A Black Civilization*, New York, Harper, 241–242.

Waxler, N. E. (1977) 'Is mental illness cured in traditional societies? A theoretical analysis', *Culture, Medicine and Psychiatry*, vol. 1, 233–253, 242.

Westermeyer, J. and Wintrob, D. (1978) '"Folk" diagnosis in rural Laos'; Westermeyer, J. and Kroll, J., 'Violence and mental illness in a peasant society: Characteristics of violent behaviors and "folk" use of restraints,' *British Journal of Psychiatry*, vol. 133, 529–541.

World Health Organization (WHO) (1979), *Schizophrenia*, Chichester, Wiley, 105.

8

Labelling mental illness

THOMAS J. SCHEFF

Suppose that in your next conversation with a stranger, instead of looking at his eyes or mouth, you scrutinize his ear. Although the deviation from ordinary behavior is slight (involving only a shifting of the direction of gaze a few degrees, from the eyes to an ear), its effects are explosive. The conversation is disrupted almost instantaneously. In some cases, the subject of this experiment will seek to save the situation by rotating to bring his eyes into your line of gaze; if you continue to gaze at his ear, he may rotate through a full 360 degrees. Most often, however, the conversation is irretrievably damaged. Shock, anger, and vertigo are experienced not only by the 'victim' but, oddly enough, by the experimenter himself. It is virtually impossible for either party to sustain the conversation, or even to think coherently, as long as the experiment continues.

The point of this experiment is to suggest the presence of a public order that is all-pervasive, yet taken almost completely for granted. During the simplest kinds of public encounter, there are myriad understandings about comportment that govern the participants' behavior – understandings governing posture, facial expression, and gestures, as well as the content and form of the language used. In speech itself, the types of conformity are extremely diverse and include pronunciation; grammar and syntax; loudness, pitch, and phrasing; and aspiration. Almost all of these elements are so taken for granted that they 'go without saying' and are more or less invisible, not only to the speakers but to society at large. These understandings constitute part of our society's assumptive world, the world that is thought of as normal, decent, and possible.

The probability that these understandings are, for the most part, arbitrary to a particular historical culture (is shaking hands or rubbing noses a better form of greeting?) is immaterial to the individual member of society whose attitude of everyday life is, *whatever is, is right*. There is a social, cultural, and interpersonal status quo whose existence is felt only when abrogated. Since violations occur infrequently, and since the

Edited from 'Schizophrenia as Ideology', *Schizophrenia Bulletin*, 1 (Fall 1970), 15–20 (reprinted in Scheff, T. (1975) *Labeling Madness*, Engelwood Cliffs, NJ, Prentice-Hall, 5–12).

culture provides no very adequate vocabulary for talking about either the presence or abuse of its invisible understandings, such deviations are considered disruptive and disturbing. The society member's loyalty to his culture's unstated conventions is unthinking but extremely intense.

Residual rule violations

It is the thesis of this paper that the concepts of mental illness in general – and schizophrenia in particular – are not neutral, value-free, scientifically precise terms but are, for the most part, the leading edge of an ideology embedded in the historical and cultural present of the white middle class of Western societies. The concept of illness and its associated vocabulary – symptoms, therapies, patients, and physicians – reify and legitimate the prevailing public order at the expense of other possible worlds. The medical model of disease refers to culture-free processes that are independent of the public order; a case of pneumonia or syphilis is pretty much the same in New York or New Caledonia.

Most of the 'symptoms' of mental illness, however, are of an entirely different nature. Far from being culture-free, such 'symptoms' are themselves offenses against implicit understandings of particular cultures. Every society provides its members with a set of explicit norms – understandings governing conduct with regard to such central institutions as the state, the family, and private property. Offenses against these norms have conventional names; for example, an offense against property is called 'theft,' and an offense against sexual propriety is called 'perversion.' As we have seen above, however, the public order also is made up of countless unnamed understandings. 'Everyone knows,' for example, that during a conversation one looks at the other's eyes or mouth, but not at his ear. For the convenience of the society, offenses against these unnamed residual understandings are usually lumped together in a miscellaneous, catchall category. If people reacting to an offense exhaust the conventional categories that might define it (e.g., theft, prostitution, and drunkenness), yet are certain that an offense has been committed, they may resort to this residual category. In earlier societies, the residual category was witchcraft, spirit possession, or possession by the devil; today, it is mental illness. The symptoms of mental illness are, therefore, violations of residual rules.

To be sure, some residual-rule violations are expressions of underlying physiological processes: the hallucinations of the toxic psychoses and the delusions associated with general paresis, for example. Perhaps future research will identify further physiological processes that lead to violations of residual rules. For the present, however, the key attributes of the medical model have yet to be established and verified for the major mental illnesses. There has been no scientific verification of the cause, course, site of pathology, uniform and invariant signs and symptoms, and treatment of choice for almost all the conventional, 'functional' diagnostic categories. Psychiatric knowledge in these matters rests almost entirely on unsystematic clinical impressions and professional lore. It is quite possible, therefore, that many psychiatrists' and other mental-health workers' 'absolute certainty' about the cause, site, course, symptoms, and treatment of mental illness represents an ideological reflex, a spirited defense of the present social order.

Residue of residues

Viewed as offenses against the public order, the symptoms of schizophrenia are particularly interesting. Of all the major diagnostic categories, the concept of schizophrenia (although widely used by psychiatrists in the United States and in those countries influenced by American psychiatric nomenclature) is the vaguest and least clearly defined. Such categories as obsession, depression, and mania at least have a vernacular meaning. Schizophrenia, however, is a broad gloss; it involves, in no very clear relationship, ideas such as 'inappropriateness of affect,' 'impoverishment of thought,' 'inability to be involved in meaningful human relationships,' 'bizarre behavior' (such as delusions and hallucinations), 'disorder of speech and communication,' and 'withdrawal.'

These very broadly defined symptoms can be redefined as offenses against implicit social understandings. The appropriateness of emotional expression is, after all, a cultural judgment. Grief is deemed appropriate in our society at a funeral, but not at a party. In other cultures, however, such judgments of propriety may be reversed. With regard to thought disorder, cultural anthropologists have long been at pains to point out that ways of thought are fundamentally different in different societies. What constitutes a meaningful human relationship, anthropologists also report, is basically different in other times and places. Likewise, behavior that is bizarre in one culture is deemed tolerable or even necessary in another. Disorders of speech and communication, again, can be seen as offenses against culturally prescribed rules of language and expression. Finally, the notion of 'withdrawal' assumes a cultural standard concerning the degree of involvement and the amount of distance between the individual and those around him.

The broadness and vagueness of the concept of schizophrenia suggest that it may serve as the residue of residues. As diagnostic categories such as hysteria and depression have become conventionalized names for residual rule breaking, a need seems to have developed for a still more generalized, miscellaneous diagnostic category. If this is true, the schizophrenic explores not only 'inner space' (Ronald Laing's phrase) but also the normative boundaries of his society.

These remarks should not be taken to suggest that there is no internal experience associated with 'symptomatic' behavior; the individual with symptoms *does* experience distress and suffering, or under some conditions, exhilaration and freedom. The point is, however, that public, consensual 'knowledge' of mental illness is based, by and large, on knowledge not of these internal states but of their overt manifestations. When a person runs down the street naked and screaming, lay and professional diagnosticians alike assume the existence of mental illness within that person – even though they have not investigated his internal state. Mental-health procedure and the conceptual apparatus of the medical model posit internal states, but the events actually observed are external.

Labeling theory

A point of view that is an alternative to the medical model, and that acknowledges the culture-bound nature of mental illness, is afforded by labeling theory in sociology (Becker, 1963). Like the evidence supporting the medical model, which is uneven and in large measure unreliable, the body of knowledge in support of the labeling theory of mental illness is by no means weighty or complete enough to prove its correctness. But even though labeling theory is hypothetical, its use may afford perspective – if only because it offers a viewpoint that, along a number of different dimensions, is diametrically opposed to the medical model.

The labeling theory of deviance, when applied to mental illness, may be presented as a series of nine hypotheses (Scheff, 1966):

1 Residual rule breaking arises from fundamentally diverse sources (that is, organic, psychological, situations of stress, volitional acts of innovation or defiance).
2 Relative to the rate of treated mental illness, the rate of unrecorded residual rule breaking is extremely high.
3 Most residual rule breaking is 'denied' and is of transitory significance.
4 Stereotyped imagery of mental disorder is learned in early childhood.
5 The stereotypes of insanity are continually reaffirmed, inadvertently, in ordinary social interaction.
6 Labeled deviants may be rewarded for playing the stereotyped deviant role.
7 Labeled deviants are punished when they attempt to return to conventional roles.

8 In the crisis occurring when a residual rule breaker is publicly labeled, the deviant is highly suggestible and may accept the label.
9   Among residual rule breakers, labeling is the single most important cause of careers of residual deviance.

According to labeling theory, the societal reaction is the key process that determines outcome in most cases of residual rule breaking. That reaction may be either denial (the most frequent reaction) or labeling. Denial is to 'normalize' the rule breaking by ignoring or rationalizing it ('boys will be boys'). The key hypothesis in labeling theory is that when residual rule breaking is denied, the rule breaking will generally be transitory (as when the stress causing rule breaking is removed: e.g., the cessation of sleep deprivation), compensated for, or channeled into some socially acceptable form. If, however, labeling occurs (that is, if the rule breaker is segregated as a stigmatized deviant), the rule breaking that would otherwise have been terminated, compensated for, or channeled may be stabilized; thus, the offender, through the agency of labeling, is launched on a career of 'chronic mental illness.' Crucial to the production of chronicity, therefore, are the contingencies (often external to the deviants) that give rise to labeling rather than denial: for instance, the visibility of the rule breaking, the power of the rule breaker relative to persons reacting to his behavior, the tolerance level of the community, and the availability in the culture of alternative channels of response other than labeling (among Indian tribes, for example, involuntary trance states may be seen as a qualification for a desirable position in the society, such as that of shaman).

'Schizophrenia' – a label

On the basis of the foregoing discussion, it seems likely that labeling theory would prove particularly strategic in investigating schizophrenia. Schizophrenia is the single most widely used diagnosis for mental illness in the United States, yet the cause, site, course, and treatment of choice are unknown, or are subjects of heated and voluminous controversy. Moreover, there is some evidence that the reliability of diagnosis of schizophrenia is quite low. Finally, there is little agreement on whether a disease entity of schizophrenia even exists, what constitutes schizophrenia's basic signs and symptoms if it *does* exist, and how these symptoms are to be reliably and positively identified in the diagnostic process. Because of the all but overwhelming uncertainties and ambiguities inherent in its definition, 'schizophrenia' is an appellation, or 'label,' which may be easily applied to those residual rule breakers whose deviant behavior is difficult to classify (see Boyle in this volume).

As I indicated earlier, in social interaction there is a public order that is continually reaffirmed. Each time a member of the society conforms to the stated or unstated cultural expectations of that society, as when he gazes at the eyes of the person with whom he is conversing, he helps maintain the social status quo. Any deviation from these expectations, however small and regardless of its motivation, may be a threat to the status quo, since most social change occurs through the gradual erosion of custom.

Since all social orders are, as far as we know, basically arbitrary, a threat to society's fundamental customs impels its conforming members to look to extrasocial sources of legitimacy for the status quo. In societies completely under the sway of a single, monolithic religion, the source of legitimacy is always supernatural. Thus, during the Middle Ages the legitimacy of the social order was maintained by reference to God's commands, as found in the Bible and interpreted by the Catholic Church. The Pope was God's deputy, the kings ruled by divine right, the particular cultural form that the family happened to take at the time – the patrilocal, monogamous, nuclear family – was sanctified by the church, and so on.

In modern societies, however, it is increasingly difficult to base legitimacy upon appeals to supernatural sources. As complete, unquestioning religious faith has weakened, one very important new source of legitimacy has emerged: in the eyes of laymen, modern science offers the kind of absolute certainty once provided by the church. The institution of medicine is in a particularly strategic position in this regard, because the physician is the only representative of science with whom the average man associates. To the extent that medical science lends its name to the labeling of nonconformity as mental illness, it is giving legitimacy to the social status quo.

References

Becker, H. (1963) *Outsiders: essays on the sociology of deviance*, New York, Free Press.
Scheff, T. (1966) *Being Mentally Ill: a sociological theory*, Chicago, Aldine.

9

On being sane in insane places

DAVID L. ROSENHAN

If sanity and insanity exist, how shall we know them?

The question is neither capricious nor itself insane. However much we may be personally convinced that we can tell the normal from the abnormal, the evidence is simply not compelling. It is commonplace, for example, to read about murder trials in which eminent psychiatrists for the defense are contradicted by equally eminent psychiatrists for the prosecution on the matter of the defendant's sanity. More generally, there is a great deal of conflicting data on the reliability, utility, and meaning of such terms as 'sanity,' 'insanity,' 'mental illness,' and 'schizophrenia'.

Finally, as early as 1934 Benedict suggested that normality and abnormality are not universal. What is viewed as normal in one culture may be seen as quite aberrant in another. Thus, notions of normality and abnormality may not be quite as accurate as people believe they are.

To raise questions regarding normality and abnormality is in no way to question the fact that some behaviors are deviant or odd. Murder is deviant. So, too, are hallucinations. Nor does raising such questions deny the existence of the personal anguish that is often associated with 'mental illness.' Anxiety and depression exist. Psychological suffering exists. But normality and abnormality, sanity and insanity, and the diagnoses that flow from them may be less substantive than many believe them to be.

At its heart, the question of whether the sane can be distinguished from the insane (and whether degrees of insanity can be distinguished from one another) is a simple matter: do the salient characteristics that lead to diagnoses reside in the patients themselves or in the environments and contexts in which observers find them? From Bleuler, through Kretchmer, through the formulators of the recently revised *Diagnostic and*

Slightly abbreviated from Rosenhan, D. L. (1975) 'On Being Sane In Insane Places', in Scheff, T. (ed.), *Labelling Madness*, Englewood Cliffs, NJ, Prentice-Hall, 54–62.

Statistical Manual of the American Psychiatric Association, the belief has been strong that patients present symptoms, that those symptoms can be categorized, and, implicitly, that the sane are distinguishable from the insane. More recently, however, this belief has been questioned. Based in part on theoretical and anthropological considerations, but also on philosophical, legal, and therapeutic ones, the view has grown that psychological categorization of mental illness is useless at best and downright harmful, misleading, and pejorative at worst. Psychiatric diagnoses, in this view, are in the minds of the observers and are not valid summaries of characteristics displayed by the observed.

Gains can be made in deciding which of these is more nearly accurate by getting normal people (that is, people who do not have, and have never suffered, symptoms of serious psychiatric disorders) admitted to psychiatric hospitals and then determining whether they were discovered to be sane and, if so, how. If the sanity of such pseudopatients were always detected, there would be prima facie evidence that a sane individual can be distinguished from the insane context in which he is found. Normality (and presumably abnormality) is distinct enough that it can be recognized wherever it occurs, for it is carried within the person. If, on the other hand, the sanity of the pseudopatients were never discovered, serious difficulties would arise for those who support traditional modes of psychiatric diagnosis. Given that the hospital staff was not incompetent, that the pseudopatient had been behaving as sanely as he had been outside of the hospital, and that it had never been previously suggested that he belonged in a psychiatric hospital, such an unlikely outcome would support the view that psychiatric diagnosis betrays little about the patient but much about the environment in which an observer finds him.

This article describes such an experiment. Eight sane people gained secret admission to twelve different hospitals. Their diagnostic experiences constitute the data of this article. Too few psychiatrists and psychologists, even those who have worked in such hospitals, know what the experience is like. They rarely talk about it with former patients, perhaps because they distrust information coming from the previously insane. Those who have worked in psychiatric hospitals are likely to have adapted so thoroughly to the settings that they are insensitive to the impact of that experience. And while there have been occasional reports of researchers who submitted themselves to psychiatric hospitalization these researchers have commonly remained in the hospitals for short periods of time, often with the knowledge of the hospital staff. It is difficult to know the extent to which they were treated as patients or as research colleagues. Nevertheless, their reports about the inside of the psychiatric hospital have been valuable. This article extends those efforts.

Pseudopatients and their settings

The eight pseudopatients were a varied group. One was a psychology graduate student in his twenties. The remaining seven were older and 'established'; among them were three psychologists, a paediatrician, a psychiatrist, a painter, and a housewife. Three pseudopatients were women, five were men. All of them used pseudonyms, lest their alleged diagnoses embarrass them later. Those who were in mental health professions alleged another occupation in order to avoid the special attentions that might be accorded by staff, as a matter of courtesy or caution, to ailing colleagues. With the exception of myself (I was the first pseudopatient, and my presence was known to the hospital administrator and chief psychologist and, as far as I can tell, to them alone), the presence of pseudopatients and the nature of the research program was not known to the hospital staffs.

The settings were similarly varied. In order to generalize the findings, admission into a variety of hospitals was sought. The twelve hospitals in the sample were located in five different states on the East and West coasts. Some were old and shabby, some were quite new. Some were research oriented, others were not. Some had good staff-patient ratios, others were quite understaffed. Only one was a strictly private hospital. All of the others were supported by state or federal funds or, in one instance, by university funds.

After calling the hospital for an appointment, the pseudopatient arrived at the admissions office complaining that he had been hearing voices. Asked what the voices said, he replied that they were often unclear, but as far as he could tell they said 'empty,' 'hollow,' and 'thud.' The voices were unfamiliar and were of the same sex as the pseudopatient. The choice of these symptoms was occasioned by their apparent similarity to existential symptoms. Such symptoms are alleged to arise from painful concerns about the perceived meaningless of one's life. It is as if the hallucinating person were saying, 'My life is empty and hollow.' The choice of these symptoms was also determined by the *absence* of a single report of existential psychoses in the literature.

Beyond alleging the symptoms and falsifying name, vocation, and employment, no further alterations of person, history, or circumstances were made. The significant events of the pseudopatient's life history were presented as they had actually occurred. Relationships with parents and siblings, with spouse and children, and with people at work and in school were described as they were or had been, consistent with the aforementioned exceptions. Frustrations and upsets were described along with joys and satisfactions. These facts are important to remember. If anything, they strongly biased the subsequent results in favor of detecting sanity, for none of their histories or current behaviors were seriously pathological in any way.

Immediately upon admission to the psychiatric ward, the pseudo-patient ceased simulating *any* symptoms of abnormality. In some cases, there was a brief period of mild nervousness and anxiety, for none of the pseudopatients really believed that they would be admitted so easily. Indeed, their shared fear was that they would be immediately exposed as frauds and greatly embarrassed. Moreover, many of them had never visited a psychiatric ward; even those who had, nevertheless had some genuine fears about what might happen to them. Their nervousness, then, was quite appropriate to the novelty of the hospital setting, and it abated rapidly.

Apart from that short-lived nervousness, the pseudopatient behaved on the ward as he behaved 'normally.' The pseudopatient spoke to patients and staff as he might ordinarily. Because there is uncommonly little to do on a psychiatric ward, he attempted to engage others in conversation. When asked by staff how he was feeling, he indicated that he was fine, that he no longer experienced symptoms. He responded to instructions from attendants, to calls for medication (which was not swallowed), and to dining-hall instructions. Beyond such activities as were available to him on the admissions ward, he spent his time writing down his observations about the ward, its patients, and the staff. Initially, these notes were written 'secretly,' but as it soon became clear that no one much cared, they were subsequently written on standard tablets of paper in such public places as the dayroom. No secret was made of these activities.

The pseudopatient, very much as a true psychiatric patient, entered a hospital with no foreknowledge of when he would be discharged. Each was told that he would have to get out by his own devices, essentially by convincing the staff that he was sane. The psychological stresses associated with hospitalization were considerable, and all but one of the pseudopatients desired to be discharged almost immediately after being admitted. They were, therefore, motivated not only to behave sanely, but to be paragons of cooperation. That their behavior was in no way disruptive is confirmed by nursing reports, which have been obtained on most of the patients. These reports uniformly indicate that the patients were 'friendly,' 'cooperative,' and 'exhibited no abnormal indications.'

The normal are not detectably sane

Despite their public 'show' of sanity, the pseudopatients were never detected. Admitted, except in one case, with a diagnosis of schizophrenia, each was discharged with a diagnosis of schizophrenia 'in remission.' The label 'in remission' should in no way be dismissed as a formality, for at no time during any hospitalization had any question been raised about any pseudopatient's simulation. Nor are there any indications in the hospital records that the pseudopatient's status was suspect. Rather, the evidence is strong that once labeled schizophrenic, the pseudopatient was stuck with that label. If the pseudopatient was to be discharged, he must naturally be 'in remission'; but he was not sane, nor, in the institution's view, had he ever been sane.

The uniform failure to recognize sanity cannot be attributed to the quality of the hospitals, for although there were considerable variations among them, several are considered excellent. Nor can it be alleged that there was simply not enough time to observe the pseudopatients. Length of hospitalization ranged from 7 to 52 days, with an average of 19 days. The pseudopatients were not, in fact, carefully observed, but this failure clearly speaks more to traditions within psychiatric hospitals than to lack of opportunity.

Finally, it cannot be said that the failure to recognize the pseudo-patients' sanity was due to the fact that they were not behaving sanely. Though there was clearly some tension present in all of them, their daily visitors could detect no serious behavioral consequences – nor, indeed, could other patients. It was quite common for the patients to 'detect' the pseudopatients' sanity. During the first three hospitalizations, when accurate counts were kept, 35 of a total of 118 patients on the admissions ward voiced their suspicions, some vigorously. 'You're not crazy. You're a journalist, or a professor [referring to the continual note taking]. You're checking up on the hospital.' While most of the patients were reassured by the pseudopatient's insistence that he had been sick before he came in but was fine now, some continued to believe that the pseudopatient was sane throughout his hospitalization. The fact that the patient often recognized normality when staff did not raises important questions.

Failure to detect sanity during the course of hospitalization may be due to the fact that physicians operate with a strong bias toward what statisticians call the type 2 error. That is, physicians are more inclined to call a healthy person sick (a false positive, type 2) than a sick person healthy (a false negative, type 1). The reasons for this are not hard to find: it is clearly more dangerous to misdiagnose illness than health. Better to err on the side of caution, to suspect illness even among the healthy.

But what holds for medicine does not hold equally well for psychiatry. Medical illnesses, though unfortunate, are not commonly pejorative. Psychiatric diagnoses, on the contrary, carry with them personal, legal, and social stigmas. It was therefore important to see whether the tendency toward diagnosing the sane insane could be reversed. The following experiment was arranged at a research and teaching hospital whose staff had heard these findings but doubted that such an error could occur in their hospital. The staff was informed that at some time during the following three months, one or more pseudopatients would attempt to be admitted into the psychiatric hospital. Each staff member was asked to rate each patient who presented himself at admissions or on the ward according to the likelihood that the patient was a pseudopatient. A 10-point scale was used, with a 1 and 2 reflecting high confidence that the patient was a pseudopatient.

Judgments were obtained on 193 patients who were admitted for psychiatric treatment. All staff who had had sustained contact with or primary responsibility for the patient – attendants, nurses, psychiatrists, physicians, and psychologists – were asked to make judgements. Forty-one patients were alleged, with high confidence, to be pseudopatients by at least one member of the staff. Twenty-three were considered suspect by at least one psychiatrist. Nineteen were suspected by one psychiatrist *and* one other staff member. Actually, no genuine pseudopatient (at least from my group) presented himself during this period.

The experiment is instructive. It indicates that the tendency to designate sane people as insane can be reversed when the stakes (in this case, prestige and diagnostic acumen) are high. But what can be said of the 19 persons who were suspected of being 'sane' by one psychiatrist and another staff member? Were these people truly 'sane,' or was it rather the case that in the course of avoiding the type 2 error the staff tended to make more errors of the first sort – calling the crazy 'sane'? There is no way of knowing. But one thing is certain: any diagnostic process that lends itself so readily to massive errors of this sort cannot be a very reliable one.

The stickiness of psychodiagnostic labels

Beyond the tendency to call the healthy sick – a tendency that accounts better for diagnostic behavior on admission than it does for such behavior after a lengthy period of exposure – the data speak to the massive role of labeling in psychiatric assessment. Having once been labeled schizophrenic, there is nothing the pseudopatient can do to overcome the tag. The tag profoundly colors others' perceptions of him and his behavior.

From one viewpoint, these data are hardly surprising, for it has long been known that elements are given meaning by the context in which they occur. Gestalt psychology made this point vigorously, and Asch (1952) demonstrated that there are 'central' personality traits (such as 'warm' versus 'cold') that are so powerful that they markedly color the meaning of other information in forming an impression of a given personality. 'Insane,' 'schizophrenic,' 'manic-depressive,' and 'crazy' are probably among the most powerful of such central traits. Once a person is designated abnormal, all of his other behaviors and characteristics are colored by that label. Indeed, that label is so powerful that many of the pseudopatients' normal behaviors were overlooked entirely or misinterpreted profoundly. Some examples may clarify this issue.

Earlier, I indicated that there were no changes in the pseudopatient's personal history and current status beyond those of name, employment, and, where necessary, vocation. Otherwise, a veridical description of personal history and circumstances was offered. Those circumstances were not psychotic. How were they made consonant with the diagnosis of psychosis? Or were those diagnoses modified in such a way as to bring them into accord with the circumstances of the pseudopatient's life, as described by him?

As far as I can determine, diagnoses were in no way affected by the relative health of the circumstances of a pseudopatient's life. Rather, the reverse occurred: the perception of his circumstances was shaped entirely by the diagnosis. A clear example of such translation is found in the case of a pseudopatient who had had a close relationship with his mother but was rather remote from his father during his early childhood. During adolescence and beyond, however, his father became a close friend, while his relationship with his mother cooled. His present relationship with his wife was characteristically close and warm. Apart from occasional angry exchanges, friction was minimal. The children had rarely been spanked. Surely there is nothing especially pathological about such a history. Indeed, many readers may see a similar pattern in their own experiences, with no markedly deleterious consequences. Observe, however, how such a history was translated in the psychopathological context, this from the case summary prepared after the patient was discharged.

> This white 39-year-old male ... manifests a long history of considerable ambivalence in close relationships, which begins in early childhood. A warm relationship with his mother cools during his adolescence. A distant relationship to his father is described as becoming very intense. Affective stability is absent. His attempts to control emotionality with his wife and children are punctuated by angry outbursts and, in the case of the children, spankings. And while he says that he has several good friends, one senses considerable ambivalence embedded in those relationships also.

The facts of the case were unintentionally distorted by the staff to achieve consistency with a popular theory of the dynamics of a schizophrenic reaction. Nothing of an ambivalent nature had been described in relations with parents, spouse, or friends. To the extent that ambivalence could be inferred, it was probably not greater than is found in all human relationships. It is true the pseudopatient's relationships with his parents changed over time, but in the ordinary context that would hardly be remarkable – indeed, it might very well be expected. Clearly, the meaning ascribed to his verbalizations (that is, ambivalence, affective instability) was determined by the diagnosis: schizophrenia. An entirely different meaning would have been ascribed if it were known that the man was 'normal.'

All pseudopatients took extensive notes publicly. Under ordinary circumstances, such behavior would have raised questions in the minds of observers, as, in fact, it did among patients. Indeed, it seemed so certain that the notes would elicit suspicion that elaborate precautions were taken to remove them from the ward each day. But the precautions proved needless. The closest any staff member came to questioning these notes occurred when one pseudopatient asked his physician what kind of medication he was receiving and began to write down the response. 'You needn't write it,' he was told gently. 'If you have trouble remembering, just ask me again.'

If no questions were asked of the pseudopatients, how was their writing interpreted? Nursing records for three patients indicate that the writing was seen as an aspect of their pathological behavior. 'Patient engages in writing behavior' was the daily nursing comment on one of the pseudopatients who was never questioned about his writing. Given that the patient is in the hospital, he must be psychologically disturbed. And given that he is disturbed, continual writing must be a behavioral manifestation of that disturbance, perhaps a subset of the compulsive behaviors that are sometimes correlated with schizophrenia.

One tacit characteristic of psychiatric diagnosis is that it locates the sources of aberration within the individual and only rarely within the complex of stimuli that surrounds him. Consequently, behaviors that are stimulated by the environment are commonly misattributed to the patient's disorder. For example, one kindly nurse found a pseudopatient pacing the long hospital corridors. 'Nervous, Mr. X?' she asked. 'No, bored,' he said.

The notes kept by pseudopatients are full of patient behaviors that were misinterpreted by well-intentioned staff. Often enough, a patient would go 'berserk' because he had, wittingly or unwittingly, been mistreated by, say, an attendant. A nurse coming upon the scene would rarely inquire even cursorily into the environmental stimuli of the patient's behavior. Rather, she assumed that his upset derived from his pathology, not from his present interactions with other staff members. Occasionally, the staff might assume that the patient's family (especially when they had recently visited) or other patients had stimulated the outburst. But never were the staff found to assume that one of themselves or the structure of the hospital had anything to do with a patient's behavior. One psychiatrist pointed to a group of patients who were sitting outside the cafeteria entrance half an hour before lunchtime. He indicated to a group of young residents that such behavior was characteristic of the oral-acquisitive nature of the syndrome. It did not seem to occur to him that there were very few things to anticipate in a psychiatric hospital besides eating.

A psychiatric label has a life and an influence of its own. Once the impression has been formed that the patient is schizophrenic, the expectation is that he will continue to be schizophrenic. When a sufficient amount of time has passed during which the patient has done nothing bizarre, he is considered to be in remission and available for discharge. But the label endures beyond discharge, with the unconfirmed expectation that he will behave as a schizophrenic again. Such labels, conferred by mental health professionals, are as influential on the patient as they are on his relatives and friends, and it should not surprise anyone that the diagnosis acts on all of them as a self-fulfilling prophecy. Eventually, the patient himself accepts the diagnosis, with all of its surplus meanings and expectations, and he behaves accordingly.

The inferences to be made from these matters are quite simple. Much as Zigler and Phillips (1956) have demonstrated that there is enormous overlap in the symptoms presented by patients who have been variously diagnosed,[1] so there is enormous overlap in the behaviors of the sane and the insane. The sane are not 'sane' all of the time. We lose our tempers 'for no good reason.' We are occasionally depressed or anxious, again for no good reason. And we may find it difficult to get along with one or another person – again for no reason that we can specify. Similarly, the insane are not always insane. Indeed, it was the impression of the pseudopatients while living with them that they were sane for long periods of time – that the bizarre behaviors upon which their diagnoses were allegedly predicated constituted only a small fraction of their total behavior. If it makes no sense to label ourselves permanently depressed on the basis of an occasional depression, then it takes better evidence than is presently available to label all patients insane or schizophrenic on the basis of bizarre behaviors or cognitions. It seems more useful, as Mischel[4] has pointed out, to limit our discussions to *behaviors*, the stimuli that provoke them, and their correlates.

Note

1. See also Freudenberg and Robertson (1956).

References

Asch, S. E. (1946) *Journal of Abnormal Social Psychology*, vol. 41, 258.
Asch, S. E. (1952) *Social Psychology*, Prentice-Hall, New York.
Benedict, R. (1934) *Journal of General Psychology* vol. 10, 59.
Freudenberg, R. K. and Robertson, J. P. (1956) *American Medical Association Archives of Neurological Psychiatry*, vol. 76, 14.
Mischel, W. (1968) *Personality and Assessment*, New York, John Wiley.
Zigler, E. and L. Phillips (1961) *Journal of Abnormal Social Psychology*, vol. 63, 69.

10

Reversing deviance

ROGER GOMM

Introduction

The article by Thomas Scheff in this volume (pp. 64–69) is one of the classic statements of the sociology of deviance as it developed in the 1960s and 1970s: a perspective which applies essentially the same set of ideas to social phenomena as apparently diverse as crime, learning disabilities, left-wing subversion, cheating in science, homosexuality, rape, drug addiction, industrial disputes and mental illness (Lemert, 1951; Becker, 1963, 1964). All of these are instances of deviance because they are regarded as objectionable by people who are powerful and credible enough to ensure that their worries influence public debates and public policy. In this framework, then, deviance derives from the moral judgements made by some people that some other person or group of persons, or some condition, is undesirable, disreputable, wicked, threatening, pathetic, contagious, outrageous or otherwise unacceptable. Since different social groups have different ideas, this notion of deviance implies disputes and struggles between people as to how life should be lived, and about how it is established – powerfully – that some are in the right and some are in the wrong.

Scheff's article deals with the labelling of behaviour as being deviance of a particular kind: mental illness. On the one hand this labelling process reflects the vested interests of those who make a living from offering to control and correct deviance. To label someone as 'mentally ill' is to label them as work for the doctor and allied occupations. Medical work in this sphere depends on public acceptance that there is a problem to be dealt with, and that it is the kind of problem for which medicine offers the solutions.

Constituting the normal

On the other hand, as Scheff argues, the definition of deviance proceeds from a definition of normality. For the sociology of deviance, what is normal is just as shifty a matter as what is deviant. Such sociologists do not define normal as what usually happens, nor do they prescribe what ought to happen and call that 'normal'. Instead, they observe that in social life people are forever treating this as normal and that as deviant; this as objectionable and that as benign; this as something to be countered and corrected, that as something to be celebrated and promoted. Different people do this differently, and often the same person labels matters deviant or normal differently on different occasions. Deviant and normal, then, have no fixed meanings; rather, they represent ways in which people give moral appraisals of social affairs.

The sociology of deviance directs attention to the question of how the normal is constituted – which is the reciprocal of the way in which deviance is constituted – and to the question of whose interests are served by particular constitutions of the normal. The history of mental illness is littered with examples of diagnostic categories which in hindsight we can read as expressions of vested interests. For example, there is the nineteenth-century diagnosis of 'nymphomania', often applied to young women of elite families sexually attracted to working class men (Ehrenreich and English, 1974). In hindsight it is easy to see how the definition of this kind of mental illness arises from a concept of normal behaviour closely related to issues of the male control of female sexuality and the management of the reputations and inheritances of elite families. Kleptomania (Abelson, 1974) began its life as a diagnosis exclusively applied to rich women who stole what they could easily afford to buy. It could hardly be normal for them to steal, but was their stealing a mark of a criminal character? Like some other mental illness diagnoses, this one developed as a label serving to divert the rich from the criminal justice system, only later being democratised. Or again, in the 1860s, Samuel Cartwright MD of Louisiana outlined the symptomatology of a mental disorder called 'Drapetomania', afflicting only Negro slaves. Its major symptom was running away from slavery (Szasz, 1972). This kind of deviance was constituted by a conception of slavery as being the only normal and healthy condition for black people.

In retrospect such diagnostic labels have a tragi-comic appearance: they seem such flimsy ways of hiding vested interests in medical jargon and medical practice. But they appeared perfectly sensible to many people at the time because they incorporated assumptions about the normal which were highly convenient to establishment groups.

Two lines of questioning in the sociology of deviance

The more orthodox and traditional approach to any kind of deviance has been to ask questions such as 'Why do they do it?' or 'Why does it happen?'. These are still important questions. They include, for example, the kinds of questions addressed in the paper called 'Mental Health and Inequality' in this volume, investigating how differences of material condition make people more or less vulnerable to experiencing mental distress. However, the sociology of deviance added another line of questioning, asking 'Why do others find this objectionable; which others; what do they do about it, and what are the consequences of that?'

There are two important implications of this line of questioning. The first is that it implies that deviance isn't so much a state of affairs as a social process: that for someone or something to be deviant, others have to make judgements, perform actions and so on. As Scheff notes, the vast majority of people who show what might be regarded as the 'symptoms' of mental illness never get labelled as mentally ill. Whether that happens or not depends on the responses of others. The processual aspects of deviance are often referred to as 'deviancy amplification'. The second implication is that, if deviance is to be reduced, there are two ways of doing it. The first and more orthodox approach is to try to stop whatever is regarded as deviant from happening, or if it does happen, to attempt to correct it. The other way of reversing deviance is to stop regarding it as deviant. Put another way, the latter means reconstituting ideas of the normal so that they become more expansive and tolerant of diversity. This is often referred to as 'deviancy normalisation', or 'normalisation' for short (Cohen, 1973a).

Deviancy amplification

The term deviancy amplification refers to the way in which attempts to correct deviance often generate more deviance (Cohen, 1973b, Chapter 1; Schatzman, 1972). In the field of mental health this possibility is widely recognised even within orthodox psychiatry. For example, the medical term 'chronicity' refers to those effects of treatment which, instead of making someone better (from the clinician's viewpoint), freeze them into a mental illness role. 'Institutionalisation' or 'institutional neurosis' (Barton, 1976; Wing and Brown, 1970) refers to the undermining of someone's capacity for individual action and the creation of 'learned helplessness'. 'Iatrogenesis' is a general term for any kind of medical treatment which makes people ill, or adds problems to their original troubles.

Chronicity and institutional neurosis are kinds of iatrogenesis. So also are the effects of drugs given to help people mend their minds which prevent them thinking clearly, create chemical addiction, promote panic attacks or produce other disabling side effects (see Gabe in this volume). And, so the argument goes, if people who have been labelled 'mentally ill' so often have a low self-esteem, this is hardly surprising when the label is so deeply stigmatising. And if they are 'lacking in confidence', this is unsurprising when being labelled as mentally ill has made relationships with others so personally undermining. And if they 'lack self-care skills', this is hardly surprising if the opportunities to practise such skills have been taken from them. And if they are distrustful of others, showing 'symptoms' of 'paranoia', then who can wonder at this, when people have been telling them half-truths, or making decisions without involving them, or speaking to them in ways they cannot understand, or promising to make them better while making them feel worse (Lemert 1973). There is perhaps nothing more conducive to paranoia than being forcibly removed from one's home by large police officers, conveyed to hospital and being injected with drugs against one's will (see Deryck Browne in this volume).

It has to be said that the sociology of deviance as a whole often gives the impression that amplification is the inevitable result of all and any attempts to control deviance. Writers like Scheff (1966 and in this volume), for example, focus almost exclusively on the iatrogenic effects of psychiatry, and their writings nearly always feature that small minority of people among those regarded as having 'mental health problems' who do end up incarcerated in hospitals and/or heavily drugged (see also Goffman, 1961; Laing and Esterson, 1969; Sarbin, 1974; Schatzman, 1972; and Szasz 1972). As a corrective it is worth pointing to the much larger number of people whose encounters with the mental health care system leave them feeling at least no worse, and sometimes better.

Nonetheless in the mental health field, the potential for turning what are relatively small problems into much worse problems seems great. When it runs its full course, deviancy amplification puts people outside of the more usual forms of social interaction and deprives them of the opportunities most people have. As Garfinkel points out, there are elaborate courtroom and medical rituals for conferring a deviant status on people and few such rituals for re-establishing those who have been labelled deviant as thoroughly rehabilitated, cured or restored to the status normal (Garfinkel, 1956; see also Rosenhan in this volume).

There are two end-points to the process. One is being isolated, at odds with the world, hurt, confused, without valued social relationships, with little self-respect: a kind of social death, and sometimes a physical death as well, since suicide is a common associate of being treated as mentally ill (see Taylor and Gabe in this volume). The other is finding common cause with other stigmatised people in a sub-culture which is regarded by wider society as deviant, but where common understandings, mutual support and valued social relationships can be found (Goffman, 1963). Such deviant sub-cultures have often been the basis for political action to redefine what is normal and acceptable. For example, gay people have managed to turn homosexuality from an illness or an abomination into a life style whose legitimacy is now more widely accepted by others, if often rather grudgingly. People with learning difficulties and people with physical disabilities have made enormous gains in rehabilitating their image and improving the opportunities available to them (Atkinson and Williams, 1990). Psychiatric survivor movements in the USA, New Zealand, Holland and Britain (see Campbell in this volume) are beginning to change the image of, and the opportunities for, such people as are regarded as having 'mental health problems'.

Deviancy normalisation

If deviancy amplification is a process which forces labelled people out of 'normal' society, then deviancy normalisation is a process which draws them back, or which, by re-drawing the lines of the 'normal', prevents people from being labelled deviant in the first place.

Within the practice literature of health and social welfare what is sometimes called 'a normalisation approach' (Flynn and Nitsch, 1980; Brown and Smith, 1992) is an offshoot from the sociology of deviance (Whitehead, 1992). For the application of this approach to the mental health field see Ramon, 1992 and Brandon in this volume. As might be expected from its origins, 'normalisation' proposes a programme with three main strands. These are, firstly, to challenge and change the conceptions of the normal against which people are judged as deviant, for example, promoting the idea that hearing voices, while unusual, is none the less part of the healthy spectrum of human experience (see Romme in this volume); secondly, ensuring that the kinds of assistance offered to people who need it do not have iatrogenic consequences; and thirdly, rectifying the adverse effects that labelling has had on the personalities and capabilities of people who have been treated as deviant. Alongside all of these is the assumption that being treated as mentally ill and deviant means being deprived of the right to control one's own life and that normalisation entails restoring these rights and developing the skills needed to exercise them.

For the first of these purposes current images of and beliefs about mental illness are particularly important targets. These are not just the public images of the 'mentally ill' which can be seen in the mass media and in popular fiction (see Philo *et al.* in this volume), though these are extremely important. They include notions underlying professional perspectives on which treatment regimes are based: notions such as that, for example, people who are mentally ill don't understand what they want, or are manipulative, or have to have decisions made for them, or present a risk to themselves and others (see Rogers and Pilgrim, and Muijen in this volume). Of course, from time to time these are accurate appraisals, but being right occasionally is not the same thing as basing a service on the assumption that these will be characteristic of all of the clients all of the time. As implied by the idea of deviancy amplification, what is assumed in labelling people as deviant has a nasty tendency to come true. David Brandon's article in this volume describes some of the ways in which professionals need to change in order to pursue a normalisation approach. Moreover, it is not just the beliefs held by other people which are important. Willy nilly these are often conveyed to the person labelled deviant to become the basis for their own feelings of worthlessness (see Philo et al. in this volume and Rotenberg, 1974).

The rectification of erroneous and prejudicial images and beliefs is an important task in normalisation. Logically the people who ought to take the leading role in image work are the labelled people themselves. To proceed otherwise would be to collude with a process which imposes identities on them.

The second strand of a normalisation approach is to 'normalise' the ways in which people in distress are treated. It is a characteristic of many treatment regimes in the mental illness field that they require the patient to learn to behave in very odd ways indeed. Most psychiatric regimes encourage the patient to believe that they are not normal people: the patient is then said to have 'gained insight'. The daily routines of the mental patient in hospital are quite different from those of everyday life. Rosenhan's pseudo-patients (see Rosenhan in this volume) found that the circumstances of the mental hospital made it quite impossible for them to behave in ways which they would regard as normal (see also Braginski et al., 1969; and Goffman, 1961). The possibilities for iatrogenesis arising from imposing an orthodox psychiatric regime on a person have been discussed above. However, perhaps the most important deviancy amplifying effect of psychiatric treatment is the way in which it excludes people from the normal opportunities of life, both at the time of treatment and thereafter because of the persistently stigmatising effect of having been mentally ill. The general thrust of a normalisation approach here is for assistance to be offered without the disempowering effect of imposition, and for such assistance as is offered to be that which is consistent with the person maintaining what they consider to be a normal and desirable life. Note that the idea of normalisation implies that people have a right to decide what is normal and desirable for them (for example, see Brown and Smith, 1992, Chapter 10; and Ferns, 1992).

Normalisation suggests responses which do not segregate and exclude, but which offer people opportunities for living in the kind of housing they want to live in, doing the kinds of jobs they want to do and participating in the leisure activities they prefer. This is an important principle in many of the papers in the third and fourth parts of this volume.

Whether because they have been labelled as mentally ill or otherwise, many such people are seriously distressed and incapacitated. They need the support and assistance of others. However, normalisation is not something which can be done to people (Perrin and Nirje, 1993). Rather, a normalisation programme is one which offers people opportunities to learn the skills they need in order to lead a satisfactory life. Self-advocacy and self-assertion skills are particularly important in this respect (see Campbell, and Williams and Watson in this volume).

The political and ethical nature of deviance and normalcy

There are of course limits to a normalisation approach. We could, for example, attempt to normalise rape by being more tolerant of rapists and offering them more opportunities for sexual aggression. The case against normalising rape is the harm this would do to those who would become rape victims. In this example it is quite clear that defining the normal and defining the deviant is a political and ethical matter of balancing the interests of different groups of people. By contrast, labelling people as 'mentally ill' evades the ethical and political issues involved. People are rarely told that they are being detained in hospital because they are annoying their relatives and frightening the neighbours. They are more usually told that they are being incarcerated because they are ill and need treatment for their own good. Thus are very real political and ethical issues hidden away and placed beyond discussion. Perhaps the most important insight deriving from the sociology of deviance is the way in which labelling is used to decide conflicts of interest, without appearing to do so, and usually at the expense of the least powerful people involved. Any proposal to pursue a normalisation approach unearths hidden political questions about conflicts of interest and puts them on the agenda for discussion.

References

Abelson, E. (1974) 'The invention of kleptomania', *Journal of Women in Culture and Society*, 123–43.

Atkinson, D. and Williams, F. (eds) (1990) *Know me as I am: an anthology of prose, poetry and art by people with learning difficulties*, London, Hodder & Stoughton.

Barton, R. (1976) *Institutional Neurosis*, Bristol, John Wright.

Becker, H. (1963) *Outsiders: essays in the sociology of deviance*, New York, Free Press.

Becker, H. (ed.) (1964) *The Other Side: perspectives on deviance*, New York, Free Press.

Braginski, B., Braginski, D. and Ring, K. (1969) *Methods of Madness: the mental hospital as a last resort*, New York, Holt, Rinehart & Winston.

Brown, H. and Smith, H. (eds) (1992) *Normalisation: a reader for the nineties*, London, Routledge.

Cohen, S. (1973a) 'Living with crime', *New Society* 8 November, 330–333.

Cohen, S. (1973b) *Folk Devils and Moral Panics*, St Albans, Granada.

Ehrenreich, B. and English, D. (1974) *Complaints and Disorders: the sexual politics of sickness, Glass Mountain Pamphlets*, No. 2, London, Compendium.

Ferns, P. (1992) 'Promoting race equality through normalisation', in Brown and Smith, *Normalisation*, 134–148.

Flynn, R. and Nitsch, K. (eds) (1980) *Normalisation, Social Integration and Community Services*, Baltimore University Park Press.

Garfinkel, H. (1956) 'Conditions of successful degradation ceremonies', *American Journal of Sociology*, vol. 61, 420–424.

Goffman, E. (1961) *Asylums: essays on the social situation of mental patients and other inmates*, New York, Anchor Books (Penguin edn 1968).

Goffman, E. (1963) *Stigma: notes on the management of a spoiled identity*, New Jersey, Prentice-Hall.

Laing, R. and Esterson, A. (1969) *Sanity, Madness and the Family*, Harmondsworth, Penguin.

Lemert, E. (1951) *Social Pathology*, New York, McGraw Hill.

Lemert, F. (1973) 'Paranoia and the dynamics of exclusion' in Rubington, E. and Weinberg, M. (eds) *Deviance: the interactionist perspective*, New York, Macmillan, 106–116.

Perrin, B. and Nirje, B. (1993) 'Setting the record straight: a critique of some frequent misunderstandings of the normalisation principle', in Brechin, A. and Walmsley, J. (eds) *Making Connections: reflecting on the lives and experiences of people with learning difficulties*, London, Hodder & Stoughton.

Ramon, S. (1992) *Beyond Community Care: normalisation and integration work*, London, MIND/Macmillan.

Rotenberg, M. (1974) 'Self-labelling: a missing link in the "societal reaction" theory of deviance', *Sociological Review*, vol. 22, 335–354.

Sarbin, T. (1974) 'Stimulus/response: schizophrenia is a myth born of metaphor: meaningless', *Psychology Today*, 6 (June), 18–27.

Schatzman, M. (1972) 'Madness and morals', in Boyers, R. and Orrill, R. (eds), *Laing and Anti-psychiatry*, Harmondsworth, Penguin, 181–208.

Scheff, T. (1966) *Being Mentally Ill: a sociological theory*, Chicago, Aldine.

Szasz, T. (1971) 'The sane slave: an historical note on the use of medical diagnosis as a justificatory rhetoric', *American Journal of Psychotherapy*, vol. 25 (2).

Szasz, T. (1972) *The Myth of Mental Illness*, London, Paladin.

Whitehead, S. (1992) 'The social origins of normalisation', in Brown and Smith, *Normalisation*, 47–59.

Wing, J. and Brown, G. (1970) *Institutionalisation and Schizophrenia: a comparative study of three mental hospitals 1960–1968*, Cambridge, Cambridge University Press.

11

Some problematic aspects of dementia

TOM KITWOOD

For some years now the category 'dementia' has maintained a relatively stable and uncontested meaning in psychiatry. The accepted general definitions[1] all point to a deterioration of an individual's mental functioning as compared to previous levels; and here they focus particularly on impairments of cognition (memory, orientation, comprehension, judgement, and so on), when there is no 'clouding of consciousness'. These definitions also specify that dementia is associated with irreversible pathological processes in nerve tissue.

The story of how dementia has come to be thus narrowed down is a convoluted one, as several historians have demonstrated (for example, Berrios and Freeman, 1991). Even in its modern and restricted meaning, however, dementia is a far more problematic category than has generally been recognized, and the much-cherished distinction between so-called 'organic' and 'functional' mental disorders is by no means clear. This can be illustrated by reference to four topics: the prevalence of dementia, research on psychological and social factors associated with the dementing process, the rise of the 'Alzheimer culture' at a popular level, and approaches to dementia care.

The prevalence of dementia

With the ageing of the populations of the more affluent societies of the world, the problem of the so-called 'rising tide' of dementia has become paramount in the planning of services for those in later life. However, despite enormous effort and expenditure, figures for both incidence and prevalence remain elusive. The most-cited of all prevalence studies is

Previously published in German Berriuf and Roy Parker (eds) (1995), *The History of Clinical Psychology*, London, Athlone Press.

that of Kay *et al.* (1964). A total of 758 persons over 65 were interviewed, and of these 6.2 per cent were assessed as having 'chronic brain syndrome': roughly equivalent, in contemporary terms, to severe and moderately severe dementia. Since that time there have been numerous surveys giving prevalence figures among the over 65's ranging from around 2 per cent to 20 per cent overall, and around 1 per cent to 8 per cent for severe dementia (Ineichen, 1987). Only one cross-cultural study with really sound methodology has been carried out thus far, that of Gurland *et al.* (1983) in New York and London. Higher prevalence rates were found in New York than in London at all ages, and for all degrees of severity.

To some extent the anomalies in reported prevalence figures can be explained in terms of a standard empiricist methodology. For example, the samples have generally been relatively small as judged by epidemiological standards, typically numbering a few hundred persons. Refusal rates tend to be rather high, of the order of 20 per cent (although some studies do not report the figure). There is also the fact that no study is truly comprehensive. Those which are based on community-survey methods leave out individuals who are already in long-stay institutional care; those which are based on medical and social provision leave out persons who have not yet made use of existing services.

Other aspects of the prevalence data, however, render the category of dementia much less robust than is commonly assumed. First and foremost there is the fact that the criteria for diagnosing a dementing illness are far from clear. Clinical judgments take many aspects into account simultaneously, and involve implicit decisions as to how to weigh cognitive and non-cognitive factors. The standardised tests which are most reliable (Blessed *et al.* 1991) deal mainly with cognitive impairments, and here much depends on which aspects of cognition are prioritised. Tests which emphasise defective memory (which is relatively simple to measure) tend to give higher prevalence rates, blurring the boundaries between dementia and 'benign senescent forgetfulness' (Kral, 1962). The problems are particularly severe in the assessment of mild dementia.

The interpretation of prevalence data is made more complex by the fact that a person who is depressed may show symptoms very like those of a dementing illness: this fact has led to the creation of the further, and even more problematic, diagnostic category of pseudodementia (Arie, 1983; Ames *et al.*, 1990). Finally, there is the unresolved question of the difference in prevalence between New York and London. Psychiatry at present accepts the possibility that the explanation might lie with environmental factors of a physical kind. It seems resolutely opposed, however, to considering that social-psychological factors might be involved. One clear hypothesis would be that stress and its accompanying biochemical and immunological changes can have a causative role in some dementing conditions. Since these ideas are entirely compatible with what is known in neurochemistry, it can only be a gross prejudice which would exclude them a priori.

Psychological and social factors in the dementing process

In the current framing of dementia there is no serious dispute over the idea that the pattern of an individual's symptoms is related to his or her pre-morbid personality, or that the existence of a dementia might be 'unmasked' by critical events. To go beyond this, however, and suggest that psychological and social factors might be causally implicated in dementia, is generally forbidden (see, for example, Gilhooly, 1984).

This has not always been the case. Not long after the research which laid the foundations for the disease categories of today, it was recognised that the correspondence between indices of dementia and those of neuropathology was not always high; the most serious anomalies were those persons who went through the entire course of a dementing illness, and yet whose brains showed no degenerative changes beyond those typical of 'normal' persons of the same age.[2] The suggestion, therefore, was made by a succession of workers in the field that neuropathology did not provide a sufficient explanation, and that other types of causative factor should also be sought. From around 1925 to 1975 various hypotheses were advanced about ways in which psychological, social and neurological factors might interact with each other in causing a dementing illness.

Probably the most important mid-century work on this topic is that of David Rothschild. Drawing on his clinical work, and particularly on 'extreme' case studies, he insisted that the dementia sufferer (like the stroke victim) should not be regarded as the passive carrier of a disease process; factors of a personal kind must be considered in every case (Rothschild and Sharpe, 1941). This theme is repeated many times in the post-war literature. Williams *et al.* (1947), for example, presented primary data which associated 'senile psychosis' with a loss or lack of social integration and the means of independence. Wilson *et al.* (1955) reviewed a considerable body of research and concluded that 'senility' has many of the features typical of a psychosomatic disorder.

Oakley (1965), using relatives' accounts, found an association between dementia and the obsessional personality (roughly equivalent, in his terms, to Freud's 'anal character'). Morgan (1965) attempted to use psychoanalytic ideas, suggesting that dementia is a form of defence against impending death. Folsom (1968), the founder of 'reality orientation', claimed that dementia can be understood as a valid reaction to such negative experiences in old age as isolation and disempowerment. Similar ideas were put forward by Meacher (1972), in his study of the institutionalization of old people. A significant conceptual development was the work of Barnes, Sack and Shore (1973), who viewed the dementing process as an iterative generation of dependency and lowered self-esteem, leading to a state of vegetation. Amster and Kraus (1974) found a high incidence of crucial life events to be associated with the onset of dementia.

Some of the work cited here is over-speculative, and some studies contain such serious methodological flaws that they can easily be dismissed (Meacher, for example, paid no attention at all to the direct effects of neurological impairment). There has, however, been no rebuttal of the case that a full account of the dementing process requires research into psychological and social factors, some of which may be causal. Recently that case has been taken up again. Gilleard (1989), for example, has shown that 'losing one's mind' takes a very different aspect if mind itself is understood as social rather than individual. Kitwood (1989, 1990) has elaborated a view of dementia as a dialectical interplay between neuro-pathology and the 'malignant social psychology' that often bears down on those who are elderly and mentally frail. Sabat and Harré (1992) have questioned the idea of 'loss of self' in dementia, on social-psychological grounds. The significant point here is that while an immensely powerful research programme grounded in the biomedical sciences has been consolidated over two decades and more, nothing comparable has developed in the psycho-social field, although the logic of evidence strongly requires it. If this fact is to be explained, we must look at the cluster of interests that gave rise to the modern conception of 'Alzheimer's Disease'.

The 'Alzheimer Culture'

Twenty-five years ago the name of Alois Alzheimer was known to only a small handful of specialists, and the main dementias of old age were generally labelled as 'senility'. Today Alzheimer is a household word, often carrying dreadful connotations; Alzheimer's Disease Societies have been established in virtually every country of the developed world, and in several of those that are fast developing. There is also a world-wide federation, Alzheimer International. How has this extraordinary transformation come about?

The crucial initiatives were taken in the USA between 1960 and 1980, at a time when there was the potential for mobilising vast resources for biomedical research (Fox, 1989). No additional information was needed to convert senility into Alzheimer's Disease, nor was there a shift from one scientific paradigm to another. It was a consequence, rather, of a set of pragmatic and political decisions, mainly in the attempt to attract funding for research in neuroscience, following the conspicuous success of lobbying around the issues of heart disease and cancer. The 'Alzheimer culture' came into being essentially out of a confluence of the interests of research groups and the concerns of family carers. This fact is reflected in the two types of discourse that still predominate in Alzheimer Society newsletters: the one technical, giving information about research findings, and the other sentimental describing carers' lived experience.

The neurological interest arose primarily from the application of electron microscopy to fresh brain tissue, where it soon became evident that much new knowledge would be forthcoming. Brain biopsies were needed, and those with a primary degenerative dementia were judged to be appropriate subjects (Terry, 1963). This and other work generated a surge of interest in the brains of dementing persons, and a new research lobby began to emerge. A key figure here was Robert Katzman, who entered this field in the early 1970s. Since there had been only limited success in making senility a focus of widespread concern, he made two crucial proposals in 1974. The first was that senility should thenceforward be re-named as Alzheimer's Disease: that is, taking over the term that had previously been applied only to certain pre-senile dementias. The second was to use epidemiological data to proclaim the disease as the fourth largest cause of death in the USA (confounding the categories 'dying from' and 'dying with' a dementia). In this way the needed focus was created, and with it a new medical–moral panic. In the same year coincidentally, the National Institute of Ageing was established. Katzman reiterated his thesis several times in subsequent years (for example, Katzman, 1976).

The recategorisation of what had been known as senility came to provide a major new resource for interpreting life history. The troubles, inadequacies and confusions of human existence, especially in later life, could now be re-framed as manifestations of the insidious advance of Alzheimer's Disease (Gubrium and Lynnott, 1985). Many biographies have been written along these lines (for example, Roach, 1985). The paradigm of this re-framing is that of the life of Rita Hayworth, the former Hollywood actress. The medicalisation of senility may have sharpened the work of medical scientists and practitioners; but it has also generated a pervasive (and perhaps unjustified) pessimism, as epitomised in such images as 'the prison that waits'[3] or 'the living death' (Woods, 1989).

Developments in dementia care

During the earlier part of this century there were no specific prescriptions for the care of those suffering from a dementing illness. When their behaviour became intolerable they, like others classed as insane, were to be taken into institutions, there to remain until their death. The creation of 'Alzheimer's Disease' as a broad category did little to promote better care practice. According to the new ideology care-givers would attend mainly to physical needs, while standing witness to the sufferer's tragic undoing, as he or she went through the 'stages of dementia' as variously described (for example, Reisberg, 1984).

Over the last 30 years or so there has accumulated a portfolio of intervention tactics viewed as appropriate to dementia care. This includes Behaviour Modification, Reality Orientation, Reminiscence, Stimulation, Validation Therapy, Resolution Therapy, Art Therapy, Music Therapy and even Psychodynamic Psychotherapy. (See, for example, Jones and Miesen, 1991.) If any historical trend can be observed here it is a move away from behavioural approaches, towards engaging with the dementing person's subjectivity, and the enhancing of his or her agency. All the most recent developments, such as Resolution Therapy (Stokes and Goudie, 1990) have been along these lines. Care-givers, for their part, often seem to live with a kind of 'doublethink'. On the one hand they hold to the disease formulation, with its dire implications. On the other, they use a different, and more optimistic, set of beliefs in their actual practice (Roth, 1980). The various intervention tactics provide rudiments of a theory appropriate to dementia care, but no more (Mace and Rabins, 1989).

The oldest, and perhaps the best known, of the interventions is Reality Orientation (Taulbee and Folsom, 1960). Its origins lie in the attempt to rehabilitate severely disturbed war veterans, not in geriatric work. The method as typically used involves the continual presentation of 'correct' information, often of a fairly banal kind, and sometimes actual classes modelled on an old-time primary school. Folsom himself, it must be said, never advocated this kind of one-sidedness, and attempted to address such issues as the patient's general well-being and self-worth. Reality Orientation has been reconstructed a number of times (Holden and Woods, 1988), and as it has moved away from a narrowly cognitive frame the earlier and more holistic emphases have begun to reappear.

Validation Therapy also is of long standing, originating in part as a reaction to some of the crudities performed in the name of Reality Orientation (Feil, 1962). The purported aim is to help the 'confused old-old' to resolve long-standing intra-psychic conflicts, rather as in psychoanalysis. To validate, literally, means to make strong or robust; thus in this context it implies accepting the subjective reality of another's experience, especially his or her feelings. Feil offers a set of specific techniques for those with various degrees of confusion. For a long time it was not clear whether the approach was deemed relevant to dementia, and although the answer is now clearly positive, there has been no attempt to take serious account of the effects of neurological impairment.

Evidence is now appearing which suggests that individuals can stabilise in a dementing illness, and that the stages of dementia are not an ineluctible path (Kitwood and Bredin, 1992); there is even some ground for claiming that a changed social-psychological environment may have benign neurological consequences (Karlsson *et al.*, 1988). However, evidence for the specific efficacy of the intervention tactics listed above is extremely slight. The methodological problems are notoriously difficult to resolve; and where effects do occur, these are attributable as much to general arousal of interest, and greater social involvement, as to the intervention *per se*. Belief in the efficacy of a specific 'therapy' may rest, to some extent, on observations too subtle to be measured by standard techniques. But also, and possibly more significant, care-givers need some framework within which to work with hope; and this is certainly not provided by the Alzheimer ideology.

The Disease, the Person and the Future

Even within its contemporary and restricted meaning, dementia is a deeply paradoxical category. The figures on prevalence show enormous variability, reflecting in part the arbitrariness with which a dementing condition is demarcated from 'normality'. Psychological and social factors which, if properly adduced, might help to rationalise some of the most serious anomalies are excluded from serious consideration. A disease, or a group of diseases, has been proclaimed to the world, but without meeting the accepted criteria of a disease entity: the symptoms are vague, the course is unclear, and the links with pathology are far less robust than medical science normally demands. Tactics in care practice have grown up largely without reference to psychiatry, and carers often hold to two sets of beliefs simultaneously, neither of them well justified by evidence. The present construction of dementia, then, is far from being the direct and logical consequence of biomedical science. It should be seen, rather, as a feat accomplished often in the face of countervailing evidence, and made possible by a unique conjuncture of social interests and economic opportunity. There is a powerful web of social forces which keeps the construction in place. The question now is whether those forces can withstand new pressures, as the century comes to its close: of disillusionment with medical science, of sharper theoretisation, of so-called care in the community, and of rising anger with inadequate provision.

Notes

1. See, for example, the definition of dementia proferred by DCM III (Lipowski, 1984) and by the NINCDS ARDRDA Working Group (McKhann et al., 1984).

2. This finding has been corroborated repeatedly, and still poses major problems for those who hold to exclusively biomedical views. (See, for example, Tomlinson et al., 1970, and Homer et al., 1988).
3. This is the title of a BBC Horizon programme first shown in 1988.

References

Ames, O., Dolan, R., and Mann, A. (1990) 'The distinction between dementia and depression in the very old', *International Journal of Geriatric Psychiatry*, vol. 5, 195–198.

Amster, L. R. and Kraus, H. H. (1974) 'The relationship between life events and mental deterioration in old age', *International Journal of Ageing and Human Development*, vol. 5, 51–55.

Arie, T. (1983), 'Pseudodementia' *British Medical Journal*, vol. 286, 1300–1302.

Barnes, E. R., Sack, A. and Shore, H. (1973) 'The cycle of dementia'. *Gerontologist*, vol. 13, 513–527.

Berrios, G. E. and Freeman, H. L. (eds) (1991) *Alzheimer and the Dementias*, London, Royal Society of Medicine.

Blessed, G., Black, S. E., Butler, T. and Kay, O. W. K. (1991) 'The diagnosis of dementia in the elderly: a comparison of CAMCOG (the cognitive section of CAMDEX), the AGECAT program, OSM-III, the mini mental state examination and some short rating scales', *British Journal of Psychiatry*, vol. 159, 193–198.

Feil, N. (1962) *Validation: the Feil Method*, Cleveland: Edward Feil Productions.

Folsom, J. C. (1968), 'Reality orientation for the elderly patient', *Journal of Geriatric Psychiatry*, vol. 1, 291–307.

Fox, P. (1989) 'From senility to Alzheimer's Disease: the rise of the Alzheimer's Disease movement', *The Millbank Quarterly*, vol. 67, 58–102.

Gilhooly, M. (1984) 'The social dimensions of senile dementia', in Hanley, I. and Hodge, J. (eds), *Psychological Approaches to the Care of the Elderly*, London: Croom Helm.

Gilleard, C. (1989) 'Losing one's mind and losing one's place: a psychosocial model of dementia', Address to the British Society of Gerontology, 10th Annual Conference.

Gurland, B., Copeland, J., Kuriansky, J., Kellerer, M., Sharpe, I. and Dean, L. L. (1983) *The Mind and Mood of Ageing: Mental Health Problems of the Community Elderly in New York and London*, London: Croom Helm.

Holden, U. and Woods, R. (1988), *Reality Orientation: psychological approaches to the confined elderly*, New York: Churchill Livingstone.

Homer, A. C., Honavar, M., Lantos, P. L., Hastie, I. R., Kellett, J. M. and Millard, P. H. (1988) 'Diagnosing dementia: do we get it right?' *British Medical Journal*, vol. 297, 894–896.

Ineichen, B. (1987) 'Measuring the rising tide: how many dementia cases will there be by 2001?' *British Journal of Psychiatry*, vol. 150, 193–200.

Jones, G. M. M. and Miesen, B. M. L. (eds) (1991) *Care-giving in Dementia: Research and Applications*, London: Routledge.

Karlsson, I., Brane, G., Melin, E., Nyth, A. L., and Rybo, E. (1988), 'Effects of environmental stimulation on biochemical and psychological variables in dementia', *Acta Psychiatrica Scandinavica*, vol. 77, 201–213.

Katzman, R. (1976) 'The prevalence and malignancy of Alzheimer's Disease', *Archives of Neurology*, vol. 33, 217–218.

Kay, O. W. K., Beamish, P. and Roth, M. (1964) 'Old age mental disorders in Newcastle upon Tyne: Part I, A Study of Prevalence', *British Journal of Psychiatry*, vol. 110, 146–158.

Kitwood, T. (1988), 'The technical, the personal and the framing of dementia', *Social Behaviour*, vol. 3, 161–179.

Kitwood, T. (1989) 'Brain, mind and dementia: with particular reference to Alzheimer's Disease', *Ageing and Society*, vol. 9, 1–15.

Kitwood, T. (1990) 'The dialectics of dementia: with particular reference to Alzheimer's Disease', *Ageing and Society*, vol. 10, 177–196.

Kitwood, T. and Bredin, K. (1992) 'Towards a theory of dementia care: personhood and well-being', *Ageing and Society*.

Kral, V. A., (1962) 'Senescent forgetfulness: benign and malignant', *Canadian Medical Assocation Journal*, vol. 86, 257–260.

Lipowski, Z. J. (1984), 'Organic mental disorders – an American perspective', *British Journal of Psychiatry*, vol. 144, 542–546.

Mace, N. and Rabins, P. V. (1989), *The 36 Hour Day: a family guide to caring for persons with Alzheimer's Disease, related illnesses and memory loss in later life*, New York: Warner Books.

McKhann, G., Drachman, D., Folstein, M., Kalzman, R., Price, D., and Stadlan, E. M. (1984) 'Clinical diagnosis of Alzheimer's disease: Report of the NINCDS–ADRDA Working Group under the auspices of the Department of Health and Social Services Task Force on Alzheimer's Disease', *Neurology*, vol. 34, 939–944.

Meacher, M. (1972) *Taken for a Ride: special residential homes for confused old people: a study of separatism in social policy*, London: Longman.

Morgan, R. F. (1965) 'Note on the psychopathology of senility: senescent defence against the threat of death', *Psychological Reports*, vol. 16, 303–306.

Oakley, D. P. (1965), 'Senile dementia: some aetiological factors', *British Journal of Psychiatry*, vol. 114, 414–419.

Reisberg, B. (1984) 'Stages of cognitive decline', *American Journal of Nursing*, vol. 84, 225–228.

Roach, M. (1985) *Another Name for Madness*, Boston: Houghton Mifflin.

Roth, M. (1980) 'Senile dementia and its borderlands', in Cole, J. O. and Barrett, M. D. (eds), *Psychopathology in the Aged*, New York: Raven Press.

Rothschild, D. and Sharpe, M. L. (1941) 'The origin of senile psychoses: neuropathological factors and factors of a more personal nature', *Diseases of the Nervous System*, vol. 2, 49–54.

Sabat, S. R. and Harré, R. (1992) 'The construction and deconstruction of self in Alzheimer's disease', *Ageing and Society*.

Stokes, G. and Goudie, F. (1991), 'Counselling confused elderly people', in Stokes, G. and Goudie, F. (eds), *Working with Dementia*, Bicester: Winslow Press.

Taulbee, L. and Folsom, J. (1960), 'Reality orientation for geriatric patients', *Hospital and Community Psychiatry*, 17, 133–135.

Terry, R. (1963), The fine structure of neurofibrillary tangles in Alzheimer's Disease', *Journal of Neuropathology and Experimental Neurology*, vol. 22, 629–642.

Tomlinson, B. E., Blessed, G. and Roth, M. (1970) 'Observations on the brains of demented old people', *Journal of Neurological Science,* vol. 11, 205–242.

Williams, H. W., Quesnel, E., Fish, V. W. and Goodman, L. (1942) 'Studies in senile and arteriosclerotic psychoses: relative significance of extrinsic factors in their development'. *American Journal of Psychiatry,* vol. 98, 712–715.

Wilson, D. C. (1955) 'The pathology of senility'. *American Journal of Psychiatry,* vol. 111, 902–906.

Woods, R. (1989) *Alzheimer's Disease: coping with a living death,* London: Souvenir Press.

12

Families and the experience of mental distress

DAVID W. JONES

Introduction

Donald Peters is in his mid-thirties; he has spent most of his life since his late teens in and out of psychiatric hospitals with a diagnosis of schizophrenia. This is an excerpt from an interview with his mother (MP) and sister Caroline (CP) in which they talk about how they have experienced this. Here they are talking about their relationship to the professional world and their knowledge of psychiatric illness.

What is apparent here is the anger, and the felt lack of communication with professionals, who apparently will not label this illness. Caroline and Donald's mother see Donald as suffering from schizophrenia and have educated themselves about this. Caroline makes a contrast with 'normal medicine', when even the distressing experience of having her father-in-law suffer from a brain tumour was eased by there being a process that was explained to them, and which they were guided through.

> *MP:* No, no-one's ever explained it. I've only learnt by going to lectures and reading books.
>
> *CP:* But they also have this idea that, as you say, they don't label the illness, they don't like to label the illness so therefore they won't tell you what it is ... that he's just ill. You see it's probably something in their training that they've got ... that it should be taken as far back as when they're all being trained for these jobs, as to how to deal with the families – I mean in normal nursing, when you're dealing with ... like my father-in-law when he had that brain tumour, the nurse was wonderful with us. She took us into a room, she explained exactly what was happening, the fact that it was malignant, what was going to happen to him, that at 76 hours he would be this – but that

during that time he wouldn't recollect anything. She went through the whole bit. Now that's what you need in mental illness ...

At first glance the demand for labelling and the association made with the brain tumour might seem hard to understand, given the great stigma attached to mental illness, and the distress and drama identified with cancer. How are we to understand this family wanting one of their number to be diagnosed as suffering from an apparently incurable and highly stigmatised illness?

Light is thrown on this conundrum by the series of questions that Caroline raises when she then explains what she needs from professionals:

> CP: You need somebody who will sit down and ... you can say – 'How can we deal with this, how are we meant to react, what do you want us to do?' We can only be there for Donald, and you go through these stages where Donald thinks he hasn't got a family, he doesn't want to know you, he'll throw you out of the place, he'll scream at you, he'll shout at you ... You need somebody, when at times like that happen, you know you're not immune to it all – it hurts. 'We know he's ill, can you explain to us what is going on in his brain that he is suddenly screaming and shouting at us, and abusing us and everything else, you know why? ... what can be done about it and what do you want us to do about it, except make nuisances of ourselves, with both them and with him' – because that's what you feel like?

Caroline is making an appeal for meaning in the face of manifestly distressing circumstances. She is having to cope with a lot of mixed emotions. There is certainly anger – her brother shouts at her and abuses her – and yet she obviously wants to be involved and cares.

Families have been subject to a great deal of research and speculation in the field of mental health. A lot of research has examined the family for the causes of mental illness (see Burnham, 1986 for review – *or other course material?*). More recently research has sought to quantify the burden experienced by families who care for someone suffering from mental illness (much of this is reviewed by Perring *et al.*) or has sought the families' views of services (Shepherd, 1994). Yet there has been very little attention paid to their point of view. This chapter is based on research interviews with people who have a close family member suffering from severe mental health problems, usually involving a diagnosis of schizophrenia.

Experience of loss

The seemingly most central and common experience was the feeling that the person the interviewees had known who had become ill had gone away: they had become like someone else. This defies our normal sense of the consistency of the self, where we see ourselves and others as, if not changing, at least as being part of a developing whole. Many of the relatives' experiences and views can be understood as an attempt to come to terms with this experience of discontinuity and loss. Mrs M., who has experience herself as a bereavement counsellor, is eloquent in talking about her difficulty in coming to terms with her husband's illness. She expresses the paradox that on the one hand she sees her husband, whom she is separated from, as having changed so fundamentally that it is as if he has gone away, but the knowledge that physically he is still around takes up mental space inside her, such that she could not have a relationship with someone else.

DJ: How do you feel about Alfred now?

Mrs M: Part of me has to see him sometimes. One thing is I can't grieve properly. If he had died you know what to do. Being a bereavement counsellor, I know what to expect, what to do and you can perhaps make a new beginning, but with Alfred he's never asked me whether I have a relationship with anybody. I haven't. But he just takes it for granted that I'm here. I don't think I have anything left for another relationship [...] If I see him, that shows me that he's not dead and that although my grieving, I know it's there, it's true. I find it difficult to know that he's in the world. He's not dead. Why's he not with us ... Because we'd like to be a normal family.

Grief and ambivalence

As Freud (1917) highlighted, there are at least two facets of loss in bereavement. Firstly there is the loss of the person that was, and secondly, and more complexly, there is the experience of the loss of the previous possibilities. When we lose someone close to us we forfeit something of our own future. For families, parents in particular, this feeling of the loss of future expectations they may have of their children can be terribly poignant.

In the following extract I ask a mother whose son was forty-five at the time of interview, and had been ill since mid-teens, whether she encounters difficulties in explaining how she feels to other people. Mrs C. makes revealing reference to her own (and her family's) appreciation of Peter in the past tense (1, 2, 3, 4 and underlined) as she jumps to talking about how Peter was some twenty-odd years earlier when he was at school (5). I think what is being expressed here is how very difficult it is for Mrs C. to reconcile those memories of her son as a successful schoolboy with the experience of him as he is now. This is an ongoing conflict, not open to easy reconciliation as she says: 'I don't think you can ever get used to it' (6). Mrs C. then says that she feels as though she is 'on the edge of the world' (8). Through this metaphor Mrs C. gives voice to the feeling that her experience has isolated her, that as others cannot comprehend her experience, she is left feeling excluded, marginalised on the edge of society. It is toward the fellow members of the National Schizophrenia Fellowship that she turns for comfort (7). It is there that she finds some common understanding of the long-term nature of the conflict.

> *DJ*: Is it something you find difficult to talk about to people because they don't understand?
>
> *Mrs C*: Sometimes yes. My sisters are very good and very concerned about Peter *(1)*, he was loved by all my family *(2)*; he was such a pleasant child *(3)*. He was never moody *(4)*, never had problems with him ... from an early age when he had homework he'd come home and start straight away *(5)*. My family do care. Some relatives I find think, I should be used to it. I don't think you can ever get used to it *(6)*, and a lot of our members [of the NSF], if you really talk to them *(7)*, they would say that you just learn to cope with the rest of the family or for each other, you learn to cope that's all you do. You're living on the edge of the world sometimes *(8)*, but people find it very hard, I think, to know how I'm feeling, I don't wear my heart on my sleeve. I tend to say 'I'm OK'.

For some time now grief has been construed as being the process through which people accommodate to loss and find fresh meaning (Murray Parkes, 1972). The grief of these relatives is complicated by certain features: as already highlighted by Mrs M., the fact that the lost person has not really gone away; the presence of strong emotions, such as anger, that can be difficult to manage; and the stigma and shame of mental illness which means communicating with others about how they feel can be difficult, hence Mrs C. feeling that she is 'on the edge of the world'.

Theories of cause

Clearly, there are complex feelings involved here. One important way that the effects of these feelings can be mediated is through making sense of what has happened. Typically families seem to seek medical help, which would suggest their perceptions of events are consistent with a medical definition of mental illness. Within this definition, however, families may have complex theories of what has caused their relatives' difficulties. In thinking about theories of cause, relatives are well aware of the moral implications of different theories.

Some people, who usually had reason to distance themselves from family disorder, did blame their own family environment. For example, a woman who had been left by her husband blamed this trauma for her daughter's illness. Another woman struggled with the idea of 'family cause' in considering her sister: if the family environment was to blame this made her partly responsible. On the other hand if it was 'organic' she felt less obliged to make sacrifices to help. Theories of organic cause were generally, from the families' point of view, the most 'morally neutral'. Of course, from a government's and service providers' point of view, the reverse is the case.

Sam (SM), in talking about his brother, is well aware of the power machinations that may lie behind the application of different diagnoses. He also feels that he is not being listened to, his contribution is not valued.

SM: ... I don't think they understand it to be honest, to be honest I don't think they really understand mental illness, because when I'm talking to some of the psychiatrists ... they are mad, really! [laughing]. They really are, they are crazy. Because you'll be telling them, you'll be, you'll be the member of the family and you'll be saying 'This person is doing this and this person is not doing this, they're not thinking in this way', and they'll be saying 'There's nothing wrong, they've just got a slight behavioural problem!' [laughing].

DJ: Why do you think they said that?

SM: Because they are mad! [laughing] ... No, I don't know how much society really wants to care for these people, and sometimes I think that they [think], 'Yeh fob them off to the family, let the family deal with them'. If they do say there is something wrong then they may feel that they have to do something about it and that may cost time and money, or whatever. And the system is not geared for that, the system is not really geared for that, so the professional people do say ... his doctor, his doctor said he had 'a slight behavioural problem' and this is after years of going in and out of hospital, after years of that doctor seeing him and giving that diagnosis that he was schizophrenic ... he's going it wasn't a behavioural problem then he said something like 'he's extrovert' [laughing], this is before the last admission into hospital! So I'm led to the conclusion that they are crazy, they're absolutely crazy!

Coming to terms: shared understandings

To accept a medical diagnosis may free people from debilitating feelings of guilt and responsibility. Such diagnoses may also give access to resources. Yet of course such diagnoses may have different implications for the person themselves. Mrs Peters and Caroline, who introduced this chapter, discuss how the development of a shared understanding of events with Donald and with the professionals they are now currently involved with is extremely important to them.

For families to reach a degree of acceptance often means accepting apparently pessimistic scenarios. Yet the belief in the idea of medical illness helps Caroline cope with stigma.

CP: I think there is another thing as well, that we've accepted that he has a mental illness, and that it has to be dealt with like any other illness or if anyone else was in hospital you go and visit them whether it's cancer or a mental illness and I think, from our point of view, yes we accept there is mental illness and we deal with that accordingly. I think we don't really differentiate between that illness or any other. I mean, you know when I'm in hospital, mummy comes to see me if there is something wrong and I think that probably is also there, apart from being her son, it is also something that you naturally do for a friend or for a brother or for anybody which you know well, you're not going to get put off by the stigma attached because that is other peoples' failings, I think. Not ours, not as a unit anyway.

During the interview I noticed how crucial to them this acceptance seemed to be. When I suggest this, the point is lukewarmly agreed with (1), but what seems to be felt as even more crucial to them is Peter's own acceptance of his illness status (2). This, in retrospect, is regarded as a crucial moment in the family finding a more even keel:

DJ: It sounds as though your accepting of him becoming somewhat different has been important in being able to cope.

CP: Um ... yeah ... I don't know, I suppose so *(1)*. Because not that it's important, I just think that is something that you have to do, it's not that it's important I just think ...

MP: It was when he accepted it, that was a great milestone, when he accepted that he was ill *(2)*, he didn't for a long-time ...

CP: Yes, that's right ... He didn't ... one's been through so much that ... there's so much that's gone through and you tend to feel ... 'well at what stage did you accept it?' and I suppose one accepted it when he accepted it. It was a big thing for all of us.

DJ: What happened then?

CP: Nothing ... he just started talking about it openly.

MP: Talking about it openly which he never did before, and he wouldn't accept that there was anything wrong with him.

CP: You know he kept saying that ...

MP: 'It was all rubbish, everyone was', ... I can't remember ... 'the doctors were making it all up' ...

CP: Yes, that 'they were victimising him and that work were victimising him, the company were victimising him, they wanted to put him somewhere where they wouldn't have to look at him', and all this sort of thing ... it was all sort ... everybody else ... And then all of a sudden he just started talking about ...

MP: I can't remember now ...

CP: the hospital and the fact that he couldn't work and he knew he couldn't work

MP: And never will work.

CP: ... and never will work and that really one's got to look at it like he were an invalid, remember that? ... When he was going on about 'really I'm an invalid' ...

MP: Mmmm ... well he does get a disability pension.

CP: yes ... he wasn't going on about the pension but he was going on about the fact that he was an invalid.

MP: Well, he was quite pleased about that!

CP: Yes, you see this is the thing, you know he's still our brother and that's all there is really.

Whilst a medical model of events is commonly accepted by virtually all the people interviewed, what I want to highlight here is that what seems to lead this to be a less troubled situation is not just acceptance of the medical model of events, nor the acceptance of the long-term nature of those difficulties, but the acceptance of the long-term nature of the changes at an everyday emotional level. Although recognising that he has changed, this family is able to accept Donald as the person that he is.

What the Peters family seem to be telling me here is that they feel they have developed a shared narrative. The family's acceptance is apparently matched by Donald's own. The understanding is shared by them, by Donald, and by the professionals they (now) have contact with. They share a discourse, which functions to explain what has happened: why Donald lives as he does and behaves as he does, and why he (to an extent) needs looking after. For the family, there is a coherent web of meaning that holds events together. They are therefore able to relate in a real emotional way (with affection for example) to the person that Donald is. To Carol Peters, Donald is someone whose sense of humour and artistic abilities, as well as occasional violence and strange ideas, are all parts of a person that is 'still our brother'.

Yet much of this is uncomfortable, perhaps particularly for professionals who work on the principle of therapeutic optimism. Construing a relative as suffering from an illness that has irreversibly changed them so that they are an invalid who will never work clearly has implications.

Nevertheless it is important that professionals understand what function this apparent 'pessimism' serves. Family members benefit from being able to develop shared understandings of what has happened with the sufferer and with professionals. This can only occur through dialogue and the sharing of information. This point again raises ethical dilemmas: who are the professionals working with, and whose confidentiality has to be respected – the identified patient or the family?

A possibly useful strategy might be to encourage involvement in mutual support groups. In this environment families may be able to support each other and realise their own experiences and feelings are not unique. However achieved, if people are able to acknowledge their own, sometimes very negative feelings, they will be better able to build meaningful relationships again with those suffering from severe mental health difficulties.

Referencess

Burnham, J. B. (1986) *Family Therapy: first steps toward a systemic approach*, London, Tavistock

Perring, C., Twigg, J. and Atkin, K. (1990) *Families Caring for People Diagnosed as Mentally Ill: the literature re-examined*, London, HMSO.

Shepherd, G., Murray, A. and Muijen, M. (1994) *Relative Values: the differing views of users, family carers and professionals on services for people with schizophrenia in the community*, London, The Sainsbury Centre for Mental Health.

Freud, S. (1917) *On Mourning and Melancholia*, The Penguin Freud Library, vol. 11. Harmondsworth, Penguin.

Murray Parkes, C. (1972) *Bereavement*, Harmondsworth, Penguin.

Part II

MENTAL HEALTH POLICY, SOCIAL INEQUALITY AND CIVIL RIGHTS

13

Introduction

In this part of the book, Roger Gomm (Article 14) gives an epidemiological survey of diagnosed mental illness. This demonstrates that 'mental illness' has essentially the same relationship with deprivation that is shown by other kinds of illness. The lesson usually drawn from mapping ill-health against social deprivation is that the appropriate policies for promoting health are economic and social rather than curative and medical. In this sense, much mental health policy has become policy only for dealing with 'mental illness'.

Most of the other articles in this part are about policies which reflect this public policy conceptualisation of 'mental health'. This is usually in a narrow sense defined as the problem of curing and controlling distressed individuals, rather than as a matter of providing populations with the resources with which they may be able to lead satisfactory lives. However, as Linda Jones shows (article ??) in her overview of the historical period up to the beginning of this century, isolating deviants (see also Scheff, Article 8) has been the predominant pre-occupation for a long time.

The nineteenth century was the age of the large lunatic asylums, when mental illness and incarceration became equated with each other. More recently there has been considerable debate considering the move away from incarceration and towards community care. Joan Busfield's article (16) gives one interpretation, emphasising the influence of organised professionals on public policy. Specifically she cites the influence of psychiatrists seeking to move out of a ghetto of mental hospitals and into the more attractive sphere of 'the community'.

The first half of Article 17 by Matt Muijen gives an account of policy initiatives leading to the run-down of mental hospitals, as well as noting the importance of scandals and inquiries which further tarnished their images. Muijen goes on to focus on the way in which current mental health policy is being driven by public and political reaction to a handful of high-profile homicides by 'mentally ill' people.

Julian Leff's article (18) takes up the theme of hospital closure. He gives a synopsis of the lessons learned for the successful resettlement in the community of long-stay mental hospital patients from research on the closure of Friern Barnet and Claybury Hospitals.

Central to Muijen's article is the image of 'mental illness' as dangerousness. The article (19) by Greg Philo and his colleagues, of the Glasgow University Media Group, investigates the way in which the mass media disseminate these images of dangerousness. In fact, people with a mental illness diagnosis are rarely a danger to others, and much more likely to be a danger to themselves through acts of self-harm or suicide. These are the topics of Article 20 by Steve Taylor and Alan Gilmour.

Risk assessment has become a key requirement of mental health work. David Pilgrim and Anne Rogers (Article 21) show that practitioners do not have the means to make accurate predictions as to who might constitute a danger to others. They remind us that mentally distressed people are also at risk from professional interventions in their lives. They argue that concentrating on the harm that mental patients might do to others or themselves reflects attention away from the harm done to people in treating them as 'mental patients'. This is an issue already touched upon by Scheff, Gomm and Rosenhan in Part I of this volume. The companion volume to this one, *Speaking Our Minds*, also contains many personal accounts of distressing experiences of psychiatric treatments.

Jonathan Gabe's article (22), on the effects of minor tranquillisers, deals in detail with one kind of damage arising from treatment, and with the vested interests of the pharmaceutical industry. The effects of another kind of damage – being stigmatised as mentally ill – are demonstrated in the article by Philo and his colleagues.

Deryck Browne (Article 23) draws attention to the apparently disproportionately coercive way in which younger African and Afro-Caribbean people are dealt with under the provisions of mental health law, paralleling their treatment by the Criminal Justice system. Here the image of dangerousness associated with mental illness is added to the more general stereotype of dangerous Black people.

Nigel Eastman's article (24) is his personal commentary on a conference on mental health law held in 1993, while the 1995 Mental Health (Patients in the Community) Bill was being discussed in Parliament. As enacted, this led to the introduction of supervised discharge in England and Wales in April 1996, with similar provisions for Scotland. Supervised discharge seems to add very little to the range of powers available to mental health workers – there are ways of accomplishing exactly the same results using pre-existing legislation. As Muijen indicates, the main effect of this legislation is to make mental health workers more accountable for the behaviour of mental patients, while giving them no adequate resources for discharging these responsibilities. However the legislation does serve to heighten concern about the coerciveness of mental health policy. This is also Eastman's concern. Like Pilgrim and Rogers, and Browne, he notes that coercion, via mental health law, often seems to have no benefits for the person coerced. Central to his thinking is the 'principle of reciprocity', recognised in several legislatures of the USA, that people with mental health problems should not be deprived of their liberty unless services can offer them corresponding benefits in the way of improvement to their lives.

14

Mental health and inequality

ROGER GOMM

Introduction

The link between health and material conditions is well-known and well-documented (for convenient summaries see Benzeval *et al.*, 1995; Townsend, Whitehead and Davidson, 1992). A very simple statement will serve to summarise all the research findings on this matter: for nearly every kind of illness, disease or disability, 'physical' and 'mental', poorer people are afflicted more than richer people: more often, more seriously and for longer – unless, of course, they die from the condition, which they do at an earlier age.

> There would be 42 000 fewer deaths per year for people aged 16–74 if the death rates for people with manual jobs were the same as for those in non-manual occupations. (Jacobson et al., 1991, cited in Benzeval *et al.*, 1995, 4)

Not only have links between material inequality and health been persistently obvious for the last 130 years (Farr, 1860) but inequalities in health have been widening recently.

Dis-ease rather than disease

'Physical' and 'mental' health both show the same kind of relationships to differences of material condition. Some obvious 'life and death' matters in the mental health field contribute directly to social class differences in life-expectancy. Suicide, which is now the second most important cause of death for younger men, shows a very strong relationship to social class and deprivation (see Taylor and Gilmour in this volume). Suicides are also closely linked to alcoholism and addiction to illicit drugs (Morgan and Owen; 1990) both officially regarded as mental health issues, and both

110

more common among the unskilled, the unemployed (Meltzer *et al.*, 1995a) and the homeless (Stark et al., 1989). Excessive use of alcohol is closely linked to both accidental and violent deaths. People who are depressed, or otherwise mentally distressed, show elevated accident proneness without medication, and an even greater proneness to life-threatening accidents when taking anti-depressant medication (Freeman and O'Hanlon, 1995). The children of mothers suffering from depression are also more vulnerable to accidents (Pound *et al.*, 1988): accidents and violence account for around 50 per cent of deaths of children under 14. Depression, which is probably the most common form of 'mental illness' in terms of the number of people affected, shows a very strong social class profile (see Brown in this volume). The poorer you are the more likely you are to be miserable enough to be regarded as clinically depressed. Seriously self-harming behaviour – at least ten times more common than suicide and often a precursor – also shows the same social class pattern (Gunnell *et al.*, 1995).

Although less closely affecting life-expectancy, neurotic states are more common lower down the social class scale (Meltzer *et al.*, 1995a). Schizophrenia is 3 times more common among those from working class origins than among those from other backgrounds (Barbigian, 1985). Its strong association with social class has led some researchers to suggest pre-natal infection (possibly rubella or influenza) or obstetric complications as causative factors (Eagles, 1991). Both are more likely in the impoverished populations of developed countries. However schizophrenia is too rare a condition to have much influence on the overall pattern. Much more important in numerical terms are the dementias of later life. Alzheimer's disease particularly, and to a lesser extent Multi-infarct dementia (MID), again show a greater prevalence among lower classes: although here years of education seem to be the associate rather than material conditions as such (Ott *et al*, 1995), and the linkage between social class and dementia tends to be less obvious since fewer lower-class people survive into their eighties, when the risk of dementia increases dramatically.

There are some 'mental illness' conditions which show a reverse pattern: anorexia nervosa – though a life-threatening condition – seems more common among well-educated, middle-class young women (Cohen and Hart, 1988:343), and there are some conditions which show no clear social-class pattern. Manic depression may be more common among those from middle-class backgrounds (Cohen and Hart, 1988, 87), but people who get diagnosed as manic depressives themselves often end up among the poorly paid or unemployed (Giggs and Cooper, 1987). So despite some contrary examples something like the earlier summary will serve: that for nearly every kind of 'mental' illness, disease or disability, and especially those which afflict large numbers of people, poorer people are afflicted more than richer people, more often, more seriously and for longer. Embroidered on this general pattern of social-class inequalities are inequalities of gender (Judge and Benzeval, 1993; Glendinning and Millar, 1992) ethnicity (Smaje, 1995) and age (Arber and Ginn, 1993).

'Mental' and 'physical' illnesses are frequently found together in the same person (Meltzer *et al.*, 1995b). Judged by their elevated rates of mortality from causes in addition to suicide, and from morbidity studies, people regarded as 'mentally ill' are often 'physically unwell' (Brugha et al., 1988), while those who are diagnosed as physically unwell or handicapped are often very distressed mentally (Broome, 1989). In later life the association of depression with physical ailments, loss of continence, loss of ambulation and hearing loss is well known (Arie, 1988), as is the association between depression, cognitive impairment and/or confusion with infection, drug side-effects, poorly controlled diabetes and dietary deficiencies (Holland and Rabbitt, 1991). Poorer sections of the elderly population are either more likely to contract the condition concerned, or less likely to receive adequate treatment for it (Arie, 1988).

The links between 'physical ailments' and 'mental' conditions are many and complex. One source of linkage should be obvious from the discussion so far – a triangular relationship between social and economic conditions, physical illness and psychological stress:

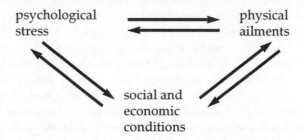

If you take any social group, or any neighbourhood with a high rate of premature death, or coronary heart disease, or long-term and limiting illness, or gastric ulcers, or childhood accidents, and so on, then the chances are that the group or the neighbourhood will be a poor one, and that it will also show a high rate of suicides and depression and anxiety states and schizophrenia. Research on local patterns of health and socioeconomic circumstances routinely demonstrates that the census wards which generate most cases for psychiatric admissions or most cases of suicide also tend to be those with the highest death rates (all causes), highest rates of low birth-weight babies, and highest rates of coronary disease (see, for instance, Townsend *et al.*, 1985; or Townsend *et al.*, 1988).

Material conditions cause ill-health: ill health causes material conditions

In this unholy triangle causality runs in all directions. Poorer people are more vulnerable to physical illnesses and physical disabilities; being physically ill or disabled is depressing and anxiety-provoking. Poorer people are more prone to 'mental illnesses': 'mental illness' lowers immunity (Lewis *et al.*, 1994) and is often associated with health-damaging behaviour such as excessive smoking and drinking, careless and risky activities, self-neglect, malnutrition and being roofless or without income. Once physically ill, psychological disturbance impedes recovery. The side effects of psychoactive drugs might be regarded as a physical ailment to which people with a diagnosis of mental illness are highly vulnerable (see Gabe in this volume). And, to complete the triangle, ill health of whatever kind not only arises from social and material deprivation, but sometimes destines people to unemployment, low incomes and poor housing or no housing at all, all of which make a negative contribution to physical and mental health.

The links between physical ill-health, mental ill-health and material deprivation are not automatic. If they were Scotland, which has the lowest average incomes and the highest death rate for the mainland UK, would also have the highest rates of mental illness, which it does not. And Greater London, where incomes are higher and death rates are lower, would not have a higher rate of mental malaise than Scotland (Lewis and Booth, 1992). Three kinds of factors mediate between material conditions and health. One of these is the 'life-style' factors so often addressed in health education programmes. Scotland does show higher rates of drinking, smoking and people with diets predisposing them to coronary disease (Scottish Office, 1992). This may explain the co-presence of high death rates with rather lower rates of mental illness than might be predicted from them. Among life-style factors are ways of coping with the stresses of life, and these do vary with social class (West, 1995). However Blaxter (1990) and others have argued that the first step towards adopting a healthier life-style is usually an improvement in material conditions.

The second set of factors is constituted by the expectations people have of life and the comparisons they make between themselves and others. Wilkinson (1992) and others have suggested that while gross material inequalities are important to health, so also are relative inequalities, perceived by people as frustrations of expectations and as indications of their own failure.

The third important kind of mediating factor is the quality of social relationships (West, 1995). These may be independent of material conditions as in a close, but poor, community, but equally, isolation and alienation and fraught relationships may be an associate of poverty as in many inner urban areas and out-of-town council estates.

Thus while there is no automatic link between material conditions and dis-ease of the physical or the mental kinds, poorer people are vulnerable in multiple ways by comparison with the better off.

Long-term health effects of material deprivation

Research seems to show that there is some very rapid feedback between the pattern of inequality and the pattern of health: for example, between 1921 and 1981, social class differentials in death rates rose and fell almost exactly in tune with rising and falling differentials in income (Wilkinson 1992). Since the war suicide among young men has shown a close relationship with rates of unemployment. But there is also evidence that past deprivation may show itself many years later in the form of ill health. For example, of men born in Hertfordshire in the 1930s, those who showed a poor weight gain in their first year of life were appreciably more likely to have committed suicide 50 years later (Barker *et al.*, 1995). Low weight-gain is a rough and ready indicator of family poverty. By 1983 the mortality pattern for men born in Sweden in 1933 appeared closely related to their experience of earlier adverse life events. Most such life events were of the kind experienced more by people from manual backgrounds. The greater the number of adverse events, the more likely were the men to be dead before the age of 50 (Rosengren *et al.*, 1993). The various cross-sectional and longitudinal child development studies in Britain show much the same, charting the effects of low birth weight, early childhood illness, family poverty, family disruption, and mentally distressed parents many years later in such multifarious ways as vulnerability to illness in general and to coronary, gastric, liver and depressive illnesses in particular, inability to cope with life crises, poor educational achievements, low levels of occupational achievement, criminality, alcoholism and difficulties in forming relationships (Wadsworth 1991; Pound *et al.*, 1988).

As an antidote to this depressing picture it is important to note that adverse circumstances in childhood merely predispose towards adverse outcomes in adulthood. Equally important is what happens to people throughout their lives, and particularly their opportunities for achieving self-esteem and making satisfactory relationships (Rutter, 1985; Rosengren *et al.*, 1993). The implication of long-term studies such as these is that adversity now breeds adversity later, sometimes much later in advanced old age, but it is never too late to attempt a remedy.

Increasing inequality in health

In Britain social-class differences in life-expectancy have been widening since the 1950s and the gap showed a marked increase in the 1970s (Townsend *et al.*, 1992; Wilkinson, 1992). Up to the end of the 1970s all social-class groups increased their life expectancy and lived longer than their equivalents at an earlier time, but the better-off increased theirs more rapidly than the poorest: everyone a winner, as it were, but some winning more than others. However the picture emerging for the 1980s seems different and more worrying. Phillimore *et al.* (1994), on the five counties of northern England for example, show that between 1981 and 1991 the very well-off increased their life-expectancy markedly, there were modest gains for people in areas of moderate affluence, but for some age groups in the poorest 10 per cent of electoral wards life expectancy was the same as, or worse than, at the end of the 1970s. For young men it was worse than it was in the 1950s.

Forwell (1992) and McCarron *et al.* (1994) on Glasgow show much the same. So does the study of the poorest post-code districts in Scotland (McLoone and Boddy, 1994). Studies from other areas in the UK, as yet unpublished, seem to show that this is a national phenomenon.

Lewis and Wilkinson (1993) show that, as measured by the General Health Questionnaire, mental health in West London deteriorated between 1977 and 1985, and make a good case for these findings to be generalised to England as a whole. While their study does not actually show increasing social-class inequality in mental health, it would be very surprising if this were not happening, since there is a widening gap for 'physical health' (see above), and the physical and mental health status of populations is usually closely related.

Widening material inequalities

It is not difficult to chart the growth of material inequality which lies behind increasing inequality in health. In Britain, throughout the 1980s and into the 1990s, the income gap between the richest and the poorest widened. There was a substantial redistribution of wealth and income in favour of the rich, which left the poorest 10 per cent of the population in 1991 with less purchasing power than the poorest 10 per cent of the population had in 1979 (DSS, 1993; Hills, 1994). Bradshaw *et al.* (1992) show that by 1991 the value of social security benefits had declined so that they were insufficient to purchase adequate food and clothing. The same period of time has shown the 'feminisation of poverty', with a large rise in female-headed single parent families (DSS, 1994) and with the growth of the population over 70, which is predominantly female and disproportionately poor.

A wide range of factors have consolidated socio-economic disadvantage. These include the effects of unemployment and of increasing insecurity of employment, which impact disproportionately on manual workers, new entrants to the labour market, older workers and ethnic minorities and, of course, create a new poor of once well-paid, but now unemployed or prematurely retired people (Warr, 1987). Unemployment rates are a good predictor of the mental health of a neighbourhood (Kammerling and O'Connor, 1993). There has been legislation to reduce the bargaining power of employees and keep down wages.

There is some evidence to suggest that working conditions have been deteriorating, as a result both of attempts to cut costs and raise productivity, and as an increasing proportion of work is provided through small and medium sized, rather than large, employers, and through self-employment. Job insecurity and casualisation have increased markedly. Job insecurity appears to effect health adversely, more so when there are unconfirmed rumours about redundancy (Ferrie *et al.*, 1995). The Demos Report of 1995 found 41 per cent of a large representative sample of employed people 'fairly or very concerned' about being made redundant over the next 12 months (*The Times*, 6 June 1995). The same survey reported considerable stress with regard to long working hours and the pressure to achieve more in less time with fewer resources. The lower the occupational category the more people are likely to experience physical strain, serious injury and high levels of chemical and noise pollution (Hasan, 1989). Research has consistently shown that those who experience most workplace stress and illness related to this are low-placed employees with little discretion over the way they work. The latter has been associated with vulnerability to coronary illness, which in turn is correlated with mental illness conditions (Karasek and Theorell, 1990).

Another increasing inequality derives from the declining service and rising prices for public transport and the movement of shopping facilities to out-of-town sites, both of which marginalise those who cannot afford to own cars.

Housing has always been shown to be a crucial determinant of health (Best, 1995). Over 16 years to 1996 the main emphases in housing policy have been to encourage owner-occupation and to restrict public expenditure. Housing was one of the two areas of government expenditure which fell substantially in real terms up to 1995; the other area was defence. There has been an increase in the number of people who cannot afford to keep up a mortgage or maintain their properties adequately, a gross shortage of affordable housing for rent, very large rent rises in the public sector and high levels of homelessness. The same housing policy has tended to concentrate the poorest and most vulnerable people into neighbourhoods of intense deprivation with high rates of vandalism, crime, drug use and racial tension (Power and Tunstall, 1995). The government's inner-city improvement programme, costing over £10bn, was described in the official evaluation report as doing much for property developers and little for the well-being, health, education and safety of inner-city populations (Robson and Parkinson, 1994).

Conclusion

In much of the literature, although not all of that cited here, and even more so in service planning and delivery, 'physical health' is divided from 'mental health' and then each category is chopped up finely into discrete diagnoses. This has the effect of tearing out the details and viewing them in isolation from the pattern as a whole. Thus studies which investigate only the social correlates of coronary heart disease fail to note the way in which the same circumstances predict high levels of mental distress. Studies which focus on the epidemiology of mental illness may miss the fact that populations vulnerable to 'depression' or 'anxiety states' are also vulnerable to 'physical' diseases and handicaps. I use the term 'population' advisedly here, because the distribution of dis-ease is a population phenomenon. It may be that someone is 'mentally ill' without impairment of physical health, but if he or she comes from a vulnerable population, there will probably be a spouse, or a parent or a child or a neighbour whose 'physical health' is indeed impaired and for the same complex of reasons.

Fragmenting dis-ease into discrete diagnoses is questionable even where treatment is the aim. It colludes with a tendency to treat diseases rather than people. In terms of health promotion or the prevention of illness, to think of communities as divided first into individuals who are sick or well, then into individuals who are physically ill or mentally ill, and then into individuals sorted into diagnostic categories is not only questionable, it is downright misleading. Its main effect is to distract attention away from the social processes which distribute the wherewithal that people need in order to lead healthy lives.

References

Arber, S. and Ginn, J. (1993) 'Gender and inequalities in health in later life' *Social Science and Medicine*, vol. 32 (4), 425–436.

Arie, T. (1988) 'Questions in the psychiatry of old age', in Evered, D. and Whelan, J. (eds), *Research and the ageing population*, Ciba Foundation Symposium, 134, Chichester, Wiley.

Baraclough, B. and Hughes, J. (1987) *Suicide: clinical and epidemiological studies* Beckenham, Croom Helm.

Barbigian, H. (1985) 'Schizophrenia epidemiology' in Kaplan, H. and Saddock, B. (eds) *Comprehensive textbook of psychiatry: IV*, Baltimore, Williams & Wilkins.

Barker, J., Osmond, C., Rodin, I., Fall, C. and Winter, P. (1995) 'Low weight gain in infancy and suicide in later life', *British Medical Journal*, vol. 331, 1203–1204.

Benzeval, M., Judge, K. and Whitehead, M. (eds) (1995) *Tackling Inequalities in Health: an agenda for action*, London, King's Fund.

Best, R. 'The Housing Dimension' in Benzeval *et al.*, *Tackling Inequalities*, 53–68.

Blaxter, M. (1990) *Health and Lifestyles*, London, Tavistock/Routledge.

Bradshaw, J., Hicks, L. and Parker, H. (1992) *Summary Budget Standards for Six Households, Working Paper*, 12, Family Budget Unit York, Department of Social Policy, University of York.

Broome, A. (ed.) (1989) *Health Psychology: processes and applications* London, Chapman Hall.

Brugha, T., Wing, J. and Smith, B. (1988) 'Physical Health of the long-term mentally ill in the community' *British Journal of Psychiatry*, vol. 155, 777–781.

Cohen, R. and Hart, T. (1988) *Student Psychiatry Today: a comprehensive textbook*, Oxford, Butterworth Heinemann.

DSS (1933) *Households below Average Income 1979–1990/91: a statistical analysis*, Government Statistical Service, London, HMSO.

DSS (1994) *Households Below Average Income 1979–1991/92: a statistical analysis*, Government Statistical Service, London, HMSO .

Eagles, J. (1991) 'The relationship between schizophrenia and immigration: are there alternatives to psychosocial hypotheses?', *British Journal of Psychiatry*, vol. 159, 783–789.

Farr, W. (1960) 'On the construction of life tables; illustrated by a new life table of the healthy districts of England' *Journal of the Institute of Actuaries* IX.

Ferrie, J., Shipley, M., Marmot, M., Stanfield, S. and Davey Smith, G. (1995) 'Health effects of anticipation of job change and non-employment: longitudinal data from the Whitehall II study', *British Medical Journal*, vol. 311, 1264–1269.

Forwell, G. (1992) *Annual Report of the Director of Public Health for Glasgow*, Glasgow, Greater Glasgow Health Board.

Freeman, H. and O'Hanlon, J. (1995) 'Acute and subacute effects of antidepressants on performance', *Journal of Drug Development and Clinical Practice*, vol. 7, 7–20.

Giggs, D. and Cooper J. (1987) 'Ecological structure and the distribution of schizophrenia and affective psychoses in Nottingham', *British Journal of Psychiatry*, vol. 151, 627–663.

Glendinning, C. and Millar, J. (eds) (1992) *Women and Poverty In Britain: the 1990s*, Hemel Hempstead, Harvester Wheatsheaf.

Gunnel, D., Peters, T., Kammerling, R. and Brooks, J. (1995) 'The relation between parasuicide, suicide and psychiatric admission and socioeconomic deprivation', *British Medical Journal*, vol. 311, 226–230.

Hasan, J. (1989) 'Way-of-life, stress and differences in morbidity between occupational classes', in Fox, J. (ed), *Health Inequalities in European Countries*, European Science Foundation, Aldershot, Gower, 372–385.

Hills, J. (1994) *Inquiry into Income and Wealth Chaired by Sir Peter Barclay*, vol. 1, York, Joseph Rowntree Foundation.

Holland, C. and Rabbitt, P. (1991) 'The course and causes of cognitive change with advancing age' *Reviews in Clinical Gerontology*, vol. 1, 79–94.

Jacobson, B., Smith, A. and Whitehead, M. (eds) (1991) *The Nation's Health: A strategy for the 1990s*, London, King Edward's Hospital Fund for London.

Judge, K. and Benzeval, M. (1993) 'Health Inequalities: new concerns about the children of single mothers', *British Medical Journal*, vol. 306, 677–680.

Kammerling, R. and O'Connor, S. (1993) 'Unemployment rate as a predictor of psychiatric admission', *British Medical Journal*, vol. 307, 1536–1539.

Karasek, J. and Theorell, T. (1990) *Healthy Work: stress, productivity and the reconstruction of working life*, New York, Basic Books.

Lewis, C., Sullivan, C. and Barraclough, J. (1994) *The Psychoimmunology of Cancer*, Oxford, Oxford University Press.

Lewis, G. and Booth, M. (1992) 'Regional Difference in Mental Health in Great Britain' *Journal of Epidemiology and Community Health*, vol. 46, 608–611.

Lewis, G. and Wilkinson, G. (1993) 'Another British disease? a recent increase in the prevalence of psychiatric morbidity', *Journal of Epidemiology and Community Health*, vol. 47, 358–361.

McCarron, P., Davey Smith, G. and Womersley, J. (1994) 'Deprivation and Mortality in Scotland: changes from 1980–1992', *British Medical Journal*, vol. 309, 1481–1482.

McLoone, P. and Boddy, F. (1984) 'Deprivation and Mortality in Scotland 1981 and 1991' *British Medical Journal*, vol. 309, 1465–1470.

Meltzer, H., Gill, B., Petticrew, M. and Hinds, K. (1995a) *The prevalence of psychiatric morbidity among adults living in private households*, OPCS Surveys of Psychiatric Morbidity in Great Britain, Report 1, London, HMSO.

Meltzer, H., Gill, B., Petticrew, M. and Hinds, K. (1995b) *Physical complaints, service use and the treatment of adults with psychiatric disorders*, OPCS Surveys of Psychiatric Morbidity in Great Britain, Report 2, London, HMSO.

Morgan, G. and Owen, J. (1990) *Persons at Risk of Suicide: guidelines on good clinical practice*, Bristol, Department of Mental Health, University of Bristol.

Ott, A., Breteler, M., van Harsksmp, F., Clans, J., van der Cammen, T., Grobbe, D. and Hoffman, A. (1995) 'Prevalence of Alzheimer's Disease and vascular dementia: association with education', *British Medical Journal*, vol. 310, 970–973.

Phillimore, P., Beattie, A. and Townsend, P. (1994) 'Widening inequality in health in Northern England 1981–91', *British Medical Journal*, vol. 308, 1125–1128.

Pound, A., Puckering, C., Mills, M. and Cox, A. (1988) 'The impact of maternal depression on young children', *British Journal of Psychotherapy*, vol. 4, 240–252.

Power, A. and Tunstall, R. (1995) *Swimming Against the Tide*, York, Joseph Rowntree Foundation.

Robson, B. and Parkinson, M. (1994) *Evaluating Government Urban Policy*, London, DOE, HMSO.

Rosengren, A., Orth-Gomer, K., Wedel, H. and Wilhelmsen, L. (1993) 'Stressful life events, social support and mortality in men born in 1933', *British Medical Journal*, vol. 307, 1102–1105.

Rutter, M. (1985) 'Resilience in the face of adversity: protective factors and resistance to psychiatric disorder', *British Journal of Psychiatry*, vol. 147, 598–611.

Scottish Office (1992) *Scotland's Health: a challenge to us all: a policy statement*, Edinburgh, HMSO.

Smaje, C. (1995) *Health, 'Race', and Ethnicity, making sense of the evidence*, London, King's Fund Institute/Share.

Stark, C., Scott, J. and Mill, M. (1989) *A Survey of the 'Long-Stay' Users of DSS Resettlement Units: a research report*, London, Department of Social Security.

Townsend, P., Simpson, P. and Tibbs, N. (1985) 'Inequalities in health in the City of Bristol: a preliminary review of the statistical evidence', *International Journal of Health Services*, vol. 15, 637–643.

Townsend, P., Phillimore, P. and Beattie, A. (1988) *Health and Deprivation: inequality and the north*, Beckenham, Croom Helm.

Townsend, P., Whitehead, M. and Davidson, N. (eds) (1992) *Inequalities in Health: the Black Report and the health divide*, Harmondsworth, Penguin.

Wadsworth, M. (1991) *The Imprint of Time: childhood history and adult life*, Oxford, Clarendon.

Warr, P. (1987) *Work, Unemployment and Mental Health*, Oxford, Oxford University Press.

West, R. (1995) 'Psychosocial health', in Health and Lifestyles Project, *A survey of the UK population: part 1*, London, Health Education Authority, 57–86.

Wilkinson, R. (1992) 'Income Distribution and Life Expectancy', *British Medical Journal*, vol. 304, 165–168.

15

George III and changing views of madness

LINDA JONES

The king is mad! The madness of George III

The madness of King George III (crowned 1761, died 1820) is a useful starting point for thinking about madness in history. Traditional ideas about madness and its treatment were being challenged in the later eighteenth century by new therapies. The treatment of the King became a battleground between the royal physicians and the mad-doctors and between politicians too, providing a snap-shot of how mental health issues were contested at this different point in time. In addition, the thoughts and feelings of the King were recorded by his family and friends, giving us a brief glimpse of what it was like to be 'mad'.

On 22 October 1788 a senior government minister wrote a 'most secret' letter to his brother about George III's sudden illness. It had begun:

> ... with a violent spasmodic attach in his stomach; and has continued with more or less violence, and with different symptoms ever since.... . He brought on this particular attack by the great imprudence of remaining the whole day in wet stockings; but, on the whole, I am afraid that his health is evidently much worse than it has been, and that is some lurking disorder in his constitution ... part of the King's disorder is an agitation and flurry of spirits, which hardly gives him any rest. (Mr W.W. Grenville, to the Marquis of Buckingham, 22 October 1788, quoted in Macalpine and Hunter, 1969, 17)

Fanny Burney, the diarist and confidante of the royal family, wrote of:

> a manner so uncommon, that a high fever alone could not account for it; a rapidity, a hoarseness of voice, a volubility, an earnestness – a vehemence rather – it startled me inexpressibly ... The Queen grows more and more uneasy. (Quoted in Macalpine and Hunter, 1969, 18)

The King himself was alarmed by what he called his 'degree of bodily agitation' and 'a desire of talking he was scarcely able to control'. One of the characteristics of his illness was uncontrollable bouts of foul language and lewd talk which his doctors found very disturbing.

After rallying a little, George III broke 'into positive delirium' one evening at dinner on November 5, and that night 'had no longer the least command over himself'. Fanny Burney reported that 'his eyes, the Queen has since told me, she could compare to nothing but black currant jelly, the veins in his face were swelled, the sound of his voice was dreadful; he spoke till he was exhausted ... while the foam ran out of his mouth' (Ibid).

Different diagnoses

His doctors, who had at first diagnosed a fever and then delirium, now spoke of derangement of his faculties, of 'water on the brain' and of 'an ossification of the membrane'. The King was moved to Kew Palace, away from prying eyes, and to the team of doctors treating him was added Dr John Willis, a mad-doctor who claimed twenty-eight years' experience and a 90 per cent success rate of 'the particular species of disorder with which His Majesty is afflicted'. He argued that the King's madness had been brought on by 'weighty business, severe exercise, and too great abstemiousness, and little rest' and gave a diagnosis of 'consequential madness', delirium or derangement, which was curable. On the other hand, Dr Warren, the King's Physician, feared that it might be 'original madness', mania or insanity, which was not generally amenable to cure.

Different treatments

To the debate about diagnosis was added a battle over treatment. The King was subjected by his physicians to blistering to create pus, and cupping and leeching to draw off bad blood, treatments which were seen as crucial to draw out physical and mental 'bad humours' from his body.

All the doctors supported 'tonic' treatment: a controlled diet, plenty of sleep and exercise. But Dr Willis and his brother the Reverend Dr Francis Willis had more faith in physical control over the patient, seclusion and silence. Physical restraint, by means of a strait-jacket, was widely used by them to cope with the King's state. John Willis also used his new restraining chair and, while the King was bound, lectured him on his lewdness and 'improper conversation' and attempted to argue him out of his madness. The King's illness struck again in 1804 and in 1810, and on both occasions battle was joined between the doctors, although the Willis brothers maintained the upper hand. Dr William Heberden, who attended the King in 1810, was appalled by the physical coercion used by the Willis brothers and argued that:

... his Majesty's mind should, if possible be roused from its disordered actions, and not suffered to degenerate into a state of habitual error ... At present there is not one moment of the day passed by his Majesty in his usual manner; scarcely one moment that he is not reminded of his unhappy situation. (Macalpine and Hunter, 1969, 162–163).

He called for friendly conversation to relieve the King's mind, amusements to lift his spirits, company and music to entertain him. If his mind was engaged in business there would be less time for delusions and fantasies. Heberden was unable to persuade the Queen or the Willis brothers to change the regime, beyond the occasional horse ride or concert,

The King's point of view

For his part, observers reported that the King seemed often 'sensible of his unhappy condition'. During his 1810 attack, when Dr Heberden's idea of activity was briefly taken up, George III was allowed to select the programme for a family concert. He chose one which 'consisted of all the finest passages ... in Handel, descriptive of madness and blindness; particularly of those in the opera 'Samson'; ... and it closed with "God Save the King".' (Macalpine and Hunter, 1969, 158). He was rebellious at his harsh treatment and angry at Dr Willis, whom he saw as the main instigator of it. 'I hate all Physicians but most of all the Willises they treat me like a Madman' he protested on 20 December 1788. On his recovery he banished Dr John Willis from his sight, although the mad-doctor returned in 1804 and 1810 when the illness struck again.

The battle over madness

The episode of the King's madness highlights several important issues. It is clear that diagnosis and categorisation of madness were of a different order to that which emerged in the nineteenth and twentieth centuries. Madness was seen as striking the whole person. For example, blisters, by drawing out bad humours, could help to heal the King's mind.

The distinctive conceptualisation of madness drove the treatment regime. Bleeding and blistering had been used for centuries to treat illness, and were based on the theory of humours (see Figure 13.1), which dates back at lest to ancient Greece. The theory has counterparts in ancient Chinese medicine and in Aruyveda, the ancient medical belief system of India. The world was conceived as being made up of four basic elements – fire, earth, air and water which have four qualities – heat, cold, dryness and dampness. These gave rise to four humours – blood, phlegm, yellow bile and black bile – and resulted in four different temperaments – sanguinity, phlegmatism, choler, and mechancholy. A balance of these humours in the individual meant that health was

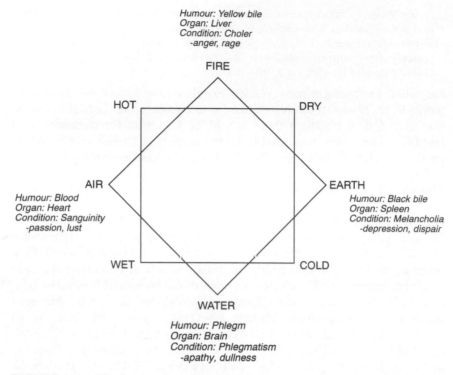

FIGURE 15.1   The humoral model of illness, showing the relationship between the four qualities, the four elements, and the four humours. Illness was attributed to an excess of one of the humours over the others.

maintained but an excess of choler or bile signalled danger and disorder. Draining the bodily fluids through blistering, purging and bleeding restored equilibrium and therefore health.

Control over diet, and a regime for exercise and sleep, also has ancient origins. In medieval times, monastic diets aimed to control the humours and passions of the monks. Diet became a more prominent part of medical practice in seventeenth- and eighteenth-century Europe. A fashionable London doctor, George Cheyne (1671–1743), linked a balanced diet with mental stability and developed a system of dietary management to cure mechancholy (Turner, 1987).

So these treatments prescribed for George III were rooted in traditional medical practices. The treatment prescribed was for the whole person, body, mind and spirit. Physicians did not examine their patients in the modern medical sense, by testing temperature, blood pressure and so on. They made a diagnosis based on a combination of theory, some assumption and the patient's view of what was wrong.

New theories of madness

The regime set up by Dr Willis, based on his experience as a mad-house keeper, demonstrates some of the newer theories about how to treat madness. The increasing use of more sophisticated devices to control the mad was one aspect of this. The mechanical age of European industrialisation was reflected in mad-house technology. As well as the 'strait waistcoat' and the restraining chair, several machines were developed around this time for immersing or douching the patient with cold water and for spinning patients in a type of gyroscope. These were claimed to be able to shock or disorient the patient back into sanity.

Perhaps the most significant break with past treatments was Dr Heberden's call for a normal routine which would enable the King to use his mind and energies, rather than to indulge his delusions. Dr Willis' emphasis on moral lectures to persuade the patient back to rationality was also distinctive; both were part of an emerging optimistic concept of madness as delusion and irrationality which were curable.

The secularisation of madness

The illness of George III was not attributed by any observers to divine intervention or explained away as 'God's will'. The explanations given were secular, rooted in a belief that disease could be understood and cured by the application of human knowledge.

Religious perspectives on madness

In medieval times, however, while doctors did attempt to relieve suffering and religious hospitals gave assistance, madness was viewed as supernatural in origin and divine in purpose. The mad were seen as possessed by demons and the Church urged prayer – to a saint such as St Bartholomew – as the cure. Poor mad people were cared for by their families or in small hospitals under the control of religious authorities, along with the sick, aged and diseased. The control of the Catholic Church and the tradition of almsgiving and care-giving provided some support to the mad, although the Church found it difficult to cope with growing numbers of poor, sick and insane people (Allderidge, 1990).

Until the seventeenth century Bethlem hospital in London was the only specialist asylum, with four patients in 1403 and 27 by 1632 (Scull, 1993). Others roamed the streets or wandered from town to town seeking charity. The state controlled the land and affairs of young wealthy lunatics through the Court of Wards and Liveries, but families generally acted as custodians of the mad (MacDonald, 1981).

Attitudes towards the mad were contradictory and ambivalent (Porter 1987a, 1987b). On the one hand lunatics were often outcasts from society, but they were also revered for their spiritual insights. There was a strong medieval tradition of creating madness – misrule – in which order was overturned and those in authority could be slighted. The dividing line between madness and evil was blurred as well. The mad could be divinely inspired but they could also be in league with the devil, trying to destroy social order: hence witch trials and tortures. It seems that people had difficulty in distinguishing one from the other.

In histories and literature mad people were often pictured as visionaries, uttering religiously inspired truths and warnings to which ordinary people were blind. In Shakespeare's plays, for example, the madman appears as a seer, predicting the future and warning of dangers and chaos. Shakespeare drew on legends and earlier histories for his complex characterisation of madness. In King Lear, the wisest characters – the fool and Edgar – use madness as a disguise and King Lear himself, through his madness, gains self-knowledge and is cleansed of guilt.

Changing views of madness

Views of madness underwent a change in the late-sixteenth and seventeenth centuries. Religious conflict and the rise of nationalism in Europe weakened the power of the Catholic Church. Its ability to control peoples' thinking and to police new ideas was undermined.

Scientific ideas, in particular, challenged religious teaching about the place of human beings in the universe. Science gave increasingly convincing explanations of the natural world and of the workings of human bodies within this world. The notion of madness as a God-given state was challenged by secular explanations of madness as a physical state.

The rise of Protestantism and its adoption in England through the establishment of the Church of England led to the closure of the monasteries and the curtailment of Church authority. Alms-giving and hospital care declined. Attitudes hardened towards mad people, who were more visible and seemed more threatening, and were no longer protected by the Church's religious teaching about madness.

The mad were increasingly seen as different and dangerous: subhuman, irrational and bestial. Through giving in to their madness they had forfeited their claims to be human and to be treated as such. Their madness was seen as their own fault, the result of self-indulgence, an excess of passion and egoism. Punishment and imprisonment were justified. States passed vagrancy acts to try and control 'study beggars' and in 1601 the English Poor Law Act made beggars and outcasts the responsibility of local parishes. Over the next century increasing numbers of poor mad people were confined in gaols, workhouses or poor-houses (Allderidge, 1990).

McDonald (1990) has argued that politics also played a key role in discrediting religious ideas about insanity. Tory Church of England politicians disliked and feared the protestant sects associated with radical politics, such as Puritanism and later Methodism. These sects continued to emphasise witchcraft and demonism, so the attack on supernatural explanations of madness was also a political attack on radicalism. Madness became redefined as the result of deluded ideas and an unrestrained and violent imagination.

Being mad in early modern Britain

It has been claimed that the eighteenth and nineteenth centuries witnessed 'a great confinement' in which mad people were rounded up by European states and put into asylums or prisons (Foucault, 1965; Scull, 1993). But while in France there was a widespread and often penal use of custody to contain the mad, with 5000 people confined in the Hospital General by the 1780s, there does not seem to have been any comparable movement by the British state. There was, however, some growth in the number of private madhouses and pauper workhouses, especially in the eighteenth century (Parry Jones, 1972).

Some of these madhouses were very grim places, even though they catered for the better-off in society. Mr Spencer's madhouse at Fonthill, Wiltshire was described in evidence to the Government Select Committee of 1815 as composed of cells nine feet by six feet, with unplastered damp greenstone walls, bare earth floors, and no light or ventilation. Inmates were chained to their beds (Allderidge, 1990). Other private madhouses were more humane and offered asylum to inmates. But the assumption that the mad were subhuman and punishable was the subtext of treatment in these institutions. This was the case in Dr Willis' treatment of George III.

What was the experience of madness itself at this time? The accounts that have survived are not those of the poorest and most vulnerable groups, who could not read or write and who left no trace of their thoughts and feelings. But we do have some first-hand evidence which helps to build up a picture. Porter (1981) notes the predominance of religious crises, 'the search for identity and respect under God', that characterises these accounts. William Cowper the poet, for example, felt abandoned by God and his own father and devastated by his sense of guilt and sin (Porter 1987b). In Dr Cotton's asylum in St Albans, Cowper sank into lunacy and despair until after many months:

> I flung myself into a chair near the window, and seeing a Bible there, ventured once more to apply to it for comfort and instruction. The first verse I saw, was

the 25th of the 3rd of Roman: 'Whom God hath set forth to be a propitiation through faith in his blood, to declare his righteousness for the remission of sins that are past, through the forebearance of God.' Immediately I received strength to believe, and the full beams of the Sun of righteousness shone upon me. I saw the sufficiency of the atonement he had made, my pardon sealed in his blood, and all the fulness and completeness of his justification. (Quoted by Porter, 1981)

Other inmates of asylums, such as Kit Smart the poet, incarcerated in Bethnal Green asylum in the 1760s, and John Perceval, confined in Ticehurst, also recorded their anguish and the gulf between their mad-doctors and themselves. Belittling Dr F's diagnosis of excessive imagination to explain his religious struggles, Perceval noted down that 'it was not likely, therefore, that I should confide the difficulties of my mind to men who, by slighting the origin of them, betrayed their presumption, whilst affecting excellent acuteness'.

The age of the asylum

In the later eighteenth and early nineteenth centuries there were increasing numbers of public subscription hospitals with wards for the insane and a small but growing number of specialist public asylums, such as those in York and Newcastle. In 1807 a government Select Committee recommended that each county should provide an asylum financed out of local rates. Eighteen counties established asylums and in 1845 the rest were compelled by legislation to follow. Within these institutions, in particular in York, innovative approaches to the treatment of madness were developed based on the view that the mad were misguided and deluded but capable of being cured. This reflected a further secularisation of the explanations and the treatment of madness.

In some ways, the new approaches seemed optimistic and enlightened. The essential humanity of the mad was emphasised; they might be irrational and childlike but they were still human beings capable of rational thought. John Locke, the influential seventeenth century philosopher, argued that madmen

[Do] not appear to have lost the faculty of reasoning, but have joined together some ideas very wrongly, they mistake them for truths; and they err as men do that argue rightly from wrong principles. (Locke, quoted in McDonald, 1990, 71)

Given the right kind of therapy, the mad could be cured of these wrong ideas and persuaded by the regime of 'moral treatment' to behave rationally again.

The York Retreat, built in 1796, by and for Quakers, emphasised moral therapy, to help the patients gain enough self-discipline to master their illness. It was built to be as home-like as possible, and all the attendants were Quakers who used a combination of firmness, kindness and physical restraint to control the patients. Inmates' delusions and hallucinations were ignored. Instead a rational regime of healthy exercise, walks, rest, reading and conversations was developed to guide patients back to health, not unlike the regime prescribed by Dr Heberden for George III.

Samuel Tuke, the medical adviser, commented that

> no advantage has been found to arise from reasoning with them on their particular hallucinations … every means is taken to seduce the mind from its favourite but unhappy musings, by bodily exercise, walks, reading and other innocent recreations. (Porter, 1981)

The recovery rates were much higher than in the traditional mad-houses and York methods were generalised across British asylums. In continental Europe Philippe Pinel, governor of the Bicêtre asylum in France, freed the prisoners from their chains and set up a similar therapeutic regime relying more on moral than physical coercion.

The application of enlightenment thought to madness

This changing approach to insanity reflected and was underpinned by the working out of new social, political and economic ideas, in what is usually termed 'the age of enlightenment'. A central tenet of enlightenment thought was belief in science and rationality as the root of all real progress, and in the ability of human individuals to control and improve society. In the sphere of physical health, for example, campaigners fought for public health legislation to improve the physical environment of the towns. Factory reformers in the early nineteenth century campaigned to curb the worst excesses of industrial capitalism.

The emerging group of medical experts in the treatment of insanity drew attention to the importance of mental hygiene and preventive psychiatry as well as specialist medical treatment for the insane. By emphasising the significance of environmental factors and pressures in mental illness, they offered a potentially powerful critique of industrial capitalism. Dr John Hawkes, assistant medical officer of Wiltshire county asylum, commented:

> Let us endeavour to promote mental sanitary reform by combining to introduce these changes in the social conditions, more especially of the working classes, by which that high pressure system, so prejudicial to the health of the mind, shall be slackened, and the strain, which it occasions, relaxed. Let the hours of labour be abridged and let childhood no longer share the curse of the Fall. Let the multitudes, who have not the means or opportunities of learning from books, be instructed by public teachers in the first principles of mental as well as physical hygiene. (quoted in Birley, 1990, 5)

The institutionalisation of madness

Unfortunately, few of the hopes of the reformers were fulfilled. Indeed, it is still a matter of fierce debate whether reform or rationalisation was the main aim. Advocates of moral therapy, such as Samuel Tuke, medical officer at the York Retreat, argued that if the insane were housed in small, homely institutions away from the bustle of everyday life, with proper supervision and a moral treatment regime, there was no doubt that most would quickly recover and become productive citizens. The 1845 Act proposed the establishment of asylums for not more than 300 patients and assumed that a regime of fresh air and therapeutic work would achieve high rates of cure. The county asylums that developed, however, grew in size during the nineteenth century to accommodate more and more seemingly insane people who, far from being cured, spent their whole life in the institution. This was justified on the grounds of economy and reflected the increasing interest of the state in controlling and segregating the 'unfit' and 'unproductive' members of society. The West Riding asylum, for example, had a population of 150 in 1818. By 1900 this had risen to 1469 inmates. Whereas in England in 1800 there were around 1000 asylum inmates, by 1900 there were 100 000.

Conclusion

This further stage in secularisation treated madness as a psychiatric disorder. It viewed the patient as the object of medical attention – to be re-educated into sanity through a programme of rational treatment if possible, and contained within the asylum if not. A century of experience in insanity had given the mad-doctors increasing authority over the domain of mental illness. From being seen as divine in origin and purpose, madness was coming to be viewed as a disorder of the brain analogous to disorders of the body.

By the mid nineteenth century mental illness was increasingly seen as another branch of medicine, where practitioners would be able to cure the mind just as surgeons cured the body.

References

Allderidge, P. (1990) 'Hospitals, madness and asylums: cycles in the care of the insane', in Murray and Turner (eds) *Lectures in the History of Psychiatry*, 28–46.
Birley, J. (1990) 'The history of psychiatry as the history of an art', in Murray and Turner, *Lectures in the History of Psychiatry*, 1–46.
Foucault, M. (1965) *Madness and Civilisation*, London, Tavistock.

Macalpine, I. and Hunter, R. (1969) *George III and the Mad-Business*, London, Allen Lane.

MacDonald, M. (1981) *Mystical Bedlam*, Cambridge, Cambridge University Press.

McDonald, M. (1990) 'Insanity and the realities of history in early modern England', in Murray and Turner, *Lectures in the History of Psychiatry*, 60–81.

Murray, R, M. and Turner, T. M. (eds), *Lectures in the History of Psychiatry*, London, Gaskell/Royal College of Psychiatrists.

Parry Jones, W.L. (1972) *The Trade in Lunacy*, London, Routledge & Kegan Paul.

Porter, R. (1981) 'Being Mad in Georgian England', *History Today*, vol. 31, 42–48.

Porter, R (1987a) *Mind-forg'd Manacles: a History of Madness in England from the Restoration to the Regency*, London, Athlone Press.

Porter, R. (1987b) *A Social History of Madness*, London, Weidenfeld & Nicolson.

Scull, A.T. (1977) *Decarceration: community treatment and the deviant: a radical view*, Cambridge, Polity Press.

Scull, A. (1993) *The Most Solitary of Afflictions: Madness and Society in Britain, 1700–1900*, Yale, Yale University Press.

Szasz, T. (1971) *The Manufacturer of Madness*. London, Routledge & Kegan Paul.

Turner, B. (1987) *Medical Power and Social Knowledge*, London, Sage.

Walton, J.N. (1985) 'Casting out and bringing back: Pauper lunatics, 1840–70', in Bynum, W.F., Porter, R. and Shepherd M. (eds) *The Anatomy of Madness: Essays in the History of Psychiatry*, London, Tavistock 132–46.

16

Professionals, the state and the development of mental health policy

JOAN BUSFIELD

Contemporary mental health services and psychiatric ideas and practices have been moulded and fashioned by the complex interplay of two inter-related but often opposing spheres of influence: medicine and the state.

Medicine

First, and most obviously, there is medicine itself. It is the ideas and practices of the medical profession that have structured and continue to structure some of the most fundamental features of contemporary ways of thinking about the phenomena that fall within the category of mental illness: the idea that there are discrete, separately identifiable mental ill-nesses with distinctive symptom syndromes and causes, as well as typi-cal modes of onset, course and prognosis; the idea that an understanding of the physical processes associated with these symptoms can offer the most satisfactory analysis of the illness in question; the idea that physical treatments can generally offer a, if not the, most valuable tool in the care and treatment of mental illnesses. It is the medical profession too, by virtue of the power, status, and authority it has achieved, that ensures that psychiatrists are at the top of the hierarchy of the occupational groups involved in the care of the mentally ill; that gives certain exclusive legal powers to psychiatrists *vis-à-vis* their patients – to prescribe drugs, to admit and discharge from hospital against their patients' wishes if necessary, to offer treatment within the NHS and so forth; and it is their identity as doctors that gives psychiatrists a key role in determining the

Edited from Busfield, J. (1986) *Managing Madness: Changing Ideas and Practice*, London: Unwin Hyman, 358–370.

content and character of mental health services, both directly via medical and psychiatric representation on key policy making bodies and indirectly via the impact of their ideas and practices. For example, the development of psychotropic drugs and their presentation as a valuable therapeutic tool for a wide range of mental disorders has undoubtedly encouraged the resort to medical practitioners for a range of phenomena which the practitioners themselves consider mental illnesses.

However, the content and character of medical ideas and practices is not itself unchanging. Much of what we now take as typical of medicine both as a body of ideas and practices and as a profession is relatively new. The immediate origins of contemporary medical ideas about illness are to be found in nineteenth-century developments: in the localization of diseases associated with the Paris school in the early decades of the nineteenth century; in the development of germ theory through the work of Koch and Pasteur in Germany and France in the middle of the nineteenth century, as well as in twentieth-century scientific and technological developments that had a marked impact on diagnosis and treatment – the introduction of X-rays around the turn of the century, the development of electro-cardiographic techniques, the chemical synthesis of drugs in the 1930s and 1940s and so forth. Institutionally it was the rapid and increasing importance of hospitals in the nineteenth century, following their emergence as distinctively medical institutions (that is, places specifically concerned with the treatment of illness by medical practitioners) that gave medicine much of its present-day character. The hospitals contributed to medicine's reliance on science and technology as the basis of its expertise and, thereby, facilitated the successful professionalization of its practice, as well as giving shape to the pattern of health service provisions.

The professional power of medicine in the clinical field is, likewise, relatively new. It was only in the nineteenth century, partly as a result of developments associated with the rise of hospital medicine, that the competing groups of authorized healers formed themselves into what can be regarded as a single profession with a single register of qualified and licensed practitioners. Divisions of course remained: between specialists of varying status and prestige, and between hospital doctors and the newly emerging general practitioners, and the continuing existence of different colleges of medicine reflects this; but there is much that is shared including many of the legal powers which registered practitioners have acquired over the last century. Over the same period the increasing role played by the state in the provision of health care has diminished the profession's direct power to control the social and economic organization of its work, and has changed the nature of the economic and organizational constraints on medical practice. However, since the development of state health services has also led to an expansion of medical activity, the impact of medicine on the ways in which people think about, make sense of, and deal with events and experiences in their lives must have also increased.

In the sphere of lunacy and mental disturbance the changing character of medical ideas and practices and the growing professionalization of medicine have had a profound effect. On the one hand they contributed to and encouraged the establishment of new hospitals for lunatics paralleling the general hospitals. In so doing they facilitated the development of a new medical specialty focused on the care of lunatics in asylums, a speciality of official medicine whose legal powers, although contested, were more firmly established in the nineteenth century. On the other hand the nineteenth-century changes in medical knowledge encouraged and strengthened the interest in the organic aspects of lunacy and mental disorder, which was reflected in the search for physical lesions and the high levels of autopsy common in the second half of the nineteenth century. These medical changes also contributed to the therapeutic nihilism of the latter part of the century as the old faith in moral treatment and the positive value of institutional care declined. And then, in turn, the development of new physical treatments for mental disorders in the twentieth century led to a new wave of therapeutic optimism which began at the end of the 1930s with the introduction of insulin therapy, ECT and psychosurgery. To describe these nineteenth- and twentieth-century changes as the medicalisation of insanity or mental disorder is to say little more than that healing activities in this sphere, like other healing activities, increasingly took on the character and shape that we now take to be distinctively medical. For medical interest in mental disorder was not new nor did it extend to entirely new territories. Rather the character of medical work in this field changed. First, it became increasingly professionalised with psychiatrists extending their legal powers, organizing themselves as a group more effectively, and developing their professional expertise. Second, medical work increasingly concentrated on physical processes and physical treatments and the former environmentalism largely disappeared. And third, the number of medical interventions increased and their balance changed, as the mechanisms for providing medical care without direct cost to the patient developed and people from all social backgrounds with all types of complaints became patients of registered psychiatric practitioners.

The state

Mental health services and psychiatric ideas and practices area also moulded and fashioned by the policies and activities of the state and its attendant bureaucratic structures. It is the state that has acceded to an licensed medical practitioners' claims for professional power and professional autonomy and given them legislative backing; it is the state that has given institutions powers to control and confine lunatics considered dangerous or in need of treatment; it is the state that has given medical practitioners, sometimes in conjunction with magistrates, sometimes without, powers to certify and decertify patients; it is the state that has first permitted and then required local authorities to make separate institutional provision for pauper lunatics; and it is the state that has developed and funded a broader range of mental health services, out-patient clinics, day hospitals, psychiatric units in general hospitals, primary care facilities, home helps, district nursing and so forth. Undoubtedly during the nineteenth and twentieth centuries, the state's involvement in this as in so many other spheres of social policy widened, although there has been some retrenchment during the past decade. The state's initial role was confined to that of statutory regulation, at first indirectly and then directly. It gave powers to a range of authorities to license healers and to attempt to suppress the activities of irregular healers; later it gave powers to physicians and magistrates to inspect private madhouses and to admit and discharge persons from them. In the nineteenth century first permissive and then mandatory legislation broadened the state's involvement from regulation to that of specific service provision, initially in relation to a narrow category of pauper lunatics and, during this century, to a wide range of persons from all social classes, for all types of mental disorder.

Indeed, the state, by virtue of the services it makes available and by its interventions and activities in the economic and social realm, plays an important part in helping to determine the use of health services. This is because by its legislative enactments and its ideological apparatuses it helps to create and structure both social dependence and psychological distress and, in consequence, the 'demand' for mental health services ... For instance, the institutional bias of the New Poor Law, and the reluctance to provide outdoor relief during much of the nineteenth century, helped to create the 'demand' for asylum care and the enormous expansion in numbers confined within public asylums during the nineteenth century. Similarly the state's reluctance since the Second World War to provide much in the way of publicly funded community services in the form of purpose-built residential homes, home helps, district nurses and so forth, and its recent willingness to see an expansion in the use of poorly regulated and inadequate private residential care while running down the old mental hospitals, is likely to affect the perception of the old mental hospitals, and we could see a new demand for greater investment in these decaying institutions on the grounds that even they are preferable to virtually non-existent public community services. In a similar fashion, decisions about issues such as the age of retirement, the level of pensions, housing, unemployment, industrial investment, all help to structure and create the sorts of social dependence or psychological difficulties that may lead to medical intervention. This is most obvious in the case of the over-65s, who can expect little in the way of employment in the labour market, material benefits or social status, and who are forced into greater social and economic dependence as well as greater psychological distress. Not surprisingly the result is high levels of 'demand' on the mental health services among this group.

Phases of development

Commercial and charitable healing

Viewed historically we can see three major phases in the development of services for those with mental disorders which highlight the complex interaction of medical ideas and practices with state policies in determining the character of psychiatry and the mental health services. First, the period of commercial and charitable healing – a period in which cures and advice for mental problems could be purchased in the market or might be given on a charitable basis to those in need. The range of conditions for which remedies were sought was as diverse as the healers themselves and the ideas and practice on which their healing art was based. Healing, like many other commercial activities, was stratified, with the most affluent receiving help from higher status, academically better educated practitioners and calling more freely on their services for more minor as well as more severe conditions. Among the poor, where lower status practitioners would find their market, a problem would generally need to be more severe to merit the cost of the services of any commercial healer, or to call upon itself a healer's charity.

It was in this commercial and charitable context that the first moves to segregate the insane began with the establishment of private madhouses on a commercial basis in the seventeenth century, followed by the charitable voluntary asylums in the middle of the eighteenth. At this point the only public institutions that confined the insane were more general places of confinement, the workhouses and common gaols, largely developed during the eighteenth century as a means of policing and containing the consequences of economic dependency, although some paupers were sent to the private madhouses under the auspices of the poor law system. While the first separate institutions for the insane were prompted by motives of profit and professional interest and played upon a family's concern to deal with a difficult family member both humanely and effectively, public asylums were developed as a solution to the problems of social and economic order created by economic, social and personal dependency and had a clear motive of collective social control.

The segregation of the insane

It was in the nineteenth century that the second phase, the segregation of the insane in separate institutions, developed fully with the establishment and rapid growth of public asylums. The factors underlying the state's funding of these new institutions were numerous. There were the difficulties that had been encountered in funding the charitable asylums which were beginning to serve as the model for institutional provision for lunatics. There were the widespread criticisms of private madhouses where the motive of profit was considered to be at the root of the problem of wrongful detention. There were the new concerns about the harsh treatment of pauper lunatics in workhouses and private madhouses that a changing consciousness about insanity, especially pauper insanity, was beginning to engender. There was the new optimism about the curability of insanity associated with the principles of moral treatment; and there was, too, the increased power of the bourgeoisie, the class especially sympathetic to the case for lunacy reforms, and willing, if necessary, to accept the necessity for central intervention.

That state intervention in respect of the problem of lunacy should have taken an institutional form was overdetermined. First, there was the keen and strengthening medical interest in hospital care. The relocation of medical practice had important professional and material advantages for medical men in terms of access to new patients, a broader range of clinical experience, more adequate training, contact with other medical men, and clinical research. Second, the elaboration of the principles of moral treatment provided a positive institutional ideology that justified and legitimated the development of separate institutional provision for lunatics. The ideology asserted both the need to remove the individual from existing environmental pressures and the merits of creating a special, well-ordered institution which would have its own therapeutic properties. Third, there was the institutional bias of the poor law system itself, heightened in the nineteenth century with the publication of the Poor Law Report in 1834. Although outdoor relief was not abolished following the 1834 Poor Law Amendment Act, the philosophy of the New Poor Law embodied a view of institutions as a convenient and appropriate way of deterring the economically dependent from seeking poor law relief, which would not only keep the numbers in receipt of relief to a minimum, but also instill proper social attitudes and values amongst inmates. Hence in the establishment of public asylums professional advantage, humanitarian concerns and the economic and political interests of the state combined.

It was, however, primarily poor law concerns, in conjunction with the power of public asylums to legally detain their inmates (first established in relation to lunatics considered dangerous), that contributed to the custodial character of the public asylums. Medical ideals of therapy, reinforced and amplified by the apparent potential of moral treatment, could not flourish in the large-scale, understaffed and poorly funded institutions, despite the presence of a medical superintendent, when saving an extra halfpenny on the weekly cost of care counted for more than rates of cure or individual attention to inmates. Public asylums for lunatics functioned, despite their original therapeutic intent, like other poor law institutions as places of last resort where those for whom little could be hoped in the way of improvement were to be found, largely because the poor law did not provide any alternative means of economic or social support.

The move to community care

During the second half of the nineteenth century, public asylums, like the private madhouses and charitable asylums that preceded them, came to be widely criticized and there were new demands for lunacy reform. The aim was to transform the asylums from custodial into therapeutic institutions by ensuring that persons disturbed in mind were given treatment in the early stage of the disease. This was to be achieved both by allowing voluntary admission and by setting up special facilities for early, acute cases. Very little reform, however, occurred in the public sector until the passing of the Mental Treatment Act in 1930. By that time the poor law system, which constituted the main obstacle to reform, had largely been supplanted by a range of welfare provisions developed outside the framework of the poor law which heralded the beginnings of the welfare state and the end of the poor law system.

The beginning of the third phase of services for those with mental disorders – of the movement away from the asylums and mental hospital and the development of a policy of community care – can be dated back to the same period. The new policy objectives were brought about as much by the changing character of welfare provisions as by changing medical ideas. The expansion in a broad range of state welfare provisions, both in the form of financial benefits and in the form of services, created a new context in which care outside the mental hospital was not only more feasible but also more acceptable. The undermining of the poor law system and the expansion of welfare provisions based largely on insurance principles meant that publicly funded care was now to be provided outside the institution, and indeed the whole institutional bias of public provision was undermined.

Changing ideas on insanity

These basic stages in the development of mental health services have been associated with and contributed to changing conceptualisations of insanity and mental disorder which highlight the way in which such concepts are socially and historically constituted. In the sixteenth, seventeenth and eighteenth centuries insanity and lunacy were narrow, primarily societal categories, albeit usually imprecisely defined, but they were only the most extreme of a range of mental states that might be considered problematic and might be brought to the attention of a healer if circumstances (particularly the family's material circumstances) allowed. When, however, institutions began to set up, and particularly when public asylums were introduced and expanded in the nineteenth century, much of the attention, especially in the discussion of public policy, focused on lunacy and insanity alone, and the less severe disorders were largely ignored. This was because the more minor problems of mind were far less likely to produce the degree of social dependency or difficulty that was held to merit the dominant form of public intervention – institutional care. In this process, however, the more restricted notions of lunacy and insanity (real madness) were themselves broadened, as concepts such as those of partial insanity and moral insanity were introduced and developed by medical men in their effort to categorise and treat the full range of cases that ended up in institutions. But the more minor problems – that did not typically fall within the compass of asylum care which provided the basic core of the work of the emergent psychiatric profession – received far less academic attention and political comment.

Towards the end of the nineteenth century and during the twentieth century, however, the milder forms of mental disorder have attracted increasing interest and new concepts, new theories and new treatments have been introduced which have modified our understanding of them. Initially this medical interest existed largely in the private sector and found its best-known examplar in the work of Freud and the development of psychoanalysis. But with the expansion of publicly funded services outside the mental hospital as welfare services were transformed, and with the development of relatively cheap forms of the treatment suitable for wide-scale therapy, interest in the realm of psychiatric activity broadened once more, and the balance of psychiatric work began to change. As a result the less severe forms of mental disorder, the neuroses and behaviour disorders, of people from all social backgrounds, not just the affluent, have become a major feature of medical and psychiatric work and of public and political concern.

This shift in the character of psychiatric work has led to new tensions and contradictions for psychiatry as a profession. In the second half of the nineteenth century the major tension was between the curative orientation of medicine and the increasing custodialism of psychiatric work in asylums. While this contradiction has not entirely disappeared during this century, and has been a major motive of endeavours to transform the character of asylums, a new contradiction has emerged. This is the contradiction between the increasing emphasis within medicine on physical causes of illness and physical treatments and the visibly social and psychological character of the neuroses and behaviour disorders which play so large a part in medical and psychiatric work.

Freud wished to establish psychoanalysis as a separate science, in practice largely directed towards the neuroses, which was to exist outside medicine, though closely linked with it. His desire to do this was but one reflection of, and attempt to resolve this new contradiction. The 1960s critiques of Szasz, Laing and Scheff were but a more recent assertion of psychiatry's neglect of the social and psychological as the new scientific medicine has taken its hold. It is not, therefore, surprising that these latter critiques attracted widespread public interest at a time when the balance of psychiatric and medical activity in relation to mental disorders was changing very markedly. For it is precisely in relation to the less severe mental disorders that questions about the nature and appropriateness of medical intervention are most obvious.

Occupations and ideologies

The response to this tension within the mental health field has been twofold. On the one hand many psychiatrists have incorporated elements of psychological and sociological thought into their ideas and practice, but they have done so in a selective manner and have transformed and moulded them into a form that is more consistent with their own ideological perspective.

On the other hand, there has been a proliferation of new professionals within the mental health services in the areas of potential expertise that fall outside the current boundaries of psychiatric knowledge and practice – a proliferation encouraged by the development of mental health services to new patient groups in new locations. In Britain these new mental health professionals, clinical psychologists, psychotherapists, counsellors and social workers, who clearly pose a potential threat to the authority and status of the psychiatric profession, have emerged historically largely within the medical ambit and have in many respects been contained as subordinate groups within the medical hierarchy.

Care rather than prevention

Neither development tackles what many critics view as the fundamental limitation of the mental health services, and of psychiatry and the other mental health professions, which is that their orientation is almost entirely curative. Psychiatry (like its parent medicine), and the mental health services in which its practice is carried out, focus on the care and treatment of specific individuals who, for whatever reason, have become psychologically disturbed. Their mode of action is almost entirely responsive.

To contend that curative individualism is the fundamental limitation of psychiatry and of the mental health services is to move beyond debates as to the precise causes of mental illness. It also calls into question the nature and form of social intervention in the complex interplay of forces that may lead to what society now regards as mental health problems. The argument is that instead of directing all our resources and attention on to disturbed individuals, important though it is to ameliorate their situation, we need also to look beyond the individual to the forms and levels of intervention which would make mental disorders less likely for the population as a whole, or for particular groups within it.

Preventive social interventions, like curative interventions, make assumptions about the causes of mental health problems. Our knowledge of these causes is, however, often fragmentary, whether we examine the casual assumptions that underpin curative, individualist intervention or those on which preventive, public intervention is founded. In the case of the latter the inadequacy of our understanding is itself in part the result of the curative bias of our contemporary orientation to health which ensures that resources largely go on research that justifies and legitimates curative interventions and not on research that would facilitate preventive interventions at the social level. More importantly, however, the absence of preventive social interventions is usually a matter of the lack of political will and not of the deficiencies of existing knowledge.

As developments in the field of public health demonstrate, successful preventive intervention requires collective action by the state and cannot usually be effected by individuals acting on their own. Nor, indeed, can we expect it to be effected by medical practitioners whose expertise and ideology are narrowly clinical and who work in services organized to deliver care and cure to individuals who are already disturbed. Without doubt the curative character of these services and of their work has served the interests of medical practitioners and has contributed to their successful professionalization. But it is the state that has ultimate responsibility for the curative character of mental health services and has the necessary powers to generate the sort of collective interventions that effective prevention requires. That it does not do so is, therefore, a political matter and requires debate and action in the political arena. It is political interests and political opposition to more radical social interventions that maintains the curative approach.

Much of the intervention that would be required to tackle health problems in a preventive manner would not be of a 'medical' character, or at least not medical in the sense that we now understand that word, as examples from the field of physical health demonstrate. Legislation requiring the wearing of seat belts, although designed to prevent death and physical injury, does not involve any provision of medical services. Nor is much in the way of medical expertise required to establish whether or not the wearing of seat belts does lower mortality and physical injury in car accidents. In contrast, in the case of smoking and lung cancer medical expertise has been involved in initial research which has established the deleterious consequences for health of smoking. But preventive intervention designed to reduce smoking does not require the expertise of medical practitioners, especially if it occurs at the societal rather than individual level. If intervention takes the form of strong restraints on the publicity and lobbying of the tobacco industry or serious use of mechanisms such as taxation to reduce tobacco consumption, then the services of medical practitioners or other caring professionals are not required. If it is restricted to exhorting individuals to stop smoking then the public may well turn to professional carers and healers in their efforts to do so, in the face of the relatively low cost, general acceptability, and high promotion of cigarette smoking.

Issues about the desirable forms of intervention to deal with the high levels of mental disorder now manifest in our society cannot, therefore, be settled simply by resort to the facts about the causes of mental disorder. For decisions as to the most appropriate forms of intervention involve matters of political judgement. Contrary to existing medical beliefs, it is not that knowledge as to the precise causes of mental disorders determines the way in which we intervene to deal with the problem of mental disorder. Rather, it would be more accurate to say that the political acceptability of particular forms of intervention determines what we choose to emphasise as *the* causes of the different types of mental disorder. Consequently, the contribution of sociology and history to our understanding of mental disorder is, above all, to challenge the belief that psychiatric interventions are value free and to illumine the way in which existing ideas and practices about mental disorder are shaped by social, political and economic forces.

17

Scare in the community: Britain in moral panic

MATT MUIJEN

Out of sight, out of mind – the age of institutions

Scandals are nothing new to mental health care, although the type and scale of scandals change. Until the early 1980s they were almost exclusively associated with hospitals. Names like Ely (1969), Whittingham (1972), Napsbury (1973), Warlingham Park (1976), Darlington Memorial (1976), St Augustine's (1976), Winterton (1979), Brookwood (1980) and Rampton (1980) are associated with institutional abuse, sometimes large-scale, and the list could go on. In the Rampton inquiry more than 800 cases of brutality affecting patients were specified, involving more than 100 nurses.

The practice in these hospitals as reported to the inquiries is strongly reminiscent of Goffmann's (1961) picture of total institutions: all aspects of patients' lives are conducted in the same place under the same authority; patients receive the same treatment, often together; days are structured with explicit rules; and activities are designed from an institutional perspective.

The lessons of these hospital inquiries are reviewed by Martin (1984), and are an indictment of institutions in general. Hospitals, wards and staff often operated in isolation, with little supervision. Sometimes the same staff were responsible for the same patients over years, even decades. Patients were dependent and vulnerable, in no position to resist or complain. Professional leadership and management were incompetent, and training inadequate. Resources were insufficient, with wards

Edited from Muijen, M. (1995) 'Scare in the Community: Part Five: Care of Mentally Ill People', *Community Care*, (7–13 September), i–viii.

severely understaffed. National policy did not offer the resources and incentives to improve quality.

Finally personal failings were a significant factor, well summarised by Martin: 'It cannot be denied that cruelty and weakness have played their parts in many incidents of ill-treatment, but always in situations where bad management has provided opportunities where an unsuitable person was given the wrong tasks, with inadequate training and leadership.'

In the 1970s, the most powerful and enduring images of the plight of people submitted to mental health 'care' were not found in the news media, but in books – fiction (Kesey, 1972; Piercy 1979), personal accounts (Chamberlin, 1977) and academic works (Wing and Brown, 1970; see also Scheft and Rosenhan in this volume)....

The common theme of all those who criticised institutions was an abhorrence of the rigid care which ignored human values, dominated by a self-satisfied medical model insensitive to patients' experiences.

The early years

It is intriguing that the inquiries throughout the 1960, 1970s and early 1980s were preoccupied with large mental hospitals, which had been expected no longer to exist. In 1959, Enoch Powell, then Secretary of State for Health, in his famous 'Watertower' speech, had already predicted the closure of mental hospitals within the next 15 years. Even the hospitals were shrinking, with patients numbers gradually declining from a peak of 155 000 in 1955.

The first day hospital opened in 1948 – the same year the National Health Service was introduced. The Mental Health Act 1959 introduced community care principles by allowing patients to choose their place of treatment, provided this would not put anyone at risk.

As early as 1962 the Hospital Plan for England and Wales proposed small-scale psychiatric units (MOH, 1962; DHSS 1972). A controversial large reduction in mental health beds was planned. It was anticipated that local authorities would provide home care by recruiting more social workers – familiar themes to the next generation.

Another familiar theme – the division between hospital and community services – was addressed a decade later (DHSS, 1972). After years of consideration, it was announced that the NHS would be administered by health authorities, not local government. Keith Joseph, the then Secretary of State for Social Services, provided a fascinating insight into the progress made over almost 25 years: 'It is well understood now, moreover, that the domiciliary and community services are underdeveloped – that there is a need for far more home helps, home nurses, hostels and day centres and other services that support people outside hospital.

Often what there is could achieve more it if were better co-ordinated with other services in and out of hospital.'

A further shift away from large mental hospitals was promoted by *Better Services for the Mentally Ill* (DHSS, 1975). Care for acutely mentally ill people was to be provided locally, mostly in district general hospitals. People with long-term illnesses were to receive asylum and rehabilitation in small-scale hostels and day centres in the community, funded by local authorities. It was fairly predictable that local authorities would be unenthusiastic about their proposed role in financing long-term 'social care' provision and support, just as they were a decade earlier. This lack of enthusiasm is reflected in the available services in 1980 (Table 17.1). Community care only took off in the early 1980s, and the mechanism was largely unintentional.

The explosion of residential care

Although the number of hospital beds had gradually decreased since 1955, this was not always the result of co-ordinated discharge planning and careful follow-up. Most had not been replaced as patients died. Some people had been discharged to live on their own, but transfers to residential care were comparatively rare: there were few local authority and registered homes available in 1980 (see Table 17.1).

Both local authorities and health authorities were reluctant to fund residential care, and the private sector was fairly insignificant at this time. Savings from the reduction in hospital beds were minimal, since staff numbers had increased in an effort to improve the quality of care (Table 17.1).

The decision that people in private residential homes could fund their places through supplementary benefit achieved unintentionally what deliberate policy had failed to do – but at price. Suddenly patient could be transferred into the community without burdening the budgets of health and local authorities. The private sector exploded. Between 1978 and 1989 total spending on supplementary benefit leapt from £80 million to £1.5 billion!

The Audit Commission (1986) raised the alarm in 1986, in a report that makes the reasons for the present system of local authority funding of residential care very obvious. A 'perverse incentive' against community care was signalled. Patients in residential homes were fully funded by social security, but those at home, requiring community care, depended upon the resources of local authorities. It was calculated that local authority expenditure had increased between 1976 and 1984 by only £20 million to £52 million, while 17 000 patients had been discharged from mental hospitals. The NHS budget for mental hospitals had increased by £60 million to £900 million, spent on more staff for fewer patients.

TABLE 17.1   Available places for adult people with mental problems in England

	1980	1990
HOSPITALS		
Beds	87 396	59 290
Admissions	121 000	119 800
OUTPATIENTS		
New patients	184 000	207 000
DAY PLACES		
Local Authority	4967	6979
Voluntary	621	2903
Health Authority	13 500	15 000
HOMES		
Local Authority	3724	4349
Registered nursing homes	2142	7377
Health Authority and joint		8000*
CPNs	1670	5000**

* Estimated from Faulkner et al. (1994)
** From White (1990)

In 1988, two reports heralded the modern era: *Community Care: Agenda for Action* by Sir Roy Griffiths (DOH. 1989a); which inspired the White Paper *Caring For People*, and *The Report of the committee of inquiry into the Care and Aftercare of Miss Sharon Campbell* (Spokes *et al.*, 1988). From now on it was to become increasingly unclear who was determining policy: government or inquiry panels. Service users, staff and the public became uncertain about responsibilities, expectations and events. The years since then could be called 'the age of instability'.

The age of instability

In 1989 two White Papers were published almost simultaneously *Caring For People* dealt with community care, and *Working For Patients* proposed a substantial review of the NHS. Mental health care was significantly affected by these changes, but not always as intended. (DOH, 1989a and 1989b).

The reasons for community care and NHS reforms were similar, and in principle timely. Demand and expenditure were out of control.

Community care had shifted money into rapidly expanding private residential care from an uncapped central government budget. In the health service, the yearly efficiency squeeze had produced bed closures and growing waiting lists. This was complicated by increasing demand because of an ageing population, the continuing introduction of new and ever costlier technology and, especially, customers' changing attitudes. No longer were people happy with treatment whenever and whatever providers were willing to offer. They expected top quality care now.

Purchaser/provider splits were created, making purchasers accountable for the availability of high quality services within a set budget. But the implications for health and social services were very different.

Caring For People established an 'enabling' role for social services departments in local authorities. Compulsory competition was introduced to create 'a level playing field' between service providers within local authorities and the private and voluntary sectors. Supplementary income support for residential care was abolished, and about £700 million was transferred through the mechanism of the special transitional grant to local government, which became responsible for purchasing care for dependent people based on individual need. However, 85 per cent of the grant had to be spent on independent sector care, which severely limited the development of innovative community care.

Joint purchasing between health and social services was strongly encouraged in rhetoric, but was not facilitated. It was suggested that social services departments would be the 'lead agency' for the community care of people with mental health problems. although they only spend about 5 per cent of the combined mental health budget (excluding the small mental health proportion of the special transitional grant). Health authorities would remain responsible for the 'health' element of care.

However, the meaning of 'lead agency' and the distinction between social care and health care were not clarified. With resources shrinking, nobody was likely to volunteer funding, and inevitable conflicts affected the care people received.

A game with no winners

Poor co-ordination had been a key concern in *Community Care: Agenda For Action* (Griffiths, 1988). This was dealt with by introducing care managers in social services and key workers in mental health services, the latter as part of the Care Programme Approach (DOH, 1990; DOH, 1995).

Both were responsible for co-ordinating care, although care managers were likely to have a budget to purchase care for individual clients, and key workers were not. The implication was that, as care managers, social workers became responsible for assessments from a purchasing perspective, while their health colleagues would provide direct care, sometimes in the same team.

In practice, market principles have been adopted in an over-enthusiastic manner by managers, breaking up fragile alliances between different services and sowing terrible confusion and duplicated effort.

In one instance, community mental health staff and social services staff worked jointly in teams until GP fundholders waded in. Social services withdrew, because they could not accept GPs determining team priorities and caseload, and since then independent teams have been working with similar client groups. Elsewhere, in joint teams, community psychiatric nurses were allowed to assess on behalf of social services. Even so, the problem of potential charging for social services, but not for health care, could not always be resolved.

Meanwhile, hospital wards are overflowing because social workers are not in a hurry to assess patients and take on responsibility for finding and funding residential care. Assessments in London boroughs can take six weeks. This obviously means health staff are either not admitting people with severe problems or discharging people early who are still at risk. Co-ordinated aftercare is an agreed principle, but rare in practice, with funding still tied up in hospital care.

A catch-22 situation exists: close hospital beds to find money for community teams or establish community teams to prevent admissions to hospital.

The onset of fear

Until 1981, inquiries had addressed scandals in hospitals on behalf of an angry public, almost invariably showing patients as helpless victims and staff as abusers. In July 1984, when the first green shoots of community care could be seen, one shocking event had major repercussions for mental health care, and eventually, public attitudes. Sharon Campbell, a former in-patient, killed Isabel Schwarz, her former social worker, at Bexley Hospital. From then on, community care became associated with danger.

The *Report of the Committee of inquiry into the Care and Aftercare of Miss Sharon Campbell* was published in July 1988 (Spokes *et al.*, 1988), before *Caring for People*. The importance of taking seriously reports of clients carrying knifes – Campbell had attacked others at least twice before she killed Schwarz – was chillingly echoed years later in the report into the care of Christopher Clunis in 1994 (Ritchie et al.), which also found that a series of assaults with a knife were not noted or communicated to other staff.

The Campbell inquiry's recommendations came to be enshrined in the Care Programme Approach: health and local authorities have a joint responsibility, along with voluntary organisations, to provide suitable aftercare for former informal patients until it is jointly agreed this is no longer necessary: before a patient is discharged a plan should be prepared setting out proposals for community care and the arrangements for reviewing them; multidisciplinary work in the community is essential. The Royal College of Psychiatrists was asked to produce a document on good practice for discharge and aftercare procedures, and the inquiry recommended registers of 'designated' patients living in the community, which are now in existence as supervision registers (DOH, 1995).

There was a cluster of catastrophic cases at the end of 1992: Michael Buchanan, Christopher Clunis and Ben Silcock. The impression of widespread danger is reinforced by tragedies such as the killing of Jonathan Newby, a care worker in a voluntary sector hostel, by a resident suffering from schizophrenia.

Such dramatic incidents have shifted the centre of the debate away from the potential benefits of community care for service users to the need for more beds in secure units, allegedly as much to protect the public as to protect service users from themselves. Many inquiry reports have stressed the need for better co-ordination, various types of registers and more secure beds. Interestingly, the Department of Health announced measures to address precisely these areas of concern around the time the Buchanan, Silcock and Clunis reports were published. No action was taken on other recommendations, such as intensive outreach teams and a review of community psychiatric nurses' caseloads. Both of these measures would, of course, have demanded greater resources.

The publicity around reports has created a sense of random and irrational danger. The public are angered by the message that it is not the fault of vulnerable patients, who are victims to, but of an inadequate system. Many people believe it is time to go back to the way it was, offering asylum.

The past vs. the present

Before 1984, problems in hospitals dominated the public agenda. Since then, attention – and blame – have shifted almost completely to community care. Even when incidents occurred in hospitals, such as in Georgina Robinson's case, care in the community seemed suddenly to play a significant part (Blom-Cooper et al., 1995).

This raises several important questions. Did hospital care before the 1980s prevent community scandals? Has mental health care deteriorated since then? Are people with mental health problems dangerous in the community? Jointly, the answer to these questions could determine the feasibility of community care.

In 1966, when 120 000 beds were available for mentally ill people in the UK, compared to 57 000 in 1990, a study in east London followed up people discharged from hospitals. (PRA, 1968). Not much information was available. Only 54 per cent of patients with schizophrenia discharged to known addresses could be traced after a year. Only about one-third of these lived in satisfactory circumstances and a third neglected themselves. Not much was known about the other 46 per cent, nor about people discharged to unknown addresses.

In the early 1990s, several studies give an idea of the state of community services. A survey in west Lambeth followed up 140 patients with schizophrenia after discharge from an acute unit (Meltzer et al., 1991). After a year, only one patient was lost to the services, but four had died, three from probable suicide. About half the patients could be diagnosed as psychotic and functioned poorly. Two-thirds had moved at least twice during the year, many lived in deprived circumstances, and only 10 per cent were employed. There was little co-ordination of services.

Studies of long-stay residents who left residential care during the 1980s showed that only very few were lost, and that their quality of life as well as satisfaction with services increased after moving to residential care (Anderson et al., 1993; Pickard et al., 1992). Evaluations of teams who care for people in the community with the aim of preventing hospital admission, show community care produces only marginally better outcomes in terms of psychiatric symptoms and social functioning (Muijen et al., 1994; Marks et al., 1994).

However, engagement with community services and satisfaction among users are consistently higher than with standard hospital care (Ford et al., 1995). Suicide rates do not seem to be affected by the type of care someone receives, although studies have been too small for a reliable estimate of risk.

The picture can be summarised as follows. Mental health care was mainly being provided in hospital settings during the 1960s, and scandals occurred exclusively in hospitals even though drop-out rates of service users following discharge were extremely high because aftercare was inadequate. In the 1990s, care has gradually shifted to the community, with vast improvements in engagement with services and satisfaction, and the scandals are associated with community care.

There are several possible interpretations, and none can be ruled out. Firstly, discharge planning and risk assessment might have been more cautious 30 years ago. The higher number of beds, especially long-stay beds, would have allowed longer stays, and people such as Campbell, Clunis or Buchanan might have been kept in hospital for a lifetime. Therefore, scandals would not have occurred in the community. The weakness of this point is that so many people were lost to services after discharge, as the east London study shows (PRA, 1968).

Second, there has been a shift in the limits of responsibility of mental health services. Scandals in the community during the 1960s, would simply not have been attributed to mental health services, since people would have been discharged from care as and when decided by the psychiatrist. From then on, mental health services would no longer be responsible for any of the former patient's actions. Only now are mental health workers considered accountable for the actions of any person suffering from mental health problems, whether in hospital or in the community, and whether receiving care at the time, in the past or not at all – a position all too familiar to social workers.

In principle this shift of responsibility is based on good principles of care, such as easy access, tailoring services to individual needs and continuity between hospital and community. In practice it puts unrealistic pressures on a service which has neither the resources nor the skills to satisfy such expectations.

Finally, the social context has changed dramatically. In the 'golden sixties' the incidence of violent crime was low, unemployment and homelessness were rare and people cared for each other. The increase in crimes associated with mental health, if real, could be a reflection of an increase in criminal behaviour in society in general. There is also the fact that both crime and mental illness are associated with similar risk factors such as inner city problems, homelessness, poverty and unemployment.

The public associates these violent incidents with the large numbers of people they hear are discharged into the community, understandably responding with fear and anger. It is assumed that reported incidents are only the tip of the iceberg.

The Confidential Inquiry into Homicides and Suicides by Mentally Ill People was set up in 1991 to investigate the relationship between care for mentally ill people and violence (Confidential Inquiry, 1996). The inquiry had received information from the Home Office of 100 cases of homicide over 18 months by people who subsequently received psychiatric care, 34 of whom had been in contact with specialist psychiatric services in the 12 months before the homicide. (The other 66 were to receive some kind of mental health care following the homicide, but no further details are known. This is obviously worrying, since some of these might have been mistakenly ignored by services.) During the same period about 900–1000 homicide cases could have been expected in England and Wales, meaning those previously in contact with services were responsible for about 3 per cent of homicides. (This 3 per cent is debatable. For example, 398 people were charged with homicide during 1992. Extrapolated to 600 over 18 months, this would produce 7 per cent.)

It is particularly important that only two victims were randomly killed by strangers; the majority were family. Equally important, in 15 of the cases the patients had refused some form of care.

None of the homicides were committed by a patient who had been discharged from a long-stay mental hospital into the community. Another study found that out of 278 people being followed up for a year after discharge, only two (0.7 per cent) were involved in violent offences (Dayson, 1993). One of these was hardly independent of the research, since the patient assaulted his father who accompanied a researcher. More typical of police involvement with this group was a woman who panicked in a shopping centre, and asked a police officer for help.

Even though the incidence of homicide may be marginally higher for people suffering from mental health problems than for the population as a whole, the risk is clearly very low, especially to non-relatives. Far from being the tip, reported incidents seem to be the whole iceberg.…

Staff under fire

Mental health staff of all disciplines have been subjected to a continual bombardment of policies and guidelines over the past five years, mostly intending to prioritise people with the most severe mental health prob lems. Many have not yet digested the full implications of the purchaser/ provider split, the formation of trusts, the care programme approach, care management, supervision registers, guidance on discharge, and now … supervised discharge (DOH, 1996). They have reached the point where another set of instructions is at best ignored and at worst reinforces a sense of being scapegoated for political convenience.

An example is central London, where the issue of discharge planning and arranging follow-up and aftercare is complicated by the enormous pressure on acute in-patient beds. Planning and co-ordinating aftercare takes time and effort, and staff are often unable to make the necessary arrangements before someone is discharged without blocking a bed.

When the 1994 guidance on discharge was issued, one consultant psychiatrist said privately that it looked very sensible and he would have liked to implement it, but that in fact if he did so the ward would instantly become closed to any further admissions, pressure would build up on neighbouring mental health units, and within weeks the whole system would grind to a halt. He continued to provide basic and sensible care, rather than following guidelines which were impossible to implement without the injection of massive resources – not only into health and social services, but also housing and a large number of support services.

However, mental health workers put themselves at increasing risk unless they adhere to the letter of guidance, rather than the spirit of good practice. An inquiry culture is dawning, with staff constantly aware that any variation from recommended perfect practice could lead to an unpleasant afternoon in front of a cynical inquiry panel and the humiliation of being named in a report, and possibly dismissed. Following the guidance to set up inquiries for every case of suicide and homicide involving mental health services, another 50 are due to be held in 1995–96.

In principle, it can only be right to question bad practice. However, the perceived message – that services are guilty unless proven innocent – is damaging even if it is unintended. Defensive practice is the result, concentrating all energy on detailed care and filling in forms, praying that no nasty events will emerge from the many patients for whom resources cannot offer the comprehensive care they need.

The effect on staff of this cumulative burden of guidance, recommendations and legislation is bewilderment and burn-out.

The future: defusing paranoia

The disarray of mental health services, with low morale from top to bottom in the NHS and local authorities, has completely overshadowed the real progress that has been made during the last decade, and has sent repeated signals to the public that community care is out of control ... The only way to reassure everyone, especially the public, is to put forward a constructive and feasible strategy for developing mental health care, without sidestepping or delegating the most difficult decisions, or denying the problems.

Providers need to be told what their objectives are. If it is decided that specialist mental health services are responsible for the high-risk group, they should be given what they need to achieve this, including 24-hour crisis services, beds, residential care and employment schemes. Either resources must increase or expectations must shrink. By international standards mental health care in the UK is underfunded, although it is expected to deliver a high quality and comprehensive service.

Policy and guidance should support the aims of the service, not contradict them. For example, GP fundholding frequently demands shifts in priorities from severely mentally ill people towards the 'worried well'. Policy should not be allowed to be construed as critical or unrealistic, as is the case with aspects of the care programme approach, care management, the supervision register and discharge guidance.

Inquiries should not be used as a constant threat, but as a corrective mechanism, encouraging better care. They should focus more on local circumstances, rather than playing to a national gallery, allowing managers to deal with care issues. Examples of good practice could be rewarded, rather than taken for granted. And it needs to be recognised that not every disaster is preventable, even if mediocre care was occasionally provided.

The duplication of services across health, social services and the independent sector is absurd and wasteful, and bewildering to service users. At a strategic level, incentives need to be created towards joint commissioning: for example, increasing the mental illness specific grant by top-slicing the NHS budget, and only allocating funding if a genuine joint management structure with clear accountability is in place. At a team level, the care programme approach and care management should be unified.

Finally, the public should know who is responsible for mental health care. Someone needs to be accountable locally, and it should be possible to contact them. Standards of care should be defined by contracts and adhered to by local services, and if not, action should be expected. This sort of management would bring mental health services closer to the public, establishing a sense of trust and ownership.

References

Anderson, J., Dayson, D., Wills, W., Gooch, C., Margolius, O., O'Driscoll, C. and Leff, J. (1993) 'The TAPS project 13: clinical and social outcomes of long-stay psychiatric patients after one year in the community' British Journal of Psychiatry, vol. 162, 45–55.

Audit Commission (1986) Making a Reality of Community Care, London, HMSO.

Blom-Cooper, L., Murphy, E. and Hally, H. (1995) The Falling Shadow, Torquay, South Devon Health-Care Trust.

Chamberlin, J. (1977) On Our Own, London, Mind.

Confidential Inquiry (1996) Report of the Confidential Inquiry into homicides and suicides by mentally ill people, London, Royal College of Psychiatrists.

Dayson, D. (1993) 'The TAPS Project 12: crime, vagrancy, death and re-admission', British Medical Journal, vol. 162 , 40–44.

Department of Health (1989a) Caring for People: community care in the next decade and beyond, London, HMSO.

Department of Health (1989b) Working for Patients, London, HMSO.

Department of Health (1990) The Care Programme Approach for People with a Mental Illness London, HC(90) 24/LASSL(90)11, DoH.

Department of Health (1995) Building Bridges: a guide to arrangements for inter-agency working for the care and protection of severely mentally ill people, London, DoH.

Department of Health (1996) Guidance on Supervised Discharge (after-care under supervision) and Related Provisions: supplement to the code of practice published August 1993 pursuant to Section 118 of the Mental Health Act 1983, London, HMSO.

Department of Health and Social Security (1972) White paper on National Health Service Reorganisations, London, HMSO.

Department of Health and Social Security (1975) Better Services for the Mentally Ill, London, HMSO.

Goffman, E. (1961) Asylums, Harmondsworth, Pelican.

Griffiths, Sir, R. (1988) Community Care; agenda for action, London, HMSO.

Faulkner, A., Vida, F. and Muijen, M. (1994) A survey of adult mental health services, London, Sainsbury Centre for Mental Health.

Ford, R., Beadsmoore, A., Ryan, P., Repper, J., Craig, T. and Muijen, M. (1995) 'Providing the safety net: case management for people with serious mental illness', Journal of Mental Health, vol. 1, 91–97.

Kesey, K. (1972) *One Flew Over the Cuckoo's Nest*, London, Marion Boyars.

Marks, I., Connolly, J., Muijen, M., Audini, B., McNamee, G. and Lawrence, R. (1994) 'Home-based versus hospital-based care for people with serious mental illness', *British Journal of Psychiatry*, vol. 165, 179–194.

Martin, J. (1984) *Hospitals in trouble*, Oxford, Blackwell.

Meltzer, D., Hale, A., Malik, S., Hogman, G. and Wood, S. (1991) 'Community care for patients with schizophrenia one year after hospital discharge', *British Medical Journal*, vol. 303, 1023–1026.

Ministry of Health (1962) *A hospital plan for England and Wales*, London, HMSO.

Muijen, M., Cooney, J., Strathdee, G., Bell, R. and Hudson, A. (1994) 'Community psychiatric nurse teams: intensive support versus generic care', *British Journal of Psychiatry*, vol. 165, 179–194.

Pickard, L., Proudfoot, R., Wolfsen, P., Clifford, P., Holloway, F. and Lindesay, J. (1992) *Evaluating the Closure of Cane Hill*, London, The Sainsbury Centre for Mental Health.

Piercy, M. (1979) *Woman on the Edge of Time*, London, Women's Press.

Psychiatric Rehabilitation Association (1968) *The mental health of East London*, London, Psychiatric Rehabilitation Association.

Ritchie, J., Dick, D. and Lingham, R (1994). *The report of the inquiry into the care and treatment of Christopher Clunis*, London, HMSO.

Spokes, J., Pare, M. and Royle, G. (1988) *The report of the committee of inquiry into the care and aftercare of Miss Sharon Campbell*, London, HMSO.

White, E. (1990) *The Third Quinquennial National Community Psychiatric Nursing Survey*, Manchester, University of Manchester.

Wing, J. and Brown, G. (1970) *Institutionalisation and Schizophrenia*, Cambridge, Cambridge University Press.

18

Beyond the asylum

JULIAN LEFF

Since the early 1950s, psychiatric hospitals in Europe and the US have been running down. It has been estimated that over half a million long-stay patients have been discharged from psychiatric hospitals in the US and UK (Lamb, 1993; see also Muijen in this volume). This massive change in psychiatric care has been endorsed by successive British governments since 1961, when then-health minister Enoch Powell predicted the closure of 75 000 beds in asylums over the next 15 years:

> There they stand, majestic, imperious, brooded over by the gigantic water-tower and chimney combined, rising unmistakable and daunting out of the countryside, the asylums which our forefathers built with such solidity. Do not for a moment underestimate their power of resistance to our assault.
>
> (Enoch Powell, 1961)

Nevertheless, only a handful of small-scale studies of the key issues have been conducted (O'Driscoll, 1993).

In 1983, North East Thames Regional Health Authority announced a 10-year programme to close two of its six psychiatric hospitals, Friern and Claybury. Friern closed in March 1993; Claybury is now expected to close in 1996. The announcement of a planned closure programme was seen as an unrivalled opportunity to mount a comprehensive evaluation of the policy and its implementation. With considerable foresight, the RHA established TAPS, the team for the assessment of psychiatric service, with funding help from the Department of Health and the King's Fund. TAPS set itself a broad agenda of research, including the three main types of patient services in the hospitals – acute, long-stay non-demented and psychogeriatric. Also covered were staff, relatives and members of the public. Close collaboration was maintained with health economists from the Personal Social Services Research Unit at Kent University, who costed the services for all groups of patients studied in the hospital and in the community. A sociological study was made of the

Edited from Leff, J. (1995) 'Beyond the Asylum', *Health Services Journal*, (22 June), 28–30.

planning process, including decision-making and implementation. This research programme was so broad it needed a multidisciplinary team, and TAPS has included social scientists, psychiatrists, psychologists, psychiatric nurses, health economists, statisticians, data managers and administrators.

We have distilled a series of recommendations which we believe will be of value to people planning community services to replace psychiatric hospitals.

Long-stay non-demented patients

Staffed homes work well, but it would be preferable to group them in core and cluster arrangements. This combats staff isolation, ensures staff support in case of crises and provides a central facility for patients to engage in structured activity. It also offers patients the chance to move between more and less highly staffed settings as their needs and preferences change.

Drop-in centres should be accessible to patients who find structured activity too demanding.

Patients with a history of vagrancy and/or absconding should be placed in staffed houses.

As the patients show little progression to greater independence, the houses should be considered their permanent homes. A range of accommodation should be provided with graded staffing levels to allow moves to more or less intensive care. Since rehabilitation in the hospitals is as effective as in the community, and since the patients with the highest readmission rate are those who are young, new, long-stay patients, this group should be the target of intensive rehabilitation before reprovision.

The most socially disabled patients should not be left until last but should be mixed with more able patients at all stages of reprovision. Selecting patients for reprovision should be flexible and not dictated by current ward composition.

Difficult-to-place patients (DTP)

DTP patients should be identified well in advance of closure by surveying the long-stay population and selecting patients with persistent special problems, such as aggression and inappropriate sexual behaviour.

These patients should be assembled in one unit and given an intensive behavioural programme tailored to their special problems. Ideally the staff in the unit will accompany the patients into their community placements.

The medication of each DTP patient should be reviewed and rigorous attempts made to rationalise high-dose, multi-neuroleptic regimes. A typical neuroleptics should be given a trial.

Any special needs unit established for DTP patients should be located in hospital grounds or nearby but should operate an open-door policy which allows patients to go out into the community. It should provide a domestic environment.

Patients in a special needs unit should have the frequency and intensity of their special problems monitored regularly with a view to transfer to a community home as soon as practicable.

Two-thirds of the DTP patients from Friern hospital were new long-stay. It must be recognised that patients of this type will continue to arise from within a community-based service, and will take up any vacated places in a special needs unit.

Psychogeriatric patients

Elderly patients with functional psychoses should be transferred from hospital to community residences as soon as possible to prevent further deterioration of behaviour and cognitive function.

Elderly patients with functional psychoses, although they may be as dependent as dementia patients in terms of physical care, function at a higher cognitive level and, for ease of nursing, should not be mixed with dementia patients.

Staff caring for dementia patients, whether in a hospital ward or a community home, should ensure that patients can have personal possessions, that there is a facility for them to make drinks for themselves and that it is easy for relatives to take them out.

Admission facilities

The closure of a psychiatric hospital with admission facilities should not be used as an opportunity to reduce the overall complement of admission beds.

In calculating the number of admission beds required for a catchment area without a psychiatric hospital, provision must be made for the needs of discharged long-stay patients for inpatient care. The number of beds required is likely to vary with the characteristics of the population of former long-stay patients.

Calculating the number of admission beds needed should take into account the accumulation of new long-stay patients, the rate of which is strongly determined by the socio-economic level of the catchment-area population.

A range of sheltered accommodation should be available to discharge patients on admission wards who are well enough to leave.

An intermediate stay rehabilitation unit is required to accommodate patients who stay for more than a few months on an admission ward. Such a unit should provide a domestic environment within the community and should have input from staff such as psychologists and occupational therapists.

Decision-making process

Some psychiatric hospital staff who are sympathetic to reprovision should be recruited as leaders of the process, or 'product champions'.

Purchasers should involve users and carers in service specifications. This is essential for the development of flexible services responsive to users' needs.

Central co-ordination is necessary to integrate planning work with providers, local authorities, users, carers, housing associations and other non-statutory agencies. Overall accountability should remain with the health authorities. Co-ordination is also required at area (strategic) and case (tactical) level.

In addition to housing needs, planning should include facilities nearby for structured and unstructured activities. More effort should be directed at developing work schemes. Clinical supervision of patients should be planned to redistribute responsibility between primary / secondary care.

There is a need for better information on providers (for example, a survey of non-statutory provision that already exists), on purchasers' intentions and on users' needs.

A poor understanding of the benefits of community care exists among members of the public and some psychiatric hospital staff. This can hamper the reprovision programme. Information campaigns are necessary, with the functions of consulting and educating as well as passing on decisions.

Purchasing property for the reprovision programme is a specialist function, not carried out well by the NHS. Purchasing, leasing property and obtaining planning permission require adequate advice from local authority housing officers and legal support. There is a place for a fully staffed, specialist legal department in reprovision.

Health economics study

Community care in general is not cheaper than hospital care, although it may be a great deal cheaper for some people.

From our various prediction analyses, we estimate that the savings from closing hospitals should be adequate – in the long term – to fund reprovision. But in the short term, additional funds are needed to allow capital investment in community facilities, and to fund the 'double running costs'.

It is important that adequate funds are moved from hospital to community to ensure good quality reprovision and that those funds are ringfenced.

Community care services need funds before hospitals can release them – for capital investment, to recruit and train staff, to build a service up to its full or most economical operating level – and this requires bridging funds or some form of 'double funding' of hospitals as they run down.

The eventual cost of providing community care may, to a certain extent, be predicted by characteristics assessed before people move out of hospital. Higher costs are associated with greater needs and appear to produce better outcomes in the first year after discharge.

There is a need to alter the funding balance, not only to move resources from hospitals to the community, but also from NHS budgets to the budgets of other agencies. The challenge for NHS purchasers and community care planners is to relocate funds between agencies, for the financial burden of reprovision is not distributed equally or in a similar way to hospital-related costs. Resource flexibility is difficult to achieve but highly desirable.

Service co-ordination is needed at both the tactical and strategic levels – for case planning as well as area planning. Care management and the care programme approach are as relevant here as anywhere. At the strategic level, the main statutory and non-statutory agencies need to develop joint long-term plans for community care development.

The costs of reprovision for DTP patients may be high because effective modes of support have yet to be fully developed. Over time, and with experience, purchasers should be encouraging providers to develop more appropriately targeted services for these people.

New service developments, particularly work schemes, domiciliary support and befriending, require additional resources, not for the health service but for local authorities or the voluntary sector.

The health and local authority, voluntary and private sectors are not equally good at providing effective and cost-effective care for every former hospital patient. Careful planning and discussion is needed to ensure the appropriate targeting of resources from across the sectors on different types and levels of need.

References

Lamb, H. R. (1993) 'Lessons learned from deinstitutionalisation in the US', *British Journal of Psychiatry*, vol. 162, 587–592.

O'Driscoll, C. (1993) 'The TAPS Project 7, Mental hospital closure: a literature review of outcome studies and evaluative techniques', in Leff, J. (ed.) 'Evaluating community placement of long-stay psychiatric patients', *British Journal of Psychiatry*, vol. 162, Supplement 19, 7–17.

Powell, E. J. (1961) Speech by the Minister of Health, the Rt. Hon. Enoch Powell, *Report of the Annual Conference of the National Association for Mental Health*, London.

19

Media images of mental distress

GREG PHILO, JENNY SECKER, STEVE PLATT,
LESLEY HENDERSON, GREG McLAUGHLIN and
JOCELYN BURNSIDE

This article reports on the findings of research on the media coverage of mental illness conducted by the Glasgow University Media Group and the Health Education Board for Scotland. The research used content analysis and an audience response study. This extract from the article begins with a description of the methods used.

Methods

Content analysis

The sample for the content analysis comprised a range of local Scottish and national media output for the month of April 1993, including factual and fictional formats aimed at both adults and children. Factual formats included news items, comment and analysis in the press; news and current affairs programmes on television; and feature items, medical columns and problem pages in magazines. Fictional formats on television included soap operas, single dramas and films. In the press and magazines they included short stories and comic strips. In some cases, for example when soap opera story lines ran before or after the sample period, it was necessary to record material from outwith the period. Otherwise relevant output was recorded and/or stored for a month.

The purpose of the content analysis was to reveal the dominant messages about mental health issues presented across the range of media examined.

Edited from Philo, G., Secker, J., Platt, S., Henderson, L., McLaughlin, G. and Burnside, J. (1994) 'The Impact of the Mass Media on Public Images of Mental Illness: Media Content And Audience Belief', *Health Education Journal*, vol. 53, 271–281.

For the initial stage of our study we established a general profile of media content examining the focus of news items, magazine features and story lines in fictional drama (including soap operas). For this initial profile, each of these was counted as a single item. Such profiles give a sense of the main areas of media interest and the general contours within which the coverage is grouped. In the second phase of the content analysis we engaged in a detailed examination of individual texts. This method of analysis had three dimensions:

(1) explanatory or interpretative themes were identified:
(2) the way in which each theme was developed in its specific context was examined;
(3) the frequency with which different themes appeared and their relative power in terms of the size of audience they can be expected to reach were assessed.

The audience reception study

The sample for the audience reception study consisted of seven groups with an average of ten people participating in each. A total of 70 people took part. Six of the groups comprised a general sample drawn from the west of Scotland and structured to be broadly representative of the area. The other group consisted of seven people who had used psychiatric services and who were working in a computer skills training programme in Edinburgh.

The sample was not large enough to make generalisations about the whole Scottish population. Instead, the intention was to explore the process by which media accounts are interpreted and contribute to the formation of beliefs. For this reason an effort was made to work with people in naturally occurring units, for example a family, couples living in the same housing block, or a group of people who worked together, in order to preserve elements of the social context within which people might read newspapers, watch television programmes, and discuss the issues raised.

The methods we used to examine the audience sample's beliefs involved three phases:

(1) A series of exercises involving sub-groups of two or three people: each sub-group was asked to write news reports prompted by copies of the original headlines; those who watched *Coronation Street* were also asked to write dialogue for an episode of the programme prompted by still photographs.
(2) Individual group members were asked to write answers to a series of nine open questions. Two questions related to the content of the exercises, while the others were intended to enable respondents to

express beliefs about mental illness and to indicate where their ideas had come from.

(3) Individual in-depth interviews designed to explore respondents' answers to the written questions.

Findings

Content analysis

The survey of media output yielded a total of 562 items relating to mental health. Five main categories of coverage emerged from the content analysis: violence to others; sympathetic coverage: harm to self; 'comic' images; and criticisms of accepted definitions of mental illness. Table 19.1 shows the distribution of the 562 items across these five categories.

Similar themes emerged from the fictional accounts in this category. For example, in films shown on television, Richard Dreyfus pursued a 'crazed killer' (*Stakeout*, ITV, 27 April), while Kurt Russell played a reporter involved with a 'psycho killer' (*The Mean Season*, BBC1, 30 April). In forging the link with violence, films of this genre also linked mental illness with the notion of 'split personality', and this theme was reproduced during the sample period by most of the main soap operas. From early 1993, for example, *Coronation Street* developed a story line concerning an angelic-looking nurse, 'Carmel', who was portrayed at first as a 'fresh-faced, home-loving Irish girl'. In later episodes, however, it emerged that behind this angelic front lurked an intensely manipulative character who would clearly stop at nothing to win the married man with whom she had become obsessed.

Such items linking violence and mental illness outweighed the second most common category, sympathetic coverage, by a ratio of almost four to one. Even these figures exaggerate the relative coverage of the two categories, because items portraying violence tended to be given a high

TABLE 19.1   Media coverage of mental health/illness, April 1993

Output category	Number of items	Percentage of total items
Violence to others	373	66
Sympathetic coverage	102	18
Harm to self	71	13
'Comic' images	12	2
Criticism of accepted definitions of mental illness	4	1
Total	*562*	*100*

profile, whereas sympathetic coverage was largely confined to back-page material in newspapers and magazines such as problem pages and health columns. Unsurprisingly, given its origin, the content of this coverage revolved mainly around the theme of how to 'cope' with mental health problems.

The majority of items in the third most common category, harm to self, were non-fictional reports of suicides or attempted suicides. For the most part these reports portrayed the events concerned as tragedies by focusing on their human context of depression and anxiety. For example, The *Sun* reported the death of a model under the headline, 'Death Leap a Cry for Help' (4 April), and the death of a Conservative Party worker was reported by the *Daily Record* under 'Secret Pain of Suicide Tory' (8 April).

In some cases, however, reports of suicides and attempted suicides emphasised a 'bizarre' aspect of the events described. Extensive coverage of this sort was given, for example, to the story of a man who had apparently jumped from a high-rise block and survived by landing on a car: 'Nissan impressed as man falls 200 ft' (The *Guardian*, 3 April). In similar vein, some newspapers focused on sexual angles which could be linked to suicides: 'Suicide of Sex Slave Nutter' (*Sunday Sport*, 24 April); 'Tragic Patient had Sex Fantasy' (The *Sun*, 20 April).

Audience reception study

In writing their own stories under the headlines they were given for the first media exercise, the audience groups demonstrated a remarkable ability to reproduce the style and language of the popular press. The first headline they were given related to a story, later proved to be completely untrue, about an arson attack on a young boy. Most of the stories produced in response to this headline included phrases like 'evil maniac', and many stories also made clear the fears which such a report can generate. This passage was written by three women whose agreement with the sentiments it expresses became apparent as they worked on it: 'Police are today looking for the evil maniac who tried to disfigure this innocent child. What kind of world do we live in when a child can't even be allowed to play in the street?' In response to a written question about the original story, 18 people stated that they remembered it, but only nine of them knew that it had proved to be untrue.

A very different note was sounded by a second headline relating to a man who had spent 33 years in hospital before receiving his Diploma in Education. The stories written by the audience groups reflected the more upbeat tone of this report, again in terms which closely paralleled the themes of the original. Some stories also developed other themes relating to mental illness and community care. Although these were not contained in the original, they clearly derived from other media accounts.

The second exercise, involving writing dialogue prompted by stills from *Coronation Street*'s 'Carmel' story line, produced a dramatic response: as soon as the pictures were shown there were murmurs of recognition and animosity. Although they were given no information at all about the plot and had to work entirely from photographs, some group members were extremely accurate in their reproduction of the dialogue. For example, a key moment occurs when 'Carmel' is put out of the house by another character, 'Sally', and this scene recurred in some of the audience groups' own work with startlingly accurate dialogue.

A question included in the second phase of the study referred to the 'Carmel' story line and was intended to tap attitudes to people portrayed as experiencing mental health problems. Group members were asked: 'How would you have reacted to Carmel, if you have been Gail?' (the wife of the man with whom Carmel was obsessed.) Two-thirds of the people in the general sample gave replies threatening aggression or violence. These varied from: 'Battered her bloody mouth in', to the more genteely expressed but no less aggressive: 'I would have been very upset and taken violent action and chucked Carmel out of the house'. However, nine respondents among the general sample were more sympathetic in that they suggested obtaining medical help. Their responses were in part related to personal experiences of mental illness, but such experience did not always lead to a sympathetic response to 'Carmel'. The issue of how media output can inter-relate with direct experience and other cultural factors is crucial, and this was the focus both of the other written questions, and of the individual interviews.

One of the key issues explored was whether mental illness was believed to be associated with violence. Almost two-thirds of the people in the general sample believed this to be so, and two-thirds of these people, or two-fifths of the whole general sample, gave the media as the source of their beliefs. In many cases they referred to a combination of factual and fictional sources. These extracts from two interviews illustrate how beliefs could be shaped through the interaction of the two formats:

A lot of things you read in the papers and they've been diagnosed as being schizophrenic. These murderers – say Donald Neilson, was he no schizophrenic? – the Yorkshire Ripper ... in Brookside that man who is the child-abuser and the wife-beater – he looks like a schizophrenic – he's like a split personality, like two different people.

I always thought mental ill people would not be able to do much for themselves and would be looked after by someone, but then you get evil people like the Kray twins who were evil, violent men who I suppose must have had some form of mental illness to do the things they have done to other people and seemed to enjoy it ... I seen the film – they looked like pretty normal people but they weren't.

Most of the people who disagreed with these views and who rejected the dominant media message made their judgement on the basis of personal experience. In particular, the group from Edinburgh cited their own experience and that of meeting other people diagnosed as suffering from illnesses such as schizophrenia. This man, for example, wrote explaining why he did not associate schizophrenia with violence:

Some of my answers came from being depressed myself and through this the people I came into contact with gave me more knowledge.... As a day-patient, out-patient, occupational therapy ... I met about five people who had schizophrenia.

In addition to the Edinburgh group, a third of the general sample cited personal or other direct experience as the key factor informing their rejection of the dominant media message. Conversely, a smaller group of ten people also had experience of mental illness and gave this as a reason for associating it with violence. Such experience could then be confirmed or developed by negative media coverage.

However, 13 people who took part in the study had non-violent experience of mental illness which was apparently overlaid by media influences. A striking illustration was given by a young woman who lived near a psychiatric hospital. She wrote that she had worked there at a jumble sale and mixed with patients. Yet she associated mental illness with violence and wrote of 'split/double personalities'. She went on to say:

The actual people I met weren't violent – that I think they are violent, that comes from television, from plays and things. That's the strange thing – the people were mainly geriatric – it wasn't the people you hear of on television. Not all of them were old, some of them were younger. None of them were violent – but I remember being scared of them, because it was a mental hospital – it's not a very good attitude to have but it is the way things come across on TV, and films – you know, mental axe murderers and plays and things – the people I met weren't like that, but that is what I associate them with.

Discussion

The relationship between different media messages, our beliefs and personal experience is extremely complex. Although there is little doubt that the mass media can exert great influence over audiences, people are not simply blank slates on which its messages are written. The media exist within developing social cultures. They do not create the whole social world or how we think about it. On the other hand, they are certainly very important sources of information and can generate strong emotional responses in their viewers and readers. The study reported here was small in scale, but the findings suggest that the media can play a significant role not only in informing the public, but also in fuelling beliefs which contribute to the stigmatisation of mental illness.

The potential for informing the public is illustrated by the accuracy with which our audience sample were able to reproduce and develop media stories with only minimal prompts. News reports and 'factual' programmes are only one part of the development of such social consciousness, but they do appear to have a significant influence.

As a corollary, however, the potential for misinformation is also great. At one level, responses to questions about the 'fire maniac' story illustrate how a completely false account can enter into the public store of memories and beliefs. At another, more subtle, level the use of loaded terms like 'madman' or 'maniac' to describe violent criminals suggests an association with mental illness which recent research in the United States indicates is minimal (Monahon, 1992). That this medium has the capacity to produce intense affective responses within its audience and to direct these against certain characters or types of behaviour is illustrated by our sample's response to writing dialogue prompted by stills from the 'Carmel' story line. The ability to reproduce dialogue and the intensity of response to the story were both sustained by group members over long periods of time, suggesting that the impact of this particular message was very strong. That the portrayal of a 'disturbed' woman should have provoked such a violent response in this 'normal' sample is not without irony.

In some respects, 'our findings in relation to the interaction of this media output linking mental illness with violence and other influences on beliefs confirm those of previous research in other areas, that personal experience, where it exists, is a much stronger influence on belief than media content (Philo, 1990). But one of the most striking findings of this study was that in several cases this pattern was reversed: in our sample several people appeared to believe media messages in preference to the evidence of their own eyes. Where mental illness is concerned it seems that some media accounts can exert exceptional power over readers and viewers.

Although less obviously problematic, the sympathetic coverage we found also raises questions about the image of mental illness presented. While this coverage was clearly well intentioned, the views presented were almost always those of medical 'experts'. Although people with direct experience of mental health problems have their own, sometimes very different views, these were rarely represented. In keeping with the resulting impression of a group of helpless 'victims' in need of 'expert' advice, coverage depicting people's ability to live relatively competent, independent lives was also very unusual. Taken together with the rarity of any questioning of accepted definitions of mental illness, an issue which continues to provoke intense debate (Johnstone, 1989), these findings suggest that even 'sympathetic' coverage can present a limited version of mental health issues.

References

Johnstone, L. (1989) *Users and Abusers of Psychiatry* London, Routledge.

Monahon, J. (1992) 'Mental disorder and violent behaviour: perceptions and evidence', *American Psychologist*, vol. 47(4), 511–521.

Philo, G. (1990) *Seeing and Believing: the influence of television*, London, Routledge.

20

Towards understanding suicide

STEVE TAYLOR and ALAN GILMOUR

Introduction

One of the major dangers of mental disorder and emotional breakdown is the risk of suicide. This article examines the incidence of suicide and deliberate self harm, discusses research into its characteristics and considers some of the issues involved in recognition and prevention.

Incidence

In recent years in Britain around 5000 deaths a year are officially recorded as suicides, a suicide rate of just over 11 per 100 000 population. Although this is comparatively low for a developed country, after illness and general accidents, suicide is the next most common cause of death in Britain and nearly as many people kill themselves as are killed in motor vehicle accidents. One of the targets of the Department of Health's policy document, *The Health of the Nation* (Department of Health, 1992) is to reduce the suicide rate by 15 per cent by the year 2000.

Researchers and policy makers use the official suicide rates not only to document the incidence of suicide, but also to help to identify some of the groups most at risk. According to the statistics, men are now three and a half times more likely to kill themselves than women (Figure 20.1), and particularly vulnerable groups include farmers, health professionals, the unemployed, those divorced or recently separated, people being treated for mental or life-threatening physical illnesses and prisoners. One of the most recent developments is the increasing suicide rate of young males (Burton et al., 1990). In 1982, 7 young men in every 100 000 killed themselves. By 1992 the figure had risen to 12 in 100 000, an increase of 71 per cent.

172

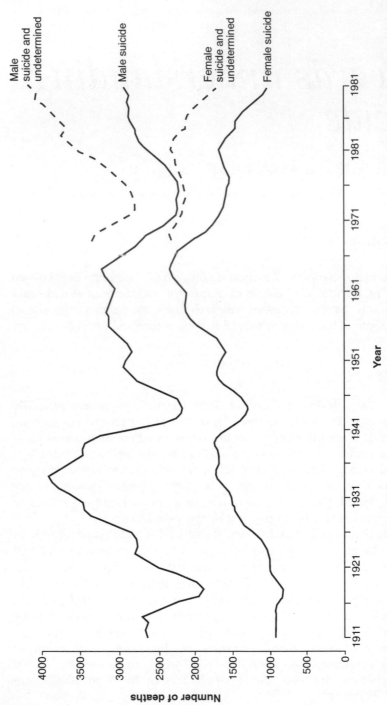

FIGURE 20.1 Number of suicide deaths according to gender, 1911–1990, plotted as a three-year moving average: England and Wales. To allow for the possibility of under-recording suicide deaths, verdicts of suicide are shown alone and shown combined with verdicts of death from undetermined causes.
Source: Charlton et al, 1992: figure 1, page 10.

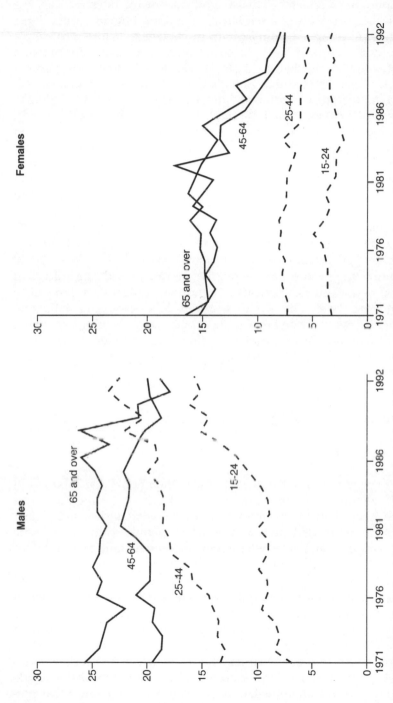

FIGURE 20.2 Suicide and age. Verdicts of suicide and deaths from undetermined causes; males and females; England and Wales 1971–82; Rates per 1000 000 of each age group. The graphs show the overall fall in suicide rates for females at all ages, and for both genders post-retirement age, and the rise in suicide rates for males under 45.
Source: Social Trends No 25 1995, Central Statistical Office, page 132.

It is important to remember that the official suicide rates are only the tip of an iceberg of deliberate self harm, injury and human misery. First, the statistics are compiled from *legal* decisions where, before recording a suicide verdict, a coroner has to be sure beyond reasonable doubt that a person intended to take their own life. Where there is doubt and particularly a lack of clear evidence of 'suicide motive' such as depression, officials tend to record a verdict of 'undetermined', or even accidental, death. Taylor (1982) examined the inquests on a number of people seen to have jumped, or fallen, in front of London Underground trains and found just over half recorded as suicides. This suggests that the number of people killing themselves may be as much as double the official rate.

Second, large numbers of people commit acts of deliberate self harm which would lead to death without medical attention. While there is no comprehensive data on 'attempted suicide', research from Oxford suggests that about 120 000 people a year, 60 per cent of whom are under 35, are admitted to hospitals in England and Wales for emergency treatment for deliberate self poisoning and self injury (Hawton and Fagg, 1992). Third, there is the problem of people having to live with the suicide of someone close to them. The suicide of a partner, relative or friend can be emotionally scarring. The usual sense of loss and grief following bereavement is compounded by feelings of guilt, anger and the recurring question of why did they do it? The suicide of an individual can devastate so many lives that The Samaritans have likened the costs of suicide to ripples on a pond.

Characteristics of suicidal behaviour

Uncertainty of outcome

The word suicide seems clear enough: it is intentional self killing and attempted suicide is where a person wants to die but fails. From this basis most people, including most health professionals, assume a clear distinction can be made between cases where people really want to die ('genuine' suicides) and cases where they are only making a 'gesture', or 'cry for help', and really want to live ('false' suicides). However detailed research into the micro social contexts of suicidal behaviour has rendered this common-sense assumption untenable. The majority of suicidal acts are undertaken with confused and ambivalent intentions. This uncertainty is reflected in the ways in which people go about harming themselves. In a pioneering study two Swedish psychiatrists made a detailed reconstruction of the circumstances in which 500 serious overdoses took place (Ettlinger and Flordah, 1955). They found that while only 7 per cent of the attempts had been 'planned' in the sense that precautions had been taken against discovery, only 4 per cent were 'harmless'. In the vast majority of cases the outcome was left to chance.

Many subsequent studies have confirmed that the majority of serious suicidal acts, including many that actually end in death, are gambles with life and death, the outcome often being determined by factors outside the individual's control.

Some researchers use the term parasuicide to describe behaviour which is something less than a wholehearted attempt to end life but much more than a suicidal gesture. Most acts of deliberate self harm are parasuicides. Even suicidal acts which are objectively lethal may be *subjectively* gambles with death.

The observation that the majority of suicidal acts are characterised by uncertainty of outcome has important implications both for research and for working with suicidal people. While more men than women commit suicide, parasuicide is more common among women (Rich et al., 1992). It is also very important for practitioners to realise that acts of deliberate self harm which do not end in death are not necessarily manipulative gestures, but may in fact have a strong suicidal component.

Process

Suicide researchers do not confine their attentions to the suicidal act itself but, as far as possible, try to reconstruct events leading up to it. We can refer to this as studying the *suicide process*. From decades of research into suicide and attempted suicide the psychiatrist Erwin Stengel reached a rather paradoxical conclusion. He argued that the suicide process, as well as being directed towards death and dying, was also in most cases directed towards life and survival. He therefore referred to the suicidal act as being 'Janus faced' – Janus being a mythological figure with two faces pointing in opposite directions (Stengel, 1973, 102–103)

Low self-esteem

Suicide is often a form of aggression turned against the self. While the motivation for suicide can be triggered by an event such as a broken relationship, the origins of suicidal feelings are often buried deep in a person's past. One of the common characteristics found in people who deliberately harm themselves is a low self image which often has its origins in loveless, traumatic or abusive childhoods.

While help and support for potentially suicidal people may be directed initially towards the short-term problem, such as helping sort out financial worries, longer-term counselling may focus on the origins of the client's feelings of worthlessness and helping to build up a more positive self image. In this context it is also worth noting that the off-hand, even hostile, treatment of suicide attempters by some hospitals staff is likely only to reinforce a negative self image and make a re-attempt more likely.

Isolation and detachment

One of the most common sentiments expressed by many of those who resort to suicidal behaviour is a sense of detachment from others. This is not so much physical isolation but refers more to a sense of moral insulation, where the individual has come to define his, or her, situation as so hopeless that others cannot help to put it right.

The vast majority of people who go on to harm themselves have tried to communicate their unhappiness and growing despair, suggesting a wish to live. Research has suggested that at least two thirds of those who deliberately harm themselves have given some direct warning of their suicidal intent. There may also be other less direct 'suicidal clues' such as a morbid preoccupation with death and dying, continual expressions of worthlessness and being a burden to people, dramatic mood swings, changes in established habits and giving away treasured possessions. Observation and understanding of these 'clues', coupled with known 'risk factors', such as depression, unemployment, and a history of a previous attempt, can help to make a reasonable evaluation of suicidal risk (Pokorny, 1993).

The fact that suicidal behaviour usually begins long before people reach for the tablets or razor blades has important implications for intervention and prevention, as many attempts to communicate suicidal thoughts are presently falling on death ears. For example, two thirds of those who kill themselves have visited their doctor in the previous month. Social workers, community and psychiatric nurses and various counselling agencies are also frequent recipients of suicidal communications. This suggests that much better training for health professionals in the recognition and management of suicidal clues should be an important component of any preventative strategy. This may be illustrated by a well-known programme in Sweden where a group of doctors were taught about identifying depression and giving appropriate treatment. In a short space of time, admission to the psychiatric services fell, the level of prescribed drugs fell sharply and the doctors' attitude to suicidal patients improved (Rutz *et al.*, 1989). When the doctors in the area retired or moved away and were replaced by new doctors who had not received the training, the suicide rate, which had fallen, began to rise again.

Methods

Well over half of the people who kill themselves in Britain do so by poisons of various sorts. Poisoning by tablets is the most common

TABLE 20.1   **Methods of suicide in men and women in England and Wales, 1992**

	Males (%)	Females (%)
Poisoning by solids or liquids	15.7	47.6
Poisoning by gases	36.1	13.1
Hanging, suffocation or strangulation	29.6	21.9
Drowning	2.1	5.3
Firearms	5.3	0.9
Other	11.2	11.2

Source: OPCS (1993) *Mortality Statistics: Accidents and Violence*, OPCS Monitors Series DH45, London: OPCS.

method of suicide for women and poisoning by gases is the most common method for men (Table 20.1). Removing or reducing the availability of common methods of suicide can reduce the suicide rate. For example, until the 1960s self poisoning by domestic coal gas was a common method of suicide in Britain. However, within a year of the introduction of natural gas, the overall suicide rate dropped significantly. In the United States, where access to guns is part of the culture, suicidal death by firearms is common. In the United Kingdom stricter gun controls mean that there is less risk of suicide by this means, a notable exception being farmers, who are more likely to have legal access to guns.

A common method of suicide in Britain since the 1980s has been the use of a hose pipe feeding exhaust fumes into the car. *The Health of the Nation* suggests that this method will become less successful with the introduction of catalytic converters into cars.

Mental Illness

There is a high risk of suicide amongst those suffering from depressive disorders, especially when a person is recovering from depression. However most people who deliberately harm themselves with suicidal intent are not mentally ill. Many are simply unable to cope with a personal crisis in their lives and have come to see suicide as the only solution. Some research has identified a lack of problem-solving skills amongst those who attempt suicide (Schotte and Clum, 1987).

While some areas have developed professional 'crisis intervention' counselling services (Cantley, 1990), most non-medical support for people who feel suicidal comes from voluntary organisations. The Samaritans was founded in 1953 to take calls from people in despair. In 1993 almost two and a half million people made contact with the Samaritans, an increase of 35 per cent from ten years previously. Three quarters of the callers discussed suicide. The Samaritans aim to reduce suicide, first by listening to people and offering further support and possibly referring them to other agencies, and second, by raising public awareness of suicide.

The causes of suicide

Sociological theories

Almost a century ago, the famous French sociologist Emile Durkheim (1897, trans. 1952) showed that suicide rates were linked to social and institutional factors. For example, Catholic districts had consistently lower suicide rates than Protestant ones, people who were married with children tended to have lower suicide rates than the unmarried and childless, and the suicide rate of a society decreased in times of war or political upheaval. Durkheim argued that the underlying cause of these correlations was *social integration*; that is, the extent to which the lives of individuals are tied in to other people and the social institutions around them. The more individuals were detached from others, the more likely they were to consider suicide.

A great deal of subsequent sociological research has tended to confirm Durkheim's observation that lack of integration and social support is not only a major cause of suicide but also of psychological and physical ill-health generally (Lester, 1992). In this context, the links between suicidal behaviour and unemployment, for example, are explained less by the economic effects and more by the fact that being in work gives people a sense of purpose and binds them to others in some common purpose.

Psychological theories

Psycho-analytical theory has focused on displaced aggression as an unconscious motive for suicide and, in a later hypothesis, Freud suggested that suicidal behaviour represents a victory of the death instinct over the life instinct. In this context, some researchers have suggested that self-harming behaviours, such as eating disorders, alcohol and drug abuse, are a form of partial or chronic suicide.

More psychological work has tended to concentrate on uncovering factors which appear to be more common amongst 'suicide prone' individuals (Leenaars, 1990). Cognitive psychology has, for example, suggested that the suicidal are characterised by more polarised, less imaginative and more constricted thought processes. A range of psychological studies has identified 'suicidal' people scoring higher on scales of 'hostility', 'hopelessness' and 'low self esteem'.

Biological theories

Biological and genetic theories offer potentially more objective indicators of suicidal risk. There is growing evidence to suggest that a range of pathologically altered states of mind are due to impaired neuro-transmitter function. For example, a number of studies have found links between suicidal behaviour and low levels of serotonin metabolite 5-hydroxin-doleacetic acid (5HIAA) in the cerebrospinal fluid. However, attempts to explain this relationship are faced with the problem of relating increasingly sophisticated biological measures to crude psychological scales (Korn et al., 1990).

Law, ethics and suicide

There are three areas where the law is relevant to suicide. First, while attempting to commit suicide has not been illegal in Britain since 1961, it is still a criminal offence under the Suicide Act 1961 to help someone commit suicide. Second, health professionals who do not take reasonable precautions to safeguard a suicidal patient who then goes on to commit suicide may be sued for negligence in the civil courts. Third, in some cases, people felt to be at grave risk of harming themselves can be detained for their own safety under the Mental Health Act 1983 (England and Wales), 1984 (Scotland), or Mental Health Order 1986 (Northern Ireland).

Most ethical issues concerning suicide revolve around individuals' rights to determine their own lives (autonomy) and the obligation on others to protect those believed to be incapable of making rational decisions for themselves (paternalism). The detention of people, sometimes against their will, has generated a great deal of controversy. Thomas Szasz, the radical American psychiatrist, has argued that suicide may be likened to emigration, and the various health professions' stance on detaining potential suicides is similar to the refusal of totalitarian countries to allow people to emigrate.

Attitudes to suicide do not remain constant but change in response to social changes. For a long time suicide was considered a mortal sin. However, in the twentieth century it has come to be seen as the result of mental and social pressures and there is more sympathy and concern for the person who commits or attempts suicide.

References

Burton, P., Lowy, A. and Briggs, A. (1990) 'Increasing Suicide Rates Among Young Men in England and Wales', *British Medical Journal*. vol. 300, 1695–1696.

Cantley, C. (1990) 'Crisis Intervention: users' views of a mental health service', *Research, Policy and Planning*, vol. 8, 1–6.

Charlton, J., Kelly, S., Evans, B., Jenkins, R. and Wallis. (1992) 'Trends in suicide deaths in England and Wales' *Population Trends*, 69, 10–16.

Department of Health (1992) *The Health of the Nation*, London, HMSO.

Durkheim, E. (1952) *Suicide: A Study in Sociology*, London, Routledge.

Ettlinger, R. and Flordah, G. (1955) 'Attempted Suicide', *Acta Psychiatrica Scandinavica*, vol. 103.

Hawton, K. and Fagg, J. (1992) 'Trends in Deliberate Self Poisoning and Self Injury in Oxford, 1976–1990', *British Medical Journal*, vol. 304, 1409–1411.

Korn, M., Brown, S., Apter, A. and Van Praag, H. (1990) 'Serotonin and Suicide: a functional/dimensional viewpoint', in Lester, D. (ed.), *Current Concepts of Suicide*, Philadelphia, Charles Press. 57–70.

Leenaars, A. (1990) 'Psychological Perspectives on Suicide', in Lester, D. (ed.), *Current Concepts of Suicide*, Philadelphia, Charles Press, 159–167.

Lester, D. (ed.) (1992) *'Le Suicide' – One Hundred Years On*, Philadelphia, Charles Press.

Pokorny, A. (1993) 'Suicide Prediction Revisited', *Suicide and Life Threatening Behaviour*, vol. 23, 1–10.

Rich, A., Kirkpatrick-Smith, J., Bonner, R. and Jans, F. (1992) 'Gender Differences in the Psychological Correlates of Suicidal Ideation Among Adolescents', *Suicide and Life Threatening Behaviour*, vol. 22, 364–373.

Rutz, W., Von Korring, L. and Walinder, J. (1989) 'Frequency of Suicide on Gotland After Systematic Postgraduate Education of General Practitioners', *Acta Psychiatrica Scandinavica*, vol. 80, 151–154.

Schotte, D. and Clum, G. (1987) 'Problem Solving Skills in Suicidal Psychiatric Patients' *Journal of Consulting and Clinical Psychology*, vol 55, 49–54.

Stengel, E. (1973) *Suicide and Attempted Suicide* Harmondsworth, Penguin.

Taylor, S. (1982) *Durkheim and the Study of Suicide*, Basingstoke, Macmillan.

21

Two notions of risk in mental health debates

DAVID PILGRIM and ANNE ROGERS

Introduction

The Anglicised meaning of the word 'risk', derived from the French 'risque' at the start of the nineteenth century, referred to a wager between individuals after taking account of the probabilities of losses and gains. More recently, and particularly in relation to aspects of public policy, the term refers only to negative outcomes or hazards (Gabe, 1995). Accordingly 'risk' and 'hazard' are conflated in contemporary debates about mental health. These now take place in a society which has become increasingly sensitive to risk, risk taking and risk management (Giddens, 1991).

The risk posed by psychiatric patients

Despite recent concerns about the relationship between mental abnormality and danger, the actual threat posed by psychiatric patients is very small. The great majority of people with a psychiatric diagnosis are never violent and most of the violence in society is not committed by people with a psychiatric diagnosis. Variables such as low social class, young age, male gender and substance or alcohol abuse are much better predictors of violence than mental state. Despite this, a disproportionate amount of media attention and public fear is focused on those patients who *do* commit violent acts (see Philo *et al.* and Muijen in this volume; and Jones and Cochrane, 1981).

Professional competence in risk assessment

Empirical data on the performance of professional risk assessors does not inspire confidence. In a review of five major studies, Monahan noted that psychiatrists and psychologists are accurate in no more than one out of three predictions of violent behaviour over a several-year period among institutionalised populations that had both committed violence in the past (and thus had high base rates for it) and who were diagnosed as mentally ill (Monahan, 1981:47–9)

When Monahan and Steadman (1994) came to review the literature on this topic, they found that only one study of clinical prediction of violence by psychiatric patients *in the community* had been published between 1979 and 1993. In that study (Lidz *et al.*, 1993) it was found that clinicians' prediction of violence was significantly greater than chance for male patients but did not exceed the chance level for female patients.

Being a psychiatric patient *per se* is not a predictor of violence in society but psychiatric diagnosis and violence are correlated. The reason for this apparently paradoxical finding is that a sub-group of patients who are violent recidivists magnify the overall group effect. This sub-group, as it were, give psychiatric patients, as a whole, a bad name.

Violence and mental disorder: ambiguity and misattribution

Artifacts may also account for a proportion of violent incidents currently attributed to mental illness, which might at other times be attributed to alternative motives, such as avarice. To an extent attributions are determined by who is responsible for labelling. When homicides and other violent crimes are committed it is commonplace for psychiatric reports to be requested. This may lead to the motives (or their apparent absence) being framed in psychiatric terms which are then recorded by the court and may influence its decisions. An example of the variable impact of this is that rapists and paedophiles committing similar offences are found in both penal and secure psychiatric settings. Another example is when a criminal act is committed by someone who has had a previous psychiatric diagnosis. Sayce (1995) cites the case in 1993 in south London of a man who had previously had psychiatric treatment. He robbed an elderly man, who had a heart attack and presently died. There was no direct evidence that the assailant's mental state was implicated at the time. The motive, as in other cases of street crime, was acquisitive. But the case was publicised widely as another killing by a psychiatric patient. Examples such as this suggest the need for a much more rigorous approach to risk assessment in the area of mental health, such as that suggested by Monahan and Steadman (1994). Until a rigorous, empirically based approach is deployed by professionals we cannot be confident that they are making fair and accurate judgements when detaining or discharging psychiatric patients.

The risk posed by psychiatric *services*

Whilst the dominant discourse about risk in the field of mental health is about patient behaviour, there is a less-publicised side to this coin: services pose a risk to their recipients. Two types of risk are noted here.

Loss of freedom and access to material and social resources

The 1983 Mental Health Act (MHA) allows professionals lawfully to detain people who are deemed to be a risk to their own health and safety or those of others. The Act does not use the term 'danger' to self or others. The much wider notion of 'health and safety' legitimises wide-ranging powers of professionals. Indeed the 'health' rather than 'danger' notion means that for a person merely to present with symptoms of mental illness can be sufficient for professionals to justify compulsory detention and treatment. For example, to hallucinate or to be deluded, even in the absence of any evidence of danger to self or others, can lead to lawful detention. The responsible psychiatrist is likely to argue in these circumstances that the untreated mental illness is putting the health of the patient at risk.

The scale of forced detention in British mental health services is difficult to estimate. Officially around 7 per cent of patients are recorded as being detained under civil sections of the 1983 MHA. This compares very favourably with other countries. For example, in some states of the USA around 25 per cent of psychiatric patients are coercively detained (Rosenstein *et al.*, 1986). However, official records disguise the number of informal admissions which are made as a forced choice. Bean (1986) describes this process as *coactus voluit* ('at his will although coerced'). Rogers (1993) examined the reports of 412 British patients who had been officially recorded as informal admissions. She found a substantial minority (44 per cent) to be in a 'pseudo-voluntary' category because they remember being pressurised by staff who were signalling the threat of forced admission. This group was more likely to be young, single and have a diagnosis of schizophrenia. Rogers suggests that the vulnerability, lack of credibility and absence of advocacy for people with these characteristics may account for their over-representation.

Wrongful detention is only one aspect of the hazards run by those entering hospital. Even for short admissions, patients run the risk of losing their homes (Bean and Mounser, 1993). Also, patients may risk the loss of a job if admitted as an inpatient. Existing employment may also be threatened by regular attendance as an outpatient for counselling or psychotherapy. The provision of professional help may also marginalise the informal support that patients receive on a daily basis from a range of sources and social networks.

Iatrogenic risk

Being damaged by what is provided as treatment is another type of risk encountered by patients (iatrogenic risk). Talking treatments can lead to 'deterioration effects' produced by abusive or incompetent practitioners. Major tranquillisers ('anti-psychotic' drugs) can cause brain damage manifesting itself in permanent and disfiguring movement disorders (tardive dyskinesia). They may also lead to sudden death due to their toxicity to the heart. Minor tranquillisers lead to sedation effects which increase the probability of both domestic and road-traffic accidents. They are ineffective after a few days and then rapidly become addictive (see Article 22 by Gabe). Anti-depressants are also associated with sedation effects and self-poisoning and suicide by depressed patients. ECT is the most controversial of the psychiatric treatments yet the one with the most contested evidence of long-term iatrogenic effects. Patients complain of memory loss and extreme fear of the treatment. In a recent survey, two thirds of 308 respondents did not regard their experience of ECT as helpful and half of these felt they had been damaged by the procedure (UKAN, 1995).

This summary of the dangers of various forms of treatment can be augmented by evidence of the mistreatment of inpatients by psychiatric staff (Martin, 1985). The two most dramatic examples in recent years were from the Special Hospitals of Rampton and Ashworth (DHSS, 1980; DoH, 1992).

Conclusion

We have provided a brief outline of two types of risk in mental-health policy debates. Policy makers need to balance these. It is no longer satisfactory for them to become preoccupied with the risk of patient behaviour whilst ignoring the risks posed by psychiatric services to their recipients.

References

Bean, P. (1986) *Mental Disorder and Legal Control*, Cambridge, Cambridge University Press.

Bean, P. and Mounser, P. (1993) *Discharged From Mental Hospitals*, London, Macmillan.

Crichton, J. (ed.) (1995) *Psychiatric Patient Violence: risk and response*, London, Duckworth.

DHSS (1980) *Report of the Review of Rampton Hospital*, London, HMSO.

DoH (1992) *Report of the Committee of Inquiry into Complaints about Ashworth Hospital*, London, HMSO.

Gabe, J. (1995) 'Health, medicine and risk: the need for a sociological approach', in J. Gabe (ed.), *Medicine, Health and Risk*, Oxford, Blackwell.

Giddens, A. (1991) *Modernity and Self-Identity*, London, Polity Press.

Jones, L. and Cochrane, R. (1981) 'Stereotypes of mental illness: a test of the labelling hypothesis', *International Journal of Social Psychiatry*, vol. 27, 99–107.

Lidz, C., Mulvey, E. and Gardner, W. (1993) 'The accuracy of predictions of violence to others', *Journal of the American Medical Association*, vol. 269, 1007–1011.

Martin, J. (1985) *Hospitals In Trouble*, Oxford, Blackwell.

Monahan, J. (1981) *Predicting Violent Behaviour*, Beverly Hills, Sage.

Monahan, J. and Steadman, H. (1994) 'Toward a rejuvenation of risk assessment research', in Monahan and Steadman (eds), *Violence and Mental Disorder*.

Monahan, J. and Steadmand, H. (eds) (1994) *Violence and Mental Disorder: developments in risk assessment*, Chicago, University of Chicago Press.

Philo, G., Secker, J., Platt, S., Henderson, L., McLaughlin, G. and Burnside, J. (1994) 'The impact of the mass media on public images of mental illness: media content and audience belief', *Health Education Journal*, vol. 53(3), 282–290.

Rogers, A. (1993) 'Coercion and "voluntary" admission: an examination of psychiatric patient views', *Behavioural Sciences and the Law*, vol. 11, 258–267.

Rosenstein, M., Steadman, H., Macaskill, R. and Manderscheid, R. (1986) *Legal status of admission to three inpatient psychiatric settings. Mental Health Statistical Note, 178*, Washington DC, National Institute of Mental Health.

Sayce, L. (1995) 'Response to violence: a framework for fair treatment', in Crichton (ed.), *Psychiatric Patient Violence*.

UKAN (1995) 'UKAN's national user survey', *Open Mind*, vol. 78, 11–14.

22

The history of
tranquilliser use

JONATHAN GABE

Anxiolytic drugs have been taken by men and women throughout history. There is evidence that opium was used by lakeside dwellers as far back as prehistoric times (Stimmel, 1983) and its medicinal use was later encouraged by Hippocrates in Greece and Pliny in Rome (Olivieri *et al.*, 1986).

It was during the nineteenth century that synthetic substances were first introduced for mental distress and insomnia to wide acclaim. The 1850s saw the marketing of morphine as a hypnotic and bromide as a sedative, followed in the 1860s by chloral hydrate which was marketed as a hypnotic and an anaesthetic. All three substances were initially well received by clinicians who saw them as a replacement for opium. However, cases of dependence soon started to be reported and disillusionment set in, prior to a more judicious evaluation of these drugs' comparative worth.

This cyclical pattern of response has been repeated with each of the tranquillisers which has been produced since that time (Cohen, 1983). For instance, the barbiturates, of which fifty compounds were marketed in the first half of the twentieth century (Lader, 1978), were initially welcomed as 'safer' than bromide and chloral hydrate. Although cases of abuse and dependence on short-acting barbiturates were reported in the 1930s, it was only in the 1950s that the risks involved were widely acknowledged and people became apprehensive (Hollister, 1983).

The barbiturates were in turn replaced by meprobamate (Miltown), again to wide acclaim. However, its reign was short-lived. In 1960 the first of the benzodiazepines, chlordiazepoxide (Librium), was introduced, followed in 1963 by diazepam (Valium), its more successful stable mate (Cohen, 1970).

During the early 1970s the number of prescriptions for benzodiazepines continued to rise in Britain and elsewhere (Marks, 1983). As a result concern started to be expressed, especially by hospital-based medical experts, about their overuse and misuse and about the possible 'total tranquillisation of society' (Tyrer, 1974). Such a warning seemed to have some effect as benzodiazepine prescriptions peaked in Britain in 1979 when 30.7 million scripts were dispensed (Taylor, 1987).

The 1980s saw a switch in concern by medical experts to the dependence potential of benzodiazepines, especially for long-term users, and the occurrence of intense physical disturbance when the drug was withdrawn (Gabe and Bury, 1988). These issues were also taken up by lay pressure groups such as Release and MIND and by the mass media, particularly consumer-oriented television programmes such as *That's Life* and *The Cook Report*. These public expressions of concern are likely to have contributed to the year-on-year decline in the prescribing of benzodiazepines during the decade, with 15 million scripts for these drugs being dispensed in 1993 (DoH, 1995), a fall of 40 per cent in 11 years. This decrease in use is attributed primarily to a decline in new prescribing (King et al., 1990), leaving a significant number of long-term users continuing to receive their benzodiazepine prescriptions. In 1989 it was estimated that there were still 1.2 million people in Britain who had been taking these drugs for a year or more (Ashton and Golding, 1989).

Twice as many women as men have been prescribed and have used benzodiazepines (Parish, 1971; Coopertock and Parnell, 1982; Ashton and Golding, 1989; Ashton 1991). It is not surprising, therefore, that women have been identified as twice as likely to be dependent on these drugs as men (Ettorre, 1985).

Experiencing tranquilliser use

Perceptions of tranquilliser use

As concern about taking tranquillisers has grown, so attitudes towards taking these drugs have changed. Early research conducted in the United States revealed a generally positive view of these drugs (Linn and Davis, 1971). More recent evidence of lay perceptions of tranquilliser use comes from small-scale qualitative studies undertaken in the UK. These paint a complex picture of the perceived risks involved in taking these drugs at a time of heightened public concern. In one such study, Gabe and Thorogood (1986) undertook semi-structured interviews about the meaning of tranquillisers with 60 middle-aged women from East London. When asked what they felt about taking these drugs a range of opinions were expressed. Some stated that they were concerned about the danger of becoming dependent or 'addicted' to tranquillisers and felt that they might be harming their body or mind by ingesting such unnatural substances.

Others said they felt these drugs were helpful in that they offered them 'peace of mind'. Yet others expressed both views. Overall a quarter emphasised only the unwanted effects and a tenth only the benefits; the remaining two-thirds expressed mixed views.

Perceptions of the risks and benefits involved in taking benzo-diazepines were influenced by experience of personal use and by knowl-edge of others' experiences, relayed first-hand or through the media. Indeed it is possible that perceptions have become even more negative since this study was undertaken, as tranquilliser use has been frequently covered by both British television and newspapers, often in a sensation-alist fashion with parallels being drawn between this medication and illicit drugs such as heroin (Bury and Gabe, 1990).

Reasons for long-term use

Despite the increasingly negative view which people have of tranquil-lisers there remains a substantial number of people who continue to take these drugs on a long-term basis. How can this be explained? Research on the reasons people give for taking tranquillisers suggests that social and material factors are of crucial importance. For example, a Canadian study by Cooperstock and Lennard (1979), based on group interviews with 68 mainly female and middle-class participants, found that the stress caused by traditional gender roles was a primary reason for con-tinued tranquilliser use. Thus, in the case of the women in the study, par-ticular emphasis was placed on the drug's role in enabling them to maintain themselves in the traditional female role of wife, mother and houseworker. Frequently, they indicated that they resented being restricted to this caring and nurturing role but said that they saw no alternative.

The men in the Cooperstock and Lennard study, on the other hand, tended to discuss occupational role strain when explaining their continu-ing tranquilliser use. The drug was seen as helping them to alleviate new strains brought about by a change of job or maintaining a stressful job.

The evidence from these studies is that people take these drugs on a long-term basis in part at least because of their roles at home and/or as paid workers and because of their position in the social structure. These social and material factors combine in a complex way to make these peoples' lives sufficiently stressful that they continue to use tranquillisers on a longer-term basis to help them get by.

The social consequences of use

The more general social consequences for society at large have been addressed by those who have argued that the prescription of benzodiazepines primarily to women represents a form of social control over such women. Two mechanisms for such control have been identified. First, it is argued that in prescribing benzodiazepines for symptoms which are generally socially induced, doctors are in effect medicalising the everyday lives of their patients; to put it another way, by prescribing they are encouraging their patients to deny or ignore the social concomitants of distress and thereby helping to minimise pressure for social change (Waldron, 1977; Koumjian, 1981). Second, it has been suggested that doctors legitimise and reinforce existing hierarchical social relations as a result of their attitudes and behaviour towards their women patients (Ettorre, 1992). To date, however, there is only limited empirical support for such claims (Gabe and Lipshitz-Phillips, 1984).

The role of general practitioners

General practitioners (GPs) play a crucial role in determining the use of tranquillisers as they are responsible for prescribing these drugs to patients in the community. Faced with the changing nature of public opinion about these drugs over the last twenty years, how have they responded?

A study by Gabe and Lipshitz-Phillips (1984), based on interviews with a small sample of GPs (n=14) in London, suggests that attitudes have hardened. All the GPs said they were aware of the dangers of patients becoming dependent on tranquillisers and seemed to be united in their concern to restrict their prescribing to the short-term management of crisis situations. All expressed doubts about prescribing although some seemed more reluctant to prescribe than others. Just under half, however, said they always warned patients, especially first-time users, of the drug's possible side effects and all of them emphasised that they did offer alternatives to benzodiazepines such as counselling and support, or recommended relaxing social activities like yoga.

In addition, just under half the doctors felt that organisational factors affected their prescribing behaviour. For these doctors a full waiting room and administrative interruptions combined to increase the likelihood that they would take the 'easy option' and prescribe a tranquilliser rather than try to reach agreement with the patient about an alternative course of action.

The activities of the pharmaceutical industry

While the pharmaceutical industry aims to meet the health needs of patients through the drugs it produces, it does so in order to maximize its profits (Ray, 1991). The manufacture of benzodiazepines for anxiety and insomnia has been extremely profitable for the small number of companies that have dominated this therapeutic sub-market. Of these Hoffman-La-Roche, the manufacturer of Valium, Librium and Mogadon, has historically been the major beneficiary. During the 1970s it secured world market dominance and produced 60 per cent of all minor tranquillisers, earning $1.5 billion from Valium and Librium alone (Lall and Parish, 1976). According to Ray (1991) the return on these products represented 100–150 per cent on capital deployed over ten years. And even when the company was subjected to an investigation by the UK Monopolies Commission and ordered by the High Court in 1975 to repay £12 million for excess profits and reduce its prices to half the 1970 levels (Braithwaite, 1984), it was still able to minimise the consequences. By trading-off currencies and repaying in sterling after it had depreciated by 30 per cent, having remitted the profits to its Basle headquarters at the higher rate of exchange, it saved one third of the cost of the fine (Ray, 1991). More recently, the issue of dependence and the industry's role in its development has become the focus of considerable attention. Medical experts, radicalised consumer groups such as TRANX and the Council for Involuntary Tranquilliser Addiction, feminist pressure groups representing drug users like DAWN, and the mental health pressure group MIND, have all been openly critical of drug companies for withholding information about the dependence potential of benzodiazepines in order to continue maximizing profits. Indeed some of these groups have been instrumental in galvanising individuals to seek legal redress from the manufacturers for allegedly supplying these drugs without due warning of the risk of dependence. The resulting class action attracted around 17 000 claimants and involved 2000 firms of solicitors (Dyer, 1994). Billed as the largest class action in British legal history it collapsed in 1994 when legal aid was withdrawn.

Of perhaps even greater significance have been the activities of the mass media. It is now widely recognised that the media play a central role in shaping public discourse about health risks (Stallings, 1990). In recent years the risk of benzodiazepine dependence and the role of pharmaceutical companies in disguising such risks has been the focus of particular attention for television and newspaper journalists (Gabe and Bury, 1988; Gabe et al., 1991). For example, in May 1988, Central Television's *The Cook Report* targeted the drug company Wyeth, the manufacturer of Ativan, for making high levels of profit from illness and suffering while remaining indifferent to the dependence resulting from taking this product. The message was graphically illustrated by showing Roger Cook confronting and being assaulted on a golf course by the Chairman of Wyeth, with the latter giving the distinct impression of having something to hide. More recently, in 1991, the BBC's current-affairs programme *Panorama* focused its sights on the Upjohn Company and its sleeping tablet, Halcion. In this case the issue was whether the company had under-reported serious side effects of the drug observed during an important clinical trial. Two of the companys' representatives, on being confronted with the evidence, were forced to admit that there had been 'transcription errors'. As they were unable to provide more robust evidence for the efficacy and safety of the drug viewers were encouraged to believe that the company had behaved unethically.

The industry has responded to these criticisms by attempting to influence public, and in particular expert, opinion about the value of their product. This has been done in part by financing researchers who might challenge the risks of benzodiazepine dependence (Gabe and Williams, 1986) and, more recently, by funding a number of symposia in Britain and elsewhere. These symposia have been organised in the hope of providing a forum for developing a more 'balanced' view of the risks of dependence which can be used to challenge an alleged 'climate of confusion' on the part of many in the medical profession, fuelled in particular by media coverage of the issue.

While there is apparent agreement within the industry about the benefits of this strategy, there is less of a consensus about how to respond to the challenges generated by such media coverage. A debate within the industry's trade journal *Scrip* in the late 1980s illustrates the point nicely (Brown, 1988; Klein, 1988). On the one side, it seems, were the companies' public relations executives and advisers, who argued that they should accept media invitations to participate in television and radio programmes in order to avoid the charge of cover-up and the associated bad publicity which this accusation generates. On the other side were the companies' senior management executives and lawyers, who argued that they should decline all contact with the media, thereby forcing programme makers to concentrate on the 'real issues – doctor diagnosis and prescribing and the cause of patients' anxiety – instead of making the companies the 'scapegoat'.

Whatever the strategy, however, what was at issue was not simply the image of the industry but how to fend off media attacks in the light of legal claims for compensation from patients dependent on tranquillisers. While the legal case subsequently collapsed, as we have already noted, the problem of how to respond to adverse media coverage remains.

The response of the state

In Britain the Committee on the Safety of Medicines (CSM), established in 1970 under the Medicines Act of 1968, is required to promote the collection and regulation of information relating to adverse drug reactions for the purpose of enabling advice on the safety, quality and efficacy of medicines (Hallstrom, 1991). If one examines the actions of the CSM and other regulatory bodies on benzodiazepines it is clear that they have been extremely cautious, preferring to acknowledge side-effects only when there is 'pretty clear evidence of actual harm' (Medawar, 1992: 195) rather than acting earlier on evidence of risk. For instance, when the Committee on the Review of Medicines (CRM) published its guidelines in 1980, after a decade of concern amongst some medical experts, it took a conservative position. While casting doubt over the efficacy of long-term use, it stated that 'on present available evidence the true addictive potential of benzodiazepines is low' (Committee on the Review of Medicines, 1980:911). Even so its conclusions about efficacy and dependence were not incorporated in most data sheets for almost five years, apparently because of lack of cooperation by the drug companies (Medawar, 1992).

It was only eight years later, in 1988, that the first definitive statement was published by the CSM (CSM, 1988). In it the Committee reacted to the problem of dependence by advising doctors against long-term prescribing. It was suggested, following closely the recommendations of a Royal College of Psychiatrists (RCP) Working Party (RCP 1988), that courses of treatment should be for no more than four weeks and only when anxiety or insomnia were disabling or severe. This advice was quickly incorporated into data sheets for doctors without the long delays that occurred in 1980 (Medawar, 1992).

Even so, the CSM still refused to take action to revoke the licence for any of the short-acting benzodiazepines which were causing particular concern among medical experts. This only occurred in 1991, when the CSM initially suspended the licence for just one of the short-acting benzodiazepines, Halcion. Subsequently it revoked the licence for this drug, despite successful appeals by its manufacturer, Upjohn, to the Medicine's Commission and to a Committee of 'persons appointed'. Subsequently Upjohn decided to challenge the validity of the government's decision in the High Court. At the time of writing the outcome of this action still has to be decided.

Though interpreting events as they unfold is difficult, this unusual decision may be explained in the following way. First, the CSM was no doubt mindful that its own actions in licensing benzodiazepine were likely to come under scrutiny from the courts as the litigation against the drug companies in the UK proceeded. Second, the CSM may have acted as it did on this occasion in order to defend itself from the intense gaze of the media, particularly the BBC's *Panorama* programme, having already been savaged by the programme on a previous occasion over the safety of Opren in 1983 (Brown, 1992). Third, the British authorities, as defenders of national pharmaceutical interests alongside their responsibilities for promoting better prescribing (Stacey, 1988; Medawar, 1992), may have felt less compunction about revoking Halcion's licence than would have been the case if the product had been made by a British rather than an overseas company. Fourth, the CSM might have been influenced by the British government's desire to have the proposed European Medicines Evaluation Agency (EMEA) located in London (Commission of the European Communities, 1993). At present no other benzodiazepine product licence has been suspended or revoked in Britain.

Conclusion

As far as patients are concerned there are clear benefits to be found in taking these drugs. Yet serious concerns have increasingly been voiced by users as they have become aware of the risks of dependence and other undesirable side effects increasingly claimed by medical experts, lay pressure groups, lawyers and the media.

The pharmaceutical companies have attempted to address these criticisms in a variety of ways, both educational and legal. However, it appears that at present they have still to find a way of answering satisfactorily the powerful criticisms expressed by the media and other interested parties. Indeed the power of the media has been such that it has actually forced the British government's hand with regard to one benzodiazepine – Halcion – leading to its product licence being withdrawn.

All the other benzodiazepines apart from Halcion are still available, either through the National Health Service or privately, resulting in their continuing use, albeit at a lower level. This suggests that they may still move beyond the current stage of 'paranoid uncertainty' about their safety and efficacy (Clare, 1991) and come to be used judiciously for a relatively discrete cluster of symptoms, with a clear series of limitations attached to their use. The prospect of this happening at present remains uncertain.

References

Ashton, H. (1991) 'Psychotropic drug prescribing for women', *British Journal of Psychiatry*, vol. 158 (Suppl. 10), 30 35.

Ashton, H. and Golding, J. (1989) 'Tranquillisers: prevalence, predictors and possible consequences: data from a large United Kingdom survey,' *British Journal of Addiction*, vol. 84, 541–546.

Braithwaite, J. (1984) *Corporate Crime in the Pharmaceutical Industry*. London, Routledge & Kegan Paul.

Brown, P. J. (1988) 'When the media attacks, what do you do?', *Scrip*, vol. 1306, 8–10.

Brown, P. (1992) 'Halcion and UK intransigence', *Scrip*, vol. 7, 3–4.

Bury, M. and Gabe, J. (1990) 'Hooked? Media response to tranquillizer dependence', in Abbott, P. and Payne, G. (eds), *New Directions in the Sociology of Health.* Basingstoke, Falmer Press.

Clare, A. (1991) 'The benzodiazepine controversy: a psychiatrist's reaction', in Gabe, J. (ed.), *Understanding Tranquilliser Use*, London, Routledge.

Cohen, I. (1970) 'The benzodiazepines', in Ayd, F. J. and Blackwell, B. (eds), *Discoveries in Biological Psychiatry*, Philadelphia, J. B. Lippincott.

Cohen, S. (1983) 'Current attitudes about the benzodiazepines: trial by media', *Journal of Psychoactive Drugs*, vol. 15, 109–113.

Commission of the European Communities (1993) *The European Medicines Evaluation Agency*, London, Commission of the European Communities.

Committee on the Review of Medicines (1980) 'Systematic review of the benzodiazepines', *British Medical Journal*, vol. 280, 910–912.

Committee on the Safety of Medicines (1988) 'Benzodiazepines, dependence and withdrawal symptoms', *Current Problems*, vol. 21, 1–2.

Cooperstock, R. and Lennard, H.L. (1979) 'Some social meanings of tranquilliser use', *Sociology of Health and Illness*, vol. 1, 331–347.

Cooperstock, R and Parnell, P. (1982) 'Research on psychotropic drugs: a review of findings and methods', *Social Science and Medicine*, vol. 16, 1179–1196.

Department of Health (DoH) (1995) Statistics Division, personal communication.

Dyer, C. (1994) 'Sad story of the happy pills', *Guardian*, 10 May, 21.

Ettorre, E. (1985) 'Psychotropics, passivity and the pharmaceutical industry', in Henman, A., Lewis, R. and Maylon, T. (eds), *Big Deal: The Politics of the Illicit Drug Business*, London, Routledge.

Ettorre, E. (1992) *Women and Substance Use*, Basingstoke, Macmillan.

Gabe, J. and Bury, M. (1988) 'Tranquillisers as a social problem', *Sociological Review*, vol. 36, 320–352.

Gabe, J., Gustafsson, U. and Bury, M. (1991) 'Mediating illness: newspaper coverage of tranquilliser dependence', *Sociology of Health and Illness*, vol. 13, 332–353.

Gabe, J. and Lipshitz-Phillips, S. (1984) 'Tranquillisers as social control?', *Sociological Review*, vol. 32, 524–546.

Gabe, J. and Thorogood, N. (1986) 'Prescribed drugs and the management of everyday life: the experience of black and white working class women', *Sociological Review*, vol. 34, 737–772.

Gabe, J. and Williams, P. (1986) 'Tranquilliser use', in Gabe, J. and Williams, P. (eds) *Tranquillisers: social, psychological and clinical perspectives*, London, Tavistock Publications.

Hallstrom, C. (1991) 'Benzodiazepine dependence: who is responsible?', *Journal of Forensic Psychiatry*, vol. 2, 5–7.

Hollister, L. E. (1983) 'The pre-benzodiazepine era', *Journal of Psychoactive Drugs*, vol. 15, 9–13.

King, M., Gabe, J., Williams, P. and Rodrigo, E. K. (1990) 'Long term use of benzodiazepines: the views of patients,' *British Journal of General Practice*, vol. 40, 194–196.

Klein, T. (1988) 'When keeping your head down can lead to decapitation', *Scrip*, vol. 1320, 20–21.

Koumjian, K.. (1981) 'The use of Valium as a form of social control', *Social Science and Medicine*, vol. 15E, 245–249.

Lader, M. (1978) 'Benzodiazepines – opium of the masses?', *Neuroscience*, vol. 3, 159–167.

Lall, S. and Parish, P. (1976) *The Price of Tranquillity – The Manufacture and Use of Psychotropic Drugs*, London, *Mind Occasional Paper*, 4.

Linn, L. S. and Davis, M. S. (1971) 'The use of psychotherapeutic drugs by middle-aged women', *Journal of Health and Social Behaviour*, vol. 12, 331–340.

Marks, J. (1983) 'The benzodiazepines: an international perspective', *Journal of Psychoactive Drugs*, vol. 15, 137–149.

Medawar, C. (1992) *Power and Dependence*, London, Social Audit.

Olivieri, S., Cantopher, T. and Edwards, J. G. (1986) 'Two hundred years of dependence on antianxiety drugs', *Human Psychopharmacology*, vol. 1, 117–123.

Parish, P. (1971) 'The prescribing of psychotropic drugs in general practice', *Journal of the Royal College of General Practitioners*, vol. 21, (Suppl. 4), 1–77.

Porter, R. (1989) *Health For Sale*, Manchester, Manchester University Press.

Ray, L. (1991) 'The political economy of long-term tranquilliser use', in Gabe, J. (ed.) *Understanding Tranquilliser Use: the role of the social sciences*, London, Routledge.

Royal College of Psychiatrists (1988) 'Benzodiazepines and dependence: A College statement', *Bulletin of the Royal College of Psychiatrists*, vol. 12, 107–109.

Stacey, M. (1988) *The Sociology of Health and Healing*, London, Unwin Hyman.

Stallings, R. (1990) 'Media discourse and the social construction of risk', *Social Problems*, vol. 37, 80–95.

Stimmel, B. (1983) *Pain, Analgesia and Addiction*, New York, Raven Press.

Taylor, D. (1987) 'Current usage of benzodiazepines in Britain', in Freeman, H. and Rue, Y. (eds), *Benzodiazepines in Current Clinical Practice*, London, Royal Society of Medicine Services.

Tyrer, P. (1974) 'The benzodiazepine bonanza', *The Lancet*, vol. 2, 709–710.

Waldron, I. (1977) 'Increased prescribing of Valium, Librium and other drugs – an example of economic and social factors in the practice of medicine', *International Journal of Health Services*, vol. 7, 37–62.

23

The black experience of mental health law

DERYCK BROWNE

When structures of domination identify a group of people (as racist ideology does black folks in this society) as mentally inferior, implying that they are more body than mind, it should come as no surprise that there is little societal concern for the mental health care of that group. (bell hooks, 1993)

Sectioning is the term commonly used to describe the process whereby an individual is *compelled* to enter a psychiatric hospital (including a secure hospital) under the Mental Health Act 1983 and detained there against his/her expressed will. Black people's experience of sectioning – and indeed of psychiatry as a whole – has been and continues to be, a negative one. Notwithstanding the above quote by bell hooks, almost all of the available research into the black experience of psychiatry, and more specifically compulsory admissions, confirms this (Fernando, 1995). Studies in Birmingham of hospital admissions between 1979 and 1984 showed that in addition to there being a general over-representation of African–Caribbean people, they were more likely to be detained under the Act (that is, sectioned), and to receive treatment in secure facilities. The proportion of black people detained tended to increase as the focus of study shifted to higher tariff points, from informal admission to the use of the civil sections and finally to the use of the forensic sections via the courts (see below). This kind of study has been replicated across the country with similar findings. The question that needs to be addressed is about the context in which compulsory detention is used resulting in what amounts to a crisis for black people with regard to mental health care. This chapter will provide an overview of the sectioning process as it affects black people, by referring to findings of a recent survey of

Edited from Browne, D. (1995) 'Sectioning: The Black Experience', in Fernando, S. (ed.), *Mental Health in a Multi-ethnic Society: A Multi-disciplinary Handbook*, London, Routledge, 62–72.

decision-making during the process of sectioning, and a study of practices at a magistrates' court, both carried out by the author. In doing so, the debate around race and compulsory detention and treatment will be pursued, placing these issues in context by exploring various related issues, such as the persistence of negative stereotyping of black people within psychiatry and related professions.

Civil sections

Although the evidence is that the criminal justice system is a major route by which black people present to psychiatry, there is a body of evidence which suggests that the forces at play within the civil sectioning process have similar effects. Not only do black people find themselves being admitted to hospital at higher rates than do white people, but many factors that accompany this 'over-admission' – the attitudes of personnel involved, the manner of admission and the kinds of diagnoses and resulting treatments – are very different for black people compared to whites.

Moodley and Perkins (1991) found that pathways to psychiatric admission, as manifested in a London borough, were different for black people to that for whites. Those under 30 years of age were usually brought into hospital by the police or presented directly to psychiatric emergency services; those over 30 tended to present via medical/surgical hospital services, domiciliary psychiatric services or psychiatric out-patients; finally, higher proportions of African–Caribbeans were given a diagnosis of schizophrenia, were compulsorily detained and, interestingly, considered themselves to have nothing wrong with them, while higher proportions of whites were diagnosed as depressed and tended to consider themselves to have physical rather than psychiatric problems. With regard to sectioning, the authors found that it was *ethnic status* rather than diagnostic category that accounted for the higher rates of compulsory detention of black people. A study in South London (Pipe *et al.*, 1991) in 1986 found an over-representation of African–Caribbean people amongst those detained (by the police) under section 136 of the Mental Health Act. This was accounted for by young men under the age of 30 (who had relatively high rates of arrest under section 136) being perceived as threatening, incoherent and disturbed but with an unclear 'diagnosis' of mental illness.

Such general research findings provided the impetus for a detailed study of the impact that the race (of the prospective patient) might have on decision-making involved in the application of the civil sections of the Mental Health Act (Browne, 1995). In this work, perceptions about race (*vis-à-vis* the sectioning process) held by professionals were explored using semi-structured interviews with a range of personnel involved, people working in relevant voluntary organisations and ex-users of the psychiatric services. We supplemented this material with an examination of case records and statistical data (although this was not our primary focus of study). The aim of the survey was to uncover any evidence of deviation from laid-out procedures, preconceived, possibly stereotypical, notions concerning black people and the general attitude to sectioning. The interview data illustrated the accuracy of our opening quote from hooks (1993), while posing further questions about perceptions of dangerousness and the role of race and culture here.

While there existed a range of written guidelines and procedures for social services staff, police officers, nurses and hospital managers in relation to their enactment of the civil sections, there was nothing similar provided for psychiatrists – whether consultants, junior doctors or general practitioners. We concluded therefore that a certain degree of latitude must exist in the way they deal with their patients, in turn allowing for differential procedures to be followed. Results of the semi-structured questionnaires (designated to determine whether or not the race of the patient was having any significant impact on decision-making during sectioning) illustrated that there was a strong association in the minds of many personnel between race and dangerousness, that is, that black people were likely to be seen as such for no other reason than their colour. In the case of 14 police officers interviewed, the answers to questionnaires showed that the ways in which this perception (of black people being inherently dangerous) manifested itself linked up with (what was seen by them as a necessary) disregard for the written guidelines available. The following were typical statements by police officers:

> The policies do not always work effectively – for example when a person is in a private place you still act for the benefit of that person. If you have to bend the rules then you have to. If a person is in a private place you might lie and say he was taken from a public place. This is in the spirit of the guidelines, and often the families are in agreement.

> If you can't understand them then they probably won't be able to understand you therefore the more likely you are to find yourself using some form of restraint. Violence is more of a factor because persuasion can't be used – and particular (racial) groups do tend to be more excitable than others.

Although many Approved Social Workers (ASWs) were aware of the dangers of racial stereotyping, some admitted to a large degree of flexibility in the way that they applied written guidelines. To quote two typical statements:

> There is a false pathology seen in some cultures by the professionals involved – yes the police are guilty but so are many GPs, psychiatrists and social workers. This is bound to affect the way they carry out their assessments.

Procedures state that you must interview in an appropriate manner but it does not define that, so I can't say what an appropriate manner is.

Such a stance, coupled with an already negative view of black people, is very likely to result in an oppressive service for black people. In fact 75 per cent of all professionals interviewed concurred that black clients were more likely than their white counterparts to be perceived as 'dangerous'.

An examination of hospital records showed that restraint was used more commonly with black patients upon or prior to admission (16 per cent versus 2 per cent), and that black patients were more likely to be administered medication as a sedative than white clients (24 per cent versus 14 per cent). Although this was a fairly small-scale study – focusing upon two urban hospitals – these findings are entirely consistent with national research. The corollary to these kinds of actions by professionals is that psychiatry is a more dangerous port of call for black people than it is for whites. Cases such as the death of Orville Blackwood at the hands of staff in Broadmoor (Special) Hospital (Prins *et al.*, 1993) serve to highlight problems at the higher end of the tariff – that is, in the Special Hospitals or Regional/Medium Secure Units. But many cases go under (or un)-reported, such as the death in North London of a young black man (Jerome Scott) uncovered by a television programme. Scott, who had previously been diagnosed as 'a schizophrenic' died after being administered lethal doses of the anti-psychotic drug haloperidol and the tranquilliser diazepam after being held down by police on a London street. In our study we found that, although ASWs were critical of a police tendency to over-react towards black people deemed to require hospital admission, ASWs also seemed to over-react in the same way in that they appeared to request the assistance of the police more readily when dealing with black clients than when dealing with white people.

As one of the agents of primary care, the role of the General Practitioner is central to the civil sectioning process. Many GPs refused to be interviewed and so the six practitioners who were interviewed, although not a representative sample, were likely to have been supportive of our study and aware of race issues. In fact, we found serious insensitivity to the danger of racial stereotyping. One GP whose patient list was 55 per cent African–Caribbean said:

It seems there is something in the physical make-up of black people which predetermines the presence of schizophrenia. They [black people] would require higher doses of sedative drugs than white people as they don't respond to normal measures.

Racist views such as these sit uneasily alongside the fact that GPs have some considerable role to play in the civil sectioning process.

An examination of case records and interviews with hospital staff suggested that the experience of people from black and ethnic minority communities is often very different from that of white people. There was little evidence that doctors considered issues of race and culture in any depth and considerable evidence that the treatment and care of black patients on one or other of the civil sections depended very much on the subjective notions of the practitioner involved. Doctors conceded that black patients were more likely to be screened for drugs use than other groups, yet hospital records showed that white patients were more likely to have a history of cannabis use than any other ethnic groups (21 per cent versus 7 per cent). Doctors concurred that they attached a good deal of weight to information contained in case files and that these notes came to be seen as medical facts. Although 'cultural factors' were noted much more frequently in the case of black patients than in the case of whites (50 per cent versus 6 per cent), 'culture' was viewed as something belonging solely to black people or applying to them only. This is all the more worrying as we found that there was a tendency in some notes for doctors to regard certain 'cultural' characteristics as pathological. One doctor noted the following about a black patient: 'She tended to talk past the point; it is difficult to tell whether this was a sign of psychosis or because of her culture.' Another said:

> The typical black admission is young, in his twenties, loud, paranoid, resisting strongly – you need to get him sedated to restrain him, and the doctors don't know what's going on – he's usually brought in by the police, therefore the doctor hasn't got a clue as to his history – and as with men generally they would be more aggressive, you would be more frightened of them and you would put them on more medication.

Although the study did not examine medication in detail, black patients were perceived as being on relatively high doses. One nurse on a locked ward said:

> Most of the white patients are on medication three times a day; all of the black patients are on (medication) four times a day. Black patients are more susceptible to PRN and restraint.

(PRN is the term used to describe a prescription of medication, the administration of which is left to the discretion of nursing staff.)

Conclusions

The way in which the thread of perceived dangerousness of black individuals runs through the civil sectioning process is not peculiar to that process. Historically, Western psychiatry, and indeed, Western culture, has portrayed black people as having some increased propensity to dangerousness and risk. Current psychiatric diagnoses carry their own images which in turn connect up with other images held in the mind (Fernando, 1991). A painting by Tam Joseph (1984) encapsulates the argument succinctly. The black youngster's passage through the British education system is punctuated by excerpts from three of his school reports. In the first caption the youngster is, stereotypically, 'good at sports'. As an adolescent the report focuses on his liking for music, and finally, as a school leaver in the third portrait the youth has made the transition to the other side of society and is deemed to 'need surveillance'.

To be young and black, particularly for males, is to be deemed a greater risk and in need of increased surveillance and greater control. This fictitious example is, of course, no different from the view of the psychiatric services which, as far as many black people are concerned, are agents of control. Our own research, as well as much of the existing body of work, tells us that when black people come into contact with the policing agents in the generic sense (thereby including psychiatry) they are seen as requiring control as opposed to care, and custody (or physical restraint) as opposed to cure; preconceptions are more often of madness than of sanity and as Joseph's picture illustrates, the focus is on the body rather than on the mind.

The link with the sectioning process is implicit. The crisis for black people with regard to mental health is not one of large numbers of black people breaking down with psychiatric disorders, but one of large numbers of black people coming to psychiatry forcibly (that is, on one or other of the sections), receiving more serious diagnoses compared to other groups, and receiving greater doses of medication and greater restraint in settings of greater security (Fernando, 1991; DoH and Home Office, 1992).

Forensic sections

The arguments around the process involved in the use of the civil sections of the Mental Health Act 1983 hold good for the forensic sections. In fact it may be the case that arguments around the inappropriate projection of that quality of dangerousness on to black patients and individuals are even more relevant here with the joining of these two formidable systems of control – criminal justice and psychiatry – with black individuals being over-represented in each.

As with civil sectioning, the evidence here too is that the experience of black people is a negative one. Black patients in a Regional Secure Unit were found to be significantly more likely to be referred from the prison system while on remand (Cope and Ndegwa, 1990), differing from white patients who were more often admitted from National Health Service and Special Hospitals. An investigation into the psychiatric remand process at magistrates' courts (Browne, 1991) carried out by the author between 1988 and 1990 gave some insight into the way in which some of the forensic sections, and processes leading up to the use of these, impacted on black defendants (although this was not a study of the forensic sectioning process *per se*). The study showed that the experience of black defendants deemed to be mentally disturbed was markedly more negative (when being passed through the courts) than their white counterparts. For example, 37 per cent of white defendants were granted bail compared with only 13 per cent of black defendants; black defendants were more likely than whites to be detained for longer periods when remanded in prison custody, and 39 per cent of black defendants were diagnosed as suffering from mental illness following remand for reports compared with 22 per cent of white defendants. When it came to disposal, 43 per cent of black defendants received psychiatric probation orders compared with 29 per cent of whites and more black defendants received hospital orders than their white counterparts.

Discussion

When considering the issue of mental health in a multi-ethnic society we cannot avoid the historical factors which have led us to the present situation. Western psychiatry and psychology were developed during an era where racist white supremacist thinking was also becoming more clearly defined. When Europe was busying itself with a colonialist agenda, accompanying racist philosophies were necessary to justify these actions. One of the ways in which this was done was quite deliberately to link concepts of madness with race, and to link irrationality with blackness. Examples of this abound in the psychological journals of the time (Fernando, 1988) and consequently any attempt to examine the current impact of sectioning on black people must be placed in its historical context. Suffice it to say here that not only was psychological sanction given to slavery but also to a range of racist practices and philosophies during more recent history. The current overadmission of black people on compulsory orders cannot be divorced from all this.

The diversion of mentally disordered offenders from the criminal justice system to supportive and caring settings provided by the health services and social services is a fashionable and apparently humane policy being pursued vigorously by the Government. But given the crisis that already exists for black people wherever they come into contact with the mental health professions – be that in hospital, in the community or other therapeutic (*sic*) settings – we have to register the question, can the equation be this simple? Is diversion from one system (that is, criminal justice) to another (that is, mental health) a justifiable and good thing when in fact both systems are negative and oppressive for black people? Clearly where people do suffer some mental health problem and are simultaneously, or consequently, caught within the coils of the criminal justice system, every effort should be made to divert them from this system. The available research (for example, Dell *et al.*, 1991) points to prison being an 'inhumane' place for the mentally disturbed. But at the same time attention must be paid to the nature of the so-called 'humanity' of the facilities to which diversion is proposed.

Our own work confirms fears that the sectioning process militates against black people, in keeping with other aspects of the psychiatric system. This encompasses both the civil and forensic sections and would appear to be linked to a persistent and unjustified view of black individuals (with or without mental health problems) as having some increased propensity to dangerousness. Although there are historical reasons for this development, its current manifestation tends to be in the ways in which sectioning procedures are carried out and the simultaneous creation of a climate within psychiatry (and related professions) which has almost come to sanction the 'need' for the *compulsory* treatment of black people. This then opens up a series of possibilities, any of which might be taken up, and most of which are disadvantageous to black people – (over)medication, increased surveillance, greater restraint, increased security and so forth.

References

Browne, D. (1991) *Black people, Mental Health and the Courts: an exploratory study into the psychiatric remand process as it affects black defendants at magistrates court*, London, National Association for the Care and Resettlement of Offenders.

Browne, D. (1995) *An Element of Compulsion*, London Commission for Racial Equality.

Cope, R. and Ndegwa, D. (1990) 'Ethnic differences in admission to a regional secure unit', *Journal of Forensic Psychiatry*, vol. 3, 343–378.

Dell, S., Grounds, A., James, K. and Robertson, G. (1991) *Mentally Disordered Remand Prisoners*, Report to Home Office.

Department of Health and the Home Office (1992) *Review of Health and Social Services for Mentally Disordered Offenders and Others Requiring Similar Services: services for people from Black and ethnic minorities groups: issues of race and culture. A discussion paper*, London, DoH/Home Office.

Fernando, S. (1988) *Race and Culture in Psychiatry*. London, Croom Helm.

Fernando, S. (1991) *Mental Health, Race and Culture*. London, Mind/Macmillan.

Fernando, S. (1995) *Mental Health in a Multi-ethnic Society: a multi-disciplinary handbook*, London, Routledge.

Hooks, B. (1993) *Sisters of the Yam: black women and self-recovery*, London, Turnaround.

Joseph, T. (1984) *UK school report* painting: acrylic on canvas, Sheffield, exhibited Sheffield City Art Galleries.

Moodley, P. and Perkins, R. (1991) 'Routes to psychiatric in-patient care in an inner London borough', *Social Psychiatry and Psychiatric Epidemiology*, vol. 26, 47–51.

Pipe, R., Bhat, A., Mathews, B. and Hampstead, J. (1991) 'Section 136 and the African/Afro Caribbean minorities', *International Journal of Social Psychiatry*, vol. 37(1), 14–23.

Prins, H., Blacker-Holst, T., Francis, E. and Keitch, I. (1993) *Report of the committee of inquiry into the death in Broadmoor Hospital of Orville Blackwood and a review of the deaths of two other Afro-Caribbean patients: big, black and dangerous?*, London, Special Hospitals Service Authority.

24

The need to change mental health law

NIGEL EASTMAN

Mental health law removes from some psychiatric patients civil liberties otherwise inherent in our legal system. Through both statute and common law it balances a patient's right to autonomy with doctors' duty of care by reference to the health and safety of the patient. It also balances the civil rights of individual patients against the right of society to protection. Does current law correctly strike these various balances?

Lack of resources

Civil rights are granted by law but effected by resources. Hence the 'principle of reciprocity' insists that restriction or removal of civil liberties for the purpose of care must be matched by adequate quality of services. This is pursued in some American states to the point where courts discharge otherwise detainable psychiatric patients because of lack of services. Even protection of the public cannot justify detention for treatment without adequate resources. Indeed, public protection is not achievable without adequate resources, as is perhaps shown by some of the recent cases of homicide in the community (Eastman, 1993). The Royal College of Psychiatrists' confidential enquiry into suicides and homicides suggests that these cases are too frequent to be explained as a minor 'incidental' problem of current community care provision (personal communication).

Professor Murphy, for the Mental Health Act Commission, has described a total of £3 billion ($4.5 billion) spent towards mental health care as 'ill-directed, uncoordinated, and inadequate'. The reduction in the numbers of acute psychiatric beds has caused a doubling of costs per

Edited from Eastman, N. (1994) 'Mental Health Law: Civil Liberties and the Principle of Reciprocity', *British Medical Journal*, vol. 308, 43–45. It is a commentary on the conference *The Mental Health Act 1983: Time for Change?* organised by the Law Society, the Mental Health Act Commission and the Institute of Psychiatry, London (November 1993).

head and a transfer of resources from long stay care. Necessary increased staffing of acute beds is reflected in the proportion of patients who are 'sectioned' having risen by 40 per cent nationally (80–90 per cent in some London districts). Hence, the process of transferring care to the community leaves a 'residue' problem of acute care which limits the resources available for community care. As a result 'resources are increasingly almost entirely reserved for those who do not want them.'

Principle rather than pragmatism

The 1959 and 1983 Mental Health Acts arose from considered pragmatism, including the experience of dealing with dangerous patients (Aarvold et al., 1973; Home Office, 1975), hospital inquiries (DHSS, 1969), pressure from civil rights groups (Gostin, 1975, 1977), and response to judgments of the European Court of Human Rights (European Rights Reports, 1981). Professor Brenda Hoggett, now Mrs Justice Hale, a law commissioner, has argued for radical legal change based not on pragmatism but on principle. Aside from reciprocity, such principles would include promotion of self determination; services designed for the individual; least restriction; close proximity of services; protection from exploitation, neglect, and abuse; and patients taking all decisions of which they are capable.

Ian Bynoe, legal director of the mental health charity Mind, has gone further by calling for reform to make mental health law congruent with common law provisions, which base non-consensual medical treatment solely on incapacity to consent. Professor Hoggett advocates separation of legal rules between provision for 'disabled' as opposed to 'disordered' people and for 'dangerous or dissenting' as opposed to 'vulnerable' people (Law Commission, 1993a, 1996b, 1993c). She also requests new legislation to fill the legal lacuna relating to treatment for physical disorders of people who are mentally incapacitated (Law Commission, 1996b).

Principles for a new mental health act

- Promotion of self determination and personal responsibility; patients taking all decisions of which they are capable
- Protection from exploitation, neglect, and abuse
- Proper consideration of views of family and carers
- Services designed for individuals
- Preference for care in the community; hospital care based on closest proximity
- Procedural safeguards consistent with European Convention on Human Rights
- Principle of reciprocity: adequacy of service to match infringement of civil rights

---

The recent case of re C (The Times, 15 October 1993) has established the principle of advance directives, although they could not be used to refuse future treatment for a mental disorder under the Mental Health Act 1983. The same case confirmed that even psychotic delusions that are specifically related to a patient's decision to accept or refuse medical treatment for a physical disorder need not necessarily remove the patient's capacity to accept or refuse such treatment. This emphasises that capacity is determined by reference to a legal test and is reliant on medicine only in an evidential way.

Compulsion in the community

Community care challenges current perceptions of mental health civil rights. The 1983 Mental Health Act is based on admission to hospital. Should any new act be based on admission to a service instead? The government's rejection of the proposal by the Royal College of Psychiatrists for a community supervision order (RCP, 1993; Bluglass, 1993a, 1993b), reflects, at least in part, recognition that such an order would probably be judged unlawful by the European Court of Human Rights. In contrast, supervised discharge orders (DoH 1993, 1996), now enacted in the Mental Health (Patients in the Community) Act 1995 use existing rules for admission under the Mental Health Act combined with a discharge package to establish, according to the government, a legal framework for community care. However, supervised discharge orders will have little influence on patients beyond that already contained in current guardianship order provisions under section 7: the only additional power is to convey a patient to a mental health care facility, where he or she might then again refuse treatment (the basis for redetention remains exactly as currently provided for in the 1983 Act).

Of course, it can be argued that a legal framework for care in the com-

---

Medical consent by mentally incompetent adult patients.

- Competence is based on a legal test – the ability to understand in broad terms the nature and purpose of the treatment
- The Mental Health Act 1983 is solely for treatment of mental disorder; patients cannot be 'sectioned' to facilitate treatment of a physical disorder
- No legal basis for proxy consent
- If a patient refusing treatment is incompetent the patient can be treated without consent if he or she would otherwise die or suffer grave harm
- For unconscious patients there is a presumption of implied consent
- Advance directives can be made but not to cover future treatment of a mental disorder

munity is already provided under section 117 of the Act, albeit grossly under-resourced and therefore ineffective. Without further resources, using the legal framework of supervised discharge orders in community care will probably do little more than identify the responsibility of the 'key worker' (who may then be made a scapegoat). Lucy Scott-Moncreiff, a lawyer, challenges the need for any extension of power over patients in the community, arguing cogently that there was no evidence to support the government's view that there are about 3000 'revolving door' patients for whom additional community legal provision is required (Dott, 1993).

Supervised discharge order

- Community treatment agreement with patient before discharge, initiated by mental health team
- Somewhat similar to guardianship order
- Identified key worker
- patient default on agreement requires key worker to call immediate case review
- Decision to redetain based on existing criteria of the Mental Health Act
- No additional powers to treat in the community

The 'compulsion in the community' debate reflects a more fundamental debate of principle underlying calls for legislative reform. Hence, it can be argued that if it is moral to have a compulsory right of treatment of some patients in hospital then improvement of their care by moving the locus of care to the community does not affect that morality. Alternatively, it can be argued that if a patient is well enough to be in the community compulsion must be wrong. Supervised discharge orders represent a fudged compromise that does not move the Mental Health Act towards greater congruence with community care. They represent tinkering with a law essentially designed for hospital care. Indeed, supervised discharge orders may have little effect other than to concentrate the allocation of scarce community resources towards the small group of patients to whom the orders are applied. Patients may be able to gain community resources only by showing reluctance to accept treatment before discharge.

Psychopathic disorder

Continued inclusion of psychopathic disorder in the Mental Health Act is controversial (Chiswick, 1993) because it defines disorder largely by behaviour and because it represents the border between 'madness' and 'badness' and between the appropriateness of treatment and of punishment. The fact that its use is almost entirely restricted to hospital orders made by courts emphasises the civil rights issue. Dr Bridget Dolan argues that there is no scientific basis either for retention or for removal of psychopathic disorder from the act on the grounds of demonstrated (un)treatability (Dolan and Coid 1994). The legal decision that a psychopathic patient must be discharged if he or she is deemed untreatable (*R. v. Cannons Park: The Times*, 24 August 1993) would have made unlawful the preventive detention of dangerous patients and focused a previously sterile debate, had it not been overturned by the Court of Appeal.

Psychopathic disorder

- A legal category not equivalent to any specific personality disorder diagnosis; defined largely by behaviour; covers borderline between 'madness' and 'badness'
- Used as a basis for 'sectioning' almost solely by courts
- Has been used as a basis for effective preventive detention in spite of untreatability; untreatability now implies tribunal must discharge
- Psychiatrists divided on continued inclusion in the Mental Health Act
- 'Hybrid hospital order' proposed for psychopathic disorder defendants

---

Forensic psychiatrists are divided on continued inclusion of psychopathic disorder in the act (Cope, 1993). Drs Coid and Chiswick propose a compromise 'hybrid hospital order', which would link a fixed punishment tariff with alternatives of hospital care or prison. The government has recently indicated its intention to introduce such an order (R. v. Cannons Park: *The Times*, 24 August 1993).

Mental health review tribunals

Mental health review tribunals offer apparent rights of redress to detained patients. Since 'due process' is not applied, however, they may sometimes be little more than legalised case conferences (Peay, 1989) applying an odd mix of investigative and adversarial approaches.

Doctors and the law

All doctors should have knowledge of general medical law. The power of doctors, especially psychiatrists, to make medical recommendations that have the effect of removing the civil liberties of psychiatric patients means that knowledge of mental health law specifically is an ethical imperative. Currently, the Royal College of Psychiatrists does not comprehensively examine in mental health law (because of a dispute with Irish candidates who have objected to examination in English law). Further, there is no requirement for demonstration of either training or competence in mental health law before recognition by regional health authorities under section 12(2) of the Act. Both inadequacies are in stark contrast with the requirement that social workers undergo approved social work training in mental health law. The lack of commitment shown by professional medical bodies in requiring doctors to acquire detailed knowledge of mental health law is likely to increase division with and criticism by civil rights pressure groups. This will be particularly damaging with regard to the reform of mental health law that is needed to make it more congruent with care in the community. The Royal College of Psychiatrists and the BMA should commit themselves to developing a principle of reciprocity which recognises that the right to infringe a patient's civil liberties must be matched by a duty to maintain detailed knowledge of the enabling law. The College has begun to approach this by making a significant commitment to legal education.

Conclusion

Legal reform should be radical. It should not only address civil detention but also introduce statute law to fill gaps relating to incapacity to consent to treatment for physical disorders and relating to patients' private property and their public protection. In relation to treatment for mental disorders, legal provisions should be designed specifically for a 'mixed economy' of care between hospital and community.

Above all, legal reform must enshrine the principle of reciprocity. Society has no right to remove civil liberties from patients for the purpose of treatment (whether in hospital or in the community) if resources for that treatment are inadequate. It has no right to legislate solely in the interests of the protection of society from nuisance or even violence. A new mental health act should continue legal provision for compulsion or persuasion of patients, whether in hospital or the community, only if the state also offers specific rights to treatment that go beyond the ineffective general rights to treatment offered by primary NHS legislation. Psychiatric patients are distinguished from all others by virtue of their condition, which potentially renders them liable to civil detention. Even if specific rights to treatment cannot, for reasons of public financial prudence, be given to all NHS patients they must be given to psychiatric patients. Infringement of individual rights requires acceptance of social duties.

References

Aarvold, C., Hill, D. and Newton, G. (1973) *Report on the review of procedures for the discharge and supervision of psychiatric patients subject to special restrictions* London, HMSO.

Bluglass, R. (1993a) 'Maintaining the treatment of mentally ill people in the community', *British Medical Journal*, vol. 306, 159–160.

Bluglass, R. (1993b) 'New powers of supervised discharge of mentally ill people', *British Medical Journal*, vol. 307, 1660.

Chiswick, D. (1993) 'Compulsory treatment of patients with psychopathic disorder; an abnormally aggressive or seriously irresponsible exercise?', *Criminal Behaviour and Mental Health*, No. 2, 106–113.

Cope, R. (1993) 'A survey of forensic psychiatrists' views on psychopathic disorder', *Journal of Forensic Psychiatry*, No. 4, 215.

Department of Health (1993) *Legal powers on the care of mentally ill people in the community. Report of the internal review*, London, DoH.

Department of Health (1996) *Guidance on supervised discharge (after under supervision) and related provisions: supplement to the Code of Practice published August 1993 pursuant to Section 118 of the Mental Health Act 1983*, London, HMSO.

Department of Health and Social Security (1969) *Report of the committee of enquiry into allegations of ill-treatment of patients and other irregularities at the Ely Hospital, Cardiff, London*, HMSO.

Dolan, B. and Coid, J. (1994) *Psychopathic and antisocial personality disorders: treatment and research issues*, London, Gaskell.

Eastman, N. (1993) 'Clunis: the wider failure', *Independent*, 23 July.

European Rights Reports (1981) *X v. United Kingdom*, European Human Rights Reports, No. 181.

Gostin, L. (1975) *A human condition: volume 1*, London, National Association for Mental Health.

Gostin, L. (1977) *A human condition: volume 2*, London, National Association for Mental Health.

Home Office (1975) *Report of the committee on mentally disordered offenders* (Butler Report), London, HMSO.

Law Commission (1993a) *Mentally incapacitated adults and decision making, a new jurisdiction*, London, HMSO.

Law Commission (1993b) *Mentally incapacitated adults and decision making, medical treatment and research*, London, HMSO.

Law Commission (1993c) *Mentally incapacitated adults and decision making, public law and protection*, London, HMSO.

Peay, J. (1989) *Tribunals on trial: study of decision making in the Mental Health Act 1983*, Oxford, Oxford University Press.

Royal College of Psychiatrists (1993) *Community supervision orders*, London, RCP.

Part III

INVOLVING USERS IN MENTAL HEALTH SERVICES

25

Introduction

'How is it that despite statements by Health and Local Authorities that they consult users, users of mental health services still feel they have little or no control over most service planning or provision?' This heartfelt plea is the opening statement in the article by Guillermo Maza (Article 28) and forms the basis of this part of the volume. Most government directives explicitly specify that local purchasers and providers should consult with service users. However, there is often a substantial gap between these good intentions and outcomes that are satisfactory to users. The articles in this part attempt to bridge this gap, firstly, by focusing on users and what they want, and then by considering ways in which they might be more effectively involved in the planning and running of services.

One of the most significant developments on the mental health scene in the recent past has been the emergence of an articulate user or survivor movement which is determined to change the way service providers (and society more generally) see users, and to improve the services provided. From a stance of personal engagement, Peter Campbell (Article 26) charts the history of the survivor movement in the United Kingdom, its strengths and weaknesses, and the challenges that now lie before it. Although the future is far from clear and certain, Campbell detects fundamental changes in society that will continue to ensure that greater attention be paid to user and survivor voices, which are becoming ever more confident.

'The Lives of "Users" ' by Peter Barham and Robert Hayward (Article 27) is based on interviews with those who have had extensive experience of psychiatric care. The authors talked to a number of people who had been discharged from institutional forms of mental health care and who are trying to re-establish themselves in the community. Some of the ways that users and survivors think about themselves and their situations are discussed, together with the services that they would like. The article shows the way that many people remain trapped in particular forms of 'ex-patient' social inter-action, isolated from the rest of society almost as effectively as if they had remained in hospital. The notion of participation in decision-making about the services that could potentially help overcome their isolation, let alone entry into meaningful planning of services at a more strategic level, seems very remote from these people.

Guillermo Maza (Article 28) discusses ideas for structuring effective user involvement and describes the detailed institutional arrangements that may be necessary in order to attempt to involve users in strategic decisions within a statutory authority. Even within an authority which is committed to helping users take part in decision-making it is apparent that the process takes considerable effort and is fraught with pitfalls. He discusses the three things that will have to change for users to be taken seriously within the planning process: the ethos of public service as a provider of services 'to' or 'for' consumers' needs must be challenged; a more democratic management structure is required within the authorities; and finally, financial investment in the user groups is needed to provide the information and training that they require.

The remaining articles in this part look at specific areas of user involvement. Jennie Williams and Gilli Watson's discussion (Article 29) of the social inequalities affecting women focuses on one aspect of the social inequalities outlined by Gomm in Part II. They suggest ways in which mental health services (specifically psychological services) for women can be developed using the skills and experiences of women who have been users of the service. Some women have developed very clear ideas of the types of services they want and the ways that the workers within the service should act. At one Women's Mental Health Forum, described in the article, the women users of the service are generous in their criticism of the current service and open about the types of help that they feel might be appropriate for them. The challenge for the services is to develop ways of responding positively to these requests and demands. They need to develop ways of seeing the users of the services as a 'source of knowledge' rather than reacting defensively or negatively to such forceful comments. Further practical examples are provided in Sue Holland's article (37) in Part IV.

In 'Asian Women Speak Out', Steve Fenton and Azra Sadiq (Article 30) describe interviews with South Asian women living in Bristol. This group of women recount the emotional and mental turmoil they were suffering. In particular, they focus upon how their problems of isolation are compounded by the specific conditions they find themselves in. Features such as language problems, ignorance about service provision, unfamiliarity with officialdom, severe economic hardship, and the constant fear of racial harassment and attack are mentioned. When individual challenges like bereavement or family disputes were added to this background, it is not surprising that many Asian women became ill with a collection of symptoms that could be described as 'depression'. In this context, services need to develop ways of listening and responding more adequately. The Inner City Health Project described in the article has developed ways of responding to these needs with workers who can speak the relevant languages.

The processes that may be necessary to make contact and engage with refugee communities are discussed in the article by My Tang and Christopher Cuninghame (Article 31). They describe the establishment of focus groups made up of members of the Vietnamese community specifically convened to discuss their concerns and needs. The groups demonstrate how disadvantaged sections of the community can be brought closer to the authorities and how positive outcomes can develop when potential service users' views are sought.

The final article in this part (Article 32) by Liz Sayce, Yvonne Christie, Mike Slade and Alison Cobb gives a picture of the views of mental health service users on current acute mental health services and outlines some suggestions for improvement in the services. Drawing on a number of surveys of the views of users, they demonstrate that many users remain unhappy with the services that are provided, especially when users are experiencing a crisis.

It may be concluded that, although there are the beginnings and ideas for more user-sensitive services, there is still a long way to go. There is much scope for more effective partnership between users and providers.

26

The history of the user movement in the United Kingdom

PETER CAMPBELL

Since 1980 the mental health system in the United Kingdom (UK) has changed dramatically. While the main types of care and treatment provided by mental health services have changed little, the location of services and mechanisms of delivery have undergone significant alteration. Simultaneously, and partly as a direct result of these transformations, those diagnosed as mentally ill, the current and former mental patients, have become more visible and more vocal within society and in the corridors of power. Government seeks their views. The service provider is obliged to consult with them. Mental health workers seek their help in professional training.

There is little published research on the history of the user/survivor movement in the UK in the twentieth century. However, the current flowering of action and activity is based on a tradition which, although previously less organised and resourced, is very long-established. There has always been protest by mad persons at their negative designation in the eyes of society and at the systems societies have set up to deal with them. Literature is filled with the personal and often self-justificatory accounts of mad individuals. An early example of collective protest is 'The Petition of the Poor Distracted People In the House of Bedlam' in 1620. In the middle and late nineteenth century, following the establishment of the asylum system, the works of the Alleged Lunatics' Friend Society and the Lunacy Law Reform Association are well documented. Protest grounded in experience of damage inflicted by care-givers and custodians continues to be a major inspiration for action by users/survivors, whether it is carried out individually or in organisations of, or for, people diagnosed as having a mental illness.

It is customary to locate the beginnings of the current user/survivor movement in the year 1985. In summer 1985, an international conference was held in Brighton at which figures from survivor movements in other countries were present and the underdeveloped, fragmented nature of user/survivor activity in the UK was noted. In autumn 1985, the National Mind Annual Conference for the first time substantially privileged the viewpoints of mental health service users in some plenary and workshop presentations. Two significant, long-lasting user/survivor organisations, Nottingham Advocacy Group and Survivors Speak Out, were set up early in 1986. Although the middle of the 1980s saw an upsurge of independent user/survivor action groups, the focus on 1985 as a new beginning is somewhat arbitrary. In particular, it encourages an underestimation of the role of action groups that were already in existence at the time and traced their origins and influences back to the Mental Patients Union (early 1970s) and the so-called 'anti-psychiatry movement'. Two of these groups were the British Network for Alternatives to Psychiatry (BNAP) and the Campaign Against Psychiatric Oppression (CAPO). Both were strong presences in the early and middle 1980s and were made up of a number of people who went on to play important parts in user/survivor action up to the present. It is not possible to elaborate on the importance of these often more ideological, more radical forerunners of the current, pragmatically-oriented user/survivor organisations. The available evidence and the author's own personal experience suggest that groups like BNAP and CAPO were crucially important. Their range of contacts, especially with individuals and groups outside the UK, may not have been matched within the existing, and much-better-resourced, UK movement until quite recently.

Leaving to one side the detailed history of the user/survivor movement and the significance of particular organisations and approaches within it, certain broader questions remain clear. In ten years we have gone from a situation where less than a dozen independent user/survivor-led action groups existed to a position (1995) where there are probably over 350 local, regional and national groups. Why did all this happen in the UK?

Certain factors have been of general importance. The movement within Western industrialised countries since the Second World War towards increasing civil rights for disadvantaged groups had to touch eventually on those diagnosed as having a mental illness. Madpersons may be mad, but they are not stupid. Inevitably, we would recognise and act upon the discrepancy between being promised that we were the same as everyone else and being treated as burdensome, dangerous, inferior aliens. By the mid-1970s there were substantial numbers of people (including the author) who had been brought up in a civil-rights climate, lived for years being thought of as psychotically ill, been enabled to live with their distress predominantly within the community, yet had been excluded from the anticipated respect and dignities. We had too little to lose not to act. Unlike fellow-psychotics of a previous era, we were living in the community. We could really taste the discrepancy and do something about it. The pace of change within mental health services has been accelerating over the last two decades. This has created uncertainties and therefore opportunities for new ideas and perspectives. Psychiatry, a profession which grew up within the nineteenth-century asylums, has been facing the loss of its favoured centres of operation. A technology and a world-view that appeared monolithic and unassailable at the close of the 1960s was much less assured by the early 1980s. In particular, mental health workers from different, less medically-orientated disciplines were starting to claim a more influential position within the multi-disciplinary pecking order. Radical workers from some of these professions began looking for new alliances. Increasingly, people who actually used mental health services began to gain their attention. It is not insignificant that radical workers, particularly psychologists, played important parts in establishing a number of the early user/survivor groups like BNAP, Survivors Speak Out and Contact in Chesterfield.

Health services and social services have been opened up to consumerist approaches over the last twenty years. This has begun to influence the way in which mental health services are planned and provided, and the way in which people on the receiving end – service users or consumers – are viewed. Although there is a considerable difference between valuing a madperson as a consumer or recipient of mental health services and valuing a madperson as a contributing and insightful member of society, the growth of a consumerist ideology has undoubtedly added to current willingness for service providers and purchasers to consider the views of people with a mental illness diagnosis. Customers may not be citizens, but they may be listened to in a way madpersons can rarely expect. Because of Government legislation, it is now necessary for users and carers to be consulted during the formulation of plans for local community care.

The UK has also witnessed the growth of self-help as a concept and the consequent establishment of a wide variety of self-help organisations. Currently, most types of distress or so-called mental illness have attracted self-help approaches. In many areas people will now be able to find support groups substantially, if not totally, made up of people with very similar difficulties and experiences to their own. While user/survivor action is often significantly different from mainstream self-help work because of its concentration on social, structural and political change rather than individual change, every action group will spend energy and time supporting members through distress. Self-help principles lie close to the heart of most user/survivor enterprises. The public acceptance of the value of self-help, the valid therapeutic contribution of the non-expert, and the centrality of personal experience as a powerful tool for change have helped create a climate in which it is increasingly possible to tolerate and respect the positive activity of madpersons.

The influence of the 'anti-psychiatry movement' on user/survivor action in the 1980s and 1990s is a subject of some interest and uncertainty. Assessment is made more difficult by the fact that 'anti-psychiatry' has become a slogan that is routinely used by traditional mental health workers to denigrate and dismiss ideas that threaten their expert world-view and status. As a result, some survivors are starting to talk of a 'post-psychiatric' approach. In general, user/survivors will often distance themselves from the 'anti-psychiatric' labelling to gain a decent hearing for their proposals. Many user/survivor activists and radical workers who were influential in setting up groups and initiatives in the mid 1980s were influenced by, and had sometimes had close contact with, people like R. D. Laing, David Cooper and Thomas Szasz. Many users and survivors support some of the general propositions put forward by 'anti-psychiatrists' like Laing and Cooper. The possible intelligibility of madpersons, the possible value of their insights and agonies when in periods usually described as psychosis, are among the respectful declarations that users and survivors warmly welcome and frequently seek to build upon. But this acceptance (and it is notable how much greater respect users and survivors pay R. D. Laing than most of his former colleagues and their followers in mainstream psychiatry ever do) should not be taken to mean that the current UK user/survivor movement proposes, underwrites or even understands the detailed positions put forward by Szasz, Cooper or Laing in their many publications. The movement's inheritance from 'anti-psychiatry' has been emotional and spiritual rather than programmatic and practical.

What is much less arguable is the influence of user/survivor movements in other countries on developments in the UK. Ideas and examples from countries where action had gone further, particularly from North America and from the Netherlands, trickled into activities here before the mid-1980s. BNAP was part of an international network which, although very loosely organised, had good contacts in some countries and brought people from Trieste (following the ideas of Psychiatria Democratica psychiatrist, Franco Basaglia) to London and elsewhere. After 1985 the knowledge that action was being carried out successfully in other countries was reinforced by regular visits from individual survivor activists to the UK. The influence of the Netherlands on the development of Patients' Councils and Advocacy in psychiatric units is notable. A Dutch survivor came to Nottingham to help establish the first Patients' Council in the UK.

When considering the development of user/survivor action in the UK over the last fifteen years, it is worth noting the following points:

- the rapidity of the growth in activity;
- the very wide variety of actions being attempted – many groups have found it difficult to reject invitations to involvement, most groups will be active in a number of areas, not all of which are directly related to mental health services;
- the fundamental pragmatism of user/survivor action in the 1980s and 1990s;
- the comparative speed with which ideas around 'user involvement' gained official approval – most of the new user/survivor groups spent only quite a short time in the wilderness;
- the tendency for many user/survivor organisations to promote and prioritise self-advocacy (people speaking out and acting for themselves) rather than to establish coherent positions on important issues or coordinate unified programmes of action.

All the above have contributed significantly to the particular character of the movement in the mid-1990s and contain clues towards the types of problems the movement might need to address in the immediate future.

There are dangers in focusing too narrowly on service use as the central concern of user/survivor action. Although negative experiences of receiving care and treatment are a primary bond and stimulus to involvement, the lives of users/survivors are not – and perhaps are increasingly not – simply about using, consuming or receiving mental health services. The activities of groups reflect this. While most groups are involved in consultation and monitoring processes around service provision and practice, and many groups are directly concerned in the creation and running of individual and collective advocacy schemes, there is also significant work being undertaken in relation to public and media misconceptions about madpersons, in the field of the creative arts and in the detailed exploration and exposition of personal encounters with madness.

The UK user/survivor movement continues to grow and diversify. However, a tension between the major concerns for action may become more apparent. While most user/survivor activists are united by common and devalued experiences and a desire to re-formulate that experience and use it to improve the status and living conditions of people diagnosed as being mentally ill, what are the primary goals? Should our experience of madness and of service use be employed to improve mental health services, or to transform the position of mad persons in society? It can be argued that these are not mutually exclusive propositions. On the other hand, the differences between an activist who is happy with the self-definition of mental patient, and someone content to call themselves a service user, or between either of them and a mental health system survivor, let alone someone who values the designation of mad person, may be important. Moreover, although it is clear that the status of the service user within mental health services has improved substantially over recent years, the status of the 'loony in the community' seems to have gone in the reverse direction.

A number of the immediate problems facing user/survivor groups are those shared by all small, underfunded voluntary organisations. These groups must learn new skills, attract and retain new members, establish and maintain better facilities, achieve demonstrable successes, perhaps create identifiable services to their communities. Many user/survivor action groups are now at the stage when organisational growth becomes problematic: for example, they may have to face the decision to employ a paid worker and see their executive group turn into a management group. In the past, user/survivor activists have not often wanted to become voluntary organisation bureaucrats. Groups continue to contend with the difficulties imposed by the likelihood that a proportion of key members will always be non-contributing because of periods of distress. While outside expectations over greater effectiveness and wider involvement increases, activists grapple with problems about supporting each other and working in ways that are empowering for group members and sensitive to the particularity of their lives. The message that 'user involvement' requires significant amounts of time and money often fails to penetrate the corridors of power. The idea that the process is not just about results but about new ways of doing things frequently gets lost altogether.

To maintain credibility and effectiveness, user/survivor organisations must now consider the issue of representativeness. This could mean confronting the current confusion over what service providers and policy-makers want when they speak sporadically and accusingly about the need for representativeness, and entering into a proper debate about the difficulties of reconciling different models of involvement – for example, representative and participative. It should certainly mean greater activity within the user/survivor movement to include the active participation of black and ethnic groups and people who experience multiple oppressions and isolations. The UK movement remains a predominantly white enterprise. Although many user/survivor organisations have been clear about who they claim to represent and not represent, the time is overdue for wider involvement and for groups to act seriously to improve the representation and feedback processes within their own organisations.

Wider questions remain. Talk about tokenism and co-option has been frequent since the end of the 1980s, yet it is only recently that groups have started to turn down invitations to involvement. While there is so much going on, and even the regional and UK-wide networking organisations are running hard just to maintain their service to the movement, it is extremely difficult to bring people together to assess what is or is not working and whether more strategic overall approaches would help. For example, there is considerable concern about whether involvement in planning and consultative mechanisms really works, yet no general positions or demands around this type of 'user involvement' have been put forward. There is widespread theoretical enthusiasm for user/survivor-led alternative services, but very few of these actually exist on the ground. The overall questions of how much is really open for change within a medically-dominated mental health system, and whether improving services without changing social and cultural attitudes towards those who use them is a satisfactory goal, persist on the horizon and are insufficiently addressed by user/survivor activists.

In the 1990s, when it is accepted that health and social service recipients and their relatives and loved ones should have some say in the planning and provision of these services, it is perhaps easy to forget how far and how fast mental health service users have advanced as stakeholders. The 1983 Mental Health Act was developed largely without the direct involvement of users/survivors. No organisation of users/survivors was significantly consulted. This neglect would now be impossible. The movement would not allow it. The Government could not get away with it and would not even try to. This is some indication of the change that the user/survivor movement has helped bring about.

Madpersons as empowered consumers of services and madpersons as equal citizens are two quite different propositions. Outside the mental health services and despite the good work of user/survivor action groups, current or former mental health service users are held in little higher regard now than they were in 1983. Arguably, they are now more likely to be seen as dangerous and inferior. There is a dilemma here that the user/survivor movement must soon address. While it may never result in a clear choice between changing services or changing society, the evidence suggests that effective change in future will demand increasingly definite choices and specialisation from local, regional and national organisations. One of the major challenges of the next few years may be to find a coherent overall philosophy that can integrate a clearer range of discrete focuses.

Suggested Reading

Beresford , P. and Campbell, J. (1990) 'Disabled People, Service Users, User Involvement and Representation', *Disability and Society*, vol. 9, no. 3, 315–325.

Brandon, D. (1990) *Innovation Without Change? – Consumer Power in Psychiatric Services*, London, Macmillan.

Plumb, A. (1994) *Distress or Disability? A Discussion Document*, Greater Manchester Coalition of Disabled People, February.

Winn, L, (ed.) (1990) *Power to the People*, King's Fund Centre Publication.

27

The lives of 'users'

PETER BARHAM and ROBERT HAYWARD

Introduction

For many people the experience of admission to hospital, and of being deemed a mental patient, delivers a considerable shock to their sense of their own powers and judgements. Though people may in an important sense feel 'better' after a period in hospital, the treatment they have received will not of itself help them to pick up the pieces of their lives again and restore their confidence in themselves as viable participants in social life. The task of reconstitution that confronts them is considerable and they may easily be relegated to the margins of social life and brought to regard themselves as some sort of social rubbish,

How then did the ex-hospital patients participating in this research set about the task of re-establishing themselves and attempt to resist demoralisation, and what assistance did they receive in doing so? A crucial dimension of the field of forces in which [they] find themselves [is] specialised psychiatric services in the community. In this chapter we shall examine how [people] assess, and negotiate, their relationships to these services. For example, what sorts of conflicts arose between [their] strategic concerns and professional ideologies about how ex-mental patients should conduct themselves? As we shall go on to explore, services proved helpful in certain respects but much less so in others. Most importantly, the majority of our participants were unwilling to be incorporated in a service system in which their social identities appeared to be defined by their psychiatric histories. Because psychiatric services were generally described by participants as a system in which they had conferred on them the status of mental patient, in what follows we shall refer at points to the 'community mental patient' or CMP system.

Originally 'Community Mental Patients?', in Barham, P. and Hayward, R. (eds) (1991) *Relocating Madness: From the Mental Patient to the Person*, London, Tavistock Routledge.

The struggle for value

In order to grasp how our participants assess services we need first to understand some of their strategic concerns a little more clearly. Bob, a former polytechnic student in his early thirties, points to an important dimension when he describes what it is like to face a new day within the protective containment that his experience of service affords him:

> It's awful. I'm very bored. I go down to the MIND centre quite a bit and help in the coffee bar. I go down there to meet other people who are schizophrenic or epileptic or problems like that. I run the shop for them on a Friday, I go and see my friends, I listen to the radio a lot and read quite a lot. I go to the Job Centre ... I find it a very inadequate existence at the moment. I could do with a sense of purpose, of actually doing something.

From accounts of boredom like this it is plain that what [people] complain about is not that they do not have anything to do but that they do not have anything *useful* to do. A characteristic statement from our participants was

> I want to feel that I'm working on something that is useful to somebody.

Our participants were persistently haunted by doubts about their worth. As Philip puts it, 'You never actually conquer this problem', and regardless of what is achieved the struggle for value always has to be renewed ... .

It is against the background of this evaluative activity of self-accounting – of what people feel able to tell themselves about what they are doing – that we need to examine the perspectives of our participants on psychiatric services. We shall now provide some examples of the difficulties that people experience in negotiating their relationships to such services and in making a stand that reflected their strategic concerns. As we shall see, [people] find themselves in a field thick with dilemmas and ambiguities in which the choice of direction is rarely obvious.

'A dead-end sort of joke'

After they were first discharged from hospital several of our participants felt that they had no alternative but to permit themselves to be conscripted into the CMP system. For example, Simon came to see the CMP system as a punishment with the threat of something worse if he failed to comply with what was required of him:

> They say 'You're going to do this if you don't do that! ... You're going to do this job if you don't do as you're told', sort of thing. They do that with the ECT in some of the hospitals. They say 'If you don't do as you're told then it's ECT for you tomorrow'. Mind you, you don't know whether they mean it but they never laugh when they say it.

After Jeffrey was discharged from hospital he was dispatched to what he was told was a sheltered work-scheme in a local factory:

> You have to abide by the rules really ... You see, when they say, 'Righto then, we want you to go down to the wire-works' ... well, I had no choice, I had to go.

He didn't like what he found there:

> I reckoned I should have had the same amount of money as the workers were getting down there ... We only got £4 a week. So I just told my psychiatrist, 'It's just sheer exploitation, you're not going to get me down there unless they pay me a decent wage'.

Jeffrey and six or seven other ex-patients used to work from nine through to four, with an hour off at dinner time. He stuck the job for two years because he was frightened that if he rebelled he would be refused his sick note. He has learned from experience:

> Now you know the ropes it's a bit different. Because you see a psychiatrist, you go to the doctor, and they automatically give you a thirteen-week sick note.

After he left the wire-works he agreed to attend the rehabilitation centre:

> The jobs there are boring to be honest with you ... There's a workroom – that's wood, making chairs – but apart from the workroom, if you're not interested in that, well what they do they have envelopes and you put them in stacks of ten and you put them to one side and you're counting them and you do it in a rotation, and that's what you're doing.

Jeffrey attended the centre for three years by which time he was

> a bit fed up with it. So I decided to leave. Then the doctor gave me the option of going to a Social Services day centre, but all they do there is drink tea and talk. That's all they do. They don't do anything constructive.

Jeffrey now feels that if you spend too much time in the CMP system you can 'become like the people there yourself'; and that 'you're better off if you're outside'.

Simon also attended the same rehabilitation centre for a period. He describes how he found himself

> taking little sticky tabs off an X-ray file over and over again all day long ... it was just soul-destroying. You could never tell yourself you were doing a job doing that ... It never felt like work ... It felt like a dead-end sort of joke, almost ... A living joke sort of thing, doing that.

The ideology of the community mental patient: 'How are you filling your time?'

In attempting to re-establish themselves in social life, participants not infrequently find themselves in conflict with the definitions and ideologies held by psychiatric professionals as to how ex-mental patients ought to think of themselves, conduct their lives and, in the jargon of psychiatric containment, 'fill their time'. For example, Jeffrey has succeeded in breaking free of the psychiatric system to a considerable extent. He goes up to the hospital every fortnight for his injection and afterwards he might spend an hour or two having a cup of coffee with other ex-patients he knows before walking home, but overall, 'I see everybody looking as miserable as sin so I'm only too glad to get away from the place.'

He receives a visit from a social worker about once every six months who gives him advice on filling in forms if he needs it and he doesn't feel that he needs more contact with official agencies. Yet despite what Jeffrey feels to be his own achievements, he continues to feel a pressure from the psychiatrist, whom he sees once every three months to redefine himself as a community mental patient and to comply with the psychiatrist's notion of how he ought to be 'filling his time'. As Jeffrey puts it, Dr Watkins tends to make him feel like a 'pawn in the psychiatric system':

> Well, I no longer live in hostels, I no longer go to these units like psychiatric workplaces but everytime I see the doctor which is about once every three months he suggests I go to this centre which is another workplace ... . Well, I've said to him that I feel I'm far better off doing what I'm doing, just going for a drink. I know I don't know a lot of people in the town but just going for a drink – all right I'm a loner ... going for a drink, having an occasional bet and filling my time either going swimming or walking. I mean the psychiatrist wants me to come under his spell.

Jeffrey resists the pressure but he clearly finds the relationship unsatisfactory: 'You're only in for five minutes with him. Well, you can't get much out of five minutes can you? And all he asks me now is the same question, 'How are you filling your time?'

For the individual to hold his own ground under these circumstances requires considerable tenacity for, as Ben remarked, people of low income and status often lack the confidence to challenge medical authority. Ian provides another example of a similar form of resilience. Ian ... is a man in his mid-thirties who has been hospitalised for a schizophrenic condition twenty-five times over the past fifteen years. Over the past five years he has become more settled and he has managed to stay out of hospital to a large extent, with only occasional admissions for a few weeks at a time when he has become severely depressed.

According to Ian, Dr Perkins told him that: 'I might never work again and I'd have to be content to be on the sick and cope and manage as best I could.' Ian is unwilling to settle for such a bleak and denigrating definition of himself and his future and he continues to make plans and to attempt to hold on to a more optimistic sense of his capabilities. So, for example, he is presently studying for physics A level. Studying gives

him some sense of purpose and value and he eschews the suggestion
that he might prefer to attend a psychiatric day centre:

> I use the physics like it was a job. I will go down to the library and work dur-
> ing the day and just come back to the flat at night. I spent about five hours in
> the library yesterday working. It gives me something to do and it might get me
> a job eventually if I did A level physics.

Resisting 'rehabilitation' but making use of it

The individual who attempts to resist incorporation into the CMP system
may therefore distance himself from a formal patient role but in doing so
he may be made to pay a heavy price for his resistance. Ben describes
how coming out to hospital:

> It was social isolation that I was aware of, and I had to find people and things
> to do. Certainly I don't think the hospital really helped in that way. I certainly
> didn't want to go to the North Road Rehabilitation Centre which was one of
> the places they suggested. I did go to the day hospital for a while but I found
> that absolutely useless really.

In attempting to go it alone and to discover alternative solutions to
social isolation, however, the individual may find that all occupational
doors are closed to him and that wherever he goes he still carries with
him the burden of his psychiatric identity. This, for example, is what
happened to Simon when he tried to go it alone. Getting himself back
into work was essential to his sense of his own worth:

> It is very important. You can call it corny and old-fashioned and Victorian
> work ethic and all that, and they're trying to tell you that there aren't going to
> be jobs for everybody, well there aren't jobs for everybody now and it's going
> to get even worse in future, but to my way of thinking I feel guilty sort of
> living off the tax payer . . . .

So:

> I went to the Job Centre and said 'Look I want a part-time job at the very least,
> can you help us?', and they said, 'Well, why aren't you signing on?' of course.
> 'Are you signing on?' and I said not, and they said 'Well, what's the problem?'
> and I told them and they said 'Oh we usually find people from "that place" ',
> as they put it – the hospital – 'can't cope with a job'. [The Job Centre then]
> wrote a letter to the psychiatrist and the psychiatrist said that I could do with
> rehabilitation.

So Simon found himself back where he had started.

Another example of someone who has tried to resist incorporation into
the CMP system and instead to negotiate use of it on terms that he
judges beneficial is Barry, a gentleman's hairdresser by training who has

been hospitalised on a number of occasions for schizophrenia. Over the past five years Barry has resisted pressure on him to affiliate himself more closely to the official spaces for ex-mental patients in the community .... A community psychiatric nurse used to visit once a week and tried to encourage Barry to attend a day centre .... He is still under pressure to attend the day centre and still he refuses. He describes his experience:

> Well I went for an hour. It was full of smoke, someone playing snooker, the radio blasting, people talking. The social worker took me down and I found out it would cost 60p for my dinner which I didn't have, so that was that. They asked why did I bugger off and I said I couldn't stand the place ... There must have been six to eight people I knew, the rest of them I didn't know. Twelve or fifteen in the room, maybe more. They were doing keep fit when I came away ... That frightened me off! What a thing to do, keep fit in all this smoke!

The staff at the centre also wanted Barry to attend occupational therapy to which he said, 'No, I'm doing my own therapy cutting hair'. The occupational therapy didn't appeal to him because the only activity on offer was 'cutting paper and colouring boxes' and Barry 'just didn't see where it was going to lead'.

Over the years since he first became ill Barry has still been able to practise his trade as a hairdresser, even if only on an informal basis – for example he describes how the previous week a community psychiatric nurse came to his flat for a haircut. On the occasions that he has been admitted to hospital he has taken his tools with him and he complains that while in hospital he did a lot of haircutting for which he didn't get paid. What he would like now is paid employment cutting hair.

For all the difficulties, Barry is intent on maintaining his identity as a hairdresser and like Simon refuses to be drawn into activities that seem to him not to lead anywhere and hold no meaning for him within the narrative of his life. It is in large part this sense of the integrity of his life project that gives him the motivation and the capacity to resist incorporation into a service-dominated form of life and instead to regard services as a resource or support on which to draw on his own terms.

Moving on from the hostel system

Another dimension of the official spaces in which people who leave hospital after a schizophrenic breakdown often have to feel their way is the 'hostel system', as Jeffrey terms it. As our discussion of Simon brought out in the case of rehabilitation, here again [people] find themselves in a field of forces, thick with tension and ambiguity, which restrict the options open to them. And whichever way they move they confront a new combination of losses and gains.

The sorts of people we are concerned with often lack the material resources to re-establish their personhood on their own terms. A significant aspect of the devastation of personhood that often results from a severe mental illness in our society is the loss, not only of a job, but also of a home. After they have been discharged from hospital, many people therefore find themselves vulnerable to co-option into the hostel system as the only alternative to the icy blasts of homelessness. As Ben describes: 'I was put into a hostel. I was homeless and so I had to go somewhere; .... Ben felt that the hostel

> would have been all right if it had acted as a community but everybody was very isolated and although we might go for a drink with somebody from time to time we didn't function as a house – it was individuals living close together but isolated.

Some people value the protection and support that the hostel affords them. This was true in Sarah's case, for example. For the first two years after leaving hospital Sarah lived in a self-care hostel where she cooked for herself but nevertheless had the support of other people in the same predicament as herself and a weekly visit from a social worker. She liked the other people she was living with but nevertheless there came a point where she felt that it was 'time she moved on', as she put it, and she has recently found a flat of her own. The formative experience for her in this respect was her contact with other people at the day centre she attends. Between them they developed a shared understanding of what it meant to 'move forward': 'I wanted to be on a par with them. That's probably why I did it. Get out of hostels and getting my own flat and everything.'

Her new flat provides her with more privacy, in the double sense of providing her with a space that is all her own and of permitting her more control over what to reveal of her private life. So, for example, when she lived in the hostel she felt awkward about inviting people she met in the pub back for coffee because of the inevitable questions that would arise. She now feels more confident: 'I'm more able to invite people back without having the problems of explaining why I'm living there.'

Going it alone

Facing life alone often places a severe strain on the person's resilience and inner resources and, at some point, he may come to find that his sense of pride is at odds with his actual felt needs. This is perhaps particularly so, as we shall see, in the case of men. Not infrequently, the person finds himself in a field of forces in which he is made to feel demoralised about his own prospects, about what he can realistically hope for, and his demoralisation may then impact upon his ability to care for himself, which in its turn may demoralise him still further. For the person to ask for help in this situation – even to admit to himself that he is in need of help – may seem to confirm the feelings of incompetence and humiliation that have already been borne upon him.

For Ben, going it alone has

> led me to something which I don't consider is my mental illness but which is part of my life and that's depression ... I wasn't depressive particularly before ... So it leads to a certain kind of depression and I can lead not a very good life on my own in the flat – you know, I won't wash up and clean up and things and I can get in a mess at times. But I don't consider that as mental illness – it might be something to do with my mental health and routines and things, but when you're unemployed you've got no external discipline to make you do anything.

Ben has befriended a number of other people in the same predicament as himself in his neighbourhood and he offers his own observations:

> ... One man, he just sits in front of the television and must have four pints of beer a night, and he's happy in a way. He's quite a good mechanic and he helps me repair my car – we repaired my car together. The other man I organised to do a garden with, an allotment – neither of us really liked gardening so it didn't really get off the ground, it was just full of weeds! – but we did do that. And I also got him involved in the MIND group, but he's interested really in playing bridge or just going for a drink. He's very isolated – I go round to his house and sometimes he gets in quite a mess, he doesn't clean up and things like that.

In Ben's experience

> There are quite a few ex-patients like that and that's why I think that community nurses should be around, really, because you do feel better – certainly I felt better – when you clear up ... but when you're living on your own – I don't think it's unique to mentally ill patients – nobody's coming round ... it's a bit of laziness really!

As Ben rightly remarks, the problems he describes are neither peculiar to ex-mental patients nor are they necessarily a product of mental illness itself. They are, instead, the kinds of problems faced by single men in general and isolated single men in particular. ... Yet if nurses were to appear uninvited at the doors of the men he has described, would they not feel humiliated?

> That's only because we've got this stigma about being helped – nurses coming in and doing it for you when you should be able to do it yourself. But I know from my own self that I need people to come and give me a bit of a push, and if you're not married and you're not working there is nobody to kind of force you to do anything.

Ben outlines a general position about the merits of professional intervention in this sphere, but when the question is brought back to his own individual case he becomes more equivocal. Nobody from the hospital has crossed Ben's door since he was discharged, but would he actually *want* someone to? In the answer that he gives, we can sense the tension between his dignity and his actual felt needs:

> Ah! ... I think perhaps they *should* have done ... but whether I *want* them or not ... I'm not sure ... I don't seem to have an opinion on that. I don't particularly not want them to, but I just feel they perhaps should have visited me, just seeing that I was coping all right really.

Relationships between ex-mental patients

In having recourse to the official spaces for ex-mental patients, the individual is, of course, brought into contact not only with psychiatric professionals but also with other ex-patients. These relationships bring out another aspect of the ambiguity and tension in the individual's dealings with the psychiatric system. Viewed and experienced in one way, such relationships may offer solidarity in a common predicament; viewed and experienced in another, they may serve to put in question the [person's] membership of a wider community. People who associate with other ex-mental patients may risk incorporation into a mental patient sub-culture and into marginalised self-definitions; the price of the integrity of going it alone, on the other hand, may be the loss of real relations of value and the risk of painful social isolation.

Ben, for example, has learned to seek out relationships, with ex-mental patients who share his concerns and to distance himself from those that embrace – albeit ironically – images of mental patienthood. He explains why he tends to avoid what he calls the 'patient sub-culture':

> It's a funny sub-culture ... There's a certain amount of mockery in it as if they say, 'Come on, take your tablets!' kind of thing ... There's a strange sort of sub-culture there where they say, 'You haven't got a job have you? How did you get a job? You're a *mental patient!*' – because I'm known as a mental patient not as a technician or a college student or something ... There is that kind of prejudice there and I do it myself. If someone's talking about making a record I say, 'You're not really going to make a record!', whereas I wouldn't really say that if I hadn't known he'd been a mental patient.

Some people manage the tension we have described by learning how to move in and out of 'mental patient' space on their own terms, without feeling contained or defined by it. One such person is Barry. As we described earlier, Barry refuses to attend the day centre but he continues to value the social contacts he made at the hospital and is a member of a support group of ex-patients.

> If I go anywhere else like a pub I don't know what is going on. If I go to the hospital I can relate to people. They tell me what they're going through and I

tell them what I'm going through ... Sometimes it works, sometimes it doesn't. That's how you meet people.

One person who has a wide network of social relationships outside the psychiatric system is Rachel, a single women in her early forties. She describes how

> I don't particularly want to make social contacts with people who I *know* have had illnesses. No doubt there are many people in the community who have had illnesses before, but I would rather be accepted as being normal and mix with normal people as far as possible.

After her first breakdown her consultant suggested that her isolated life-style had been a contributory factor in her illness and encouraged her to expand her social life. She is now active in a wide variety of groups.

None the less, at a deeper level she feels that something is missing. She describes how she was disappointed to find that

> it was mostly older people that belonged to these societies. I seem to belong to a missing age group. Probably because people of my age group have got young families and tend to be married to a large extent.

Much as many of the people we are concerned with may want to make relationships outside the psychiatric system, doing so is not easy, not least because of the heavy burden of self-accounting that is involved. People may then find themselves thrown back to relationships within the system as the only alternative to isolation. Sidney describes one aspect of his experience:

> I do lack confidence in mixing with groups of people. Not so long ago since I went to college to study a foundation course in music and I found that I couldn't cope with just mixing with people in the class ... The work I could cope with but I couldn't cope with mixing with groups of people.

At the day hospital he attends once a week by contrast:

> There it's different, we all have the same problems, there's very little cynicism there, I mean everybody's in the same boat, it's a lot easier to mix socially there. There's no competition at the day centre between people, nobody's trying to prove anything there.

Sidney now has a girl-friend whom he met in a night-club but with whom he had been acquainted previously when she was a patient in the same hospital as him. He would prefer to strike up a relationship with a woman who had no connection with the psychiatric system but his experience has brought him to the reluctant conclusion that this avenue is not open to him. In his description of what might happen in such an encounter we can readily appreciate the painful constraints that the burden of accounting for a psychiatric history imposes:

> I might have got talking to her but I would have had to cut off at a certain point because I knew that once she found out about the background she'd lose interest ... It would have been more difficult to carry on with the relationship ... I have actually avoided getting into relationships because of the difficulty I would have in explaining what I'd been through, and what it all means, to someone.

One effect of his psychiatric history, he goes on to say, has been to 'stop a few relationships before they'd started'.

Medication? 'Yes, but it's not the *only thing*'

The pressure to accept a definition of oneself as a community mental patient emerges most forcibly in relation to medication. It is here that awkward questions about the capacities of people with a history of mental illness to exert control over their own lives, and to make rational judgements about matters that concern their own well-being, are brought into sharp relief. One view on the benefits of medication is given by Steve;

> What it's like is you get your injection and it's just like having a bottle of whisky – not that I've ever had a bottle of whisky! – but you feel calm. You know it's a funny thing you feel knocked out really. The stuff you were worried about before your injection just flows over you.

Were he to give up the injections, Steve says:

> I'd probably get into trouble. I'd probably get a bit excited and trouble would develop. I like to be kept tranquillised after all this time ... I'm not violent, I've never been violent in all my record, but I act on impulse and I think I'm better on injections.

The majority of our participants, however, were much less enamoured of the virtues of being 'knocked out' or 'kept tranquillised' than Steve. Most of them did not object – in the short term at least – to taking medication, and indeed many of them found it beneficial, but they were intent upon ensuring that medication did not interfere with what they held to be the main priorities in their lives. And it was here that conflicts with the medical profession arose. What participants looked for from psychiatrists was an approach in which – to put the matter in more formal terms – the prescription of antipsychotic drugs was an adjunct to a psychosocial understanding of their predicaments rather than a substitute for such understanding.

A number of participants felt that doctors took inadequate account of the tendency of psychotropic drugs, particularly in high dosages, to reduce the efficiency and stamina of the person in performing day-to-day tasks.

The shock of the first breakdown often brings about a collapse of confidence in which patients take on trust what medical authorities declare to be best for them. Simon describes how over the years he has learned to hold firm on his own judgements:

> At first I just accepted everything but I think when I had that bad side-effect I realised all was not well. Then they put me on something else that didn't have the same side-effects so I thought maybe I was just unlucky and then I started getting Modecate shuffle where I could hardly move my legs and just couldn't get out of the chair. They thought I was playing silly buggers but I wasn't. Now I am adamant that I stay on this 25 mg and there is no way they are going to increase it or alter it in any way. I tell them what I want. I don't let them tell me.

Dr Wilkins told him he might need to take medication for the rest of his life and compared the long-term treatment for schizophrenia with that for diabetes. However, Simon was unimpressed by this effort at the medical normalisation of his fate: 'I've heard of diabetics who are able to stop using insulin so to my way of thinking, even using the comparison the doctor used, it ought to be possible for me one day to say "I don't need this thing".'

Conclusion

Perhaps the most consistent message that comes through here is that in participants' experience of services the significant questions that concerned them about the value and direction of their lives were left unaddressed or obscured. Medication was judged to be beneficial but delivered crudely: as the primary form of intervention, it became a currency that devalued participants' efforts to re-establish their personhood and resist entrapment in an unremitting state of mental patienthood. For the most part services appeared to offer participants a form of protective containment within the identity of a community mental patient. As our participants saw it, such containment did not generally provide a means or stepping-stone to a more meaningful form of inclusion in social life as much as confirmation of their marginalisation and lack of social worth. An image that was often borne upon us when participants discussed their experiences of services was of containment in a confined space cut off from where they and others judged the significant action to be. Most of all, agents wanted purposive activity that, as Barry expressed it, appeared to be leading somewhere. Only then could they be said to be doing something more than 'passing the day' or 'filling their time'.

28

Structuring effective user involvement

GUILLERMO GARCIA MAZA

How is it that despite statements by Health and Local Authorities that they consult users, users of mental health services still feel they have little or no control over most of service planning or provision?

As a result of the National Health Service reforms and the introduction of the Community Care Act, mental health services in the Chesterfield area are now provided by a range of organisations. The Hospital Trust provides a hospital ward with 25 beds for acute admissions and an acute therapy and day service, which functions from 9 am to 5 pm and is shared with other areas in north Derbyshire. Private residential homes cover some of the residential shortfall created by the closure of the big psychiatric hospitals. The Community Health Trust manages the multi-disciplinary Chesterfield Community Mental Health Team (CMHT), who are responsible for mental health assessment and co-ordinating individual care plans. The CMHT in turn also manages the Walton Villas rehabilitation team, offering a range of temporary accommodation and rehabilitation options. Social Services manage the Chesterfield Support Network (CSN), which provides individual support and advisors to self-help groups. The National Schizophrenia Fellowship runs a drop-in centre and a range of employment schemes, and participates in the provision of individual support. As a result of community care funding, a number of private agencies are also providing individual support. All mental health services are co-ordinated through the Local Planning Group and the Operational Management Group for Chesterfield.

At first sight, there is an impressive range of structures for user-participation in mental health services in Chesterfield, of which some are long-standing, some are new and experimental, and others are just being created. Nevertheless, the history of user participation is a chequered and confusing one, mainly as a result of its *ad hoc* development, which involves two aspects: on the one hand, the involvement of users in their own care whether as individuals or as groups, and, on the other, the involvement of users in planning services.

Users' participation in their own care

Three main developments have taken place in users' participation in their own care since the Chesterfield Support Network (CSN) began.

First, CSN started from a model of partnership with users, based on supporting self-help groups. CSN provides advisors to the groups, as requested. Advisors have no voting or vetoing powers, but support the groups on a number of issues mutually agreed with the groups. These may range from financial advice to dealing with issues of conflict between members. Their support helps the group welcome the most disabled mental health service users, who might otherwise be effectively excluded. There are guidelines written between users and workers, setting limits and boundaries to the partnership. Each group writes its own constitution, elects a committee and receives funding from Health and Social Services through the Local Planning Group.

Secondly, it has been the practice of CSN to involve users in drawing up and reviewing their own care plan, which has become one of the key elements of community care and of the Care Programme Approach. While there are users who choose not to participate in the formal process of review meetings, their presence in the meetings shifts power from professionals to users almost automatically. Users can use the multi-disciplinary, multi-agency meetings to their advantage, by getting some workers to advocate for them and gain improvements in the support they receive.

Thirdly, the Acute Day Therapy Service at the hospital started a patients' forum, where patients discuss issues affecting their attendance at the unit with the members of staff.

It is at these levels of direct participation that users generally claim to be satisfied with how they exercise their own power. This is not to deny that often problems exist and that considerable time and effort is sometimes required to solve them. Some of the problems stem from differences between users and workers on the result of assessments and access to appropriate support. As always, there is a background of lack of resources available to meet all their needs.

User participation in planning services

User participation in planning services in Chesterfield has developed as a result of two historical factors. The first and most important one is the partnership model between the support network (CSN) and the self-help groups, which gave rise to the creation of The North East Derbyshire and Chesterfield Association of Self-Help groups (NEDCASH). NEDCASH represents around 25 groups with about 250 members. The groups have a number of functions: for example, gardening, social, a group for African–Caribbean people, survivors of sexual abuse, women. Some of them meet for a session a week, while Contact, the largest of all with about a hundred members, meets six days a week. The role of NEDCASH is to safeguard the interests of the self-help groups and of their individual members, to advocate on their behalf and to campaign for the improvement of services from statutory, voluntary and private organisations as well as trying to secure funding for the groups.

The Chesterfield Local Planning Group (LPG), made up of representatives of Health, Social Services and voluntary organisations, was the first planning structure where users became involved. However, this did not happen without problems. Most professionals at one point disengaged themselves from the Local Planning Group, because they felt these meetings became mostly talking shops for individual users. For their part, users felt frustrated with the lack of willingness of professionals to share power with them. They felt it gave them little opportunity to influence the many changes that were taking place in Health and Social Services, that is, the implementation of the Care Programme Approach. However, NEDCASH itself was able to gain control of the process of distribution of grants to the self-help groups. Participation at this level gave users the opportunity to further their political skills, by focusing their efforts on establishing NEDCASH as the organisation representing users.

Furthermore, users took crucial roles within the LPG structure that eventually took them to the higher planning structure of the LPG Chairs' meetings. Participations in the LPG Chairs' meetings provided a wealth of information about the planning processes of senior management, though only a slightly improved chance of influencing it. After three years of participation, NEDCASH still report that their comments are not listened to and that they still are having to repeat the same arguments without making a noticeable impact on service delivery.

The second factor in the development of user participation in planning has been the mental health guidelines from the Department of Health (*The Health of the Nation, Mental Illness Key Area Handbook*, 2nd edition, 1994), which clearly recommend the involvement of users in the planning and reviewing of services. As a result of this, Trent Regional Health Authority has also developed a policy for user participation (NHS Executive, Trent Regional Health Authority, *Focus on User Empowerment in Mental Health Care*, 1994).

The two factors have combined to become an impressive force in the development of user participation. Two other crucial developments have taken place. First is the inclusion of user organisations in the joint commissioning group for north Derbyshire. They will implement the recommendations of the Audit Commission and monitoring services. This will include user-defined quality standards as a key feature. Secondly, user inclusion in the review of the Care Programme Approach has been negotiated by CSN staff on behalf of user organisations.

The way ahead

For user participation to function properly it has to be about power sharing, about recognising and treating user organisations as stakeholders and power sharers and not as consumers. For this to happen, three things must change.

One is the ethos of public service itself, which is based on the provision of service for or to people. This needs to change to the provision of service with the involvement of the people it affects. The Care Programme Approach has brought about a negotiated process with users, so at the level of direct contact the ethos is changing rapidly. It is, however, at the level of senior management and planners, where there is little contact with users, that the change in attitudes needs to permeate deeper.

Two, following from this, a more democratic structure of management is needed. The statutory agencies are hierarchical organisations with complicated decision-making processes and a top-down management style. Understandably, they have only taken tentative steps to share power with users, and have not consciously reappraised their ethos so far. There has been an *ad hoc* and reluctant participation process with very little consideration given to the implications of power-sharing, its limits, strengths and effectiveness (see King, 1996).

The third issue is that survivors and user groups have not yet developed into successful political organisations, able to participate in full and set the agenda within decision-making structures. For this to happen statutory agencies need to invest in them. That would involve the provision of the necessary finances, information, space, trust and – most importantly – the right type of processes of decision-making. This is something that user organisations, both locally and nationally, have been seeking for a long time.

Reference

King, S. (1996) 'We can still find a way to be heard', in Read, J. and Reynolds, J. (eds), *Speaking our Minds: personal experience of mental distress and its consequences*, London, Macmillan.

29

Mental health services that empower women

JENNIE WILLIAMS and GILLI WATSON

Introduction

Mental health workers continue to be reluctant to identify social inequalities as a major cause of the despair, distress and confusion that is named mental illness and mental disorder (Williams *et al.*, 1993; Fernando, 1996). As mental health practitioners we find it difficult to think constructively about the effects of structural inequalities on the lives of our clients and on our practice as clinicians/workers, and to be aware of the many ways that mental health services and organisations are shaped by these inequalities. This huge deficit in our knowledge and understanding is partly due to mental health training which typically ignores, denies and obscures connections between social inequalities and psychological distress (see Williams and Watson, 1991; Phillipson, 1992). This chapter addresses this gap and Figure 29.1 identifies knowledge bases that we can draw upon to transform mental health services for women.

Women users and ex-users of mental health services

Women recipients of mental health services are an important source of knowledge. We illustrate this point in the following account of a recent meeting between one of us (JW) and a Womens' Mental Health Forum – a community group for women users and ex-users of mental health services.

The Women's Mental Health Forum

That evening there were 15 women in the group. The group was racially mixed, and the age range must have been from early twenties to late

---

FIGURE 29.1   Sources of knowledge: *Where, and how, we can learn about the.*
effects of social inequalities on the lives of women

- Women users and ex-users of mental health services
- Theory and research about inequality and mental health
- Women-centred mental health projects
- Our own experience of power and powerlessness

---

sixties. Most of the women lived in poverty and one woman had a serious physical disability. After we had introduced ourselves, they asked me if I would tell them about women and mental health. I suggested that it would be more informative if we talked about their knowledge of women and mental health. They talked readily though often with pain, anger and despair. First, they described what had brought them, and sometimes other women they knew, to the mental health services. They identified sexual, physical and emotional abuse they had experienced as children and as adults; they talked about poverty; about loss, and not feeling loved, liked or valued; of feeling responsible, blamed, and power-less for what happened in their families; several women also recounted difficult experiences around childbirth. When the conversation moved to their experience of mental health services the talk was of widespread dissatisfaction mixed with some recognition and gratitude for the efforts – however misguided – of individual mental health professionals. They described: not being listened to; therapists who were incompetent and sometimes damaging; inappropriate routinised treatment including those provided by rehabilitation and day services; the enormous amounts of psychotropic drugs they had taken; their courses of ECT; multiple admissions to psychiatric hospitals; their children being taken into care; and sexual and physical abuse in services.

We do not think this group of women is unusual. Their stories, pre-occupations and observations were strikingly similar to those offered by other groups of women service users we have known and worked with since the 1970s. The only significant change over the years has been the increasing possibility of talking about experiences of physical and sexual abuse. The women in these groups – including those who have lived for many years on the long-stay wards of psychiatric hospitals – have taught us an immense amount about the experiences that bring women to mental health services and the grim reality of using these services. In all these groups we have also been taught about, and reminded about, the ways that women together can empower and enable each other.

We have much to learn from those who use our mental health services, and it is essential that our theory and practice is shaped by their knowledge. This can happen in a range of ways. For example, service users can be centrally involved in training mental health workers, they can be partners in research, and directly and indirectly involved in shaping service provision (for example, Sassoon and Lindow, 1995; Williams and Lindley, 1996). However, it is very important that we collaborate with service users in ways that are not exploitative, abusive, or simply ineffective, and that we are attentive to factors that influence the willingness of women service users to share their views and experiences.

Theory and research about inequality and mental health

The experiences, observations and theories of women who use the mental health services are now echoed and supported by work published in the extensive field of women and mental health. The work that is most valuable is that which helps us name, and take action about, the effects of inequalities on women's mental health and the mental health services they receive.

FIGURE 29.2   Mental health consequences of women's everyday life

- **Marriage** is more likely to be beneficial to the psychological well-being of men, and detrimental to the psychological well-being of women[1]
- **Childbirth** is linked with depression for a significant number (estimates vary between 10%–30%) of women[2]
- **Caring** for children and dependant relatives carried high costs when associated with isolation, low social value and a lack of resources[3]
- **Poverty** and levels of deprivation are much higher amongst women than men; the links between poverty and psychological distress and disturbance are well documented[4]; poverty amongst women is strongly associated with:
    - being a single parent[5]
    - being divorced[6]
    - being old[4]
    - being Black or a member of an ethnic minority group[7]
- **Domestic violence** is estimated as occurring in 1 to 4 households[8], largely to women by male partners; links between battery and long term mental health difficulties are now well established[9]
- **Childhood sexual abuse** is estimated as occurring to between 1 in 3 and 1 in 10 girl children[10] and can have profound mental health consequences[11]
- **Sexual violence:** 1 in 4 women have experienced rape or attempted rape[12], victimisation is a powerful predictor of psychiatric treatment[13]
- **Being female** is a risk factor: women's feelings, thoughts and behaviours are more likely to be defined as madness than men's[14]

[1]McRae and Brody (1989); [2]Nicolson (1989); [3]Smith (1991); Platt *et al.*, (1990); [4]Belle, (1990); [5]Platt *et al.*, (1990); [6]Day and Bahr (1986); [7]EOC (1992); [8]Smith (1989); Mooney, (1993); [9]Rosewater (1985); [10]Whitewell, (1990); [11]Polusny and Follette (1995); see also Gardner, this volume; [12]Painter, (1991); [13]Goodman *et al.* (1993); [14]Williams (1984); Ussher (1991).

In Figure 29.2 there are some points about the psychological costs for women of normal everyday life in a society structured by gender. We include the material in Figures 29.3, 29.4 and 29.5 as a reminder that the everyday life of women is not only shaped by gender. It is particularly important that mental health workers develop understandings of the ways in which multiple discrimination and disadvantage exacts costs from the mental health of women. It is also important to be aware – as the findings below indicate – that multiple discrimination and disadvantage *decrease* the likelihood that a woman's needs will be met by mental health services. We also note that research in the field of women and mental health has serious omissions. Comparatively little research, for example, has been directed to women with long-term mental health problems.

From this belief overview of the research literature it is evident that socially structured differences between women have very important, though frequently ignored, mental health consequences. Attention has also been drawn to research documenting the mental health consequences of the sexual, physical and emotional abuse of women. These socially unacceptable, though not unusual, ways in which men maintain and use their power have profound implications. For example, there is evidence that at least 50 per cent of women who use community- and hospital-based mental health services have been sexually and/or physically abused as children and as adults (Williams et al., 1993); that sexual and physical abuse are major factors in the mental health difficulties women experience, including difficulties diagnosed as severe mental illness (Goodman et al., 1993; Polusny and Follette, 1995; Bryer et al., 1987) and that abuse is strongly linked with high service use (Potier, 1993). Given what is now known about the incidence of abuse and its implications for women's mental health, neglecting to inquire about a history of abuse constitutes malpractice. When we fail to provide or find sensitive help for women in dealing with abuse, we replicate processes which damage psychological integrity. It is crucially important that we are trained to intervene effectively around sexual and physical violence, and that this is part of a broad-based training designed to familiarise us with the dynamics of power relations, and the mental health implications of the uses and abuses of power in a wide variety of contexts. It is fortunate that there is a growing body of material that can be used to inform practice (e.g. Briere, 1992; Burstow, 1992; Watson and Williams, 1992; Gender Working Group 1993; Wooley, 1994).

In a context of growing evidence about the prevalence of power abuses in therapy and counselling (e.g. Edwardes and Fasal, 1992; Russell, 1993) each of us, regardless of our backgrounds and allegiances, needs training and support to ensure that we use our own power ethically. We also need knowledge that sensitises us to, and enables us to take action about,

---

FIGURE 29.3   Black and ethnic minority women

- Racism, poverty and isolation shape the lives, and experiences of Black and ethnic minority women who use mental health services;[1] these hidden causes are not being acknowledged within mental health services
- Black women are more likely to receive physical treatments of drugs or ECT than counselling[2]
- The specific needs of Black and ethnic minority women are only just beginning to surface, in the literature advocating the provision of racially sensitive mental health services
- Racism also impacts on the lives of Black and ethnic minority women who provide mental health services[3]
- Access to Black and ethnic minority counsellors and interpreters is important[4] but very limited[2]

[1] Holland (1995); Fernando (1995); [2]MIND (1992); [3]Sayal-Bennett (1991); [4]Webb-Johnson and Nadirshaw (1993).

FIGURE 29.4   Older women

- Over 60 per cent of women over 65 years old live below the official poverty line in Britain;[1] the effects of poverty on mental health are well documented[2]
- Older women with mental health difficulties are the age/gender group least likely to be offered counselling and therapy[3]
- Psychotropic drugs are prescribed more often to older women than any other age/gender group[4]
- Older women are the age/sex group most at risk from side effects of medication[5]

[1]Titley *et al.* (1992); [2]Bruce *et al.* (1991); [3]Wallen *et al.* (1987); [4]Catalan *et al.* (1988); [5]Woerner *et al.* (1991).

FIGURE 29.5   Lesbian women

- Lesbian women service users and workers are invisible within most mental health services[1]
- Mental health services commonly assume heterosexuality which intensifies the stress of homophobia for lesbian services users and workers[1]
- Mental health services for women who are openly lesbian are often based on the assumption that their sexual orientation and life style is the cause of their difficulties[2]
- Mental health services often minimise major life events experienced by lesbian women, e.g., loss/death of partner; relationship difficulties[2]

[1]Rothblum (1990, 1994)' [2]Perkins (1991); Greene, (1994).

the abuses of power within services, including sexual abuse of clients by mental health workers (e.g. Crossmaker, 1991; Potier, 1993). Additionally, working in mental health services means being confronted daily with the use of psychotropic medications and ECT to silence women and manage their distress. It is important therefore, that we inform ourselves about the effects of these treatments on women (Hamilton et al., 1995a, 1995b; Wade and Wade, 1995), and find ways to use this knowledge to support women service users make informed choices about their treatment, and to promote the development of accessible and effective talking treatments.

Women-centred mental health projects and services

In the UK expertise in working from a social-inequalities perspective is with some exceptions (eg. Holland, 1995; Watson, 1993) developing outside of statutory mental health services, in contexts which support equality-based management, supervision, and theorising. Examples include: Shanti (see Williams et al., 1993), Threshold (Davis, 1993), the White City Project (Holland, this volume) NEWPIN (NEWPIN, 1992), POPAN (Edwards and Fasal, 1992), and SAGE, Sexual Abuse Groups, Exeter (Watson, 1993). There are also countless examples of good practice to be found in self-help groups, mutual-aid groups, help lines, and the Women's Aid Movement (see GPMH, 1983; Lindow, 1994). The work involved, while innovative, inspirational and significant, is rarely well documented in the academic literature. Those involved are often too busy, developing and providing services and fund raising, to write about their work with clients. However, there are other ways in which we can learn from these developments. Project workers often provide training, and can be bought in as trainers: they may also be able to offer direct training in the form of supervised placements. Many projects also welcome the involvement of researchers, particularly those trained to use participative and empowering methodologies.

Our own experience of power and powerlessness

Finally, we can learn from our own experience of power and powerlessness. Many of these experiences are rooted in our social position defined by our gender, race and other parameters of social difference and status. For feminist mental health workers, making sense of our own experiences of oppression, and making connections with the experiences of other women, has been a major route to developing feminist clinical theory and practice (see Burstow, 1992, Watson and Williams, 1992; Holland, this volume). However, this route can be legitimately used by all mental health workers to develop a greater sensitivity to the dynamics and consequences of power abuses. Training for mental health workers needs to include safe and respectful opportunities for people to learn from their own experiences of oppression, privilege, and empowerment.

Conclusion

In this paper we have argued for fundamental changes in the training of mental health professionals and practitioners. The case for making these changes is now very strong and supported by an extensive and legitimate body of knowledge. At the same time it is essential for our own safety, and the safety of the women we work with, that we appreciate the deep-rooted resistance to these changes that exists within mental health services and the broader social context.

References

Belle, D. (1990) 'Poverty and women's mental health'. *American Psychologist*, vol. 45(3), 385–389.

Briere, J. N. (1992) *Child Abuse Trauma*, London, Sage.

Bruce, M. L., Takeuchi, D. T. and Leaf, P. J. (1991) 'Poverty and psychiatric status: longitudinal evidence from the New Haven Epidemiologic Catchment Area Study', *Archives of General Psychiatry*, vol. 48(5), 470–474.

Bryer, J. B., Nelson, B. A., Miller, J. B. and Krol, P. A. (1987) 'Childhood sexual and physical abuse as factors in adult psychiatric illness', *American Journal of Psychiatry*, vol. 144(11), 1426–1431.

Burstow, B. (1992) *Radical Feminist Therapy: Working in the Context of Violence*, London, Sage.

Catalan, J., Gath, D. H., Bond, A., Edmonds, G., *et al.* (1988) 'General practice patients on long-term psychotropic drugs: A controlled investigation', *British Journal of Psychiatry*, vol. 152, 262–263.

Crossmaker, M. (1991) 'Behind locked doors – institutional sexual abuse', *Sexuality and Disability*, vol. 9(3), 201–219.

Davis, S. (1993) 'Threshold – a local initiative for women and mental health', *SPRING*, vol. 10, 4–5.

Day, R. D. and Bahr, S. J. (1986) 'Income changes following divorce and remarriage', *Journal of Divorce*, vol. 9(3), 75–88.

Edwardes, M and Fasal, J. (1992) 'Keeping an intimate relationship professional', *Openmind*, vol. 57(June/July), 10–11.

EOC (1992) *Some Facts About women 1992*, Manchester, Equal Opportunities Commission.

Fernando, S. (1995) 'Social realities and mental health', in Fernando, S. (ed.), *Mental health in a Multi-ethnic Society*, London, Routledge.

Gender Working Group (1993) *The Gender Resource Pack: A Training and Practice*

Resource for Clinical Psychology, Exeter, Department of Clinical and Community Psychology.

Goodman, L. A., Koss, M. P. and Russo, N. F. (1993) 'Violence against women: physical and mental health effects. Part 1: research findings', *Applied and Preventative Psychology*, vol. 2, 79–89.

GPMH (1993) *An Information Pack of Mental Health Services for Women in the United Kingdom*, London, Good Practices in Mental Health.

Greene, B. (1994a) 'Ethnic-minority lesbians and gay men: Mental health and treatment issues. Special Section: Mental health of lesbians and gay men', *Journal of Consulting and Clinical Psychology*, vol. 62(2), 243–251.

Greene, B. (1994b) 'Lesbian and gay sexual orientations: implications for clinical training, practice, and research', in Green, B. and Herek, G. M. (eds), *Lesbian and Gay Psychology: theory research, and clinical applications*, London, Sage.

Hamilton, J. A. and Jensvold, M. F. (1995a) 'Introduction: Feminist psychopharmacology', Special Issue: Psychopharmacology from a feminist perspective, *Women and Therapy*, vol. 16(1), 1–7.

Hamilton, J. A. and Jensvold, M. F. (1995b) 'Sex and gender as critical variables in feminist psychopharmacology research and pharmacotherapy'. Special Issue: Psychopharmacology from a feminist perspective, *Women and Therapy*, vol. 16(1), 9–30.

Holland, S. (1995). 'Interaction in women's mental health and neighbourhood development', in Fernando, S. (ed.) *Mental Health in a Multi-ethnic Society*, London, Routledge.

Lindow, V. (1994) *Self-Help Alternatives to Mental Health Services*, London, Mind Publications.

McRae, J. A. and Brody, C. J. (1989) 'The differential importance of marital experiences for the well-being of women and men: a research note', *Social Science Research*, vol. 18(3), 237–248.

Mind (1992) *The Hidden Majority*, London, Mind Publications.

Mooney, J. (1993) *The Hidden Figure: domestic violence in North London*, London, Islington Council.

NEWPIN (1992) *Annual Report 1992*, National Newpin, London.

Nicolson, P. (1989) 'Counselling women with post-natal depression: implications from recent qualitative research', *Counselling Psychology Quarterly*, vol. 2(2), 123–132.

Painter, K. (1991) *Wife Rape, Marriage and Law: survey report, key findings and recommendations*, Manchester, Manchester University, Department of Social Policy and Social Work.

Penfold, P. S. and Walker, G. A. (1984) *Women and the Psychiatric Paradox*, Milton Keynes, Open University Press.

Perkins, R. (1991) 'Therapy for lesbians? the case against', *Feminism and Psychology*, vol. 1(3), 325–338.

Phillipson, J. (1992) *Practising Equality: women, men and social work*, London, Central Council for Education and Training in Social Work.

Platt, S., Martin, C., Hunt, S. and Tantam, D. (1990) 'The mental health of women with children living in deprived areas of Great Britain: the role of living conditions, poverty and unemployment', in Goldberg, D. (ed), *The Public Health Impact of Mental Disorder*, Gottingen, Hogrefe & Huber.

Polusny, M. A. and Follette, V. M. (1995) 'Long-term correlates of child sexual abuse: theory and review of the empirical literature', *Applied and Preventative Psychology*, vol. 4, 143–166.

Potier, M. (1993) 'Giving evidence: women's lives in Ashworth maximum security psychiatric hospital', *Feminism and Psychology*, vol. 3(3), 335–347.

Rothblum, E. D. (1990) 'Depression among lesbians: an invisible and unresearched phenomenon', *Journal of Gay and Lesbian Psychotherapy*, vol. 1(3), 67–87.

Rothblum, E. D. (1994) ' "I only read about myself on bathroom walls": the need for research on the mental health of lesbians and gay men', Special Section: Mental health of lesbians and gay men, *Journal of Consulting and Clinical Psychology*, vol. 62(2), 213–220.

Russell, J. (1993) *Out of Bounds: sexual exploitation in counselling and therapy*, London, Sage.

Sassoon, M. and Lindow, V. (1995) 'Consulting and empowering Black mental health system users', in Fernando, S. (ed.), *Mental Health in a Multi-ethnic Society*, London, Routledge.

Sayal-Bennett, A. (1991) 'Equal opportunities – empty rhetoric?', *Feminism and Psychology*, vol. 1(1), 74–77.

Smith, H. (1991) 'Caring for everyone? the implications for women of the changes in community care services', *Feminism and Psychology*, vol. 1(2), 279–292.

Smith, L. J. F. (1989) *Domestic Violence: An Overview of the Literature. A Home Office Research and Planning Report*, London, HMSO.

Titley, M., Watson, G. and Williams, J. (1992) 'Working with women, including older women, in the mental health services', paper presented at the *British Psychological Society (London Conference)*, London:

Ussher, J. M. (1991) *Women's Madness: misogyny or mental illness?*, Amherst, MA, University of Massachusetts Press.

Wade, T. C. and Wade, D. K. (1995) 'Biopsychiatric attacks on women: an aberration or a predictable outcome of biopolitics?', Special Issue: Psychopharmacology from a feminist perspective, *Women and Therapy*, vol. 16(1), 143–161.

Wallen, J., Pincus, H. A., Goldman, H. H. and Marcus, S. E. (1987) 'Psychiatric consultations in short-term general hospitals', *Archives of General Psychiatry*, vol. 44(2), 163–168.

Watson, G. (1993) 'Mental health services for women who have been sexually abused in childhood', in Williams, J. *et al.* (eds), *Purchasing Effective Mental Health Services for Women: a framework for action*, Canterbury, University of Kent/Mind Publications.

Watson, G. and Williams, J. (1992) 'Feminist practice in therapy', in Ussher, J. and Nicholson, P. (eds), *Gender Issues in Clinical Psychology*, London, Routledge, 212–236.

Webb-Johnson, A. and Nadirshaw, Z. (1993) 'Good practice in transcultural counselling: an Asian perspective', *British Journal of Guidance and Counselling*, vol. 21(1), 20–29.

Whitewell, D. (1990) 'The significance of childhood sexual abuse for adult psychiatry', *British Journal of Hospital Medicine*, vol. 43(May), 346–352.

Williams, J. and Lindley, P. (1996) 'Working with mental health service users to change mental health services', *Journal of Community and Applied Social Psychology*, vol. 6(1), 1–14.

Williams, J., Watson, G., Smith, H., Copperman, J. and Wood, D. (1993) *Purchasing Effective Mental Health Service for Women: a framework for action*, Centre for the Applied Psychology of Social Care, University of Kent at Canterbury.

Williams, J. A. (1984) 'Women and mental illness', in Nicholson, J. and Beloff, H. (eds), *Psychological Survey*, vol. 5 Leicester: British Psychological Society.

Williams, J. A., Watson, G. (Eds) (1991) 'Clinical psychology training in oppression?', *Feminism and Psychology*, vol. 1(1), 55–101.

Woerner, M. G., Kane, J. M., Lieberman, J. A. and Alvir, J. (1991) 'The prevalence of tardive dyskinesia', *Journal of Clinical Psychopharmacology*, vol. 11(1), 34–42.

Wooley, S. (1994). 'Sexual abuse and eating disorders; the concealed debate', in Fallon, P. (ed.), *Feminist Perspectives on Eating Disorders*, New York: Guilford.

30

Asian women speak out

STEVE FENTON and AZRA SADIQ

In the language of western psychiatric medicine, the term 'depression' is used to describe a specific mental and emotional state with particular symptoms, and is treated, in part, by anti-depressant drugs. This does not mean, however, that there is absolute clarity about what exactly depression is, or what causes it, or the best way of treating it. Indeed, it is often argued that there is no real distinction between someone who cannot deal with life's pressures and someone who is ill with depression, between 'minor' and 'major' depression. No-one is sure when, or whether, this condition can be called an illness – at least not without clearly defining what they mean by illness. If the specialists are in disarray about this, it seems pointless to argue about whether the women in the study were truly suffering from depression. Hardly any of them used the word, and most were probably unaware of it.

But we can be sure of one thing; the women whom Azra Sadiq spoke to described symptoms which, if they had been described to their GPs by English speakers, would have been diagnosed as 'depression' – the women spoke of weakness, listlessness, and tearfulness; they were unable to sleep; they could not cope with the simplest things; they lost their sense of self-confidence and the meaning of life; and they contemplated suicide as a way out. All eight symptoms were mentioned by a majority of the women, most mentioned six or more symptoms, and none mentioned fewer than five.

While their suffering always began as a completely natural and normal response to some terrible personal shock, like the death of a close relative back in India or Pakistan, this turned into a virtually unmanageable state of mental and emotional distress which the women invariably described

Edited from Fenton, S. and Sadiq, A. (1993) 'Sixteen Asian Women Speak Out About Depression', in *The Sorrow in My Heart: Asian Women Speak Out About Depression*, London: Commission for Racial Equality', 8–36.

as a kind of illness, not ordinary illness like rheumatism or arthritis, but an illness which took over their lives and trapped them, a kind of 'thought sickness' – *soochne ke bimaari*. As one of them said, 'My sorrow has become my illness'.

It all began when ...

All the women could point to the exact moment when their lives began to unravel. Often, 'it all began when ... ' a close member of their family died back home, or when there were problems with their husbands or in-laws.

A death in the family

The pain and grief of losing someone you love can seem like the end of the world when it happens. For some of the women, it was as though they had lost their loved ones twice over, once when they left their countries, and then when their mothers, fathers, sisters or brothers died. This was a different sort of parting, though, and they had to come to terms with it on their own, far away from home.

For many, the news usually came too late to travel back, even if they could afford this. The shock was made worse by the helplessness the women felt being miles away; it was as if things might have been different, somehow, if they'd been there. For some, not knowing what was actually happening to their dying relatives, moment by moment, became a terrible ache.

Most of the women turned to their faith for help in accepting what had happened, but even this solace was only partial, cut off as they were from the customary rites and mourning ceremonies that help those who are left behind to reconcile themselves to their loss.

Very few people here, even the women's husbands and children, knew their families back home; there might be sympathy, but they could never fully understand. Talking about the past, about their loss, was the only way they might learn to remember differently, but for many of the women this simple remedy was not available. Torn between their need to remember, and the mounting pressures on them to forget, some just stopped trying to juggle between the two and lost themselves in their own world, brooding endlessly. Three of the women could not "pull themselves together' enough to cope, and had to be hospitalised.

All the women were intensely aware of their distress, and the way it consumed their lives, leaving them almost paralysed in the face of urgent responsibilities and claims.

Mounira was older than the other women, and this made her feel less able to cope with shock. She was with her husband in Pakistan when he

died, but it was the more recent loss of her nephew in a tragic accident
that devastated her. The boy had been like a son to her, and she was here
in England when it happened.

> It was God's will. The roof of the cave just fell in on him. Someone else also
> died with him. I can't forget his death, even now. I feel it in my heart (*dil*).
> When you're young you can take these things more easily, but not when your
> blood gets weak from the fears (*fikar*) and illness. There's ordinary illness, and
> there's the illness of sorrow. I don't sleep very well. I can't forget everything
> that's happened, even though I want to ... I keep telling myself, and everyone
> else also tells me, to forget the past. But some things are beyond your control.
> That's why I can say that I am old now.

Unhappy relationships

While western cultures give a lot of importance to people as individuals,
Asian traditions set greater store by people's relationships with others,
and mostly those with their family and community. A breakdown in
family relationships is a threat to the roles by which individuals define
themselves, and can cause tremendous mental and emotional turmoil.

Jaswinder was 56 years old. Her ordeal began when her husband told
her to take the four children and leave the house. It was like the end
of her world; everyone she knew seemed to be avoiding her, all her
husband's relatives, and everyone who knew her story. As a single
mother, she felt it was always her morality that was being questioned by
people, never mind how much her husband was to blame.

> It was a time of sorrow and anxiety. I thought I was going mad (*paagl*) ... that's
> why they sent me to hospital ... I've been in this country for nearly sixteen
> years now, and to be honest, it's only now that there's a little happiness in my
> heart. When I came here from Pakistan it was hard, what with young children
> and not knowing anybody or anything. My husband wasn't too bad then, but
> suddenly one day he just told me to leave his house ... I was stunned. I didn't
> know what to say, or where to go. I have no relatives in England, let alone here
> in Bristol. His relatives all took his side and didn't even want to see us.

Jaswinder finally turned to someone from her village in Pakistan. The
man took her and the children in, in spite of the inconvenience, and went
out of his way to help them.

Amina's depression began when her husband threw her out and she
had to rent a room as a single Asian woman. In a community where
living in a family is the norm, a single, married woman living on her
own can all too easily become an object of malicious gossip. Amina was
renting a room from an (Asian) family, and used to spend some time
after work in the communal lounge watching TV or talking before going
to her bedroom. But the family's attitude suddenly changed, and Amina
could find no explanation for it.

The common stereotype about Asian families is that they "look after their own'. This is largely because of the extended family structure prevalent in the Indian subcontinent. Seen as tightly-knit and unfailing support structures, Asian households have been overlooked by health and social services departments when planning their services.

In the first place, only 16 per cent of Asian households in Britain are extended families. Secondly, they are far from being as close-knit and trouble-free as they are thought to be. As in any family, there can be tensions and stresses, with everyone taking one side or another in an argument or dispute. Occasionally, individual members of the family can become very isolated, and feel victimised and shunned.

Migration necessarily divides families, but this has been exacerbated in Britain by the increasingly restrictive immigration controls applied to particular ethnic groups. It can take many heart-breaking years before a family is allowed to be reunited – or not, as the case may be. Moreover, since the introduction of visas for all countries in the Indian subcontinent, and an ever-increasing number of African countries, even holiday visits have been made more difficult for nationals of these countries.

Many of the women had few family members nearby – some had none. Also, they did not have much chance of making friends with other South Asian women, and those who did not speak English were even more socially isolated.

Shahnaz's problems went back to the time her father-in-law remarried. She had the feeling that her new relatives did not really accept her, that they thought she was unworthy. She felt desperately lonely, and longed for support.

> I get frightened, and it feels as though there's a ball rolling round in my stomach. I'm terrified that something bad is going to happen. I think constantly in my heart. Sometimes it all gets too much, and my heart sinks and my stomach surges, and my head feels as though it will burst from the pressure. I'm always apologising to my husband for being like this, but I can't help it, it just happens ... I start crying, and cannot stop myself, even when the children are there, and they get upset – they're still very young. I keep telling myself that I must live for them, that if I don't, there's no-one else to look after them.

Shahida told us that her illness began when her husband and his brother had a row about the ownership of their house. They had bought the place jointly, both putting in a lot of hard-earned money for the deposit. The brother took care of the business side, as he had more experience in these matters. Shahida and her husband then moved into the house, and put in a lot of work to improve it. A year or two later, her husband decided to sell the house and recoup his share. But his brother disputed his stake, saying the house was his – legally, it was in his name.

The two brothers almost had a fight. It was a very bad time for us, and that's when I started to get ill. In those six months I lost a lot of weight and became very weak. I used to worry a lot, and spent days and nights just thinking and thinking in my heart about it all, about what would happen. Sometimes they'd have big arguments in the house, and I'd feel even worse. The life would just go out of my heart.

What made it worse ...

Racial hostility

All the women had experienced racial hostility and abuse, from snubs and sneers to open aggression by neighbours and others in public places. Suffering from an illness whose basic symptoms are insecurity and withdrawal, the fear of racial violence further intensified the women's feelings of persecution.

Their greatest anxiety was for their children's safety. Most of the women recounted incidents of racial bullying, abuse or attack, in parks, in the street and elsewhere, and they spoke of the dread they lived in all the time of 'what might happen'.

... Shahnaz's anxieties about what her new in-laws might do to discredit her were compounded by her fear of being attacked or robbed.

I'm afraid that something bad is going to happen, so I don't go out on my own at all ... If I have to go to the clinic, I wait for my husband. He works in a restaurant and this means he is home most mornings. But he's away from the house from half-past-two in the afternoon till two or three in the morning. I'm in a state of terror all that time. I lock all the doors and the windows when he leaves – there are a lot of people around here prepared to do that, you know – English and black people. We're always afraid of them. We rarely take the children to the park. We just go into the garden when the weather is fine.

Kuldip had been living in a small council flat with her young son and elderly mother ever since her separation from her husband. Racial taunts and harassment were daily events in her life.

There are some in the block who don't look kindly on our sort. It wasn't very long ago that I bought my son this bike, and he went out on it in the passage – I thought it was alright if he just went up and down a little. Then I heard him screaming. I ran out to find all these English children around him, hitting him, and they'd thrown his bike over the side of the flats. When they saw me, they just started calling me nasty names. This went on for about five or ten minutes, when a black lady who lives in the end flat came to my rescue and told them all off. It was only then that they went away. She is good to me, and helps me a lot. She then went down and got the bike and said that if there was any more trouble I was to call her.

Looking after the children

Many of the women had young children to look after. What should have been a joy became a terrible responsibility because of their poor health. Almost all the women reported symptoms of general weakness and tiredness, and described how impossible it was to cope with the everyday business of living. Not surprisingly, the problem of looking after children was more acute when the women had sole responsibility for the children because their husbands had walked out on them.

Women with older children also talked about the worries they had about their children's education, and their prospects of finding a good job. All the women were keen to see their children do well in their studies. Bringing up children is a big responsibility for all parents, but these women were unwell and needed help themselves. Some were able to turn to their husbands, but those who did not have this option found themselves in a desperate situation.

Housing problems

About half the women had problems with their housing. They did not talk about their housing conditions as the primary factor in their 'illness', but various aspects of their housing situation affected their well-being – run-down, dilapidated houses lacking basic amenities like running hot and cold water, and in urgent need of repair; the hopeless task of getting the council or landlord to make improvements; difficult stairs for women who had asthma or arthritis; lack of a garden when the public park was a dangerous place; and unpleasant encounters with neighbours who did nothing to conceal their hostility towards Asian.

Salma had been widowed twice, and now believed that people were plotting against her. At the same time, she was in desperate straits, living with her four children in a small, crumbling, two-bedroomed terraced house. She had no money for repairs, and no husband to support her or help put things right. 'When we moved here – from one pit to another – I left all my furniture behind. We live like animals here.' Salma slept with her daughter in one of the bedrooms and her three sons slept in the other. One of the downstairs rooms could not be used because it needed replastering, and the floor boards were dangerous in another.

> Yes, we've applied for a repair grant, but that was about a year and a half ago. They came and took pictures and didn't do anything about it. You know what these people are like, they just see us and turn away. I don't understand it. We've also applied for a council house, but they say it will take a long time ... You asked about my health, where do I start? There's nothing wrong with me, just nerves ... I feel like my life is being squeezed out of me.

Many of the women mentioned difficulties with speaking and reading English, especially if their husbands had left them and they had to sort out their housing problems on their own. Dealing with complex and usually unsympathetic bureaucracies is a problem for everyone, but the women who had been left by their husbands or forced out of their homes were confronted with it all of a sudden, as a matter of extreme urgency. Few of them had dealt with the authorities before, they were unfamiliar with the system, and, worst of all, they did not understand English well enough.

Someone to turn to ...

GPs and social services

All but one of the 16 women had visited their GP. They did not necessarily talk about their problem in psychological terms, but concentrated on symptoms such as sleeplessness, loss of weight, and aches and pains. It was very clear that the women knew exactly what their problem was, but in most cases they were reluctant to talk about it openly to their GP – they did not think a doctor could do much to help. Kuldip thought that she would do better to go back home and see a religious healer – 'If you ask me truthfully, I don't know what to do ... What should I do? Go on taking anti-depressants or go to Pakistan?' Ayesha was the only one who did not see her GP –

> I didn't go to the doctor about my trouble, because there was nothing she could do for me. I get a lot of different ailments – aches and pains, and tiredness – and I just take Panadols. I don't think I'll get well till my husband stops seeing that other woman. Hopefully, the move to Leicester will be better for us all.

Salma saw her GP with a linkworker, because she did not speak much English. She told him about the aches in her body and a sense of constriction all over, which made her feel very tired and unable to do her usual chores in the house. But it was after her consultation with the GP that she told the linkworker that she thought people were conspiring against her, and that this was upsetting her dreadfully, and making it impossible to cope, especially with looking after her four children properly. She found it easier to pour out her problems to the linkworker because they both spoke the same language, and the linkworker was able to understand, certainly more than the doctor.

Jaswinder saw her GP with her daughter, who explained that her mother's bad health had begun since her father threw them out of the house, and that she was very depressed. She spent so much time thinking about it all and crying all the time that she got terrible headaches, and became desperately tired. Jaswinder said that she couldn't stop herself from imagining all that could happen to her and her children. The doctor told her not to worry so much, and that thinking about it all would only make her feel worse.

> I explained I was feeling very weak and had dizzy spells. So the doctor gave me some iron tablets and said that I must take care that I ate enough. But I still felt weak, as if all the strength had left my body. And I wasn't sleeping very well at night. I stopped going to the doctor because I didn't think he could give me a cure for my worries. He just said I mustn't worry so much.

There was no doubt that the women were more direct and forthright about their problems when they spoke to an ICMHP worker, because they were able to talk in much greater depth. They were quite clear what their problem was, and realised that it helped to talk to someone about it.

> It's good to talk to her (the ICMHP worker). It's good to talk about my family in Pakistan to another woman. She comes and sees me at home, which is nice – to have another woman in the house. I like women's company.

Seven of the women consulted the Inner City Mental Health Protect (ICMHP) team about their depression or other disorders. Most were getting help for the first time.

Other sources of help were staff from the social services department and social workers who helped to sort out the women's housing problems and chronic money difficulties. Three of the women mentioned getting help from a social worker.

Religious faith

All the women turned to their religion for help and comfort in various ways. This was plain in the way they expressed their understanding of illness and misfortune, and by the fact that they might decide to visit a religious healer.

Most of the women mentioned 'the will of Allah' or 'Rab's will' when talking about illness, death, or something they saw as fated or predestined, and said that they prayed regularly. They understood their illness or predicament as having been sent by Allah, and expressed no resentment about it. But this does not mean they had reconciled themselves to what had happened, and to the fact that their fate was to be 'a life of tears'.

31

Focusing on health: focus groups for consulting about health needs

TANG MY and CHRISTOPHER CUNINGHAME

Where refugees are concerned, it is often not helpful to distinguish mental health needs from needs for basic primary health care. Most Vietnamese refugees escaped from their country by boat, either at the fall of Saigon in 1975, or the subsequent Chinese invasion in 1979 (Todd and Gelbier, 1988). They share common refugee experiences of intense and multiplying trauma: from the devastating effects of war, from feelings of great shock and anger about leaving their land and families, and from the terrible circumstances of their voyages (Karmi, 1992). They have experienced an invidious bond of forced migration and, often, long and difficult stays in refugee camps (Bell and Clinton, 1993). The problems facing refugees on their arrival in the UK are wide-ranging, and include political, social and structural dimensions. Their needs for health care can usefully be linked with their need to build or rebuild community networks.

Too often, health authorities regard gathering information for local needs assessment as being of benefit to themselves alone, and as an end in itself (Smithies and Webster, 1993). The contract culture, demanding concentration of time and resources internally, and a main focus elsewhere on provider issues, may be part of the explanation.

But where planners are confident in the means of achieving greater community participation, assessment becomes a much more effective and dynamic interpretation of local needs. This is demonstrated by a

Originally Tang, M. and Cuninghame, C. (1994) 'Ways of Saying', *Health Service Journal* (15 September), 28–30.

project established in the inner-London area of Lewisham by Save the Children in 1992, in partnership with Optimum Health Services and Lewisham social services, and with two-year funding from City Challenge. Its main concern is with the health status of Vietnamese refugees and their underuse of local health services.

The 25 000 refugees who settled here in the UK are mainly from farming and fishing backgrounds, are from North Vietnam, and are of Vietnamese or ethnic Chinese (Cantonese-speaking) origin (Todd and Gelbier, 1988).

More than half the UK Vietnamese population now live in London and some 1000 individuals – 300 families – live in Deptford. Some arrived in the early 1980s, others very recently. Compared with other London boroughs, Lewisham has the highest number of Vietnamese people registered with mental health problems, and these are mainly from Deptford (Vietnamese Mental Health Project, 1992).

The proposal to use focus groups within the project arose from concerns that a basic understanding of Vietnamese health needs was missing in local service planning and provision.

For many Vietnamese people, expression of views as service users would be virtually unknown to them before they left Vietnam, and the refugee experience has resulted in further mistrust of authority. Language and cultural barriers mean that the community prefers to use local resources run by Vietnamese and Chinese speakers. Focus groups provide opportunities to develop structured discussions on a given topic in a group setting. Typically, a group may consist of six to 12 members. They are often created near the start of a community development process.

Focus groups can promote discussion of shared agendas, address common needs, and consider broader social, environmental and cultural determinants, such as those of health.

Most importantly, the focus groups are built on existing community networks with which the health adviser (My Tang) had established a good rapport over more than a year. Familiarity helped in developing feelings of trust and safety about the groups. First meetings in which Save the Children staff and community workers participated were held in July 1993. Aims and objectives of the focus groups, and participant recruitment methods, were agreed. Besides those of group members themselves, additional roles were defined for facilitator, record keeper and observers.

Discussion questions fell into three broad categories:

- factual information about details such as length of time in the UK, English language ability;
- awareness of relevant health services and experience of using health services;
- specific needs and ideas for improving services.

Each focus group aimed to recruit representative members of the Vietnamese community in terms of:

- origin in Vietnam;
- length of time in the UK;
- linguistic ability.

Two existing community groups – of women and elders – were identified. Members were invited to participate in focus groups, and timing of two-hour sessions were planned with them. These took place in September to October 1993, at midday, after community group activities had ended.

Vietnamese lunches were provided, but no payment. Cash payment would have led to uneasy relationships within the community groups from which focus group members came.

Two sessions were run with each focus group, in successive weeks, to provide continuity and aid recall. They were conducted in Vietnamese. For both groups, the idea and aims of the sessions were introduced, roles clarified, confidentiality explained, and agreement sought to use tape-recording, which would help with writing-up.

Women's voices

Twelve women participated in the first session. Nine of them, plus two new members, came to the second.

All had children. All but one needed interpreters when seeing GPs, and some when making appointments. All lived within the family health services authority interpreting-service catchment area, but none had ever self-referred. About half had been referred by health visitors or hospitals.

The women wanted a 24-hour telephone interpreting service to be provided. They were particularly concerned that when they, or their children, fell sick during evenings, or at weekends, there was no access at all to Vietnamese-speaking interpreters:

> What I fear most is my children getting sick in the night, because I don't speak English. I have no transport. I just don't know who to call for help.

Some women found illnesses were hard to explain to GPs, even with an interpreter. This was a main reason for preferring Chinese medicine and other traditional treatments. But these are expensive, with diagnosis and travel adding to the total cost. Women who speak only Vietnamese also need a Chinese interpreter. Provision of a locally based Vietnamese speaking GP was strongly supported.

Women were aware of family planning, but not breast screening, or HIV/AIDS services. Those with young children also expressed concerns abut childhood illnesses, and how to recognise them and what to do about them.

Many women were unhappy in their marital relationships and some suffered from domestic violence.

Elders' concerns

Eight elders – seven male, one female – participated. All needed language help when accessing health services, but only one had self-referred to family health services authority interpreters.

Elders often asked children or friends to interpret for them, and one 75-year-old, unable to find anyone else, had to communicate with his GP through gesture.

Elders saw British weather as a main cause of illness, including serious problems, such as arthritis, bronchitis, dizziness, fainting and stroke. To explain this, they used Vietnamese expressions such as '*trung gio*' meaning 'apoplexy'. Bad weather left elders feeling very vulnerable, through 'apoplexy after being caught by the wind', or because 'our blood can't stand this cold weather'.

Elders regularly visited Chinese doctors and used traditional medicines. One man explained the virtues of Chinese medicines, obtained from Hong Kong, as follows: 'The cold wind makes me vomit if I eat. I have been to my GP but it was not good. I now just use Chinese medicine ... I have recommended it to many friends'.

The quality of service experienced at GPs and elsewhere was also a major deterrent, for example, for the wife of one man who went to hospital 'two months ago for a (tuberculosis) test but has not been told of the result and we don't know who to ask: our GP or the hospital'.

Early results

The project has completed two evaluation phases. These were: short post-sessional discussions, each of about 20 minutes, for those running the sessions, and a later, combined meeting of Save the Children staff, health authority managers and community workers.

Aims of the post-sessional evaluations were to reflect on:

- what had been discussed;
- practical management issues;
- the content and style of future sessions.

All sessions seemed lively and committed and they contained much relevant discussion.

There was excitement among members in being able to share opinions in a group setting. Participants appeared to share factual, personal information readily

The women's group moved freely on to its own agendas. Concerns included isolation and loneliness, marital relationships, unemployment and childcare.

Elders followed the prepared agenda more strictly. They were also more compliant in their expectations of health services, seeing themselves as recipients, and finding 'consumer rights' concepts especially hard to entertain.

Cultural influences on group dynamics were observed. Vietnamese people expect the young to defer to elders, so the facilitator sometimes felt inhibited when older speakers dominated discussions.

There were external interruptions, for example from young children playing nearby. There was an on-site creche, but mothers and children chose to be together during sessions.

The main purpose of the combined evaluation meeting was to discuss initial outcomes. Pilot focus groups appeared to have helped to assess effectively the shared health needs and priorities of the community. They should be used with other Vietnamese groups. Health commissioning representatives agreed that they ought to carry out this kind of task routinely, and that they should feed the pilot results back to local health and other authorities.

In informal discussions after the sessions, members said that they would welcome opportunities for more shared discussion, but felt groups should include the statutory authorities, whom they perceive to have shown little interest in their health needs so far.

It is important for health authorities to take up such opportunities as an essential element of their own needs-assessment processes.

Lessons learned

What lessons have we learned from the project that we can pass on? Focus groups worked well, but need to be based on proper community involvement, and sufficient development and planning time must be allowed.

Focus groups are only one kind of contribution to needs assessment and when used they should not be isolated from related activities. Furthermore, focus groups provide insights into groups interactions and cultural norms that should be built into the whole needs-assessment process.

The greatest benefit we ascribe to the project is its demonstration of how authorities and the local community can be brought closer together.

The project found means for Vietnamese refugees in Deptford both to express and to address long-running, unmet health needs. We hope that, when the health authority puts its new understanding into action, the community may relax a little its suspicion of all authority.

The project discovered an opportunity for health and other local authorities to work directly with the community. This has already revealed the quality of information that is available within the community and the potential for further development.

It also shows how sought-after service improvements can begin at basic levels, such as developing better communication, not only between providers and purchasers and among themselves, but, above all, between them and users.

References

Bell, J. and Clinton, L. (1993) *The Unheard Community*, Derby, Refugee Action.

Deptford Vietnamese Health Project (1993) *Report to City Challenge Ltd re Project's Developments and Outcomes to Date*, London, DVHP.

Karmi, G. (1992) 'Refugee health requires a comprehensive strategy', *British Medical Journal*, vol. 305, 205–206.

Smithies, J. and Webster, G. (1993) *Responding to Local Voices: an Overview of the Implications for Purchasing Organisations*. Haworth, Labyrynth Training for NHS management executive.

Todd, R. and Gelbier, S. (1988) 'Eat More Food, Get More Health attitudes and food habits of a group of Vietnamese refugees'. *Health Education Journal*, vol. 47(4), 149–153.

Vietnamese Mental Health Project (1992) *Annual Report*, London, Vietnamese Mental Health Project.

32

Developing crisis services

LIZ SAYCE, YVONNE CHRISTIE, MIKE SLADE and
ALISON COBB

Existing alternatives to acute hospital care

Despite a general picture of a total imbalance between hospital and community expenditure, there are some excellent examples of non-hospital
crisis services in place or under development, as well as individual
practitioners working in more imaginative ways:

> When I escaped from the hospital I went to the other end of the country where
> I met a GP who ... refused to enforce the section [authority to admit someone
> to hospital against their will] and treated me herself. She supported me on
> withdrawing from the medication and came to see me every day – talking to
> me and reinforcing my sense of self. She gave me practical tasks by which I
> could measure my progress. (*Mindlink, 1993.*)

There are a small but growing number of alternatives to acute units, both
in the statutory and the voluntary sectors. Some psychiatrists are coming
out of their hospitals and treating people at home, retaining just a small
number of beds. The home treatment service in West Birmingham
explains its success by its comprehensive approach to meeting needs (not
overly medical), its non-coercive nature and round the clock availability.
It is able to respond flexibly to people's needs rather than slot them into
an existing system: for instance, if someone in crisis needs to be able to
contact the team, they can be lent a mobile phone; if someone needs
immediate help with getting the electricity put back on or the children
cared for, the team will assist people to ensure it happens; if someone
needs to talk to someone in a language other than English it can be

Edited from Sayce, L., Christie, Y., Slade, M. and Cobb, A. (1995) 'Users' Perspective on
Emergency Needs', in Phelan, M., Strathdee, G. and Thornicroft, G. (eds), *Emergency Mental
Health Services in the Community*, Cambridge, Cambridge University Press, 39–60.

arranged. These creative solutions were often absent in the old institutions – and absent also in acute units.

In Newcastle the acute unit is a large house, offering residential and day care and a range of treatment approaches. It is less stigmatising than a hospital and has close links with the community mental health team. In Barnet, London, there is a long established crisis intervention service in which a three person multidisciplinary team visits the person or family to assist in working through the crisis and in enabling them to learn coping strategies for the future. Redbridge, Essex, crisis team has cards in their local accident and emergency unit to encourage those who have attempted suicide to use their services. Southampton MIND's Crisis Point provides someone to talk to by telephone or in person between 10 p.m. Saturday and 8.30 a.m. Sunday – 'they have said that they left feeling more relaxed and able to sleep, and a few have said that it saved calling out the duty doctor'. 'Choices' in Cambridge, a project for women and children who are being or have been sexually abused, provides a crisis refuge and counselling.

Where these newer services have been evaluated the results are promising. A 24-hour helpline in Mid-Downs (southern England) was found to have reduced admissions (Health Advisory Service, 1990). An unstaffed flat in Bassetlaw, Nottinghamshire resulted in improved symptoms and praise for the 'peace and quiet of the flat' (Turkington, 1991). Home treatment is preferred to the traditional acute hospital service by both users and carers and can reduce admission by over 50 per cent (Dean and Gadd, 1990).

The main key to positive development is that users are involved from the outset in running or planning the projects. It is important that different groups of users have a say, as relevant – for example, black people as well as white. According to the 1991 Census there are over three million black people in Britain, 5.5 per cent of the total population (Centre for Research in Ethnic Relations, 1992). Issues that have been identified as relevant to black people using mental health service are increases in: diagnosis of schizophrenia (Harrison *et al.*, 1988; Knowles, 1992); use of medication and ECT (Sashidiran and Francis, 1993); forced custodial treatment (Browne, this volume); involuntary admissions (Moodley and Perkins, 1990); and likelihood of transfer whilst on remand (Cope and Ndegwa, 1990). The response of service developers to these issues has been inadequate (Jones, 1991). Good practice guidelines have been developed (Wilson, 1993), yet culturally inappropriate concepts of normality are still applied, such as the nuclear family (Webb-Johnson, 1991).

The dissatisfaction expressed by black people results in not willingly engaging with services before a crisis, having negative experiences of interventions and dropping out of care rapidly afterwards. Problems with these three stages of care were addressed in the development of the

Sanctuary Project. The idea for this project initially evolved at the King's Fund Centre in London, a service development centre which promotes improvements in health and social care. Local consultation exercises were then undertaken. The philosophy and consultation process undertaken in developing the Project are discussed.

The Sanctuary Project is envisaged as a community mental health development, which will be more appropriate to the needs of black people than existing services. It is intended to help black people with serious mental health problems. Black users, carers and other community members will be centrally involved in the planning, running and evaluating of the service, with the aim of encouraging participation and a sense of ownership by the black community as a whole. An accessible 24-hour, seven-days-a-week service is envisaged, that offers short-stay facilities. People will be seen in their home or in the Sanctuary. Child-care will be a central component, since the experience of black women with psychiatry has been particularly negative – one research study found that Afro–Caribbean women are twice as likely as Afro–Caribbean men to be diagnosed as schizophrenic, and 13 times as likely as white women (Knowles, 1992). The aim will be for early engagement between the persons in crisis and the service, hence avoiding the involuntary and negative contact with police and psychiatrist that arises from high use of Section 136 of the Mental Health Act on black people (Ferguson, 1992).

Assessment will be holistic, rather than based on the medical model. It will avoid early diagnosis and use of drugs, emphasising instead information, counselling and other therapies. Counselling will be appropriate, meaning more than just by another black person, but also considering class, race, religion and gender issues (Bhugra, 1993). Any staff employed will be sensitive to these concerns. Complementary therapies, such as dance, drama, music, acupuncture and herbalism will be available. A time-out room will allow safe emotional expression. The option of medication will be available, but not as the central therapeutic approach. In particular, there will be no medical or nursing staff based at the Sanctuary. The therapy process will therefore have a different orientation to most existing psychiatric services, and the intention is that the experience of therapy be positive for the black person in crisis.

After the initial crisis contact, seeing the person soon (for example, next morning) will allow on-going relationships and support to be established. Flexible and assertive outreach practices will be used, to work with black people whose previous experience has left them feeling marginalised by psychiatry (Phaure, 1991).

Consultation took place at a local level in the two London districts selected as sites for Sanctuary projects, under the direction of a steering group for each district. The purpose of consultation was to encourage involvement in and ownership of the Sanctuary Project by the local community, and for the project to be seen as complementary to, rather than in competition with, existing services. The key issues addressed were similarity in values and culture of agencies, agreement on roles and responsibilities, network awareness, all parties gaining from working together, and the absence of alternative resources (Smith *et al.*, 1993).

The steering groups comprised users, carers, advocacy workers, voluntary sector staff, mental health and social services managers, and a King's Fund representative. Public consultation meetings were then held at neutral settings, which is important if people are not to be disadvantaged (Christie and Blunden, 1991). The exercise generated new ideas, such as the need for two venues, one for peace and tranquillity, the other for help and information. The issue of feeling safe at the Sanctuary was highlighted by users, as was the need to educate GPs about cultural needs and alternative therapeutic options. Practical suggestions for involving black women in the project were made. Local issues were also raised, such as a request for the involvement of a particular worker. The consultation therefore was of practical benefit in operationalising the Sanctuary idea. However, the problems that occurred during the process may inform future consultation exercises.

It was difficult to encourage service users to attend these meetings. A danger of this is that poorly-attended stake-holder meetings can be used (wrongly) to justify decision-making as being based on users' views. Even when people do attend, while having had bad experiences of a system they may not know what they would like to see change, with the possibility that whatever suggestions are being discussed are approved in the absence of anything better. Furthermore, few people will have the confidence to speak, typically without training, in a public forum to people who they do not know or (necessarily) trust. The steering group therefore held meetings in various day centres, and liaised with organisations that work with black people.

A second issue in the consultation process was the difficulty in involving local service providers, such as GPs, who would ensure high visibility for the Sanctuary Project. This was where the lack of clinical staff on the steering group was a drawback. However, there is a trade-off between the number of people on the steering group and how quickly progress is made. The King's Fund representative had links with staff in one of the two districts, and it was noticeably easier to develop a vision for the Sanctuary in that district. This suggests that, even where there is a wish to avoid operating under the control of health or social services, it is still helpful to have informal links with practitioners during the consultation process. The ideal configuration is a small steering group, which has links with both community-level and formal services, and whose members have a vested interest in the development of new models of care for black people.

Several principles emerge from the consultation process. Discussions should flexibly encourage service user participation, and should go out to them, instead of expecting users to attend organised meetings. Training for service users in public speaking can increase participation. A steering group should be small but representative, and can facilitate both mental health staff and user group involvement. Individual workers can be used to gain access to the local networks of black people. If services are to address the needs of black people, or indeed any marginalised group within society, then meaningful local consultation and community involvement must take place.

The way forward for British crisis services

From experience with existing projects and consultation with service users, we can make recommendations for new services:

- As a first principle that it does not do harm.
- That it is as effective as possible.
- That it offers autonomy, including a choice of treatments, a choice of worker and a right not to be intruded upon.
- That it gives opportunities to talk through underlying causes of distress, and adopts a holistic, rather than mainly medical, approach.
- That it helps people with their problems where they arise, and does not unnecessarily cut people off from normal life.
- That it offers safety, including safety from abuse for women, black people and others; and including some genuinely safe containment in crisis houses – or in hospital-type care for the minority who want or need that.
- That it takes place in people's own environments or in environments geared towards mental health needs.
- That it manages people's dependence on others when in a crisis with sensitivity – for instance, by ensuring people can withdraw from the service gradually and be linked into other help, such as self-help or longer term professional support.
- That it can offer anonymity.
- That it safeguards resources from the hospitals for support for mental health service users.
- That it respects the support that users offer each other.

The potential pitfalls of non-hospital crisis services identified by users could be tackled by following the above principles – and by ensuring that the new services are developed by or with users.

Many respondents to MIND's consultation on crisis and acute services were attracted to the idea of a safe house (sanctuary, crisis house). Safety can mean safety to be out of control without coming to harm – for example, harming yourself or being oblivious to danger (such as traffic), or being emotionally destructive to those nearest to you. Safety can mean separate space for women, or black women, or black men and women. It can mean 'safety from psychiatry' to those who have had bad experiences of hospital, or safety from drugs. Someone commented that 'you don't want to be "quizzed to death" before being accepted – or rejected'.

Crisis services could be preventive – a place that people can turn to when they anticipate a crisis. For some people it is important that support should come from other users of mental health services – people who can offer peer support, 'fellow travellers'. Diverse users have diverse needs – for example as women, or Afro-Caribbean or Asian people. People using crisis houses may need to have their child(ren) with them or benefit from having their partner stay over. They may need the service to value their spiritual beliefs. Or to respect a lesbian or gay relationship. Or understand money worries.

There is demand both for residential services and support services which come to individuals at home. This is not preference alone – some people have to get away from their home when in a crisis, others are unable to. For other people it is having a complete rest that matters. Befriending, or telephone contact may be what is needed. One suggestion for people in rural areas was for the kind of phone-based alarm system developed for elderly and disabled people living alone to be adapted for those vulnerable to a 'mental health emergency'. A 'sitting service' was mentioned in the survey, where someone stays with the person in distress. In Prato, Italy, a co-operative organisation provides a rota of people to stay with the person in crisis round the clock until they are able gradually to withdraw support.

There is no doubt that a partial revolution has been brought about by user involvement. For the first time users in some areas have begun to have some real say in what is – and is not – developed.

> 250 people turned up to the public meeting and told the District General Manager where to put his DGH unit and his concrete jungle. (Personal communication, on how Nottingham's plans to create another acute unit were foiled.)

However, in some quarters a backlash is occurring, fuelled by concern about the management of risk in the community, especially since the Clunis inquiry. The Royal College of Psychiatrists in 1992 argued that the way to improve the nation's mental health under the Government's Health of the Nation programme was through more psychiatrists and more beds (Royal College of Psychiatrists, 1992). Old arguments resurfaced about the need for psychiatric wards to be physically near to general medicine – arguments that were forcefully demolished, including by some psychiatrists, many years ago. There is another concern: that progressive ideas will be co-opted by professionals such that we end up with community crisis services that are coercive, intrusive and drugs-based. If the Royal College of Psychiatrists takes up the Ritchie Report's suggestion that it develops a crisis card, the result could be a less empowering version of the card already developed by Survivors Speak Out. Or co-option could be by the managers, who might introduce community crisis services as a cost-cutting measure, providing minimal service only to those at 'risk' and consigning users' hopes for early intervention and holistic care to the scrap-heap.

The next few years demand that professionals, managers and the voluntary sector support users in setting up the services which users state that they need. There could be broad public and professional support for this. A survey of the general public found that 76 per cent believe that people with mental health problems should have a legal right to 24-hour crisis services (MIND/RSGB, 1994), as do 89 per cent of Directors of Social Services (MIND/Community Care, 1994). As one person in the consultation put it: 'Services must be run on the lines of a co-operative effect between users, professionals and voluntary agencies.' Exactly which agencies will fulfil which roles in the purchase and provision of these services remains to be established. But what is clear is that change of substance will only occur if users call the tune.

References

Bhugra, D. (1993) 'Setting up services for ethnic minorities', in Weller, M. and Muijen, M. (eds), *Dimensions of Community Mental Health Care*, London, W. B. Saunders.

Centre for Research in Ethnic Relations (University of Warwick) (1992) *1991 Census Statistical Paper*, No. 1. Centre for Research in Ethnic Relations.

Christie, Y. and Blunden, R. (1991) *Is Race on your Agenda?*, London, King's Fund Centre.

Cope, R. and Ndegwa, D. (1990) 'Ethnic differences in admission to a regional secure unit', *Journal of Forensic Psychiatry*, vol. 1(3), 365–376.

Dean, C. & Gadd, E. M. (1990) 'Home Treatment for Acute Psychiatric Illness', *British Medical Journal*, vol. 301, 1021–1023.

Ferguson, G. (1992) 'Race and mental health', *Community Psychiatric Nursing Journal*, vol. 12(6), 11.

Harrison, G., Owens, D., Holton, A., Neilson, D. and Boot, D. (1988) 'A prospective study of severe mental disorder in Afro-Caribbean patients', *Psychological Medicine*, vol. 18, 643–657.

Health Advisory Service (1990) *Report on Services for Mentally Ill People and Elderly People in the Mid-Downs Health District*, NHS Health Advisory Service, Department of Health Social Services Inspectorate.

Jones, A. (1991) *Black Communities Care: Report of the Black Communities Care Project*, Leeds, NISW.

Knowles, C. (1992) 'Afro-Caribbeans and schizophrenia: how does psychiatry deal with issues of race, culture and ethnicity?', *Journal of Social Policy*, vol. 20(2), 173–190.

Mind/Community Care (1994) Survey reported in *Community Care Magazine*, 30 April.

Mindlink (1993) 'Treatments consultations' (written communications in consultation with Mindlink members on treatment, unpublished).

Mind/RSGB (1994) *The Public's View of Mental Health Services*, Mind/RSGB Survey quoted in *Community Care Magazine*, 10 March.

Moodley, P. and Perkins, R. (1990) 'Routes to psychiatric inpatient care in an inner London borough', *Social Psychiatry and Psychiatric Epidemiology*, vol. 26, 47–51.

Murphy, E., Jenkins, J., Scott, J. and Rooney, P. (1991) *Proceedings of Community Mental Health Service – Models for the Future*, Conference Series, No. 11, University of Newcastle Upon Tyne.

Phaure, S. (1991) *Who Really Cares?: Models of Voluntary Sector Community Care and Black Communities*, London, London Voluntary Service Council.

Ritchie, J. H., Dick, D. and Lingham, R. (1994) *Report of Inquiry into the Care and Treatment of Christopher Clunis*, London, HMSO.

Royal College of Psychiatrists (1992) *Mental Health of the Nation: the contribution of psychiatry*. Council Report, R16.

Sashidharan, S. and Francis, E. (1993) 'Epidemiology, ethnicity and schizophrenia', in Ahmed, W. (ed.), *Race and Health in Contemporary Britain*, Milton Kaynes, Open University Press.

Smith, R., Gaster, L., Harrison, L., Martin, L., Means, R. and Thistlethwaite, P. (1993) *Working Together for Better Community Care*, Bristol, SAUS Publications.

Turkington, D. (1991) 'The use of an unstaffed flat for crisis intervention and rehabilitation', *Psychiatric Bulletin*, vol. 15, 13–14.

Webb-Johnson, A. (1991) *A Cry for Change: An Asian Perspective on Developing Quality Mental Health Care*, London, Confederation of Indian Organisations.

Wilson, M. (1993) *Mental Health and Britain's Black Communities*, London, King's Fund Centre.

Part IV

EXAMINING PRACTICE

33

Introduction

This part of the book, looks at innovative and reflective practice. The articles reflect debates and disputes outlined in other parts of the Reader, and consider their implications for practice. The contributions draw on understanding gained through evaluative studies as well as more personal reflection on practice and experience.

In her article Glenys Parry (Article 34) explores some of the misapprehensions concerning the research credentials of psychological treatments. Often counselling in primary care is depicted as an under-researched indulgence, and compared unfavourably with 'tried and tested psychiatric services'. Parry outlines the considerable difficulties in comparing, for instance, the outcome in treatment of depression with cognitive behaviour therapy, and with drugs, through randomised control trials. Parry advocates instead a more descriptive and evaluative audit of counselling and psychotherapeutic services, preferably conducted by counsellors and mental health professionals themselves. She suggests ways in which research studies can be made more useful to practitioners, and ways in which practitioners can inform themselves more fully about research relevant to their practice. (Julian's Leff's article 'Beyond the Asylum' in Part II is an example of a summary of research specifically designed to influence practice.) However, the possibilities for funding research and service evaluation of psychotherapies are discouraging in comparison with the apparent opportunities for drug research.

Rosalind Ramsay and Tom Fahy in 'Reviewing Advances in Psychiatry' (Article 35), from their background as psychiatrists, acknowledge the complexity of studies needed to understand 'psychiatric illness' and treatment. They point to the range of factors involved in mental health problems, and note recent developments of both drug-based and psychological interventions. For instance, the effects of cognitive behaviour therapy in helping people to monitor their own symptoms of schizophrenia are discussed alongside the effects of neuroleptic drugs on the same condition. On the whole however, their review is weighted towards drug-based advances, and in contrast to Parry's article, there is only brief mention here of the relevance of psychosocial factors in the treatment of depression.

The article by David Brandon (36) moves discussion away from individual pathology to changes in services. He argues that mental health work tends to lead to devalued roles for users and for staff involved in their care, and looks at the implications of 'normalisation' for professional skills. Brandon writes on the themes of good relationships, maximising choices, effective participation, personal development and greater mixing. He gives examples of practice which breaks through barriers between professional and client, and makes suggestions for training.

Brandon seeks to normalise mental distress and to help service users find more valued roles, taking part themselves in service provision where possible. Sue Holland in Article 37 describes an approach which challenges existing professional boundaries and engages local women in 'social action psychotherapy'. She outlines an innovative model of therapeutic work with depressed women in their own community setting of a working-class inner-London housing estate. This project has led to a thriving and independent self-help programme which offers a 'Centre for Women's Emotional Wellbeing'. Workers are members of the community and the centre is a resource for all those women who may distrust 'the professional'.

The next four articles focus on specific forms of mental distress and offer new insights or different approaches. Fiona Gardner (38) considers the implications for psychotherapeutic work of the effects in adult survivors of child sexual abuse. She introduces into her classical psycho-dynamic framework the issues of the language used, questions of fantasy and reality, power and control and the implications of gender. By using the language of sexual abuse, rather than that of seduction, therapists can acknowledge the harm that has been done. Even though patients may imbue real-life abuse with aspects of fantasy in their memory, it is important for the therapist to hold on to the reality of actual abuse, and to help patients to confront their trauma. Power and control are always issues in therapeutic relationships. In work with survivors of child sexual abuse the issues are around the potential abuse of power. This can happen through the imposition of theoretical frameworks which are not congenial to the patient, as well as through overt re-enactment of sexual abuse. The gender of the therapist is relevant for the patient, and it is important, even in work with adults, to remember that the majority of those abused are girls and the majority of abusers are men.

Janice Russell's article 'Treating Anorexia Nervosa' (39) first appeared in the *British Medical Journal* editorial and prefaces an anonymous pair of letters which were published in the 'personal view' column of the same issue. Russell highlights the difficulties for doctors in determining what action is most ethical in the treatment of a condition where the patient's interpretation often differs greatly from their own. She reminds us that anorexia sometimes develops in people who have experienced sexual abuse in childhood. Published accounts by people who have been subjected to refeeding programmes often describe the trauma they experienced at such coercive measures (see for instance Caplin, 1996). Yet the physical effects of starvation on the brain mean that a decision to continue with starvation is made under the constraint of altered thinking. Improvements in nutrition can lead a patient to a change of mind and acceptance of weight gain.

The letters which accompany this article give, respectively, the perspective of a colleague, who has difficulty in confronting a friend with the evidence of her anorexic behaviour, and the perspective of a sufferer who found that humiliating treatment made the illness worse rather than better. There are no easy answers to the dilemmas posed by anorexia and bulimia. The letter from the anorexic patient emphasises the importance of help which treats the person as an individual and builds on the motive to seek help. Russell affirms the need to offer treatment which helps rather than hinders the sufferer's process of achieving a secure sense of self.

Marius Romme (Article 40), a psychiatrist from the Netherlands, provides a guide to the reintegration of voice-hearers into their social environment. Hearing voices is usually taken to be a sign of serious disturbance and is alarming for the voice-hearer and for family, friends and helpers. Romme stresses the importance of good quality, honest relationships with helpers; careful reflection on the nature of the voices and their meaning to the hearer; receptivity to the emotions engendered in hearer and helper; social autonomy and trusting relationships with significant friends, family or partner. This guidance is in contrast to the failure of many psychiatrists to pay attention to content, criticised by Mary Boyle back in Article 4. Self-determination is at the heart of Romme's approach, which is basically a normalising one, aiming at removal of stigma. This does not mean collusion with the voice-hearer, but may mean acceptance if voices continue to be heard. 'The ultimate objective is the full development of a personal identity as one who happens to hear voices.'

John Killick's work as a writer-in-residence with people with dementia offers an inspirational approach which can help people to understand what it is like to suffer such neurological devastation. Killick (Article 41) describes how he works with people with the aim of producing a written piece which expresses something that the person has wished to communicate. It involves painstaking attention, willingness to be mistaken for a doctor, a visitor or another resident, compassionate and loving communication, and preparedness to accept rebuffs.

Many of articles so far discussed have involved a broadly psychosocial approach which requires helpers to enter into the world of the person they are trying to help. Sourangshu Acharyya (Article 42) writes of the dilemma facing doctors in working with people from cultural minorities. Western systems of classification dominate psychiatry, yet may seem sadly irrelevant to a person whose presentation of their distress takes a form unaccounted for in these systems. People from cultural or ethnic minorities may be labelled inaccurately, sometimes with disastrous consequences. This is also noted by Browne in his article in Part II. Acharyya argues for recognition that psychological distress in all societies is equally a mixture of cultural and biological factors, in order to avoid the notion that only people from 'other cultures' have cultural problems. Like Romme and Killick, he advocates greater concentration on what the person is saying in their own terms – whether verbally or non-verbally – and to take special note of the cultural dimensions that make up personhood.

The last three articles in this part turn to issues from practitioners in dealing with demands for their services. Peter Tyrer, in Article 43, provides a description and evaluation of a community-based emergency mental health service. Although he says it is not a crisis service as it does not offer 24-hour cover, the aim of providing a rapid response in a community setting means that the service has something in common with the services discussed by Sayce *et al.* in Article 32. Tyrer's article discusses the opportunities and pitfalls in setting up such services. Tyrer attributes the strength of the team and the maintenance of motivation to the model of 'skill-sharing' used. Team members are encouraged to work closely with each other and learn from regular review discussions. They contribute their specialist knowledge, and may provide direct involvement, but overall the person who acts as the co-ordinator of the case takes on a far broader role than might normally be expected from members of the discipline concerned.

Jane Shackman and Jill Reynolds (Article 44) consider a different type of emergency: that created by war-like or politically repressive situations and resulting in the displacement of large numbers of people who become asylum-seekers and refugees. The stress involved in such upheavals is well documented, but there has been less attention to the stress for people working in organisations set up to help refugees in the UK. While stress is felt personally, it is often generated by structural and organisational factors. Shackman and Reynolds look at ways in which services, which often have a sense of 'emergency' built into their hastily-put-together structure, can provide more supportive and nurturing environments for their workers.

The final article by Tom Heller (45) is a personal statement on the dilemmas which face health workers in 'being there' for their patients, which can lead to neglect of their own emotional needs and professional development. He draws on resources and strategies which he has found helpful in his work as a GP. Heller's article completes the circle of breaking through the barriers between professional and service user as he considers his own need for regeneration, professional development and therapy in undertaking this demanding work.

Reference

Caplin, R. (1996) 'Time, faith and encouragement', in Reed, J. and Reynolds, J. (eds), *Speaking Our Minds: an authology of personal experiences of mental distress and its consequences*, London, Macmillan.

34

Using research to change practice

GLENYS PARRY

Research is often invoked as a method of establishing whether or not to provide a service, or which service to provide. For instance, it is sometimes argued that counselling in primary care is popular with GPs and patients, and competes for funds with tried and tested psychiatric services, yet its popularity is not based on sound research evidence of its effectiveness. Conversely it may be argued that many psychiatric services and other professional practices are unevaluated or without research justification, and that an increase in primary care counselling may reflect the failure of services to meet people's needs.

There are problems inherent in the link between research and practice. This article outlines some of the pitfalls in undertaking outcome research in relation to counselling, using research on depression as an example. Research has not had the impact we might hope for on the design of services or treatments. Two complementary issues are how to make research more relevant to practitioners, and how to help practitioners to become more aware of and more interested in research.

Counselling and outcome research

The question 'Does primary care counselling work?' is actually unanswerable. For a start, when people talk about 'counselling' in primary care, they almost always use the term in the broadest sense. Any psychological intervention or treatment in primary care is misleadingly referred to as 'counselling', whether or not it is based on humanistic, client-centred counselling derived from Rogers, Egan and others of that school. I use the term counselling in the narrower sense of skilled therapeutic

Edited from a paper presented at the conference *Collaboration in Care*, St George's Hospital, London (1994).

listening to facilitate self knowledge, emotional acceptance and personal resources, and I distinguish it from other forms of psychotherapy.

A more useful question is: 'Which *type of therapy* produces which specific *effects* with which *patients* compared to which *alternatives*?' In the most general sense, we already know that humanistic counselling 'works', in that meta-analysis shows superiority of treated over untreated groups (Greenberg *et al.*, 1994). We can also deduce from process analysis that it is unlikely, unless modified, to be helpful to people with severe difficulties in regulating emotion and major repetitive problems in interpersonal relationships. So it would be meaningful, for example, to say that one wished to evaluate the relative effectiveness of humanistic client-centred counselling compared with cognitive behaviour therapy and with the GP's usual practice. We could define effectiveness in terms of reducing symptoms and improving coping skills and social support mobilisation. The patients could be people suffering depression and anxiety linked with major life events such as divorce, bereavement or job loss, but no evidence of personality disorder or severe mental illness, in a primary-care setting.

A case example: therapy for depression

Taking studies of major depressive disorder as a starting point, results have recently been reviewed by Roth and Fonagy (1996) and show that psychological treatment is superior to no treatment, with statistical and clinical significance. Results on superiority to placebo treatments are more equivocal, but generally positive. Cognitive behaviour therapy has the edge over less structured forms of therapy. There is little consistent evidence of advantage in combining drug and psychological treatments, and cognitive behaviour therapy is as helpful as drugs for moderate depression. There is some evidence, not decisive, that the very high relapse rates following drug treatments can be improved by using cognitive behaviour therapy. In general, it is better to understand depression as a chronic disorder, with a fluctuating course of episodes, remission and relapse. Research which only examines the treatment of a single episode but does not consider longer-term outcomes is flawed.

Even this simple summary of a field of research hides a number of problems in interpreting the results, problems which lead to a number of caveats. For example, placebo blindness is not equivalent in psychotherapy and drug treatments. Studies of drug treatments using 'active' placebos (which mimic side-effects of anti-depressants) get far less impressive results. On the other hand, studies which find therapy as helpful or more helpful than medication are often found to be flawed, in that they either give low dosages of the drug or they fail to monitor plasma drug levels. Therapy and drug comparisons mask a very high attrition rate in the patients recruited to the drug treatment group, and many people seem to find therapy more acceptable than drugs. This reduces the effectiveness of the drug treatment as delivered in every-day practice. Another problem with interpretation is that some studies include patients with a depressive episode overlaid on a long-standing dysthymic disorder, and these people are less responsive to treatment. This hidden variable is confounded between groups and across studies. Studies which demonstrate better relapse prevention in cognitive behaviour therapy tend to have been conducted by enthusiasts for this mode of treatment, whereas those who are not primary proponents of the approach achieve less impressive results. It is hard to know if this is because they are less skilful or more objective. The natural history of depression follows a fluctuating course, so most follow-up periods are too short to distinguish therapy effects from this. Furthermore, evaluating outcome at follow-up in a study designed to look at the effects of treatment is problematic, since only those who improved originally are able to relapse, and the improvers in one treatment can differ from those in another, so that a systematic bias is introduced and the original randomisation is lost. You either have to re-randomise or, better still, design a study specifically to examine relapse.

What is the role of randomised controlled trials?

The 'gold standard' for medical research is the randomised controlled trial (RCT), because it is only if you are sure there are no systematic differences between the treatment and control groups before therapy that you can attribute change to the therapy. The idea of a randomised trial of counselling is therefore attractive, but it is important to understand what such trials can do and what they can't. There are both practical problems and intrinsic limitations.

Randomisation is very difficult to achieve in practice, especially when the GP makes the allocation and believes that counselling would be of benefit (King, 1994). Even when achieved through rigorous quality control during the research, one often finds, especially with small sample sizes, that the pre-therapy groups differ systematically despite randomisation. For example, the treated group ends up as younger or less severely unwell. There is a further problem with differential attrition between groups, where people leave the trial for different reasons in each group, leading to systematic but hidden bias. People who have a strong preference for one treatment or another often refuse to join the trial if they are assigned to the non-preferred option. It is possible to overcome these problems to some extent: by stratified sampling, by reporting on the 'intention to treat' sample, by studying the characteristics of those who left the trial compared to those who stayed, by considering a patient-focused partial randomisation design (Brewin & Bradley, 1989).

When a well-designed trial overcomes these problems it will be able to tell us whether a particular form of counselling intervention can achieve better outcomes than something else with certain patients. But even then, it will be fundamentally unable to evaluate the effectiveness of services *as delivered*. This is because by guarding against threats to internal validity, the results do not generalise easily to ordinary settings where people are not randomly assigned, patient preferences are taken into account when planning treatments, patients are not so highly selected, counsellors are less adherent to a particular model and the results obtained depend on many other sources of variance.

We need to see clinical controlled trials as part of a general Research and Development strategy for psychological treatments in primary care. The first steps are clinical development of promising approaches, theoretical development and descriptive studies of case series. Then, but not before then, RCTs are necessary to establish that an approach has the necessary 'active ingredient' to produce therapeutic change under controlled conditions. Finally there is a need to disseminate the methods, to ensure that people on the ground are delivering them skilfully.

The specific effects of counselling in relation to other forms of psychotherapy for different general practice patients is a valid question for trials-based research. However, counselling in general practice is not a coherent or homogenous therapeutic approach, but a mix of more-or-less established techniques being implemented with more-or-less competence. To answer the general question posed by Wessely (1994) 'Why do 31 per cent of general practitioners now employ counsellors, although no-one knows what they do, let alone if it works?' we require an entirely different approach. This sort of question must be addressed by good descriptive and evaluative audit, preferably conducted by counsellors and fellow mental health professionals themselves. Thus I see experimental and quasi-experimental research, qualitative and process research, audit, quality assurance and service evaluation as members of a family of techniques which complement each other and are informative in different ways (Parry, 1992). All are needed at different points in developing, testing and implementing effective therapeutic interventions.

Weak impact of research on services and treatments

The huge amount of research conducted into psychological therapy over the last 40 years has, to a large extent, failed to influence the design of either services or treatments. In fact, it is hard to point to an example of major changes in service delivery attributable to research. I am thinking of something like the impact over the last 40 years of Bowlby's and others' attachment research on hospital management of children separated from their parents.

On the other hand, there are many instances of services ignoring both process and outcome research. As an obvious example, it is a good bet that many people with phobic anxiety states and related disorders will benefit from a form of treatment, competently carried out, which includes exposure to the feared stimulus, preferably combined with some attention to the role of catastrophic cognitions in stoking up anxiety. This is not a panacea for all such problems, but available research clearly points to this as a good treatment of choice as a first intervention (Roth and Fonagy, 1996). Yet on the whole, NHS service providers have not ensured the skill base within their organisations to deliver this relatively simple requirement. Health Authorities in their role of purchasers of services have not specified the amount or the quality of such work they want to buy, based on any estimate of the population need.

A second example is the strong evidence that a good working alliance between patient and therapist is the *sine qua non* of successful therapy. This applies to all therapies, but particularly those approaches, such as humanistic counselling, which depend on using the relationship between the therapist and the patient in a systematic way, as a vehicle of change. The effect of a good working relationship is at least as strong as most demonstrated differences between therapy types, yet generally speaking, our approach to fostering this alliance and to correcting failures in establishing the alliance, is haphazard.

Another related example is the demonstrated attenuation of the effect of any therapy if the patient has what is sometimes termed 'personality disorder'. We know that these individuals are high consumers of health resources and that the existing mental health services, whether at primary or secondary levels of care, are providing neither efficient nor effective help for them. Individual client-centred counselling is certainly not the treatment of choice. Yet part-time primary care counsellors, often without much support or many links to more specialist services, are still referred these extremely challenging patients, *without anyone even making an informed decision to do so*. The fact that many struggle valiantly to help, often at great personal cost, does not make the situation any better. There has been very little transfer of knowledge and skill from the research base to the field.

Indeed, it is not only with regard to personality disorders that research seems to be disregarded. More generally, GPs and Family Health Services Authorities (FHSAs) seem to ignore research evidence that, for people in poor mental health, there is little value in 'generic' counselling which is not supported by a rationale of treatment related to psychological models of mental disorder. The balance of evidence suggests that 'generic' counselling is only appropriate to the least impaired of the practice population (Roth and Fonagy, 1996).

Designing practitioner-friendly research

There are several ways in which research could become more useful and accessible to clinical practitioners, health authorities and GPs.

- take more active care over generalisability and take steps to reduce threats to external validity, for example, by better tracking of patients lost to research by attrition, reporting the characteristics of the treated in comparison with the 'intention to treat' sample, using more clinically realistic treatments and measures;
- establish clinical as well as statistical significance of change (Jacobson and Truax, 1991);
- give greater emphasis to dissemination of research results in non-specialist, practitioners' journals. Spend time transferring the research group's skills and knowledge to the field. Include costs for these activities in research funding;
- encourage those involved in service audit to link to psychotherapy research groups and vice versa. Join with a group of counsellors to set up descriptive, naturalistic studies and process studies;
- ensure that a good range of appropriate research methods, including single case, qualitative and process methods, are commissioned.

Becoming a research-friendly practitioner

There is much that can be done by practitioners, and those training them, to improve the impact of research on practice. A good part of the problem is about attitudes – believing that research is difficult, boring or irrelevant. Of course, such a point of view is self-fulfilling. Practitioners can bridge the research-to-practice gap in a number of ways:

- they can work at overcoming attitudinal barriers to research, for example, the idea that it is incompatible with humanist values;
- pick up a research journal in the library and browse through it from time to time;
- join with others to share information about current reports in the journals and to plan simple collaborative projects, in-service evaluation and therapy-process analysis;

- read around a specific topic in relation to a patient/client. For example, do a library search to find recent review articles on, say, treatment models in bulimia;
- learn to use basic audit tools, such as simple process monitoring, outcome measurement, as part of one's practice (see Firth-Cozens 1993);
- trainers should ensure that their curricula include opportunities for trainee counsellors to become familiar with applied research methods so that they can read research reports and gain basic skills in the evaluation of their own practice.

Concluding comment

There is no doubt that research benefits practitioners in many ways, not least through people being active consumers of research and undertaking audit and service evaluation. It breeds a healthy scepticism and is a corrective for dogmatism and complacency. Researchers, too, need strong links with practice, otherwise their work becomes less and less applicable.

However, there are great difficulties in ensuring adequate funding for research and service evaluation in the psychotherapies compared with, for example, testing and marketing a new drug. All the research and development costs of the drug research are loaded on to the prices of drugs in a multi-billion-dollar industry. There are no analogous commercial incentives to develop a better therapy. There is no way that the average counsellor or psychotherapist can put a ten per cent surcharge on to the price of their service to fund research. We therefore continue to need well-designed protocols in this field to be submitted to government and charitable agencies which fund research. Only in this way can we increase the proportion of research and evaluation funding the psychological treatments receive.

References

Brewin, C. R. and Bradley, C. (1989) 'Patient preferences and randomised clinical trials'. *British Medical Journal*, vol. 299, 313–315.
Firth-Cozens, J. (1993) *Audit in Mental Health Services*, Hove, Lawrence Erlbaum.
Greenberg, L. S., Elliott, R. K. and Lietaer, G. (1994) 'Research on experiential psychotherapies', in Bergin, A. E. and Garfield, S. L. (eds), *Handbook of Psychotherapy and Behaviour Change*, 4th edn, New York, Wiley.
Jacobson, N. S. and Truax P. (1991) 'Clinical significance: a statistical approach to defining meaningful change in psychotherapy research', *Journal of Counselling and Clinical Psychology*, vol. 59, 12–19.
King, M. B. (1994) 'Counselling services in general practice. The need for evaluation'. *Psychiatric Bulletin*, vol. 18, 65–67.

Parry, G. (1992) 'Improving psychotherapy services: applications of research, audit and evaluation', *British Journal of Clinical Psychology*, vol. 31, 3–19.

Roth, A. D. and Fonagy, P. (1996) 'What works for whom? a critical review of psychotherapy research', New York, Guilford Press. Report to Department of Health strategic policy review on psychotherapy services.

Wessely, S. (1994) 'Research foundations for psychotherapy'. *Psychiatric Bulletin*, vol. 18, 167–168.

35

Reviewing advances in psychiatry

ROSALIND RAMSAY and TOM FAHY

Psychiatric illness stands out as having a rich multi-factorial aetiology. This makes it unreasonable to expect massive advances because of the complexity of studies needed to understand it. Acknowledging the multifactorial nature of psychiatric illness, over the past year researchers have tried to consolidate information about different disorders – for example, by studying environmental aspects of the more biological conditions and biological aspects of the traditionally more environmental disorders – and in drawing up treatment protocols.

Schizophrenia

There have been some promising developments in the treatment of schizophrenia, with both pharmacological and psychological interventions. It is now recognised that the atypical neuroleptic drug clozapine is effective in 30 to 60 per cent of patients with schizophrenia who do not respond to conventional neuroleptics (Kerwin, 1994; Clozapine Study Group, 1993). As up to 2 per cent of patients treated with clozapine develop agranulocytosis, obligatory blood monitoring has been introduced, which can be a considerable obstacle to compliance for many chronically psychotic patients.

The introduction of clozapine in Britain in 1990 was followed by the release of another atypical neuroleptic, risperidone, in 1993. The clinical effects of these drugs have stimulated investigations into the neurochemistry of schizophrenia and are leading to the formulation and testing of different hypotheses to explain the disorder's complex phenomenology.

Conventional neuroleptics are presumed to exert their effect through

Edited from Ramsay, R. and Fahy, T. (1995) 'Recent Advances on Psychiatry', British Medical Journal, vol. 311, 167–170.

blockade of dopamine D_2 receptors. Clozapine has a more complex pharmacology, with a lower occupancy of D_2 receptors and higher occupancy of D_1 receptors than conventional neuroleptics (Pilowsky *et al.*, 1992; Farde *et al.*, 1992), in addition to potent blockade of $5HT_2$ receptors. The finding that clozapine also selectively binds to the D_4 receptor led to speculation about the role of this receptor in the aetiology of schizophrenia. There is, however, no difference between patients who respond to clozapine and those who do not respond in the distribution of alleles for the D_4 gene (Shaikh *et al.*, 1993), which suggests that clozapine's antipsychotic activity is not merely a result of its activity at the D_4 receptor. These findings reveal the inadequacy of a simplistic dopamine hypothesis to explain the aetiology of schizophrenia and the therapeutic effect of neuroleptic drugs.

Many patients with schizophrenia exhibit delusional beliefs, often with a paranoid theme, which can lead them to engage in irrational or self destructive behaviour. In most cases delusions response to antipsychotic drugs, but some patients remain chronically deluded despite intensive pharmacotherapy. It is well recognised that psychosocial and family interventions can reduce the rate of relapse for schizophrenia (Bebbington and Kuipers, 1994), but it now seems that individual psychotherapy may also help patients to modify or cope with delusional beliefs. The techniques used in this type of work are derived from cognitive behaviour therapy. Over a series of sessions patients are encouraged to systemmatically review the start of their symptoms and are questioned in detail to identify faulty reasoning (Kingdon *et al.*, 1994). Patients learn to monitor their psychotic symptoms and to develop coping strategies, including changing attention, relaxation techniques, and modifying behaviours which exacerbate psychotic symptoms. Efforts are also made to educate patients about their illness and to provide them with an acceptable explanation about the origin of their symptoms, with the aim of destigmatising the illness and improving compliance.

Recent advances in psychiatry

- New atypical neuroleptics clozapine and risperidone are useful in treating schizophrenia.
- Cognitive behaviour therapy techniques help patients cope with delusional beliefs.
- Serotinergic neurotransmitters are important in the aetiology of obsessive-compulsive disorder.
- Education programmes for professionals and public have been developed about depression and a consensus statement for treatment has been introduced.
- Second line strategies for 'treatment resistant depression' include lithium augmentation and electroconvulsive therapy.

These techniques have been evaluated in a study that compared a cognitive behaviour therapy intervention designed specifically for psychotic patients with a non-specific problem solving therapy (Tarrier *et al.*, 1993). Both treatments were effective in reducing psychotic symptoms (compared with the control period, when the patients were on a waiting list), but the cognitive behaviour therapy produced the greater reduction in symptoms. The patients had maintained the improvement six months later. The treatment has a drop out rate of up to half, but if the results of this study are replicated at other centres individual psychotherapy will probably, resources permitting, become part of the standard treatment for patients with chronic psychotic symptoms.

Obsessive-compulsive disorder

Obsessive-compulsive disorder classically presents as a combination of senseless repetitive rituals (compulsions) and intrusive repetitive thoughts (obsessions). The importance of neurobiological factors in the aetiology of this disorder is suggested by the observed increase in obsessive-compulsive symptoms after head injury and encephalitis. Structural brain imaging has failed to reveal consistent abnormalities in patients with the disorder, but recent functional imaging studies show a consistency that is unusual in the functional imaging studies of other psychiatric disorders. Positron emission tomography in patients with obsessive-compulsive disorder who were not receiving drugs showed increased cerebral blood flow in the orbitofrontal cortex and the dorsal parietal cortex and decreased cerebral blood flow to the caudate nucleus compared with controls (Rubin *et al.*, 1992). These findings are consistent with previous studies with positron emission tomography, which reported increased metabolic activity in the frontal cortex. Relevance to the psychopathology of obsessive-compulsive disorder has been shown in studies that induced obsessive-compulsive symptoms in patients undergoing positron emission tomography. It has been possible to show a close relation between severity of symptoms and cerebral blood flow in the orbitofrontal cortex, basal ganglia, hippocampus, and cingulate gyrus (McGuire et al. 1994; Raunch *et al.*, 1994). Two elegant studies have compared the results of positron emission tomography before and after treatment. One of these found that improvement in obsessive-compulsive symptoms correlated with changes in caudate metabolism and that these changes were present in patients treated with either drugs or behaviour therapy (Baxter *et al.*, 1992). The other study also showed a correlation between reduction in symptoms and decrease in orbitofrontal metabolism (Swedo *et al.*, 1992). One interpretation of these findings is that obsessive-compulsive symptoms may be the result of an abnormally functioning neurological circuit encompassing the orbitofrontal cortex, cingulate gyrus, and caudate nucleus (Insel, 1992). This hypothesis is open to testing and refinement with further functional imaging and neuropsychological studies.

Investigations of the neurochemistry of obsessive-compulsive disorder highlight the role of abnormalities in serotoninergic neurotransmitters (Lucey, 1994). Controversy over the superiority of antidepressants with specific serotoninergic agonist activity over other antidepressants has now concluded in favour of serotoninergic drugs. Some of the evidence that supports this conclusion comes from an unlikely source: a study of the treatment of a type of obsessive-compulsive disorder in dogs known as canine acral lick (excessive licking of paws or flanks, which can produce ulcers and infection) showed clear superiority in serotoninergic drugs (clomipramine and fluoxetine) over non-serotoninergic antidepressants (Rapoport *et al.*, 1992). A meta-analysis of studies of treatment for obsessive-compulsive disorder concluded that behaviour therapy, clomipramine, and fluoxetine were all effective (Cox *et al.*, 1993). In view of the lower risk of relapse after effective behaviour therapy, however, this should still be considered the first line treatment.

Depression

Attempts to raise the profile of depression among professionals and the public have resulted in consensus statements, and public education programmes such as the 'defeat depression campaign' run by the Royal College of Psychiatrists in Britain (Paykel and Priest, 1992).

General practitioners still do not diagnose up to half of cases of depression, and, of those patients who receive treatment, 80 per cent have ineffective doses for less than the minimum required course while 60 per cent of patients who recover do not receive preventive drug treatment (Fishman, 1994). These findings prompted the International Committee for the Advancement of Neuroscience and Psychiatry to issue a consensus statement on the long term management of depression. The key message is that, with a first episode of depression, a course of antidepressants should continue for six months and stopping the treatment sooner is likely to lead to a return of depression in up to 60 per cent of cases.

Over the past year researchers have attempted to consolidate knowledge about other aspects of the management of affective disorder. Goodwin (1994), analysing the results of studies on bipolar illness, found that over half of new episodes, especially of mania, occurred within three months of stopping treatment with lithium. He concluded that psychiatrists should not introduce lithium for the prophylactic treatment of bipolar illness until both the doctor and the patient understand that the drug must be used for a minimum of two years.

Treatment of depression may be helped by knowledge of the course of the illness in subjects with different temperaments. Double depression (acute illness superimposed on background depression) is associated with a dysthymic temperament, while cyclothymic depression (depression fluctuating with a normal or elated mood) is associated with a cyclothymic temperament (Akiskal, 1995). It seems that these two types of depression respond to different treatments, traditional antidepressants being suitable for double depression and mood stabilisers such as lithium being particularly useful in cyclothymic depression.

Another example in which knowledge about the nature of a depressive illness might affect the treatment is the maternity blues. Interest is now focused on progesterone, Harris *et al.* (1994) having shown that maternal mood in the days immediately after delivery is related to the withdrawal of naturally occurring progesterone. In future this might mean that treatment with progesterone would decrease the severity of the blues.

In spite of the increased range of antidepressant drugs available, patients with 'treatment resistant depression' continue to challenge clinicians. Evidence has been accumulating in support of alternative second line treatment strategies – for example, the addition of lithium to an antidepressant. Results from the first large controlled trial of lithium augmentation of fluoxetine or lofepramine showed a significant increase in response to treatment in the lithium augmentation group compared with the group given antidepressant only, provided subjects received adequate doses of lithium (Katona *et al.*, 1995). Another option is to use electroconvulsive therapy, to which at least half of patients with treatment resistant depression will respond (Prudic *et al.*, 1994).

Understanding depression requires knowledge of psychosocial factors too. Kupfer, arguing that anti-depressant drugs had long term benefit in preventing recurrent episodes of depression, also found that monthly psychotherapy had a modest but significant prophylactic effect (Kupfer, 1993).

References

Akiskal, H. S. (1994) Dysthymic and cyclothmic depressions: therapeutic considerations, *Journal of Clinical Psychiatry*, vol. 55(supplement), 46–52.

Baxter, L., Schwartz, J. M., Bergman, K. S., Szuba, M. P., Guze H., Mazziotta J. C., *et al.* (1992) 'Caudate glucose metabolic rate changes with both drug and behaviour therapy for obsessive-compulsive disorder', *Archives of General Psychiatry*, vol. 49, 681–689.

Bebbington, P. and Kuipers, L. (1994) 'The predictive utility of expressed emotion in schizophrenia: an aggregate analysis', *Psychological Medicine* vol. 24, 707–718.

Clozapine Study Group (1993) 'The safety and efficacy of clozapine in severe treatment-resistant schizophrenic patients in the UK', *British Journal of Psychiatry*, vol. 163, 150–154.

Cox, B. J., Swinson, R. P., Morrison, B. and Lee, P. S. (1993) 'Clomipramine, fluoxetine and behavior therapy in the treatment of obsessive-compulsive disorder: a meta-analysis', *Journal of Behaviour Therapy and Experimental Psychiatry*, vol. 24, 149–153.

Farde, L., Nordstrom, A. L., Wiesel, F. A., Pauli, S., Haldin, C. and Sedvall, G. (1994) 'Positron emission tomographic analysis of central D_1 and D_2 receptor occupancy in patients treated with classical neuroleptics and clozapine. Relation in extrapyramidal side effects', *Archives of General Psychiatry*, vol. 49, 538–544.

Fishman, R. H. B. (1994) 'Treating depression', *Lancet*, vol. 344, 1291.

Goodwin, G. M. (1994) 'Recurrence of mania after lithium withdrawal: implications for the use of lithium in the treatment of bipolar affective disorder', *British Journal of Psychiatry*, vol. 164, 149–152.

Harris, B., Lovett, L., Newcombe, R. G., Read, G. F., Walker, R. and Riad-Fahmy, D. 'Maternity blues and major endocrine changes: Cardiff puerperal mood and hormone study II', *British Medical Journal*, vol. 308, 949–953.

Insel, T. R. (1992) 'Towards a neuroanatomy of obsessive-compulsive disorder', *Archives of General Psychiatry*, vol. 49, 739–744.

Katona, C. L. E., Abou-Saleh, M. T., Harrison, D. A., Nairac, B. A., Edwards, D. R. I., Lock, T. et al. (1995) 'Placebo-controlled trial of lithium augmentation of fluoxetine and lofepramine', *British Journal of Psychiatry*, vol. 166, 80–86.

Kerwin, R. W. (1994) 'The new atypical antipsychotics', *British Journal of Psychiatry*, vol. 164, 141–148.

Kingdon, D., Turkington, D. and John, C. (1994) 'Cognitive behaviour therapy of schizophrenia', *British Journal of Psychiatry*, vol. 164, 581–587.

Kupfer, D. J. (1993) 'Management of recurrent depression', *Journal of Clinical Psychiatry*, vol. 54(supplement), 29–33.

Lucey, J. V. (1994) 'Towards a neuroendocrinology of obsessive-compulsive disorder', *Journal of Psychopharmacology*, vol. 8, 250–257.

McGuire, P. K., Bench, C. J., Frith, C. D., Marks, I. M., Frackowiak, R. S. J. and Dolan, R. J. (1994) 'Functional anatomy of obsessive-compulsive phenomena.' *British Journal of Psychiatry*, vol. 164, 459–468.

Paykel, E. S. and Priest, R. G. (1992) 'Recognition and management of depression in general practice: consensus statement', *British Medical Journal*, vol. 305, 1198–1202.

Pilowsky, L., Costa, D. C., Ell, P. J., Murray, R. M. and Kerwin, R. W. (1992) 'Clozapine, single photon emission tomography and D_2 receptor blockade hypothesis of schizophrenia', *Lancet*, vol. 340, 199–202.

Prudic, J. M., Sackeim, H. A. and Rifas, S. (1994) 'Medication resistance, response to ECT, and prevention of relapse', *Psychiatric Annals*, vol. 24, 228–231.

Rapoport, J. L., Ryland, D. H. and Kreite, M. (1992) 'Drug treatment of canine acral lick. An animal model of obsessive–compulsive disorder', *Archives of General Psychiatry*, vol. 49, 517–521.

Raunch, S. I., Jenike, M. A., Alpert, N. M., Baeur, K., Breiter, H. C., Savage, C. R. et al. (1992) 'Regional cerebral blood flow measured during symptom provocation in obsessive-compulsive disorder using oxygen 15-labelled carbon dioxide and positron emission tomography', *Archives of General Psychiatry* , vol. 51, 62–70.

Rubin, R. T., Villanueva-Meyer, J., Anath, J., Trajmar, P. G. and Menon, I. (1992) 'Regional xenon 133 blood flow and cerebral technetium 99 m HMPAO uptake in unmedicated patients with obsessive-compulsive disorder and matched control subjects', *Archives of General Psychiatry*, vol. 49, 695–702.

Shaikh, S., Collier, D., Kerwin, R. W., Pilowsky, L. S., Gill, M., Xu, W.-M. *et al.* (1993) 'Dopamine D_4 receptor subtypes and response to clozapine', *Lancet*, vol. 341, 116.

Swedo, S. E., Pietrini, P., Leonard, L. H., Schapiro, M. B., Rettwe, D. C., Goldberger, E. L. *et al.* (1992) 'Cerebral glucose metabolism in childhood-onset obsessive–compulsive disorder. Revisualisation during pharmacotherapy', *Archives of General Psychiatry*, vol. 49, 690–694.

Tarrier, N., Beckett, R., Harwood, S., Baker, A., Yusupoff, L. and Ugarteburu, I. (1993) 'A trial of two cognitive-behavioural methods of treating drug-resistant residual psychotic symptoms in schizophrenia patients. I: Outcome', *British Journal of Psychiatry*, vol. 162, 524–532.

36

Normalising professional skills

DAVID BRANDON

Normalisation invites us to focus on the nature of change in services rather than on changing the pathology of the devalued individual. Five major themes – *good relationships; maximising choices; effective participation; personal development and greater mixing* (Brandon and Brandon, 1988) – present a challenge for user-based services with a coherent philosophical base measured against a clearly outlined outcome. Wolfenberger calls this 'Model Coherency'. There is very little material which tries to sketch out their necessary implications for the training and skills development of professionals.

Presently, we take student professionals right away from their customers for large chunks of their education and training. Their direct experiences are often devalued in training sessions. They are trained in an elitism based on an undercurrent that professionals really know best. Promotion and increased salaries mean greater social distance from the customers. Frontline staff are usually the worst paid and the least qualified and experienced.

Good relationships

We have to provide settings which nourish and support the depth of contact between paid staff and consumers. Professionals are often trained to feel guilty about their relationships with users. Over-emotionalism is suspect. Relationships can become emotionally constipated – cool and detached, never offering real friendship, which is considered dangerous and uncool.

Extracted from Brandon, D. (1991) 'The Implication of Normalisation Work for Professional Skills', in Ramon, S. (ed.), *Beyond Community Care: Normalisation and Integration Work*, London, Mind/Macmillan.

The implications for training are immense. We need to explore and share our vulnerability – to move away from the macho traditions of our professions. We are not aiming for independent heroes and heroines but for interdependence, with people working in a nourishing environment and trusting more in each others' wisdom and skills. That must include more appreciation of users' attainments and more diffusive relationships which allow consumers to make a greater contribution to the quality of services. That involves a 180° turn in the existing training.

The emphasis needs to be on the skills of openness and flexibility. We need to have access to support and use it. We need to express our difficulties with others and feel confident in sharing. That means working more closely as a team and not being afraid of ignorance and personal limitations. Often ignorance is the only wisdom. There are no real answers to many areas of human suffering except that we eventually and unavoidably fall ill and die. Being strong has to take on a new meaning in the professions. There are none so weak as those who *must* cope with everything.

What kind of support structures will work? Recently, in Holland, I saw some former service users, women who had been depressed, working with professionals in helping others. This helps fuzz the boundaries between helpers and helped. Boundaries become more diffusive. Professionals also felt more able to talk about their distress.

Maximising choices

Our systems are frequently monopolistic and autocratic. People need to have real and increasing choices as well as meaningful options – increased control over – food, drink, patterns of the day like mealtimes and bedtimes, holidays, sexuality, relationships. All that involves access to resources like money, telephones, transport, friends to share with and help, adequate knowledge and information, social skills and experience, competence with tools, musical instruments, gardening … above all, surrounded by people who believe you can do it and what Bob Perske calls the 'dignity of risk'.

People start off by being devalued and labelled 'mentally ill' or 'wheelchair case'. They are shunted off into devalued services which amplify their deviance, herding people with handicaps together in one place with staff who are poorly paid, trained and supported.

Staff need to communicate *realistic optimism*. We must select innovative people, full of ideas and possibilities. Practical dreamers. They need training in practical visions, constructive options and how to work towards them. The services must give the staff a direct feeling of making choices, of having control over their lives, instead of feeling like a private in the Red Army.

Choices are widened through us having respect for others so that they can learn and expand. Staff also need help in widening people's experience and possibilities for ordinary living and independence. If they are to model good practice, they need to experience it. Users are usually going to need real work and real wages to have effective choices in our desperately materialistic society.

Our concept of choice is widened further by the Avoider movement amongst people with disabilities (Chamberlin, 1987). These are pioneering individuals and groups of people who distance themselves from services run by professionals. They want cash not staff. They want their own services to free themselves of the colonialism of able-bodied people. It has some parallels with the separatist branch of the Feminist Movement. It doubts fundamentally that professionals can ever engage in a genuine partnership with people who have disabilities.

Service brokerage, developed in Canada, particularly in British Columbia and Alberta, gives devalued people the opportunity to finance their own services. As in the advocacy movement, it produces a new kind of professional – the service broker – who fixes deals for people with disabilities (Brandon and Towe, 1989).

Brian Salisbury (Broker):

> As a Broker, I'm an agent for the person whom I act on behalf of. I don't do things to the person. My whole role is designed to on the one hand identify needs and on the other hand take the vision of those needs and translate it into what's appropriate and obtainable in the community. What I really do is walk through the community system ... The greatest sense of fulfilment or reward that I get as a Broker is the empowerment of other people (National Symposium on Brokerage, 1987).

As I saw in Canada recently, some users are rapidly acquiring the skills of service providers. They are able to hire and sometimes to fire staff.

Effective participation

We have to see the development of real democracy within human services so that users exercise their rights as citizens even when entering an old people's home or a psychiatric hospital. Too many devalued people go from relative democracy into services which are the political equivalents of Central American republics and lose any rights they formerly held.

The tide slowly turns as some professionals learn skills of sharing power (see Holland, this volume).

Harper (1988, 18) writes of the involvement of people with learning difficulties in service planning groups in Clwyd. 'Initially, staff were challenged. It was all new and strange. Staff were having to acquire new skills in self advocacy, reorganising traditional committee structures, new approaches to participating in meetings so that consumers were not dominated and had time to express their points of view.'

Devalued people are being encouraged and trained to speak out for themselves – self-advocacy. Current professional training offers little guidance and some barriers. Student professionals can be trained to be open to radical innovation. Services need to respect users' gifts and wisdom and provide adequate information; machinery for consultation; for the development of both staff and users' skills in areas like self-advocacy and assertion; access to independent representation as in citizen advocacy for example.

Professionals are often fearful of power sharing. They came in to social work, nursing or medicine to take up quasi-parental positions. They may not know how to cope with greater equality, more brotherly-sisterly relations. Such relationships may seem threatening and less controllable. They lose influence. Users will probably make decisions that are seemingly foolish and risky when the role of professionals, for so long, has been to control, to protect and to cosset. The structures in which they operate hold them directly accountable for things which go wrong. Given no radical changes in organisational structures, reasonable risks going wrong may lead to blaming the professionals.

Personal development

The personal development of both users and workers is among the most difficult tasks for our services. We are moving very slowly away from negative 'you can't' philosophies to see our customers as achievers. All kinds of packages like individualised programme planning and case management can assist but packages are not enough. What is required is a substantial attitude and skill shift.

What makes us feel worthwhile and valued is a complex jigsaw. Perhaps we have firm convictions about religion and/or politics. 'Being a Christian or a Muslim' gives a sense of identity and common faith with others. We are part of a group of fellow believers, attending services, worship, study groups. Perhaps we gain confidence from skills in woodwork, car maintenance, gardening, running, cooking. Some people find confidence in their knowledge: they know about geography, German, maths, chemistry.

Most find confidence in relationships. We are loved by others; feel warm about close friends; feel nourished; feel attractive. Most of us own cherished possessions – things which have a sentimental value – an engagement ring, a favourite pen given as a present by Mum, photos, records. We feel we have personal attributes.

Normalisation reminds us that the blossoming of individual selves is difficult in the 'special' services. People usually don't feel 'special'. Both users and staff are used to marginalised groups. Devalued people are looked after often by devalued and poorly paid staff. Talent and initiative are knocked out of them by an extensive and brutal system. So-called helping cultures can be extremely oppressive, rendering most people passive and ritualised. Not only users get infantilised and institutionalised, it happens to staff as well.

In exploring the normalisation principle, we need skills which take us away from the demands of services towards seeing the individuality of Alice and Fred (see, for instance, Killick, this volume). That means moving towards a view of people as lovers, tenants, house owners, workers rather than as patients, bedwetters, deviants, self-mutilators. These skills help the emergence of individuality and personal development and achievements.

That emergence needs core counselling training for all staff. We need to listen and respect individual wishes and differences in ways which have not so far been evident. *Good listening,* helping people to construct their visions for their own increasingly independent existence, are vital components in professional help. It needs an ability to see far beyond the wheelchair, the crutches and the dark glasses into the person's true nature and potential.

Greater mixing

This is about integration with non-paid and non-devalued people. Most existing services encourage segregation and congregation so that disabled people feel weird to 'ordinary' people who have had little contact with them. Ordinary people also experience a segregated education system: segregated away from people with disabilities.

Mixing means access to buildings through necessary ramps and wide doors at times when others, who are not handicapped, are using the facilities, with understanding staff to help if necessary. It means receiving respect, good manners and esteem from others within a society which has a positive image of handicapped people; having the opportunity for a satisfying private life with meaningful relationships; being a citizen with legal status and opportunities for growth, maturing and self-fulfilment; living in a society with structures and forms that assist all kinds of integration, primarily through ordinary and valued services (Flynn and Nitsch, 1980).

Mixing involves narrowing the social distance between staff and consumers. It means changing congregated provision for devalued people into more general use for a wider variety of 'ordinary' and valued people. It means encouraging users to use unsegregated facilities like art galleries, ordinary buses and taxis, snooker halls and moving away from services which only provide for 'special groups' like Gateway clubs, Special Olympics, psychiatric clubs. For staff it will involve major changes, often helping people to make use of their networks – painting clubs, community contacts, even friends and relatives.

Staff must learn skills to help access. They learn to be effective bridges into wider neighbourhood and community facilities. They are less central to the support of the devalued person and more facilitating of other relationships, which are more important over the long term. Instead of being omnipotent providers, they make ordinary contacts with valued facilities from libraries to community centres.

Martin and Parrott saw that many young people with learning difficulties lived very isolated lives in Sheffield. They started a project to use the city's recreation facilities which had far-reaching benefits. They produced pen profiles of all young adult people known to workers in north-west Sheffield. Many wanted to learn to play snooker so eight young men joined a class with an adult education tutor and social worker. With the skills learned, most people joined snooker clubs from this class and played with ordinary people. The development of this system has led to a wide range of people exploring different leisure pursuits – sewing, car maintenance, adult literacy, cookery.

> We have come to recognise the value of local group working in supporting individuals who have needs in common. From initial contact with three young women who had low self confidence and lacked friends, we identified eight young women in the area with similar needs. We responded by running a short life assertion group to discuss common problems, finding new and more constructive ways of emotional expression and supported friendships. (Martin and Parrott, 1989).

All this material has fundamental consequences for the skills required for the managing of services. In many ways it increases the extent of chaos. Creativity comes from a mess where unexpected and inspirational connections can be made. Managers need to welcome chaos and live comfortably with it rather than attempting a clean-up. Instead of traditionally setting clear guidelines, and developing rather rigid categories and working out prescribed goals, it throws the windows and doors wide open.

All of us, as doctors, psychologists, social workers, nurses, occupational therapists, boldly claim to pursue userist ends. 'The patient/client/resident/consumer comes first in this profession.' All of us know in our hearts that is untrue. We know it from our stay as patients in hospitals; from the experiences of our elderly relatives. Users are usually last in the queue. They don't ever have reserved car park places. No one really wants it that way. We are all userists now but there are just too many meetings, too many vested interests which stand between good intentions and good effects.

We will need completely new forms of professionalism in our diligent pursuit of normalisation. They must be more feminist in structure and operation, and increasingly seek more cooperation and equity. We need to move towards new values, towards mutually humanising and liberating practices, rather than the traditional ignorance. It is essential to model and communicate the value and goodness inherent in ordinary human relationships.

We need new skills involving softer and more vulnerable parts of ourselves, less controlled and controlling, less manipulative and more loving. It means turning aside from the traditional hard shell of the paid workers. These new people will need substantial support and warmth to avoid the worst aspects of traumas.

We need increased flexibility based on the nature and needs of particular persons rather than on some generalised prescription for so called client groups. Person feeling rather than group think. General prescriptions are of limited help and techniques can be inherently dangerous.

We need more softness and gentleness to see who is really there. Our brothers and sisters everywhere suffer in poverty. Close contact can lead us naturally towards anger and indignation rather than to control. We must avoid further oppressing our own brothers and sisters.

References

Brandon, A. and Brandon, D. (1988) *Putting People First*, London, Good Impressions.

Brandon, D. and Towe, N. (1989) *Free to Choose: An Introduction to Service Brokerage*, London, Good Impressions.

Chamberlin, J. (1988) *On Our Own*, London, Mind Publications.

Harper, G. (1988) 'Consumer-led service planning', *Community Living*, September, 18.

National Symposium on Brokerage (1987) British Columbia.

37

Developing a bridge to women's social action

SUE HOLLAND

This article describes a method of working with depressed women on a multiracial inner city estate in such a way as to make connections to the wider community. The social and political factors involved often remain invisible, although contributing massively to mental distress. The work is described more fully elsewhere (Holland, 1988, 1990, 1992).

Essential to the principles of empowerment and liberation which inform mental health promotion is the idea of beneficial movement or change. In this case, the aim was to bring together a combination of techniques drawn from sociological, psychological and therapeutic sources, known therefore as 'sociopsychotherapy' or 'social action psychotherapy', in order to facilitate change. The theory and method described here led to the development of an innovative model of psychotherapeutic work with depressed women in their own community setting. The women were residents of a large working-class, multiracial inner-London housing estate. The psychological and inter-personal problems they experienced were usually compounded by inequalities and deprivations: poverty, racism, unemployment, lack of childcare facilities, and limited educational and recreational choices, to name a few.

In order to address not only their personal psychic pain – which shows itself in the form of depression, anxiety, or phobias – but also their social suffering, a model of mental health intervention was used which confronted both psychic depression and social oppression. The women were encouraged to move through a series of perceptions of themselves, ranging from passive victim/patient to active participant in their own and their neighbours' well-being. In practice this involved a series of steps from brief focal psychotherapy to educational discussions to social

Edited from Holland, S. (1995) 'Interaction in Women's Mental Health and Neighbourhood Development', in Fernando, S. (ed.) *Mental Health in a Multi-Ethnic Society: A Multidisciplinary Handbook,*, London, Routledge, 139–147.

action around specific local issues or demands. The model of intervention is *dynamic* in that the woman moved through a series of psychic and social 'spaces', each more socially complex than the last. This has proved to be a particularly effective method of therapy with women whose depression stems from childhood abuse, and who later experience the added abuse of economic exploitation, racism and sexism.

Four steps in social action psychotherapy

The steps in social action psychotherapy are described in non-technical terms so that educational deprivations and inequalities are not allowed to rob people of a means of liberation. A leaflet outlines the 'Four Steps in Social Action Psychotherapy'.

Step one: patients on pills

Women sometimes feel so bad about themselves that they can't face their everyday life. We go to the doctor complaining of 'nerves' and get given pills to calm us down (tranquillisers) or cheer us up (anti-depressants). We then see ourselves as having a 'medical' problem. Sometimes the doctor will send us to see a psychiatrist who continues the regime of, usually stronger, mood-changing drugs. We now see ourselves as a 'psychiatric case', passively expecting to be cured.

Step two: person-to-person psychotherapy

White City Mental Health Project offers women an alternative to pills. Talking to a woman therapist helps us to explore the meanings of our depression and so reveals our buried feelings, such as anger and guilt. We can then take charge of all our painful 'ghosts' from the past.

Step three: talking in groups

Now, freed from our personal ghosts, we can get together in groups and discover that we share a common history (HER-STORY) of abuse, misuse and exploitation of ourselves as infants, as girls, as women, as working-class women, as black women .... Now we can see, and say together, what we really want!

Step four: taking action

Having changed ourselves from patient to person, from a state of depression to self and self–other awareness, we can now use our collective voice to demand changes outside in our community ... in our schools, our health centres, our community centres, our housing, transport, and in anything else that affects our lives.

Issues and obstacles

By using this model of therapeutic intervention, depressed women can move through psychic space into social space and so into political space. 'Finding a space for oneself' thus becomes a series of options, each more socially connected than the last, in a progression from private symptom to public action. This therapeutic practice grew out of the theoretical foundations laid down by Burrell and Morgan's 'sociological paradigms' (1979) and further developed as social work theory by Whittington and Holland (1985).

It is a *reflexive* model of practice and teaching which assumes that what is so for the client is so for the practitioner. For example, the clinical psychologist who is more comfortable desensitising a woman's agoraphobia than exploring her unconscious desires will have more 'success' with the woman 'patient' who has also internalised a functionalist/mechanistic view of herself and the world. Both practitioner and 'patient' are capable of changing and moving on to other modes of intervention – for example, a consciousness-raising group based on Paolo Freire's (1972) concept of 'conscientisation', which has a significant collective and socio-historical meaning – but both will require retraining in order to move them into this alternative radical humanist mode.

Emergent issues

Since professional people of many kinds are charged with, or lay claim to, responsibility for community mental health, it is necessary to say something about the issues of identifying who can do what to help facilitate this kind of movement.

The work is disrespectful of existing professional boundaries in three senses. First, it challenges the vested, and anxiously protected, interests of the many distinct professional groups involved in community work. Second, it has been necessary to demystify the conceptual language of psychology and sociology, and it is no easy task to put rigorous theory into the local vernacular, without posturing as a member of the working class. Finally, in attempting to use available services it has been evident that the welfare system condones the passive victim but criminalises those who are angry and assertive. It has been necessary to address not only personal loss but also expropriation (having been robbed) and the desire for justice. It challenges the very notion that professional intervention is necessarily better informed and more beneficial than lay action. It challenges the notion of a hierarchy of professional legitimacy and power, usually referred to as 'medical dominance'. This challenge does not obliterate the idea that specialists of various kinds may provide essential services. It objects only to the claim by any group that they always, exclusively, know best. There are some people in all these professions with sufficient flexibility to make a valuable contribution, and part of the work is to identify and relate to these helpers. But what a challenge it is to all professions that there has emerged from this work a completely autonomous group of local women with their own funds and premises, known as WAMH.

Women's Action for Mental Health (WAMH)

WAMH is a resource for all those women who distrust 'the professional', who might or might not take the prescribed pills, who would like to work but have no confidence or opportunity, as well as those who need advice on how to prevent themselves being robbed, raped, exploited and mistreated in their personal relationships and in their dealings with social institutions and authorities of various kinds. For those who have already suffered some kind of abuse it is also a refuge. WAMH is on the spot and easily contractable during unsocial hours in an emergency. (Ironically their most social hours – evenings, weekends, bank holidays.)

This unit is able to provide services that are acceptable to local people because its workers are members of the community. But acceptability does not merely depend on the location of buildings (Community Mental Health Centres) or with proclaimed 'interest' in a community approach. It is the common bond forged through sharing both the pride and the stigma of the 'council tenant'.

From 'fringe' to 'mainstream'

The White City Mental Health Project started in October 1980 and is still in existence as a mainstream social-services resource 14 years later. The ex-client/users' organisation which sprang from it, Women's Action for Mental Health (WAMH), still thrives, and received its grant from Hammersmith and Fulham Council for the twelfth year running. This durability and tenacity is a highly unusual feature in the present political climate in which small, innovative and radical multiracial, black and working-class projects have not survived longer than two or three years.

Unless such experimental projects can fight their way into the mainstream there will be little prospect of progressive multi-ethnic resources within our community health systems. The fact that the White City Project and WAMH have been successful in surviving and influencing other programmes outside their immediate neighbourhood is due directly to a number of features, the most crucial being its coherent therapeutic theory and practice of 'social action psychotherapy'. Its other features show that the project had incorporated in a pilot form almost every characteristic which was later drawn up by social services in 1990 as a blueprint for community care. So by its tenth birthday the project could demonstrate a twenty-point quality audit (see Holland, 1995 pp. 144–6).

Now into its fourteenth year and firmly rooted in its working-class and multiracial catchment area, the project can afford to be less exclusive and separatist. Renamed 'Bridge' and shedding the clinical associations with the word 'mental', the service now projects itself as a 'Centre for Women's Emotional Wellbeing', offering counselling, psychotherapy, group-work, family-work, workshops, training, consultancy and a specialist library. The group-work will open its membership to a wider catchment area, forging links between women living on the estate and those outside. The title 'Bridge' conveys the links to be made between women, and also echoes the core philosophy, which is to make bridges between the private/personal and the public/social lives of women.

References

Burrell, G. and Morgan, G. (1979) *Social Paradigms and Organisation analysis*, London, Heinemann.

Freire, P. (1972) *Pedagogy of the Oppressed*, Harmondsworth, Penguin.

Holland, S. (1988) 'Defining and experimenting with prevention', in Ramon, S. and Giannichedda, M. D. (eds), *Psychiatry in Transition: the British and Italian experiences*, London, Pluto.

Holland, S. (1990) 'Psychotherapy, oppression and social action: Gender, race, and class in Black women's depression', in Perelberg, R. J. and Miller, A. C. (eds), *Gender and Power in Families*, London, Routledge, 256–269.

Holland, S. (1992) 'From social abuse to social action', in Ussher, J. and Nicholson, P. (eds), *Gender Issues in Clinical Psychology*, London, Routledge, 68–77.

Holland, S. (1995) 'Interaction in women's mental health and neighbourhood development', in Fernando, S. (ed.), *Mental Health in a Multi-ethnic Society: A Multidisciplinary Handbook*, London, Routledge, 139–147.

Whittington, C. and Holland, R. (1985) 'A framework for theory in social work', *Issues in Social Work Education*, vol. 5(1), 25–50.

38

Working psychotherapeutically with adult survivors of child sexual abuse

FIONA GARDNER

Introduction

A wide definition of sexual abuse is taken for the purposes of this study: 'The actual molestation of a child by any person whom that child sees as a figure of trust or authority ... we see the questions of age, blood relationship and taboo as red herrings which obscure the central issue: the irresponsible exploitation of children's ignorance, trust and obedience' (Nelson, 1987, 14.)

This definition highlights three key elements: the betrayal of trust and responsibility, the inability of children to consent, and the abuse of power. It is suggested in this paper that in the therapeutic work with adult survivors of child sexual abuse these three elements may emerge in the re-enactment in the transference relationship. These aspects need careful assessment by the therapist in the light of her or his theoretical stance.

In this country two surveys, the Mori poll (Baker and Duncan, 1985) and a study by Nash and West (1985), concluded that between one in ten and one in three girls have been sexually abused in the wider sense of the word. Studies that have included boys as well as girls found that the proportion of girl to boy is about 4:1 in all cases of sexual abuse (Bentovim and Tranter, 1994).

All the studies agree that the overwhelming majority of abusers are

Edited and updated from Gardner, F. (1990) 'Psychotherapy with Adult Survivors of Child Sexual Abuse', *British Journal of Psychotherapy*, vol. 6(3), 285–294.

men, and many men who have been imprisoned for sexual offences have themselves been victims of child sexual abuse. Although girls are more often the victims of sexual abuse, they rarely become sexual offenders against children. Many women rationalise and repress the incidents of child sexual abuse which then become an internal preoccupation which may emerge in a different configuration and symptomatology. For many adult survivors of child sexual abuse the emotional work in resolving this violation and preoccupation was not and could not be done. In turning to therapy in adulthood there is a second chance to alleviate some of the pain.

Therapists need some guidelines about how to work with understanding the abuse and how it is manifested in the transference. This paper is an initial exploration of the work with adult survivors within psychotherapeutic practice. Psycho-analytic psychotherapy is defined partly by an emphasis on helping the client understand their own unconscious patterns as they are re-enacted in the transference relationship. This is a repetition of past reactions to significant people in early childhood. The psychotherapist will also use their own emotional responses to the client, called countertransference, to further the work (see Dallos, this volume, Article 2, for further discussion). I look first at the psychological effect of child sexual abuse.

Psychological effects of child sexual abuse

Why do therapists encounter adult clients often many years after their sexual abuse? Lindberg and Distad (1985, 332) suggest that 'Following a severe trauma a pattern of repression, denial and emotional avoidance emerges. This denial-numbing phase can last days or decades, then is followed by an intrusive-repetitive phase, in which disquieting symptoms such as nightmares or guilt re-occur'.

The child is overwhelmed by fear, and because he or she cannot retain any control or give informed consent the experience becomes an assault. Detachment then acts as a survival technique. The dissociation, or cutting-off may be complete so that neither the event nor the emotions are recollected. The patient may later recollect the actual sexual abuse but have no appropriate emotional responses.

A patient in her late thirties entered therapy following relationship difficulties at work. Her presenting complaints were of depression and a feeling of hopeless futility, emotions which were quickly felt in the therapy sessions. After six months of twice weekly therapy the patient revealed that while away at the week-end she had realised that between the ages of eight and twelve she had been regularly sexually abused by her uncle, where she had stayed most weekends as a child. Although she had known that this had happened, she felt she had accepted it as part of her 'special' relationship with him that involved secrets. Once she realised she had been abused the patient insisted that she felt nothing about it and for some time did not speak again about the memories.

Assessing the long-term effects

Research based primarily on female clients already in psychotherapy is beginning to show that there are devastating long-term effects of child sexual abuse. Symptoms and behaviours manifested by abused victims include a high likelihood of depression, suicide attempts, low self-esteem, alienation, distrust, sexual acting-out, problems with sex and men, and self-destructive behaviours (Herman and Hirschman, 1981; Haugard and Repucci, 1988; Bentovim and Tranter, 1994). Some research has focused on survivors of incest, abuse within the child's immediate family and, when compared with other psychotherapy clients or with women who had seductive relationships with their fathers, the incest victims' disturbances have appeared the most severe.

It would appear that the trauma of child abuse is exacerbated by life circumstances that are deprived, inconsistent and lonely, and that the degree of trauma correlates with the degree of powerlessness experienced by the child during the sexual acts. A comprehensive summary and estimation of the long-term effects of sexual abuse is an extremely complex task because of the enormous variety of factors involved. For example, a very young child may be more deeply traumatised than an eight-year-old, but so many other variables such as frequency and aggression and degree of physical damage have to be considered. Undoubtedly for those children able to confide in an adult the adult's response affects the way the event is assimilated and the reaction of the rest of the family and outside agencies. In situations of incest placing children in care may prevent further abuse but can leave the child confused about who is the 'guilty' one, as removal from the rest of the family (not from the abuser) may feel like punishment.

The psychic dilemma

The sexual relationship is often explained to the child as a 'special' relationship. This creates a psychic dilemma when the child then simultaneously perceives and confuses the 'good' adult, sometimes referred to as 'object' in analytic writing, with the 'bad' behaviour. This unbearable confusion leads to a fragmented development of the self with continuous splitting or separating into dual aspects of bad-v-good, abusive-v-non-abusive, affect-v-experience and mind-v-body.

'The rage and horror of victimisation are internalised as parts of the "bad self" thus allowing the victim to maintain a fantasy of the object as "good" ' (Kilgore, 1988, p. 226). In this way the child protects the vulnerable part of the self but at the cost of internal polarisation. The internalisation of bad self and good object, a process where intersubjective relations become changed into intrasubjective ones, affects reality testing and judgement, and the abusive relationship is re-enacted in an actual or idealised way in other relationships and experiences.

In the continuing therapeutic work with the patient described above emotional aspects around the child sexual abuse did emerge. However the form they took was of blame and guilt. The patient felt that if she had really wanted to she could have stopped the abuse, therefore she must have wanted it to happen in the first place. She felt further guilt because she became 'the other woman' in the relationship between her aunt and uncle and so betrayed her much loved aunt. Both aunt and uncle gave her many presents and treats and this compounded the blame she felt. If they were 'good' and generous she then became 'bad' and ungrateful. This was the psychic dilemma brought to the sessions, and in turn re-enacted in the transference and counter-transference.

The thought processes seemed to be 'stuck' around configurations of the trauma and it is to this fixation that we now turn.

Unconscious Preoccupations in the Inner Space

In order to understand and work with survivors of child sexual abuse there needs to be a conceptual understanding of the psychic dilemma described above. This would be the inner psychological representation of the external event of the abuse and the relationship with the abuser. So the abuse may be internalised in unconscious form and evoked though not necessarily recollected in random images, symbols and certain relationships and so on.

The process of dissociation and splitting is a form of internal preoccupation with the abuse. The preoccupation blocks internal space preventing thoughts and feelings from being assimilated. As in an occupied zone, gaining entrance is perceived as attack and assault. The abused child may formulate internal concepts of blame, guilt and punishment to explain what happened, while at the same time repressing the hurt feelings and damaged sense of self. The rational feelings remain conscious, the repressed feelings are unconscious and not accessible to the child.

The language of internalisation is metaphorical, and the danger (highlighted by Meissner, 1980) is that the metaphor is lost sight of and we then begin to interpret such metaphors literally. In the work with survivors of sexual abuse this language is used as an attempt to understand what aspects of the external event and the abusive relationship become internalised, and so form structures and unconscious psychic organisations. It is this process and formation that has to be understood and worked with in the context of the external happenings. The process is one of working with the dissociated feelings belonging in the emotional space in order to affect and reformulate aspects of the rational space, thus constituting a less punishing and deprecating sense of self.

Child sexual abuse indicates that the child's body has been violated and professional procedures are accepted to protect the child and deal with the abuser. Psychotherapists need to explore and develop conceptual understanding to work with the emotional legacy of the violation. This understanding can have direct influence on our part in the transference relationship and the working alliance.

In the therapy between therapist and the patient who is a survivor of child sexual abuse, four aspects are pertinent to the transference: the language available, theoretical pressures around fantasy and reality, the issues of power and control, and the implications of the gender of the therapist.

The language we use – seduction or sexual abuse?

In Freud's theory of seduction he was clearly referring to the child's sexual seduction by the father. The theory of infantile sexuality which Freud put in place of the seduction theory attributes unconscious drives and desires to small children, whom he claimed became increasingly sexually interested in the parent of the opposite sex. The combined effect of the theories in semantic terms is of seduction as a sexual advance implying suggestibility and willingness – that was almost asked for, albeit unconsciously. In contrast, the starkness of the term sexual abuse is a clearer concept reflecting the betrayal of trust, the lack of consent, and emphasising the abuse of power.

A year on in the therapy with the patient sexually abused by her uncle, we entered a difficult phase where it felt as if everything about the therapy was very wrong. The patient felt therapy itself was no good, that the comments and interpretations I gave made no sense and that she was not the right kind of person to be a patient. While I listened carefully to what she said, the patient felt the need to act out her feelings by angry silences, fierce retorts and questioning of me followed by a general 'rubbishing' of the whole thing. She also experimented with the idea of stopping, missing several sessions and then reporting how much better she felt. In reviewing what was going on it felt possible to take two views and use two languages.

One possible approach could have been to analyse the patient's behaviour as 'provocative', 'tantalising', 'seductive' and even that she was 'asking for it'. A process of almost adult justification for the abuse could have been made. Taking this view and using this language would have diverted away from what the patient might be revealing about her own experiences and how she herself had been made to feel in the abusive relationship. With this approach it might have felt better and safer being the therapist. With this attitude the theories of infantile sexuality appeared seductive and acceptable. However the second approach involved using the language of sexual abuse to analyse the behaviour. It therefore involved a conscious siding with the child as a sufferer of sexual abuse. It also involved a more careful scrutiny of what were the feelings that the patient was trying to express. Using this approach and language gave insight and helped me understand the confusion, anger and powerlessness the patient had been made to feel in the relationship with her uncle. If she participated in the relationship she was rewarded with approval and gifts, if she protested the uncle and ultimately the aunt would be angry and withdraw.

Fantasy or reality

If a form of re-enactment of the abuse is at times present in the therapeutic relationship then issues of fantasy and reality are pertinent. For if the patient has split off and dissociated herself from the event it may be tempting for the therapist to acknowledge the re-enactment in the transference and counter-transference as indicating fantasy rather than a 'real' event.

One effect of the 'unbalanced' feel of the inner space is that there can be a blurring between the actual abuse and the evocation through image and symbol. The distinction between fantasy and reality is also blurred in that real-life situations become sexualised. The confusion and sexualisation are therefore repeated in the transference and the fantasies may in turn confirm, comfort and console the patient and preoccupy her away from taking part in external reality. The working alliance is dependent on the patient's ability to sustain a realistic tie to the therapist and the therapist's task would appear to be holding the actual abuse as the 'real' event and trying to understand the function of the fantasies within the transference.

Miller reminds us that: 'If we become practised in overlooking the sexual abuse that actually took place we will call our patient's complaints excessive and abandon them to their trauma. Since they cannot confront their trauma without assistance, they must struggle to keep the cause from becoming conscious' (1985, p. 32).

It is important that we try to differentiate between fantasy and reality as it does matter for our patients and will affect the relationship. It appears that the truth of child sexual abuse as with all extreme violence against the person is hard to assimilate. Our theoretical knowledge can help a great deal and accepted technical practices are a necessary comfort to fall back on, but sometimes it seems we can use both of these to protect ourselves from 'unbearable' issues. It may be easier to see only the fantasy than the reality. Child sexual abuse is also only one aspect of a wider form of sexual abuse that is present in our wider society. This too is sometimes hard to recognise and then to acknowledge.

Issues of power and control

Issues of power and control are fundamental in the abusive relationship and as pertinent in the therapeutic relationship. Technical issues of lying on the couch and payment always require awareness but take on further implications in this situation. The way interpretations are heard and incorporated may also reflect the effect of the abuse. Dorey (1986) suggests that in relationships the victim feels that he is really threatened in his autonomy and that his very identity is being denied. This re-enacted in the transference may lead to the therapist experiencing in the countertransference the anxiety of the threat of obliteration, as well as the experience of both helplessness and violence commensurate with the degree felt and inflicted on the child. The therapist may feel uneasy that her/his interpretations are probing and penetrating. Issues around structure and boundaries may feel harsh and unyielding and have meaning beyond that of accepted technical practice.

Power and control are issues always present in the relationship between patient and therapist but in the work with survivors of child sexual abuse the issues are more around the potential *abuse* of power. In some situations overt acting-out of the misuse of power and trust takes place (Chesler, 1974). Such crude re-enactments can neither help women into self-definition, self-esteem or independence, nor provide protection for their conditioned helplessness.

More insidious and covert re-enactments of the abuse of power and trust through theoretical attitudes and beliefs can be as damaging to the patient. Miller (1985) discusses the experience of an analyst in training who worked under supervision with a woman raped in adolescence. During the analysis the patient found it hard to approach the experience on an emotional level. The supervisor explained the patient's attitude as '... an expression of her guilt feelings at the fulfilment of her libidinous desires, which was forbidden by her superego' (p. 39). Despite disapproval at his rejection of this interpretation the analyst in training worked on imagining through reading and then empathising with what the patient had undergone, and talked with her about the trauma not about libidinous feelings. He was taking the rape so seriously that the patient felt able to get in touch with a whole range of emotions about the abuse, which eventually relieved her depression.

Until recently many children have been condemned by disbelief and fear. A more subtle hostility is the notion of 'willing victim' and the 'flirtatious nature' of young children, especially little girls. The notion that on some level the control lies with the child is erroneous.

Such beliefs need careful scrutiny and reassessment in the context of the therapist's own recollections of helplessness and fear as a child. For the patient, issues of control may be further confused if at a physical level they may have felt pleasure in the sexual act. So it is necessary to differentiate between physical sensation and emotional response. The therapist has to be clear that the experience of pleasurable body sensations does not imply collusion. The acknowledgement of the inability to consent clarifies the issues of control.

Aspects of gender

The final aspect to be explored is that of awareness of gender issues. The statistics confirm the vast majority of abused are girls and the vast majority of abusers are men. There seems a reluctance generally to accept this imbalance. Press reports invariably refer to 'parents' and 'children', and the family therapy model which focuses on the role of the mother either as the colluder in the abuse or as responsible for family dysfunction further confuses the issue. Reasons for men to abuse are worth greater attention although findings on the limiting of emotional expression in boys would appear to play a part (Gardner, 1987).

Should male therapists work with survivors of child sexual abuse? In many cases it may be well into the therapy that abuse is recollected and emerges as a trauma. Frosh (1987) argues that: 'if there are systematic factors that make men more likely to abuse children sexually then those factors will be present more or less strongly, in all men', 335. In his paper he suggests that the associative link between children and sex is so uncomfortable as to encourage an exploration of the therapist's own personality.

In work with adult survivors of child sexual abuse sexual stereotypes increasingly confuse that association. In the work with sexually abused children the association is clear and immediate – there are children who have recently been sexually abused, they are aware of this and so are the clinic staff working with them. Frosh found that some abused children were upset by the presence of a male worker during the disclosure interviewing. The team subsequently agreed that the first interview should always be carried out by women. However in work with adults the distress is initially repressed and the emotions and memories less immediate. Cultural and sexist stereotypes have undoubtedly affected both female patient and male therapist. It is suggested that the association and link is not the uncomfortable one between children and sex, but the comfortable and accepted notion of women and sex. The fantasies in turn are not of reparation towards and protection of children but rather of penetration and conquest of women.

Recognition of sexism and cultural conditioning is as problematical as the recognition Frosh advocates of the male workers' own abusive possibilities. The recognition of sexism involves for male therapists a radical challenge to their self-perception. Furthermore it is too easy to assume that women therapists are above all this by nature of their biology, and for some women recognition of their own oppression appears too hard to acknowledge. However the fantasies of penetration and conquest are not generally present and it may be easier for women therapists to understand the fear of male violence and sexual assault as it universally affects women and children.

The emotions produced in the transference and counter-transference between therapist and survivor of child sexual abuse need careful accounting and more indepth exploration. This is especially important for male therapists working in this situation.

The relationship between patient and therapist is always a 'special' relationship. We have to be careful that the trust placed in us is not abused, that the inherent power is understood and that the treatment is at the deepest level with the consent of the patient. In this way the abuse is not re-enacted as abuse but the transference offers a way of understanding if we can align ourselves alongside the abused child.

References

Baker, A. and Duncan, S. (1985) 'Child sexual abuse: a study of prevalence in Great Britain', *Child Abuse and Neglect*, vol. 9, 457–467.

Bentovim, A. and Tranter, M. (1994) 'Psychotherapeutic work with adult survivors of child sexual abuse', in Clarkson, P. and Pokorny, M. (eds), *The Handbook of Psychotherapy*, London, Routledge.

Chesler, P. (1974) *Women and Madness*, London, Allen Lane.

Dallos, R. (1996) 'Psychological approaches to mental health and distress', this volume.

Dorey, R. (1986) 'The relationship of mastery', *International Review of Psychoanalysis*, vol. 13, 323.

Frosh, S. (1987) 'Issues for men working with sexually abused children', *British Journal of Psychotherapy*, vol. 3(4), 332–339.

Gardner, F. (1987) 'Gender issues in child guidance', *British Journal of Social Work*, vol. 17, 187–198.

Houqard, J. and Repucci, N. (1988) *The Sexual Abuse of Children*, London: Jossey-Bass.

Herman, J. and Hirschman, L. (1981) 'Incest between fathers and daughters', in Howell, E. and Bayes, M. (eds), *Women and Mental Health*, New York, Basic Books.

Kilgore, L. (1988) 'Effect of early childhood sexual abuse on self and ego development', *Social Casework*, vol. 69(4), 224.

Lindberg, F. and Distad, L. (1985) 'Post-traumatic stress disorders in women who experienced childhood incest', *Child Abuse and Neglect*, vol. 9, 329–334.

Meissner, W. (1980) 'The problem of internalisation and structure formation', *International Journal of Psycho-Analysis*, vol. 61, 237–248.

Miller, A. (1985) *Thou Shalt Not Be Aware*, London, Pluto Press.

Nash, C. and West, D. (1985) 'Sexual molestation of young girls: a retrospective survey', in West, D. (ed.), *Sexual Victimization*, Aldershot, Gower.

Nelson, S. (1987) *Incest: Fact and Myth*, Edinburgh, Stramullion.

39

Treating anorexia nervosa

JANICE RUSSELL

'None of these cases, however exhausted, are really hopeless as long as life exists,' commented William Gull about anorexia nervosa in 1874. Doctors have up to now been exhorted never to give up in the attempt to induce anorectic patients to eat and restore weight even if, as Gull also said, this might entail the need to 'fight for every mouthful'. Two personal views below, and a recent paper on treating anorexia nervosa as a terminal disease (O'Neill *et al.*, 1994), raise disturbing questions for those who battle with this common and tragic illness.

Treatment of patients with anorexia nervosa must be constantly reviewed to ascertain that it accords with the five ethical principles of beneficence, autonomy, non-maleficence, justice, and utility. All of these, however, are open to interpretation, and our patients' interpretations often differ greatly from out own. Because we know the effects of starvation on the brain, is it ever ethical to act in accordance with a severely emaciated patient's decision to reject active treatment (Newman *et al.*, 1995; Tiller *et al.*, 1993)? Yet are some patients truly incurable, and is palliative care then more appropriate? Conversely, given that many of our patients are both contrary and ambivalent, can the clinician's acceptance of palliation as a treatment goal even have the paradoxical effect of motivating recovery?

We must also question exactly whom we are treating and whether we are colluding with the denial of the nihilism of the patient, her family, or the funding authority. Are we being moved to inappropriate action or excessive restraint by the impotence and guilt that anorexia nervosa can so powerfully engender? Although draconian in-patient regimens

Edited and adapted from Russell, J. (1995) 'Treating Anorexia Nervosa: Clinical Concerns, Personal Views', *British Medical Journal*, vol. 311, 584 and 635, where it appeared as an editorial and an anonymous pair of letters which were published in the 'personal view' column.

achieve weight gain, lenient treatment works as well for most patients (Touyz *et al.*, 1984). But what do we do with a patient who repeatedly rejects or fails in our user-friendly refeeding programme? Even if we do not believe everything our patients tell us, we must be sensitive to the trauma described in the published accounts (see for example Caplin, 1996) and mindful of the potential sequelae, particularly in a condition in which abusive antecedents may have occurred (Waller, 1991).

Improving a patient's nutrition by whatever means might result in a change of mind, even to the point of accepting weight gain and eventual return to normal life. Even when restoration of weight is only partial or intermittent the quality of life might still be deemed acceptable, and who are we to argue? Doctors never cease to be surprised by patients with seemingly chronic anorexia nervosa who recover after 10 years or more (Ratnisuriya *et al.*, 1991). After all, in a condition in which the average duration of illness is around five years (Ratnisuriya *et al.*, 1991; Hsu, 1988) what should we regard as chronic disease? Like the patient whose personal view appears below, recovered or recovering patients often extol the virtues of a particular clinician's psychotherapy or a treatment programme in which their individuality was respected and they at last felt safe to be themselves. If at the deepest level anorexia nervosa signifies that the sufferer is having difficulty achieving a secure sense of self (Bruch, 1982; Goodsitt, 1985) we as clinicians must take care that our treatment is a help, not a destructive hindrance in this process.

None the less, our patients often feel desperately out of control, and our firm limits would seem to be the only effective antidote. Most of us can recall a patient treated coercively who made a full recovery and later sincerely thanked us for doing what was experienced at the time as totally reprehensible (Tiller *et al.* 1993). Unfortunately, not all our patients do as well or are as happy with their treatment. Anorexia nervosa is a humbling condition, and it behoves ethical clinicians not only to accept this with good grace but to appreciate just how fine is the line they walk.

My friend

I have a friend. She is intelligent, witty, well read, extrovert, charming, and beautiful. She has anorexia nervosa. My friend has not told me this – it is entirely supposition. The evidence, however, is overwhelming.

She is underweight and seems to be losing weight all the time. She has never been seen to eat anything and has never accepted any offer of food of any sort from anyone. She talks about food far more frequently than is normal. She spends large amounts of time in the staff toilet where she keeps several different types of mouthwash, toothpaste, and toothbrush. Retching has been heard on many occasions.

My friend is a doctor.

I appreciate that environmentally imposed eating disorders are common in junior doctors, but the circumstances under which my friend works are conducive to eating at least one meal a day with several coffee breaks.

The fact that my friend has anorexia nervosa confuses me more than I think it should.

Am I perplexed because she is a doctor? No. I appreciate that the medical profession has an above-average incidence and prevalence of maladaptive behaviour and I know that overachieving young women are the epidemiological stereotypes for anorexia nervosa.

Am I bewildered to be faced in reality with a condition about which I learnt in books? No. Every day I deal with both the textbook defying and the textbook defining. Do I believe that doctors have greater personal insight than lay people and should therefore be better able to diagnose self-destructive behaviour? No. From personal experience the medical profession achieves at best introspective parity with lay people.

I have concluded that the reason my friend's illness confuses me is that she is my friend. I have met many people with mental illnesses but they have been strangers. I have known other people with anorexia nervosa, but they have been cured and openly confess to their illness. I have never known anyone whose illness, obvious to the outside world, is either veiled from their perception or cannot, they believe, be disclosed.

My other dilemma is what I can do to help my friend. I am reticent to confront her as I am well aware that until the sufferer independently makes a commitment to change little can be done by way of treatment. I do not know my friend's family, so I am unaware if they suspect her condition. If they do then I am sure that they are doing all in their power and would perhaps not take kindly to a meddling stranger. If they are ignorant of her illness, and assuming they believed me, we would again by hampered by my friend's denial. It seems, therefore, that all any of us can do until she decides to seek help, is wait and give generously of our companionship. I hope then that my friend will feel that she is surrounded by people to whom she can turn for support.

I am unsure as to my reasons for writing this. I do not believe that my friend will read it, recognise herself, and in a moment of therapeutic clarity renounce her ways and seek help. I do not believe that others in a similar position will recognise cognate signs in one of their colleagues and be able to offer timely support. Perhaps I was hoping to convince myself that we have all been joyously wrong about my friend. This has not happened.

Which option would you take?

Dear Doctor,

I would like to give you my views about certain regimens used in treating people with anorexia nervosa. I have been a sufferer for over seven years, since I was 13. Ironically, both my parents are general practitioners and their concern and worry were exacerbated by a feeling of inadequacy at being unable to cure me themselves. Our plight as a family (my elder and younger brother included) was not improved after I agreed to undergo one particular treatment when I was 15. In fact, it has taken both me and my parents many years to put that destructive episode behind us. It is only now, five years later, that I am able to express my anger at what amounted to a kind of brain washing. With hindsight I can question the reasons behind this demeaning regimen without having to accept the word of a professor.

You may be curious as to what this treatment involved. Initially the programme began by shutting you in one room (a cubicle) and feeding you on 3000 calories a day until 'target weight' was reached. 'Target weight' was a source of great resentment by all those undergoing this treatment. It was simply the average weight of anyone your height and age. Because I was 1.7 m, my target weight was deemed to be 62.5 kg – a daunting prospect from the word go. Other factors, such as build, family tendencies, and previous experiences were ignored.

My home for the next few months was a room with windows all down one side, giving the effect of a museum exhibit case. The only way of escaping gaping stares was to draw the curtains, but as these were on the outside of the room it would cause other more deranged patients great amusement to whip these open. Besides, we were permitted to draw them only when we washed or used the much loathed commodes. Visits to the bathroom were forbidden (save twice a week) and we were confined to wash bowls, brought three times a day, and the use of a commode, which was emptied only when the student nurses failed to find more pleasant tasks. The fact that getting dressed each day was also forbidden only added to the humiliation.

If you had to leave the room for any reason, such as to use the telephone, it was necessary to shout for someone to push you, while you sat on the covered commode to prevent your legs from so much as taking one step. Unfortunately, if the commode had not been emptied recently the contents would splash on to your feet, adding an extreme feeling of revulsion to the degradation and shame already being experienced as this pyjama clad spectacle was wheeled to its destination. What possible purpose could this cruel and humiliating treatment serve? Surely walking a few steps to a therapy session or telephone would have no effect on subsequent weight gain?

The theory behind this extreme lack of independence was that we had to suppress any anorexic tendencies. Unfortunately, this seemed to include a total disregard for our basic human instincts and an obliteration of our privacy and pride. Ironically, the fact that we underwent this treatment in a room with an open window and a rubbish bin seemed to go overlooked. In preventing the patients from using any sophisticated techniques for the removal of food, the most obvious means were left unguarded.

After a four-week period of such treatment a few therapy sessions were tentatively included. These consisted of group therapy as well as individual sessions once a week. Family therapy was delivered if it was deemed appropriate. In the meantime, meals were delivered punctually on a tray, weighing was carried out twice a week, and any activity, such as making your bed or standing at the window (as opposed to sitting or lying), was forbidden. To say that the days crawled by would be an extreme understatement.

The moment you weighed in at your target weight, however, you were moved on to the psychiatric ward for a further eight-week reintroduction to the outside world. After a hellish period of private metamorphosis, this emergence was terrifying. To exhibit and clothe this uncomfortable, unfamiliar new body was very traumatic.

I did reach my target weight under this regimen, but at what cost? The anger and bitterness I felt after this humiliating treatment exacerbated rather than lessened my illness over the years to follow.

I found the ferocious generalisation as to the cause of the problem frightening. We were persuaded to believe in supposedly brilliant new theories without being given any credit for being individuals with differing anxieties and ambitions. Thus the process of delving into your family life and childhood was moulded to fit this idea of a common cause or causes of the problem. Much time was spent looking into the past and criticising the whole family set up, hence destroying any lingering self esteem; moving forward became confusing and impossible. I now see the psychological explanations for my problem to be a complete sham. I feel angry at being persuaded to believe such unhelpful complicated nonsense at a time when I was vulnerable and begging for help. The most hurtful part of the treatment was the fact that I was stripped of all my better points along with what was perceived to be the problem. It seemed to be considered that any part of the treatment, which would surely have been unbearable to any sane human being, seemed intolerable to us only because we were ill, and it was the 'illness' that made us wish to rebel. Because we were so desperate to escape the grip of the anorexia, we silently but miserably stayed in this harsh regimen, believing that what were in fact perfectly normal and understandable emotions were wrong and must therefore be repressed.

How is the success of a treatment measured? Is it weight gained at the end of the hospital period, the length of time for which weight is maintained, or the patient's mental wellbeing?

I am writing this letter five years later in the Russell Unit of the Royal Free Hospital. Here I am treated as an individual. Each programme for recovery is tailor-made after discussions with the patient. Help is designed to enhance your good points and to provide support in challenging the destructive manifestations of the illness. The aim is to build on what has led you to seek help in the first place, rather than to strip you of your self esteem and remould you to fit a psychiatric model.

It is difficult to judge progress in this rather complicated disease, but I do feel that for the first time I am in partnership with people who are working with me rather than against me. Weight gain is taken one step at a time without setting ambitious target weights. You are fully aware of the minimum medically safe weight which you should be aiming for. How this is achieved is up to you but advice and encouragement are freely available from those around the unit. Once forward motion is being achieved every effort is made to maintain that momentum.

The comparison I have made between these two approaches can be likened to a dentist examining teeth. On the one hand, I have all my teeth removed and replaced with an ill fitting set of dentures. On the other hand I have a caring and detailed examination of each tooth, with the aim of preserving the good teeth while attempting to stop the decay in the bad teeth.

This letter is to ask those of you who may have patients like me to fully consider the long term consequences of the treatment options available to you. After all, which option would you take?

Yours faithfully,
An anorexic patient

References

Bruch, H. (1982) 'Anorexia nervosa: therapy and theory', *American Journal of Psychiatry*, vol. 139, 1531–1538.

Caplin, R. (1996) 'Time, Faith and Encouragement', in Reed, J. and Reynolds, J. (eds), *Speaking Our Minds*, Basingstoke, Macmillan.

Goodsitt, A. (1985) 'Self psychology and the treatment of anorexia nervosa', in Garner, D. and Garfinkel, P. (eds) *Handbook of Psychotherapy of the Eating Disorders*, New York, Guilford Press.

Gull, W. W. (1894) 'Anorexia nervosa (apepsia hysterica, anorexia hysterical)', *Clinical Society's Transactions*, vol. 7, 22, reprinted in Ackland, T. D. (ed.) *A Collection of the published writings of William Withey Gull*, London, New Sydenham Society, 305–314.

Hsu, L. K. G. (1988) 'The outcome of anorexia nervosa: a reappraisal', *Psychol Medical*, vol. 18, 807–812.

Newman, L., Russell, J. and Beumont, P. (1995) 'Legal and ethical issues in the treatment of very low weight anorexics', in: Kenny, D. and Job, S. (eds), *Current Issues in Adolescent Health and Lifestyle: an Australian perspective*, Sydney, Faculty of Health Sciences, University of Sydney.

O'Neill, J., Crowther, T. and Sampson, G. (1994) 'Anorexia nervosa: palliative care of terminal psychiatric disease', *American Journal of Hospice and Palliative Care*, Nov/Dec, 36–38.

Ratnisuriya, R. H., Eisler, I., Szmukler, G. I. and Russell, G. F. M. (1991) 'Anorexia nervosa outcome and prognostic factors after 20 years', *British Journal of Psychiatry*, vol. 158, 495–512.

Tiller, J., Schmidt, U. and Treasure, J. (1993) 'Compulsory treatment for anorexia nervosa: compassion or coercion', *British Journal of Psychiatry*, vol. 162, 679–680.

Touyz, S. W., Beumont, P. J. V., Glaun, D., Phillips, T. and Cowie, I. (1984) 'Comparison of lenient and strict operant conditioning programmes in refeeding patients with anorexia nervosa', *British Journal of Psychiatry*, vol. 144, 512–520.

Waller, G. (1991) 'Sexual abuse as a factor in the eating disorders', *British Journal of Psychiatry*, vol. 159, 664–671.

40

Rehabilitating voice-hearers

MARIUS ROMME

Hearing voices is not a purely personal phenomenon: it is intimately bound up with the hearer's social environment. His or her place in this environment is often negatively affected by the experience, and this means that due attention must be paid to the process of rehabilitation. This and other social interventions are essential to provide the proper milieu in which helpful strategies and activities can flourish. The aim of rehabilitation is to ensure the best possible circumstances for the individual's development, thus allowing the therapies employed to bear fruit. The essential ingredients in this process are:

• quality relationships (for example, with the social worker);
• quality information;
• receptivity to emotions;
• social autonomy;
• trusting relationships (significant others).

Quality relationships

As we have already observed, the hearers of voices are understandably averse to interpretations which deny their experiences and perceptions. It is important to them that others listen to them with genuine interest and concern, rather than patronisingly or purely for the purpose of acquiring data. The relationship must be dependable, and founded on equality.

To illustrate the centrality of these qualities to a good therapeutic relationship, I would like to quote an ex-patient (Romme and Escher, 1993, p. 141):

Edited and revised from Romme, M. (1993) 'Rehabilitation', in Romme, M. and Escher, S. (eds), *Accepting Voices,,* London, Mind, 227–235.

Only once in 15 years of psychiatric intervention, and at the age of 36, was I able to find someone who was willing to listen. This proved a turning point for me, and from this I was able to break out of being a victim and start owning my experience. This nurse actually found time to listen to my experiences and feelings. She always made me feel welcome, and would make arrangements so we would not be disturbed. She would switch off her bleeper and take her phone off the hook, and sometimes, as there were people outside her room, she would close the blinds. These actions made me feel at ease. She would sit to one side of me instead of across a desk. She told me that what we said was confidential, but that there were some exceptions, so I could decide what to reveal. Slowly, as trust grew between us, I was able to tell her about the abuse, but also about the voices. Sometimes when I was describing what happened to me, she would tell me that it was hurting her and she needed a break. At last, I had found someone who recognised the pain I was feeling. She helped me realise that my voices were part of me, and had a purpose and validity. Over a six-month period, I was able to develop a basic strategy for coping. The most important thing she did was that she was honest – honest in her motivations and in her responses to what I told her.

I would like to say that maybe other mental health/social work professionals could learn something from her approach:

- be honest about your motivation and the reasons for your intervention;
- establish ground rules at the start;
- provide a safe environment, and keep it safe;
- don't force the agenda – provide people with a breathing space so they can decide what to bring and what not to bring;
- be honest about your own feelings – this is not rejection, but a sign that you are also alive;
- let the person decide what the goals should be and say if they want changes.

Quality information

In order to understand and help someone, and to know what that person wants, it is essential for both parties to be fully informed. It is therefore a good idea to begin with a comprehensive inventory of the number of voices, their gender, age and characteristics, to whom they belong, how they are organised, what influence they have on the hearer, what they say, how the hearer reacts to this, what has happened since they were first heard, and so on.

The supposed identity and the way of speaking of the voices might reflect the persons represented by the voices. The content mostly indicates the themes which the person has difficulty expressing in daily life. The triggers might point to troublesome emotions or situations the voice-hearer cannot handle. A belief in almightiness of the voices with regard to life experiences and traumas might make the person's reaction to the voice more understandable. And the time when the voices were heard for the first time often shows traumas or interactions with others that made the voice-hearer quite powerless. This gives both the social worker and the voice-hearer something to think about, and this process of reflection is a vital precursor to the formation of an approach to the voices.

The aim of the approach is to understand. The first thing to understand is the connection between the negativity of the voices and the anxiety or resistance towards the voices. This understanding brings a sense of relief. The relationship is also oriented toward equality, so the emphasis should be given to freedom to withdraw from therapy and freedom to continue one's own beliefs about the voices. Short-term goals are generally particularly concerned with anxiety management. The techniques used are, for example, that the voice-hearer gives to the voices a special hour in the day or evening to talk with them, or the voice-hearer asks the voices to stay away the next hour because the person has a task to fulfil and cannot be disturbed but after the hour he then will listen again.

Another possibility is to bring the voice-hearer into contact with another voice-hearer to show that others have similar problems. Reducing anxiety and gaining even a small degree of control obviously makes thinking considerably easier. Other aids to facilitate thinking may also be used, such as writing assignments, the keeping of a diary, and focusing exercises.

This process of thought and reflection is designed, slowly but surely, to improve the quality of insight into the significance of the voices in the hearer's life. It is therefore important to identify the situations, emotions and persons associated with the occurrence or intensification of voices – the so-called triggers. Triggers represent the circumstances which encapsulate the voices. The voices seem to react to something recognisable: they may recall a traumatic experience (flashbacks), interfere in a situation (assignments) or protect the hearer (prohibitions). Triggers can often be difficult to identify. Many hearers have found it a great help to compose a life-history in which the voices are cast as the leading characters (an ego-document).

It is particularly important, in these explorations, to recognise metaphoric or symbolic meaning. Consider, for example, the case of a voice that speaks like a robot. A robot is suggestive of strength without emotion, a force that is difficult to resist or overcome, and may therefore be the symbol of a person without fear. The robot becomes a metaphor for dealing with emotions; if it is a frightening figure to the hearer, it may also be a metaphor for coping with fear. In such cases, it is vital that the fear of fear be focused upon in the discussions, and the hearer must then be taught to cope with this fear. Another example might be the voice of a small child: this may signify being treated like a child, or suggest a traumatic childhood. In the first instance the voices would be likely to assign childish tasks, while in the second, the voice might be expected to represent difficult experience in the hearer's own childhood.

Receptivity to emotions

In their dealings with the hearers of voices, family, friends and helpers alike must be receptive, and prepared to acknowledge their own emotions and the ways in which they go about coping with them. This receptivity entails a sensitivity to the following issues:

Empathy

This involves listening with appropriate emotion, to show that the fear of the voices is felt and appreciated, but that this is no reason to run away. This is not to say that the helper should encourage anxiety but nor should he or she avoid the display of any emotion when extremely unpleasant experiences are recounted. Appropriate human reactions on the helper's part allow the voice-hearer to admit more easily to the severity of the emotions involved, and to feel that his or her turmoil is quite understandable under the circumstances.

When the voices are very threatening, the helper should still try to elicit a description of the experience. Talking things over can help establish which aspects of the phenomenon are grounded in reality and which are imaginary; remaining silent can only serve to reinforce the idea that the problem is entirely the hearer's problem. Proper discussion can also enable the hearer to gain insight into the power structure of the voices; this power exists only to the extent that the hearer attributes it to the voices, however persuasively he or she may insist that it is unlimited.

Confrontation

Voices can serve to protect the hearer against certain emotions inspiring avoidance and flight; under these circumstances, it is clearly not possible for them to learn to cope with such emotions. A helper must therefore appreciate that confrontation with these emotions may be essential to progress, and must not be side-stepped, even if this seems to run the risk of precipitating a violent reaction. There should, of course, be proper provision of emergency measures in case of such a crisis. For example, a female student first started to hear voices at the age of 23. During the previous year she had undergone hazing, an initiation ceremony for first-year students. (In the Netherlands, some student organisations have a custom of seeking to shock new students over sexual matters during the induction period.) She had been thrown into confusion by the confrontations involved in this ritual, and had avoided sexuality ever since.

Even after several years of help, including a couple of years of out-patient treatment with us, she continued with this evasion.

Ever since the onset of this problem, she had consistently embraced the role of patient, and saw this as her future. She could not see herself finding a partner, having children or holding down a job. This was such a severe handicap that we confronted her with the lifestyle she had chosen, and with what we felt to be the underlying problem: her fear of her own sexuality. This confrontation triggered absolute uproar from the voices; the women became psychotic, and fled to her parents' home, as though she might somehow find protection there from her own emotions. This spontaneous reaction confirmed our suspicions of a sexual problem.

A violent reaction like this calls for special support and guidance. Clearly, these confrontations should not be undertaken without meticulous preparation, but it is vitally important that helpers should not collude with their patient in evading a problem out of fear of psychosis. Indeed, a psychotic reaction may prove to be extremely instructive for both patient and helper.

Acknowledgement

Voices may refer directly to situations and experiences, past or present, which inspire shame, fear or horror; they may, for example, be associated with aggressive sexual abuse which is too painful to recall.

Unfortunately, painful events do not always reside in the past: sometimes they persist into the present, and then they are even more difficult to inquire into. In the process of exploring these, the helper will occasionally find him or herself reluctantly involved in uncharted territory, and must therefore be prepared for any eventuality in daring to feel the voice-hearer's emotions as well as his or her own reactions to these.

It can be difficult not to flinch from this challenge, but full acknowledgement of the hearer's experience is all-important. When a situation is glossed over, the threatening emotions will disappear from consciousness, but will continue to be verbalised by heavily-charged voices.

We know, for example, a mother whose son had hanged himself. After his death, the son regularly called out to her, asking her to join him. In this situation, if the help offered did not fully acknowledge the content of the voice's message, there might have been great danger of an eventual suicide attempt by the mother. Consequently, we explored the nature of this message before tackling the mourning process.

Social autonomy

Hearing voices is a very penetrating experience, and it demands considerable strength and stamina for the hearers to retain control over their conduct.┆This is scarcely helped by the fact that society often shows them little understanding or tolerance, so that they are all-too-prone to isolation. ┆The maintenance or development of personal power requires proper social autonomy: in other words, the granting of a place within society that will help to foster identity and independence. Amongst other things, this entails the full provision of social services such as independent housing, some kind of occupation, and a degree of financial independence. These are the elements that enable individuals to structure their daily lives and build relationships, and are the foundations of social identity.

An obstacle to the acquisition of social autonomy may be seen in those who remain living in the parental home for too long. This is as true for voice-hearers as it is for any other adolescent; when one's dependence on others is excessive, it is difficult to learn to make decisions independently, and thus to take personal responsibility and develop own's own identity. The parents of voice-hearers are sometimes heard to observe that their children's lack of any real sense of responsibility is due to an incapacity for independence, but our experience is that they are very often capable of leading quite normal, independent lives, even during so-called psychotic episodes.

The privacy offered by having one's own house or flat can also be an important element in the development of a sense of social autonomy, although it is essential to distinguish clearly between privacy and isolation. As is the case for us all, it is crucial for people who hear voices to have relationships, to maintain contact with others and to engage in social activities. What we are speaking of here is the ability to be master or mistress of one's own living-space without interference from others. It may, however, be wise in the first instance to give priority to those things likely to avoid any danger of isolation: for some, this may mean training in social skills, while others may need to develop an active interest in expressive pursuits such as dance, music, and drawing, or to engage in some kind of study.

Given the way our society is organised, any sense of social autonomy is particularly dependent on having some kind of employment; this makes it especially important for the hearers of voices to be able to retain their jobs or find other suitable occupation. The evidence suggests that the job or activity concerned must allow scope for exercising some responsibility, in accordance with the temperament of the individual. Hearing voices can be debilitating, and often leaves the hearer with only limited energy to create such opportunities for him or herself. Given the right circumstances, many voice-hearers show themselves to be willing and able to adjust imaginatively to their working conditions when these are flexible and allow room for their creativity.

Trusting relationships

Almost all those voice-hearers who have learned to live with the experience describe how important it was for them to have a friend, partner or family member who listened to them, accepted them, and made them feel safe. What matters here is not the number of these relationships, but the quality: even one such can offer a real sense of security during periods when the voices are especially overwhelming.

Finding appropriate support during difficult phases can be something of an art. Even those voice-hearers who manage to avoid recourse to psychiatric treatment of any kind report that there are times when they are overwhelmed by voices and run the risk of hospital admission and long-term treatment. In London, the organisation Lambeth Link offers the facility of a rented flat where voice-hearers can stay a night or a couple of days and be given support by other members of the network who themselves hear voices.

Not everyone, however, is fortunate enough to have such a network at their disposal; in any case, it is perhaps just as important to find ways of feeling safe in one's own milieu. Any campaign of rehabilitation must therefore include adequate information for voice-hearers' families, partners, friends, acquaintances, and anyone else with whom they may have significant contact.

It will be clear that the rehabilitative measures discussed here are all based on principles of equality and on the involvement of the voice-hearer's own experiences in the helping process. In this way, any support that may be offered is always designed to avoid encouraging dependence, and to enhance the possibility of a growth in self-awareness and self-determination.

Hearing voices represents an enormous challenge: a challenge which can be regarded either as a threat which renders one powerless or as a teacher capable of empowering one to withstand the trials of life. However, the development of personal power can only take place when the immediate surroundings provide the appropriate stimulation and real opportunities within society at large. The ultimate objective is the full development of a personal identity as one who happens to hear voices.

Reference

Romme, M. and Escher, S. (1993) *Accepting Voices*, London, Mind.

41

Communicating as if your life depended upon it: life history work with people with dementia

JOHN KILLICK

Phyllis says to me 'You can climb up the family tree and see the vista from there', thus encapsulating in a sentence the theory behind life history work with elderly people. It is an activity which can be life-enhancing, capable of offering confirmation of self-worth to subjects. It enables them to gain new perspectives, forming part of a re-evaluation and a coming-to-terms with their having entered the last stage of life. Having success-fully completed the process, Enid is able to proclaim proudly 'Age and I are great friends together.'

But life history work with people with dementia? Surely it is a contra-diction in terms to attempt to compile the memories of those for whom loss of that faculty is a daily reality? Well, in choosing to work with such subjects one is certainly not taking the easy option'. The literature of life story work with people who do not have dementia is a growing one whilst the study of writing projects with memory-loss subjects is in its infancy. Yet Oliver Sacks (1985) has written:

> In Korsakoff's, or dementia, or other such catastrophes, however great the
> organic damage and human dissolution, there remains the undiminished

This article describes work done by the author for Westminster Health Care and is extracted from Killick, J. (1994) 'Please Give Me Back My Personality!' in *Writing and Dementia*, Stirling: The Dementia Services Development Centre, University of Stirling, 6–10.

possibility of reintegration by art, by communication, by touching the human spirit: and this can be preserved in what seems at first a hopeless state of neurological devastation.

I do not see it as part of my brief here to theorise about the process in which I am engaged. None of the observations that I might make, in any case, could at present be backed up by research evidence. I would prefer to state my convictions in the following terms:

(1) That it is possible to enter the minds of people with dementia at every stage of their illness and to be vouchsafed insights, however fragmentary, of their lives and experiences.
(2) That these insights are, by any standards, an achievement in their own right.
(3) That, when written down and shared with carers, relatives and friends, they often constitute communications valued by those persons.
(4) That, disseminated to a wider audience, they can contribute to a greater understanding of what it is like to suffer from these illnesses.

A moment of truth from Joanne: after struggling for over an hour to discover where she was, and, indeed, to articulate at all clearly, she came up with the following: 'Most of the people here need looking after. And if I was truthful I would say that that includes me.'

'This writing it down': the practicalities

I have developed basically three methods of working.

The first is for the subject to write down what he or she wishes to share with me, and I can then discuss with them the shaping or extension of it. This process is extremely rare with people with dementia. Only someone in the very early stages would be capable of the manual skill and mental organisation involved. Sometimes, however, there is material which the subject has written earlier in their life, which they may have with them or a relative may supply. This is not a substitute for one-to-one contact, and such material would need to be supplemented by work with the subject as he or she is now.

The second, and most common, method, is for the writer to write down what the subject says. Of course most sessions are conversations, and it has to be your judgement whether it is necessary for you to record your question to the subject, and your response to the answer received. When the level of confusion is high in what the subject says, it may well be necessary for you to read to the subject a transcript of what they have said in order to gain approval of the text.

The third method is tape-recording. This is useful where the subject speaks rather fast, or their use of language is difficult to interpret. Playing it back at leisure when you can pause the narrative is essential in these circumstances.

It is as well to remember that when you are working with people with dementia you can rarely ask them to repeat what they have said, nor can you often cross-question them about the meaning of an utterance.

The use of shorthand in transcribing the speech of people with dementia would not be appropriate. Because of the many unusual uses of language which occur it is necessary to produce a text as near to the actual sounds used as possible.

In choosing a notebook or cassette-recorder it is advisable to take as small a version as possible in both cases. Large notebooks appear portentous, and subjects can be alarmed by technological equipment. In any case, you will be asking the subject's permission to use the tools of the trade. There is a further problem connected with the cassette-recorder. If a cassette-recorder is to be used in a public area then it is inevitable that, as well as that of the subject, you will be recording the voices of other residents, staff and visitors who have not given their permission. Wherever possible you should aim to use the cassette-recorder in a resident's room or an office or somewhere private, and employ the pause button if someone else enters.

Sometimes it is possible to work with a subject with the assistance of a relative or a member of staff, either of whom will be able to supply valuable added information as well as capitalising on their own relationship with the subject in conversations.

Occasionally relatives may wish to work with a writer on their own behalf. It can be of therapeutic value of them to unburden themselves of the circumstances of the onset of the disease and the ways in which the carer was (or was not) able to cope. Many relatives have strong guilt feelings to be assuaged.

It is possible to use certain visual or aural aids to stimulate remembrance. Of these photographs (particularly of family and friends) and recordings of music can be the most potent. But one must beware of putting too much store by them. The relationship of writer and subject is what matters most, and such aids should only be used within the context of that developing bond.

I do not find it necessary or desirable to be briefed by staff before working with anybody. Sometimes, of course, staff recommend particular subjects, either because they think they would benefit from the activity, or because a request to engage in such work has come from an individual. Experience has taught me, however, that subjects recommended by staff are often the most socially communicative, and that other residents who are more withdrawn often have a greater need for talking and resolving issues. Such subjects are capable of identifying themselves to the writer if given the opportunity. Often the most successful practice is for me to sit in a social area and wait for people to make contact. Subjects who make the choice themselves are obviously the most motivated, and such work often proves the most rewarding.

Sometimes I am mistaken for a visitor, perhaps a doctor, and this provides me with an opening to explain what I do and ask the potential subject if they would like to participate. Sometimes I am mistaken for a new resident, and this allows for introductions to be made, the routines of the establishment to be explained, and even sometimes for a tour of inspection to be made. The more spontaneously the contact occurs, the more uninhibited is likely to be the interaction.

I do not want to know about people's medical condition prior to working with them; it would get in the way of my responding to them as individuals. Many of the commonest symptoms are obvious to me as I increase the tally of subjects with whom I have worked. What must be resisted is the temptation to recognise and classify according to either clinical theory or uninformed observation: they are both equally destructive of the instinctive response. I am convinced that if one is to engage in writing activities with people with dementia one must keep as open a mind as possible and let the subjects set the agenda.

A lot of time working with an individual is spent waiting, and then the insights often come in a rush. You may not be able to keep up on transcribing what is said, or you may not have the cassette-recorder switched on at the precise moment the material is presented. It can be a frustrating experience. If you are to work with the same subject on subsequent occasions it is a good idea to alert staff to the fact, so that they may be prepared for more material to surface in the subject's conversations over the period. The stimulus provided by your presence can often be quite lasting in its effects, and the subject may not choose (or be able to choose) to wait for your next appearance.

Sometimes a session may prove quite distressing to the subject. Memories can be stirred which have long remained hidden, and they may be accompanied by strong negative emotions such as guilt or loss. Or the very effort of communication which ultimately proved self-defeating can provoke tears of frustration. It is most important in these instances for you to inform staff and relatives of what has occurred (not necessarily the substance of what has been vouchsafed; here you must respect the subjects' confidences) so that they may exercise their sensitivity if the reaction is prolonged or recurs.

The quality of attention

You must have respect and curiosity and love in the presence of your subject. At the time of speaking to them you must believe that at that moment it is the most important activity in which you could possibly be engaged, and not be distracted from it by others or deflected from it by any other consideration. You are truly communicating as if your life depended upon it. Is it not a positive activity that you are engaged in? You are using all your powers of empathetic attention. If, at the end of a session, you feel physically and emotionally drained, then, though this may not be a sign of success, it is an indication that you have given what is necessary for success, which is as much as you can ask of yourself.

These powers of attention include:

(1) Watching the eyes and face for expressive indicators, and the body language.
(2) Listening to the tone and inflection of any language used; this can sometimes tell you as much as the actual words employed; indeed when the sentences come out 'scrambled' these may be your only guides to interpretation.
(3) Responding quickly to changes of attention and mood in the subject, going along with these wherever possible; you may not be able to keep one step ahead of your subject, but you will certainly lose them if you once allow yourself to fall behind. This is literally true if your subject perambulates ceaselessly – you must move with them in order to develop the interaction.
(4) Providing physical reassurance where necessary: this will very frequently involve holding hands; not only does this provide demonstrable proof of your concern, but much can be told by the intensity of pressure exerted in return. At times it may also be appropriate to offer hugs and kisses of fellow-feeling and solidarity. (A senior sister said to me 'I give more hugs and kisses in the course of one day than you would think humanly possible'.)

I must emphasise that the amount of planning which should go into the use of these powers is minimal. Through practice you can develop a confidence and a spontaneity which allows you to perform instinctively. That is the ideal to aim at.

Of course there are individuals who will reject your advances. They may misinterpret your motives, or not be feeling well enough to make the effort involved in a social activity. Or they may be afraid of letting their emotions become engaged in a relationship, as in the following example:

GO AWAY!

You have hurt me.
You have hurt me deeply.
Because you will go away.

> I want to be looking
> for somewhere else.
> I didn't choose this place.
> I don't like this area.
> And oh the noise!
> I can't stand the noise!
> What was that sound?
> It really upsets me.
>
> I don't want to talk to you.
> I don't want to see you again.
> Because you will go away.

These occasions will put your sensitivity to the test. You will have to learn from experience when to accept a rebuff, and to persevere, trying a different approach, or the same approach after some time has elapsed. Whatever form the rejection takes two principles should be paramount:

(1) The right of the individual not to co-operate with you in your strategy if they so choose.
(2) The lack of condemnation implied by this rebuff; where people with dementia are concerned such motivation is unlikely to be present; the rejection may indeed be symptomatic of their condition. You must learn not to take such experiences to heart. On no account should you display the slightest sense of grievance to your subject.

Of course it has to be faced that many of the subjects you will encounter display anti-social characteristics to varying degrees: some will be mis-shapen, some noisy, some incontinent, some have developed tics or other involuntary physical or verbal mannerisms which you may find repellent at close quarters. It is because of these features that I believe so many people tend to reject elderly people with mental illness and to want them out of sight and mind. You have to overcome these natural aversions and penetrate the disguises of dementia to the common humanity beneath.

The Quakers have a phrase about seeking for 'that of God in every man'. Without necessarily subscribing to a fully religious philosophy we can surely assert that in carrying out this, and other work, with and on behalf of elderly people with a mental illness, we should be actively searching for qualities which command assent. And that is another way of saying that we should adopt a compassionate and tender approach to those who have been dealt such a cruel and undeserved blow by fate.

Reference

Sacks, O. (1985) *The Man Who Mistook his Wife for a Hat*, London, Duckworth.

42

Practising cultural psychiatry: the doctor's dilemma

SOURANGSHU ACHARYYA

In the West, psychiatrists have tended, covertly or overtly, to play the 'colonialist game' that their own elected politicians have developed. This process has consisted of devising systems of classification of psychiatric disorders which are essentially home-based. If such categories do not embrace migrant peoples (who do not readily fit in the classification system), these people tend to be segregated into something else. Thus they are seen as suffering from an 'exotic' condition – 'West Indian Psychosis', or 'the Begum Syndrome', for example. This problem, I would suggest, has been amplified by the trend in the former colonies to perceive as the root of the power of their erstwhile colonial masters a new religion called 'science and technology'. This has led these nations to send their brightest young people to the West to be trained in this new craft. Psychiatry has been no exception.

The most distinguished of Third World psychiatrists, the doyens of their profession, are mainly (if not wholly) trained by Western psychiatrists in the prestigious teaching centres of the West. The psychiatric elite of the New Commonwealth has been trained by British psychiatrists in Britain. Therefore, these people on their subsequent return to their own cultures, perceive mental illness along the 'British model', and find difficulty in evolving new ways or methods of examining psychiatric disorders within their own cultures (Chakraborty, 1990).

Hughes (1985) suggests that 'both the labels "atypical psychoses" and "exotic syndrome" imply deviance from some "standard" diagnostic base'. If something is 'atypical' it obviously falls outside the range of

Edited from Acharyya, S. (1992) 'The Doctor's Dilemma: The Practice of Cultural Psychiatry in Multicultural Britain', in Kareem, J. and Littlewood, R. (eds), *Intercultural Therapy: Themes, Interpretations and Practice*, Oxford, Blackwell, 74–82.

what is 'typical'. Similarly, 'exotic' means 'foreign', 'different', or 'deviant' from something that is not exotic, that is something that is familiar and well known. So to use these two categorical labels is to imply that we are dealing with psychiatric (or imputed psychiatric) syndromes that fall outside the scope of known and described (and thus legitimate) phenomena.

The doctor's dilemma

What constitutes the 'scope of known and described phenomena'? The frame of reference might be that embodied in the most recent edition of the International Classification of Diseases (currently the 10th edn), the latest edition of the Diagnostic and Statistical Manual of Disorders of the American Psychiatric Association (currently DSM IV), or any of the contemporary versions of numerous other formal medical diagnostic systems. Herein lies a problem for the doctor diagnosing and treating mental illness in people from cultural minorities: the doctor's dilemma. One is dealing with a person who comes from a different culture, with an attitude to well-being or illness that is perhaps quite different from that of the examining psychiatrist. In addition, the patient might speak a language other than English as a first language. Even if English is the mother-tongue, the language used may be semantically and syntactically different from that of the examining psychiatrist. Therefore, the dilemma which arises for our examining doctor is whether this distress, as presented by the patient, is (a) one of the familiar entities according to standard Western classification systems, or (b) if not, then what it is (Littlewood, 1992). The doctor's dilemma is in deciding where, in a broad group of symptomatological classification it can be included, or indeed (c) whether this condition truly constitutes a 'psychiatric illness' at all.

What is seen by the individual (or by the individual's family or community) as complex personal distress has, of course, to be slotted into a neat diagnostic box by the doctor, a personage seen by them as the professional expert who can alleviate this distress. Both sides, the patient and the doctor alike, bring their own preconceptions about the other to the encounter. The patient walks through the door of the surgery or consulting room with assumptions about the doctor's work: these may include a range of qualities or qualifications which doctors may feel are hardly their role – perhaps detailed advice on social and financial matters, for example. Similarly, the doctor may have preconceptions of what constitutes the 'patient-ness' of a person. And it may be difficult to redefine this 'patient-ness' in terms of how the patient – this particular patient – presents themselves.

One of the problems for a patient from a minority cultural group may be that what the patient presents as a problem, and the way it is presented, does not fit easily into the standard European text-book classification that the doctor has so meticulously absorbed, categories which are themselves taken to be culture-free. It has been suggested by Littlewood (1990; Littlewood and Lipsedge, 1989), that while people from 'other cultures' are very often slotted into the existing categories of mental disorders developed in the West, the reverse does not happen. The categories therefore remain sacrosanct; the patient either has to be shoe-horned in, or left in some marginal 'exotic' or 'culture-bound' category.

In the UK there have been some recent studies which show marked differences in psychiatric morbidity and treatment between the cultural majority and cultural minorities. One study points out that there is a marked increase in the incidence of diagnosed schizophrenia amongst British-born people of West Indian descent compared with the native white population in Britain (Harrison *et al.*, 1988; McGovern and Cope, 1987) found higher rates of compulsory psychiatric admissions of people under the Mental Health Act for immigrants from the West Indies, although British-born West Indians were also over-represented compared with the native white population and South Asians.

Can the high rate of detained West Indian patients be explained simply by an increase in the incidence of schizophrenia? It would be comforting to put these findings together in a neat little box and conclude that a genetic predisposition amongst this ethnic group accounts both for the high rate of schizophrenia and the consequent detention. This hypothesis does not stand up to close scrutiny. Littlewood and Lipsedge (1989) found that excessive detention of West Indians under the Mental Health Act was independent of diagnoses. Ineichen and his colleagues (1984) suggested that no significant differences in voluntary admissions of West Indians are found compared with the white population. When one looks at how the population was diagnosed in McGovern and Cope's paper (1987), there is little detailed description of the process of how the diagnosis was reached except to say that they followed the 'final case note diagnoses'.

The pattern of schizophrenia across the world – in various centres in Europe, North America, Asia and Africa – is extremely similar (Sartorius *et al.*, 1977). However, when it comes to the outcome of 'schizophrenic illness' across the world, there are differences. In the Third World countries the prognosis appears to be markedly better than in the West. Now, in medicine, 'a disease entity' is marked not only by common aetiology (causes), pathogenesis (disease process) and symptoms, but also by prognosis. This is particularly so for schizophrenia. However, in the case of schizophrenia, although the International Pilot Study finds markedly different prognoses in different societies, it still adheres to the diagnosis of 'schizophrenia' for these conditions in the Third World societies rather than to question the soundness of the diagnosis of schizophrenia (Kleinman, 1987).

There is some evidence to suggest that when mental illnesses are taken out of the hospital-derived context (which is where the classificatory systems were conceived), identification of symptoms in the case of people from cultural minorities does not follow the same trend as has been found by Harrison and his colleagues (1988), or McGovern and Cope (Chakraborty, 1990). Recent community-based studies, carried out at the Nafsiyat Intercultural Therapy Centre found that: (a) the largest single group of the patient population consisted of people who referred themselves to the Centre rather than those referred through the usual statutory services such as general practitioners or hospital psychiatric emergency clinics; and (b) that contrary to the findings of other community 'walk-in' mental health centres (Bouras et al., 1982; 1983; Lim 1983; Hutton 1985), the majority of patients were suffering from severe affective disorders which, according to the usual classification systems, would warrant categorisation as 'affective (depressive) disorder'. About 19 per cent of their population were suffering from what would be described as psychoses (Acharyya et al., 1989; Moorhouse et al., 1989), but a high proportion of these people had a series of adverse life events during their lives such as separation from parents (especially the mother) at an early age, sexual and physical abuse, and emotional conflicts within their own families.

It has been suggested both by McGovern and Cope (1987) and by Littlewood and Lipsedge (1989) that what causes minority patients to be labelled as suffering from schizophrenia might have something to do with the manner in which they present their distress. Behaviour which perhaps appears to the observer as 'bizarre', involving verbal or non-verbal patterns which are interpreted as aggressive, implies to an observer from another culture that not only is this behaviour abnormal, but it also indicates schizophrenia. Particularly in the West, since the notion was first introduced by Kraepelin and Bleuler, 'schizophrenia' has signified to lay people, as well as the medical profession, mental disorder of a very profound degree.

We might wonder whether the observer, even the trained observer, observing behaviour in another person from a different culture which is 'bizarre', perhaps talking with them through an interpreter, could be tempted into diagnosing 'schizophrenia' more readily than, say, affective (emotional) disorder. Since the international systems of classification are vague about defining and classifying affective disorder, but, on the other hand, certain when describing and defining schizophrenia, it is perhaps small wonder that, when the patient is not understandable, the doctor may turn more readily to the box marked 'schizophrenia' into which to fit the patient.

If this is so, how does one escape this stereotyping which I have suggested is the neo-colonial practice of psychiatry? It might prove to be an appropriate starting point to view the problem no longer as one of 'transcultural psychiatry' but simply as one of 'cultural psychiatry', in that psychological distress in all societies and groups is equally cultural, equally biological. Perhaps one could also concentrate more on how to help alleviate the distress rather than on wondering what exactly 'it' is.

On the question of epidemiology, we need to take a more open-ended view of concepts, to allow more elbow room for concepts to develop, rather than smaller boxes for them to shrink into. The newer multi-axial classification, in which diagnosis is considered together with the person's personality, physical health and preceding life stresses, has been widely welcomed by psychiatrists and other mental health professionals all over the world (Littlewood, 1992).

Multi-axial classifications still lack, however, the acceptance of cultural variation within societies as well as across societies. Our world has become much more multicultural than it ever has been, due to faster communication and transport, an international economic system, and the consequent migration of people. Perhaps it is important for us, when attempting to understand 'the patient', to concentrate harder on actually listening to what the patient is saying in their own terms – whether verbally or non verbally – and to take special note of the particular cultural dimensions that make up personhood. The preconceptions about each other that both doctor and patient bring to their consultation will have to be examined and a dialogue can then take place which may go towards alleviating the patient's distress. In mutually divesting the other of their preconceptions, the doctor has a special role in allowing the patient to feel that they have the power and the right to present their own view of what is wrong. Although the doctors may or may not agree with the patient's constructs, they need to accept such constructs as legitimate, worthy of being given a place in the clinical dialogue. It is the doctor who needs to initiate such a sense of dialogue rather than of confrontation with the patient.

A major difficulty in continuing any such dialogue may be what I would describe as the 'cultural divergence' between the professional and the patient. Every human being, when suffering distress, initially attempts to define the nature of this distress and what may have caused it, and to make sense of it. Once the person has reached a decision (whether this is medically 'correct' or 'incorrect' is immaterial), then the person actively chooses to find appropriate means to alleviate the distress. If, for example, a person with abdominal pain attempts to define the cause of the distress and concludes that because it comes in spasms, in periodic bursts, it is caused by the attacks of a malevolent spirit, then he or she might choose a priest or an exorcist as the most appropriate agent to alleviate this distress. On the other hand, the medical professional may conclude that this is a colic, caused by an inflammation of the gut, and simply prescribe an anti-inflammatory drug. On biomedical criteria the professional may be 'right' and the patient may be 'wrong', but, unless the professional can initiate a dialogue, no joint collaboration in attempting to alleviate the distress can take place. The patient may instead go off to find an exorcist while, in terms of the therapeutic programme, the professional complains about the 'non-compliance' of the patient.

With the power that is invested generally by society (which ultimately includes all of us as potential 'patients') in the professional, and that invested by the specific professional training, it is only too comfortable for the professional to remain in the position of power and to retain a position of supremacy, of prescribing the forms of understanding and treatment to the patient. As this reprehensible, if not iniquitous, distribution of power remains, then so does the 'doctor's dilemma'.

References

Acharyya, S., Moorhouse, S., Kareem, J. and Littlewood, R. (1989) 'Nafsiyat: a psychotherapy centre for ethnic minorities', *Psychiatric Bulletin*, vol. 13, 358–360.

Bouras, N., Brough, D. I. and Watson, P. J. (1982) *Mental Health Advice Centre: Three Years of Experience*, Research Report No 1, London, Guys' Hospital (ms).

Bouras, N., Tufnell, G., Brough, D. I., and Watson, P. J. (1983) *Mental Health Advice Centre, The Crisis Intervention Team*, Research Report No. 2, London, Guys' Hospital (ms).

Chakraborty, A. (1990) *Social Stress and Mental Health: A Social Psychiatric Field Study of Calcutta*, New Delhi, Sage.

Harrison, G., Owens, D., Holton, A., Neilon, D. and Boot, D. (1988) 'A prospective study of severe mental disorder in Afro-Caribbean patients', *Psychological Medicine*, vol. 18, 643–658.

Hughes, C. C. (1985) 'Culture-bound or construct-bound? the syndromes and DSM–III' in Simons, R. C. and Hughes, C. C. (eds) *The Culture-bound Syndromes: Folk Illnesses of Psychiatric and Anthropological Interest*, Dordrecht, Reidel.

Hutton, F. (1985) 'Self referrals to a community mental health centre: a three year study', *British Journal of Psychiatry*, vol. 147, 540–544.

Ineichen, B., Harrison, G. and Morgan, M. S. (1984) 'Psychiatric hospital admissions in Bristol', *British Journal of Psychiatry*, vol. 145, 600–611.

Kleinman, A. (1987) *Rethinking Psychiatry: from Cultural Category to Personal Experience*, New York, Free Press.

Lim, M. H. (1983) 'A psychiatric emergency clinic: a study of attendances over six months', *British Journal of Psychiatry*, vol. 143, 460–466.

Littlewood, R. (1990) 'From categories to contexts: a decade of the "New Cross-Cultural Psychiatry" ', *British Journal of Psychiatry*, vol. 156, 308–327.

Littlewood, R. (1992) 'Symptoms, struggles and functions: what does overdose represent?', in McDonals, M. (ed.), *Women and Drugs*, London, Berg.

Littlewood, R. and Lipsedge, M. (1989) *Aliens and Alienists: ethnic minorities and psychiatry*, 2nd rev. edn, London, Unwin Hyman (originally published 1982).

McGovern, D. and Cope, R. V. (1987) 'The compulsory detention of males of different ethnic groups', *British Journal of Psychiatry*, vol. 150, 505–512.

Moorhouse, S., Acharyya, S., Littlewood, R. and Kareem, J. (1989) *An Evaluation of the Nafsiyat Psychotherapy Centre for Ethnic and Cultural Minorities*, Report of a Three-Year Study funded by the Department of Health and Social Security (unpublished ms).

Sartorius, N., Jablensky, A. and Shapiro, R. (1977) "Two year follow-up of the patients included in the WHO International Pilot Study of Schizophrenia', *Psychological Medicine*, vol. 7, 529–541.

43

Maintaining an emergency service

PETER TYRER

Although there is now overwhelming evidence that psychiatric services concentrating on rapid response in community settings are superior to more conventional services (Stein & Test, 1980; Dean & Gadd, 1990; Merson et al., 1992; Burns et al., 1993), there is still scepticism about the feasibility of organising rapid services nationally.

This article is based on work done in the Early Intervention Service (EIS), a community-based service set up as a demonstration project by the Department of Health in 1987 to provide a rapid response service for patients with severe mental illness. It is now integrated into the mental health services for the area. The service sees patients in their homes or at other appropriate settings, including general practices, the community team base, day centres or, relatively rarely, hospitals.

The EIS was established in 1987 in Paddington and North Kensington in the western part of central London. The district was the fourth most deprived area in England using the Jarman scores for underprivileged areas derived from the 1981 UK census (Jarman, 1983, 1984). It is particularly noted for its large number of temporary residents, many of whom are placed in bed-and-breakfast hotels in the district, and it ranks second in the Jarman indices for mobility of population.

The impetus for setting up the EIS was the need to develop a community arm of the existing hospital service, particularly the one based in Paddington. However, it was appreciated that many community services become involved with the less severely mentally ill, and as the area concerned was such a deprived one, it was felt that the numbers of patients with severe mental illness in the community were sufficiently high to accord their needs priority.

Edited from Tyrer, P. (1994) 'Maintaining an Emergency Service', in Phelan, M., Strathdee, G. and Thornicroft, G. (eds), *Emergency Mental Health Services in the Community*, Cambridge, Cambridge University Press, 917–1212.

The community service

The EIS is an eight-strong multidisciplinary team of two community psychiatric nurses, two social workers, a clinical psychologist, occupational therapist, senior psychiatrist (P.T.) and an administrator. Recently some of the staff have been working part-time and sessional senior registrar sessions have also been attached. Despite this, the team remains a small one with only the equivalent of 6.5 full-time members. The average caseload varies from 20 to 25. The EIS takes patients from a defined catchment area (Paddington and North Kensington) and has an open referral, rapid response system with assessment of all referrals, usually at the patient's home, within a few days; however, it is not a crisis intervention service and does not have 24-hour cover (Onyett *et al.*, 1990). A case manager system is used, with clinical decisions reached by consensus at regular team meetings and implemented by a named key worker who can be reinforced by other team members when needed. The overall philosophy is to try and treat all mental disorders outside hospital in the first instance, with particular emphasis on joint working with other agencies, home treatment when necessary, and a collaborative approach to care that involves the patient as an active participant in all treatment decisions.

This is a more radical approach than that commonly employed in most psychiatric services and could be regarded as too idealistic. In theory, the EIS model could become badly unstuck when admission to hospital, particularly compulsory admission, is necessary, because it appears to contradict the tenets of the community-based treatment ethos of the service. In practice, however, this has not caused any serious difficulties, and the strong orientation of team members to community treatment wherever possible has cemented working relationships and has had some surprising gains. For example, the EIS has had great success with keeping Schizophrenic patients out of hospital, and the impact has been greater for this diagnosis than for any other (Tyrer, 1992).

Because it is appreciated that the EIS will be likely to receive a large number of referrals it was important not to have too great a case-load of continuing care. The early policy of the service was to see patients for no longer than six months and to concentrate on liaison with other services as part of its clinical work. In practice, a significant number of patients have no appropriate agency to refer to because of the needs that still remain after early intervention. Nevertheless, most of the patients can be referred to another agency and, as the EIS has a policy of taking re-referrals at any time, there is no reason why such patients cannot come back to the EIS at times of crisis or of deterioration in their conditions.

Management of the service

The EIS is a relatively cheap service with, for example, a total annual budget in 1992 of around £180 000. This is partly because management costs are kept to a minimum. Most decisions are reached by the team as a whole and this includes both clinical and business decisions. There is a team co-ordinator, currently a social worker, who represents the team in its negotiations with management and with other agencies. Line management for each of the disciplines within the team is maintained by senior professionals from the relevant disciplines, although in practice this occupies a very small amount of time and is largely concerned with career developments and other long-term aims, because almost all clinical and professional support is given from within the team.

In this respect the EIS differs from many other services in which there is a clear management who gives the team direction and, at least to some extent, imposes control. The mistake is often made by those who visit the service in thinking that I, as the senior psychiatrist, am the team manager and the controlling influence. This misperception is resented by other members of the team, not least because I am a part-time member of the service, and such views give the impression that the team has much less autonomy than is indeed the case. However, I do have an important role in being able to link the EIS to other parts of the mental health services in the district. The team now has acute beds in the psychiatric unit and so when patients are admitted, key workers in the EIS can continue to retain contact and are involved with planning discharge and after-care (and often retaining their role as a key worker throughout). These links to other parts of the service help to maintain the profile of the EIS and prevent it from becoming isolated at the periphery of the mental health services at a time of rapid reform.

The patients

Patients can be referred to the EIS from any source, including self-referral. The advantages of this approach are mentioned below. The disadvantage of such an approach is that any service which is seen to be successful in meeting need tends to get flooded with increasing numbers of patients from a wide variety of sources and is unable to cope with the demand. From time to time we have to remind our referrers, particularly GPs, that we remain a service that is concentrating on treating mental illness that is both severe and acute, and that many patients referred do not meet the criteria for our service because their complaints are relatively minor or chronic.

There have been some alterations in the profile of referrals over the years of operation of the service but there has always been a rough separation into three diagnostic groups of equal size: mood disorders, adjustment and stress disorders, and schizophrenia. Although the team does not treat primary drug and alcohol problems, a small group of patients does have these problems to a significant extent.

Good diagnostic comparison has been possible because the EIS has from the beginning adopted the policy of giving a formal diagnosis to all patients taken on by the service using the ICD–10 notation. Comparison of referrals early in its history and those recently show relatively little change in the type of patient being taken on by the service and, in particular, the phenomenon of 'up-market drift' so often found with community mental health teams (Dowell and Ciarlo, 1983) has not been shown; our referrals are not getting any easier!

Patients are almost invariably seen as a joint assessment on the first occasion, at which attempts are made initially to carry out the assessment with the referrer. If this is not possible then two members of the EIS see the patient. Clinical review meetings are held twice weekly, in which all members of the team take part. The first part of these meetings involves allocation of new referrals and feed-back from assessments. In many cases it is possible after the end of the first assessment to plan further care but if necessary, this decision is delayed until the next clinical review meeting, in which the decision about the best form of further care is made and then communicated to the patient. Most patients are at home in the first instance, although subsequently care may take place at any setting, including the team's base at the EIS, which has an interview room.

Maintaining expertise and morale: the skill-share model

The EIS operates a model of care that is somewhat unusual. It is often aspired to in theory but seldom achieved in practice. We call it the skill-share model. In this model of care all disciplines contribute to each other's skills through close working arrangements and open discussion at clinical review meetings. Every patient seen by the service is reviewed weekly at the two clinical review meetings. As the average case-load is around 140 this requires economy and efficiency and this is imposed by the Chair of the meeting (who rotates every two months). Team members are encouraged to contribute their specialist knowledge at all times so that the key worker for the patient is equipped to deal with all relevant aspects of the clinical problem.

The consequence of this is that, almost imperceptibly, each team member loses the professional label that was originally attached. Team members now find it somewhat embarrassing to be introduced to visitors with the epithet of their parent discipline because in their work with the service they are not really acting as a member of that discipline. Clearly some responsibilities remain restricted to certain disciplines. (for example, injections of drugs to medical and nursing staff; assessments for compulsory admission by social workers and doctors). Many others can be shared and, for example, one of our social workers is now our main cognitive therapist, both of our community psychiatric nurses have skills in family therapy, and an occupational therapist has become expert in monitoring drug treatment.

There are important aspects of the skill-share model:

- The case manager can be of any discipline and responsibilities for co-ordinating care do not change when the patient is admitted to hospital or moves to other centres.
- The problems that are beyond the individual case manager's expertise are tackled within the team by joint working.
- The case manager takes on a much greater role than is normally expected of the relevant discipline.
- When consultation is sought from other disciplines it can be provided indirectly through a clinical review meeting or directly by involvement of additional members of the team. There is no armchair advice.

The skill-share model: budget

Because all members of the team are clinically active the functioning of the service is economical. In strict terms there are no management charges apart from a small sum given to the team co-ordinator for additional responsibilities.

In the formal comparison of the EIS and the standard hospital service, described in more detail below, confirmation of the relatively low cost of the EIS was demonstrated. Despite the fact that each visit from the service cost £132, the total cost was 2.5 times greater in the hospital service when all aspects of health care were taken into account (Merson *et al.*, 1995).

Integration with psychiatric services

One of the major problems attending the introduction of community mental health centres in the United States was the relative isolation of staff, leading to a loss of morale and subsequently to a mass exodus of staff to other more attractive work settings. The EIS has tried to maintain close contact with other services, although this has not always been easy. Initially the service was restricted in its access to referrers and there was also the suspicion that reduction in services elsewhere were a direct consequence of the introduction of the EIS. These difficulties have naturally been more marked with other community services and still there are some difficulties outstanding.

This has disadvantages but also has the bonus of helping the team. One of the consequences of this is that staff turnover in the EIS has been remarkably low. Only six changes have occurred in five years; four of these have been promotions, and four staff have stayed in post since the inception of the service. One problem this has highlighted is the difficulty of establishing an adequate career structure in a multidisciplinary team. It is difficult for team members to leave for other posts where they can continue to act in the same way as in the EIS; to some extent all those who have left for promotions have had to become more restrictive in their clinical work.

As the EIS has become established close links have been established to important parts of the service. This includes the day hospital (where a former member of the EIS is not the clinical manager), in-patient wards (improved now the EIS has access to its own beds), general practices (particularly those with many practitioners who are also fundholders) and voluntary organisations such as Mind. The policy of joint working means that members of the service are always in contact with their colleagues from other parts of the psychiatric network.

Problems with a mobile population

One of the main problems of a mobile population is the difficulties that many people have in integrating into the local community services. These include, in particular, primary care and social support networks. The consequence of this is that many psychiatric problems bypass the filters that normally prevent preventable psychiatric admissions.

An open referral system tends to obviate these problems. If referrals are accepted from patients themselves, their neighbours, voluntary agencies and other health professionals apart from doctors, then they are likely to be seen earlier than if they are seen in the sometimes bureaucratic statutory systems. In the EIS the policy of open referral has been operated from its beginning and has worked well. There is no evidence that those who are not medically qualified refer more inappropriate cases and in practice the more severely mentally ill patients are referred from non-medical sources.

Evaluation and audit, including comparison with other services

After the first year of its operation the EIS circulated all those agencies who had referred patients to ask their opinions. These were presented in the form of positive and negative statements about the service, to which respondents indicated whether they agreed or disagreed to varying levels of intensity. At this stage the EIS was still a demonstration project and its future was uncertain, so one of the statements referred to the possibility of permanent status in the area. 61 per cent (62 of 102) of the referrers returned their forms and their results indicated a strong preference for the EIS procedures of joint working, visiting patients at their homes, referral arrangements and need for permanent status. Further audit of referrers has continued; although there is still general approval there has been concern expressed in the last year, particularly by GPs, about the priority we give to referrals of major mental illness, or to be more precise, the relative lower priority given to less severe illness.

We also need to determine whether our service was different from others dealing with the same population. This was possible because the EIS, although dealing with a specified catchment area, was only one of several service providers.

A formal study was set up to evaluate the EIS by comparing its efficacy, impact on patients' views and use of resources by comparison with the parallel hospital service. Using a randomised controlled design, patients presenting as emergencies to either the hospital doctors or psychiatric social work department (who could equally have been referred to the hospital service or the EIS) were allocated to their service by the referring doctor or social worker opening a sealed envelope.

Selection of treatment in both services was clinically determined without any restrictions imposed by the study design, but in practice more EIS patients received psychological intervention than in the hospital

TABLE 43.1 Summary of findings of randomised controlled trial of Early Intervention Service and standard hospital service in treatment of psychiatric emergencies over a period of 12 weeks

	Early Intervention Service	Standard service
Number allocated	48	52
Number seen and engaged by services	47	37
% reduction in symptoms:	36.4	16.9
in personality disorder	1.6	25.6
in non-personality disorder	34.0	17.2
% improvement in social function overall	15.4	18.8
Mean psychiatric bed days	1.2	9.3
% satisfied with care	83	54
Mean cost of care/patient	£1160	£2502

Source: Merson *et al.*, 1992, 1995; Tyrer *et al.*, 1994.

service. Most EIS referrals were seen at home initially, with referrals to the hospital service seen mainly in psychiatric out-patient clinics.

After randomisation, patients were assessed by a psychiatrist blind to service allocation, which included demographic and clinical information (symptoms, social function, premorbid personality). Further assessment and scoring of symptoms and social function were carried out two, four and twelve weeks after randomisation, either at home or hospital depending upon patients' preference.

Of 100 patients consenting to inclusion in the study, 52 were randomly allocated to the control and 48 to the experimental group; 95 patients completed the research assessments at 12 weeks. The results (Merson *et al.*, 1992) are summarised in Table 43.1 and showed greater take-up of services with the EIS together with greater satisfaction with care, reduced in-patient bed use and greater reduction in symptoms but not social functioning.

Difficulties in developing skill-share services

Although we feel confident that we have identified an efficient and effective model of practice that reinforces practitioners' skills and avoids burn-out, we are well aware of the difficulties involved in implementing this approach elsewhere.

Until we have a consistent core of training for all mental health professionals we are bound to have continuing conflict in community psychiatric teams. The past arguments between professional groups, still unnecessarily rehearsed and repeated in formal teaching, build up the ammunition of resentment and antagonism ready to be exploded when trainees are released into clinical practice. As a consequence early conflict in community teams is the norm and many respond by retreating into uni-disciplinary purdah, only venturing out to converse with other disciplines on formal occasions when rules of etiquette are strictly observed.

Inter-professional rivalry follows from inter-professional conflict and should also be reduced by common training. The skill-share approach removes rivalry and aids communication. Similarly, arguments over who is clinically responsible for patients should not be used as an excuse to create an unnecessary hierarchy in community teams. The doctor is medically responsible, the social worker is similarly responsible for specific social work tasks, and so on. Even medical students when attached to the team carry some responsibility for their behaviour and actions that can in no way be transferred to others. In the skill-share approach the clinical team becomes a responsible body in its own right and much of this unnecessary debate disappears.

If each member of the team feels their particular skills and potential are being exercised and acknowledged this is an excellent reinforcer of self-esteem. This is particularly important for new team members. We can usually detect when such members have achieved the necessary empowerment for good functioning in the team; they have the confidence to express their disagreements with others openly! A strong team is one that can express its disagreements without rancour and when these are genuine ones involving opinions about difficult clinical decisions rather than the phoney exercise of pre-existing prejudices they ultimately improve the team's function.

References

Burns, T., Beadsmoore, A., Bhat, A. V., Oliver, A. and Mathers, C. (1993) 'A controlled trial of home-based acute psychiatric services. I. Clinical and social outcome', *British Journal of Psychiatry*, vol. 163, 49–54.

Dean, C. and Gadd, E. M. (1990) 'Home treatment for acute psychiatric illness', *British Medical Journal*, vol. 301, 1021–1023.

Dowell, D. A. and Ciarlo, J. A. (1983) 'Overview of the community mental health centers program from an evaluation perspective', *Community Mental Health Journal*, vol. 19, 95–125.

Jarman, B. (1983) 'Identification of underprivileged areas', *British Medical Journal*, vol. 286, 1705–1709.

Jarman, B. (1984) 'Validation and distribution of scores', *British Medical Journal*, vol. 289, 1587–1592.

Merson, S., Tyrer, P., Onyett, S., Lack, S., Birkett, P., Lynch, S. and Johnson, T. (1992) 'Early intervention in psychiatric emergencies: a controlled clinical trial', *Lancet*, vol. 339, 1311–1314.

Merson, S., Tyrer, P., Carlen, D. and Johnson, A. L. (1995) 'The cost of treatment of psychiatric emergencies: a comparison of hospital and community services', *Psychological Medicine*, vol. 26(4), 727–734.

Onyett, S., Tyrer, P., Connolly, J., Malone, S., Rennison, J., Parslow, S., Shia, N., Davey, T., Lynch, S. and Merson, S. (1990) 'The Early Intervention Service: the first eighteen months of an inner London demonstration project', *Psychiatric Bulletin*, vol. 14, 267–269.

Stein, L. I. and Test, M. A. (1980) 'Alternative to mental hospital treatment. 1. Conceptual model, treatment program and clinical evaluation', *Archives of General Psychiatry*, vol. 36, 1073–1079.

Tyrer, P. (1992) 'Schizophrenia: early detection, early intervention', in Jenkins, R., Field, V. and Young, R. (eds), *The Primary Care of Schizophrenia*, London, HMSO.

Tyrer, P., Merson, S., Onyett, S. and Johnson, T. (1994). 'The effect of personality disorder on clinical outcome, social networks and adjustment: a controlled clinical trial of psychiatric emergencies', *Psychological Medicine*, vol. 24, 731–740.

44

Working with refugees and torture survivors: help for the helpers

JANE SHACKMAN and JILL REYNOLDS

That violence, war and torture may have psychological effects on refugees is now well established and reflected internationally in services which have words like 'torture' or 'trauma' in their names. Attention is increasingly being paid to the demands and stresses of working with refugees overseas (Stearns, 1993; van der Veer, 1992; Macnair, 1995). We look in this article at the needs of people working with asylum-seekers and refugees in the UK, drawing on our experience with different refugee agencies over the last 15 years in reception-centre work, training, supervision and programme evaluation. We examine the nature of work with refugees, and what kinds of stresses workers experience. We suggest that workers need some balance between giving and getting support and consider how organisations can help.

What kind of organisations work with refugees?

Organisations specialising in work with asylum-seekers and refugees in the UK vary in size and function. They are voluntary organisations, and may receive some government or local authority funding, but most have to be active in fund-raising. They range from agencies like the Refugee Council, the largest provider of services, through to refugee community organisations set up by local groups, sometimes with just one worker. There are also services such as the Medical Foundation for the Care of Victims of Torture and Nafsiyat, the intercultural therapy centre, which because of the nature of their work see large numbers of asylum-seekers and refugees.

The nature of the work: stresses and strains

In contrast to other mental health work, many refugees are basically mentally healthy, although going through distressing crises. Workers can see positive changes in the lives of their refugee clients as they adjust, are reunited with families, and take up educational and employment opportunities.

However, working with refugees can also be draining. At times, workers are likely to have feelings of helplessness, hopelessness and frustration: unable to do enough, unable to control, change or influence the factors and situations causing refugees to flee. There is rarely the gratification of seeing real change in the world of repression and human-rights abuse.

Workers from the same community as their clients have specific strengths in shared language, culture, and experiences. They are also subject to very particular pressures from their compatriots: envy, or expectations that workers will do everything for them, and be constantly available. Such expectations may be held at the same time as underlying beliefs that non-British workers are less effective in manipulating the system in their clients' favour. They are constantly confronted with problems which may be unresolved for themselves, or may stimulate painful memories. A Vietnamese worker in the early days of resettlement work expressed the feeling:

> Sometimes I think I have done so much for these people and I cannot do much for myself. That makes me sad. It should be my father on that plane [arriving in the UK]. But you learn to take it. (Bang, 1983)

The divisions which caused people to leave home do not disappear on arrival in the UK. Workers may find that they are subject to fears and threats from former compatriots. For instance, one Serbian worker described how: 'I was threatened for four months by one (a Bosnian Muslim) – "I killed 1000 Serbs – you will be the 1001st".' People working for refugee community organisations are especially vulnerable: they have the greatest difficulty in attracting sufficient funds and resources, and the demands on them from refugees and asylum-seekers are often the heaviest. Even when mental health work is not their main activity, they receive the full force of community members' distress and frustrations. Community workers feel a sense of guilt when a client commits suicide (Mohammed, 1995). They may feel that they should have been able to do something to prevent this: one worker was angry because his advice of such a risk was ignored by health workers.

Workers from outside the refugee community have different stresses. Not understanding the language or cultural issues makes it difficult to pick up and respond to behaviour and clues. The use of Western

Eurocentric models of intervention may feel inappropriate or have limited effectiveness. An idealistic desire to help may have set a tone and expectations which cannot be met. One non-refugee worker recalled:

[At first we were] undisciplined with the Bosnian clients, allowing them to come and abuse and threaten us in a confined space. This was very stressful. ... We all get the impression that their expectations cannot be met by us, whether those expectations were formed in Croatia before transit to the UK, or by our lack of discipline or professionalism at the beginning of the project.

Workers from all backgrounds are exposed to a high and intense degree of distress, grief and anguish, as clients face the deaths of family and comrades, or the long enforced separations from partners, children or parents left at home. They are immersed in hearing stories of atrocities and abuses that many of us would rather not have to think about. How do workers bear to hear stories of horror, repeatedly? What do they have to do, in order to be able to listen? What kind of accumulated stress is there in hearing these accounts? A non-refugee working with torture survivors wrote:

I can remember feeling sick typing reports and on occasions having to leave the office ... I feel I carry a great weight inside me that sometimes feels quite unbearable. I can feel guilty about being born in a safer, more secure environment, and about spending money and enjoying myself, and going on holiday. (Smith, 1993)

Someone who is herself a refugee working in the Bosnia programme expressed the impact on her: "We just weren't prepared – the first month I was living on adrenalin – I'd hear people's stories and be really shocked.'

It has been suggested that people working with survivors of torture and trauma are at risk of being traumatised themselves. a new term has been coined of 'vicarious traumatisation' which involves symptoms of nightmares, intrusive memories, and avoidance behaviour (Lansen, 1993; van der Veer, 1992). Like 'post-traumatic stress disorder' this may have limited use as a diagnostic concept since it does not necessarily imply a need for psychological treatment. Its value may be in drawing attention to the impact of the work for different individuals, and the need for workers to monitor their own reactions, and their agencies to provide opportunities for them to express their feelings:

The first year almost everyone on our staff suffered from nightmares. This was particularly prevalent after Tuesday, which is the day when we see a lot of patients. No one on the staff could get up Wednesday morning. Then, gradually, we began to realise that this is a reality of the service. Not only are we a family clinic to the community, but we also have to be a family clinic to the staff, to ourselves. (Mollica, quoted by Stearns, 1993)

People are affected by, and respond to, stress very differently. A challenge to one can be an impossible pressure for another. It is important for workers to recognise and acknowledge their own limitations and vulnerabilities. Burnout is a term frequently used when workers are no longer able to function effectively. Some drive themselves ever harder, others succumb to illness and take periods off sick, others become distanced from the work, apathetic, irritable with colleagues or inefficient.

Organisational hindrances to coping with stress

Culture and society shape an individual's response to stress. The culture of an organisation will not necessarily be prescribed in policy documents, but is nevertheless a potent and dynamic force within any organisation. The organisational culture will help or hinder workers in acknowledging and coping with stress, and can indeed be a source of stress.

Tensions can arise if assumptions and models of work are not shared by members of the same work team. Is the refugee a survivor or a victim? Should the approach used be an individual therapeutic or a community-based one? Fundamental disagreements, if not resolved, can lead to confusion, frustration and demoralisation. This is not to suggest that everyone must work in the same way, but different approaches need to be compatible and able to co-exist, and differences need to be out in the open.

This can be particularly difficult when the majority of staff are white and not themselves refugees, and disagreements involve ethnic or racial divisions. As members of the dominant culture in which the agency is operating, white, non-refugee workers may have the advantage in settling disputes, and this can appear racist or unfair. Conversely, non-refugees may hold back from expressing their views and give undue power to less experienced workers who are then burdened with the expectation that as members of the refugee community being worked with, they hold the answers.

Such inertia can apply to working practices too, as one Bosnian worker related:

> If someone is abusive the others in the office will wait for me to do something
> – because I've been in the war they think I have the right to throw them out.

In the smaller organisations set up by refugee community groups organisational problems are likely to be of a different nature. With sometimes only one or two paid workers, the conflicts or confusions may be between the paid workers and their volunteer management committees. Workers can feel that in addition to the main task of their project, they have to induct their management committees into good management practices, and that they do not really get the support that they need in return. The problems cut both ways: a management committee may be concerned by the workers' practices, but have difficulty (because of lack of time, or detailed knowledge) in getting them to be fully accountable.

Any refugee agency may find that just because the problems seem so urgent and endless, a culture develops in which people are expected, and expect themselves, to cope with everything. A worker asking for extra support may be regarded as self-indulgent. It can be particularly hard for people working with their own communities to draw a boundary to protect their time off, or simply to say there are limits to what they can do. Studies of Vietnamese workers have noted that their connection and commitment to the people they are working with are amongst the characteristics which make them so effective in their work (Bang, 1983; Reynolds, 1995).

Refugee crises are by nature chaotic and sudden. Lack of preparedness from agencies can mean that they also respond in a chaotic manner: sometimes this becomes part of the organisational culture. Clients then experience this chaos too.

The *ad hoc* nature of government decisions and refugee agencies' response to the arrival of refugees from Bosnia led to extra stress for workers (Graessle and Gawlinski, 1996). The need for co-ordination between the five different agencies involved was not effectively fulfilled. This resulted in mixed messages, confusion and inconsistency with clients and staff. Decisions did not get communicated within agencies, let alone between them. Co-ordination and communication improved following discussions between the agencies and a process of ongoing evaluation. Some problems remained. Staff were hastily appointed without sufficient induction or training: assumptions were made that to be Bosnian was enough of a qualification. Contracts were short-term, and some staff remained on three-month contracts over a two-year period, unable to plan for their clients or themselves. They were unclear about their roles, responsibilities and boundaries, and often felt quite overwhelmed. One team leader reported:

> Prior to my appointment the team had existed on good will. There were too many martyrs! The workplace was grossly understaffed, policies and procedures were undefined, staff were exhausted. I took some time to reverse this process.

People were unprepared for the stressful nature of working with clients who had been in concentration camps, or who were expressing extreme distress or anger. They often felt unsupported by their head office in dealing with these problems.

An organisation which helps the helpers?

An organisation can help its workers cope with stress by recognising the demanding nature of the work and developing a culture which allows stress and feelings to be expressed. Key values, such as the willingness to listen and learn from staff at all levels, or to risk admitting uncertainty, are brought to life in organisational attitudes. Organisations can also avoid creating stress, by establishing good management structures. Staff selection, clear job descriptions, induction procedures, staff supervision, training, workload management, on-going monitoring and evaluation are components of good management.

Helping organisations to manage

Refugee community organisations are often fighting for their very existence, and even in more established organisations, the crisis nature of assistance programmes means that systems are being set up with little time to plan. How feasible is it for agencies to create the framework of good management which workers and their clients need? The crisis nature of refugee emergencies will always involve reception of unexpected numbers of people from time to time. It would be helpful if there could be some acknowledgement by agencies and by government of this likelihood of need for emergency provision. One recommendation to come from the evaluation of the Bosnia project is that agencies work together to maintain and review their plans for responding to humanitarian emergencies. Although it may not always be possible to address specific details, it is important to recognise good practice in how crisis are dealt with, and to develop methods for setting up effective management structures. The larger agencies which offer reception services need organisational structures which can allow them to expand and contract as required. Agencies need to develop and monitor their own 'organisational good health', for instance in their staffing procedures, in order to respond quickly and qualitatively to a humanitarian emergency.

There are some developments which can help refugee community organisations. Some larger agencies offer advice to community groups on management, finance, fundraising and developing (Reynolds, 1994). Through skill-sharing sessions which bring different groups together, committee members and workers can share their views on how to work well as a team, and groups can learn from each other. Some refugee community projects offer training to health authorities on cultural awareness and different perceptions of mental health. Such initiatives suggest there is great potential for organisations to help each other in developing good work and management practices, shared training, and creating informal links which allow learning from different kinds of special expertise.

Supervision, training and support

The pressure on the physical and emotional resources of workers with refugees means that formal and informal opportunities are essential to off-load and share casework difficulties. Training helps workers to develop and practise skills, and supervision provides a time for reflection upon ongoing work. There are different aspects to supervision which can be distinguished as educative, supportive and managerial (Hawkins and Shohet, 1991). These aspects are all-important if organisations are to provide the kind of working environment in which workers feel confident and valued in their work.

Some organisations working with refugees are using group supervision sessions, and find this particularly valuable for the educative and supportive aspects of the process. By establishing the existence of shared issues within individual caseloads, support networks are established, and a 'good enough' ethos can be tolerated, in place of a sense of guilt and inadequacy. Supervision needs to be regular and planned, rather than an emergency response when something has gone wrong. Staff need to know that they will get supervision, and that they are expected to make use of it.

Sensitive supervision can help protect workers from over-immersion in the problems and concerns of the work. A team co-ordinator used supervision sessions to help workers in the Vietnamese community-development programme look at whether their out-of-hours involvement on the committee of a community group was part of their job. Some workers particularly value the chance in supervision to look at the impact of the work on themselves. Supervision should also address issues of power, racism, and different aspects of cross-cultural work with clients and between workers.

Training and supervision should be geared to the kind of work which an agency is undertaking, for instance whether the emphasis is on therapeutic work or community building. However, since bilingual staff are most likely to be receiving the brunt of refugees' experiences, whatever assistance is being offered, there are arguments for always providing new bilingual staff with counselling, communications and assertiveness training, which can then also help them in dealing with their own feelings. A worker from former Yugoslavia reported on how she had gained from training:

> The first training was two days on loss and bereavement – I recognised myself in that, realised I was quite normal myself, and then I recognised my clients.

When workers are dispersed, working in quite small projects, as is often the case, training which brings people together is also a means of facilitating support and creating the sense that there is a wider 'team'. A useful by-product of the first training sessions for workers in the Bosnia programme was that people were able to set up peer support amongst themselves.

Regular staff meetings can also have a supportive function. Interpreters at the Medical Foundation have valued having regular group meetings together:

> The Interpreters' Group is useful because we can share all the difficulties and the nice things, like when a client achieves something. We share happiness and sadness with each other: once an interpreter talked about a boy who died, and another time an interpreter talked about a family reunion. (Rahimi, 1995)

Of course, problems do not need to wait for meetings. Messages conveyed in the mother-tongue have an immediate impact on the interpreter. Another interpreter finds that talking to the professional worker soon after the interview can help:

> Sometimes a case touches you so badly. We are supposed to control our feelings in the session and we do, but how can we overcome these feelings? To express the feelings afterwards makes me stronger. The hopelessness is something we carry, we are so powerless, but if I know someone else is feeling the same, that helps.

The provision of support is, however, a sensitive issue, and attention needs to be given to what kind of support people from different backgrounds, and using different approaches to helping, find most useful. When people have been pitched into the work, perhaps chiefly for their language ability, without adequate preparation, they are not unnaturally suspicious of supervision and support. They may fear criticism. Some agencies have found that those meetings which are labelled 'support' are rarely attended by the workers who are refugees, and that it is easier to use meetings in which workers review their work on cases as an opportunity to give informal support. There is a balance to be struck, so that people know that it is acceptable, and indeed expected, that they will want help, but do not feel that this help is being imposed on them, or that their raw feelings will be exposed in front of others before they themselves are ready to acknowledge them.

Agencies offering psychotherapeutic help to refugees usually require that workers have appropriate professional training and some prefer that their workers are in personal therapy themselves. Nafsiyat emphasises the importance of colleague one-to-one support as well as more formally structured supervision or external consultation (Thomas, 1995). Workers are encouraged to recognise their own vulnerabilities and where possible plan ahead for support when they have a potentially stressful interview coming up, by fixing a time to see a colleague later. A particularly valuable feature of this is that people can take some responsibility themselves for monitoring their own needs.

Conclusion

People are bound to experience stress when they are working with the effects of war, torture, violence and disruption of family life. For some people the risks are greater because they can be reminded of how their own lives have been disrupted. The pressure to keep going can be intense, and agencies working with refugees have a responsibility to ensure that their employees are given enough time to stand aside from the demands and to reflect, with supervision and support, on the impact of the work on themselves. They also have a responsibility to develop organisational and management practices that do not add to the stress of the work itself.

References

Bang, S. (1983) *We Come as a Friend: towards a Vietnamese model of social work,* Derby, Refugee Action.

Graessle, L. and Gawlinski, G. (1996) *Responding to a Humanitarian Emergency: an Evaluation of the UK's Bosnia Project to Offer 'Temporary Protection' to People from Former Yugoslavia 1992–1995,* Planning Together Associates, King's Lynn.

Hawkins, P. and Shohet, R. (1991) *Supervision in the Helping Professions: an individual, group and organizational approach,* Buckingham, Open University Press.

Lansen, J. (1993) 'Vicarious traumatization in therapists treating victims of torture and persecution', *TORTURE,* vol. 3(4), 138–140.

Macnair, R. (1995) *Room for Improvement: the management and support of relief and development workers,* London, Overseas Development Institute.

Mohammed, F. (1995) Personal communication.

Rahimi, Z. (1995) Personal communication.

Reynolds, J. (1994) 'New Society, new networks: community development work with Vietnamese people in the UK', *Talking Point,* Association of Community Workers, 153.

Reynolds, J. (1995) ' "The work chose us": community development work with Vietnamese people settled in the UK', *Practice,* vol. 7(3), 19–26.

Smith, P. (1993) Personal communication.

Stearns, S. (1993) 'Psychological distress and relief work: who helps the helpers?', *Refugee Participation Network,* vol. 15, 3–8.

Thomas, L. (1995) Personal communication.

van der Veer, G. (1992) *Counselling and Therapy with Refugees: psychological problems of victims of war, torture and repression,* Chichester, Wiley.

45

Doing being human: reflective practice in mental health work

TOM HELLER

Julia knew she was dying. When her time came she called for the people she wanted to be around her. I got the phone call about half way through that evening's surgery. I explained to the receptionist and to the line of waiting people that I might well be gone for quite a long time. When I arrived at her house Julia was short of breath and with her deep sunken eyes she acknowledged me. 'I'm dying', she said. I think I said something like, 'Yes'. She asked to use the commode and afterwards settled back on to her sofa, closed her eyes and died. Up to the very end she had been in control of her final illness. A remarkable, calm woman who attracted love because she had given so much love during her short life. I wanted to be there with her as she died, and she wanted me to be there. We had arranged that previously.

On the way back to the surgery I felt the calmness she had been able to transmit to me and to those she had touched with her life and with her death. I felt something behind my eyes which was probably the start of tears, but also deep thoughts I couldn't place or explain. The line of people waiting to see me had shrunk to just a few people. I sat in my surgery for a moment and then rose and asked the first person to come in. 'Can you just have a look at this wart on my foot, Doctor Tom?'

This incident happened to me fifteen years ago, and I am still furious with the woman who had waited all that time to show me her trivial complaint which interrupted my own inner journey. How should I behave in this situation? Should I calmly explain to the woman that I had just witnessed the most remarkable human event? Should I show my anger at having my philosophical and emotional space invaded? Should

I make this woman pay in some way for the pain I was myself going through – give her inappropriate advice, trivialise her complaint? In the event my professional auto-pilot took over. I doubt if she ever knew what was going on in my head as I examined the offending lesion, offered comfort to her for having to put up with it in such a painful place, prescribed the correct treatment and offered her a follow-up appointment at the next wart clinic.

Is there any point in asking whether my way of coping was being a good doctor or a bad doctor? Should I have shown my emotions? Why do health workers continually put themselves in these sorts of positions, for although this incident sticks in my mind and has a significance in my life, it is not really all that exceptional. We are continually being asked to put on a professional show, when behind the façade all sorts of personal turmoil might be underway.

We open our doors and ask people to come in and tell us what is worrying them – why should we be angry with them when they do just that? What sort of service could we possibly run if we allowed ourselves sufficient time to feel our own feelings? But there is something rather odd about trying to get help from health workers who have not worked out their own feelings, or who deny them to themselves and others. Where do all those spontaneous feelings go, and who is to say what damage they might be doing to the delicate internal workings of our minds if we continue to repress and suppress them? You callous bastard, here was a major life event for you and for others, and you carried on as if nothing had happened.

In mental health work rather similar things are happening all the time. Being in contact with people who are journeying through their own 'dark night of the soul' is a constant theme for people who set themselves up as health workers. How is it possible to cope with such depth of human pain? 'Oh, here's a person who has just had her children taken away from her, and here is someone who has cancer; oh well it's lunch time, let's see who's in the refectory.'

Even for the most well-defended health workers, there are always a few people who slip through the defensive guard and who the worker will take home with them in their thoughts at the end of the day. I remember with guilt that woman who had her children taken away from her. I tried to fight on her behalf against the social workers, but lost the battle. I thought she was being made ill by the threats that the social workers were making; they thought she was too disturbed to have any more responsibility over her own children. Why did this scenario affect me so deeply? What parts of my own experience and psyche did it touch? Did this battle take place at a time when I felt vulnerable about my own parenting skills? Or was it just to do with power? We doctors don't like to be beaten in power struggles, you know, or to have our judgement questioned.

There is a continual philosophical problem that health workers struggle with. In the world of mental health this is to do with notions of continuity or discontinuity. Are there some people who are 'well', who are in positions of authority and who can make judgements over those with a 'disease'? Or is 'mental health' a spectrum along which we are all arranged according to certain factors? Some people, when sufficiently stressed, may find themselves at the other end, and be labelled as suffering from a mental disorder. Either we are all in this human soup together, sculling around for answers to the meaning of our existence, or 'we' are here to help 'them'. Flawed and imperfect, 'they' need to be examined, diagnosed, categorised and treated. 'We', of course, need none of those things; we just need to get better at diagnosing and finding out more about brain biochemistry and the categorisation of diseases.

A middle position might recognise that it doesn't help if the people who are given the status and facilities to do health care work feel too guilty about the 'medical model' and their own sins. There are skills to be learnt in this work and it is important to develop some understanding about the things, including physical types of treatment, that may be effective.

I have recently come into contact with many people who would describe themselves as 'users and survivors'. It appears to me that the quality of the supportive interactions that occur *between* some of the 'survivors' is of a depth and directness that is simply not matched by the relationship between 'health workers' and 'clients'. Having been through the experience of mental distress provides an immediate link which appears qualitatively different from that between worker and those being 'worked on'. Maybe all health workers should learn from this and encourage and support initiatives which enable this to happen.

Even those workers who subscribe to the theory of continuity of mental distress – that we would all develop depression or even psychotic illness if we were pushed or stressed enough – are challenged to maintain this stance consistently. In the middle of the night, or at the end of the long day, it can be hard to retain one's liberal credentials. Rather like the parent faced with a trying child who eventually resorts to saying, 'Just do it, because I tell you to', the power remains with the health workers. We do have the power to diagnose, write reports on, medicate and deprive of their freedom people whom it can be difficult to place on the same human spectrum as ourselves. How can I learn to tell myself that I am on the same spectrum as this person who is starving herself to death? Are my own problems with control of food, body image and self esteem really the same, but just in different amounts – or is this to diminish the horror of the experience she is going through?

Here is my personal exploration of some of the things that I find helpful to get me through my work. You may want to compile your own list.

Reading and Writing

Writings by mental health system survivors offer vital insights. Many of these writings are forceful and contain angry sections describing how poorly they feel they have been treated by the health care system (Read and Reynolds, 1996). I have found it hard not to react defensively, but to try to open myself up to what they are saying and acknowledge that in these accounts are also many generous suggestions about the way that health workers should behave. The task is fundamentally a human task. The person who comes to you for help is doing just that. Whatever the other responsibilities and pressures that might tend to deflect or distort the task, before you is another human being. How would it be if we changed seats for a while? In my own work, I try to imagine for at least one moment during each contact what it must be like to feel so low, so rejected, so frightened. In the writings survivors have shared, it is often the simple human gestures that have made all the difference to their recovery and to their perceptions of themselves. The person who helped them the most was not always the person with encyclopaedic knowledge of the pharmacopoeia (although this is handy sometimes), but the nurse who chatted, the 'fellow inmate' who had the energy to challenge, or the doctor who remembered they were human.

For me the process of writing is also important. I usually get my mind clearer when I try to write things down, and I think other people may be like this. For some people the process of writing brings a clarity and precision that isn't possible in spoken words. Editing and revising your own words can add an extra dimension to your thoughts on a subject. (Thank God for word processors.)

How about taking some of my own medicine?

By this I don't mean gulping down the Prozac, or sharing out the tranquillisers. I often find myself giving advice to others while behaving in a reckless way with my own health. In the field of mental health, especially in primary care work, it is usually the areas of stress that we focus on pretty soon during any encounter. 'Take a few days off, let someone else take the strain, you aren't indispensable you know', runs the external, human/caring dialogue – while inside my head a different tape might be running: 'Why doesn't anyone ever say that to me?' or, 'I've got more reasons to be stressed than this bloke, but if I took a few days off, the place here really would crumble.'

Therapy for the therapist

Many health workers might consider that psychotherapy or personal psychological work is for others. How damaged do you have to be to be in need of psychological help? What comfort is there to be had from spending even more of your precious time sorting out your own head? Wouldn't it be more effective to blot out the day with a stiff Martini, a slump in front of the TV, another visit to the fridge, or whatever other narcotic anaesthetises you to the cumulative horrors that the day has wound up for you?

Here again it is important not to be prescriptive, and I know of no studies which have led us to think that health workers who undertake therapy on their own behalf are 'better' than others who do no aspect of 'self-gardening'. I can only report that for myself some form of supportive 'therapy' has been a consistent and persistent thread through my professional life. Over the last twenty years, the key insights and changes in the way I view myself and my professional work have come through self-reflective work.

Doing things that work

I can hear my tutors sitting on my shoulder in their white coats saying, 'Well all this being human is alright, but you still need to know what works, for whom and when.' And of course they are right. It is easy to get sucked down into a depressive mood when people who could be described as 'depressed' come along. It is possible to agree with them that the world is a hopeless place, that they had been done down by a series of horrible events and that there isn't much prospect of resolution. Together we could wallow in this mutual self pity and my affirmation of their plight might be a human response, but it isn't a very useful one. It is also possible to see how this sort of 'hopeless' consultation can lead to inappropriate prescribing of psycho-active medication. My more recent training has given me some simple tools which have been shown to be effective in helping people climb out of their low mood. Using these techniques has helped me get through heavy surgeries without being personally so devastated at the end. Knowledge-based work which is effective helps to make people feel better and seems to be less draining for the practitioner also.

Making it all better

There is a temptation to come to conclusions and to make things end neatly. This article could end that way: 'it would all be OK if doctors did a bit more sensitivity training', or some similar sentiment. But the reality is that mental health work is often messy, difficult and unresolved. Major challenges remain and have been pointed out to me by a friend who was a former 'mental patient' and who commented on a previous draft (Read, 1995). He suggests the following issues are highlighted by the things I have written:

> It needs to be OK for mental health workers to have strong feelings about the situations they are in and to have the time and permission to express them directly to colleagues.

> The current mental health system has been shaped by a white, upper-class, male-dominated medical establishment. This group of people has been bullied and cajoled into becoming terrified of their and other people's feelings, and desperate to be in control. If the mental health system was taken away from that influence, how would it change?

However, these views should be treated as a stimulus for change rather than as a threat. There is a clear need for health workers to work to make things better for themselves in order to survive and remain human for the people they set out to help. The need for supportive structures within the health services is obvious, but often the mechanisms for seeking or providing such help are underdeveloped. People working in the mental health world can be very good or very bad in the way they engage with each other and with the people who come for assistance, and this difference can be absolutely crucial. We all need support, encouragement and continuing reflective training. Mental health is not something you either have or do not have, but something that it is necessary to continually engage with.

References

Read, J. (1995) Personal communication.
Read, J. and Reynolds, J. (1996) *Speaking Our Minds, Personal Accounts of Mental Distress and its Consequences*, London, Macmillan.

Index

abuse *see* drugs, abuse/dependence;
    physical abuse; sexual abuse
Acute Day Therapy Service,
    Chesterfield 239
acute hospital care 205 6, 266–70
    see also crisis services
admission facilities 159–60
aftercare 149–50, 226–37
    see also community mental patient
    (CMP) system; hostel system
age
    and suicide 173(figs), 174
    see also elderly
agency 13
aggression 49–51
agranulocytosis 290
alcoholism 14–15, 22, 110–11
Alzheimer's disease 24, 87, 90–1,
    92–3, 111
American Psychiatric Association 20
    Diagnostic and Statistical Manual 24,
    70–1, 340
anorexia nervosa 22, 111, 319–24
anti-psychotic drugs 26, 199, 291,
    293–4
antidepressant drugs 7, 23, 184,
    293–4
    long-term use of 7, 188, 237, 293–4
    serotoninergic 293, 294
    tricyclic 26
anxiety-related disorders 10
approved social workers (ASWs)
    198–9
Asian people 246(fig), 252–9
    see also ethnic minorities
asylums *see* mental hospitals and
    asylums
Ativan 190

Australia 60–1
Avoider movement 299

barbiturates 186
Beck, A. T. 9
beds
    acute 205–6
    admission 159–60
    mental health 144–5, 147, 149–50
behaviour modification 8–10, 92
behavioural cycles 14–15
benzodiazepines 186–7, 190 1, 192–3,
    199
Better Services for the Mentally Ill, DHSS
    145
Bicêtre asylum, France 19, 129
biological approach to mental illness
    6–7, 178
black people 80, 196–203, 246(fig),
    267–70, 339–44
    and crisis services 267–70
    and cultural psychiatry 80, 339–44
    medication 199–200
    mental health law and 196–203,
    341
    schizophrenia 267, 268, 341–2
    stereotypes 198–201, 203
    women 246(fig), 268
    see also ethnic minorities; racial
    inequality
Bleuler, E. 31–2
brain 20–2, 24–5
Bridge 308
British Network for Alternatives to
    Psychiatry (BNAP) 219, 220,
    221
bromide 186
Buchanan, Michael 149, 150

Campaign against Psychiatrist
    Oppression (CAPO) 219
Campbell, Sharon 148, 150
Canada 299
care managers 147–8
care (mental illness), or prevention
    140–2
Care Programme Approach 147, 148,
    154, 161, 240–1
Caring for People White Paper 146–7,
    148
cerebral pathology 24–5
charities 136
Chesterfield 220, 238–41
Chesterfield Community Mental
    Health Team (CMHT) 238
Chesterfield Support Network 238–41
child sexual abuse *see* sexual abuse
childbirth 244(fig)
childcare 244(fig), 257
childhood 10–11, 49–50
    see also sexual abuse
Chinese medicine 262, 263
chloral hydrate 186
chlordiazepoxide (Librium) 186, 190
choice maximisation 298–9
chronicity 81–2
civil rights 205–10
    see also coercion; detention;
        repression
class inequalities 110–17
clozapine 290–1
Clunis, Christopher 148, 149, 150, 271
coercion 122–3, 183, 207–8
    see also repression
cognitive behaviour therapy 9–10,
    283–4, 291–2
Colombia 58
Committee on the Safety of Medicines
    (CSM) 191–3
communication
    client/therapist 15
    families' lack of 97–8
    through life history work 333–8
Community Care: Agenda for Action
    146, 147
community care 138, 143–54, 157–61,
    226–37, 266–70, 346–54
    community mental patient (CMP)
        system 226–37
    cost of 160–1
    crisis and emergency services
        266–70, 346–54
    and mental health law 144, 207–8,
        210
    reprovision programme 158, 160
    scandals in 148–52
    TAPS research 157–61
community mental patient (CMP)
    system 226–37
compulsion *see* coercion
Confidential Inquiry into Homicides
    and Suicides by Mentally Ill
    People 151, 205
Contact, Chesterfield 220, 240
Cooper, David 221
cost, of mental health care 145–6, 147,
    160–1, 205–6, 348, 350
Council for Involuntary Tranquilliser
    Addiction 190
counselling 189, 282–7, 301
Cowper, William 127–8
criminal justice system 201–3
crisis services 266–72
    see also acute hospital care; Early
        Intervention Service (EIS)
culture, mental illness in context of
    30–4, 54–61, 64–7, 200–1, 268–9,
    339–44

DAWN 190
dehumanisation 47
delusions 27–33, 291
    see also hallucinations; voice-hearing
dementia 87–94, 111, 159, 333–8
    care of 91–3
    life history work and 333–8
    see also Alzheimer's disease
dependence
    drug 22, 26; tranquillisers 186–7,
        190–3
    social 135–6, 137–8, 139
    see also alcoholism
depression 36–43, 66–7, 252–6, 283–4,
    293–4
    Beck's cognitive model 9–10
    as diagnostic category 66–7
    psychosocial factors 41–2, 43,
        110–11, 294
    role of life events and loss 36–43
    treatment of 283–4, 293–4
    in women 36–43, 49–52, 252–6,
        294
deprivation 113–17
desensitisation, systematic 8

detention 183
    and principle of reciprocity 205, 210
    see also civil rights; coercion;
        sectioning
determinism 13
deviance 67–9, 79–85
    deviancy amplification 81–3
    deviancy normalisation 83–5
    labelling theory of 67–9, 79
diagnosis
    historically 122, 123–4
    problems of 54–8, 70–1, 340–4
    self-fulfilling 77–8
Diagnostic and Statistical Manual,
        American Psychiatric Association
        24, 70–1, 340
diazepam (Valium) 186, 190, 199
difficult-to-place patients (DTP)
        158–9, 161
discharge orders 207–8
disempowerment 81–3, 89–90,
        299–300
    and empowerment 242–8
    see also helplessness
dissociation 310–11, 312–13, 314–15
drapetomania 80
drop-in centres 158
drugs
    abuse/dependence 22, 26, 186–7,
        190–3
    anti-psychotic 26, 199, 291, 293–4
    antidepressants 7, 23, 26, 184, 293–4
    neuroleptics 26, 290–1
    psychotropic 7, 236–7, 243, 247
    tranquillisers 7, 184, 186–93
    treatment by 7, 26, 184, 186–93,
        283–4, 290–1; ex-patients attitude
        to 236–7; iatrogenic 81, 184;
        rationalisation 159

Early Intervention Service (EIS)
        346–54
ECT *see* electroconvulsive therapy
        (ECT)
elderly
    dementia in 87–94
    life history work 333
    women 246(fig)
    see also senility
electroconvulsive therapy (ECT) 184,
        227, 243, 247, 294
emergency services 346–54
    see also crisis services

empathy 14, 329, 336–7
empowerment of women 242–8,
        304–8
    see also disempowerment
entrapment 39–41
Ethiopia 60
ethnic minorities 252–9, 260–5,
        266–72, 339–44, 355–63
    Asian 246(fig), 252–9
    Black 196–203, 267–70, 339–44
    crisis services 266–72
    focus groups 260–5
    language problems 255, 258, 261–3,
        266–7, 340
    mental health of women 246(fig),
        252–9, 262–3, 268
    Vietnamese 260–5
    and Western culture 30–4, 339–44
    see also racial inequality; refugees
European Court of Human Rights
        206, 207
'exotic syndrome' 339–40, 341

families
    effect of mental illness on 97–104,
        174
    family therapy 6, 11, 13
    role in mental illness 6, 12–13,
        100–1, 113–14, 254–6
feminisation of poverty 115–16
feminism 247–8
fluoxetine (Prozac) 294
focus groups 260–5, 304–8
France 19, 127, 129
Freud, S. 10–11, 139, 140, 313

Galen 18
gender
    of asylum patients 49–51
    of child abusers and abused
        244(fig), 309–10, 316
    childhood behaviour differentials
        49–50
    inequality 112, 115–16, 188–9,
        243(fig), 244–7
    of mental health workers 316–17
    of suicides 50, 110, 171, 172–3(figs)
    tranquilliser use by 187–9
    see also women
general practitioners (GPs) 148, 153,
        189, 269, 287–8
    and ethnic minorities 199, 258–9,
        262, 263

George III 18–19, 121–5
gestalt psychology 75
Griesinger 19, 20–1
groups
  focus 260–5, 304–8
  group therapy 11, 12
  local planning 240–1
  mutual support 103–4
  in non-industrial cultures 54–61
  self-help 12, 220–1, 238–41
Gull, William 319

Halcion 190–1, 192–3
hallucinations 27–34, 54–6
  see also delusions; voice-hearing
haloperidol 199
healing ceremonies 60
health authorities 144–5, 147
health economics 160–1
health, and inequality 110–17
Heberden, Dr William 122–3, 125, 129
helplessness 39–41, 311
  learned 9, 81
  see also disempowerment
Hippocrates 18
Hoffman-La-Roche 190
Hoggett, Prof. Brenda 206
homelessness 116–17, 252
homicide 151–2, 182
homosexuality 82–3
hospitals 133, 143–4, 207–8
  mental 46–52, 143–5, 157, 160–1
  see also acute hospital care; mental
    hospitals
hostel system 231–2
housing 116–17, 257–8
humanistic approach to mental illness
  11–12
humiliation 39–41
humoral model of illness 123–4

iatrogenesis 81–2, 84, 184
identity
  defined by psychiatric history 226
  maintaining 231, 232
immigration 255, 260–1
  see also refugees
incest 311
incompetence, mental 206–7
individuality 6, 300–1
inequality 110–17, 196–203, 244–7
  gender 112, 115–16, 188–9, 243(fig),
    244–7

racial 112, 196–203
socioeconomic 110–17
Inner City Mental Health Project
  (ICMHP) 259
insanity 70–8, 138–40
  see also madness; mental illness
institutional neurosis 81–2
institutionalisation 81–2, 130, 133–4
institutions *see* asylums; hospitals;
  mental hospitals
irrationality 30
isolation
  in ethnic communities 254–5
  of ex-mental patients 230, 232–3,
    235, 301–3
  and mental illness 176, 235, 331

Jaynes, J. 33
job security 116

Katzman, R. 91
Kenya 57
King's Fund 269
kleptomania 80

labelling, mental illness 54–8, 64–9,
  75–8, 85, 97–8
  cultural context 54–8
  family need for 97–8
  labelling theory of deviance 67–9,
    79
  stickiness of 75–8
Lader, M. 29
Laing, R. D. 221
Lambeth Link 332
language
  as insight into sexual abuse 313–14
  problems for ethnic minorities 255,
    258, 261–3, 266–7, 340
  use by dementia patients 335
Laos 56
law, mental health
  and civil rights 205–10
  establishment of asylums 128, 130,
    135
  Mental Health Act (1959) 144, 206
  Mental Health Act (1983) 179, 183,
    196–203, 206–10
  Mental Health (Patients in the
    Community) Act (1995) 207
  Mental Treatment Act (1930) 138
  Poor Law 126, 135, 137–8
  and principle of reciprocity 205, 210

psychiatrists' knowledge of 209–10
Suicide Act (1961) 179
learned helplessness 9, 81
lesbian women 246(fig)
Librium *see* chlordiazepoxide
Life Events and Difficulties Schedule
(LEDS) 37, 38–9, 43
life events, and mental health 25,
36–43, 244(fig), 245, 252–4
life history work 333–8
life-expectancy 115
life-style 113
Lindsley, Ogden 28
Lipsedge, M. *see* Littlewood, R. and
Lipsedge, M.
lithium 26, 291, 293–4
Littlewood, R. and Lipsedge, M.
29–30, 32, 341–2
local authorities 144–5, 147, 153
local planning groups 240–1
lofepramine 294
long-stay non-demented patients 158
loss 38–9, 51–2, 98–9, 252–4

Madagascar 60
madness
historical background to 17–20,
121–30
institutionalisation 130, 133–4
secularisation of 125–8
women and 46–52
see also insanity; mental illness
Malaya 56
manic depression 111
marriage 244(fig)
maternity blues 294
media coverage
content analysis 163–4, 165–7
health risks from tranquillisers 187,
190–1, 192–3
of mental illness 163–70
medical approach to mental illness
6–7
Medical Foundation for the Care of
Victims of Torture 355
medication
of black patients 199–200
ex-patients' view of 236–7
see also drugs; treatment of mental
illness
medicine
Chinese 262, 263
and humoral model of illness 123–4

professionalisation of 133–4
and treatment of mental illness 130,
132–4
memory 21–2
mental disorder *see* mental illness
mental distress 54–9, 165–6, 340–4
mental health 196, 206–10, 224
Mental Health Act (1959) 144, 206–7
Mental Health Act (1983) 179, 183,
196–203, 206–10, 224
dangerous patients 206–7
designed for hospital care 207–8
experience of black people under
196–203
lack of user involvement 224
and need for change 206–10
psychopathic disorder 208–9
sectioning 196–203, 206
Mental Health (Patients in the
Community) Act (1995) 207
mental health services 134–42,
144–54, 201–3, 242–8, 297–303
care or prevention 140–2
changing 132–42, 297–303
co-ordination 147–8, 150, 153–4, 161
and criminal justice system 201–3
crisis/emergency services 266–72,
346–54
and empowerment of women
242–8, 304–8
role of state 134–6
shift of responsibility 150–1
under fire 152–3
see also community care; community
mental patient (CMP) system
mental health workers
feminism 247–8
normalisation and being human
297–303, 364–9
proliferation of skills 140
racial stereotyping 198–9
skill-sharing 349–50, 353–4
under stress 356–63, 367–8
mental hospitals and asylums 46–52,
143–5, 157, 160–1
closure 144–5, 157
cost of care 160–1
historically 19, 125, 127–30, 135,
136–8
legal requirement for 128, 130, 135
scandals in 143–4, 150
women in 46–52
see also hospitals

mental illness
  associated with violence 165–6,
    167–8, 181
  cultural background to 54–9, 60,
    340–4
  and gender inequality 112, 115–16,
    188–9, 243(fig), 244–7
  media coverage 163–70
  milder forms 139–40
  and physical illness 20–1, 22–5,
    112, 115–17
  problems of diagnosis and
    classification 54–8, 70–1, 340–4
  and social order 60–1, 64–9
  supernatural explanations 54–5
  vested interests in 79–80, 306–7
  see also insanity; madness
Mental Patients Union 219
Mental Treatment Act (1930) 138
metaphor 31–2, 328–9
Mind 187, 190, 206, 219, 270–1, 351
Mogadon 190
moral treatment of madness 19,
    128–9, 134, 137–8
multi-infarct dementia (MID) 111
mutual support groups 103–4

Nafsiyat Intercultural Therapy Centre
    342, 355, 362
National Health Service (NHS) 146, 153
National Schizophrenia Fellowship
    100, 238
Navajo Indians 58
Netherlands 221
neuroleptics 26, 290–1
neurotransmitters 7, 291, 293
NEWPIN 247
Nigeria 55–6, 60
normalisation 70–1, 79–80, 83–5,
    297–303
normality 70–1, 79–80, 83–5
North East Derbyshire and
    Chesterfield Association of Self-
    Help Groups (NEDCASH) 240
Nottingham Advocacy Group 219
nymphomania 80

obsessive-compulsive disorder 292–3
Opren 192

Parkinson's disease 22–3
patients
  aftercare 149–50, 226–37

difficult-to-place 158–9
disabled or disordered 206–7
gender differentials 49–51
long-stay non-demented 158
pseudopatients 71–8
psychogeriatric 159
relationships between 233, 234–6
relationships with staff 286–7, 297,
    301–2
'revolving door' 208
risk from 148–52, 181–2, 206–7
risk to 143–4, 183–4, 206–7
  see also users and ex-users
patients' forums 239
pharmaceutical industry 189–91, 288
phobias 8
physical abuse 243, 244(fig), 245
physical aspects of mental health 6–7,
    110–17
physical disease 22–5, 110–12, 115–17
  treating in mentallly incompetent
    206–7
Pinel, Philippe 19, 129
Plato 17–18
Poor Law 126, 135, 137–8
POPAN 247
positron emission tomography 292
poverty 115–16, 244(fig)
Powell, Ernest 144, 157
power
  male abuse of 50, 80, 245, 315–16
  power sharing 299–300
  of professionals 189, 344, 366
  and racism 202–3
powerlessness see helplessness
prediction of depression 41–2
pregnancy 49
prevention of mental illness 140–2
private sector care 145, 146–7
problem-solving therapy 292
professionalism 133–4, 140
  normalisation of 297–303
Prozac see fluoxetine
pseudodementia 88
psychiatric disease 22–5
psychiatry
  advances in 290–4
  anti-psychiatry movement 219, 221
  cultural 339–44
  curative not preventive 140–2
  diagnosis and classification
    problems 54–8, 70–8, 340–4
  and mental health law 209–10

and minor forms of mental disorder
139–40
professionalisation of 133–4
proliferation of experts 140
psycho-education 7
psychoanalysis 11, 139, 140
psychodynamic psychotherapy
10–11, 92
psychogeriatric patients 159
psychological frameworks 6–15
behavioural 8–10
biological and medical 6–7
humanistic 11–12
psychodynamic 10–11
systemic 12–13
psychopathic disorder 208–9
psychosocial factors, in depression
41–2, 294
psychotherapy
for child sexual abuse 310–11, 313
psychodynamic 10–11, 92
research into 282–7
social action 304–8
psychotropic drugs 7, 236–7, 243, 247
public perception
of mental illness 148–9, 163–70
of use of tranquillisers 187–8
Puerto Rica 58–9
punishment, for mental illness 126,
127, 227–8

racial harassment 257
racial inequality 112, 196–203
see also ethnic minorities
Rampton 143, 184
randomised controlled trials (RCT)
284–5, 352–3
Ray, L. 189–90
reality orientation 92
reciprocity, principle of 205, 210
Refugee Council 355
refugees 260–5, 335–63
`vicarious traumatisation' of
workers 357–8
Vietnamese 260–5
working with 355–63
see also ethnic minorities
rehabilitation centres 228, 230
relationships
between child abuser and abused
310–12
between ex-mental patients 233,
234–6

between users and staff 286–7, 297,
301–2
family 6, 12–13, 100–1, 113–14,
254–6
therapeutic 10–11, 14, 326–7, 332
religion 259
representativeness 223
repression, institutional
George III 122–3
historically 18–19, 122–3, 125,
126–8
of sexuality 48–9
see also coercion
reprovision programme 158, 160, 161
research
funded by pharmaceutical industry
191, 288
into effectiveness of psychotherapies
282–7
and practice 282–8
randomised controlled trials (RCT)
284–5, 352–3
residential care 150
in private sector 145, 146–7
resolution therapy 92
Rhodesia 54–5
risk
assessment 150–1, 181–2
and community care 150–1
in deviance 83–4
from patients 148–52, 181–2, 206–7
iatrogenic 184
to patients 143–4, 183–4, 206–7
in tranquilliser usage 184, 187–8
risperidone 290
Robinson, Georgina 149
Rogers, C. 12
Rothschild, D. 89
Royal College of Psychiatrists 151,
205, 210, 272, 273, 293
Rush, Benjamin 19

SAGE (Sexual Abuse Groups, Exeter)
247
Sanctuary Project 268–70
sanity 70–8
failure to detect 73–5
Save the Children, Vietnamese refugee
project 261–5
Scheff, Thomas 79, 81, 82
schizophrenia 7, 22–3, 111
among black people 267, 268, 341–2
associated with violence 167–8

and community care 149–50, 347
concepts of 66–7
family experiences of 97–104
and family relationships 25, 100–1
hallucinations and delusions 27–33
labelling 68–9, 75–8
treatment 237, 290–2
sectioning 196–203, 206
segregation of mentally ill 136–8, 301
self-advocacy 299–300
self-esteem 175–6, 226–7, 301–2
self-harm 165–6
   see also suicide
self-help groups 12, 220–1, 238–41
   see also user movements
Seligman, M. 9
Senegal 55, 61
senility 90–1
   see also Alzheimer's disease
serotoninergic drugs 293, 294
service brokerage 299
service users *see* users
sexual abuse 243, 244(fig), 245
by mental health workers 48–9,
   247
of children 243, 244(fig), 245,
   309–17
sexuality 10–11, 80
shamans 59, 60–1, 68
Shanti 247
sheltered employment 228
Silcock, Ben 149
skills
normalisation of professional
   297–303
proliferation 140
skill-sharing 349–50, 353–4
slavery 80, 202
social action psychotherapy 304–8
social autonomy 330–1
social constructionism 31
social dependency, role of state
   135–6, 137–8, 139
social order, and mental illness 60–1,
   64–9
social services 147, 148, 258–9
social workers 147, 198–9, 210, 247–8
   see also mental health workers
socioeconomic class, and mental health
   65, 110–17
sociology
of deviance 79–85
of suicide 178

special needs units 159
state
curative or preventive intervention
   141–2
inaction on tranquilliser dependence
   191–3
and mental health services 134–6
role in health care 133–6
and social dependence 135–6,
   137–8, 139
stigma of mental illness 57–8, 74,
   81–2, 98, 101–2
stress 7, 116
in mental health workers 355–63,
   367–8
suicide 82, 110–11, 150, 151–2,
   171–9
and gender 50, 110, 171, 172–3(figs)
incidence 171–4
media coverage 165–6
methods used 176–7
and social class 110–11
and type of care 150, 151–2
Suicide Act (1961) 179
supernatural 54–5, 69, 125–6
supervised discharge orders 207–8
supervision of staff 360–2
supplementary benefit 145, 147
survivor movements 83, 218–24,
   241
Survivors Speak Out 219, 220
symbols 31–2
systematic desensitisation 8
systemic approach to mental illness
   12–13
systemic family therapy 6, 13
Szasz, Thomas 50, 179, 221

Tanzania 57
team for assessment of psychiatric
   service (TAPS) 157–8
*Health of the Nation, Mental Illness Key
   Area Handbook* 240
*Report of the Committee of Inquiry into
   the Care and Aftercare of Miss
   Sharon Campbell* 146, 148
therapies
behaviour modification 92
cognitive behaviour 9–10, 283–4,
   291–2
electroconvulsive (ECT) 184, 227,
   243, 247, 294
family 6, 11, 13

group 11
non-specific problem solving 292
psychodynamic psychotherapy 92
reality orientation 92
research into psycho- 282–7
resolution 92
social action psychotherapy 304–8
validation 92
see also treatment of mental illness
Threshold 247
torture victims 355–63
see also refugees
training 298, 300, 301, 353–4, 360–2
for counselling 301
and empowerment of women
242–8
tranquillisers 7, 184, 186–93
CSM regulation of 191–3
and pharmaceutical industry
189–91
risks from 184, 186–8, 190–3
use of 7, 187–9, 237
women and 187, 188–9
transference 10–11, 309–10, 312–15,
317
TRANX 190
treatment of mental illness
as branch of medicine 130, 132–4
charitable 136
coercive 122–3, 183, 207–8
historical 18–19, 122–5, 127–30,
136–8
institutional 19, 125–30
moral 19, 128–9, 134, 137–8
political aspects 127, 141–2
private sector 136, 145, 146–7
recent advances 290–4
research into 282–8
see also community care; drugs;
therapies
tricyclic antidepressants 26
Trinidad 60
tuberculosis 58
Tuke, Samuel 129, 130
Tuke, William 19
Turkey 59

Uganda 57
unemployment 116
United States 157, 183, 350
Upjohn 191, 192
user involvement 223–4, 238–41,
267–70, 271–2, 299–300

in Chesterfield 238–41
and crisis services 267–70, 271–2
user movements 83, 218–24, 241
users and ex-users
and crisis services 266–72
going it alone 232–4
lives of 226–37
maximising choices 298–9
relationships 233, 234–6
views on CMP system 227–31
women as 242–4
see also patients

Valentine, E. 28
validation therapy 92
Valium *see* diazepam
'vicarious traumatisation' 357–8
Vietnamese refugees
focus groups 261–5
health status 260–2
language problems 261–3
violence
by women 51–2
and community care 151–2
mental illness associated with
165–6, 167–8, 181–2
to women 245
see also physical abuse; sexual abuse
voice-hearing 33, 83, 326–32
see also delusions; hallucinations
vulnerability (to depression) 41–2,
43

White City Project 247, 305, 307–8
Willis, Dr John 122–3, 125
Willis, Rev. Francis 18–19, 122
Wing, J. 33–4
withdrawal 66
women
effect of life events 36–43, 51,
244(fig), 245
empowerment of 242–8, 304–8
ethnic minorities 246(fig), 252–9,
262–3, 268
focus groups 262–3, 264, 304–8
male control over 50, 80, 245,
315–16
and mental asylums 46–52
multiple discrimination 245–7,
246(fig)
psychiatric careers 47–9
and use of tranquillisers 187, 188–9
see also gender

Women's action for mental health
    307–8
Women's Aid Movement 247
Women's Mental Health Forum 242–4
Working for Patients White Paper 146

Wyeth 190

York Retreat 19, 128–9, 130

Zimbabwe 36